STUDIES IN ENVIRONMENT AND HISTORY

Americans and their forests

STUDIES IN ENVIRONMENT AND HISTORY

Other Books in the Series

Donald Worster *Nature's Economy: A History of Ecological Ideas*
Kenneth F. Kiple *The Caribbean Slave: A Biological History*
Alfred W. Crosby *Ecological Imperialism: The Biological Expansion of Europe, 900–1900*
Arthur F. McEvoy *The Fisherman's Problem: Ecology and Law in the California Fisheries, 1850–1980*
Robert Harms *Games Against Nature: An Eco-Cultural History of the Nunu of Equatorial Africa*
Warren Dean *Brazil and the Struggle for Rubber: A Study in Environmental History*
Samuel P. Hays *Beauty, Health, and Permanence: Environmental Politics in the United States, 1955–1985*
Donald Worster *The Ends of the Earth: Perspectives on Modern Environmental History*

Americans and their forests

A historical geography

MICHAEL WILLIAMS
University of Oxford

Published by the Press Syndicate of the University of Cambridge
The Pitt Building, Trumpington Street, Cambridge CB2 1RP
40 West 20th Street, New York, NY 10011-4211, USA
10 Stamford Road, Oakleigh, Victoria 3166, Australia

First published 1989
Reprinted 1990
First paperback edition 1992

Library of Congress Cataloging-in-Publication Data

Williams, Michael, 1935–
Americans and their forests.
Bibliography: p.
1. Forests and forestry – United States – History.
2. Forest products industry – United States – History.
3. Clearing of land – United States – History. 4. United States – Historical geography. I. Title.
SD143.W48 1988 333.75'0973 87–15828
ISBN 0-521-33247-8 hardback

British Library Cataloguing-in-Publication applied for

ISBN 0-521-33247-8 hardback
ISBN 0-521-42837-8 paperback

Transferred to digital printing 2004

For Eleanore, Cathy, and Tess

Contents

List of illustrations page x
List of tables xv
Preface xvii
A note on tree, lumber, and plant association names xx
Abbreviations xxii

Part I The forest before 1600

1 The forest in American life 3
American geography 3
The American economy 5
American culture 9
National symbols 19
2 The forest and the Indian 22
The forest 22
The Indian 32

Part II Change in the forest, 1600–1859

3 The forest and pioneer life, 1600–1810 53
Pioneer life in the forest 55
Clearing 60
Stock in the forest 67
Fences and houses 69
Potash and fuel 74
4 Two centuries of change: the commercial uses of the forest 82
The demands of the sea 82
Timber products and lumber 94
Charcoal and iron making 104
5 The quickening pace: agricultural clearing, 1810–1860 111
Clearing the forest and making a farm 112
The amount of land cleared 118
"A new creation": the landscape of clearing 120
The forest edge 128
Domestic fuel supplies 133

Minor pioneer products 139
"Another landscape": environmental concern 144
6 The quickening pace: the industrial impact, 1810–1860 146
Charcoal and the iron industry 147
Fueling transport and manufacturing 152
Naval stores 157
The lumber industry 160

Part III Regional and national impacts, 1860–1920

7 The lumberman's assault on the forests of the Lake States, 1860–1890 193
The Great Lakes setting 197
Technical changes 201
The social and economic organization of the lumber business 216
Production 222
The depletion of the forests 228
The aftermath of depletion 230
8 The lumberman's assault on the southern forest, 1880–1920 238
The northern invasion 239
Technological changes 244
Business and social organization 263
Production 271
Depletion 277
Reforestation 285
9 The last lumber frontier: the rise of the Pacific Northwest, 1880–1940 289
Experiment and change 289
Tidewater lumbering, 1850–83 291
Transition, 1883–99 296
The long boom, 1900–40 309
A new approach, 1933–45 328
10 Industrial impacts on the forest, 1860–1920 331
Changes in fuel use 332
Charcoal and the iron industry 337
The railroads 344
11 Agricultural impacts on the forest, 1860–1920 353
Clearing and making a farm 353
The locale and extent of clearing 357
Ohio: microcosm of forest clearing 361
Early warnings 368
Planting in the Plains 379
Shelterbelts 386

Part IV Inquiry and concern: questions about the forest, 1870–1933

12 Preservation and management, 1870–1910 393
"The day is coming" 393

Are the forests worth preserving? 1870–90 395
How should the forests be managed? 1890–1910 411
13 Ownership, supply, protection, and use, 1900–1933 425
Who owns the forest? 425
How much forest is left? 430
How should the forest be protected? 447
How should the forest be used? 456
Transition 459
14 The rebirth of the forest, 1933 and after 466
The rebirth of the forest 467
Epilogue: gains and losses 489

Notes 495
References 543
Index 585

Illustrations

Figures

1.1 Wood as a percentage of industrial raw materials, 1900–80 *page* 8
2.1 "The forest and prairie lands of the United States" (1858), by Joseph Henry 24
2.2 "The forests and trees of North America" (1860), by James Cooper 26
2.3 Density of woodland in the United States, 1873 28
2.4 The density of existing forests, 1883 29
2.5 Natural vegetation of the United States, 1924 31
2.6 Dominant types of Indian subsistence 34
2.7 Native population density 35
2.8 Schematic representation of Indian forest use 40
3.1 Expansion of settlement, 1600–1809 54
3.2 The export of naval stores and potash, 1769–72 76
4.1 Sawmills and navigable waterways in northern New York, c. 1810 97
4.2 Early (eotechnic) forest exploitation 98
4.3 The export of staves, shingles, planks, 1768–72 103
5.1 The amount of land cleared, by state, before 1850 119
5.2 The amount of land cleared, by state, 1850–9 120
5.3 The prairie/forest edge in the northern United States 129
5.4 Entry dates and vegetation boundaries, Township TIN R6W, St. Clair County, Illinois 131
5.5 Cords of wood sold in the United States, 1840 135
5.6 Tons of potash and pearl ash produced in the United States, 1840 140
5.7 Value of skins and furs caught in the United States, 1840 142
5.8 Value of ginseng and all other forest products in the United States, 1840 143
6.1 Abandoned charcoal-fired blast furnaces, 1859 149
6.2 Charcoal and anthracite blast furnaces, 1859 151
6.3 Production of barrels of pitch, resin, and turpentine in the United States, 1840 159
6.4 Lumber cut and per capita consumption, 1800–1960 161
6.5 Lumber production in the United States, 1839, 1849, 1859 162
6.6 Number of sawmills in the United States, 1840 164
6.7 Number of men employed in sawmills in the United States, 1840 165

6.8 Value of lumber products, United States, 1840 166
6.9 The continental log transport system 177
6.10 Tons of goods moved on New York canals, 1837–1906 180
6.11 The Albany–Erie Canal waterfront, 1888 181
6.12 The Tonawandas: lumber receipts and shipments, 1873–1906 184
6.13 Chicago: receipts and shipments of lumber by lake and rail, 1847–1907 185
7.1 Lumber production, 1869, in m.b.f. by state 195
7.2 Lumber production, 1889, in m.b.f. by state 195
7.3 Lumber production, 1909, in m.b.f. by state 196
7.4 Lumber production, 1929, in m.b.f. by state 196
7.5 Production of lumber by major regions of the United States, 1869–1959 197
7.6 The rivers of the Lake States and estimates of standing white pine in major watersheds, 1880 199
7.7 Forest conditions in the Lake States, 1881 200
7.8 Logs moved, Tittabawassee River, Flint and Père Marquette Railroad, and Michigan Central Railroad, 1864–92 205
7.9 The Beef Slough boom, open and shut 205
7.10 Paleotechnic forest exploitation 209
7.11 The Bonita line, Wisconsin, c. 1919 213
7.12 Logging railroads and sawmills, Wolf River basin, Wisconsin, 1857, 1885, 1898, and 1921 215
7.13 White pine lumber production, Lake States, 1873–1920 223
7.14 Production of lumber by value, Lake States, 1860 225
7.15 Production of lumber by value, Lake States, 1870 225
7.16 Production of lumber by value, Lake States, 1880 226
7.17 Great Lakes and Mississippi River: lumber production, 1876 227
7.18 Lake States: cutover lands, 1920 233
8.1 Forest conditions in the South, c. 1880 246
8.2 Pull-boat logging 250
8.3 Main-line carriers and spurs, Calcasieu basin, Louisiana, c. 1920 257
8.4 The Fisher–Victoria tramway system in Sabine and Natchitoches parishes, Louisiana, c. 1920 259
8.5 Railroad logging sequence, in Tangipahoa Parish, Louisiana, 1936–55 260
8.6 Large timber holdings in western Louisiana, c. 1914 266
8.7 Production of lumber in the South, by state, 1869–1940 273
8.8 Production of lumber in the South, by value, 1860 274
8.9 Production of lumber in the South, by value, 1870 275
8.10 Production of lumber in the South, by value, 1880 276
8.11 Value of lumber and timber products per square mile, 1900 278
8.12 The birth and death of mill towns in the Calcasieu basin, Louisiana 282
9.1 Pacific Coast: production of lumber by value, 1860, 1870, and 1880 296
9.2 Lumber cargoes received in San Francisco, by species, 1863–1900 297

9.3 Classification of land in western Washington, 1902 306
9.4 Lumber production on the Pacific Coast, by state, 1869–1949 309
9.5 Land held by F. K. Weyerhaeuser and other larger landholders, northwestern Washington, c. 1913 311
9.6 Western Washington: forest reserves and cutover land, 1909 326
9.7 Installed daily capacity of lumber centers in Washington and Oregon, 1924 and 1947 327
10.1 Utilization of wood grown in the United States, by roundwood equivalent, 1800–1975 332
10.2 Energy sources in the United States as a percentage of total energy consumption, by five-year periods, 1850–1960 333
10.3 Type of fuel used in the United States, c. 1880 335
10.4 Proportion of farms in the eastern United States covered by woodlot, 1910 337
10.5 Charcoal pig iron production, 1854–1911 338
10.6 Location of charcoal blast furnaces, 1859, 1876, and 1890 340
10.7 The range of oaks and chestnuts, the hardwood tie market (1882), and railroad plantations 348
11.1 The amount cleared, by state, 1860–9 358
11.2 The amount cleared, by state, 1870–9 358
11.3 The amount cleared, by state, 1880–9 359
11.4 The amount cleared, by state, 1890–9 359
11.5 The amount cleared, by state, 1900–9 360
11.6 Ohio: percentage improved, by county, 1850 363
11.7 Ohio: percentage improved, by county, 1880 364
11.8 Ohio: the causes of forest destruction, 1885 366
11.9 Ohio: percentage improved, by county, 1890 368
11.10 Woodland cover, Cadiz Township, Green County, Wisconsin, 1831, 1882, 1902, and 1950 369
11.11 "Area of shelterbelt planting to date, Jan. 1941" 389
12.1 The national forests of the United States, 1898, 1907, and 1980 408
12.2 The area of national forest and other lands, 1890–1960 412
13.1 "Area of virgin forest": 1620, 1850, and 1920 436
13.2 Lumber surpluses and shortages, 1920 437
13.3 Location of some major forest fires in the United States since 1800 449
13.4 Proportion of woodland burned in the settled parts of the United States, 1880 451
13.5 Fires of 1910 in the western United States 452
13.6 Organized protection against forest fires; funds spent in relation to funds received 453
13.7 Visits to national forests, national parks, museums, and allied areas, 1923–70 459

13.8 The decrease in lumber production in the United States, 1932, as a percentage of 1929 462
14.1 Woodland as a percentage of total farmland, 1959 470
14.2 Change in cleared farmland acreage, 1910–59 474
14.3 The percentage decrease in agricultural land in the eastern United States from its peak to 1930 475
14.4 Conversion of forest to agricultural land and abandonment of agricultural land to forest, Carroll County, Georgia, 1937–74 476
14.5 Direct planting and seeding of forest, 1943–78 478
14.6 Replanting and reseeding of land to forest in 13 southern states, 1925–80 480
14.7 Annual extent of forest fires in the United States, amounts spent on fire control, and average size of area burned per fire, 1910–70 482
14.8 Per capita consumption of timber products, 1900–80 488
14.9 Annual net growth and use, 1800–1977, with projections of growth and demand to c. 2030 492

Plates

1.1 The beginning 13
2.1 The town of Secota, Virginia 36
3.1 Life in New England, 1770 62
3.2 Types of fencing in eighteenth-century America 70
3.3 Log house in the forest of Georgia 73
4.1 Cutting a box for turpentine 85
4.2 Collecting the resin 86
4.3 Transporting the resin by "rolling it to market" 86
4.4 Distilling the resin to make spirits of turpentine 87
4.5 Gathering the scrape 87
4.6 Burning old boxed wood in a forest kiln to make tar and pitch 88
5.1 Embryo town of Columbus on the Chattahoochee 121
5.2 "Newly cleared land in America" 122
5.3 Contrasting landscapes in rural Pennsylvania 124
5.4 The first six months 126
5.5 The second year 126
5.6 Ten years later 127
5.7 The work of a lifetime 127
6.1 From forest to market: felling and sawing logs 171
6.2 Loading a sled 171
6.3 Floating logs in the drive 172
6.4 A logjam 172
6.5 Loading the timber on a ship 173

7.1 Rafts of Lake States logs being pushed into position in readiness for their voyage downstream, upper Mississippi 207
7.2 Horse-drawn load on an ice road at an exhibition at Bemidji, Minnesota, c. 1890 210
7.3 The Lombard log hauler on an ice road, Minnesota, c. 1908 210
7.4 A Shay locomotive drawing 47 cars, with 393 logs totaling 61,000 b.f., through a logging settlement near Cadillac, Michigan, 1904 212
7.5 The charred stumplands of the cutovers of northern Wisconsin 237
8.1 A two-drum pull boat in a set in a canal cut in the cypress swamps of Louisiana 249
8.2 A slip-tongue cart in the forests of Louisiana, c. 1910 251
8.3 The bringer of change in the southern forests: a logging railroad near Oscilla, Irwin County, Georgia, 1903 254
8.4 A fairly large modern mill in the South, Huttig, Arkansas 288
9.1 A doghole on the rocky north California coast 295
9.2 The B. F. Brook logging camp, Cowlitz County, Washington, c. 1892 301
9.3 A Dolbeer donkey engine and skid road in the redwood country near Fort Bragg, Mendocino County 302
9.4 High-lead yarding somewhere in the Pacific Northwest 316
9.5 Revolution in harvesting the timber of the forest: trucks, tractors, and hoists 318
9.6 Part of the area burned in Tillamook, Oregon, 1932 324
10.1 Baird Creek, near Longview, southern Washington: the impact of the railroad on the forest 345
11.1 "The war on the woods" 355
11.2 Shelterbelt planting in Oklahoma 390
12.1 Forest Service Chief Gifford Pinchot with President Theodore Roosevelt during an Inland Waterways Commission excursion on the Mississippi River, 1907 417
13.1 A fairly confined fire in the Olympic Peninsula, western Washington 448
13.2 The automobile completely altered the concept of how the forest should be used 457
13.3 "Time to take an inventory of our Pantry," by J. N. "Ding" Darling 464
14.1 The ability of the forest to regenerate naturally after a time 469
14.2 Cotton gives way to pine: reforestation in the South 479
14.3 New forestry practices become environmental issues: block cutting 493
14.4 Another environmental issue: aerial spraying of herbicides and pesticides 493

Tables

1.1	Manufacturing by major industry, 1850–1920: industries ranked according to percentage of total value of manufacturing	*page* 6
1.2	Estimated value added and employment attributed to timber in timber-based economic activities, 1972	9
2.1	Area of commercial timberland by forest type, 1977	32
3.1	Estimates of the cost of clearing land, 1688–1851	64
4.1	Imports of masts into Britain for selected years	91
4.2	New York: settlement and sawmills (lags in years between first settlement and establishment of first sawmills)	94
4.3	Acres of forest cleared per 1,000 tons pig iron produced, under various assumptions	106
4.4	Labor requirements, Sterling Ironworks, New York, 1777	108
5.1	Relative advantages and costs of clearing one acre at different times of the year (mid-nineteenth century)	116
5.2	Fuel used in Philadelphia, March 1826–March 1827	137
6.1	Costs and profits ($) in producing a ton of pig iron, c. 1850	150
6.2	Cordwood consumption on steamboats, 1800–60	154
6.3	Fuel wood consumption, 1879	157
6.4	Production of naval stores, 1840–1919	160
6.5	Increasing productivity of saws, seventeenth to nineteenth centuries	167
7.1	Estimated value of forest products, 1880	194
7.2	Lumber cut in the Lake States at the census, 1869–1939	224
7.3	Population of four western Michigan counties, 1880–1930	231
8.1	Purchases of 5,000 acres or more of federal lands by northerners and southerners in five southern states, 1880–8	242
8.2	Mississippi lumber shipped by water and rail, 1891–2	258
8.3	Texas longleaf pine shipped by water and rail, 1882	258
8.4	Concentration of ownership in the southern pine region, c. 1913, by size of holding	264
8.5	Holders of over 300,000 acres of timberland in the South, c. 1914	265
8.6	Timber acreage and timber stands of 74 owners in western Louisiana, c. 1913	267

8.7	Acreage of timber (including timber rights and additional land) owned in Louisiana and in the southern pine region, by the 14 largest landholders plotted on Fig. 8.6	267
8.8	Total cutover land, restocking and not restocking, 1907 and 1920	284
9.1	Classification of land in the Pacific Northwest, 1903	305
9.2	Size and standing timber of the holdings of the three largest companies in the Pacific Northwest, c. 1913	312
9.3	Concentration of timber ownership in the Pacific Northwest, c. 1913, by size of holding	313
10.1	Charcoal pig iron production, by state, 1866–1910	341
10.2	Charcoal iron: bushels of wood used per ton produced, 1880–1900	343
10.3	Estimates of crossties used and acres of forest cleared, 1870–1910	352
11.1	Amount of land cleared and man-years expended in forested and nonforested areas, 1850–1909	354
11.2	Number of acres (millions) cleared from previously abandoned land	357
11.3	Amount of forest clearing each decade, by major region, 1850–1909	361
11.4	Reasons given, by county, for forest destruction in Ohio, c. 1884	365
13.1	Concentration of ownership of standing timber, Pacific Northwest, South, and Lake States, c. 1913	427
13.2	The wider Weyerhaeuser interests, c. 1913	427
13.3	Estimated area of commerical and noncommercial forest, and total standing sawtimber volume	433
13.4	Area of commercial and noncommercial forest, and growth potential of commercial forest lands, 1907–77	434
13.5	Estimated volume and annual growth and drain, 1909–77	439
14.1	Cleared farmland in the United States, 1910–79	473
14.2	Major land-use changes in Carroll County, Georgia, 1937–74	477
14.3	Ownership of commercial forests, 1977	489

Preface

Other than the creation of cities, possibly the greatest single factor in the evolution of the American landscape has been the clearing of the forests that covered nearly half of the country. The existence of the forest and the effort either to use or to subdue it have been a constant theme in American geography, economy, and history until the opening decades of this century. Geographically, clearing was the first step in the alteration of the visual landscape and the creation of farmland. Economically, the forest had no rival in the provision of fuel, building and construction materials, and a host of other products, and it was a major contributor to the economic growth of the United States until perhaps the end of the nineteenth century. Culturally, the forest provided symbols, themes, and concepts, such as pioneer woodsman, lumberjack, wilderness (and, conversely, the conservation of the environment), that have a special meaning in the thought and feelings of a large number of people in the United States.

The forest – or lack of it – has been, and still is, a part of the everyday experience of most Americans. Few know, however, what happened to it in the past, or what is happening to it today. Therefore, the object of this study is to describe and analyze the clearing of the forest from pre-European times to the present and to trace its subsequent regrowth since the middle years of this century.

Primarily, the book is a synthesis of topics that have been dealt with before, but in isolation and with little awareness that they all contribute to one great story. For example, lumbering, agricultural expansion, the growth of industry and spread of transport, the construction of houses and other buildings, and the supply of fuel wood all caused a demand for timber and a decrease in the forest cover. Similarly, awareness of the problems of soil conservation and of vegetational hydrological relationships, the growing national awareness of environmental values and beauty, and the growth of recreation and leisure have all had their effect on forest use and perception of the forest as a resource. In addition to the diversity of topics, there is the quantity of material. During the early 1960s it was calculated that subjects dealing *directly* with the forest were generating over 500 articles, pamphlets, and books a month, that is, over 6,000 items a year, an amount that is vastly greater now. The sheer mass of this material and its diversity and complexity are daunting, and therefore it is understandable that the usual solution has been to concentrate on one topic, one place, or one period. But in doing that there is a very real danger of missing a total picture of Americans and their changing relationship with the forest through time, of not seeing – if one may be permitted the phrase – the wood for the trees.

Before about 1840, information is tantalizingly slight, and we have to rely on literary and topographical descriptions as they occur here and there. After 1840, statistical and cartographical information increases so that by about 1880 it becomes an indigestible mass that needs a different type of analysis. Consequently, the focus of this book shifts from the individual and the local at the beginning to the aggregate and the regional or continental at the end.

In summary, therefore, this book is not a history of the lumbering and pulp industries, although they figure largely in the story; it is not a history of land settlement, although that is an essential element of the story; it is not a history of industrial and domestic fuel use, although it enters into the picture; nor is it a history of conservation and of the Forest Service, although these were important outcomes of the clearing of the forest. Though I have included some accounts of local areas, companies, and the colorful individuals involved in logging, lumbering, and conservation, and have touched on a host of other topics from forest taxation and forest technology to forest products and the forestry profession, I have focused on the relationship of Americans with their forest and how that major, visible biotic feature of the American landscape changed with the demands made on it and the uses to which it was put. Such interest in man–land relationships is the concern of the geographer and environmental scientist, but because those relationships have changed from place to place and from time to time, they are the very particular concern of the historical geographer.

Many persons have assisted in the making of this book since the idea behind it was first formulated over 12 years ago. I was encouraged to develop my original idea of the historical geography of forest clearing by the late Andrew H. Clark of the University of Wisconsin, Madison, when I visited the Department of Geography there on a sabbatical leave. At that time the idea of dealing with forest clearing was relatively novel, and there was not the wealth of material or interest then that there is now. Therefore, his encouragement and suggestions were doubly important. To him, David Ward, George Dury, and the entire Department of Geography at Madison, which extended such a warm welcome to me and my family, I am grateful.

Others have contributed in a variety of ways. Professor Paul W. Gates, Professor Norman Schmaltz, Marion Clawson and the officers of the Forest History Society, particularly the director, Harold K. Steen, and the former editor of the *Journal of Forest History*, Ronald H. Fahl, have all shared their knowledge and enthusiasm about various aspects of the topic. I owe thanks to the helpfulness of the staffs of the National Archives, Washington, D.C., and the State Historical Society Library in Madison. The Australian Research Grants Committee, the Oxford University Faculty Travel Fund, and the Provost and Fellows of Oriel College, Oxford, made it possible for me to visit the archives and libraries of the United States. I am also indebted to those who have typed various drafts of this manuscript: Sandra Stevenson, Moira Wise, and Elena Tiffert. Max Foale and Angela Newman skillfully drew many of the maps and diagrams in the book. The staff of the Press has shown unfailing courtesy and care, and I am particularly indebted to Barbara Palmer and Janis Bolster, who made this a better manuscript than it was.

I wish to thank the following organizations for permission to reproduce illustrations or extended passages of text:

British Library for Plates 3.2, 3.3, and 5.3

Chatto and Windus, Ltd., for quotations from "The Bear," from *Go Down Moses and Other Stories* by William Faulkner

Economic Geography for Figure 13.1

Forest History Society photography collection for Plates 7.2, 7.3, 9.1–9.3, 9.5, 10.1, and 14.3; and the following organizations for illustrations held in the Forest History Society collection: American Forest Institute, Plates 7.1 and 11.2; American Forest Products Institute, Plates 8.2, 8.4, 13.1, 14.1, 14.2, and 14.4; J. N. "Ding" Darling Foundation, Des Moines, Iowa, Plate 13.3; Lycoming County Historical Museum, Williamsport, Pennsylvania, Plate 11.1; Oregon Historical Society, Plate 9.4; U.S. Forest Service, Plates 7.4, 7.5, 8.1, 8.3, 9.6, 12.1, and 13.2

State Historical Society of Wisconsin for Plates 2.1 and 6.1–6.5

University of Wisconsin for Plates 1.1, 3.1, 5.1, 5.2, and 5.4–5.7

Finally, to Eleanore, Cathy, and Tess, who have had to live with this book for far too long, I give many thanks for their patience and encouragement. I only hope that the result has been worth the wait.

M.W.
Oxford

A note on tree, lumber, and plant association names

Throughout this book there are many references to tree and lumber names. The nomenclature adopted is the common or English plant name – for example Douglas-fir, ponderosa pine, hemlock, hickory, and so on – but occasionally the scientific or Latin name is also given, for example, white pine (*Pinus strobus*), longleaf pine (*P. palustris*), or baldcypress (*Taxodium distichum*).

It should be borne in mind, however, that many common tree names have other variants by which they are known locally. Thus the baldcypress is also known as the cypress, southern-cypress, swampy-cypress, red-cypress, yellow-cypress, white-cypress, tidewater red-cypress, and gulf-cypress. More simply, the red pine (*P. resinosis*) of the northern forests is also known as Norway pine, and Douglas-fir (*Pseudotsuga menziesii*) of the western forest is also known as Oregon-pine. An added complication is that lumbermen had commercial names for lumber derived from several species, and the lumber could be from any of them. At the simplest level, eastern hemlock included only two species: eastern hemlock (*Tsuga canadensis*) and tamarack (*Larix laricina*). More complicated and far more common was southern pine, which included nine species: loblolly pine (*P. taeda*), longleaf pine (*P. palustris*), pitch pine (*P. rigida*), pond pine (*P. serotina*), sand pine (*P. clausa*), shortleaf pine (*P. echinata*), slash pine (*P. elliottii*), Table Mountain pine (*P. pungens*), and Virginia pine (*P. virginiana*). Similarly, Red Oak included 16 varieties, and White Oak included 15 varieties.

Because the discussion in this book revolves more around the end products of the destruction of the forest – lumber – than around the trees themselves, the commercial lumber names occur frequently in quotations and comments. Thus the term "pines" (or "yellow pines" or "piney woods") when referring to the South can include any of the tree varieties and their lumber listed above. Similarly, "pines" or "pineries," when referring to the North, can include any of the three varieties in the lumber classification Northern Pine (e.g., jack pine, *P. banksiana*; red or Norway pine, *P. resinosa*; and pitch pine, *P. rigida*) and particularly the eastern white pine (*P. strobus*), which held such a prominent place in the commercial assessment of the forest.

For further details, see Elbert L. Little, *Checklist of United States Trees*, Agriculture Handbook no. 541, Forest Service, USDA (Washington, D.C.: GPO, 1979).

In addition to tree and lumber names, broader plant association names to describe the forest types are used, such as oak–hickory forest, loblolly–shortleaf forest, hemlock–Sitka spruce forest, and so on. These terms are derived from A. W. Küchler, *Manual to Accompany the Map of Potential Natural Vegetation of the Coterminous United States,*

American Geographical Society, Special Publication no. 36 (New York, 1964). A condensed commentary and smaller-scale map appear in *The National Atlas of the United States of America,* plate 89, U.S. Department of the Interior, Geological Survey (Washington, D.C.: GPO, 1970).

Abbreviations

b.f.	Board feet, a common measure for lumber; a board foot measures 1 ft. × 1 ft. × 1 in.
cu. ft.	Cubic feet (0.02832 meters); 12 b.f. equal 1 cu. ft.
m.b.f.	Thousand board feet
USDA	United States Department of Agriculture
USFS	United States Forest Service

Note: "Billion" throughout is the American billion (thousand million).

PART I

The forest before 1600

1

The forest in American life

> **No other economic and geographical factor has so profoundly affected the development of the country as the forest. It forms the background of our early history. . . . it enters into the everyday life of every American citizen.**
>
> **Raphael Zon, "The Vanishing Heritage," National Archives, GRG 95**

The immensity of the subject of the Americans and their forests owes much to the central role that the forest plays in American geography, economy, and culture. It is as large and complex a subject as the nation itself. The area originally occupied by the forest and the changes wrought by its removal play a large part in the evolution of the visual landscape and of the geographical organization of the country. For the development of the economy through the provision of building material and fuel and in the creation of agricultural land, the forest had no rival until the end of the twentieth century. The forest must also be accorded a high place in American cultural history as the words and phrases "woodsman," "frontier," "outdoor recreation," "environmental awareness," and "conservation" all bear out.

Each of these themes is explored briefly in this chapter, but it is the evolution of the space content, the creation of the visual scene, that is the main focus of what follows. It is well to remember, however, that in no place on earth can the geography be separated from the economy and ethos of the society that produced it. The hands and the minds of people have made the geography of any place as surely as have climate and relief and soils and vegetation. Therefore, the role of the forest in American economy and American culture cannot be overlooked.

American geography

Possibly the greatest single activity in the evolution of the rural landscape of the United States has been the clearing of the forest. In a subject that is bedeviled by a multitude of statistics that commonly run into millions and billions, and therefore are difficult to appreciate and compare, a few figures are basic. The land area of the coterminous United States is 1903 million acres. Of that land area it is calculated that between 822 and 850 million acres, or about 45 percent, were covered originally by well-formed "commercial" forests, which made the forest the most important, and certainly the most visually dominant, vegetation on the continent. Approximately one-fifth of that forest cover was located west of the Great Plains in the Pacific coastal states of Washington, Oregon, and California and those portions of the Rocky Mountain states not too arid to support trees, such as upland Colorado, Wyoming, and Idaho. Over four-fifths, the greatest bulk of the forest, lay east of the Great Plains. It stretched from eastern Texas through the South, to a lesser extent from the extreme eastern portions of Kansas and Nebraska through the center of the continent, and in great profusion and density from mid-Minnesota through the Lake States to the Atlantic Coast and New England. In fact, the land east of the Mississippi to the Atlantic Ocean was an almost unbroken expanse of forest, although, as will be pointed out later, local studies reveal a greater variation in the forest than has been appreciated up to now.

The existence of that original forest is striking enough, but its gradual and then rapidly accelerating denudation is equally impressive. Clearing had far-reaching results. One vegetation was replaced by another – usually cropland – but often by weeds and different trees. Sometimes clearing produced permanent changes in the soils, runoff, hydrology, wildlife, and a multitude of other ecological characteristics of the landscape. New patterns of settlement and transport arose in the cleared areas. New geographies were created, starting with the geography of Indian occupation and alteration of the forest and moving through the various geographies of European occupation, all of which created the landscape of much of the United States today.[1]

As a result of the agricultural clearing, lumbering, and other industrial and domestic impacts on the forests, it is estimated that by 1920 the original cover had been reduced to approximately 470 million acres. Of that remainder, only 138 million acres were original forest; 250 million acres were radically disturbed through grazing, cutting, and burning and did not sustain second growth; and 81 million acres were "wasted," that is to say, were nonrenewable and nonrestoring. By 1977, however, the story of destruction had changed to one of regrowth and birth. Trees have a remarkable capacity to grow if left alone. The commercial forest had grown to 483 million acres in extent and was still growing slightly in size and vigorously in volume. In addition, 250 million acres of noncommercial forest, that is, forest where the timber stand per acre is too low or its potential for growth is too small, also existed. In all, forests of some kind still occupy one-third of the land in the country.[2]

Precision in these sorts of figures is difficult. Initially, the forest was so common that no one bothered to write about it, let alone collect statistics, and the extent of the forest and the amount of clearing went unrecorded. The very use of the word "lumber" in North America for rough-cut wood and felled trees rather than "timber" was indicative of attitudes, for these materials were something that "lumbered the landscape"; they were useless and cumbrous. Standing timber was regarded as a waste material, of which there was an overabundance; in many places it had a value of less than zero. Bare land was worth more than land and trees. Even when statistics become increasingly refined, from about 1870 onward, the problem of determining regrowth and accompanying loss of cleared land that offered the opportunity for a second cutting (to say nothing of deciding what density of trees constituted a forest on the grassland edges of the prairies and elsewhere) adds to the complication of precise computation. The drawing up of a budget of the forests with debits and credits is fraught with all sorts of accounting difficulties.

In view of the importance of the effort to use and subdue the forest in the evolution of the landscape of the United States, it is surprising that it has not commanded more attention. But while visual change is going on, it is difficult to comprehend and appreciate: "Man gets accustomed to everything," wrote Tocqueville in 1831.

> He gets used to every sight. . . . [He] fells the forests and drains the marshes. . . . The wilds become villages, and the villages towns. The American, the daily witness of such wonders, does not see anything astonishing in all this. This incredible destruction, this even more surprising growth, seem to him the usual progress of things in this world. He gets accustomed to it as to the unalterable order of nature.[3]

The same sort of insensitivity to the imperceptible and gradual changes affecting the forest has typified the years since Tocqueville wrote. Barely anyone has attempted either to quantify or to synthesize the various impacts that have been made on the forest.[4]

The American economy

From the early seventeenth century to the early twentieth century the trees of the forests produced the most valuable raw material in American life and livelihood: wood. Unlike the western European littoral from where most settlers had come, America had enormous forests. Wood was abundant, it was ubiquitous, and consequently it was used prodigally in a multitude of ways. It entered into every aspect of life, quite literally from cradle to coffin, and formed an essential element in human needs that is difficult either to ascertain from statistics or to express with figures. When James Hall said in 1836, "Well may ours be called a *wooden country*; not merely from the extent of its forests but because in common use wood has been substituted for a number of most necessary and common articles – such as stone, iron and even leather," he was thinking of the multifarious human needs in the life and livelihood of the citizens of the country that were supplied by wood.[5]

A partial measure of wood in the economy is provided by the values of the products of major manufacturing industries provided by the census between 1850 and 1920 (Table 1.1). The lumber industry consistently outstripped almost all forms of manufactures other than meat packing and slaughtering, iron and steel, and flour and grist mills. In 1850 lumber was the second largest industry after the manufacture of flour and grist mills. By 1860 cotton goods had ousted lumber from second place, but in 1870 lumber rose again to be the second-ranking industry, accounting for 6 percent of national production of manufacturing, and was second only to the manufacture of flour and grist mills, which accounted for 10 percent of the national product of industry. By 1880 lumber had fallen to fourth place (5 percent) and by 1890 to fifth place (4.3 percent); it rose again to fourth place in 1900 (4.4 percent), then to third place in 1910 (5.6 percent), but plummeted to ninth place in 1920 when a major change became apparent in the manufacturing of the United States. For the first time shipbuilding, petroleum refining, and automobile manufacture and assembly entered into the industrial picture, and from then on all these exceeded lumber manufacture by value. The economy had changed markedly from the processing of raw materials to the fabrication and assembly of complex machinery. At

Table 1.1. *Manufacturing by major industry, 1850–1920: industries ranked according to percentage of total value of manufacturing*

Rank	1850	1860	1870	1880	1890	1900	1910	1920
1	Flour & grist (14%)	Flour & grist (13.3%)	Flour & grist (10.0%)	Flour & grist (10.5%)	Flour & grist (5.5%)	Iron & steel (6.2%)	Slaughter & meat pack. (6.6%)	Slaughter & meat pack. (6.8%)
2	LUMBER (5.9%)	Cotton goods (5.6%)	LUMBER (6.0%)	Slaughter & meat pack. (5.6%)	Slaughter & meat pack. (4.6%)	Slaughter & meat pack. (6.1%)	Foundry & mach. (5.9%)	Iron & steel (4.5%)
3	Cotton goods (5.9%)	LUMBER (5.3%)	Iron & steel (5.0%)	Iron & steel (5.5%)	Iron & steel (4.5%)	Foundry & mach. (5.0%)	LUMBER (5.6%)	Automobiles (3.8%)
4	Clothing (4.8%)	Clothing (4.3%)	Clothing (4.7%)	LUMBER (5.0%)	Foundry & mach. (4.4%)	LUMBER (4.4%)	Iron & steel (4.8%)	Foundry & mach. (3.1%)
5	Woolen goods (4.8%)	Leather goods (4.0%)	Boots & shoes (4.3%)	Foundry & mach. (4.0%)	LUMBER (4.3%)	Flour & grist (4.3%)	Flour & grist (4.3%)	Cotton goods (3.4%)
6	Leather (4.4%)	Woolen goods (3.7%)	Cotton goods (4.0%)	Cotton goods (3.9%)	Clothing (3.9%)	Clothing (men's) (3.2%)	Printing & publ. (3.6%)	Flour & grist (3.3%)
7	Liquor (2.3%)	Liquor (2.7%)	Woolen goods (3.6%)	Clothing (men's) (3.9%)	Carpentering (3.0%)	Printing & publ. (2.7%)	Cotton goods (3.0%)	Petroleums (2.6%)
8	Slaughter & meat pack. (1.0%)	Slaughter & meat pack. (1.6%)	Carpentering (3.1%)	Boots & shoes (3.7%)	Cotton goods (2.8%)	Cotton goods (2.6%)	Clothing (men's) (2.7%)	Shipbuilding (2.3%)
9	Agric. impl. (0.8%)	Agric. impl. (1.1%)	Foundry & mach. (3.0%)	Woolen goods (3.0%)	Boots & shoes (2.3%)	Carpentering (2.4%)	Boots & shoes (2.5%)	LUMBER (2.2%)
10	Paper & pulp (0.3%)	Paper & pulp (1.0%)	Sugar & molasses (2.6%)	Sugar & molasses (2.9%)	Lumber (planing, sashes, doors) (2.0%)	Woolen goods (2.3%)	Woolen goods (2.1%)	Railroad cars (2.1%)
Total value, all manuf. (s billion)	1.01	1.82	4.23	5.36	9.37	13.01	20.67	62.41

Source: U.S. Bureau of the Census, *Manufactures*, 1850–1920.

every census year the value of lumber produced rose, but the value of all other manufacturing industries rose even faster, thus relegating lumber to a position of less significance. In every case the number of wage earners engaged in the manufacturing industries bore a rough relationship to the importance of the industry by value of production, and sawmills and logging establishments were among the largest employers in the country, rising from an average number of 55,810 persons employed in 1849 to 547,178 in 1909.[6]

The census estimates of the value of manufacturing industry do not convey the importance of wood in the economy of the country. A whole technology based on wood existed in America from earliest times, as did a "society pervasively conditioned by wood," so that wood and wood products, as Hall observed correctly, entered into every walk of life.[7] To the census estimates of manufacturing, then, should be added the products of, for example, the wood planers, the packing-box manufacturers, the coopers, the tanners, the carriage makers, and the furniture makers, whose production during the late nineteenth century was at least as valuable as the products of the sawmills. The output of the shipbuilders and house builders who also used lumber has not been calculated. The vast majority of buildings (houses, mills, barns, warehouses) were made of either logs or cut planks, and wood was an essential element even of those made of brick and stone, to say nothing of the furniture within the house. Wood was the principal material in the transportation system, being essential in the majority of ships, riverboats and barges, carriages and railcars, bridges and railroad ties, in plank and corduroy roads, and even in road surface blocks and canal locks. It was the major material used in household, industrial, and agricultural implements and machines. In rural areas it was not replaced as the principal fencing material until the latter part of the 1890s.

Wood was the household fuel for at least two-thirds of the country as late as 1880, and for a quarter of households as late as 1920. It made the steam that drove the engines of the railroads, steamers, and factories well into the 1880s, and it continued to make the charcoal for a surprisingly large quantity of pig iron smelted in this century. The consumption of wood for fuel probably reached its peak in the decade of the 1880s, and even today the immense quantities of wood cut for fuel probably exceed the amount of wood cut for all other purposes.

Wood was also the source of many important chemicals such as potash, an early industrial alkali, and naval stores, which included pitch, tar, and turpentine, all essential ingredients in maritime transport. The bark was a source of tannin for the leather industry, and the sugar maple produced a sweetener.

Of course, nearly all these uses have substitutes today; concrete, steel, aluminum, and plastic have taken over in construction, transport, and utensils, and coal, oil, and electricity for energy and heat. But one has only to look around to realize how that first energy-producing and construction material of mankind still enters fully into our life, even in this last quarter of the twentieth century, and has even become the source of completely new uses such as pulp, plyboard, laminated boards, and panels. In the closing years of the twentieth century nearly as much wood is being used than was ever used before. Wood still accounts for one-quarter of all the industrial raw materials used in the country (Fig. 1.1). In 1972 3.3 million people were employed in its growing, processing,

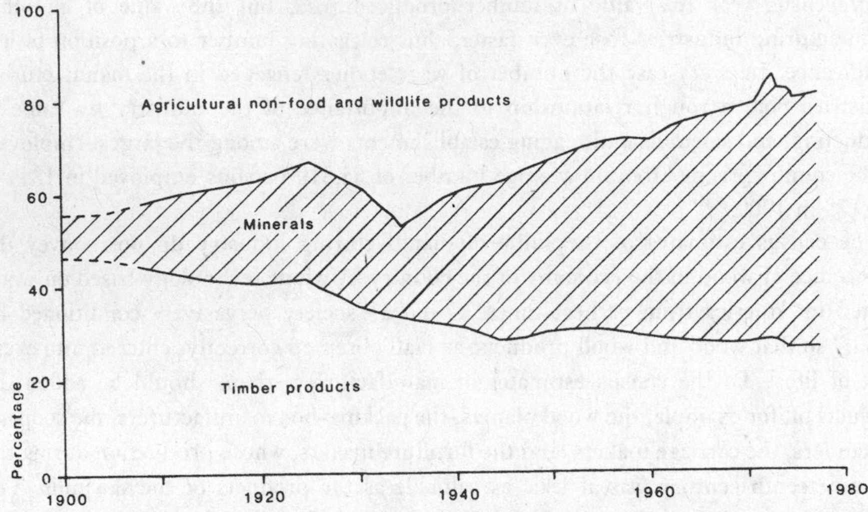

Figure 1.1. Wood as a percentage of industrial raw materials, 1900–80. (Based on USFS, *The Economic Importance of Timber in the United States*; and USFS, *An Analysis of the Timber Situation in the United States, 1952–2030*, fig. 1.1, p. 3.)

and manufacturing, and the value added through manufacturing amounted to $48.5 million (Table 1.2).[8] About 4.1 percent of the gross national product of $1,171.1 billion was attributable to timber-based activities.

Besides these material, fuel, and chemical uses, the forested lands had another value: When the forest was cleared, ground was left that was fit for agriculture and the grazing of livestock. Until about 1840 the agricultural prosperity of the United States and the expansion of the most widespread and common of all occupations – farming – depended upon the ability of the pioneer farmer to sustain a steady rate of forest clearing. After the mid-nineteenth century accurate statistics became available, and it is possible to calculate the increment of improved land created from forested land and added to the national total. During the decade 1850–9 it was nearly nine-tenths of 44.8 million acres, and the comparable figure for the next decade, 1860–9, was still over one-half of 39.6 million acres added.[9] The backbreaking task of clearing the land and making a farm was central to the geography and economy of the country until the later years of the nineteenth century.

In addition, the forest had other uses, less easily calculable in a material or monetary sense but no less valuable for that. As a watershed cover it evened out, equalized, and otherwise diminished extremes of runoff and so protected steep slopes against erosion and, by providing an even flow of moisture, facilitated irrigation, domestic water supply, and hydroelectricity production. It was the abode of wildlife, and it was, above all, the favored environment for recreation.[10]

No one has yet examined adequately the role of wood in promoting the rapid industrialization of the country, but it could probably be argued convincingly that the

Table 1.2. *Estimated value added and employment attributed to timber in timber-based economic activities, 1972*

	Value added		Employment	
Economic Activity	$ billion	%	No.	%
Timber management	2.9	6.0	117.2	4.0
Timber harvesting	3.1	6.4	190.4	6.0
Primary manufacturing	8.8	18.1	426.5	13.0
Secondary manufacturing	12.5	25.8	900.4	27.0
Construction	11.9	24.5	795.3	24.0
Transport and marketing	9.3	19.1	835.2	26.0
Total	48.5	99.9	3,265.0	100.0

Source: USFS, *Timber in the United States Economy, 1963, 1967, and 1972,* 5.

abundance of wood and the land that clearing created were the starting points for many important economic and social changes. Brooke Hindle hints at this when he says that it was during America's "wonderful Wooden Age" that the country began the "assent to industrial primacy and to the highest standard of living in the world." Lewis Mumford in his global and generic view of world technology distinguishes three ages of technology: the eotechnic, the paleotechnic, and the neotechnic.[11] The eotechnic was the dawn age of technology when water and wind and wood and stone were the motive and material components of life and industry. Certainly, in the emergent and underdeveloped economy of the United States during the late eighteenth and early nineteenth centuries, the country had an abundance of the rarest of those primary resources. Moreover, wood was visible, easy to work, and easy to exploit, unlike minerals, which were underground and needed much more complex machinery and organization for their exploitation. There was never any thought that the supply of wood was likely to run out, for in most minds, and in actual fact at that time, wood was an inexhaustible resource of unprecedented magnitude.

American culture

The forest environment created a variety of complex, and at times contradictory, myths and symbols, general ideas and emotions, which have fused to become part of the collective American moral and cultural imagination. Despite the often amorphous and imaginary nature of some of these ideas and symbols, they have been perceived to exist and thus have had an influence on practical affairs. Most important, they help to explain attitudes toward the forest at various times and also the changing relationships of Americans with their forest, which are underlying and basic themes running through the more detailed chapters of this book.[12] Some of the attitudes and ideas still have significance today, although the direct connection with the forest has long since been severed.

Attitudes toward the forest: to the late eighteenth century

Redemption. To the early American pioneers the forest was repugnant, forbidding, and repulsive. Some of those feelings and reactions had roots that went back a long way, into the culture of their ancestors in Europe, but were to be reenacted in a dialogue between the European pioneer's mind and the American environment. In Europe from classical times to postmedieval times society had developed a mythology and set of cultural mores in relation to the extensive forests of the Continent. The forests were the wild areas, alien to man and in need to felling, firing, grazing, and cultivating so that they could become civilized abodes. The forests were dark and horrible, places where there were very real dangers from wild animals, particularly bears and wolves. The word "wilderness" was almost synonymous with forest; etymologically, it was the "place of wild beasts." In addition, the forests were places of terrifying eeriness, awe, and horror, where the imagination played tricks and the limbs of the trees looked like the limbs of people, especially if animated by the wind. In that chaos the hapless peasant was first *be-wildered* and eventually succumbed to license and sin.

According to the folklore and tradition of European peoples, the forests were the abode of the supernatural and the fantastic. For the Greeks there were the genial dryads, but there were also the malevolent satyrs and centaurs; Pan was the Lord of the Woods, and his approach produced *panic* in unwary travelers. In medieval northern and central Europe the popular folk culture contained trolls, sprites, dwarfs, ogres, witches, werewolves, child-eating monsters, and demons of all descriptions. The fears embodied in these creatures have been handed down to us in the terrifying "fairy tales" of, for example, Hansel and Gretel, Red Riding Hood, Sleeping Beauty, Tom Thumb, the Three Little Pigs, and Snow White, all set in the environment of the dark, somber forest.[13] Early Christianity extolled the virtue of clearing the forest, which was equated with the devil and was the abode of natural sin. Yet if the forest was to be feared, its products were to be valued; piety and economic progress went hand in hand in the creation of "new" land that was akin to Paradise.[14]

As with so many cultural myths, there was a curious contradiction, even paradox, in these attitudes, because many holy men retired to the uninhabited solitude of an island or the depths of the forest to escape the corruption of society and to purify their souls. This tradition had strong roots in the Judeo-Christian experience of the wilderness, from the Exodus of the Israelites from Egypt, to Moses on Mount Sinai, to Christ's temptation on the Mount – all were exercises in freedom and the purification of faith. It was with such a varied background of cultural traditions, which had evolved from antiquity and were laced with biblical metaphors, that the first settlers arrived in the New World, which afforded such an abundant opportunity to extirpate the forest, reclaim the "waste," and create some sort of new Eden, Paradise, a New Canaan.

The immigrant Anglo-Saxon farmer of England, the ancestor of so many Americans, was described in the chronicles of about A.D. 800 as "the grey enemy of the wood," and within another 800 years it was the American pioneer on the eastern seaboard who reenacted this role, conquering nature, civilizing the landscape, and eliminating the abode of evil. On a physical level the forest was an insecure and uncomfortable environment that was a threat to his survival; on a symbolic level it was repugnant, being dark, sinister, and devoid of order. The settler had little time or value for the beauty and novelty of the untouched forest, which was worth contemplating only when it lay felled by the ax. His was a utilitarian view, and the forest was good only inasmuch as it became improved land or lumber or the site of settlements; "the forest stood in the way of progress."[15] As Alexis de Tocqueville wrote when describing frontier life in the Michigan forests in 1831, the pioneer "living in the wilds . . . only prizes the works of man. He will gladly send you off to see a road, a bridge, or a fine village. But that one should appreciate great trees and the beauties of solitude, that possibility completely passes him by." Generally, Americans were "insensible to the wonders of inanimate nature, and they may be said not to perceive the mighty forests that surround them till they fall beneath the hatchet. Their eyes are fixed upon another sight . . . peopling solitudes and subduing nature."[16] The Promised Land or Paradise that the Puritans had hoped to find was seen as covered by an obstacle that needed backbreaking work in order to convert it to the merest essential of living. The cleared patch, the made ground neatly fenced off, became a symbol of order and civilization.

The sinister and frightening image of the forest meant that it was portrayed as the "enemy" that needed to be "conquered." However, the dangers of the forest were real enough. It was the haunt of wild animals; few matters took up so much space in the records of the inland towns of New England during the early seventeenth century as the killing and maiming of livestock by wild animals. The terror of the forest was reinforced during King Philip's uprising when nature itself seemed to contrive to help the Indians in their attacks on the colonists so that, said Increase Mather, "Our men when in that hideous place, if they but see a Bush stir, would fire presently, whereby it is verily feared they did sometimes unhappily shoot Men instead of Indians." The Indians seemed to appear from behind trees to attack and then melt away into the darkness of the forest.[17] But the forest also had a symbolic meaning. It was the dark and sinister symbol of man's evil, where one was beyond the reach of redemption and where even a civilized man could revert to savagery if left too long. The fact that the Indian inhabited the forest and did not appear to clear it seemed proof enough of that.[18]

Social order and the Christian concept of morality seemed to stop on the edge of the clearing, and at that margin, without the ameliorating regulations of organized communities, social cohesion, and ties of positive Christianity, frontier man could become less civilized, degenerate into license, even into savagery. Hector St. John Crèvecoeur in his celebrated essay "What Is an American?" was quite explicit. If one wished to see America in "its proper light," one had to pass through the eastern coastal cities and the cultivated "middle settlements" to the "great woods, near the last inhabited districts." Here were the "off-casts" of society, where men "appear to be no better than carnivorous animals of a superior rank. . . . There, remote from the power of example and check of shame, many families exhibit the most hideous part of our society." In time that "hitherto barbarous country" would be "purged" by the next wave of decent settlers and changed into a "fine, fertile, and well regulated district."[19] The new immigrants to the American shore understood that

> There was only the dark forest, the secret home of the unknown beings. What was fact and what was legend in the minds of Europeans for whom the forests had been the forbidden place in a remote distance inhabited by the witches and ogres of fireside tales? They themselves were now to be swallowed up in the darkness to become themselves the beings of the woods. The awesome thought came to those who were alone; no reckoning of right or wrong could find them out here. . . . That was the horror.[20]

Clearing, therefore, was a form of redemption and, as it was for their European forebears, the destruction of the forest was uppermost in the minds of the first American pioneers in the making of their common wealth. In the clearings God could look down benevolently on their efforts to reestablish the order and morality that had gone out of life on departure from their native land and to counteract the evil that always threatened on the forest edge.[21] In emphasizing the hardships and temptation of the forest environment, the pioneers achieved two aims: First, the struggle with the forest tempered the spiritual quality of those who faced it, and they were better, humbler, men for the trial of clearing; second, it was a reminder to later generations of the magnitude of the accomplishment and the character of those who achieved it.

Above all, it was the size of the forest that astonished and frustrated the New World pioneers, especially after the experience of the transatlantic journey, which was often terrifying enough. The forest was impersonal and lonely in its endlessness; consequently, clearing the forest was likened to a battle or struggle between the individual and the immense obstacle that had to be overcome in order to create a new life and a new society. The image of the heroic struggle to subdue the sullen and unyielding forest by the hand of man, and to make it something better than it was, was a legacy of feeling, thought, and imagery that was handed down over the centuries. For Francis Parkman in 1885 the forest was "an enemy to be overcome by any means, fair or foul," and Frederick Jackson Turner, the inheritor and brilliantly successful interpreter of these deeply held images, echoed these ideas and attitudes in his account of early pioneer life. The frontier was the ultimate American symbol and the place where the pioneer created the world anew (Plate 1.1).[22]

Progress. Though the clearing of the forest had an almost sacred motive, increasingly it came to have a secular one. The concepts of progress, development, and ultimately civilization itself had never been far beneath the surface, even from the beginning of settlement, because they were the logical outcomes of the fight for survival. From the eighteenth century onward, however, the concept of controlling nature and of making it more useful gained strength. The ideal was the rural, domesticated, agrarian landscape. It was a landscape, said Crèvecoeur, that had been converted from forest to "fair cities, substantial villages, extensive fields, an immense country filled with decent houses, good roads, orchards, meadows, and bridges where a hundred years ago it was wild, woody,

Plate 1.1. The beginning. (*One Hundred Years of Progress in the United States.*)

and uncultivated.[23] Man was the agent of change, and as his settlement spread across the country, the fusion of industry and natural processes produced the increasingly sought-after "made" landscape.

Writing about his life in the northwestern portion of New York State in the late eighteenth century, William Cooper declared that his "great primary aim" was "to cause the wilderness to bloom and fructify."[24] Because of these and similar attitudes, the landscape of "unimproved land" was a desirable aim. From 1860 onward and without interruption to the present day, these two categories of land have been recorded in the censuses, the amount of "improved land" being one of the main criteria of material progress in rural America. To fell the forest was almost to enter the kingdom of heaven on earth, as the making of new land seemed to demonstrate the direct causal relationship between moral effort, sobriety, frugality, and industry and material reward. Increasingly, American nature seemed full of implications for ethical and material betterment. Very early in the eighteenth century, Franklin identified the frontier of cultivation with opportunity and "tended to measure moral and spiritual progress by progress in converting the wilderness into a paradise of material plenty." Half-playfully, he even attributed a cosmic influence to the clearing of the forest: "by *clearing America* of Woods" Americans were "*Scouring* our Planet . . . and so making this Side of our Globe reflect a brighter Light to the Eyes of the Inhabitants of *Mars* or *Venus*." The link with virtue was pointed out explicitly in Andrew Jackson's second annual address, in which he asked rhetorically,

> What good man would prefer a country covered with forests, and ranged by a few thousand savages to our extensive Republic, studded with cities, towns, and prosperous farms, embellished with all the improvements which art can devise or industry execute, occupied by more than 12,000,000 happy people, and filled with all the blessings of liberty, civilization, and religion?[25]

If the forest and the Indian were swept aside in the process, then so be it. Not only progress was good; it was also inevitable.

Attitudes toward the forest: early nineteenth century

Though piety and activity remained important elements in man's relationship with the forests of America, new themes began to gain ground from the end of the eighteenth century onward. These themes can best be described as the romantic and the patriotic, and from these new points it became increasingly possible to praise the forest, which formerly had been held in such disgust, and even to advocate its preservation for future generations.

The romantic. Generally, hostility and repugnance toward the forest remained because the bulk of the population was still confronting it and clearing it, but the forest began to find new champions among those who found aesthetic values in it and even associated its primitive, primordial condition with the works of God, which, if we consider all that had gone before, was an intellectual revolution. Like most other intellectual revolutions, this changed attitude was not a sudden eruption of ideas; rather, it was a gradual fermentation of ideas that had been current for a long time but were now formulated and articulated with greater clarity.

The medieval monks had found freedom and achieved purification from the corrupt world in the depths of the forests, and religious significance had been attributed to groves by their pagan predecessors. Therefore, it was not surprising that deism, the association of God with nature, should find its advocates. The advances of eighteenth-century science had produced a new wonder at and awe of Nature, and the feeling that God showed his power and excellence in such untouched environments as the forest was an important change in attitude, particularly in American pioneer society in the late eighteenth century where the achievement of an ordered landscape was both a duty and a source of pride. Moreover, as the proponents of the picturesque began to suggest, nature's roughness had a certain pleasing quality about it.[26]

Others went further: Wild nature was sublime, and disordered, chaotic scenes, be they mountains or forests, could please and exalt just as easily as the comfortable, well-ordered landscapes of the cleared and made land. If God and beauty could be found in the forest, then would not man be more perfect if he were in touch with that environment? From Daniel Defoe's *Robinson Crusoe* (1719) to Jean-Jacques Rousseau's *Emile* (1762) there had been the suggestion that the primitive life, despite its known disadvantages, produced a happiness and wholesomeness that were not to be found in the cities, or even in the rural, agricultural landscapes, which, when all was said and done, were still man-made landscapes.[27]

Deism, sublimity, the picturesque, and the primitive fused in the romantic movement with its emphasis on the strange, the solitary, and the melancholy. Romanticism initially found its greatest and most articulate champions in the poets Byron and Wordsworth, who exercised a stimulating influence on their American contemporaries.[28] An early expression of romanticism in America was in the writing of Chateaubriand, a European, who was seized with "a sort of delirium" when he found so little evidence of civilization in his travels during the winter months of 1791–2. In contrast to the situation in Europe, the imagination "could roam . . . in this deserted region, the soul delights to busy and lose itself amidst the boundless forests . . . to mix and confound . . . with the wild sublimities of Nature." But there were indigenous evocations of the forest; successive topographical writers, diarists, and natural scientists, such as Bartram, Michaux, Flint, Audubon, and Kalm, began to discover that the forest could be regarded with something other than hostility and that its vastness and variety were sources of wonder and awe.[29]

As Roderick Nash has pointed out, before the end of the eighteenth century, as few Americans had discovered primitivism. In 1781–2 Philip Freneau had published *The Philosopher of the Forest*, and his hermit was able to say, "There is something in the woods and solitudes congenial with my nature"; and in 1800 Benjamin Rush wrote in much the same vein. Both extolled the simple, moral life of the forests, which acted as a corrective and a balance to the distorted, crowded, unsympathetic, and unfeeling life of the city dweller. That most of the "primitives" were city dwellers was significant for they did not face the forest from the pioneer's viewpoint. Their experience was divorced from the harsh realities of the backbreaking struggle of the pioneer's life, and their sorties into the forests were calculated, limited, and therefore regarded as pleasant interludes in an otherwise civilized urban existence.[30] Increasingly, primitivism and romanticism became a slightly decadent cult, the hallmark of the well-educated gentleman. Francis Parkman, the historian, said that from his early youth "his thoughts were always in the forest, whose features possessed his waking and sleeping dreams, filling him with vague cravings impossible to satisfy." Certainly, by the 1830s this gloomy sentimentalism was a literary genre.[31]

The evocation of the simple and the solitary was too much for most early-nineteenth-century Americans, however, and they were content for virtue to arise out of the rural landscape, admittedly a man-made landscape but at least one that was halfway along some continuum between city and savagery and one that achieved a balance between industry and agriculture. The "middle landscape," as Leo Marx calls it, occupied a peculiarly important position in the American heart and mind because it was the visible evidence of progress, expansion, moral effort, and industry – it was the "Garden" and the "Paradise" – and yet it was still natural enough to produce feelings of spiritual renewal and simplicity. When William Penn wrote in 1688, "The country life is to be preferred for there we see the works of God, but in the Cities little else but the works of man," he was extolling the virtues of the made landscape and the rural pioneer life – its simplicity, sobriety, frugality, tranquillity, freshness, and the regenerative powers of the cycles of nature; he was not extolling the qualities of the pristine forest.[32] It was no accident that two other major figures in American history and pioneer life expounded similar ideas about the agrarian ideal. In America, Thomas Jefferson wrote, "We have an immensity of land courting the industry of the husbandman." Why, then, turn to manufacturing, especially since "Those who labor in the earth are the chosen people of God, if ever He had a chosen people, whose breasts he has made his peculiar deposit for substantial and genuine virtue"? For Benjamin Franklin, agriculture was the only "honest" way of acquiring wealth, and, moreover, it kept people virtuous and morally righteous. Undoubtedly, the spectre of the growing power and complexity of industry and mechanization, and the intrusion of the urban into the rural during the opening years of the nineteenth century, perturbed those who saw the rural ideal being destroyed and the beginning of disharmony between man and nature. Washington Irving summed up the feeling well in *The Legend of Sleepy Hollow*, written in 1820:

> I mention this peaceful spot with all possible laud; for it is in such little retired . . . valleys that population, manners, and customs remain fixed; while the great torrent of migration and improvement, which is making such incessant change in other parts of this restless country, sweeps by them unobserved. They are little nooks of still water which border a rapid stream.[33]

The patriotic and aesthetic. The romantic movement gave the forests a new meaning for some people, and this admiration for what had once been rejected was bolstered by yet another change of attitude, which can best be called the "patriotic." After the War of Independence the question was continually asked, "What was it in this new country that was distinctively American?" The continent, with its short history and ill-formed traditions, could not produce anything like the rich cultural heritage and the antiquities of Europe. One thing that America had, however, was vast areas of untouched land – forest, prairie, and mountain – and these seemingly unending wild areas were perceived by nineteenth-century naturalists, poets, writers, and artists as something uniquely American and something about which to be proud. Chateaubriand touched upon this feeling when he said, "There is nothing old in America excepting the woods. . . . they are certainly the equivalent for monuments and ancestors." The naturalists and diarists had led the way in spreading "a proper feeling of nationality," but the "boundless," "trackless," "incomparable," and "fresh" forests also became the object of literary and artistic attention and pride. The writing of Washington Irving, the poems of William Cullen Bryant ("A Forest Hymn," 1825) and James Kirk Paulding ("The Backwoodsman," 1818), and above all, the Leatherstocking novels of James Fenimore Cooper, written between 1823 and 1841, produced a new appreciation of the forest landscapes of the country and a pride and confidence in the qualities of the American scene.[34] The literary appreciation was paralleled by the visual appreciation, and Thomas Cole, in particular, the leader of the Hudson River school of painting that flourished during the 1830s and 1840s, reveled in the wildness of American scenery (of the forested Catskills in particular) and indicated it as a subject worthy of study and reproduction.[35]

The new appreciation, let it be admitted, was not wholly affirmative. Feelings of disgust and fear of the forest still come through, and feelings of pride and satisfaction in the pioneer endeavor to clear and settle the forests were also strong. After all, the process of subjecting the wild landscape to the plow produced the self-dependence, simplicity, manliness, and neighborliness of the American people. This ambivalence about the forests was rarely absent. Even Emerson, 11 years after he had written, "in the woods we return to reason and faith," could still say:

> This great savage country should be furrowed by the plough and combed by the harrow . . . these rough Alleganies should know their master, these foaming torrents should be bestridden by proud arches of stone; these wild prairies should be loaded with wheat; these swamps with rice; the hill tops should pasture sheep and cattle; the interminable forests should be graceful parks for use and delight.[36]

By the mid-nineteenth century the significance of the uniquely American character of the continent's scenery took another twist with the writing of Emerson and particularly of Thoreau. They both expounded the transcendentalist philosophy that the experience of nature in general, but of the forests in particular, produced a higher awareness and sense of reality than did one's physical surroundings, which were dominated by expansion and exploitation, particularly in the cities, which were "great conspiracies." In other words, nature mirrored God's higher meaning and was not only aesthetically pleasing but positively and rationally beneficial. Unlike the Puritan pioneers who thought that morality stopped on the edge of the clearing, the transcendentalists thought it began there, for man was inherently good and not evil, and perfection could be maximized on entering the forests. By halting stages the argument went further. If the forest and the other wilderness areas were uniquely American, and if God's purpose was made more manifest in such places, then the very spirit of America and its creativity could be found in the forests, from whence came, said Thoreau, "the tonics and barks that brace mankind."[37] Here, perhaps, were some of the intellectual seeds of Turner's frontier hypothesis, that the "frontier," the "cutting edge" of civilization as it faced the "untouched forest," prairie, or mountain, was the crucible in which were forged the American traits and institutions of self-reliance, industry, and democracy, as people fled from the uniformity and degeneration of the urban areas. Quite simply, for Turner, "American democracy was no theorist's dream . . . it came out of the American Forest."[38]

Thoreau's philosophy had one other effect. He believed that wilderness without the benefits of civilization was detrimental to mankind; therefore, the optimum condition for man was one in which he could participate and keep in contact with both ends of the spectrum. With this view of the binary requirements of man, Thoreau gave the traditional American idealization of the "rural" or pastoral environment a new respectability.[39]

Attitudes toward the forest: after the mid-nineteenth century

Economic and environmental concern. Most Americans thought that the forest and its timber were limitless, but even as early as 1745 Benjamin Franklin was lamenting the scarcity of fuel that had formerly been "at any man's door." Now it had to be brought nearly 100 miles to the large coastal towns, and at considerable cost. By the late eighteenth century the situation was acute, particularly in the urban areas, if the winter was excessively prolonged and cold. By 1840 deficiencies in constructional timber also became evident on the eastern seaboard, and by 1870 New York State was importing from beyond its boundaries over 1 million tons of timber annually.[40]

After the mid-nineteenth century, answers to the problems of deficiencies and high costs were sought in the development of better communications, in order to bring timber to the areas of consumption from the untouched forests farther afield, and in mass production methods to keep down costs. But by the latter years of the century there was also a realization that some form of silviculture was desirable, that is to say, management of the existing forest stand to maximize yield.[41]

Concern over the growing scarcity of timber and fuel was coupled with another worry: What was clearing doing to the land itself? Many late-eighteenth- and early-nineteenth-century writers regarded the New World as a "great outdoor laboratory" for observing scientific changes, for controlling nature and for making it useful.[42] But it remained for George Perkins Marsh to combine the experience of his youth in Vermont, where he saw forest clearing and erosion at first hand, with the well-documented observations of Europe (where he lived in later years) to produce his brilliant synthesis, *Man and Nature; or, Physical Geography as Modified by Human Action*, in 1864, more than one-third of which was devoted to forests and forest influences.[43]

Earlier, in 1847, Marsh had addressed the Rutland agricultural fair. After praising the effort of the pioneer farmers in changing the forest to farmland, he went on to say: "the increasing value of timber and fuel ought to teach us, that the trees are no longer what they were in our fathers' time, an encumberance." But he went beyond economic concerns to elaborate environmental concerns. Already too much of Vermont had been cleared, and every middle-aged man who returned to his birthplace after the interval of a few years would see that "the signs of artificial improvement are mingled with the tokens of improvident waste," such as erosion, ravines, and dried-up streams.[44]

Marsh went beyond the concerns of his contemporaries about conserving wood and increasing yield to consider the wider implications of clearing and of the influence of the forest. He emphasized the destructive powers of man's everyday activities, which upset the "harmonies of nature," and he suggested that wise management could mitigate some of these problems. It was a big step in American and, indeed, world thinking.

The idea that nature was an organic Whole and that the forest influenced other facets of nature had far-reaching effects. The settlers moving into the treeless plains of states such as Kansas and Nebraska during the 1870s wanted to plant trees and increase rainfall. The settlers and irrigators of the arid West, the flood controllers of the East, and those concerned with erosion everywhere wanted positive forest policies. Increasingly, people regretted and resented the devastation wrought by mass production, large-scale lumbering, particularly in the forests of the Great Lakes, which left worthless cutovers. The disquiet about the probable effects of human interference on the influence of the forest on the environment as a whole became a major factor in the public acceptance of forestry.[45]

Preservation and management. In his voluminous writing Thoreau constantly decried the felling of the forests, and in *The Maine Woods*, written in 1858, he talked of "our natural preserves" and of the necessity of retaining a portion of that natural environment for posterity. But the concept of preservation was taken one step further by Marsh, who in a later edition of *Man and Nature* in 1874 saw preservation not only as a benefit by nature to man but also as a benefit by man to nature. It was a wider concept of preservation than had been held before, and it emphasized the reciprocal bonds that tied mankind to the environment. To that end he advocated that a "large and easily accessible region should remain, as far as is possible, in its primitive condition." It was the beginning of conservation, and it started in the forest.[46]

From then on the movement to protect that national heritage became intense. There was a growing commitment by the government to preserve and manage forested catchments and other forest areas, with the creation, in time, of federal forest reserves, not only for economic purposes to ensure a supply of timber but also to provide areas of recreational, scenic, and aesthetic value and, later, even to preserve areas of "wilderness." By about 1890 the appreciation of the forest and other wild places had passed from being the concern of a small and articulate group of writers, artists, poets, and politicians to becoming a national cult. For many Americans, who were now urban dwellers, the primitive conditions of nature no longer impinged on their lives and were no longer to be feared. The frontier had passed, and the wild landscape no longer repelled. Such landscapes were now sought out actively, and they could be seen in comparative comfort by the vacationer. In the popular imagination they were now imbued with attributes all of which were good.

Paradoxically, however, the horror of the forest was transferred to the ever-expanding urban areas, which became "concrete jungles" and "asphalt jungles," godless and physically dangerous places where it was possible for one to lose one's identity, morality, sensitivity, and even one's life.[47] An intellectual process was occurring in which the values attributed to the forest and to civilized places were being exchanged. The poles had been reversed.

National symbols

The intellectual and philosophical attitudes toward the forest were based very largely on the type of person that that environment might nurture and produce. In time, that person became an emotionally powerful symbol, embodying the qualities the nation thought it had or, at least, hoped it had. It would probably be fair to say that the pioneer who hacked out his holding from the forests for himself and his family gradually began to epitomize the ideal from at least revolutionary times onward. He was hardworking, frugal, sober, independent, resourceful, God-fearing, and democratic. He was simple in his wants and demanded only a moderate prosperity, free from oppression. He was the very stuff from which the nation was built. All his characteristics were nurtured and encouraged by his forest environment, which was God-given: For Thoreau, "Adam in Paradise was not so favourably situated on the whole as is the backwoodsman in America."[48]

Not untypical was the description of the "noble backwoodsman" by Anthony Trollope, the author of such genteel novels as *Barchester Towers* and the like. It was a composite picture of people he had met many times on his journey from Niagara west to Wisconsin. The pioneer, he said, "goes to work upon his land amidst all the wilderness of nature. He levels and burns the first trees, and raises his first crop of corn amidst stumps still standing four or five feet above the soil." He was "rough," but he was fiercely independent; he had "no love for his own soil" but would sell and move on if the price was right.[49]

The ax became an emblem of independence and sturdiness just as much as did the rifle, and the backwoodsmen became a national symbol, even a stereotype. It did not matter who the backwoodsmen were in real life; all that mattered was what they became in the imaginary life and collective fantasy of the people who cherished them.

Yet, the sober, responsible, and hardworking characteristics of the backwoodsman's life, though appealing to the patriotic and moral emotions of the nation, could in no way be regarded as "popular," with the exceptions of the characterization of Natty Bumppo in the Leatherstocking novels of James Fenimore Cooper, the accounts of the archetypal pioneer Daniel Boone (and of his popularized counterpart Davy Crockett), and some of the novels of Nathanial Hawthorne set in the New England forests.[50] Perhaps people knew the reality of the hardship of backwoods life well enough for the myth not to become popular; nevertheless, by the time the pioneer woodsman and his forest-bred virtues could be popularized in mass literature for the enjoyment of the ever-growing urban society of the East during the second half of the nineteenth century, general attitudes toward the forest were undergoing radical change. Partly as a result of the westward movement of settlement out of the eastern forests and into the Plains, and partially because of the widespread commercial exploitations of the forest for mass-production lumber, two new national characters, or symbolic stereotypes, emerged: the cowboy and the lumberman.

Both the cowboy and the lumberman were as resourceful and individualistic as the backwoodsman, but both were more colorful. Obviously, the forest environment of Natty Bumppo was doomed to reduction or even to elimination with the expansion of agricultural clearing, farming, and industry. The simple egalitarian code of the backwoodsman could not survive these advances, or the restrictions, wealth, and organized institutions that the new society imposed. Perhaps it was because of the inevitable demise of the backwoodsman that the cowboy and the lumberman emerged as popular figures. The cowboy was by far the most important and popular. He was still something of a man in nature, preserved from the corrupting influences of the big centers of population. He moved freely over large distances, and "coupled with a handy skill with a gun, and an innate sense of common justice . . . was a natural equivalent of the early man in nature. [The cowboy was] the last symbol of a passing geographical and consciously felt frontier." Nevertheless, independent, free, and relatively uncorrupted as the cowboy was, he lacked the sacred qualities of the backwoodsman. The cowboy was a secularized hero.[51]

If there was any romance in lumbering, the population had an ambivalent attitude toward it. Rugged and tough as he was, the lumberjack was rarely seen as an individual in tune with nature or as one pitting himself heroically against nature. He was usually the employee of a large and growing capital-intensive industry, a mere cog in the new industrial system of ever-expanding production and consumption, and that was hardly heroic.[52] Moreover, the lumberjacks and their bosses – the "lumber barons" – were perceived to be stripping the country of a moral resource and an object of national pride, and that caused animosity.[53] But the products of lumbering were so necessary for everyday living that, paradoxically, the lumbermen and their activities had to be tolerated by the public and federal government alike, and frequently they were praised for acting in the public interest. There was no better illustration of this tolerant and even friendly relationship between government and industry than the working relationship that sprang up between the two because of shared objectives "to improve the effectiveness, efficiency and stability of the existing order: or, in the rhetoric of the industry spokesmen, to work in the national and public interest."[54] The dilemma of the "machine in the garden," so ably explored by Leo Marx, was abroad in the country in a new way: Steam and steel in the form of logging railroads, skidders, and high-line yarders were making massive inroads into the forests, and more than anything else it was the action of the lumbermen that fostered concern about the dearth of supplies and the destruction of the forest environment, with all its influences, real and imagined.[55]

In their different ways, the backwoodsman with his ax and the lumberjack with his saw were cultural symbols, and both were characters who were heavily charged with sentiment and emotion as well as loaded with fallacy and idealization. The imagery that they and the forests evoke goes deep in American culture and pervades and underlies the whole story of man's changing relationship with the forest. It is also an imagery that politicians and public speakers still use when they invoke and exhort the desirable aspects of national character, cohesiveness, and enterprise. To imagine an America without its forests is to imagine another world; the forest is inextricably a part of American life, livelihood, and landscape, in the past, in the present, and almost certainly in the future.

2

The forest and the Indian

> ... these nations [Indians] have long practised agriculture, and are not to be charged with an incapacity of providing for the future or with an absolute carelessness of their posterity.
>
> Johann David Schoepf, *Travels in the Confederation (1783–1784)*, 183

> Our eastern woodlands, at the time of white settlement, seem largely to have been in process of change to park lands.
>
> Carl O. Sauer, "The Agency of Man on Earth," 55

The plants and people of the New World were objects of curiosity and awe to the explorers and to the settlers who came after them. The curiosity arose because both were different from what was known in Europe; the awe, because they were evidence of the creativity of God in providing a varied, rich, and vastly more habitable world than had hitherto been suspected.[1]

Gradually, however, quasi-theological wonder at the new discoveries, and the "mythical atmosphere" of the New World engendered by thoughts of inexhaustible treasures,[2] gave way to the realities of the land and to the patient observation and recording of the obvious contrasts offered by the New World and the Old. From about the early seventeenth century onward, knowledge built up slowly about the extent and composition of the forest and of the life and economy of its earliest inhabitants. Curiosity, romance, and theology gave way to botany and ethnography as the Europeans learned about the forest and the Indians.

The forest

Nearly every traveler and explorer commented on the strange vegetation encountered in the New World and sent back tedious lists of plants and trees to which they gave British names. For example, Capt. John Smith in his *Description of New England* (1616) commented on the prevalence of "foure sorts of okes," and Thomas Morton in his *New English Canaan* (1637) found

> Oakes ... of two sorts, white and redd; excellent tymber for building both of howses and shipping.... Ashe ... very good for staves, oares or pikes.... Elme, ... Beech ... of two sorts, redd and white; very excellent for trenchers or chairs.... Wallnutt ... an excellent wood, for many uses approoved.

His list went on to include pine, cedar, cypress, spruce, alder, birch, maple, hawthorn, and many more.[3]

The work of observation, identification, and recording was placed on a new footing after the mid-eighteenth century when Linnaean taxonomy enabled vegetation to be identified precisely. Pehr Kalm, a Swedish pupil and protégé of Linnaeus, traveled extensively in northeastern America between 1748 and 1751 and identified upward of 50 previously unknown trees.[4] A more comprehensive knowledge of American flora was built up as a result of the extensive travels of William Bartram and his son John during the last quarter of the eighteenth century. They found and identified seeds, plants, and cuttings and exchanged them with English correspondents. The process was capped brilliantly by the exploits of André Michaux and his son, François André Michaux, who in 1785 were sent to America by Louis XVI in order to discover which trees could be shipped back to France to grow profitably in its forest-depleted areas. For seven years both assiduously collected seeds and plants, most of which never reached Europe, due to a shipwreck. However, André's more lasting contribution to botanical knowledge was the publication of his *Histoire des chênes de l'Amérique septentrionale* (1801) and his *Flora Boreali-Americana* (1803), the first systematic study of American flora, published posthumously. After the death of his father, François André returned to North America, continued his work, and in 1810 published his massive and authoritative *North American Sylva*, which remained the most comprehensive account of the forests east of the Mississippi for more than a half-century.[5] In later years Thomas Nuttall compiled a supplement to François André Michaux's *Sylva* (1842–9), which included the results of his own investigations in the western forest and those of other explorers and botanists, such as David Douglas.[6]

In contrast to the slow but steady growth of knowledge about individual species, early ideas about the composition and extent of the forest as opposed to the trees were vague and generalized. Typical was Count Volney's concept in 1807 of America as an "almost universal forest" divided into "northern," "middle," and "southern."[7] However, with the accumulation of botanical knowledge about the interior and the West by explorers, settlers, and government surveyors during the years up until the middle of the century, a picture began to emerge of American forests and their vegetation as a whole. Important in the expanding interest in plant assemblages, their relationship to the environment, and the suitability of the land they covered for cultivation was the encounter for the first time in Illinois of extensive prairie lands. It was not that prairies were a new thing to settlers who had experienced the "openings" and "savannas" of Ohio, Indiana, and Kentucky, but clearly, the Big Prairie and its climatic environment were something very different.[8] Thus, in 1858, Professor Joseph Henry, secretary to the Smithsonian Institution, published an article in the report of the commissioner of patents, "Meteorology and Its Connection with Agriculture," which included the first comprehensive map of the forest and prairie lands of the United States (Fig. 2.1). It was a fairly simple delineation of two types of prairie – dry (short grass) and arable (tall grass) – and three types of forest – foliaceous, coniferous, and deciduous – but it was remarkable in that it was probably the first attempt to map the vegetation of the continent as a whole. Unfortunately, no comments were made by Henry about the map so that contemporary opinion about the distribution shown is not available to us.[9]

Two years later, the Smithsonian Institution provided the backing for the publication of another and more ambitious map, this time compiled by James G. Cooper. Cooper had been an active natural history observer and collector for most of his

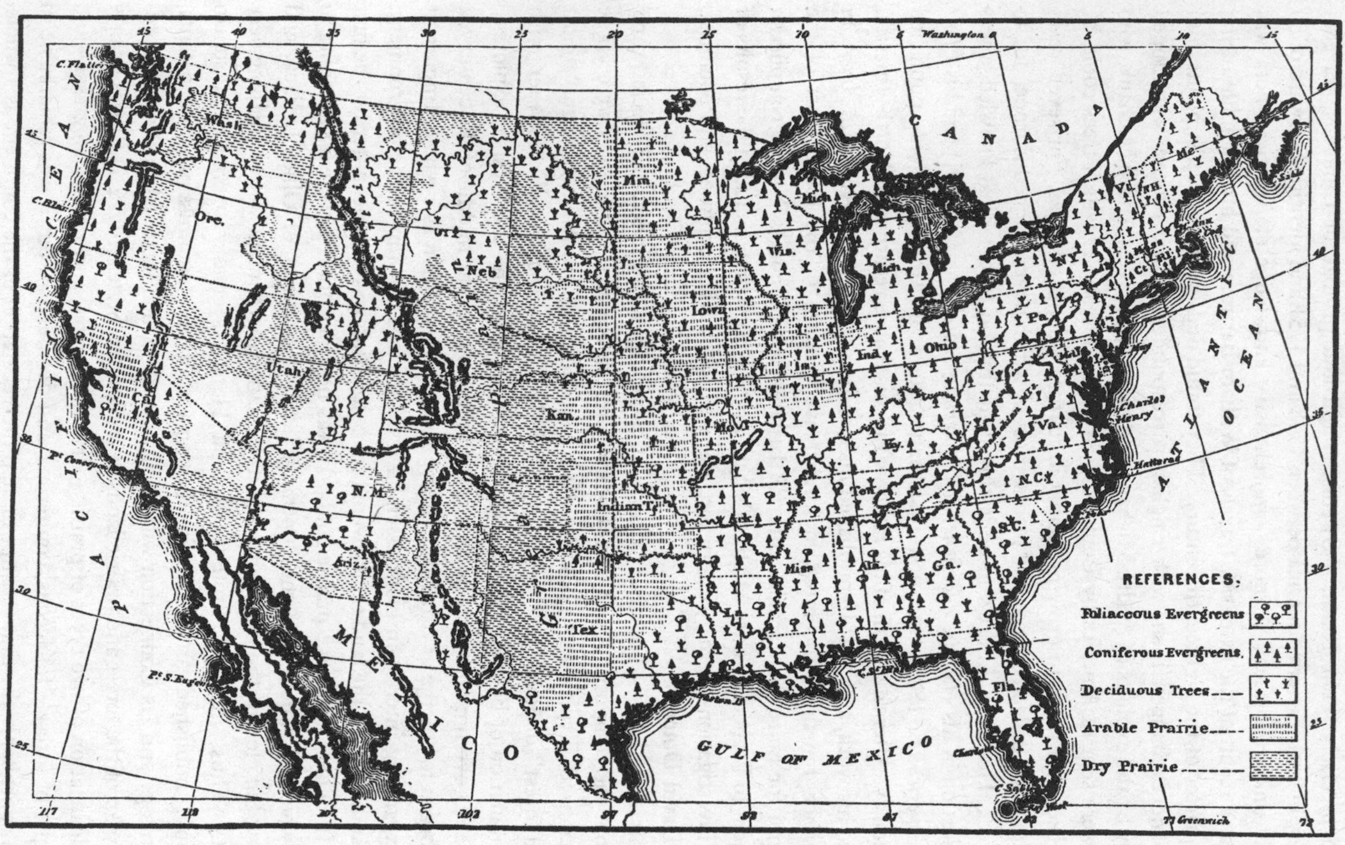

Figure 2.1. "The forest and prairie lands of the United States" (1858), by Joseph Henry. (Joseph Henry, "Meteorology and its connection with Agriculture," plate 6.)

life and had been employed in a number of expeditions to the West. Back in Washington in 1855, he began to contemplate the making of "a complete map" of "botanical and zoological regions" of the Pacific Northwest based upon his copious notes and observations. Whether that particular map was compiled is not known, but on his return to the East Coast in 1858, and under the auspices of the Smithsonian, he prepared a paper, "On the Distribution of the Forests and Trees of North America." Its 34 pages contained not only a catalogue of native American trees but, more importantly, Cooper's discussion about the "range" of individual trees and of "the extreme points to which each species extends in every direction," which were, in his opinion, controlled largely by temperature and rainfall. He thus summarized the current understanding of the distribution of the forest and its relationship to the continental environment as it stood at about midcentury, and his catalogue was, as he intended, a review and up-to-date version of the work of the Michaux. Two years later Cooper completed another paper, "The Forests and Trees of North America, as Connected with Climate and Agriculture," which included much of the old material but incorporated new findings into an amended version of the map (Fig. 2.2). It had an extended discussion about the "capacity of the regions" for the cultivation of particular crops and contained a speculative discussion about the problem of forest succession in the various provinces.[10]

By the late 1860s the massive destruction of the eastern and Great Lakes forests for lumber and the rapidly accelerating rate of agricultural clearing made people concerned about the *amount* and density of the forest. The question "Will the land grow food?" was now accompanied by the question "How much forest is left?" In addition, the possibility that humans could modify and even destroy their "God-given" habitat seemed starkly real after George Perkins Marsh had put man firmly into the center of factors influencing the environment with his book, *Man and Nature; or, Physical Geography as Modified by Human Action*, published in 1864. The long-suspected influences of the forest environment on rainfall, runoff, shelter, temperatures, and erosion were no longer in doubt.

The Bureau of the Census took seriously the concern about the depletion of the country's major resource and allocated a section on "Woodland Forest Systems" in Francis A. Walker's *Statistical Atlas of the United States*, which came out as a companion volume to the 1870 census. The map was compiled by William H. Brewer, a Yale botanist, who in his accompanying commentary noted that "no published map of any considerable area in our country is known to us in which the woodlands are laid down from actual surveys." The land office plats were numerous and extensive, but they were scattered and never assembled into a map of one region, let alone one map of the country, as had been, for example, the information on mineral deposits. The resultant map was clearly more detailed than the efforts of Henry or Cooper a few decades before, and it was an ambitious attempt to understand not only the distribution but also the density of the forest for the country as a whole. Brewer was well aware of its limitations, conceding that the amount of woodland on the map was "understated east of the Mississippi River and overstated west of it." The resultant map was "a compromise," he said, "on which I have tried to show as far as possible what is known of the woodlands." As it was the first map of woodlands ever attempted, he hoped that ultimately a more

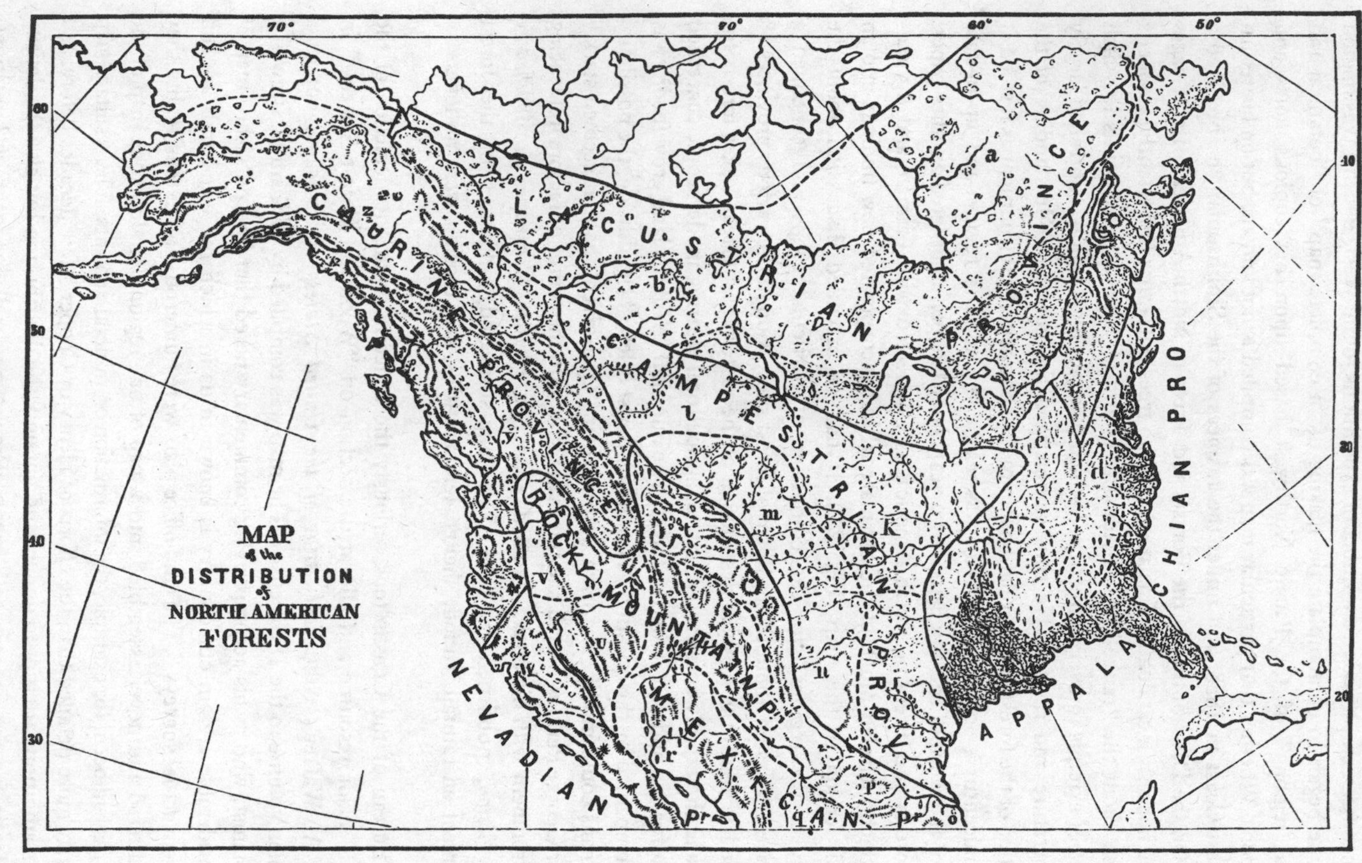

Figure 2.2. "The forests and trees of North America" (1860), by James Cooper. (James G. Cooper, "The Forests and Trees of North America, as Connected with Climate and Agriculture," frontispiece.)

satisfactory series of maps would be compiled – "perhaps by some future United States Commission of Forests" – to illustrate the distribution of specific forest types, which he could describe only in a cursory fashion in his commentary.[11]

Brewer's map (Fig. 2.3) consisted of six degrees of density of woodland, expressed in terms of acres of woodland per square mile of land. The distribution was complex because it incorporated the densities of the natural woodland and the densities of the woodland that remained after agricultural clearing. What people thought of the picture it portrayed we do not know. One can only deduce that it confirmed precisely that near total woodland cover was confined to a few places only on the continent (the Pacific Northwest coast and Idaho, eastern Minnesota and northern Wisconsin, eastern Maine and the Adirondacks, the summits of the Appalachians, southern Georgia and Alabama, and the coasts of Florida); that the woodland thinned out very rapidly between the Mississippi and the ninety-seventh meridian; and that vast areas of the West had virtually no woodland at all. Moreover, it suggested that agricultural expansion, particularly through western New York, Ohio, Indiana, Illinois, and portions of states adjacent to this core of ever-burgeoning farm-making activity, had reduced the original cover to a third, or even less than a sixth, over much of these states. Brewer himself attempted in a general way to ascertain and list the causes of the destruction, but the task was beyond his interests and knowledge, and certainly beyond his commission. Of one thing he was certain, however; the destruction was the result of "prodigal use and needless waste."

In 1880 Charles Sargent was asked to prepare a report for the Tenth Census on the distribution and value of the forests. His massive *Report on the Forests of North America* appeared in 1884, together with a *Folio Atlas of Forest Trees of North America*. Like Brewer before him, but working on a much surer basis, he attempted to summarize the status of clearing by providing a map of "average forest density" in order to illustrate "the present productive capacity of the forest."[12] The resultant map was necessarily generalized and took the form of plotting the distribution of the number of cords of wood per acre, divided into nine geometrically graded categories (simplified to six in Fig. 2.4). The map is difficult to read because of the unusual measure used, but if we ignore the absolute amounts and concentrate on the relative distributions it can be said to tell the same story as Brewer's map, although in more detail, and it gives a good idea of the location of the major forested areas.

These efforts to depict the forest as a whole during the last quarter of the nineteenth century were largely utilitarian and were influenced by concern for the destruction of the forest cover. While these concerns continued unabated, the next major step in the acquisition of knowledge of the forest as a whole came through the work of forest biologists or, more correctly, plant ecologists, who, dissatisfied with simply identifying individual species, attempted instead to identify characteristic groups or associations of plants and to investigate their interrelationship with their environment. In addition, they were aware that forests were not static entities and that changes were at work within them.[13] For example, in 1899 Henry Graves in his *Practical Forestry in the Adirondacks* used the term "forest type" to distinguish forest stands that were characterized by a plurality of trees and that were essentially similar in their composition, development,

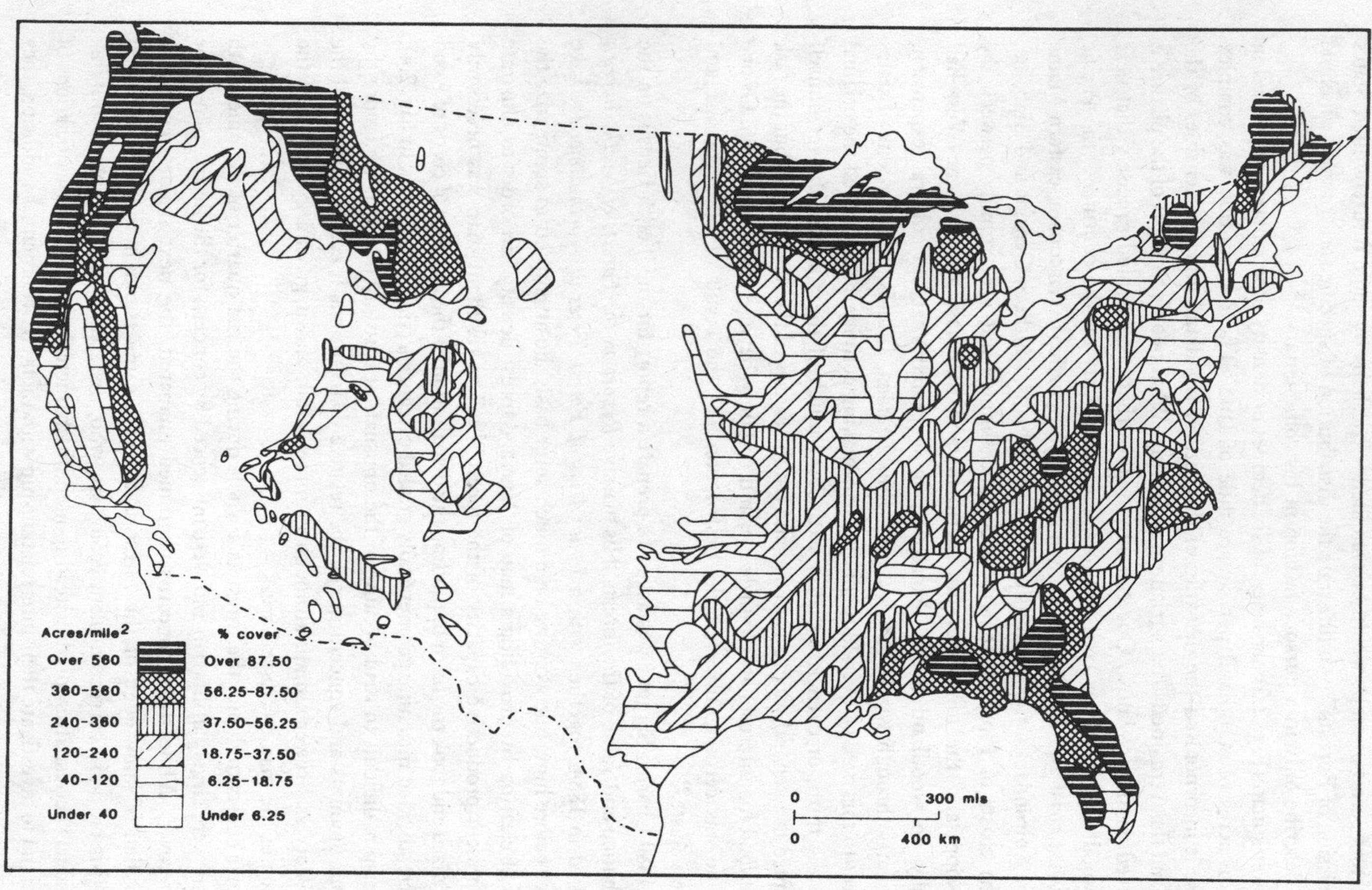

Figure 2.3. Density of woodland in the United States, 1873. (Simplified from William H. Brewer, "The Woodland and Forest Systems of the United States.")

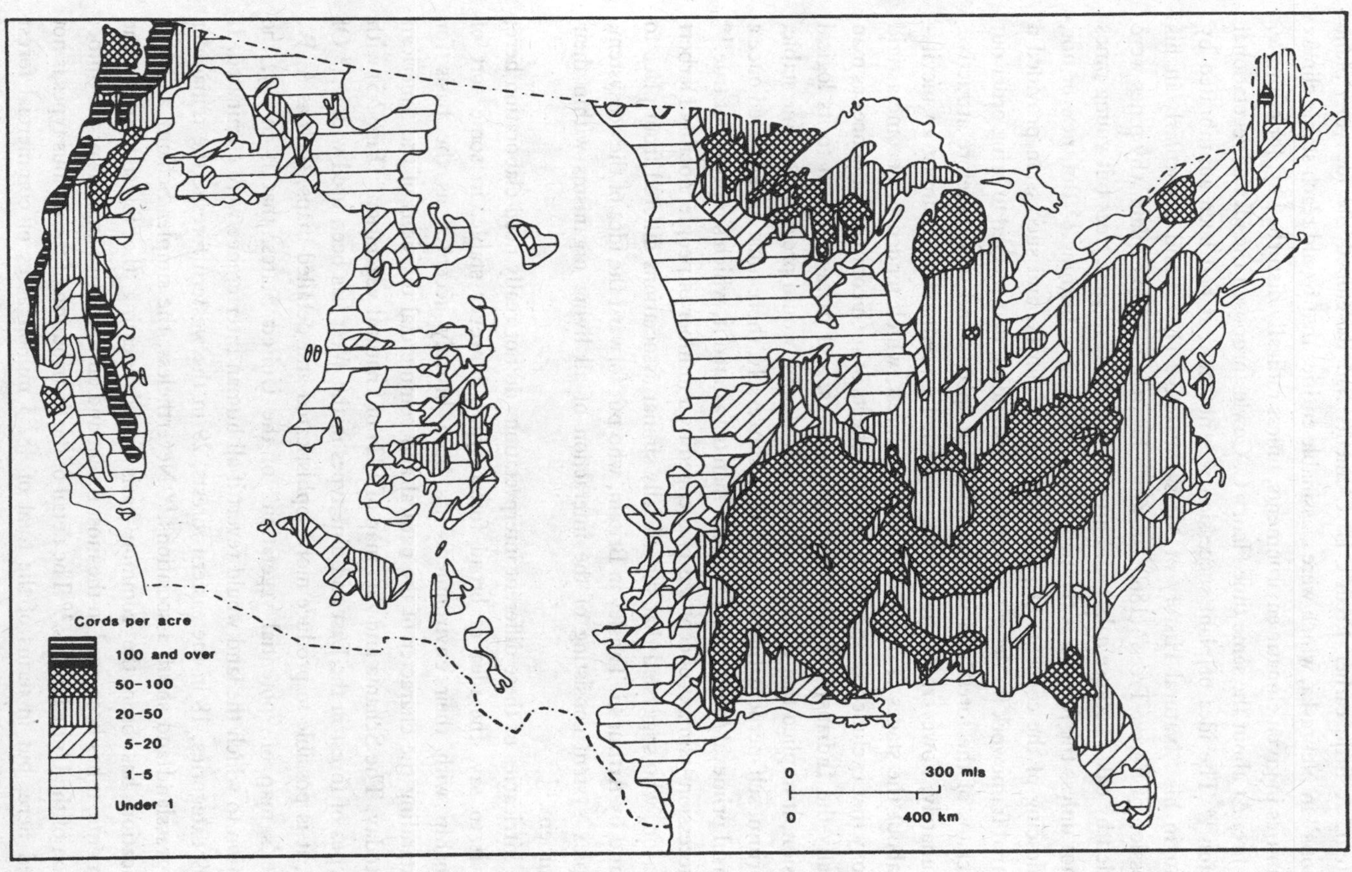

Figure 2.4. The density of existing forests, 1883. (Simplified from Charles S. Sargent, *Report on the Forests of North America*, vol. 9, and its *Folio Atlas of Forest Trees on North America*.)

and extent. A little earlier Fredric E. Clements had developed the idea of "plant formations" in Nebraska, which were classifiable entities that would reach stable climax communities in given climatic environments, unless radically disturbed by, for example, man or fire. At about the same time Henry C. Cowles proposed the idea of vegetational succession.[14] The idea of plant succession was not new; it had been adumbrated by Thoreau in his "Natural History of Massachussetts" (1842) and particularly in his "Succession of Forest Trees" (1860), in which he attempted to explain why pines were succeeded by oaks, because oaks could grow in the shade of pine trees but young pines could not unless the forest was thinned through felling or burning.[15] But, new or not, the coalescing of the concepts of plant association, climax, and succession provided a conceptual framework for description and understanding that seemed to bring order out of the chaos of the forest and its multiplicity of species and was therefore attractive. Unfortunately, however, the framework suggested a predictability, almost an inevitability, about the stages of development of the forest, which tended to become a rigid orthodoxy that excluded an appreciation of the effects of such complex influences as man (especially the Indians), animals, wind, fire, insects, and disease. Taken to its logical conclusion, it implied that if the forest were left alone it would progress to some stable, equilibrium, self-perpetuating state and that, conversely, there must have existed once a wonderful primeval forest until European man devastated it, which simply is not true.[16] The succession–climax concept was challenged by a number of people, notably Herbert A. Gleason, who suggested that superficially similar associations differed from place to place, and by Arthur G. Tansley in Britain, who put forward the idea of the ecosystem, a complex system consisting of the interaction of all living organisms within their environment.[17]

The intricacies of these different interpretations do not really need elaboration here; suffice it to say, the idea of broad forest types, however stable, in some sort of relationship with their environment has been broadly accepted as the basis for understanding the character of the natural and commercial divisions of the American forest today. The Schantz and Zon map of 1924 of "natural vegetation" (Fig. 2.5) with nine types of forest in the East and nine types in the West has been used widely.[18] Of course, it is possible to produce more sophisticated and detailed maps, such as A. W. Küchler's map of "potential vegetation" of the United States, that is to say, the vegetation to which the land would revert if all human interference were eliminated. It has 109 categories, 18 in the eastern forests, 29 in the western forests, the remainder being grassland and shrub associations.[19] Nevertheless, the simpler Schantz and Zon classification has been the common basis for delineating the forest. With some refinements, it has been used in the most recent and authoritative analysis of the timber situation of the United States.[20] The detail of the distribution of the various types is not pursued here, but in terms of the total of 482.5 million acres of commercial forest standing in 1977 nearly three-quarters are in the eastern third of the country. Of the total acreage of all forests, 109 million acres or nearly one-quarter are oak–hickory, and 50 million acres or a tenth loblolly–shortleaf pine. No other forest type accounts for more than 36 million acres (Table 2.1).

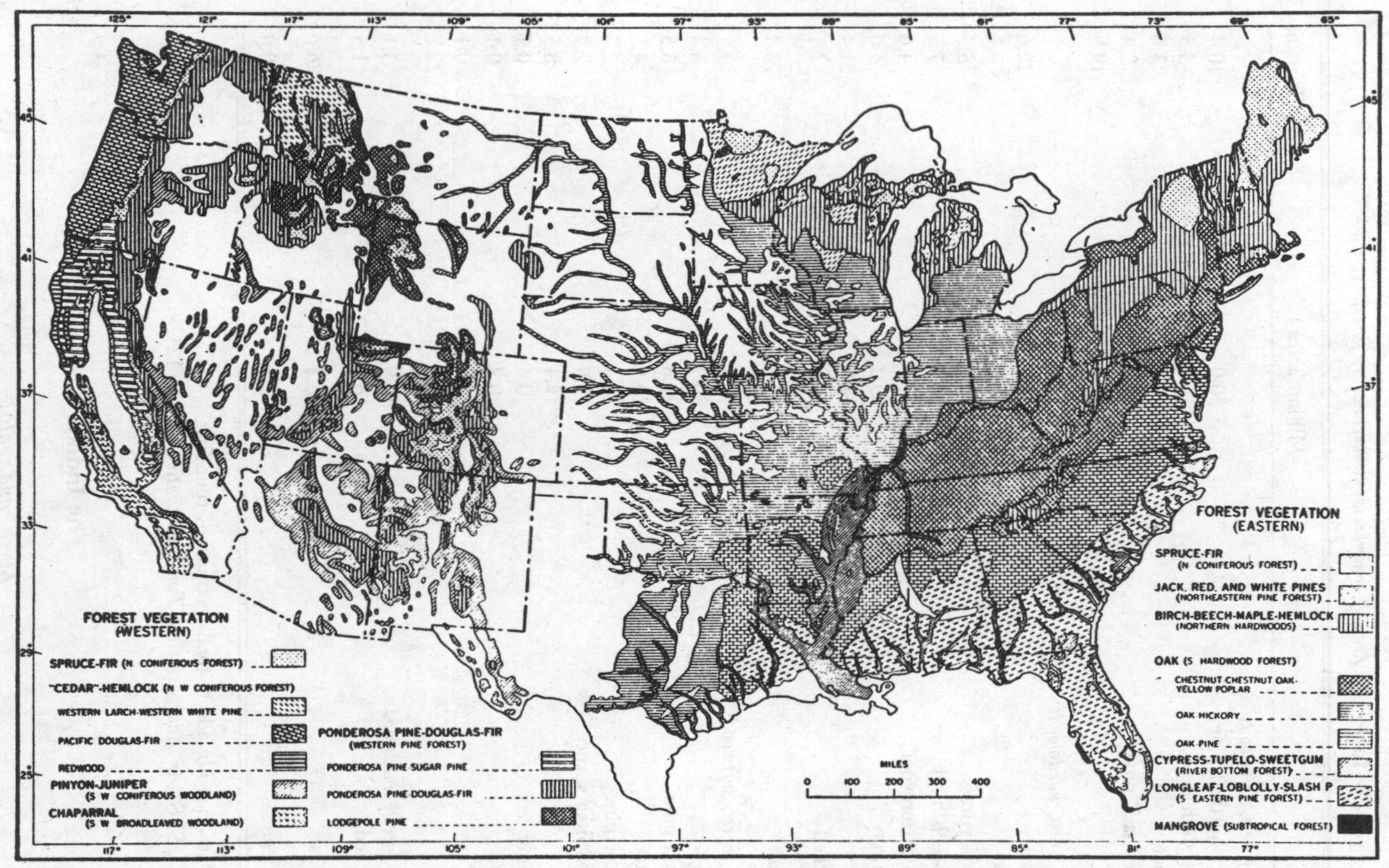

Figure 2.5. Natural vegetation of the United States, 1924. (Homer L. Schantz and Raphael Zon, "Natural Vegetation," in USDA, *Atlas of American Agriculture*, pt. 1, sec. E.)

Table 2.1. *Area of commercial timberland by forest type, 1977*

	Million acres	Percent
Eastern forest, softwoods		
Loblolly/shortleaf pine	50.0	10.4
Longleaf/slash pine	16.8	3.5
Fir/spruce	17.6	3.6
White/red/jack pine	11.8	2.4
Total	96.1	19.9
Eastern forest, hardwoods		
Oak/hickory	108.9	22.6
Oak/pine	34.6	7.2
Oak/gum/cypress	26.7	5.5
Maple/beech/birch	36.2	7.5
Elm/ash/cottonwood	22.3	4.6
Aspen/birch	19.2	4.0
Total	248.0	51.4
Nonstocked	10.0	2.1
Total East	354.2	73.4
Western forest, softwoods		
Douglas-fir	30.9	6.4
Ponderosa pine	26.6	5.5
Fir/spruce	19.9	4.1
Lodgepole/pine	12.7	2.7
Hemlock/Sitka spruce	12.9	2.7
Larch	2.4	0.5
White pine	0.4	0.1
Redwood	0.7	0.1
Others	0.5	0.1
Total	107.0	22.2
Western forest, hardwoods	14.9	3.1
Nonstocked	6.4	1.3
Total West	128.3	26.6
United States	482.5	100.0

Note: Data may not add to totals because of rounding.
Source: USFS, *An Analysis of the Timber Situation* (1982), 121.

The Indian

Myths and realities

By the time European man landed on the eastern shores of America, portions of the woodlands were in the process of being changed to a more open, parklike vegetation, largely through the agency of Indian agriculture and the use of fire for clearing and hunting. Much of the "natural" forest remained, but the forest was not the vast, silent, unbroken, impenetrable and dense tangle of trees beloved by many writers in their romantic accounts of the forest wilderness.[21] The European did not find an America that was all trees; there were extensive, naturally clear areas in the salt marshes and near some of the rivers, and the woodlands were open for miles on end where the natural forces of fire, disease, and wind had affected them. Nor did the Europeans find an uninhabited land; the Indians had been there in large numbers for millennia before them, and the Indians had been burning, clearing and collecting, and otherwise changing the forest.

The impact of the Indian on the forest has not been admitted readily by historians, in much the same way as it has been ignored by plant ecologists of the past. More often than not the Indian has been depicted as the uncivilized inhabitant of an uncivilized environment – even a product of it – a migratory hunter devoid of the ability to clear the forest and cultivate the land. However, these views are difficult to reconcile with the numerous early accounts of substantial and well-cultivated crops of tobacco, sweet potatoes, tomatoes, squash, watermelons, kidney beans, sunflowers, and, of course, maize, most of which originated in North America and all of which grew profusely in the favorable environment of the eastern forests. It was a rich biological store, which indicated a varied agricultural system.[22] The work of contemporary anthropologists suggests fairly conclusively that throughout the whole of the eastern seaboard – the theater of the first century and a half of northwestern European colonization and settlement – a complex subsistence agricultural economy prevailed, except for the country north of around Massachusetts, where a game economy predominated (Fig. 2.6).

The question of numbers, and hence the Indian impact on the forest, has similarly been misjudged and underestimated. The conventional wisdom has been that there were, perhaps, between 1 million and 1.15 million people north of the Rio Grande, and a mere 8–14 million throughout the Americas.[23] Since the mid-1960s, however, a completely new scenario of pre-Columbian America has emerged. Foremost among the revisionists has been Henry Dobyns, who has built on the previous pioneering work of distinguished Berkeley scholars such as Cook, Simpson, and Borah to calculate population figures by determining the demographic nadir of Indian population in various regions during modern times and multiplying exponentially that figure by a number representing the measure of known Indian population loss from disease, warfare, famine, and other factors. The population of the Americas *might* have been 90 million or 112.5 million, depending upon the historical depopulation ratio used, and North America alone could have had between 9.8 and 12.25 million.[24] Consequently, the scenario for the pre-Columbian forest changes radically. When America was "discovered" the population density of the forest was probably even greater than that of densely settled parts of western Europe, and there is the strong possibility that in the late fifteenth century the Western Hemisphere may have had a greater total population than western Europe.[25] The implications of these figures for forest disturbance and destruction are enormous.

The exact distribution and density of that vastly enlarged population are difficult to reconstruct with any accuracy. Harold E. Driver's attempts (Fig. 2.7) show an interesting correlation with types of subsistence and forest or other vegetation types.

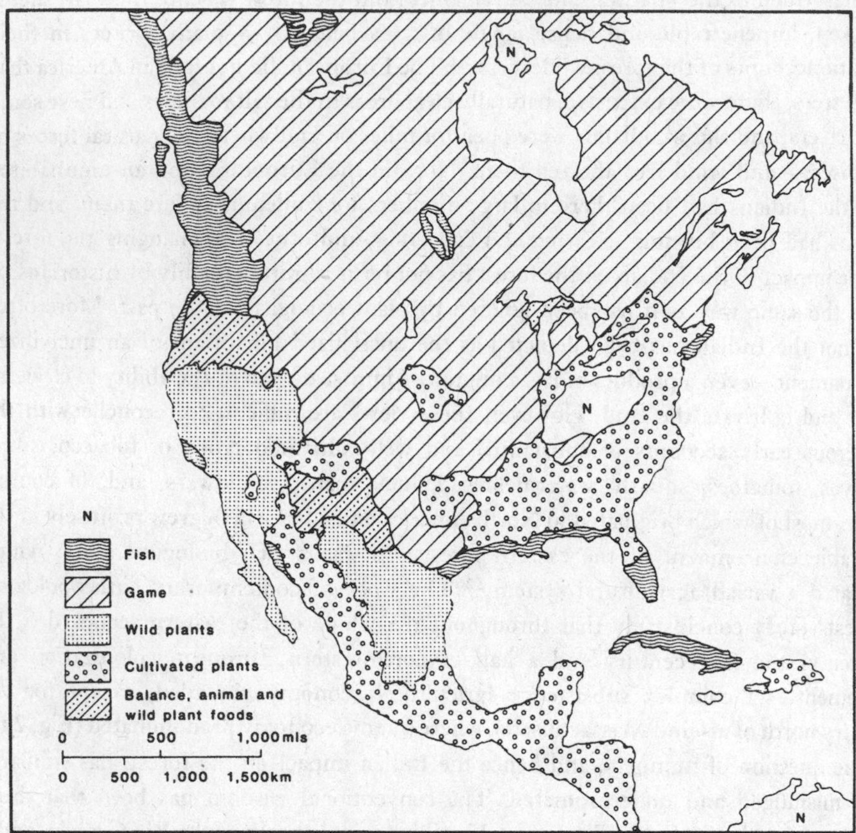

Figure 2.6. Dominant types of Indian subsistence. (Based on Harold E. Driver, *The Indians of North America*, map 3, p. 571.)

Even though the absolute figures are uncertain (and it should be noted that Driver roughly halves Dobyns's estimates though he decreases them less for North America), the relative density is probably still correct. Several generalizations are apparent. Population density was greater in the South than in the North, greater on the coast than inland, and greater in the West Coast nonfarming cultures than in the East Coast farming cultures, though nowhere in the East did the cultivation of plants seem to provide much more than half the total diet; collecting from the forest and hunting game provided the rest.[26]

Whatever the uncertainties and questions about the distribution and density of the pre-conquest Indian population, there is little doubt about the details of their economic life. Hardly any colonial explorers or articulate and literate settlers failed to comment on the inhabitants that they found. They were impressed by the number, size, and good order of the Indian settlements and by the appearance, even of the civility, of the people. These Indians were seen not as "untutored savages, but as people living in a society of appreciated values."[27] The accounts not only enable one to reconstruct a picture of the

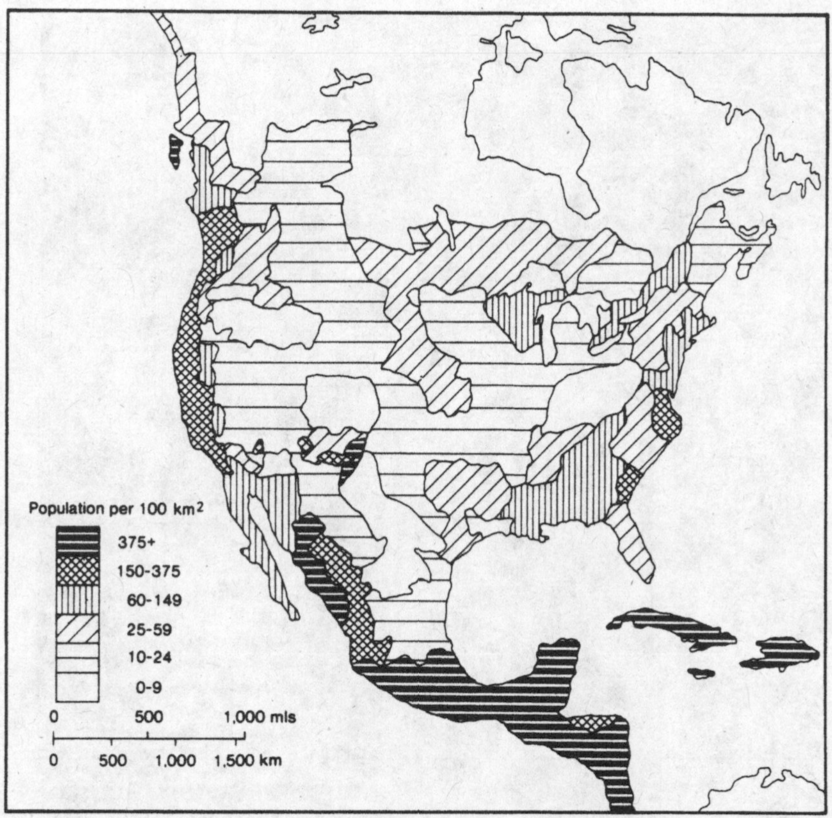

Figure 2.7. Native population density. (Based on Harold E. Driver, *The Indians of North America*, map 6, p. 574.)

effect of the Indian on the forest but also support the suggestion that the destruction and alteration of the forest were far greater than thought and greatest in the vicinity of the greatest population density.

Indian clearing

Throughout the eastern forests, from Maine to the Gulf, there is sufficient evidence to be sure that nearly all Indians lived in villages, surrounded by fields in which they grew a great variety of crops, particularly corn; wherever explorers went, mention was made of cornfields (Plate 2.1). More often than not the ground for both village and fields was hacked out of the forest and maintained by fire. In order to make a clearing or "opening," small bushes were uprooted and burned. The bark of the trees was bruised and peeled away with stone axes or burned off. Either way, this was the equivalent of ring-barking or girdling the tree, and in time the tree died, withered, and it too could be burned.

Occasionally, trees were felled with stone axes. Sometimes a few trees were left in the

Plate 2.1. The town of Secota, Virginia. The sketch depicts various aspects of life in an actual Indian village. *H* (far right center) denotes corn in various stages of cultivation and maturity. *I* (next to the corn) depicts pumpkins, and in the top left-hand corner there is a hunting scene in the woods. (John Smith, *Narrative of the First Plantation in Virginia* [1620], reproduced in Thomas Hariot's *Narrative of the First English Plantation of Virginia* [1893].)

cultivated area, or the larger stumps were left in the ground as their removal was not worth the energy expended, although smaller stumps were prized out with large crooked poles. As the clearing was enlarged, the women broke up the soil with digging sticks or light hoes, and mixed crops were planted, which always included maize. The plots varied in size from a few square yards to 200 square yards, and the crops were grown in mounds, sometimes in rows. The villages contained between 50 and 1,000 persons living in a loose agglomeration of sturdy, defensible, and weatherproof wooden houses, sometimes located roughly along a street, sometimes very scattered with a few houses on the very edges of the clearing. The Huron and Iroquois, in particular, relied upon finding a supply of conveniently sized timber for the construction of their longhouses and palisades. Conrad Heidenreich's detailed study of the Huron in Ontario suggests that a village of 1,000 people on a site of six acres needed 36 longhouses and that the houses required 16,000 poles of 4–5 inches in diameter and 10–30 feet long, 250 interior posts 10 inches in diameter and 10–30 feet long, and 162,000 square yards of American elm or eastern redcedar bark for covering. The palisade would need 3,600 stakes, 5 inches in diameter and 15–30 feet long. Such timber sizes were most commonly found in areas of secondary forest growing on abandoned cropland and within a roughly 50-acre forest zone surrounding the village. The total area of a clearing could be between 150 acres and 600 acres (about the size of from one to four quarter-sections), and it would usually be surrounded by a palisade. If the density of settlement was great enough, it was possible for clearings to merge one into the other to make large open areas, but this was unusual unless along a riverfront.[28]

The availability of fresh water was a prerequisite for the choice of a village site and ground to clear, and the quality of the ground must have also been important, its friability being essential for hoe cultivation and its fertility for sustained yields. These were points well attested to by the European colonists, who in Virginia said that the Indians "fix upon the richest ground to build their wooden houses. . . . Wherever we meet with an old Indian field, or place where they have lived, we are sure of the best ground."[29]

But the Indian drew upon the forest for a great variety of products other than food crops planted in cleared land. There were such things as fibers, trees, foods, medicines, and, above all, there was fuel wood. Beyond the fields and in the forest, fallen or dead wood was collected for fuel, mainly by the women, so that it was generally known as "squaw wood." In some groups fuel gathering was of such significance that it was an integral part of ritual activities, such as puberty rites. Many observers commented on the prodigious consumption of fuel wood in the villages, and it is not surprising, therefore, that the search for fuel went on in an ever-increasing radius – even up to three miles – and that in the last resort it became reasonable to fell trees in close proximity to the village rather than walk the long distance to collect such a bulky item. The older the village, the more likely it was that the forest was encroached on more and more and that its edge receded from the village.[30]

The zone of fuel collection was also the area of foraging for a great variety of other products. There were wild fruits (persimmon, wild plums, papaws), berries, leaves, roots, and nuts, particularly acorns (which in California were ground into a paste and baked to make a sort of bread) but also chestnuts, pecan nuts, and nuts from related hickory trees. There was maple syrup; a host of products with medicinal properties such as juniper berries and chokeberries; and trees such as American elm for making utensils like bowls and cups. So important were these products that it is thought that the present unusually high proportion of nut and fruit trees in some localities might denote purposeful promotion of favored species or, at least, the natural propagation on nutrient-deficient soils of sun-loving, aggressive, colonizing trees in clearings. Sumac and sassafras grew so abundantly on the abandoned clearings that they were commonly known as "old field trees." In addition to the products of these naturally growing trees, the edge of the forest was a favored spot for hunting birds, squirrels, rabbits, and other small game.

The cultivation of crops helped the Indians to overcome periods of famine by building up food stores, and the accumulation of surplus food allowed them to engage in joint undertakings, such as warfare and long ceremonies, and to build permanent settlements. However, cultivation was an auxiliary activity, albeit an important one, to hunting and fishing. Hunting, in particular, was an important way of gaining food, and beyond the radius of frequent human interference around the village the Indians gathered periodically in groups in search of game. But even this activity did not leave the forest undisturbed. The paucity of good forage plants made the burning off of the undergrowth in fall and spring a reasonable practice so that the new and succulent growth – the "fire grass" – would be an attraction to the deer, buffalo, bear, elk, and turkey, which were then killed. The Indian was well aware that fire-disturbed sites provided more food than mature wood vegetation, that seed, berry, and nut growth was stimulated, and that game yields were greatest in areas of constant vegetational succession. Also, the forest was sometimes burned as a means of rounding up the game in enclosures for easy slaughter. Accidentally or on purpose, these fires must have destroyed hundreds of square miles of forest every year, for there was no way of putting them out. Generally, the fires were never lit near the village because of the danger of causing destruction to the houses and crops and also of destroying the small game, wild fruits, and berries, although there is evidence that fire was sometimes used to promote the growth of berries.

Though some tribes, such as the Abnaki of coastal New England, went through an annual cycle of migration, moving their village from the woods where hunting was undertaken during the winter to the seashore for fishing during the summer, even the more sedentary tribes were not stable. Continuous occupation localized and intensified disturbances in the forest, and in time such things as decreasing yields, the sheer accumulation of filth and refuse, the infestation of vermin and weeds, and the ever-increasing distance that needed to be traversed in order to collect fuel, game, and fruits led either to the making of new openings in the forest within easy reach or, more likely, to complete abandonment of the village for a new site. It is variously estimated that such a move could occur every 10 to 15 years, or twice every generation. One authentic record of the timing of moves is that of the important settlement of Onondaga, near present-day Syracuse. As the capital of the Iroquois League it was visited often and was found in nine different locations between 1610 and 1780, or an average life span of close to 20 years for each site. One of the moves was definitely made to renew exhausted resources, for Father Jean de Lamberville wrote of the village in 1682:

> I found on my arrival the Iroquois of this town transporting their corn, their effects, and their lodges to a situation 2 leagues from their former dwelling place, where they have been for 19 years. They made this change in order to have nearer to them the convenience of firewood and fields more fertile than those which they abandoned.

This was not a small settlement and in all probability was the same town that the explorer Greenhalgh visited in 1677 which then consisted of 140 houses and land cleared and planted with corn that stretched from four to six miles away from the town.[31]

If a village was located near river deposits or "bottomlands" that had a periodic addition of new silt from floods, then there was the possibility of permanent cultivation. The need to resite such a village did not occur, a fact attested to by artificial mounds on which some villages stood and which indicated the continued attractiveness of the site. However, even these river-located towns sometimes suffered the same disabilities as the purely forest ones. Writing in 1798 on the numerous Creek settlements in Alabama, Benjamin Hawkins described Tookabatche on the Tallapoosa River as a town "on the decline. . . . The land is much exhausted with continued culture, and the wood for fuel is at a great and inconvenient distance, unless boats and land carriages were in use."[32]

It seems likely that the direction of the moves in those towns that were relocated periodically was not arbitrary, bearing in mind the requirements of water and soil, and that the villagers went to a predetermined location in an old cleared area or even relocated in an old village site that had been abandoned up to a half-century before. The annually fired game areas were too large to contemplate their being relocated, and in any case there was no advantage in doing that, so the new village site would be within 5 to 20 miles of the previous one and within striking distance of the game area. From the Mid-Atlantic Coast northward, abandoned sites needed about 20 years in order to regain their vegetation, and the interval was somewhat less toward the South. This explains the large number of "old fields," flats, "openings," "meadows," and "maize lands" that are referred to constantly in early descriptions and that always seemed to be far greater in number than the needs of the population would suggest. Old fields and villages that were abandoned by the Creek tribes in what are now Alabama, Georgia, and northern Florida were known as "tallahassees," a term that has been adopted in modern place-names, including the capital of Florida.[33]

In many ways the Indians' use of the temperate forests in the past resembled the shifting agriculture of the tropical forest of the world today. Given the population density, it was a reasonable and delicate use of environmental resources which was based on the sound economic principles of sustaining crop yields and minimizing labor inputs (Fig. 2.8). The intensity of land use and, hence, attention decreased with an increasing distance from the center, graduating from complete clearing through partial and even sporadic clearing, from thinning to periodic burning. In time the village lands became exhausted, and without manuring a rotation needed to be undertaken. Also, the gathering of wood and game had to be pushed farther afield, and in terms of labor inputs the point must have come when it was worth abandoning the village and selecting a new site. A variation on this arrangement is apparent on the river-fronting villages of the South. Although it is admitted that the fields were probably more fertile than those of the eastern woodlands because of the nearly annual deposition of silt on them by floodwater, the ease of movement by canoe and the effective minimization of distance that this provided must have extended enormously the range of cultivation and fuel gathering. Fields could be relocated up and down the river without moving the village. Yet, even here, soil and fuel exhaustion could be so great that relocation might have to

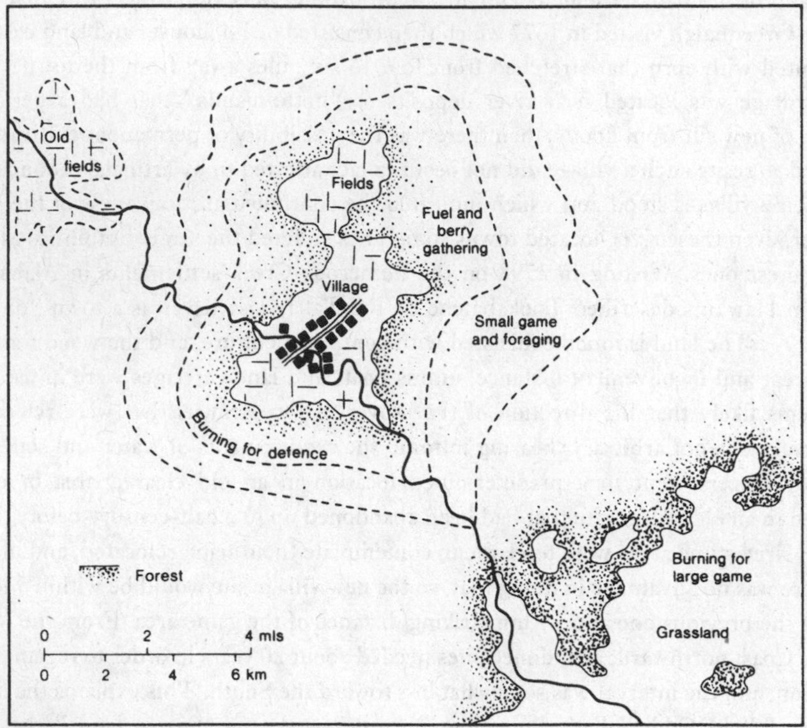

Figure 2.8. Schematic representation of Indian forest use.

occur after perhaps a hundred years. Hawkins noted that the fields of Cussetuh village on the Chattahoochee had been "cultivated beyond the memory of the oldest man" but that yields were declining so rapidly that a move was imminent.

The contemporary record

Of the dozens of existing contemporary descriptions of the Indian villages and clearings in every state in the eastern half of the continent, only a few must suffice to give a picture of conditions and locations. In Virginia, John Smith described the village of the Pawhatans as follows:

> Their houses are in the midst of their fields or gardens, which are small plots of ground. Some 20 acres, some 40, some 100, some 200, some more, some less. In some places from 20 to 50 of those houses together or but little separated by groves of trees. Near their habitations is little small wood or old trees on the ground by reason of them burning of them by fire. So that a man may gallop a horse amongst the woods any way.

Smith estimated that there were some 5,000 Indians within a 60-mile radius of Jamestown, living in villages consisting of between 30 and 200 people.[34] In 1620 William Strachey described the country around the present site of Hampton, Virginia, as

> ample and faire countrie indeed, an admirable porcion of land, comparatively high, wholesome and fruietfull; the seat sometyme of a thowsand Indians and three hundred houses, as it may well appeare better husbands [farmers] than in any part ells that we have observed which is the reason that so much ground is there cliered and opened, enough already prepared to receive corne and make viniards of two or three thowsand acres.[35]

The Indians were driven inland after 1622, and the value of the abandoned clearings was not lost on the settlers. There was so much cleared land that some colonists thought that

> the objection that the country is overgrown by woods, and consequently not in many years to be penetrable for the plow, carried a great feebleness with it for there is an immense quantity of Indian fields cleared ready to hand by the natives, which, till we are grown over-populous, may be every way abundantly sufficient.[36]

These examples refer only to the tidewater area of Virginia during the early decades of the seventeenth century, but even in the backcountry, which was penetrated a half-century later, there is evidence of clearings – in the Roanoake Valley, along the Virginia–North Carolina border, and along the border with West Virginia, which was described as "by the industry of these Indians . . . very open and clear of woods" and "open in spacious plains."[37] Probably many more clearings had existed farther west, but because of the depopulation of the area between the Alleghenies and the Ohio River caused by the forays and plundering of the Iroquois Indians in what is now New York State from the 1650s to 1670, the fields had been abandoned for so long that when Europeans first penetrated the area in about 1750 the traces of most openings had been obliterated. A few were found here and there: One traveler in 1752 found remnants of fields covered with white clover and stands of trees uniform in growth and about 50 years old, some of which were growing up through the cobblestone floors used by the Indians as places for drying nuts, fruits, and game. Farther west on the Kanawha River, where Indians were known to exist, there is more evidence of openings, for when George Washington wanted to sell 20,000 acres of land there he extolled "the excellent meadows," which could only have been the remnants of fields abandoned years before.[38]

Throughout New England similar accounts exist, particularly in southern New England. The Narragansetts had cleared land for 8 to 10 miles from the coast, and the only wood found by the colonists was in the fire-protected low and damp ground. Edward Winslow and Stephen Hopkins, who rode along the banks of the Taunton River in 1621, commented on the towns and fields deserted because of disease, which had decimated the Indians. The land "was very good on both sides, it being for the most part cleared. . . . As we passed we observed that there were few places by the river but has been inhabited; by reason whereof much ground was clear, some of the weeds which grew higher than our heads."[39]

The Reverend Francis Higginson, writing about the country around Salem in 1630, said that most of the land was

> fit for Pasture, or the Plow or Meddow ground, as men please to employ it; though all the Countrey be as it were a thicke wood for the generall, yet in diverse places there is much ground cleared by the Indians, and especially about the Plantations; and I am

> told that about three miles from us a man may stand on a little hilly place and see diverse thousands of acres of ground as good as need to be, and not a Tree on the same.[40]

The evidence of clearings is not confined to the densely populated regions of the Mid-Atlantic and southern New England coasts. In the North, Cartier had visited Hochelaga (modern Montreal) in 1534, which stood in "land cultivated and beautiful, large champaigns full of the corn of their country," and Champlain in the beginning of the next century saw or heard of similar sights in Vermont and on the shores of Lake Champlain.[41]

In the South a similarly early, extensive account comes from the plundering travels of the De Soto expedition, which landed in Tampa Bay, Florida, in 1538 and marched for the next three years in a great M-shaped journey through eastern Florida, across central Georgia, touching the edges of North and South Carolina and Tennessee, then turning south through central Alabama to around Mobile, north again into Mississippi and western Arkansas, and finally south along the Ouachita River to Louisiana and the Mississippi delta. The descriptions of the land en route were copious, the details of the alternating forest, savanna, fields, and burning and cultivation filling pages. For example, between what is now Lake City and Tallahassee in northern Florida there was an alternation of open forest, treeless land, cleared land, plains, and cultivated fields, which extended "over the plains as far as the eye could see." Everywhere the forest was open enough for this huge expedition of about 600 men and its ambulant supply of meat in the form of herds of swine to proceed without any difficulty, except for the wetness of the swamps. Fields were found alongside many of the rivers, particularly the Coosa, Tombigbee, Alabama, Savannah, Ouachita, and Mississippi, and all their tributaries. For example, along the Coosa the land was "thickly settled in numerous towns with fields extending from one to the other, a pleasant place with fertile soil and good meadows along the river . . . many corn fields and an abundance of grain."[42] The descriptions of these settlements are confirmed by later reports. Particularly noteworthy are Bartram's journal of 1791 and Benjamin Hawkins's *Sketch of the Creek Country* of 1798. Hawkins describes in detail the cultivation of 50 towns along the Coosa, Tallapoosa, and Chattahoochee rivers. One had fields extending "one mile and a half," another "four miles down the river, from one hundred to two hundred yards wide," another "three thousand acres . . . and one third in cultivation," and so on.[43]

In the Midwest similar accounts exist, although for a much later period because it was not penetrated by Europeans until the latter half of the eighteenth century. In 1724 the whole of the Virginia claim extended as far west as the Ohio River and was described as "a perfect forest except where the woods are cleared for plantations and old fields and where have been formerly Indian towns and poisoned fields and meadows where timber has been burnt down in fire in hunting or otherwise."[44] The density of clearings and towns was very great in the western extremity of New York State in the Iroquois country. The Indian town of Onondaga near Syracuse and its immense surrounding fields have been mentioned already in connection with the shifting of settlements, but farther west was the village of Genesee near present-day Rochester. Genesee consisted of about 128 houses and was said in 1779 to be "almost encircled with a cleared flat which extended for a number of miles covered by the most extensive fields of corn and every kind of vegetables that can be conceived," and there were other towns in the vicinity with cornfields of 200 acres and more. Farther west again in Indiana and Ohio Wayne's army in 1792 found settlement and cultivation along the entire length (approximately 100 miles) of the Maumee River from Fort Wayne to Lake Erie.[45]

Indian fires

These contemporary accounts leave little room for doubt that the Indians made an impact on the forest by clearing it periodically for cultivation, but in all probability they made an even greater impact with the widespread use of fire. Fire was used not only for cooking and heating but also for opening up the forest to provide land for cultivation and also to provide additional fertilizer by recycling debris and its stored nutrients. Fire was used to maintain and even extend already open grassland by preventing forest growth and thus encouraging game; to promote the growth of fresh grass in already open areas; and to encourage the growth of plants bearing nuts and berries. Finally, fire was used in forest and grassland to hunt the game, herding it together for slaughter by encircling large areas with fire, a practice known as fire hunting. Fire created a favorable environment for the Indian; the mixed grassland–forest edge in particular maximized returns in game and cultivated food. Far from being incapable of modifying his environment, the Indian created it, gradually replacing dense forest with thinner forest, thinner forest with grassland, and changing the composition of the standing forest. Other uses of fire had far less certain effects, but they were none the less real for that: to encourage rain, to drive off mosquitoes and flies, and as a means of defense against rival tribes.[46]

All the evidence points to the fact that the Indians were careful users of fire as a tool to manage the land and to promote their welfare. Cereal grasses were fired annually, basket grasses and nuts about every three years, brush and undergrowth in the forest about every 7 to 10 years, large timber in the swidden rotation every 15 to 30 years or more; and broadcast fire in the fields on an annual basis got rid of vermin, disease, weeds, and regrowth.[47] Fire for defense and hunting occurred as needed. As in most primitive societies, fire was a natural and integral – even sacred – part of the Indian landscape and livelihood in contrast to more civilized societies where fire is abhorred and suppressed, nowhere more so than in nineteenth-century North America.[48]

As in the case of clearing and cultivation, the contemporary record of the nature, purpose, and effect of Indian fires is rich, varied, and quite conclusive, and only a fraction of the evidence can be selected.[49] It does seem as though fire produced a sort of hierarchy of effects and changes in the forest. First, repeated burning retarded the reestablishment of trees by destroying the understory so as to leave the forest open or in a parklike state. Second, in dry locations, if the burning was frequent and hot enough all the seedlings and sprouts would be eliminated and only the older trees would be left. In time they died, leaving only grasslands or prairies. Third, in the more humid areas, extensive and frequent burning altered the composition of the forest stands: Fire-hardy or tolerant species invaded the old forest lands and, if the process of burning went on long enough, to such an extent that they became the dominant species. Simply, the Indian made an indelible impact on the forest.

The opening of the forest. The repeated use of broadcast fire in the forest to free it from the underbrush had its effect upon the density of the trees. Many Europeans approaching the eastern coast had seen the evidence of great conflagrations even before they landed, but few made the connection with Indian burning of the woods. However, Thomas Morton, who had lived in New England for several years by 1637, had no difficulty in explaining the openness of the forest:

> The Salvages are accustomed to set fire of the Country in all places where they come, and to burne it twize, in the yeare, viz: as the Spring and the fall of the leafe. The reason that mooves them to doe so, is because it would other wise be so overgrowne with underweedes that it would be all coppice wood, and the people would not be able in any wise to passe through the Country out of a beaten path.

This repeated broadcast firing of the undergrowth

> destroyes the underwoods and so scorcheth the elder trees that it shrinkes them, and hindreds their growth very much; so that hee that will looke to finde large trees and good tymber, must not depend upon the help of a woodden prospect to finde them on the upland ground; but must seeke for them . . . in the lower grounds, where the grounds are wett.

The result was to open out the forest and to create a sort of ecological secondary fire association: "The trees growe here and there as in our parks; and makes the Country very beautifull and commodious," he wrote.[50]

So open and parklike did the landscape become that Andrew White, on an expedition along the Potomac in 1633, said that the forest was "not choked up with an undergrowth of brambles and bushes, but as if laid out by hand in a manner so open, that you might freely drive a four horse chariot in the midst of the trees," and a little earlier John Smith in Virginia had commented that "a man may gallop a horse amongst these woods any waie, but where the creekes and Rivers shall hinder."[51] The ability to ride a horse or drive a horse and carriage between and under the trees became a favorite way to describe the open nature of the forest. If the opening out went far enough, then small meadows or prairies, as they were called later, resulted, something that evoked another typical set of descriptive comments. As early as 1654 Edward Johnson had observed that the thinness of the timber in parts of New England made the forest "like our Parkes in England," and Adam Hodgson, on a journey from Natchez through central Mississippi, found that the forest was "delightful, open and interspersed with occasional small prairies and had the appearance of an English park."[52] This type of vegetational landscape was noticed particularly by the British travelers and settlers, not only because of its utility and the ease of movement in it but also because of its aesthetic qualities. Such a landscape was comforting and pleasing, as it suggested domestication and a touch of art and intimacy, and this was a preference deeply rooted in British landscape aesthetics, the very antithesis of American native landscape taste with its stress upon the wild and the absence of artificiality.[53] Nevertheless, all we know about the frequency and extent of Indian burning suggests that the early observations were correct and that the open forest situation was more characteristic of the forest in general than the picture of the dense, dark woods of the romantic imagination.

The creation of the grasslands. Wherever fire was more frequent and devastating, completely open ground was the result. The sparsely growing, parklike forest gave way to what were variously called plains, barrens, openings, deserts, prairies, or, particularly in the South, savannas, all individual patches of open grassland varying in size from a few acres to many thousands of square miles. Their location and extent varied, but the farther west one moved the more frequent the clearings became, so that in a string of states from Texas on the Gulf through Missouri, Iowa, Illinois, and Wisconsin in the North, individual, isolated "prairies" coalesced and became larger, and the forest became less. What forest there was stretched out westward in long peninsulas along the wetter lowlands beside the river courses until it broke up into individual patches, like islands in the sea of grass that made up the prairies. When European travelers and explorers encountered the openings and later the prairies, they made copious comments and observations about them. The absence of trees in the small openings was equated with the sterility of the soil – hence, barrens and deserts – and the vast, open, larger prairies, encountered frequently after the beginning of the nineteenth century, were such an unfamiliar environment that the travelers were not sure how settlement should proceed in them.[54]

Some of the first major prairies encountered were on the Rappahannock River in Virginia, and particularly along the Potomac, in an area known as the "barrens," which extended for over 1,000 square miles through Pennsylvania and Maryland in the lower Shenandoah Valley. Traveler after traveler through the valley commented on the "large Level plains," the "large spots of meadows and savannahs wherein are hundreds of acres without any trees at all," and the "firing of the woods by the Indians" (subsequent analysis of burned tissue and tree rings shows that the forest in this area has been subject repeatedly to fire over the last thousand years). Similarly, the enormous and anomalous area of open ground in western Kentucky, later known as the bluegrass country when invaded by the European plant *Poa pratensis*, excited much comment.[55] Less spectacular in size but equally as important locally were the numerous openings of a few hundred acres in size in eastern New York, an area of intensive settlement by the Iroquois, which were kept open by repeated firing in order to promote young grass growth to entice deer and other game.[56]

The prairie phenomenon was not confined to the northern part of the country. Rostlund cites dozens of descriptions of openings when examining the evidence for the "prairie belt" in Alabama and adjacent parts of Missouri. For Bartram, for example, the origin of the Alabama prairies was clear: They were "expansive illumined grassy plains, or native fields, about 20 miles in length and in width 8 or 9." In the Midwest, on the road from Terre Haute to St. Louis, Baird encountered prairies varying in size from "several miles in circumference, to those which contain only a few acres," and some "became very extensive and stretch[ed] out in some directions so that the eye perceives with difficulty the bounding forests." Whatever their size, however, there was little doubt in his mind that they were "formed, as they are perpetuated, by annual fires, which sweep over them every autumn. . . . Wherever the fire is kept out of the prairies, they soon become covered with a dense and rapidly growing forest."[57]

The destruction of the forest by fire and the creation of individual openings in the forest, whether large or small, did not approach in impact or extent the destruction of the forest along its western edge. From Wisconsin in the North to Texas in the South repeated firing by the nomadic hunting cultures of the Plains Indians sustained the grassland vegetation against forest encroachment and the gradual extension of grassland at the expense of the forest. Their object was to extend the range of game by continually burning the forest edge and burning abandoned fields to prevent forest regeneration. Archaeological evidence suggests that burning could have started in early postglacial times before the forest became established, so that one can view Indian incendiarism as either an attempt to maintain the grassland and the game on it as the forest advanced with the improvement of the climate and the retreat of the glaciers or an attempt to remove the forest and extend the grasslands. Either way, fire was of critical significance in altering the vegetation, and it is thought that the buffalo, for example, spread throughout the continent as the forests were burned, crossing the Mississippi in about A.D. 1000, entering the South by the fifteenth century, and penetrating as far east as Pennsylvania and Massachusetts by the seventeenth century.[58]

Climatic conditions alone cannot be evoked to explain the origin and peculiar purity of the grasslands as a climax vegetation, just as it cannot be evoked to explain the expansion of the buffalo.[59] Extensive grassland areas extend across humid and semiarid environments alike, and the latter are not devoid of timber, as is shown by the extensive woodland of the cross timbers of Texas and Oklahoma, which grow vigorously and extend if fire is suppressed. Similarly, the mesquite has established itself over 40 to 50 million acres of Texas, and the chaparral over thousands of square miles of California. Evidence of recolonization of grassland by the forest and its advance westward is abundant. In the "oak openings" or prairies of southwestern Wisconsin the amount of grassland was calculated to decrease by nearly 60 percent between 1829 and 1854 following the cessation of fire. Gleason has calculated that the oak–hickory forest advanced westward into the Illinois grasslands at the rate of between one and two miles in 30 years after fire ceased and that portions of the grassland in the driftless area of northwestern Illinois and adjacent portions of Wisconsin, which are now heavily forested and where no cultivation is carried out, were 80 to 90 percent grassland when first seen by Europeans in 1823. They presented "the waved appearance of a somewhat ruffled ocean; it is covered with a dry short grass."[60] Similar accounts of oak–hickory forest advance in eastern Kansas and Nebraska were the subject of comments by Bayard Taylor in 1866. Near Omaha, on the Missouri River in northeastern Nebraska, he saw

> a phenomenon of which I had often heard – the spontaneous production of forests from prairie land. Hundreds of acres, which the cultivated fields beyond had protected against the annual inundation of fire, were completely covered with young oaks and hickory trees, from four to six feet in height. In twenty years more these thickets will be forests.

Not until there was a complete occupation of the prairies for agriculture and the suppression of wildfires did the advance of the forest stop, but even then, within a short time of cropland becoming abandoned, forest growth began.[61]

One should be careful of syllogistic reasoning from these examples. Nevertheless, the same inferences can be drawn as from the example of the southeastern pine forests outlined later in this chapter. There seems to be sufficient evidence to suggest that keeping down the growth of trees by burning, possibly started by lightning but mainly by Indians, produced most of the eastern prairies and certainly the prairie openings and stunted oak barrens that interdigitated with the forest edge. The seasonally dry plains on which the grassland flourished provided no barriers of relief but only a few barriers of water bodies, so that once a grassland fire started it would race away unchecked. The repeated firing of the prairies by the Europeans after the Indians had been removed was never so effective because grazing cattle left insufficient grass to produce a blaze hot enough to destroy the trees, and the forest advanced. In any case, the advent of permanent settlement, and particularly wooden fence posts and crops, put an end to discriminate burning to improve the range.[62]

The composition of the forest. Frequent burning had one other permanent and far-reaching effect: It altered the composition of the forest. Repeated burning and clearing tend to reduce the number of species and to simplify the biological components of the forest. For example, the extensive, open loblolly (*Pinus taeda*), longleaf (*P. palustris*), and slash pine (*P. elliottii*) forests of the Southeast are thought to be a man-induced fire subclimax within the general deciduous region of the eastern woodlands, the repeated firing in the past tending to propagate pyrophytic deformations with their characteristic needle-leafed adaptation. Some of the accounts of travelers through this region between the seventeenth and nineteenth centuries have been mentioned already, and many more are scrutinized in K. H. Garren's review of fire as a factor in the origin of the forests and in Elsie Quarterman and Catherine Keever's study of the concept of climax in the forests. All make it clear that Indian cultivation and burning were widespread and that pines were the first woody dominants to colonize abandoned land. Certainly, since European settlement, the southeastern forest has been repeatedly cleared and abandoned, usually resulting in the growth of new forests of pine.[63]

But the maintenance of that forest composition by burning is more complex than that. If the burning is frequent and fierce enough on the western extremity of the forest (i.e., eastern Texas), the forest is eliminated and grassland results. In the eastern portions of the forest, however, fire promotes the permanent establishment of the pine by removing the vegetation and debris that surround and shade the slow-growing longleaf seedlings and by eliminating brown-spot needle disease. But the ecological response is yet more complicated, for although fire is essential for the perpetuation of the forest through the clearing of the accumulation of litter, it must occur in a proper ratio of winter fires to fire-free years; that is, there must be nonfire years properly spaced so that the seedlings can pass through their critical first and sixth to eighth years of growth without burning and being killed. The suppression of burning and the reduction in the incidence of widespread fire during the present century have led to the invasion of the southern pine forests by hardwoods, such as oak and hickory, to such an extent that in recent decades the pine forests have reached the late stages of their succession, have declined, and have become relatively stable hardwood communities. Inferentially, therefore, contemporary trends in forest composition point to widespread Indian burning, which, on balance, must have favored the pinewoods and prevented the hardwoods from growing.[64]

To pursue this further takes one down many interesting sidetracks of cultural history and paleobotany, which is not our purpose, but it should be noted that similar arguments about the positive role of fire can be made for other forest types. Sequoia, Douglas-fir (*Pseudotsuga menziesii*), western white pine (*Pinus monticola*), and parts of the lodgepole pine (*P. contorta*) and ponderosa pine (*P. ponderosa*) forests of the West are subclimax fire types; fire is crucial to maintain their state. In the North, red and eastern white pine (*P. resinosa* and *P. strobus*) and, in the west, pure ponderosa, lodgepole, and Engelmann spruce (*Picea engelmannii*), though not fire types, required fire to hasten and ensure reproduction. At the other end of the scale, the extensive areas of aspen (*Populus grandidentata* and *P. tremuloides*) and jack pine (*P. banksiana*) varieties throughout the Lake States are not natural types and have been successful only because of major fire disasters during the past, just as the hardwoods have invaded the southern pine forests as a result of fire suppression and partial cutting in areas where wildfires previously kept hardwoods at bay.[65]

Conclusion

One is left with an impressive array of evidence about the impact of the Indian on the forest. The aggregate extent of the accumulated modifications is difficult to determine accurately, as the patches of clearing and burning were widely scattered and in various stages of regrowth. Also, no one has left a comprehensive record of the destruction of any considerable part of the pre-European forest to enable a calculation to be made, although the number of reports by explorers and colonists and the possibility that they saw only a small portion of Indian activity suggest that the total area involved might well be greater than the individual reports describe.

On the basis of population numbers and field sizes in selected examples, Kroeber suggested that one acre of cleared land on fertile soil, with a mixed planting and a long growing season, might sustain one person per annum. More recently, Heidenreich has calculated that the Hurons in Ontario needed 6,500 acres of cropland to support 21,000 people – or about one-third of an acre per person – but that when the extended fallowing cycle is taken into account the amount of land needed might reach 2.3 acres per person.[66] Thus, even if only a half of the estimated pre-conquest population of between 9.8 and 12.25 million for North America cleared and cultivated forest land, then between 19.8 and 24.50 million acres of forest would have been affected (and that would not have included abandoned clearings), which is approximately 10 percent of a total of 278.6 million acres of land in crops in the 31 easternmost states today. In his study of Virginia, Maxwell suggests that, if all the clearings that were cultivated and abandoned and all other burned areas were aggregated, a "conservative" estimate of treeless land would be between 30 and 40 acres per person. Maxwell's suggestion that another 500 years of

Indian settlement would have reduced the eastern forests to negligible proportions is provocative but probably exaggerated.[67] Unless population densities had grown to be so great that rotating agriculture was abandoned with the consequence that some sort of permanent cultivation and settlement occurred everywhere, then the prodigious power of the forest to regrow would have kept a large part of the country covered in forest. The greatest modifications would have occurred, as all botanical evidence suggests, in the area of the most frequent burning in the prairie ecotone, so that the grasslands, together with the changed composition of the forests over large areas, may be regarded as the greatest and most enduring cultural contribution of the Indians to the continent as a whole. Their numbers were sufficient and their agriculture extensive enough to achieve these changes.[68]

But, speculation aside, the Indians were a potent, if not crucial ecological factor in the distribution and composition of the forest. Their activities through millennia make the concept of "natural vegetation" a difficult one to uphold.[69] This does not mean that there was no untouched forest, or even fluctuations of climate, but the idea of the forest as being in some pristine state of equilibrium with nature, awaiting the arrival of the transforming hand of the Europeans, has been all too readily accepted as a comforting generalization and as a benchmark against which to measure all subsequent change. When the Europeans came to North America the forest had already been changed radically. Their coming did not alter the processes at work; it was merely their subsequent superior numbers and advanced technology that accomplished that. Paradoxically, their arrival may have been instrumental in causing forests to grow more rapidly and extensively in some places than before as fire was suppressed and controlled.

PART II

Change in the forest, 1600–1859

3

The forest and pioneer life, 1600–1810

Such are the means by which North-America, which one hundred years ago was nothing but a vast forest, is peopled with three million of inhabitants.

Marquis de Chastellux, *Travels in North America* (1789), 1:47

For approximately two centuries between 1600 and 1810 European settlers edged forward from various "hearths" along the 1,500 miles of coast from northern Florida to northern Maine, transforming the landscape into agricultural land and, by selectively removing and utilizing trees for domestic, semiindustrial, and commercial purposes, thinning out the forest and changing its composition. The map (Fig. 3.1) of the points of entry along the coast of some of the main migrant groups (and the subsequent spread of settlement inland) is a measure of the impact of man on the forest. The various frontiers of settlement in 1700, 1760, 1790, and 1810, when the total population of the country reached about 0.25, 1.6, 3.9, and 7 million, respectively, can be regarded as stages in the diffusion of the destruction of the forest. Lumbering was an important adjunct to agriculture for nearly the whole of the period, but after the midpoint of the eighteenth century it became pronounced in certain localities, particularly those where agriculture was either difficult or impossible, such as in New England.

In this process of change, the Indians were pushed out of their habitations ahead of the slow westward advance of the Europeans who disrupted their lives, either purposefully or accidentally by land alienation, a new social system, or above all, disease. Nevertheless, the initial form of European farming was not markedly different in kind from that of the Indian; it differed only in degree. The rotational clearing and widespread burning practiced by the Indians were not eliminated but were often replaced by more extensive and thorough clearing and burning, so that areas never touched by fire – on wet bottomlands or on the hilltops – were now affected. The mosaic of openings in the forest, which facilitated the use of the natural forest products, periodically recycled as the Indian moved from site to site, was replaced by a more continuous "opening," the object of which was to encourage permanent and stable settlement based on domesticated grasses, and imported grains and livestock. In addition, new demands were made on the forests for timber and new products, such as pearl ash and potash. Where Indians were eliminated ahead of settlement and abandoned fields were recolonized, the forest grew thicker than before. But the imprint of Indian occupation was never erased entirely from the landscape. When the land was cultivated

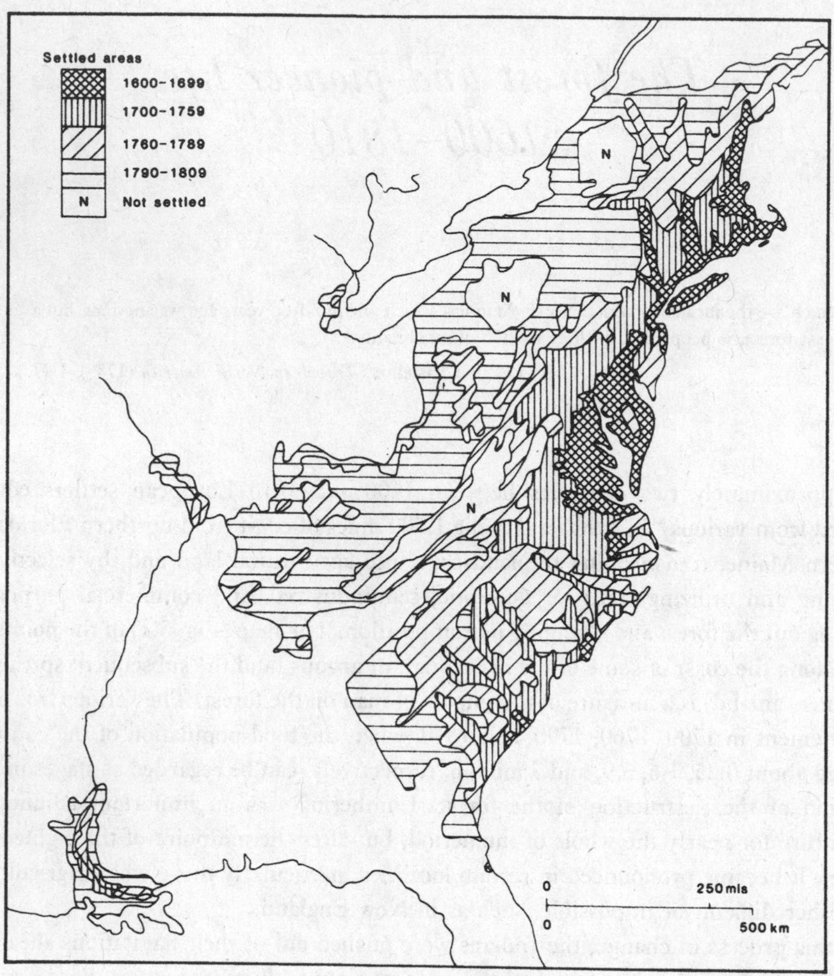

Figure 3.1. Expansion of settlement, 1600–1809. (Based on H. R. Friis, "A series of population maps of the Colonies of the United States, 1625–1790"; and 1810 census.)

continuously it was like an open wound in the flesh of the forest; when it was abandoned and allowed to regrow it was like a scar that was never exactly the same as the surrounding area because the density and composition of the forest changed. It was the beginning of a massive environmental change that had few counterparts in seventeenth- and eighteenth-century world history.

Sometime between 1710 and 1720 the European population along the East Coast reached about 300,000, which was well under the total Indian population throughout the eastern woodlands in the pre-conquest era of the previous century. Yet, curiously, it is almost more difficult to understand the European impact on the forest than the Indian impact of a century before. There is neither a body of evidence that gives a national picture nor a comprehensive account that tells all. The trees were so universal that they were barely noticed; to most pioneer farmers they were something to be gotten rid of, an impediment to the essential task of growing crops and feeding oneself and family. Consequently, the landscape changed imperceptibly from forested wilderness to fields, from fields to villages, and from villages to towns, but as Tocqueville observed even a century later, the American pioneer farmer, "the daily witness of such wonders [,] does [not] see anything astonishing in all this. This incredible destruction, this even more surprising growth, seem to him the usual progress of things in the world."[1] The destruction of the forest went almost unrecorded.

Perhaps one of the best ways to approach the topic is through a consideration of pioneer life, its characteristics and necessities, for the forest and its exploitation were a critical and basic part of that life. The tamed and domesticated landscape of open, arable fields, meadows, and intervening patches of woodland that the European had left behind appeared to be the natural order of things – as if it had existed forever. But it was different in North America where the forest presented an obstacle that had to be cleared in order to make a landscape. Clearing was a crucial experience that was multiplied many millions of times and that helped the transplanted European become an American.

Pioneer life in the forest

By the middle of the eighteenth century the forest experience of the pioneer had become a basic element in American geography and history. Chastellux, who traveled extensively in America during the late eighteenth century, said that the sight of the "work of a single man who in the space of a year" had cut down several acres of woodland and built himself a house in the resultant clearing was something he had observed "a hundred times. . . . I have never travelled three miles without meeting a new settlement either beginning to take form or already in cultivation."[2] The task of making a farm, of clearing trees, of building a house and fences, and even of making a road to his clearing was the greatest and the most labor-demanding task of the pioneer farmer and his family. Unless he could find a small Indian clearing, the making of his farm could take his lifetime, during which "he was less a farmer and more a builder."[3]

Throughout the colonial period the farmers of the northern and middle colonies built their economic basis by aiming at self-sufficiency in as many ways as possible, providing practically all the food, fuel, and clothing that was needed by their families.[4] It was a style of life that is commonly supposed to have developed traits of thrift, resourcefulness, and independence. But, laudable as these traits were and are, they have been heavily romanticized. Isolation and unremitting toil must have made pioneer life a combination of drudgery, worry, and limited cultural outlooks. As one of the ablest of pioneers' sons, Horace Greeley, said many generations later, "our farmers' sons escape from the fathers' calling whenever they can, because it is made a mindless, monotonous drudgery, instead of an ennobling, liberalizing, intellectual pursuit."[5]

The pioneer farmer could secure greater income and comforts only through his labor and energy. By clearing the forest, making buildings and fences, he was creating more capital, and even if his farm did not appreciate in value during his lifetime it probably would during that of his children. His cheap land was of little advantage to him because of the high cost of labor in colonial America, and if he had a large holding the ineffectual and primitive farming machinery available did not allow him to make many economies of scale.

But it would be wrong to think of the forest as a challenge and obstacle that needed extirpation in order to raise the level of his income and comfort. The pioneer farmer was well aware that the trees supplied many other needs and opportunities and that, if they were not there, the hardship of his life would have increased in this era of attempted self-sufficiency. Like the Indian before him, he knew that the forest was a source of wood for his home and for his fences. It was a source of much needed fuel during the cold winter months. It not only was a source of edible and nutritious wild berries, nuts, and products like maple sugar but also provided forage for cattle and immense herds of hogs, which became a mainstay of his diet. It was a source of wood for utensils, tools, and furniture and when the wood was burned, a source of potash and fertilizer. The products of the forest formed a valuable resource, and their exploitation could best be described as part of a backwoods system that was complementary to the task of clearing the land. However, there was a disadvantage, for in order to make the most of this resource the pioneer had to be a jack-of-all-trades and master of all – a carpenter, lumberman, mason, toolmaker, and blacksmith as well as a farmer – and all this added to the hardships of his life.

Sometimes the job was only poorly and half done, as the aim of the pioneer was merely to begin clearing, building, and cultivating in order to sell off the land to the next wave of more permanent farmers.[6] For the pioneer farmer who had more permanent settlement in mind, there could be a large nonsubsistence element in his activities. If he was lucky enough to have relatively clear land and to be located near the coast or a navigable river or in old, settled land within striking distance of a large town or city, he could put away his ax and use his plow to grow wheat or other food crops that could be sold. But even then he did not cease to be a pioneer making his living from the forest, for the demand in the towns was not only for food but also for lumber, fuel, and potash, the provision of these bringing in essential cash. Indeed, accessibility to markets and ease of transport were probably more essential for success in the initial stages of pioneering than they were after the land was cleared.

This outline of pioneer life was true for the majority of farmers in New England, in the northern United States, and for those German and Scots-Irish farmers of the middle colonies who subsequently traversed the Alleghenies, moved along the Appalachian valleys into western Virginia and eastern North Carolina, and settled in Tennessee and Kentucky by the end of the eighteenth century. The coastal South, however, was unaffected by this movement of people, and different conditions prevailed. Large-scale plantations growing one crop, mainly tobacco and later cotton, were always market-oriented and rarely self-sufficient. These were highly marketable crops, always in demand in Europe, light to ship, of high value, and not perishable. Slaves took the place of pioneer farmers as farm builders and toilers. Despite these differences, however, the forests of the South were affected by clearing because of the spectacular shifting of cultivation from the coast and the river valleys into the piedmont of Virginia and South Carolina during the mid-eighteenth century in order to overcome the soil exhaustion caused by the monoculture of tobacco or cotton. In addition, the pine forests yielded wood and naval supplies of pitch, tar, and turpentine, and consequently the interfluvial areas of the forests away from the plantations were severely affected.

Initial penetration

The pioneer settler with his limited resources of time soon learned that the less land he had to clear initially the more time he could devote to the growing of food and that this could possibly mean the difference between survival and death. Speed was the essence of success, and he knew that he should "begin where most ground may be easiest and speediest cleared."[7] There were very few supply ships bringing provisions to sustain him during his first few years in his new setting, and he had few goods and little money to trade for or to buy food, even if he could get it. The less dense the trees the greater the chance of success, and therefore the pioneer chose the existing clearing, particularly the old fields of the Indians, as a place to start.

In the first settlement in Jamestown, Virginia, in 1607 there had been no recent epidemic or disastrous war to cause the Indians to abandon their fields, and the English tried to clear the forest. Consequently, the crops were planted too late, and the "starving time" was the result. The lesson was not lost on the colonists. John Smith seized 300 acres "readie to plant" at Richmond, and Lord Delaware grasped the significance of the value of Indian fields quickly when he wrote, "No countrie yealdeth goodlier corne or more manifold increase, large feildes we have as prospects houerly before us of the same, and those not many miles from our quarter . . . [which] our purpose is to be masters of ere long, and to thresh it out on the flores of our barnes when the time shall serve."[8]

But in the succeeding years plague and harassment provided the colonists throughout the eastern woodlands with land enough. By 1646 the first Dutch settlers in the Lower Hudson region were growing wheat and oats for beer on land they found "fit for use, deserted by the savages who formerly had fields there." These old fields were scattered, which led to dispersed settlements and the need to clear the woods around in order to improve communication as well as to get new land.[9] A little later another Dutch observer, Andriaes Van der Donck, described some of these fields as "large extensive plains containing several hundred morgens" (about two acres per morgen), which were "very convenient for plantations, villages and towns."[10] The same was true in Pennsylvania, New Jersey, and New England; detailed descriptions of some of the old fields have been given already, and they could be multiplied by many others. Plymouth was in fact established in an old field, as were many other New England towns.[11] In New England the abandoned clearings were so extensive that one settler said in 1630 that they would provide enough "void ground" to receive all the people that the region could absorb for many years to come. The experience was repeated in Virginia, where there were said to be "abundantly sufficient" fields "ready to our hand."[12]

The Indian contribution to the early success of the European colonies was not limited to the provision of open ground. There was the gathering of wild berries and nuts, and the hunting of game. There was the planting of the quick-maturing crops like squash, pumpkins, corn, and beans, which grew easily in small mounds or on patches of irregularly and partially cleared ground. Above all, there was the shortcut to clearing by killing the trees through girdling. This method contrasted with the European method of cutting the trees and grubbing out the roots so that the ground could be plowed and grains sowed either broadcast or in rows. In the Indian system of hoe agriculture stumps could be avoided, and the fields did not have to be large or regular. A varied and bountiful supply of food could be harvested quickly without heavy equipment, draft animals, or massive inputs of labor.

The Indian crops and methods of tillage were not only used in the earliest days on the eastern seaboard but continued throughout most of the colonial period, particularly in the upland areas of the southern states of Kentucky, Tennessee, South Carolina, and Georgia. Obviously, productivity per person could be, and was, raised with draft animals and the introduction of the plow, but the solution to opening up the hardwoods country with a minimum of tools and effort was still an adaptation of an Indian method.

Acquiring land and learning about the forest

It was not long before the farmers had to face the task of clearing the forest to prepare new land for cultivation. The population soon grew beyond the numbers of its Indian predecessors, and the sedentary nature of European occupation meant that the scattered Indian fields were of less use than first might be supposed. In any case, many of the Indian fields had been cultivated for a long time before they were abandoned, and their fertility was badly depleted.

Before the pioneer could clear a single acre he had to acquire some land. In New England during the seventeenth century settlement proceeded by the creation of numerous tightly knit village communities within townships, which achieved the desired aims of religious and defensive coherence. The villagers had home lots and more distant farming lots. During the early eighteenth century, as settlement spread along the Connecticut and Merrimac river valleys, the compactness of the villages began to break down as the settlements tended to straggle alongside the river or the road, and by the time of the great expansion into New Hampshire, Vermont, and northern Maine, after the removal of Indian harassment in 1760, townships of 36 miles square were surveyed and cut up into rectangular lots of 50 to 100 acres. It was up to the individual settler to choose and buy the lot he wanted and to place his house on it, if he wished.[13]

Throughout the change in New England from a controlled, nucleated settlement to a freer, loosely aggregated settlement, pioneering was still reasonably orderly and controlled. But elsewhere on the frontier, in western Pennsylvania, Tennessee, Kentucky, Virginia, and the parts of New York not held in manors, settlement was a more haphazard affair. It was a relatively simple matter for an enterprising man to move into the forest beyond his neighbor, squat without legal title or right, sometimes blazing a tree and establishing his claim by "tomahawk right." Erecting a cabin and clearing a few acres were sufficient to establish title for the land, which could be paid for later.[14] After 1790, regular surveys by government and individuals in western New York and

Ohio resulted in a more orderly and less haphazard process, and so it had to be, for the new pioneer in these areas was likely to be not a near resident making a short-distance migration but a distant resident whose home was located somewhere on the eastern seaboard, more often than not in New England. The distant settler got information from newspapers and propaganda broadsheets; he even went to look for the land he wanted, possibly buying it, going home, and returning the next winter with "axe, gun, blanket, provisions and ammunition." He would then start to clear the land in preparation for his family, who might follow one or a few years later.

Although individuals pioneered in the traditional way in the South during the late seventeenth and most of the eighteenth centuries, the rise of the plantation system gradually forced all other considerations into the background. Large blocks of land of 250 to 1,000 acres were needed for cotton and tobacco growing, and the additional need for widespread shifting of cultivation meant that much land was disposed of by grants.

The locational criteria involved in choosing a particular piece of land in the forest were many. A near supply of water was essential as there were no pipes and few wells. Defense diminished as a consideration after 1760 with the defeat and removal of the troublesome Iroquois Indians, but the practical advantages of community cohesiveness, of neighbors and facilities like church, school, mill, and store – or at least the ability to utilize them by being located on or near a road – were rarely abandoned. In the expansion of New England during the eighteenth century and in Pennsylvania, settlement moved along the river valleys and their tributaries, which supplied water and power, required a minimum of clearing, and often had some natural meadowland. Settlement was selective, and the pioneers were now willing to bypass indifferent territory in search of the best land. This process continued during the 1770s, leaving the less desirable slopes and uplands to the latecomers during the 1780s.[15]

In the expansion of settlement during the latter part of the eighteenth century there were some significant changes in the pioneers' appraisal of the forest compared to what had gone before. Low forested ground was avoided because of the prevalence of malaria, which killed or at least so debilitated a man that he could not continue the task of clearing and making the farm. Malaria had not been entirely free from the river valleys of New England during the eighteenth century, but its prevalence in the Genesee country of western New York and in the Ohio Valley ensured its being considered seriously when settlers reached these areas after about 1790. It was widely thought that malaria (Italian *mal'aria*, literally "bad air") came from the vapors given off by the rotting vegetation of the forest during the initial stages of clearing.[16]

The practical pioneer knew that he should clear as much forest as possible in order to dry out the land and get rid of the "vegetable putridity" and that higher and sloping ground was more desirable despite its thinner soils and inherent infertility, which were, it must be admitted, well hidden under a deep humus accumulation. Sloping ground did not require draining, it was slightly easier to clear because of the lesser density of the forest, and there was no disadvantage because of machinery; none was used, and sloping land could be worked just as well as flat ground with scythe and sickle. Moreover, Yankee immigrants from New England who colonized the new lands of western New York and adjacent areas had usually entered from New Hampshire and Vermont, where they were accustomed to hill farming, and headed instinctively, so it seemed in many cases, for the uplands.

The condition of that unseen environmental element, the soil itself, was much more difficult to judge. There were no soil surveys or soil augers and no knowledge of soil chemistry, although some traditional folklore based on color and texture had been brought over from Europe. Therefore, the trees were used, as in all pioneer societies around the world at this time, as the gauge of soil quality. No trees meant infertility; hence the early avoidance of and skepticism about the fertility of the openings in the forest and the patches of natural grassland, which were quickly dubbed "deserts" or "barrens." There was a curious paradox about this, however, because Indian clearings were highly valued, and this can be explained only because pioneers knew that they had once been forest-covered. The larger the trees and the more luxuriant the growth, the better the soil. More complex than that, hardwood species, particularly oak, chestnut, walnut, and hickory, were commonly held to indicate good soil adapted to agriculture, and pitch pine and other pines to indicate dry, sandy, and infertile soils.[17] Every region had its plant indicators of soil fertility, which the locals knew and quickly pointed out to newcomers. The concept became so widespread by the early nineteenth century that vegetation was used as the basis for land valuation in Kentucky, Illinois, Pennsylvania, and North Carolina.[18]

Clearing

Northern and middle colonies

The pioneer could use either of two methods of removing the timber. He could either chop down the trees or girdle them, that is to say, hack around them deeply enough to cut off the flow of sap, causing the leaves and ultimately the branches to fall, so that the surrounding land, no longer shaded, would dry out and crops could be planted. The evidence suggests that, generally, cutting was more common in New England and Pennsylvania and in those places subsequently populated from these two regions (New York and Ohio, and Tennessee and Kentucky) than it was in the South. But girdling was by no means absent in the northern states, and the two methods mingled and were used according to the predilection and necessity of the individual settler. In northern New England cutting was common enough for people to call it the "Yankee" method of clearing, and its prevalence might well be associated with the high density of population and the need and market for logs, lumber, fuel, and masts. Certainly late-seventeenth-century leases sometimes had provisions to "stub, clear and plough," and it was believed by some that no crops could be grown until this was done.[19] It was Timothy Dwight's view that girdling had gone "much into disuse" throughout New England during the late eighteenth century, and that felling had become the universal practice. Jeremy Belknap said that upcountry areas of Vermont, New Hampshire, and Maine settled after 1760 were wholly clear-cut.[20] In Pennsylvania cutting was associated with the German immigrants, who did not "girdle the trees simply and leave them perish in the open as is the custom of their English and Irish neighbors." In New Sweden, along the Delaware

River, land was said to be completely cleared of trees and stumps by Swedish settlers, and in the New Netherlands the trees cut by the Dutch were "usually felled from the stump, cut up and burnt in the field, unless such as are suitable for building, for palisades, posts and rails."[21] But although these tendencies were observed, such ethnically based explanations are notoriously difficult to substantiate.

One of the earliest accounts of clear-cutting comes from Pennsylvania in 1692 in the form of a simple but powerful poem, which summed up the main characteristics of this form of clearing.[22] It ran:

When we began to clear the Land,
For room to sow our Seed,
And that our Corn might grow and stand,
For Food in time of Need,

Then with the Ax, with Might and Strength,
The Trees so thick and strong,
Yet on each side, such strokes at length
We laid them all along.

So when the Trees, that grew so high
Were fallen to the ground
Which we with Fire, most furiously
To Ashes did Confound.

Such clearing was a combination of "sweat, skill and strength."[23] First the settler cut away the undergrowth and grubbed out small trees and gathered them together in small piles for burning. Next he chopped down the large trees in autumn and early spring, and if he was skillful he would know how to make the trees fall not only so as to avert danger to himself but also so that they fell in parallel lines to facilitate plowing, and so that it was easy for the oxen to drag away the best logs for house building and other constructional purposes. The really experienced axman could "drive" the trees, that is, partially fell a large number so that when the last large one was cut it crashed into the rest and they fell like dominoes against each other. When the weather was dry enough the trunks were burned, and, if successful, "the greater part of them . . . consumed in the conflagration." He soon learned that to pile up the trunks for burning created an overly potash-enriched patch of ground that caused the grain to grow tall with long stalks and little ear. Such piling up for burning made sense only if he wanted to collect large quantities of potash to sell to nearby urban markets. However, if the trunks were burned where they fell the potash was spread reasonably evenly over the ground and the burning trunk helped to consume the adjacent stump.

The time when the clearing was done was important. If it was done in June when the leaves were on the trees the dead foliage would help in getting a "good burn" during the next dry period, usually the following spring. Other settlers preferred felling the trees in winter or in very early spring before the snow melted because there were fewer shoots from the remaining stumps and because this was the "off" season from their other agricultural activities. But weighed against this was the disadvantage of the timber not drying out enough for a spring or early summer burn.[24] Either way, clear cutting was an arduous task, but it left the field relatively free for agriculture after two years, and it

Plate 3.1. Life in New England, 1770. (*One Hundred Years of Progress.*)

remained clear forever provided the field was cultivated regularly to stop the regrowth of the forest. However, the problem of stumps remained, the gravity of the problem bearing an obvious relationship to the original density of the trees. Generally it was thought that the value of the land gained would not pay for the expense of grubbing out the stumps, a task that was probably more arduous than felling the trees. Only a few instances of immediate grubbing are known, and usually the stumps were left to rot until they were easy to lever out or pull with oxen. Side roots rotted in about two years, but stumps in the hardwoods in the northern and middle colonies took up to 10 years to rot. In Virginia and Maryland stumps took about seven years and in the South a little less.[25]

Stumps were no problem as long as grass was being grown because it could be browsed or the hay could be harvested with a scythe. Maize and tobacco cultivation were also no problem, nor were potatoes and flax in New England (Plate 3.1). In particular, corn required little preparation and yielded well enough, providing immediate food for family and livestock. It was common enough to see the tall stems "standing among the recently decapitated stumps of trees." If the cultivation of wheat, barley, or oats were contemplated, however, it was a different matter. Generally, a plow could not be used because it could not pass over the holes and between the stumps, the unburned logs, and the boulders. Usually, ground was broken up with either a hoe or a drag (often a log with wooden spikes driven into it) or a triangular harrow, and a crop of corn was put in.[26] If the settler could remove a few stumps, he could attempt to get a crop of buckwheat, wheat, or, in northern New England, winter rye. The yields of the first few crops were often prolific, reaching 40 to 50 bushels for wheat and 30 to 40 bushels for corn in good river valley locations, but only half that amount on upland and stony areas. But these yields were a sort of temporary bonus for initial effort, and most farmers were content to take moderate yields over larger areas. They needed a number of fields rapidly and partially cleared for different purposes – meadow hay, pasture, and corn – not one perfectly cleared plot. Farmers, such as those in New England who were near to markets, dispensed with crops and instead sowed grass only, on which cattle were grazed for a few years and walked out to the markets when sufficiently fattened.[27]

In contrast to clear-cutting, girdling was certainly more economical of labor. The advice given by John Smith in Virginia in 1625 was very sound: "The best way we found . . . to spoil the woods was first to cut a notch in the bark a hand broad about the tree, which pull off and the tree will sprout no more and all the small boughs in a year or two will decay."[28] Then the pioneer merely piled the undergrowth, small trees, and fallen boughs against the "mortally wounded" trees and let "the flames consume what the iron was unable to destroy." Finally, he planted corn between the skeletons of the deadened and half-burned trunks.[29] Girdling had its disadvantages, however. It produced less potash than clear-cutting, little good timber, and it was dangerous when the branches rotted and fell on people and livestock. "The falling branches incommode us for years, covering our grain every winter and causing a great deal of labor in picking them up," complained a settler from Pennsylvania. "The trees fall over our fences and demolish them; sometimes they fall on horses and cattle, killing or maiming them, and not infrequently men and boys have been killed."[30] In addition, the toppling trees threw up large expanses of inferior subsoil. Yet, for all these disadvantages, girdling was an effective pioneer expedient that solved many of the problems of rationing out those limited commodities of time and energy.

The clearing process was long, hard, and gradual. One writer estimated that it took the typical pioneer in western New York 10 years to clear between 30 and 40 acres. At the other extreme there were those who averred that a skilled axman could clear an acre a day and those who said that it took from 7 to 10 days to do the same. Gabriel Thomas in Pennsylvania in 1698 said that two men could clear between 20 and 30 acres, which were then "fit for the plough" in one year. These claims seem doubtful, however; a more common figure was something like 10 acres or 15 acres in one year, which was La Rochefoucauld's estimate, or 12 acres a year, which was Wansey's estimate. In South Carolina, Adam Hodgson noted a clearing of one acre now occupied by a log house and stable and granary whereas "thirteen days previously this was in the middle of a wood and not a tree was cut down."[31]

There could be no norm, for what was accomplished depended upon so many variables. There were the original density and size of the trees, the strength of the axman, the type and thoroughness of the clearing undertaken, the range of alternative tasks the pioneer had to attend to, like building a cabin, fencing, making a road, making furniture, and so on, and also how much land the pioneer had to clear. Unless he had a market for his product, there was little point in clearing more ground than he needed to raise the foodstuff for his family. It was just not necessary to clear the whole of the farm lot: La Rochefoucauld thought that few men cleared more than 30 acres, and later

Table 3.1. *Estimates of the cost of clearing land, 1688–1851*

	Cost per acre	Additional detail	Location	Source
1688–90	35–53 s.	+ fencing, stubbing, plowing	New England	Judd 1863, 432
1791	$6	+ fencing	New York	Steuben, in Ellis 1946, 75
1794	20 s.	+ fencing	New York	Wansey 1796, 191
1785–95	$4–$6	+ fencing	N. America generally	La Rochefoucauld 1799, 1:222
1791	$8 $16–$24	light timber heavy timber	New York & Penn.	T. Cooper 1774, 119
1811	$10	+ fencing	New York	Ellis 1946, 75
1823	$10–$15	+ grubbing	Virginia	Hodgson 1824, 294
1830	$10	—	New York	Fowler 1831, 79
1830	$10 or less	—	New York	Ellicot *Reports*, 2:340
1851	$3–$5	—	New York	N.Y. State Ag. Soc., *Trans.* 1851, 683–4

Beaufoy thought that 25 acres was adequate to support a large family and that 100 acres was the mark of a thriving farmer but that its achievement was "an affair of years."[32]

Sometimes the task was helped along by a "logging bee" when several neighbors came to chop the trees, burn the brush, and engage in logrolling in an attempt to help the newcomer to prepare several acres of ground for immediate cultivation. The copious quantities of rum and the games and talk afterward lightened the toil and enlivened the drudgery. The degree of help depended upon the stage of clearing and establishment that the neighbors themselves were in. If a newcomer was a later arrival than the rest then he was fortunate, but as Chastellux observed, on the pioneer forest edge "a man is never alone, never an isolated being"; help was always to hand.[33]

For the majority of pioneer farmers the cost of clearing the land was measured solely in terms of the labor and time they expended, but for those who did have money and could hire help the cost was high. It is difficult to be sure of the costs of clearing for, just as the estimates of the amount of land a man could clear in a given time varied enormously, so did the costs of clearing. Some of the cost estimates, the work done, and the location of the work are listed chronologically in Table 3.1. They cover a wide span of years, from 1688 to 1851, which is beyond the period of our immediate concern; nevertheless, the figures are included for the sake of comparison. All that one can say is that costs rose from somewhere around $5 to $6 per acre in the late eighteenth century to $10 during the early nineteenth century, when they stayed fairly stable for many years. The stability was the equivalent of a fall in real costs and can be explained only by the rise of a specialist group of clearers, of which there is some evidence, and by the increasing value and demand for the potash from the burned trees, which was variously estimated to repay the clearer half to two-thirds of his costs. Certainly, steady costs had nothing to do with improved techniques, for nothing replaced the ax until the power saw of the early twentieth century. Attempts were made in the late eighteenth and early nineteenth centuries to use pulleys and ropes to haul men to the tops of trees to lop off branches, and also to chop the roots of the trees and pull them over with ropes and windlass. The first experiment merely left gaunt poles that still needed cutting; the second left great stumps that would not burn immediately, and large areas of earth were thrown up when the stump toppled. Both experiments were so expensive as to be prohibitive to the average settlers. It took five times the labor to chop the roots as it did to cut a tree conventionally. All in all, the pioneer had the right idea in chopping down the trees in his traditional way, and if he girdled them it only cost him a tenth of clear-cutting them, although he was merely deferring his expenses of time and labor for some future period.[34]

The South

In the commercially oriented plantations in the pine forests of the South, all forms of clearing were employed, but girdling was most common. As William Byrd said in 1623, "the English have up to now with little difference imitated the Indians in this."[35] Girdling was common and popular because of the necessity for haste in tobacco and cotton growing and to a lesser extent in indigo and rice. However, as girdling did not provide the large quantities of potash needed for these demanding crops, partial clearing by cutting the trees about a yard from the ground and burning the trunks in situ became increasingly common. The land between the stumps and trunks was then hoed up and crops, particularly tobacco and cotton, planted.[36]

Because tobacco was such a heavy consumer of plant nutrients and because its growth encouraged soil toxicity it was inevitable that the land would become exhausted and "sour" in time and would then be abandoned. Therefore, grubbing up the stumps was never worthwhile. Even when the stumps rotted in about seven years it was usually calculated that the land was already "worn out." Plowing replaced hoeing only when local conditions caused a check to expansion, and the "old" lands were used again after 20 years or so.[37] Planters calculated on getting three to four crops of tobacco from a field before it was turned over to a few crops of corn or wheat, and when the yield of the latter dropped to below five bushels an acre the land was abandoned to weeds and the regrowth of the pine forests, which occurred rapidly. "After twenty or thirty years," commented one observer, "the same land would be cleared and put under a similar scourging tillage."[38]

Obviously, the shifting of fields from time to time would be economical only because new land could be acquired and cleared as cheaply as manuring the old to maintain its fertility: "We can buy an acre of new land cheaper than we can manure an old one" said Thomas Jefferson; dunging was "regarded as more irksome than cutting down trees," said Johann Schoepf. Yet hardly any cattle were kept for their manure; most were roaming wild in the forests. When a man's labor on tobacco yielded six times as great a return as any other crop, and certainly more than any stock, keeping stock for manure was not a paying proposition. (In any case, manured tobacco plants gradually began to taste of manure.) The high-paying returns on tobacco could best be achieved by planting new ground, skimming off the inherent fertility of the soil, and then moving on to lands newly cleared from the forest. It was known by some as the "Virginian mode of cultivation." The abandoned ground "when left to itself is gradually clothed with wood again," observed Hodgson, but he was perceptive enough to realize that it was "seldom of so large a growth as the original trees."[39]

To accommodate these constant shifts, estates had to be large, at least 1,000 acres in size by 1700, and many were up to 10 times that size. Inevitably, therefore, these estates were widely spaced, and coupled with the constant regrowth of abandoned fields, the impression was that they were like islands of cultivation in a sea of forest. Thomas Anburey, who traveled through the South in 1789, said:

> The house we reside in is situated upon an eminence commanding a prospect of nearly thirty miles around it, and the face of the country appears an immense forest, interspersed with various plantations, four or five miles distant from each other: on these there is a dwelling house in the centre with kitchen, smoke house, and out-houses and from the various buildings each plantation has the appearance of a small village.[40]

The investment in buildings, slaves, quarters, and curing houses was high, and few plantation owners wanted to move if they didn't have to. However, the smaller owners were forced to abandon their holdings permanently, and they moved west and south from coastal Virginia and Maryland into Kentucky, Tennessee, and North Carolina during the eighteenth century.

For the planter in the South, just as for the pioneer in the North, the amount of land cleared and the cost of clearing it depended upon a great number of variables. It was commonly thought that one slave could clear about three acres during the fall, split the timber for fencing and huts, and then prepare the ground for planting in March. The cycle of slave labor was minutely itemized by De Brahm for a large plantation in eighteenth-century South Carolina. The cutting of trees on one acre was considered "a day's task of eight slaves"; lopping and burning carried out during the evening was extra, and the "weak hands" – the women and boys – cleared and burned the bushes and shrubs and scattered the ashes over the ground. The cultivation of a cleared acre of corn or potatoes or peas and four acres of rice or indigo was considered the task of one slave for a year. When the maintenance of the cleared plot become too great the ground was abandoned, and the cycle of clearing and cultivation began again elsewhere. Although most of the new fields remained "lumbered with the bodies of the trees" for a number of years this did not "hinder the planters from cultivating the clear spots," and bit by bit the trunks were split into rails and constructional timber or collected as fuel.[41]

How much clearing was actually done depended very much on the state of the tobacco or cotton market, more being cleared when demand and prices were high, less when there was a downswing in trade. Clearing was a far more commercially oriented and influenced process than in the North.

Even though clearing was done by slaves there is no reason to think that it was done without expense, as each slave cost about £40 to buy in the early eighteenth century and had to be housed and fed. In 1710 Thomas Nairne calculated the costs of establishing various-sized planations. The plantation of 1,000 acres needed 30 slaves and cost at least £1,500 to establish, of which £1,200 was spent on buying slaves, who "beginning to work in *September* or *October* will clear 90 acres of land, plant it and hoe it."[42] What the northern farmer paid for with hard labor the southern planter paid for with hard cash.

Whether in the North, the middle colonies, or the South, the process of clearing and farm making in the forest was widespread, universal, and an integral part of rural life, which meant the life of the greater proportion of the population. "Such are the means," marveled Chastellux, "by which North-America, which one hundred years ago was nothing but a vast forest, is peopled with three million of inhabitants. . . . Four years ago, one might have travelled ten miles in the woods I traversed, without seeing a single habitation."[43]

Stock in the forest

The contribution of the forest to the pioneer farming economy did not end with the creation of land for crops by the felling of trees. Beyond the clearing lay the unfenced range of the forest, which provided free feed for stock – beechnuts, acorns, chestnuts, together with young shoots and roots (all known as "mast") for the swine and grasses for the cattle, the growth of the latter encouraged by frequent light burning in the Indian fashion.[44]

Probably the most common stock was the razorback hog, which roamed the forest in great numbers. It was a hardy and fierce animal that thrived on the abundant mast of the forest. The swine provided a semiwild and inexpensive source of protein for the pioneer family from the very beginning of settlement, and pork was the most widely used meat in the United States until the later part of the nineteenth century.[45] Either as fresh meat or salted down and packed in barrels, pork became so universal that the full pork barrel became a symbol of plenty, but when its bottom could be seen it was a symbol of starvation – one was simply "scraping the bottom of the barrel."

The utilization of the forest for swine was not a new practice that arose in America. It went back to biblical and classical times in the Middle East and Europe. In medieval Europe woodland was often measured according to the number of swine it could support rather than according to its acreage.[46] The right of pasture for swine – or pannage, as it was called – was a privilege that the medieval peasant valued highly, and in a like manner it was a highly prized resource for the American pioneer farmer during the seventeenth and eighteenth centuries and, indeed, even much later in the South.

The source of the wild swine in America was probably the De Soto expedition that had threaded its way through the South between 1538 and 1541. The absence of wild animals in the accounts of early settlers in Virginia and Maryland suggests that the stock had not got this far north by the early seventeenth century, but within a few decades of settlement the forests of all the colonies from the northern Atlantic coast to Florida were swarming with swine as well as feral horses and cattle. Stock was particularly prevalent in wild and rugged country that did not warrant clearing, and what seems to be a heavier concentration in the South probably had something to do with the warmer weather, which meant that winter shelter was not a necessity for survival.[47]

In addition to the swine, there were herds of cattle, both feral and semidomesticated. Herds of from 200 to 1,000 head were not uncommon in the Carolinas by the late seventeenth century, and some were much larger. The massive increase in their number had less to do with the warm winters than with the ready markets for hides and meat in Europe, the West Indies, and even New England and with the organization of wild grazing by the Carolinians into a fairly efficient system in tune with the forest environment. Cow pens were created, fenced enclosures of various degrees of permanence in clearings in the forest in which the best stock was kept in winter and hand-fed and in which the young stock was penned every night. The rest of the stock roamed the surrounding forest during the day, foraging for grazing, particularly in the canebrakes for winter feed and in the savannas, which were constantly fired to provide fresh grass growth in spring and summer and to keep the trees from growing back. At night the cattle returned to the cow pen to suckle the calves, and they were milked before being let out again into the forest range. With the building of rough barns and dwelling houses beside the cow pens, and with the emergence of rustling, branding, and roundup by black "cattle hunters" or "cowboys," the cow-pen system in the southern forests was the prototype of the western range of later centuries. The cattlemen were a mobile group, not only in driving their stock to distant urban markets, such as Baltimore and New York, but also in order to achieve flexibility and avoid conflict in their operations. For example, the ranges in South Carolina were so overstocked by the mid-eighteenth century that herds were being driven south into Georgia to the unsettled country between the Savannah and Ogeechee rivers. They were "kept in granges under the auspices of cowpen keepers, which move (like unto the ancient Patriarchs, or the modern Bedewins in Arabia) from forest to forest in a measure as the grass wears out, or the planters approach them, whose small stock of cattle are prejudicial to the great stocks."[48]

Cattle herding in the forest was not always as specialized as described, as some westward-moving tobacco planters incorporated parts of the stock economy into their crop economy and kept large herds of swine or cattle in the uncleared portions of their estates for little expense other than the attention of a couple of superannuated slaves.[49] In this way, pork and corn became the staple diet of many in the South.

Initially, cattle and swine either were branded or had their ears notched for identification at the subsequent annual roundup, the unmarked and illegally grazing stock being sold or killed for winter meat. There was hardly a town in eighteenth-century New England without its town pound, which was a measure of the extent and importance of woods pasturage.[50]

As with so many matters concerned with the utilization of the forests during the seventeenth and eighteenth centuries, we have a general picture of what went on, built up of isolated instances and examples, but no comprehensive or quantitative view of what was happening in a small part of the country, let alone the whole – mainly because of a lack of census data. One hint is given in the tax returns of Connecticut in 1796 where 579,847 acres, or 29.1 percent of the state, were classed as "brush pasture," which was cutover land and old fields with forest regrowth used exclusively as pasture. It was not to be confused with the 425,595 acres, or 21.4 percent of surface area, of "clear pasture." There was also another 606,573 acres of uninhabited woodland, some of which must also have held wild stock. In Massachusetts, where it is known that hog running went on in the woodlands, 59 percent of six counties was woodland-covered.[51]

Regulation of free grazing within an agricultural system worked well as long as the crops were fenced in and the animals fenced out. Inasmuch as fencing required timber, the system of wild grazing could last as long as the wood supplies were abundant. A time would come when a critical balance would be reached between the amount of woodland for free grazing and the amount of timber available for fencing. That was to come in the 1830s in areas of little woodland on the prairie edges of the eastern forests.

Fences and houses

In emphasizing the destruction and removal of the forest in the frontier areas and their use for wild grazing, it should not be forgotten that the timber was also a resource of the highest value. The pioneer farmer used the wood for his own fuel, selling it for cordwood to commercial users, making potash and pearl ash, collecting bark, turpentine, and pitch, and even selling lumber. But during the earliest pioneering days these were all incidental activities which, though they aided financially in the task of establishing the farm, were not integral parts of making the farm as were fencing and the building of a house.

Fencing

The moment the trees were felled the material for fencing was available. Fences could be crude and makeshift to begin with; they could be brush fences or piled-up stumps, or a combination of different types. All were excellent pioneer expedients. When time and money permitted, poles were selected and straight-grained logs stored for splitting into rails during the winter when the sap was less.

The purpose of the fencing was to keep the animals out of the fields, not in them, quite the reversal of the European practice in the postmedieval period. Because the forest was used as a range for hogs, cattle, and horses it was logical that, instead of fencing in the animals and thereby failing to take advantage of the natural assets of the forest, settlers chose to fence only their small, cultivated fields. The need for fences, therefore, became enforced by a variety of legislation in the different states, but the common purpose was to indemnify the owners of wild stock from claims against them, and it put the onus of responsibility on the cultivator.[52] So long as that was the case, the sort of clashes associated with frontier behavior could be expected. People soon learned what Robert Frost immortalized three centuries later in his poem "Mending Wall," that "Good fences make good neighbours" and that they contributed positively to maintaining community order.[53]

There were many sorts of fences, but the Virginia fence, or worm, snake, or zigzag fence, was probably the most innovative and certainly the most widely adopted fence of

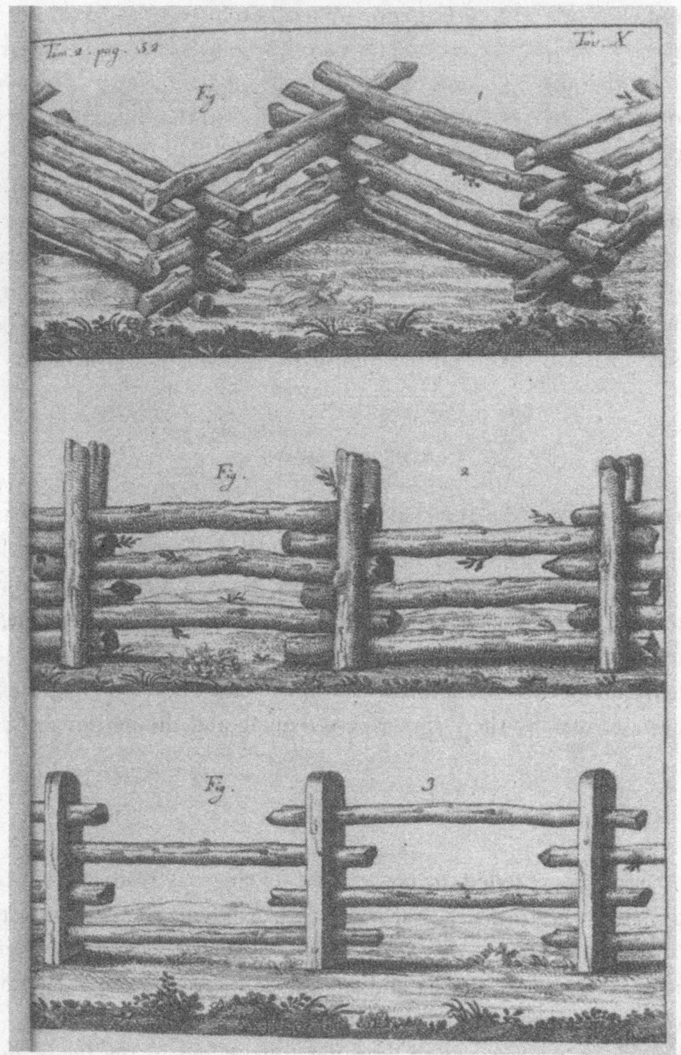

Plate 3.2. Types of fencing in eighteenth-century America. *Fig. 1* shows the worm or zigzag fence, which required enormous amounts of timber but little labor to erect. *Fig. 3.* illustrates the post-and-hole fence, which required far less wood than the worm fence but considerable labor in digging post holes and making holes through the uprights. The double-post-hole fence shown in *Fig. 2*, in which the rails were slotted between the posts in alternate layers, was somewhere between the other two; it was not in common use. (Luigi Castiglioni, *Viaggio negli Stati Uniti dell' America.*)

all (Plate 3.2). The fence was constructed by laying horizontally a set of between 6 and 10 slender 12-foot-long poles or rails in such a way as to interlock with one another at right angles in a zigzag pattern, sometimes with heavy bracing logs to lock the intersecting angles. When split wood was used rather than young trees or branches, the preference was for clear, easily split timbers, such as oak, black locust, walnut, or American chestnut. "At first sight the worm fence appears very inefficient," observed William Oliver in the middle of the last century, "but on closer inspection and more intimate acquaintance with its qualities it improves in one's estimation and would certainly be difficult for the pioneer settler to substitute anything so efficient and at the same time so easily got." He had a point: It certainly required enormous amounts of wood and took up large areas of land, which were uncultivable and became weed- and vermin-infested, but it did have the advantages of being self-supporting, requiring no post holes, no nails or pegs, no mortises or ties, and it was easy to repair and to remove to a new location, the latter being an important consideration in the incremental enlargement of clearings and in the shifting cultivation due to soil exhaustion in the tobacco fields of the South. The absence of the need for post holes meant that the fence did not rot easily (it was commonly said to last for between 20 and 35 years), and because it was not fixed into the ground no special timber was required, only what was abundant and at hand. Moreover, because of its heavy construction it was a good answer to marauding hogs, which could jump over low fences and knock down or root under flimsy ones. At its very worst, the fence was a store of firewood for the years ahead when local supplies of wood had diminished with forest clearing.[54]

The problem with the post-and-hole fence (see Plate 3.2) was that the digging of the holes was laborious; for example, a common slave task was for four slaves to dig and place posts for the erection of 35 to 40 panels a day, a target the solitary pioneer never achieved, even assuming he felt the need to work as hard as the slaves were made to work. The invention of the spiral auger after 1800 for post-hole digging caused a great rise in the popularity of the post-and-rail fence, but a side result was that the farmer had to be far more selective in the type of timber he used, and wood like eastern redcedar and American chestnut, which did not rot so easily, might not be available locally. In addition, the mortising of the uprights was both a time-consuming and a moderately skilled job. Plank fences were generally less common.[55]

In New England there was an abundance of timber, but the boulders in the glacial till of the fields and the mountainsides had to be removed before cultivation could begin, and in many places fences of piled-up boulders took the place of the wooden fences. Where settlement penetrated treeless areas, such as the bluegrass region of Kentucky, middle Tennessee, and Missouri and the oak openings of Ohio and New York, ditches, stone walls, plank fences, and embankments were resorted to. But, of course, the purpose of fencing was reversed in such areas; with no forests there were no wild swine to keep out, only domesticated livestock to keep in. The need then was for "live" hedges that grew, not "dead" hedges that had to be built from costly and scarce timber imported into the region, and the problem was not solved satisfactorily until the invention of barbed wire. Nevertheless, even as late as 1850, it is estimated that worm fences accounted for 79 percent of the 3.4 million miles of fencing in the United States, and that post-and-rail and plank fences accounted for another 14 percent; the remaining fences were of stone. Wood was the overwhelming fencing material well into the nineteenth century.[56]

Fencing was as cheap as the timber was available and as the time was free to cut and prepare it. An experienced farmer could split 50 to 100 rails a day, and a few strong men were known to split 175 to 200. Another measure was that slaves, either men or women, were commonly set the task of splitting a hundred 12-foot rails in a day or building a hundred panels of a worm fence in a day; or four slaves might erect 35 to 40 panels of a post-and-rail fence in a day, digging holes two to three feet deep and nine feet apart. If labor was hired, it cost between 50¢ and $1 per rod (16½ feet) to split timber and erect worm fences during the early nineteenth century.[57]

The smaller the field the more fencing it needed relative to a large field,[58] but whatever the size the amount of wood needed was enormous. For example, a square field of 160 acres required half a mile of fencing on each side if fenced in a straight line (a total of two miles of fencing), but nearly half as much again if fenced in right-angled zigzags. Therefore, a 10-rail, 10-foot-length zigzag fence required nearly 15,000 rails, not including overlaps at the ends, but a straight post-and-rail fence required only 8,800 rails and about 200 posts; less, but still a considerable amount of timber.

The large quantities of wood required for fencing did not matter when the wood was going to be felled anyway as in the interior, but in densely settled areas like eastern Pennsylvania and coastal Virginia, and around all urban areas where thorough clearing and stripping for firewood had occurred, post-and-rail fences became more common than zigzag fences because they required less wood in their construction. This situation became more common as the population became more urban during the late eighteenth century. As early as 1750 Pehr Kalm made the dire prediction that the forest in Pennsylvania would last only another 40 to 50 years if the zigzag fence continued to be built. That, of course, never came about, but by the end of the century La Rochefoucauld was convinced that with rising prices "sooner or later this useless waste will certainly be regretted," something Michaux echoed later.[59]

Houses

Besides fencing, the pioneer farmer in the forest had to provide a shelter for himself and his family. The first shelter was often no more than a lean-to of boughs with a grass covering in which the family huddled in blankets for the first few months until a more substantial log cabin could be built. The log cabin was the archetypal symbol of the pioneer and the frontier (Plate 3.3). The log cabin is regarded as typically American, and so it was if measured by its prevalence throughout the country well into the nineteenth century; yet it seems likely that the idea was introduced into the Delaware area by the Swedes during the early seventeenth century.[60] The log cabin had much in common with the zigzag fence. It was extravagant in its use of wood, but because it required no nails or holes it was easily and speedily constructed. All that was needed were 80 logs of between 20 and 30 feet in length, split logs for gables and roof, and a few helpful neighbors. After the first four logs had been laid to form the rectangular foundation, axmen stood at each of the corners and notched each new log as it was put in place on top of the other, the axmen rising with the walls. The gaps between the logs were filled with a clay and straw mixture or even with moss, but often so carelessly that light could be seen through the walls. The floor was made of split planks and the roof made variously from thin logs or branches. Depending upon the amount of help received from

Plate 3.3. Log house in the forest of Georgia. (Basil Hall, *Forty Etchings*, plate XXII.)

neighbors, a cabin could be completed in one to three days. A rough and wide stone fireplace was essential, for here was the source of warmth and even of light during the long winter months; the chimney was often made of clay-daubed boards, which served until such time as a stone flue could be built. The draftiness of the cabins meant that the fire was kept burning day and night, and "amazing quantities" of fuel wood were consumed in the grate, said La Rochefoucauld. "A little further off," he continued, "appears a small shed, like a sentry box, which is the necessary or privy." The cabins rarely had windows, and the door was added last, hung on wooden hinges and fastened with a wooden latch. Not a nail was used in the construction of the cabin. If rough-hewn planks were available, the floor was raised a foot or so to guard against the damp and cold. Furniture and utensils of wood were added as time permitted and need dictated, the most important of which were the two great beds that seemed to dominate most cabins and that "receive[d] the whole family."[61]

The variations in design were numerous, the most common plan being a single square or rectangle, sometimes divided by an interior partition (a hung quilt or blanket), which dominated the forested areas of Upper New York, central and western Pennsylvania, the settled areas in the West and Appalachia and Mississippi. In the southern central states a more common plan was one of two square "pens" or rooms separated by a passage and covered by a single roof. Later, as planters became more affluent, they added more "pens," cut doors through the intervening walls, and sometimes surrounded the structure with a shady porch; thus the spacious southern house evolved. Other variations of these wooden folk structures were more subtle and seemed to have regional characteristics that are difficult to generalize about or to classify.[62]

In eastern New York, eastern New England, and in the tidewater settlements, frame houses sheathed in sawed clapboards predominated from the beginning of settlement. Where log cabins and particularly blockhouses – so-called because they were built from squared logs, with small slit windows, a defensive overhanging second story, and mortised corner joints – had been built on the frontier for defensive purposes during the seventeenth century, they were slowly replaced by clapboard houses as the Indian threat diminished and sawmills were erected. For example, during a lull in the local Indian menace in New Marblehead in 1740 a sawmill was erected, and with peace a spate of frame house building began in about 1750. It was the same in all the other New England frontier towns, so that two-and-a-half-story frame houses with pitched and gabled roofs appeared throughout the countryside, especially after 1760.[63]

The altogether more "English" form of construction of the frame house was thought to be more elegant and a step above crude forest pioneering, and it became a model that most wished to copy. "For most people," said Timothy Flint, "the log cabin is only a temporary structure until a house of frame and cut boards could be built." That, of course, depended upon a suitable supply of milled timber, but so powerful were the "elements of rivalry and emulation" that many people were tempted to build "larger and more showy houses than were called for either for comfort or conformity to the circumstances of the builder." Many owners merely nailed clapboards over their log cabins to produce the required effect once the initial rigors of pioneering were behind. Although in time the log cabin was replaced by more elaborate and elegant structures, a surprising number remained. In New York there were 33,092 as late as 1855, and they must have housed nearly one-fifth of all farm families; the number was reduced only to 20,245 by 1865. Whatever the form of structure or appearance, however, over 80 percent of all dwellings were made of timber construction at the turn of the nineteenth century.[64]

Potash and fuel

In making a farm out of the "immense magazine of wooden materials" in the forest, the pioneer created two incidental but valuable by-products: potash and fuel wood.[65] The potash was not used on the farm but sold commercially; the fuel was used on the farm, but there was usually such a surplus of timber from clearing that there was plenty to spare for sale in more distant markets.

Potash

Besides fertilizing the ground *in situ*, the value of burning the felled timber lay in its production of ashes, a commodity that could be sold readily to local asheries, which then transformed it into potash, which was used for many industrial processes. The best ash was produced from hardwood timber, particularly oak and sugar maple, which yielded the highest amount of potash. American elm, hickory, and beech yielded less. The logs were burned in the open and the ashes collected and sent to the ashery where they were leached by pouring water over them in large kilns. Repeated leaching with alternate boiling produced a somewhat purer, creamy white ash, known as pearlash, which was anywhere from three to five times more concentrated and therefore more valuable. In a liquid form these solutions produced lye, which was boiled in iron kettles until the water evaporated leaving a hard brown residue. Lye was an alkali that was essential in the manufacture of soap and also for glass making, tanning, bleaching, cleaning greasy wool, calico printing, saltpeter for gunpowder, medicines, and a number of other chemical operations in which potassium carbonate compounds were needed. Lye sold readily on the East Coast and in the South , where pine trees made poor and little ash. There was also a ready market in industrial Britain, where traditional supplies from the Baltic had been interrupted by conflicts in Europe during the early eighteenth century. The demand for potash was so great that Britain, through the Society of Arts in London, offered premiums for its production in America.[66]

Although only a few cents (between 5¢ and 12¢, depending on its degree of concentration) were paid for every bushel of potash produced, the return was in hard cash, which enabled the farmer to concentrate on his main task of clearing the land and making a farm instead of wasting his time producing other goods that he could trade for necessary supplies of food, clothing, and implements. In Thomas Cooper's view, "One-half or two-thirds of the expense of clearing land in New York is repaid by the potash obtained from burning the wood," although in Pennsylvania and the southern states the "back settlers" were "not so much in the practice of this useful method." (He did not seem to realize that the soft and resinous timbers of the southern forest yielded very little potash.) Potash even bought the hired labor necessary for improvements on the farm and even for initial clearing.[67] If there were no asheries nearby, the farmer could sell his ash at a much farther distance, as its high value compared to its bulk and its ready sale at the ports made it capable of being transported great distances, the only impediment to its sale being the state of the roads. Nearly every major American port north of Philadelphia exported potash and pearl ash. Boston handled nearly half of the 10,600 tons exported during the five years for which customs returns are available, from 1768 to 1772, and New York exported nearly a quarter of the total (Fig. 3.2). Some measure of its value is given by its ranking as sixth in value among American exports. However, asheries sprang up everywhere, beside local stores and at crossroads, as part of the local pattern of trade and barter, and some farmers invested in their own kettles in order to cut out the middlemen. In one New York community, Beekmantown on the western shore of Lake Champlain, there were 22 asheries in 1821 as opposed to one grist mill, one fulling mill, and no sawmill, and Beekmantown still had 21 in 1825, although the number declined markedly after that.[68]

Fuel

Perhaps less important immediately than potash as a profitable sideline, but ultimately of far greater significance, was the fuel wood gained from clearing. It seems likely that the use of wood for fuel has exceeded any other demand on the American forest, and even now in the closing years of the twentieth century it is doubtful if the total value of timber cut for commercial purposes has yet reached the total cut for fuel since the beginning of settlement. Fuel was basic and indispensable for the very preservation of life. The normal winter climate of the northern half of the United States was such that life without shelter and warmth was impossible for seven or eight months; "so much of the comfort and convenience of our life," said Benjamin Franklin, "depends on the article of *fire*." The remedy, said one early settler, was "not to spare the wood of which there is enough," and great blazing fires halfway up the chimney were a common sight in the homes of colonial America.[69] Even in the warmer South, the winters were still cold

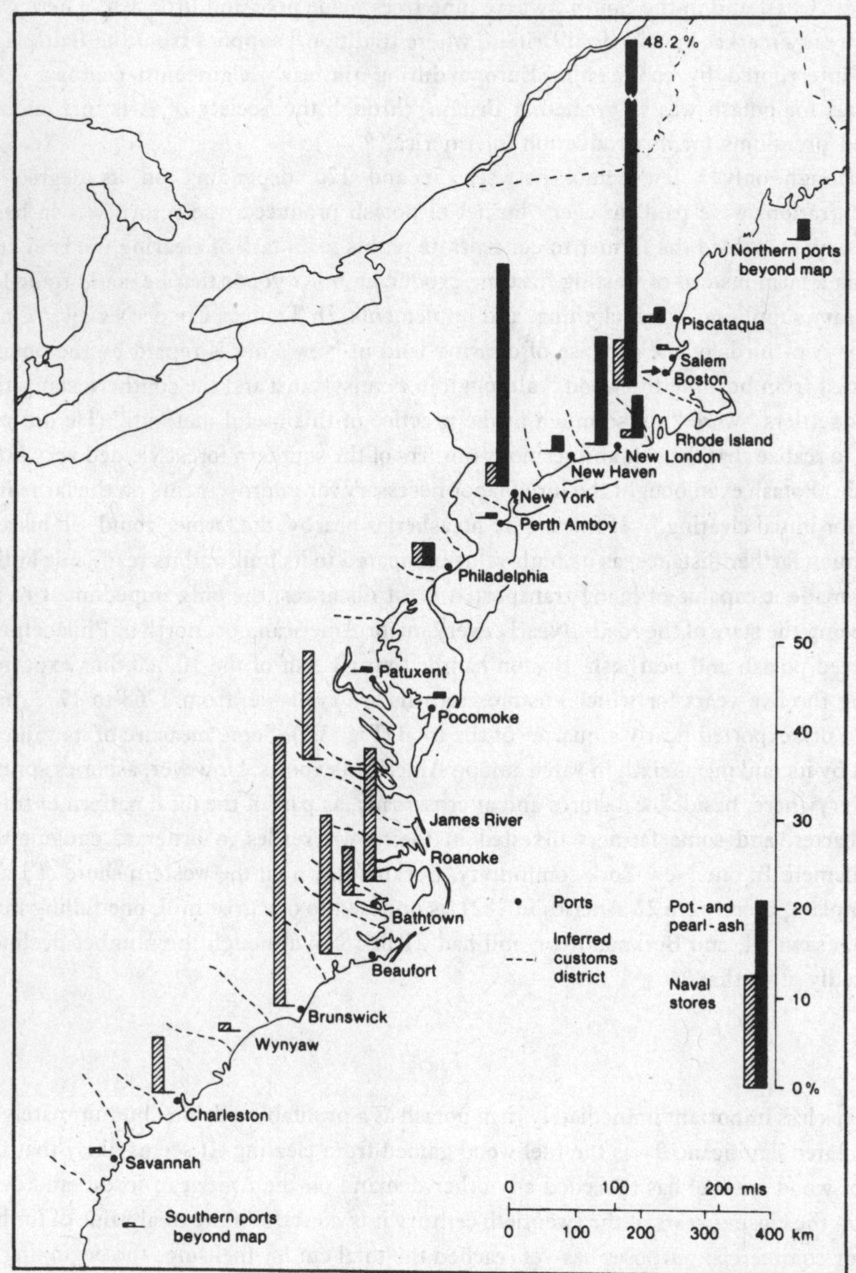

Figure 3.2. The export of naval stores and potash, 1769–72. (London: Public Record Office, Customs, 16/1.)

enough for houses and huts to require heating; the slaves never dared return from the fields "without bringing a load of firewood on their shoulders."[70]

For the pioneer farmer, fuel was plentiful and cheap. It was an incidental by-product of clearing land or building a cabin, and if the trees were cut especially to provide fuel, the gain in newly cleared land meant more food eventually. Either way, fuel gathering and making a farm went hand in hand, and as a consequence, like clearing during these years, little statistical evidence is available about the impact of fuel gathering on the forest. Little thought was given to the use of alternative forms of fuel; said one visitor to Virginia in 1705:

> And as for Coals, it is not likely they should ever be used there in anything but Forges and great Towns, if ever they happen to have any; for, in their Country Plantations, the wood grows at every Man's Door so fast that after it has been cut down, it will in seven Years time grow up again from seed to substantial Fire Wood.[71]

Therefore, fuel was available for the asking – or, perhaps more correctly, for the cutting – because the forest surrounded and was part of all the new farms that had to be hacked out of the forest.

But the task of collecting fuel was not without its cost as a great deal of hard labor had to be expended in chopping, splitting, and carting the bulky, heavy, hard commodity. Anything that could lighten that task and at the same time provide a fire with the most heat, the least smoke, and the longest-lasting flame was eagerly seized upon. The lone cutter selected the smaller trees, well under two feet in diameter, and avoided twisted and knotty ones. Chopping was often done in winter when other farm tasks could not be undertaken, but more especially because logs split more easily during freezing weather and, when there was snow, they could be dragged out on sledges. Above all other considerations, the trees were selected for their inherent qualities so that the best-burning wood was cut first. Hardwoods have the best percentage value of heat as related to one short ton of coal. For example, hickory is preeminent with 96 percent, oak has 86 percent, sugar maple 84 percent, American beech and birch both 80 percent. Chestnut, on the other hand, has a value as low as 60 percent. Most softwoods have a low heat value, pines ranging between 70 and 77 percent, tamarack being 75 percent, Douglas-fir 68 to 60 percent, and lodgepole pine 58 percent. There were considerations other than heat value. Hemlock and chestnut split badly and sent sparks flying up chimneys. American elm rarely dried enough to give good heat, and most pines were so full of pitch that, though they ignited quickly, and burned well in an open grate with an immediate and intense heat, they caused great smoke and soot accumulations in narrow chimneys and, later, in the pipes of stoves and cooking ranges. Consequently, the hardwood forests of the Northeast, where the concentration of population was greatest anyway, received the greatest impact from cutting for fuel wood of any part of the United States, and the forests were cut over repeatedly whenever regrowth occurred.[72]

Generally, it was considered that a skilled axman could cut, split, and stack one to one-and-a-half cords of wood a day, a cord a day being the amount that was commonly set for slaves as a task norm.[73] A cord was a cubic measure of 4 by 4 by 8 feet or a total of 128 cubic feet. Some evidence of the amount of wood cut and the time taken to do it does exist, although never for periods before the late eighteenth century. For example,

Crèvecoeur said in 1782 that "One year with another I burn seventy loads, that is, pretty nearly so many cords. Judge of the time and trouble it requires to fell it in the woods; to haul it home either in wagons or sleigh, besides cutting it at the wood-pile fit for the length of each chimney."[74] Crèvecoeur's consumption of about 9,000 cubic feet of wood for one family per annum seems excessive, but he certainly had more than one chimney, and to judge from the prevalence of large open fireplaces capable of taking logs of eight feet or at the very least of four feet, it is not impossible. Another estimate of the "time and trouble" that Crèvecoeur spoke of was that it took between 10 and 20 acres of woodland to supply the fuel burned by one fire annually. Be that as it may, it is certain that the open fires were very inefficient, about 80 percent of the heat going up the chimney, so that the rate of fuel consumption was not surprising. Benjamin Franklin thought that a large house with a few fireplaces needed one man full-time to cut, split, and haul fuel wood in order to keep supplies at the required level. A more common figure for annual consumption for a typical rural household was 20 to 30 cords, of which at least half would be consumed, though some were sold. This amount bears some relationship to estimates of four-and-a-half cords per capita suggested by Reynolds and Pierson as being the average figure during the period of maximum consumption in the Middle Atlantic States during the late eighteenth and early nineteenth centuries.[75]

The chance for profit and the heavy commitment to cutting and preparing forest lumber for firewood became greater the nearer the pioneer farmer was located to towns or metalworking industries, and his commitment became greater as the years progressed. Throughout the length of the eastern seaboard towns were increasing in number and in size. The total number of sizable urban centers rose from 205 in 1700 to 435 in 1750 to 730 in 1775 and to over 1,100 in 1810, with a very pronounced concentration in New England and to a lesser degree in the Mid-Atlantic coastal region.[76]

Along the coast the population of a few major cities such as Boston, Philadelphia, New York, and Baltimore was growing rapidly to over 50,000; and an adequate and inexpensive supply of firewood was indispensable for their very existence. Wherever possible, cordwood was obtained from local farmers in order to offset the high costs of transport, but as the towns grew the woodland close at hand became scarce, and an ever-increasing radius was scoured for suitable supplies. Pioneers creating and clearing farms up to 100 miles away were now drawn into the new system of trade and supply.

The problem was obviously more severe in the northern cities than in those in the South. As early as 1637 immediate supplies were exhausted around Boston and wood was brought by boat from Cape Ann. New York's supply on Manhattan Island had become depleted sufficiently by 1680 for space to be set aside for a wharf where the wood, brought by boat from Long Island and New Jersey, could be stacked and measured. The country surrounding Newport had been denuded enough by 1713 for wood to be brought by boat from Shoreham and Narragansett, and 20 years later plans were discussed, but never implemented, to reafforest the area surrounding the town. At an early date Philadelphia's supply of wood was brought easily and cheaply up the Delaware, but by 1750 that situation had changed. Pehr Kalm thought that the then prevailing high prices could be attributed to the competition from brick kilns and iron smelters. In Philadelphia 25 years later, fuel was said to be as dear as in forest-denuded Britain. By 1783, Schoepf said simply, "the forests are everywhere thin," and he blamed the ironworks, which "could not but ravage the woods." In Philadelphia "iron or tin-plate draught stoves" (or "Franklins," as they were known after their inventor) were becoming common "as a result of the increasing dearness of wood." Farther south, the gentler climate and the numerous creeks and rivers across the coastal plains made the matter of fuel supply far less perplexing for urban centers.[77]

Even as early as the last quarter of the seventeenth century, the urban trade had become so large and so complex that farmers no longer marketed their wood themselves in the large towns, and fuel dealers proliferated. A measure of the problem of supply during these years was the appointment of official corders of wood, whose duty it was to ensure that unscrupulous dealers did not short-measure the cords. The corders were appointed in New York in 1680, in Boston a little later, in Newport in 1697, in Philadelphia in 1701, and in New Hampshire in 1714, and they were common in most towns in Massachusetts by the end of the eighteenth century. In another move standard cords were established in the states from Massachusetts to Georgia between 1640 and 1780.

As the eighteenth century progressed, towns grew in size and complexity, and the intensity and extent of agriculture increased so that the timber from which fuel was cut was available only at great distances from the cities. Consequently, the fuel situation became critical, especially during particularly severe winters. In Boston in 1726, 24,000 sled loads of wood were moved across Boston Neck into the town to supplement the normally adequate summer-shipped supplies. The price of wood rose to 51 shillings per cord. In New York during the winter of 1741 the price reached 50 shillings per cord. It seemed common for the selling price to quadruple or quintuple to 40 or 50 shillings, particularly in bad weather, so that there were continual complaints that only the rich could afford to purchase wood whereas the poor experienced real hardship and literally froze unless the city authorities doled out charity supplies, as they did from time to time. Increasingly, those who could afford to relied on coal, not only the western Virginia coal discovered during the early 1700s but cheaper, less smelly British coal. Somewhere between 5,000 and 10,000 tons were imported annually toward the end of the eighteenth century, mainly into Boston, New York, Philadelphia, and Charleston.[78] By 1745 Benjamin Franklin lamented the scarcity of wood, which had formerly been "at any man's door" but which had now to be fetched nearly a hundred miles to some of the coastal towns, causing its price to rise beyond the reach of the poor. Although coal was now a cheaper fuel than wood, few people could be weaned from their traditional source of heat.[79] By the end of the century the environs of the large towns were denuded, causing collectors to move their operations even farther from the point of consumption, and the by-product of the cultivator was increasingly drawn into the urban market and traveled even farther than before.

The bulkiness and relatively low value of the wood tended to limit its haulage by cart to somewhere between 25 and 35 miles, except in times of scarcity and emergency when prices rose markedly, but whenever water transport could be used much greater distances were overcome. Small sloops of 50–60 tons, carrying anywhere between 25 and 50 cords of wood, plied between the Maine and New Hampshire coasts and Boston,[80] went 200 miles up and down the Hudson from the Palisades to New York, and also negotiated the rivers and coast around the pine barrens of New Jersey, which became a major source of fuel after 1775. The ships were so small they could negotiate the many shallow creeks, estuaries, and rivers that penetrated the coast, particularly the Egg, Mullisca, and Oswego, which crossed the New Jersey sand plain. Woodcutting on the barrens and the transport of the wood became an increasingly complex business as commercial elements began to dominate the scene, there being few farmers in this area. The whole business became integrated vertically, woodcutters owning land, ships, and wharves, and horizontally, as they became involved in charcoal burning and pitch extraction. Activity was concentrated in the summer months, and large piles were accumulated to tide the towns over the December-to-March freeze. Fuel wood collection and cutting on the barrens rose to a peak during the early nineteenth century, declining only sometime between 1850 and 1860 (one cannot be more precise) when large areas had been cut so frequently and devastated by fire so badly that they yielded no more wood and when the increasing use of anthracite in stoves decreased demand. Despite intensive study of this important fuel-producing area of the New Jersey barrens, the actions were individual and unrecorded so that one is still hazy about the details of fuel gathering.[81]

As a result of the scarcity of wood around areas of consumption and the need for a more complex organization to gather it from greater distances, the price of wood rose steadily and then dramatically between about 1785 and 1795. William Strickland, who was a keen observer and had a mania for finding out current prices of every commodity, said that "everyone complains of the scarcity and increasing prices" and quoted again and again the assertion that the average price of wood had doubled in the 10 years after 1785 from its long-established price of about 10 shillings per cord and that this was not just a seasonal variation. He attributed the increased price to the fact that trees were cleared and destroyed wantonly by the pioneer agriculturalist: "what is not wanted for any present purpose is set fire to," he said; "if care be not taken it will soon be very scarce . . . unless coal mines be discovered and worked." Of course, coal was being mined in Virginia as early as 1700, and Franklin had invented a stove to use coal, but his stove and others made little impact on the fuel-using public, except in the large eastern coastal cities where the fuel could be imported from Britain.[82]

Strickland's warning about the scarcity of fuel was echoed by Tench Coxe in 1793, and others, one writer saying as early as 1775 that the "ravaging, rather than the cutting down," of the forests in southern New England had led to the timber shortage.[83]

Scarcity was relative, however; there was certainly plenty of timber left, although the price had risen because it had to be hauled from farther afield. There were no railways and no canals to reduce costs, only waterways and the coast, and they did not always lie close to the timber sources. The rise in the price of the fuel was a measure of forest destruction.

The paucity of statistics on the destruction of the forest for fuel can be rectified, to some extent, by looking at the total consumption by decades based upon a variety of sources and statistical transformations. Crude estimates of consumption can be made based on the size and distribution of the population, the climate, housing conditions, and availability of wood, making allowance for the slow shift from open fireplaces to stoves and the gradual replacement of wood by coal. These data are analyzed more fully in

Chapter 6 where they are more central to the discussion, but we can note here that up until 1810, probably 1.08 billion cords of wood were cut and used. The total amount increased steadily each decade as the population increased, so that by the latter part of the eighteenth century, in round figures, 100 million cords were consumed between 1760 and 1769; 125 million from 1770 to 1779; 155 million from 1780 to 1789; 201 million from 1790 to 1799; and 268 million between 1800 and 1809. The greater part of the wood was, of course, consumed in the seaboard states, the area of the greatest population density; up to 1810, 30 percent of the total consumed was in New England, 38.5 percent in the Mid-Atlantic States, and 25.5 percent in the South Atlantic States.[84]

But what did this mean in terms of acres of forest destroyed? It is difficult to convert cords into stands of trees, an exercise that demands an analysis of yield, and these vary enormously under different environmental conditions of soil, moisture, climate, the degree of destruction by fire and disease, and the species of tree. Pitch pines, for example, have yields that vary from 5 to 50 cords per acre, which could give us a figure of between about 2.16 million and 21.6 million acres of land denuded by 1810, the latter figure being far in excess of the amount of land in agriculture at that time. Another complicating factor is that cordwood was a by-product of clearing and that cleared land could regrow successfully within about 20 years, so that with careful husbandry the same ground could be cleared possibly four or five times in a century. Again, some trees were "cropped" for cordwood every few years, the slender branches from the pollarded stem being much easier to cut than large trees.

Regrettably, we shall never get an index of the amount of land cleared purely for fuel; all we know is that it was an immense quantity. Certainly, if the 1.08 billion cords that it is suggested were cut by 1810 are converted to billions of board feet (b.f.), then they represent about 265 billion b.f., which far exceeds the total amount of lumber cut by 1810 and was an amount probably not reached by the lumber cutter until 1870. Like the land cleared for agriculture during these centuries, the wood used for fuel was as plentiful as air – and who wrote about that or recorded statistics about it?

4

Two centuries of change: the commercial uses of the forest

> In a very little space everything in the country proved a staple commodity. . . . Wheat, rye, oats . . . timber, masts, tar, sope, plank-board, frames of houses, clabboard, and pipestave . . . many a fair ship had her framing and finishing here . . . nor could it be imagined that this Wilderness should turn a mart for Merchants in so short a space.
>
> Edward Johnson, *Johnson's Wonder-Working Providence* (1654), 246–7

To make a distinction between the impact of agricultural clearing in the forest and the impact of commercial cutting and exploitation is difficult, as the two overlapped in so many ways. On the one hand, the infant settlements needed to clear the land in order to grow food to live; on the other hand, they also needed to export surplus products to exchange for essential manufactured goods not available in the new location. Either way, there was no hyperbole in the assertions of various writers that the harvesting of the riches of the forest was "fundamental to the developing economy," that it was "the mainstay of the agricultural and commercial economy," that life was dominated by "The Forest and the Sea," or that the economy of seventeenth- and eighteenth-century New England was basically a "timber economy."[1] Thus, we know that potash and fuel wood were incidental products of clearing and burning, but we also know that they were also crucial to the profitability of farm making. Similarly, it is clear that the agriculturalist supplied much of the wood that went into the making of timber products like staves and shingles and much of the wood for general lumber, all of which were exported worldwide. Once established, the agriculturalist also supplied bark for tanning and syrup as a sweetener. In the South, the production of tar, pitch, and turpentine, again overwhelmingly for overseas markets, often went hand in hand with agricultural clearing and activity. Like potash in the North, tar was basic to farm profitability. Nevertheless, there was an indeterminate point where the agricultural settler began to arrange his life and activities in such a way that he purposely supplied an outside, even distant market rather than his own and local needs. At that point the pioneer economy shifted from self-sufficiency, and the exploitation of the forest began to have a distinctly commercial orientation.

The demands of the sea

The story of the needs of the British navy, of the supply of naval stores and masts, and of the working, challenge, and decline of the old colonial trading system or mercantilism, has been told often and well,[2] but the effect of these activities and policies on the changing geography of the forest of North America has not been looked at closely.

In truth, these activities constituted a minor use of the forest, and their impact was a mere pinprick compared to the gashes made by agricultural clearing, fuelgathering, and lumbering. Exploitation was fairly unsystematic, scattered, and tied in with agriculture, and therefore no one area was preeminently dominant. Nevertheless, these forest-based activities cannot be neglected, if only because they helped to create important elements in the economies and trading patterns at the two extremes of the eastern seaboard, in the South and in northern New England, and because they had a strategic and political significance, together with a role in the formation of national attitudes, which gave them a prominence (and visibility in the contemporary record) out of all proportion to their economic importance.

Naval stores

"Naval stores" was a composite term that covered masts, spars, and planking but in time was reserved for such commodities as hemp for ropes, flax for sails, and the many products obtained from the sap of the coniferous trees, such as pitch, tar, turpentine, and resins for waterproofing hulls and decks and preserving rigging. The demand for naval stores came primarily from Britain and its navy, which connected a vast empire. The supply of timber from British-grown oaks and beech had been depleted after centuries of colonizing the waste and because of the climbing demands for constructional timber and for fuel for iron smelting and domestic purposes from the late sixteenth century onward. This left little timber for shipbuilding,[3] and by 1600 it was said that the forests had been "reduced to such a sicknesse and wasting consumption, as all the physick in England cannot cure."[4] From even before the time of the Armada in 1588, which marked British sea supremacy and the need for the maintenance of a navy, forests of only a few thousand acres were reserved and protected for shipbuilding by royal decree. However, masts and spars, which demanded length, cylindrical straightness, strength, durability, yet a certain elasticity, could not be grown in Britain where there were no native conifers except for the Scots pine, which was neither large enough nor of sufficient quality for the purpose. Supplies of tar, pitch, resins, and flax were not available, and Britain relied on the countries around the Baltic Sea for these supplies.

Yet Baltic supplies were precarious. At various times Prussia, Sweden, Denmark, and Russia levied custom duties on these commodities. In addition, the entrance to the Baltic might be closed off, as had happened during the First Dutch war in 1652, and there was constant warfare between Spain, France, and Holland, on the one hand, and Sweden and Russia on the other during the late seventeenth and early eighteenth centuries. In addition, the home consumption of naval stores increased in all producing countries, which led to scarcities, and their purchase from foreign powers by Britain left an adverse trade balance. Not unnaturally, therefore, Britain did not want to be dependent on foreign powers for indispensable ship commodities and looked elsewhere for supplies.

The new colonies in America seemed a natural and promising possibility, and almost every British explorer from Raleigh onward had pointed out the likely trade. But the switch to alternative sources of supply was not that simple. In the South, although tobacco farmers had attempted large-scale production of pitch and tar from time to time when the tobacco industry suffered a depression, they did not achieve much success. They were not situated in the best-yielding pine areas, and the high cost of labor and the cost of transport relative to the Baltic countries (about three times greater) made competition difficult. The sap-derived products were also said to be of inferior quality. By the closing years of the seventeenth century, domestic American needs were satisfied, but only a few thousand barrels of pitch and tar were exported annually.[5]

In New England, similar disadvantages of cost and quality worked against the large-scale production of tar and pitch from the pitch pine (*Pinus rigida*), but another factor at work here was more directly related to the other branch of the naval stores supplies – the mast and timber trade. Britain's policy under the Navigation and Trade Acts of utilizing the colonies as a source of raw materials and as a market for its manufactured products worked reasonably well in the South but not in New England. Increasingly, Yankees were building boats, producing woolen goods, engaging in trade, in fact, achieving a sort of economic independence. From about the 1660s onward true conflict began to arise over colonial and mother country demands for the timber for shipbuilding, over the right of Britain to reserve the timber for its own naval use, and the demand of the New Englanders to saw it up for the valuable West Indies and "Wine Islands" markets, as well as for their own shipbuilding and domestic uses. The situation was highly political and held all the ingredients of the discontent and ultimate revolt of 100 years later.

Britain sent a fact-finding commission to America in 1698 to investigate the possibility of obtaining naval stores and in 1704 proposed bounties to stimulate production after the Swedish government required that all naval stores from that country be sent in their own ships at the prices and quantities they determined. The Naval Stores Act of 1705 paid a bounty of £1 per ton on masts and spars, £6 per ton on hemp, £4 per ton (about eight barrels) on pitch and tar, and £4 per ton on turpentine and resin, roughly equal to the cost of transport from America to England. The law was in operation until 1776 with only a brief interlude between 1724 and 1729. Colonial adminstrations also passed legislation to stimulate production, such as measures to standardize the size of pitch and tar barrels and to initiate a system of inspection. The Virginia government passed acts in 1722 and 1748 that emphasized the uselessness of the pine forests for tobacco cultivation and their desirability for timber and naval stores production, and it added another bounty of two shillings per barrel to all tar made according to British standards. Almost every colony between 1722 and 1769 had rewards for growing hemp.[6]

Pitch, tar, and turpentine

Whereas the colonial hemp system was a failure – the special culture and techniques for hemp were never mastered – the production of tar and pitch was a success. During the first couple of decades of the eighteenth century, an average of about 40,000 barrels of pitch and tar were imported annually into Britain, but the American colonies, or "Plantations," as they were sometimes called, supplied almost none. After the 1705 subsidies, however, shipments from the Plantations rose sharply, and by 1718 they supplied 82,084 barrels out of 92,344 barrels imported, or about 89 percent. This almost

Plate 4.1. Cutting a box for turpentine.

complete saturation of the British market fluctuated with the maintenance or otherwise of the subsidy during the 1720s and 1730s, but during the 1760s (until 1776) British imports were running between 110,000 and 135,000 barrels per annum, of which the greatest bulk was coming from the North Carolina longleaf pine (*Pinus palustris*), along with some from South Carolina.[7]

There was little variation in space and time in the methods of production. The first step was to collect the natural resins from the longleaf pines. Periodically, the bark of the tree was hacked off and the tree grooved with a downward-pointing V-shaped incision, or "box," on the sunny side of the tree, as near the ground as possible (Plate 4.1). This cut stimulated the flow of resin, which collected in a cup placed at the apex of the incision. When the box filled, the resin was taken out with a ladle and poured into barrels located at intervals in the forests, which ingeniously were "rolled to market" (Plates 4.2 and 4.3). This could be done every 10 to 15 days during the summer or over a longer period during the winter months when the flow lessened. One slave could collect about two-and-a-half barrels a day, which was the product of about 1,000 trees. The clear turpentine used for paint was obtained by the distillation of the crude thin resins in large copper vats (Plate 4.4). The heavy resin (or rosin) was obtained from the residue after distillation and also from the "scrape" which was the gum that gathered on the face of the tree (Plate 4.5). Both of these could be used for glue.

The tar for preserving ropes and timber, and the pitch for caulking ships' timbers were made from burning freshly split fire-deadened pine wood, particularly the dried-out old "boxed" trees, which produced the best-quality tar that commanded the premium prices. The logs were usually piled in a cone-shaped heap over a sloping floor of clay,

Plate 4.2. Collecting the resin.

Plate 4.3. Transporting the resin by "rolling it to market."

Plate 4.4. Distilling the resin to make spirits of turpentine.

Plate 4.5. Gathering the scrape.

Plate 4.6. Burning old boxed wood in a forest kiln to make tar and pitch.

which had a wooden pipe or channel leading from the floor to barrels or a pit where the tar could be collected (Plate 4.6). Sometimes the whole pile was covered with earth, as in charcoal production, the pile lit from a hole in the top and then covered over so that the fire would work downward to the base, forcing the tar down with it. A cord of wood could make between 40 and 50 gallons of tar. Pitch was made by boiling down the tar in large vats or "kettles," or even in clay-lined pits, in order to get rid of the volatile material. Because of this extra processing, pitch cost about twice as much as tar.[8]

It was a crude form of manufacturing, and because it did not require great expertise or capital it was widely carried out by farmers and dispersed throughout the forests. Most of the work was time-consuming and laborious and, therefore, was done by slaves. Where the slaves were thickest on the ground in longleaf pine areas, there was the greatest production. Pitch and tar manufacturing represented an efficient use of the slave labor force, for when other seasonal tasks lagged it could be assigned to this profitable sideline.[9]

Contrary to British hopes, the major part of the pitch and tar came from the Carolinas and not from New England, so that that part of the colonies was not converted to more "suitable" economic pursuits. The reason was simple: The southeastern longleaf forest on the inner coastal plains away from the "sounds" and on the inland sand hills yielded better products more prolifically and cheaply than did any northern conifers. As a result, production was centered in the South wherever the longleaf pine grew, and it provided a livelihood and cash crop for small farmers. But the Yankee shippers, ever industrious and ingenious, imported southern tar and pitch and then reexported it to Britain so that they "absorbed" the naval stores premium (which totaled nearly £1 million before the Revolution) that should have gone to the Carolina producers. Less than one-fifth to one-sixth of the total pitch and tar came from New England ports, and much of the export that is recorded was not from New England forests but was transshipped Carolina produce picked up by New England traders in their all-purpose "coasters" on the return run from the West Indies.[10]

In addition to gaining financially in the transshipment of the produce, the New Englanders needed tar for their own ships. It is calculated that between 1674 and 1714 alone, Massachusetts had built at least 1,257 new vessels totaling 75,267 tons, and other places in New England had built another 364 vessels of 21,236 tons, most of which were registered in Massachusetts. These totals are much greater than supposed, and some of these were large transatlantic vessels, which took the stores to Britain. The fact that this capital investment in shipping and related services of shipbuilding and provisioning went almost unnoticed was the supreme example of Britain's unawareness of colonial realities.[11]

Although the complexities of the tar and pitch trade are not at the center of this discussion of the destruction of the forests, it is important that the sources, the ports, and the flows of trade are understood because they help one to pinpoint production. Figure 3.2 shows the exports of tar, pitch, and turpentine by barrels for a four-year period between 1769 and 1772 at the height of production, and it shows quite clearly the preeminence of North Carolina, with its ports of Roanoke, Bathtown, Beaufort, and particularly Brunswick exporting about 60 percent of the national total. Other ports of significance were Charleston in South Carolina, and the Lower James River customs district in Virginia, but a large proportion of the exports of the northern ports of Boston and New York in particular consisted of shipments of southern produce.[12]

Other states, for example Virginia and Maryland, attempted to manufacture pitch, tar, and turpentine, but despite their extensive forests and the bounties, they produced very little in the way of stores until the 1770s, by which time the bottom had fallen out of the market. Georgia and Florida tried, too, but they were minor producers and were more concerned with lumber production. Georgia's total export was only a few hundred barrels annually, but Florida's rose to about 8,000 because of the direct trade with the nearby West Indies. In Louisiana production was centered around Mobile and Pensacola, where the longleaf pines could be reached from the coast, and it was part of the general trade with the West Indies and Havana, especially when Cuba was taken over by the Spanish.[13]

With the Revolution the trade to Britain declined to almost nothing, but it did pick up again to run at about 40,000–50,000 barrels per annum during the opening years of the nineteenth century, only to go through further fluctuation during the Napoleonic wars. Britain had turned back to Baltic supplies, and North America had to find new markets for its annual export of about 140,000 barrels. As ever, the West Indies trade was reliable, and new markets were opened in Canada and Portugal. Naval stores remained a profitable sideline for planters and agriculturalists, particularly on the coastal plain of North Carolina, where production was firmly established but where the lack of readily harnessable water power left it relatively untouched by the industrial experiments that moved through the South during the early nineteenth century and disrupted production elsewhere. The growth and maintenance of pitch, tar, and turpentine exports were very closely related to foreign demand and to competitive conditions in the transatlantic market. These fluctuations could be eliminated only with the development of a strong home demand, which came in time.[14]

Masts

For a variety of interrelated reasons the extraction of masts from the New England forests was a far more complex process than the extraction of pitch and tar from the pine forests of the South. First, masts were of a greater strategic importance than naval stores, and therefore they were subject to radically different policies. Bounties were supplied to encourage the manufacture of wood-derived products from an abundant source, products that could be used only in ships, but restrictions were imposed to limit the cutting of the uniquely large trees, which could be used in many other sorts of timber construction. Attempts to exploit the forest were replaced by attempts to conserve the forest.

Second, the fact that the timber could be used for planks, clapboards, and other contructional lumber meant that the mast trade became bound up with the rapidly expanding New England lumber industry, especially after the introduction and widespread use of the water-driven reciprocal gang saw, which could cope with large logs. The first sawmill is said to have been built in 1623 near York, in Maine, and from then on sawmills sprang up everywhere and were an indispensable accompaniment of frontier settlement. Lumbering, of course, was not absent from the southern pine forests, but it was a complementary activity to pitch and tar extraction, not a competitive one.

Third, and closely related to the above because the supply of masts came from New England and not from the South, the problems were increased by what John Evelyn, the English diarist, called "the touchy humour" of those Yankee colonies, which made the exploitation of the source difficult, mainly because this meant the diversion into British markets of lumber, which along with fish was a staple export. Colonial New England had developed a complex lumbering, shipbuilding, fishing, whaling, and trading structure with little or no interference from Britain, and any attempt to restrict this with imperial edicts was bound to have political repercussions.

Though mast getting was merely an adjunct to this general trade and commerce, it probably could not have existed efficiently and economically without the organization and general cutting and transport in the woods that had developed over the years.[15] From 1652 onward the navy sent ships to Portsmouth on the Piscataqua inlet for remasting. By the end of the century, an almost monthly cargo of masts was sent to Britain, and by 1727 Falmouth on Casco Bay had become the center of the trade. The strategic importance of the New England masts was simple: Any large or small sailing ship needed timber for at least 23 masts, cross yards, and a bowsprit. Large ships needed a main mast that could be as large as 40 inches in diameter and up to 120 feet in length. Trees of these dimensions were not readily available in the German and Russian forests

Table 4.1. *Imports of masts into Britain for selected years*

	Source	Diameter Great (18″+)	Middling (12″–18″)	Small (8″–12″)	Source as % of total
1706–7	Europe	1,168	1,606	579	98.6
	N. America	48	—	—	1.4
1718–27	Europe	6,028	8,879	20,056	84.3
	N. America	2,615	1,148	2,748	15.7
1744–5	Europe	1,595	999	1,124	89.9
	N. America	419	—		10.1
1759–60	Europe	1,309	1,876	2,648	88.0
	N. America	603	127	58	12.0

Sources: Based on J. Malone, *Pine Trees and Politics*, 55–6; and E. Lord, *Industrial Experiments in the British Colonies*, App. B.

where continual cutting over the centuries in the accessible areas had left trees over 27 inches in diameter very scarce, and the superior advantages of the New England "big sticks" over the European-made masts was obvious.

Paradoxically, despite the necessity for Britain to obtain masts from New England, relatively few were shipped, which made the subsequent acrimony and conflict so unfortunate and unnecessary. Even during the European conflicts of the early and mid-eighteenth century when one might expect New Enlgand masts to be used more, the Baltic still dominated (Table 4.1). Simply, distance added enormously to the cost. Although the New England proportion increased a little in later years, there is no evidence that these supplies were very great. For example, Albion shows that British naval mast reserves in 1770 consisted of only 673 New England masts compared with 3,398 European. What was important, however, was that nearly all the New England masts were in excess of 20 inches and 397 were in excess of 27 inches and that, although 1,500 European masts were over 20 inches, none was over 27 inches in diameter. Size was the key to this trade; it was the mere skimming of the cream of the milk. The total was certainly not enough to warrant the reservation for naval needs of all white pines on the northeastern coast, as happened after 1722.[16]

The process of getting masts was far more complex and costly than was ordinary tree getting, and the whole business was handled by a few politically well-connected contractors who specialized in the business. First, many trees that appeared to be suitable for masts were not so. In 1771 George Wentworth, one of the contractors, said, "one hundred and two out of one hundred and six proved rotten in the heart and not worth a shilling," although they were all right for local timber use. Clearly, getting suitable masts was highly speculative, which accounts for the price of between £100 and £150 for a "good stick." A contributory factor to the cost was the fact that masts could not be felled in just any spot in the forests, nor could they be allowed to crash to the ground and become damaged. The trees were selected carefully, and then a path or road was cut through the forest in the direction of the required fall and for the ultimate transport of the log. The ground had to be flat to prevent damage; if the cutting was done in winter, as it often was, the ground could be packed with snow to even out the bumps and break the impact of the fall. If the ground was clear the log could be dragged to the water's edge, but if not it had to be slung by chains under two axles, each connecting two great wheels about 16–18 feet in diameter and dragged by dozens of oxen. Judge Samuel Sewell records a visit to Salmon Falls, on the Salmon River, just before it reached the Piscataqua inlet, "to see a Mast drawn of about 26 Inches or 28; about two and thirty yoke of Oxen before, and about four yoke by the side of the Mast between the fore and hinder wheels. 'Twas a very notable sight." Once in the river, the masts were floated down to the loading pond where timber ships, often specially constructed 400–500 tonners, waited to take them overseas.[17]

Enough has been said to indicate the heavy dependence of local life on timber and timber products and the small part masts played in the total picture. Therefore, the British attempt to restrict the felling of timber was as stupid as it was reckless. In 1691 the Massachusetts Bay Company charter (which included the Plymouth Colony and Maine) had restricted the cutting of all white pines over 25 inches growing on unoccupied lands outside declared and surveyed townships, and after 1695 a surveyor of pines and timbers had been in charge of New England forests to supervise the selection of suitable trees for masts, marking them with three cuts of a hatchet to make the broad arrow signs, which meant that the tree was henceforth reserved for the navy. After several preliminary acts, the White Pine Act of 1722 stipulated that all white pines, regardless of size and provided they were not growing in townships, in the forests from New Jersey to New England were reserved for the navy. Resentment ran high at these unworkable blanket restrictions; at the high prices the contractors received and the low prices the woodsmen got; at the speculation in land prices that the high value of the pines helped to promote; and at the many "paper townships" in Maine that were created by speculators in order to avoid the law and to get the pines. The years between 1722 and 1776 are confused, with illegal cutting on the part of the woodsmen, intrigue and duplicity on the part of the Crown agents, and conniving on the part of the contractors. While Britain tried to enforce the unenforceable, New England merchants carried on a good trade in masts and timber with Spain and Portugal where prices were better than in Britain, although those countries were officially at war with Britain. Crown agents attemped to seize the timber and destory illegally cut white pine logs at the mills. More often than not, the local surveyor came to a working agreement with the mill owners and lumbermen, reserving only a portion of the pines in return for nonprosecution for the illegal cutting of the smaller pines (not that a charge could be made to stick). The "woodland rebellion" against these restrictions had all the ingredients of the conflict that 50 years later was to bring about American Independence.[18]

After Independence, Britain did not buy American masts unless absolutely necessary, although masts were sold to other European countries. During the Revolution and the Napoleonic wars between Britain and France, from 80 to 90 percent of all masts were supplied by Canada, particularly by New Brunswick.[19]

Shipbuilding

After listing the staple commodities of the colony in his *Wonder-Working Providence of Sion's Saviour in New England*, Edward Johnson was not slow to point out that "many a fair ship had her frame and finishing here" – but how many ships were built and where they were built are often difficult to ascertain during these early years. It is probable that shipbuilding began as early as 1629 with the construction of fishing shallops at Salem, and the occasional 100-ton oceangoing vessel was built after the mid-1630s. But all the evidence suggests that contrary to expectations the average size of the vessels did not increase markedly during the seventeenth century in response to the lure of profits from trade with distant markets; rather, it decreased, although the number of vessels increased dramatically. According to the calculations of Charles Carroll, vessels cleared in Boston Harbor and with New England registration fell from 61 tons in 1661–2 to 43 tons in 1687, to rise slightly to 49 tons in 1698, the number registered during the same years rising from 13 to 112 to 163, respectively.[20]

The small size of the vessels was not due to a lack of skills or wood for their construction. The building of vessels of this size was fairly simple and required no great expertise. There were ample supplies of eastern redcedar, white pine, and a variety of spruce and oak along the New England coast or just inland, all timber-proven for their water resistance, strength, and durability, and there was a plentiful supply of the highly prized knees and futtocks, the naturally bent parts of the trees that went into ship construction. Masts and bowsprits were available in abundance. Rather, it was considered more efficient and profitable to use small vessels in great numbers for the West Indies trade because of its seasonality, the lack of bulky return cargoes, and the inability of the small West Indies ports to absorb much cargo. In addition, the ocean voyages of bigger vessels had proved hazardous with disproportionately large losses. It was not until British merchants began to place orders for vessels that were clearly going to be used on the transatlantic trade that the average size of ships rose. For example, between 1698 and 1714, 187 vessels were purchased in New England, with an average displacement of 110 tons. The only comprehensive view one gets of shipbuilding during these years is the Bailyns' detailed reconstruction, namely, that in the 40 years after 1674 Massachusetts had built 1,257 new vessels totaling 75,267 tons and that other places in New England had constructed 364 vessels of 21,236 tons.[21]

The shipbuilding yards were located throughout New England, and by the 1670s there were over two dozen located on creeks and inlets from New Haven in the west to Piscataqua in the north. However, four centers began to dominate the industry, and by the late seventeenth century Boston, Charlestown, Salem, and Scituate accounted for nearly three-quarters of all construction and remained dominant throughout the next century. After 1700, other centers rose in prominence, particularly Philadelphia, and Charleston and Savannah in the South. Herndon calculates that between 1722 and 1755 nearly 2,000 vessels of just over 100,000 tons were built in Georgia alone.[22]

These few isloated figures must stand as examples of the scope and magnitude of shipbuilding in the country. It is reasonably certain that on the eve of the Revolution about 40 percent of all British tonnage was built in American yards. Beyond that we

Table 4.2. *New York: settlement and sawmills (lags in years between first settlement and establishment of first sawmills)*

	Sawmill before settlement	Sawmill after settlement (years)					
		0–4	5–9	10–19	20–29	+30	Total
1640–49	—	—	—	1	3	—	4
1650–99	2	—	3	1	1	6	13
1700–49	2	2	—	4	4	15	27
1750–74	1	10	9	11	6	3	40
1775–99	—	75	45	13	3	—	136
1800–09	1	7	1	—	—	—	9
Total	6	94	58	30	17	24	229

Source: James E. Defebaugh, *History of the Lumber Industry of America*, 2:307–12.

cannot be more precise until the records of the nineteenth century become available. Ships certainly consumed large quantities of timber, though much less than houses or fences, for example. Therefore, their impact on the forest was not great. What was important, however, was that shipbuilding was an integral and fundamental part of the whole timber and trade economy of New England, and as such it assumed great importance during these years. It was, said Tench Coxe in 1790, "an art for which the United States are peculiarly qualified, by their skill in the construction, and by the materials with which the country abounds."[23]

Timber products and lumber

To trace the impact of the getting of timber products and of lumbering on the forest during the seventeenth and eighteenth centuries would be nearly as detailed and complex a task as that of tracing the origin and spread of agricultural settlement from township to township during the same period. The similarity of the task lies in the simple fact that the two activities were closely related. The making of hand-riven staves and shingles was always an adjunct to agriculture, but it would also be true to say that up until the first quarter of the nineteenth century the distribution of lumbering and the distribution of settlement were almost coincident, particularly in the northern half of the country. The establishment of lumber mills usually came a few years later than initial settlement, the lag getting less and less as time progressed. This sequence is suggested by Table 4.2, which shows the interval between the date of first settlement in a township and the establishment of the first sawmill in that township in the state of New York for a span of time straddling the seventeenth to early nineteenth centuries. Just as the establishment of the sawmill was a function of agricultural settlement, so was the number of sawmills. The number in any particular county at this time was, more than anything else, a reflection of the length of time the area had been settled and the number of people in that area; no mills meant no people.[24]

With the flour or grist mill, the sawmill was the first local industry established in the semisubsistence economies of the newly settled areas. As new ground was cleared, some of the timber (usually pine) was sent to the mill to be made into clapboards, flooring planks, beams, and other house and general constructional materials. Smaller sawed pieces (usually hardwoods) were hand-riven with axes and wedges into shingles and staves. Usually, this all occurred after the first burst of pioneer activity, which was aimed at clearing a few acres and putting up log house, but even when these essentials for existence were achieved farming still went hand in hand with more specialized sawing. In New England and New York it was only where agriculture failed to develop in the uplands that it was possible for large-scale lumbering to emerge as a commercial industry.

Sawmills

Most sawmills were small speculative enterprises run by individual farmers or families with no more than one or two workers. Because the sawmill was so essential to pioneer life, towns made grants to and townsfolk held shares in what was, in reality, a cooperative enterprise. Anyone was invited to set up a mill within a stipulated time (usually between 6 and 12 months) at a predetermined source of power, on condition that he cut the timber of the townsfolk before handling his own commercial cutting and that he cut it at a cost below that of the current price, the concession usually being between 5 and 10 percent. Maj. John Pynchon's mill on Stony River, Suffield, Connecticut, was built along these principles in 1673, the townspeople giving him

> The Privilidge of Stony River, and the streams belonging to it for the advantage of building a Saw Mill . . . and for a Corn-Mill, if he shall set up one, & also the free use of what timber may be Needful for Board Planks, or the like which is growing on the Land, Intended for Lots, which he may fetch for that purpose . . . till Persons that take up their Lots come to make use of them.

Alternatively, the intending mill owner could accept payment in kind, the farmers offering a proportion of their lumber. Sometimes grants of forest land of between 100 and 200 acres were offered to individuals "for their encouragement." Later, during the eighteenth century, it became common for these fairly simply organized sawmills to be joined by more commercial enterprises in which people had sunk shares of capital, often in bizarre fractions like thirty-sixths and forty-eighths, depending on the number of participants.[25]

As settlement fanned out into western New York after 1780, mills were established immediately, but in the new frontier areas there were often difficulties in getting the metal equipment needed. For example, John Stone established a mill in or near the town of Pittsfield, Massachusetts, in about 1790, and his saw had to be made from old scythes by the local blacksmith. Others came to the Genesee country with only an ax and provisions. They cleared land and built the mill in one year, went back to Connecticut or Massachusetts (usually by foot) where they bought their saws, and returned the next year with equipment, ox teams, and families.[26]

With the exception of a few wind-powered mills built by the Dutch in New

Amsterdam during the early seventeenth century, nearly all the mills were located on suitable water-power sites as near as possible to the woods in order to minimize haulage. The primitive manual operation of two men in a sawpit was mechanized by placing a single saw in a frame that was moved up and down by a crank attached to a large, overshot waterwheel. The log moved against the saw, either being pushed by hand or being drawn by a pawl and ratchet gear worked off the wheel. The sash saw was a crude and inefficient piece of machinery, but compared with hand sawing it achieved outstanding success. It could cut about 3,000 feet of lumber in a working day, which was 20 to 25 times the output of two men working in a pit. With an eight-month working season (from approximately late March to early December, but varying according to location) there were about 250 working days during the year. If demand warranted, the saw could be worked day and night through the year until the water froze. A substantial increase in output came with the invention of the muley saw by Israel Johnson of Schroon in Essex County New York, sometime in the late eighteenth century. The muley was a long, single-blade saw, capable of being run at very high speed because it was taken out of the cumbersome frame and merely had tension applied to top and bottom. Portability was thus an advantage, but functional limitations and other factors were such that the muley was not in widespread use. When in use it was capable of cutting up to 8,000 b.f. per 10-hour working day. Another development, the gang saw, was not a special type of saw but a system of several (eventually 20 to 30) parallel blades mounted in the same frame so that the whole log could be sawed at once into boards. Although known from the early seventeenth century, and used occasionally, the gang saw made little impact because it required much more power. It was a more appropriate feature of mills after 1850 when waterwheels were made more efficient and steam power came into widespread use. Despite the increased efficiency of the saws in the mills, the carriages did not carry the logs past the saws, their movement being limited to about 12 feet. Therefore, pit sawing and whipsawing were still employed to cut long timber and planks.[27]

Almost without exception, mills were located on rivers as running water was the source of power, but as the energy requirement for each mill was small, power sites tended not to be important considerations. The number of mills in any area, therefore, tended to be a reflection of the length of settlement and the number of people there. The advantage of a river location was strengthened by the need to transport the bulky raw material and finished product easily and cheaply, which could best be done by water, and a six-mile overland haul to a navigable waterway seemed to be a limiting factor in most cases, which is illustrated well in the case of mill location in the early nineteenth century in the lowlands surrounding the Adirondacks in northern New York (Fig. 4.1).[28]

Moving the logs

The demand for timber was still relatively small during the seventeenth and eighteenth centuries, and logging did not usually extend for more than five or six miles from the mill. The trees were felled in winter with simple tools – axes, wedges, and mauls. Crosscut saws, although known, were rarely used. Initially, there was probably no need

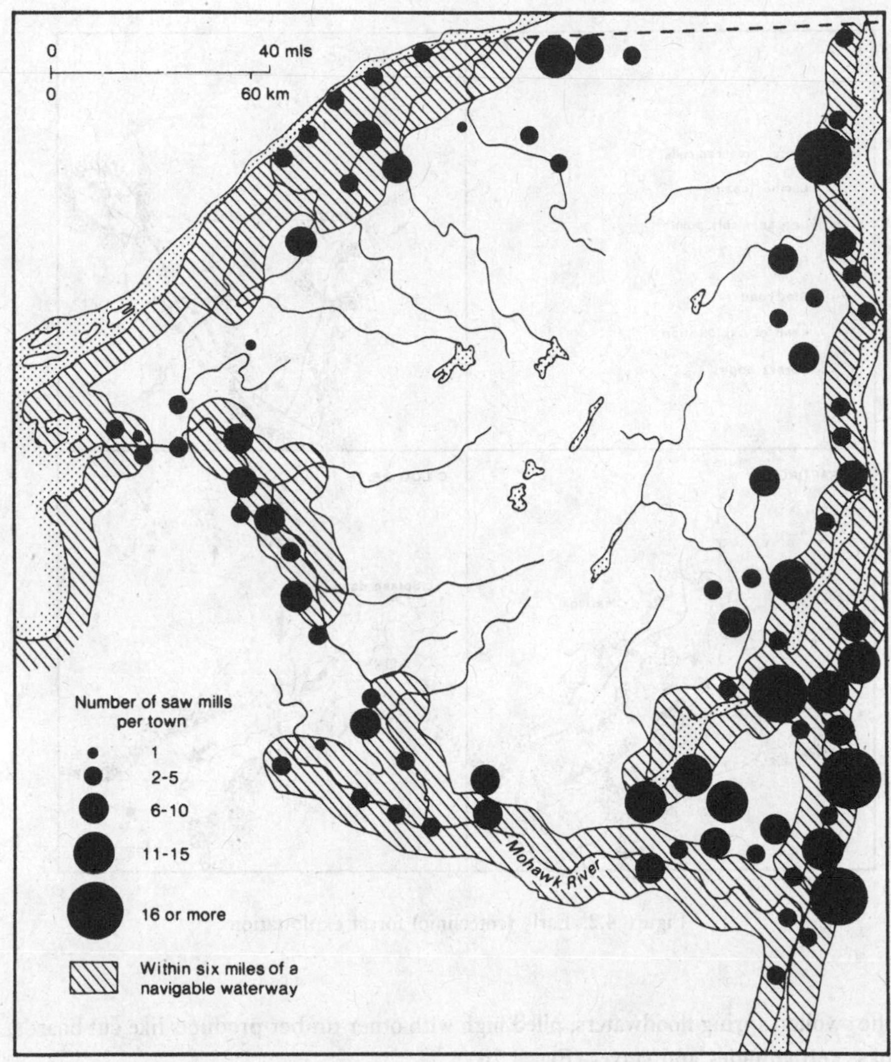

Figure 4.1. Sawmills and navigable waterways in northern New York, c. 1810. (Based on E. M. Dinsdale, "Spatial patterns of technological change," 267.)

to employ draft animals, as the nearest logs could be dragged or rolled to the mill, but as the edge of the forest receded draft animals were used to "snake" out the logs to the mill and to pull sleds during the winter (Fig. 4.2A). Oxen were preferred as they were cheaper to keep than horses and accustomed to rough work. However, as many mill owners were also part-time farmers they probably already had oxen for plowing. Eventually, if the distances from stump to mill became too great, the mill was dismantled and moved nearer to the timber. However, if the mill was on a river, as most were, and one that was navigable and flowed down to a major market, then an alternative solution was possible. The logs could be lashed together into rafts and sent downstream

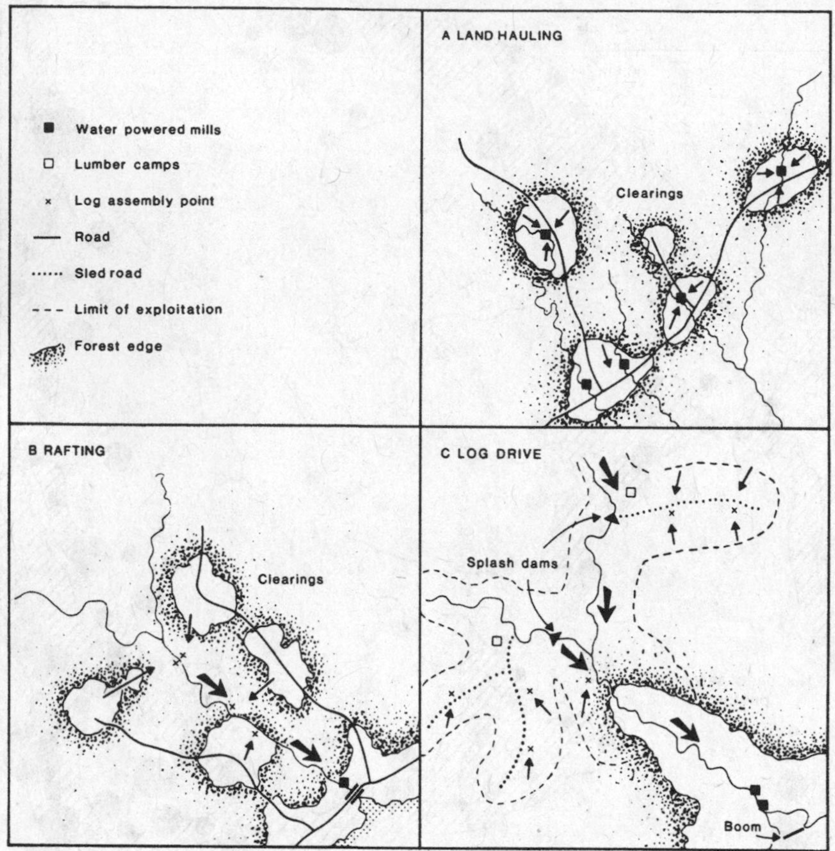

Figure 4.2. Early (eotechnic) forest exploitation

on the swollen spring floodwaters, piled high with other timber products like cut boards, planks, and shingles and staves (Fig. 4.2B).

The origin of rafting is obscure, although it is known that from at least the mid-seventeenth century onward masts and large constructional timbers had been floated carefully down the larger New England rivers to be loaded on waiting ships that took them across the Atlantic. But the idea of extending the radius of supply of coastally located mills in areas where demand had outstripped local supplies does not seem to have been developed until the second half of the eighteenth century, and then it was more the innovation of enterprising farmers than a response to organized demand by dealers. It is thought to have been started by Daniel Skinner on the Delaware during the 1750s, the idea rapidly gaining in popularity so that rafts were a common sight on other Pennsylvania rivers, first on the North Branch and then on the West Branch of the Susquehanna, the Allegheny, and rivers in neighboring New York and New Jersey by the end of the century. Despite the complex structure and often very large size of these rafts, rafting was basically a simpler operation than log driving, which, about 60 years later, became the normal method of transporting logs by water. With rafting, logs were cut by one man, usually a farmer, or perhaps by a group and then taken downstream as a whole to be sold as one lot at the mill or a timber yard. With log driving, on the other hand, possibly hundreds of individual loggers pitched tens of thousands of logs into the river and drove them downstream, then sorted them out at the boom of the sawmill (Fig. 4.2C), which demanded a complex cooperative arrangement, particularly over payments, that was simply not necessary with rafting.[29]

The sight of the immense rafts was something that foreign travelers and native Americans alike never ceased to comment upon, as it was not only spectacular but also indicative of the ingenuity and new scale of operations for which Americans were becoming known. During the 1750s Anne McVickar Grant witnessed some "very amusing scenes" from the banks of the Hudson, where the new settlers had established mills on every side stream. The cut planks were drawn in sledges down to the river edge,

> and when the season arrived that swelled the stream to its greatest height, a whole neighbourhood assembled, and made their joint stock into a large raft, which was floated down the river. . . . There is something serenely majestic in the easy progress of these large bodies on the full stream of this copious river. Some times one can see a whole family transported on this simple conveyance; the mother calmly spinning, the children sporting about her, and the father fishing at one end and watching its safety at the same time. These rafts are taken down to Albany, and put on board vessels there for conveyance to New York.[30]

Some years later, Henry Wansey saw similar rafts at Burlington, New Jersey, on the Delaware, some of which were a quarter of a mile long.

> On one of them I observed a hut erected for a family to lodge in, and a stable with a horse and cow at its entrance. This float of timber was probably framed together two hundred miles further up river by some settlers who were clearing the land and were now carrying some of the finest timber for shipbuilders and architects, down to Philadelphia, in the cheapest way imaginable to convert it into money, and therewith to purchase ironmongery, woollens, implements of husbandry, and whatever other articles may be wanting to improve the comfort of their new settlement.[31]

The practice spread; from about 1790 onward, but particularly after 1800, enormous areas of southern New York were settled and logged over and the timber rafted down the Conaunga, Conestigo, and Tioga rivers into the Susquehanna system to end up eventually at Baltimore; down the Allegheny and Ohio to Pittsburgh and Cincinnati and even, in 1821, as far as New Orleans; and down the Delaware, where besides the usual demands for wood for fuel and house construction, large timbers for shipbuilding were also in demand in Philadelphia. Another outlet for northern New York lumber was via Lake Champlain and the Richelieu River to the St. Lawrence, but this route declined in importance and popularity after the War of 1812 and the opening of the Champlain Canal in 1828, which funneled the cut timber down to New York City by barge.[32]

A particularly ingenious, one might say American, variation on the transport of timber by water was the ark. It is said that the first of these was launched on the Upper Susquehanna at Fredrickstown on the Cohocton in 1800. Arks were square, flatbottomed boats of thick-planked sides, which were caulked to stop leaks. They were up to 100 feet long and 20 feet wide with a two- to four-foot draft. They were constructed as cheaply as possible, and it was hoped they would hold together until they had made the 200-odd-mile journey to market with their cargo of shingles, potash, or wheat. Because it was impossible to return against the current and because the market value of the timber was more than the original cost of its cutting, the arks were broken up and sold. It was a cheap and effective expedient. After 1800, when sawmills became established in the headwaters of the rivers, more sawed lumber became available, and it was said that the ark was supplanting the raft as a means of transporting timber on the Delaware and the Susquehanna.[33]

It is impossible to estimate how much timber came down the rivers by raft and ark, but some clues are given in the claim that, in 1820, 1,638 rafts containing 25 million b.f. were brought down the Susquehanna to Baltimore alone. In 1825 and in 1828 estimates of 900 and 1,000 rafts were made for the Delaware River, and in Baltimore in 1812 it was claimed that seven-eighths of the timber came by raft or ark from Brown and Chanago counties in southern New York.[34]

In general, rafting allowed farmers to participate directly in the timber trade in a way they could not in later years, and it is an excellent example of the blurred distinction between the commercial and noncommercial impacts on the forest during these early years. Nevertheless, navigating the frailly constructed arks and rafts around bends and mudbanks, under bridges and over rapids, such as on the Susquehanna, did require a special expertise, and eventually a skilled group of river pilots emerged "who," as Anne Grant said, "with long poles were always ready to steer it clear of those islands or shallows that impede its course." As the trade became more complex, legislation to regulate the rivers became more common. The piling up of broken rafts was prohibited, and riparian owners were forbidden to make dams on the waterways. Between 1791 and 1792 some of the major river courses, such as the Delaware, Susquehanna, and Hudson, were declared public highways. The transition later to log driving meant that even more freedom of action was necessary on the rivers; in 1806 the relatively minor stream of the Salmon River in Franklin County, New York, was declared a public highway, and within the next half-century nearly every river that flowed through a forest was similarly designated.[35]

Finally, what of the thousands of farmers and their families who went downstream with their load of logs in the rafts? Some had brought down lightweight keel boats or "Durham" boats on the rafts or arks, which could be punted back upstream with 20 to 40 tons of essential provisions purchased in the city with the proceeds from the sale of the lumber. But the vast majority trudged back on foot, the riverbanks swarming with back traffic in the return season and the inns bursting to capacity while as many as 200 to 300 people tried to find a bed on any given night.[36] The distances and the mobility necessary to negotiate them were measures of the demand for timber and the readiness of the part-time farmer to be a part-time lumberman as well.

Production

Throughout these two long centuries there is no precise information about the location of the production of lumber and timber products, and almost no statistics about production. The use of wood was widespread and ubiquitous, and its production was regarded as a necessary part of everyday living. Consequently, during the years before the beginning of the nineteenth century almost the only time we get a hint of the concentration and magnitude of production is when the unusual happened, that is to say, when forest products entered into overseas trade and were thus subject to customs returns or were of such magnitude as to be worthy of comment by travelers.

It is clear that during the seventeenth and well into the eighteenth centuries the settlers along the entire eastern seaboard needed to export surplus products in exchange for essential manufactured goods. Nearly every port participated in this trade after 1640, but particularly those in New England. Falmouth, Portsmouth (on the Piscataqua inlet), Salem, Marblehead, Boston, and later Providence, Fall River, New London, and New Haven, developed a brisk trade with the West Indies, Spain and Portugal, the Wine Islands of the Azores, Madeira and the Canaries, and, of course, Britain. The dependence of the early economy on this trade was basic.

Only premium-quality timber products were exported: masts, potash, and naval stores to Britain, White Oak barrel and pipestaves to the Wine Islands, with dried fish as a subsidiary cargo to all. Although not unknown, it was rare that shingles and pine boards were sent across the Atlantic as their value was not great enough, and Baltic timber, although dearer, cost less to freight to England. White Oak staves for rum casks and the more porous Red Oak for sugar and molasses barrels and hogsheads went to the West Indies, it being calculated that every 100 acres of sugar needed 80 1,000-pound hogsheads and another 20 for the molasses. However, as the economy of Jamaica and the Leeward Islands expanded after the mid-seventeenth century and the indigenous supplies of timber were quickly cut out, the export to the West Indies of more general timber, such as planks, boards, clapboards, and shingles, became profitable and reasonable.

The return cargoes were the tropical products of sugar, molasses, rum, and cotton from the West Indies, wine from the Mediterranean areas, slaves from Guinea and even from as far away as Madagascar. But as timber accounted for well over half of the outward tonnage, filling the ships on their return journeys was a problem, and anything was picked up as long as it made money. Hence the attempt to promote the sale and import to New England of tropical hardwoods like mahogany and brazilwood from the Caribbean.[37]

All this commercial activity promoted shipbuilding and further consumption of timber, particularly around the Piscataqua inlet, which was said to be lined with shipbuilding yards; there were about 20 sawmills on the river and its tributaries by 1665. As the hinterland was settled and the overseas trade grew, the amount of wood exported was enormous. In an account of the exports from New Hampshire during the single year from mid-1718 to mid-1719, among other things, just over a million feet of boards, planks, and joists were exported, 2,149 tons of planks, 614,950 shingles, 252,760 hogshead- and barrelstaves, 1,000 clapboards, 6,000 hoops, 199 masts, 520 spars, and 151 bowsprits. Other settlements dotted the coast at Saco, Black Point, Casco Bay, Kittery, York, Wells, Cape Providence, and the Penobscot, each with its mills and markets.[38] Extension inland was severely limited during the seventeenth and early eighteenth centuries because of the constant threat of Indian attacks. Woodcutters, particularly the masting gangs that roamed the forest widely in search of suitable trees, needed military protection. Despite these problems, however, trade in forest products was energetic and competitive, and the power of the New England merchants and entrepreneurs over the forests of the region was great. It was little wonder that the British restrictions on the felling of white pine were strenuously resisted and ultimately defeated.[39]

The importance of the general trading of timber products to New England and, indeed, to the rest of the East Coast is placed beyond doubt by the survival of a five-year run of customs returns for 1768–72, which throw considerable light on the regional aspects of timber exports, particularly sawed planks and boards, staves, and shingles.[40] The distribution of the ports engaged in the trade of these timber commodities is shown in Figure 4.3.

In round figures, 233 million feet of boards and planks were exported during the five-year period, nearly every port participating in the trade. But, of all the ports or customs districts, that of Piscataqua with its hinterland of productive mills and active logging consistently headed the list in any year and cleared exactly one-third of all the country's exports during these five years. Boston with 12.2 percent of the total, Falmouth with 11.4 percent, and Salem and Marblehead with 9.4 percent helped establish the predominance of New England in this trade, these four ports alone accounting for exactly two-thirds of all boards and planks exported. Only Brunswick in the South, with its developed sawmills in the Cape Fear Valley (5.4 percent), and Savannah (4.8 percent) came anywhere near the totals of any of the New England ports.[41] Of all the boards and planks cleared by the northern ports, between two-thirds and three-quarters were pine, the remainder being oak; in the South it was all pine.

Staves were not a particular speciality of the northern New England ports at the end of the eighteenth century, for out of 161 million exported during the five-year period only 29 percent were cleared by ports to the north of New York. However, New York itself, with 9.3 percent of the total, and Philadelphia, with 20.9 percent, between them equaled the New England total. In general, staves were more conspicuous in the southern colonial commerce than were boards or planks, and the Patuxent and James River customs districts cleared a sizable 14 percent between them.

The export of shingles was much more evenly distributed among all ports along the coast. The need was for wood that was light, soft, easily worked, and resistant to frequent wetting and drying. Redcedar and baldcypress answered well in the South, as did various species of oak and eastern white pine in the North. Of the 215 million shingles exported, Piscataqua with 14.9 percent of the total and the James River district with 14.3 percent stood head and shoulders above all other ports. The stark contrast between the different types of timber exports cleared by one port and another, perceived by Merrens when looking at only the small sample of five North Carolina ports, is not substantiated when we examine the returns for the whole of the eastern seaboard.[42]

After 1775 sawmills spread with settlement everywhere, but particularly northeastward along the New England coast into Maine, which was destined to become one of the major timber-producing areas during the early part of the nineteenth century. First settlement and production centered on the Saco, which had 18 mills in operation above the falls by 1800. The Penobscot was exploited next, followed in 1810 by the

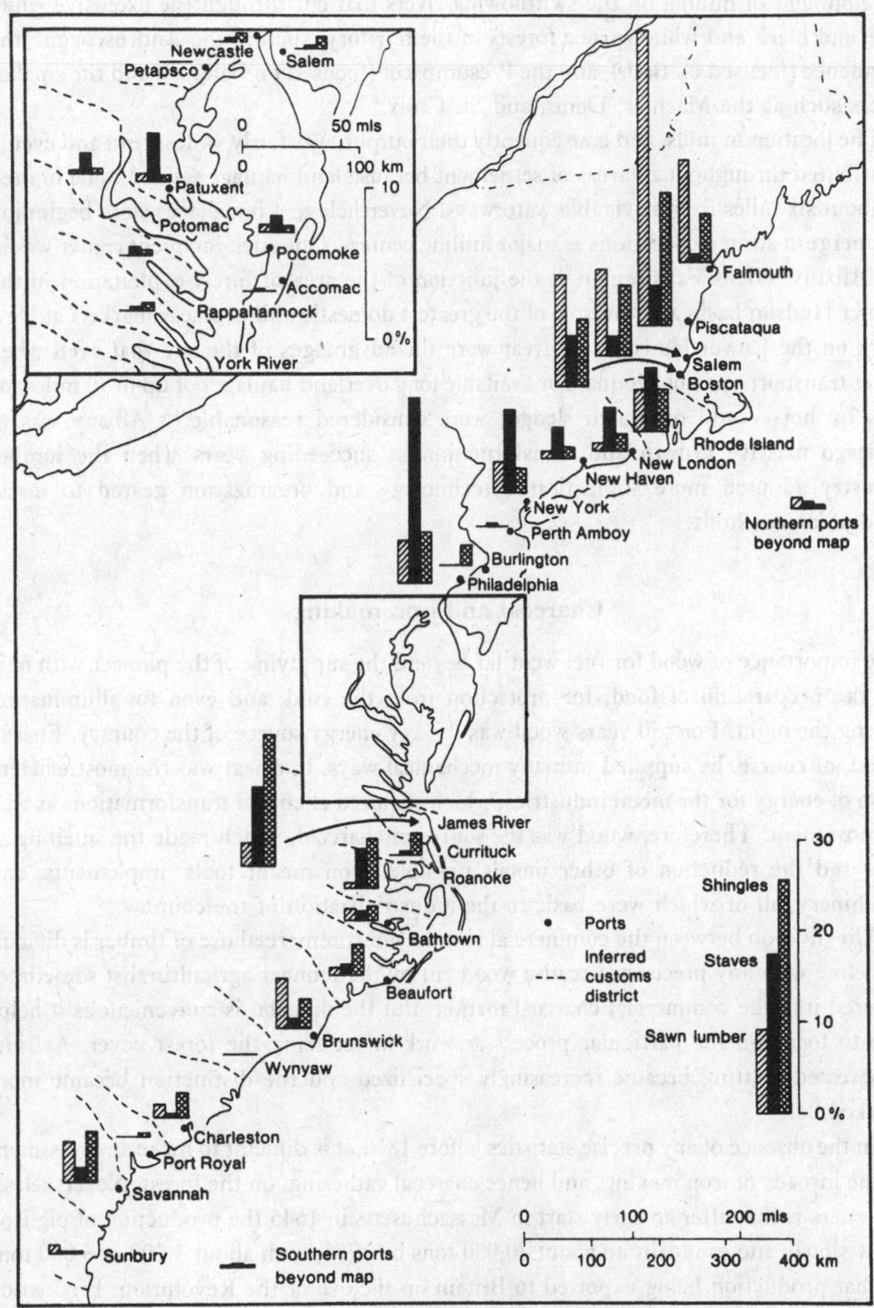

Figure 4.3. The export of staves, shingles, planks, 1768–72. (London: Public Record Office, Customs, 16/1.)

development of milling on the swiftflowing rivers that cut through the extensive white pine and black and white spruce forests of the territory, such as the Androscoggin, the Kennebec (focused on Bath), and the Presumpscot (focused on Bangor), and the smaller rivers such as the Machias, Denis, and St. Croix.[43]

The location of mills, and consequently their output, was fairly widespread and evenly distributed throughout all areas of settlement because land haulage tended to be limited to about six miles from navigable waterways. Nevertheless, a few places were beginning to emerge at strategic locations as major milling centers. One such incipient center was in the Albany–Glens Falls region at the junction of the area of forest exploitation in the Upper Hudson basin and the area of the greatest domestic and overseas markets at New York on the Lower Hudson. So great were the advantages of the site that even when water transport was not adequate or available long overland haulages of up to 70 miles for logs by horse- and ox-drawn sledges were considered reasonable.[44] Albany was to undergo massive growth and transformation in succeeding years when the lumber industry adopted more sophisticated technology and organization geared to mass-production methods.

Charcoal and iron making

The importance of wood for fuel went far beyond the supplying of the pioneer with heat for the preparation of food, for protection from the cold, and even for illumination during the night. For 150 years wood was the key energy source of the country. Energy could, of course, be supplied in many mechanical ways, but heat was the most efficient form of energy for the metal industries, which required chemical transformations as well as movement. Therefore, wood was the source of charcoal, which made the smelting of iron and the reduction of other metals possible. Iron meant tools, implements, and machinery, all of which were basic to the industrialization of the country.[45]

The division between the commercial and the noncommercial use of timber is difficult to define with any precision because wood cut by the pioneer agriculturalist sometimes entered into the commercial charcoal market. But the division is convenient as it helps one to focus on the particular process at work in reducing the forest cover. As time progressed, cutting became increasingly specialized and the distinction became more marked.

In the absence of any precise statistics before 1810, it is difficult to make an assessment of the inroads of iron making, and hence charcoal gathering, on the forest. Nevertheless, the guess is that after an early start in Massachusetts in 1645 the production of pig iron grew slowly and gradually to about 30,000 tons by 1776, with about 3,000 to 5,000 tons of that production being exported to Britain on the eve of the Revolution. Production probably exceeded that of Britain at this time. Certainly by 1810 production had reached 54,000 tons, expansion just about keeping pace with the population increase. British restrictions on the iron industry, like those on mast gathering and shipbuilding, had little practical effect, nor did they raise much ill will because iron was less basic than wood to the economy; it was not a strategic commodity, and it was rarely fabricated into large, complex, and important structures, such as ships.[46]

There were furnaces and forges in all the Atlantic Coast states with the possible exception of Georgia, and there were major concentrations in the oak forests of northern New Jersey and adjacent southern New York, in western Connecticut and Massachusetts, over most of Pennsylvania, and in the pine barrens of southern New Jersey. But generally furnaces were scattered everywhere throughout the rural areas, their production being indispensable to the farmers.

When the Mount Holly furnace and forge near the Delaware River in southern New Jersey were sold in 1743, the advertisement in the *Pennsylvania Journal* listed the equipment and the locational advantages of the furnace. They sum up most of the factors involved in siting an early ironworks. It was built

> On a good Constant Stream of water 27 miles from *Philadelphia*, and is Water Carriage within 3 or 4 Rods of the Furnace and the Forge, with plenty of wood at a small distance for making Charcoal.... Also a good Iron Mine which makes the Toughest and best Bar-Iron, about 18 Miles Distance, all water carriage except one Mile and a Quarter ... a good road without hill in the way.[47]

Most furnaces were located in proximity to either exposed or easily quarried or stripped deposits of magnetite and hematite ore that outcropped on the sides of valleys, or near to the limonite or bog ore of the lowland coastal region in New Jersey and Maryland. Water was necessary for cooling and, if the stream was navigable, for the transport of goods. Water was particularly important in providing the power to drive the bellows and the trip hammers. Wood for charçoal fuel was usually found in abundance and did not exert a strong locational influence initially, but later, when the surrounding forests were stripped and cutters had to go farther afield, the availability of trees became a paramount consideration. Poor roads limited the radius in which fuel and ore could be obtained, and this served as a check to the scale of operations, unless water transport extended the scale of operations, as in the case of the works mentioned in the advertisement. The essential character of the industry was one of many small plants producing between 300 and 800 tons of pig iron per annum. Periodically, the abandonment of furnaces was forced on operators as accessible raw materials became exhausted, and temporary shutdowns were common in order to gather stocks or because of the lack of water during the driest or coldest parts of the year.[48]

The technology of production was simple. The furnaces were usually stone structures, pyramidically shaped and about 25 feet high. They were loaded from the top through a hole into which layers of fuel, ore, and limestone for flux were emptied. Once the furnace was lit, the intense heat melted the ore, which, being heavier than the residue or slag, sank to the bottom of the furnace to await tapping and running into pigs. Additional heat was given by the bellows, which were worked by a waterwheel. The pig iron was pounded by water-driven trip hammers to remove impurities so that the amount of carbon was reduced. Finally, the pig was reheated (often in forges or "bloomeries") and beaten into the required shapes of rods or bars of malleable iron, easily adapted by the farmer to his varied needs and capable of being welded easily.[49]

Essential to all these processes was the production of charcoal for smelting and the generation of heat. Gangs of cutters, or colliers, roamed the surrounding woods

Table 4.3. *Acres of forest cleared per 1,000 tons pig iron produced, under various assumptions*

Yield/consumption ratio	Yield cords per acre	Bushels charcoal per cord	Bushels charcoal per acre	Bushels required per ton pig iron produced	Tons pig iron per acre of woodland	Acres of woodland felled per 1,000 tons pig iron
High yield/ low consumption	30	40	1,200	180	6.66	150
Medium yield/ medium consumption	20	30	600	200	3.00	333
Low yield/ high consumption	10	20	200	300	0.66	1,515
Medium yield/ high consumption	20	30	600	300	2.00	500

collecting and chopping wood. Because any kind of wood could be used for charcoal, their activity was a less exacting drain on the forest than was that of the more discriminating domestic fuel cutter. Nevertheless, large areas were affected as colliers cut suitable 6–10-foot lengths, which were then piled up into a cone or dome-shaped structure called a pit and covered with damp leaves and turf, with an open chimney on the top from where the fire was lit. After anywhere from 3 to 10 days, depending upon the quality of the wood and the weather, the earth was removed and there remained almost pure carbon, which could be carried to the furnaces. Generally, the longer and slower the burning the better the quality of the charcoal. Charcoal burning required constant attention in order to keep the fire going but suitably dampened, and also in order to prevent it from causing forest fires, as it often did. Therefore, cutters built huts near to the pits, moving them as needed when new areas of wood were to be cut. The life was almost nomadic, and it was certainly primitive, and charcoal burning was considered a dirty and lowly occupation that few liked. Always the total operations of burning and living took place as near to the forest as possible so as to reduce the bulk of the commodity to be hauled to the furnace.[50]

It is estimated that a cord of wood would produce between 20 and 40 bushels of charcoal and that most pits burned about eight cords at a time. Furnaces, however, required differing amounts of charcoal, and their consumption per ton of pig iron ranged from 125 to 400 bushels according to the quality of the fuel, the quality of the ore, and the degree of preservation of heat with blasts of air.[51] These and other figures allow one to estimate the amount of woodland required to supply an ironworks. If one acre of woodland produced 30 cords of wood and each acre yielded 40 bushels of charcoal, then an acre of timber would supply 1,200 bushels (Table 4.3). If the furnace was efficient and consumed only 180 bushels to produce one ton of pig iron, then an acre of woodland would make 6.66 tons of iron. By calcuation, therefore, 150 acres of woodland would have to be felled to produce 1,000 tons of pig iron. One can move from this high timber yield/low fuel consumption situation through an intermediate situation of medium yield/medium consumption to the other extreme where timber yields were low due to type and age of tree, soil condition, and past mismanagement and where the furnace was inefficient and required 300 bushels to produce one ton of pig iron. In this situation a mere 0.66 tons of pig iron would require 1,515 acres of forest to be cleared.

If these ratios are now multiplied by the known amount of charcoal consumed to fuel the 54,000 tons of pig iron produced in 1810, then in the high yield/low consumption situation a mere 8,100 acres of woodland would be affected, but in the low yield/high consumption situation it would rise to 81,810 acres per annum. Assuming that a 20-year period before regrowth was sufficient for more cutting, then the result would have been that 162,000 acres were felled at the best estimate, or 1.636 million acres at the worst estimate. During the eighteenth and early nineteenth centuries nearer 300 bushels were used to produce a ton of pig iron (and sometimes as much as 400 bushels) rather than the 180 suggested by Temin.[52] Assuming a medium yield from the timber, 500 acres of forest would have been affected per 1,000 tons produced. Multiplied as before, this would produce 540,000 acres felled over 20 years.

All the indications are that this medium estimate would probably be the most likely one, and about half a million acres of forest affected seems credible when measured against contemporary accounts.[53] To all this destruction one must add the demands of the forges, which often took a great deal of charcoal too. And although forge consumption may not have been as great as that of the furnaces, the larger number of forges may have compensated for this so that their demand could have been anywhere from 50 to 100 percent that of the furnaces.

These figures on the amount of woodland affected are difficult to substantiate, but at least they give some idea of the magnitude of the inroads of industry on the forests. There is some confirmatory evidence of the importance of wood in the number of men employed in various tasks and the actual size of the iron "plantations." In 1777 Noble and Townsend, the owners of the Sterling Ironworks in New York, petitioned the Council of Representatives of New York for exemption from military duties of essential personnel for their operations.[54] Their labor requirements are set out in Table 4.4, which shows that 30 percent of the labor force was engaged in woodcutting and a further 38.8 percent in burning, hauling, and stacking the wood and charcoal; in other words, more than two-thirds of the labor force were getting fuel. Elsewhere, 12 men to serve a furnace was common.

Large areas of wooded land were either leased or bought to supply the fuel. If our hypothetical output of 1,000 tons of pig iron a year was produced by efficient furnaces, and if a 20-year rotation was employed on high-yielding woodland, then about 3,000 acres of land would have been needed to keep production going. If, at the other extreme, the furnace was an inefficient producer and the yield of wood per acre was very low, then nearly 30,000 acres would be needed. The evidence is that estates of between 2,000 and 5,000 acres were common, and anything less than 1,000 acres usually meant that wood was bought from surrounding farmers, the price of between 2s. 9d. and 3s. 9d. per cord cut making it a profitable sideline for the pioneer agriculturalists. By the late eighteenth and early nineteenth centuries larger furnaces and more frequent cutting necessitated

Table 4.4 *Labor requirements, Sterling Ironworks, New York, 1777*

	Fuel-wood getting			Iron making				
	Cutters	Burners and stackers	Haulers	Ore diggers and carters	Skilled opera-tives	Ancillary trade	Manage-ment & clerical	Total
Furnace	20	22	7	7	8	2	2	68
Forge and anchory	20	17	5	—	20	2	2	66
Steelworks	15	16	4	—	11	2	1	49
Total	55	55	16	7	39	6	5	183

Source: J. M. Ransom, *The Vanishing Ironworks of the Ramapos*, 186–7.

larger estates, and 8,000–20,000-acre holdings were not uncommon. In southern New Jersey it was estimated that a successful bog-ore furnace needed the wood of at least 20,000 acres, and these tracts were divided into sections of 1,000 acres, one of which was cut each year. Cutting, however, often exceeded this allotment. Estates in the 8,000–20,000-acre category are, incidentally, confirmatory evidence of our assumptions of medium yield/high consumption in most iron-making areas.

From a sample of 47 estates in New Jersey and Pennsylvania listed by Boyer and Bining, only 5 were below 1,000 acres in size, but 6 were over 10,000 acres, the largest, the Great Egg Harbor estate in southern New Jersey, being 78,060 acres in size in 1800 with a furnace producing 900 tons of pig iron per annum and the forges casting 200 tons per annum. The median size of these estates was exactly 4,000 acres, the average size was 5,400 acres.[55] The amounts of charcoal stacked at the furnaces also suggest heavy consumption; for example, 70,000 bushels and several thousand cords of wood at the Monmouth Furnace at the northern edge of the New Jersey pine barrens in 1817, and 107,000 bushels at the Pompton Furnace in northern New Jersey in 1840.[56]

The size of the holdings and the nature and organization of the enterprise meant that the ironworks of the forests of the North resembled the agricultural plantations of the forests of the South, and indeed the name "plantation" was applied to both. The fairly lengthy sale description of the Martic Furnace and its forge, Pennsylvania, in 1769 gives a good idea of the size and complexity of the establishment.[57] After itemizing its 3,400 acres of woodland, the advertisement went on:

> At the furnace, a good dwelling-house, stores, and compting-house, a large coal-house, with eight dwelling-houses for the labourers, a good grist-mill, Smith's and Carpenter's shops, 6 good log stables . . . a number of pot patterns, and . . . stove moulds etc., etc., a good mine bank, abounding with plenty of ore, so convenient that one team can haul three loads a day; about 15 acres of good watered meadow, and as much adjoining may be made.

The forge was about four miles away and, besides its equipment, it too had its range of other buildings:

> a dwelling-house, store, compting-house, with six dwelling-houses for the labourers, two very good coal houses, large enough to contain six months' stock, three stables, Smith's and Carpenter's shops, two acres of meadow made, and about 1,500 cords of wood, cut in the woods at both places.

The sale package also included horses, wagons, and two slaves.

The variety of activities, the self-sufficiency of the economy, the "owning of labour, the high social position of the owner," his large house and patriarchal attitude were the hallmarks of this form of organization. James Swank, who lived nearer the age and conditions than later commentators, summed it up well when he said that the owners were

> almost feudal lords, to whom their workmen and their workmen's families looked for counsel and guidance in all affairs of life as well as for employment; whose word is law, who often literally owned black labourers, and to whom white "redemptioners" were frequently bound for a term of years to pay the cost of their passage across the ocean; who cultivated farms as well as made iron; who controlled the politics and largely maintained the churches and the schools in their several neighbourhoods; who were captains and colonels of military organisations; whose wives and daughters were grand ladies in the eyes of the simple people around them; whose dwellings were usually substantial structures, which were well furnished for that day and ordered in a style of liberality and hospitality.[58]

Not until there was a shift from charcoal to mineral fuel did the plantation and the patriarchal life-style end.

Because of the huge demands of the forest for fuel the supplies were often exhausted, and the plantation moved on to a new location – another similarity with the southern plantation after soil exhaustion. J. D. Shoepf, who traveled through Pennsylvania in 1783, commented on the lack of timber supplies in the country around the Schuylkill, where the forests were "everywhere thin" and the ironworks "could not but ravage the woods and to their own hurt." The Union Furnace in New Jersey had "exhausted a forest of nearly 20,000 acres in about twelve to fifteen years" and had had to be abandoned. At the Sterling Ironworks in southern New York 23,000 acres of woodland were used to supply the furnaces, but near to the town of Sterling itself "for miles there is scarcely stick as thick as a man's arm but has been scraped off and consumed in the insatiable maw of the furnace."[59] Examples of the abandonment of furnaces because of the exhaustion of fuel supplies are found everywhere, for example, at Stockbridge and Pittsfield in Massachusetts and in the area of the Housatonic River Valley in western Connecticut. In northern New Jersey the Union Furnace was abandoned in the 1780s after 20,000 acres had been cleared off in the space of 15 years, and the Clinton Furnace built not far away was closed for the same reasons in 1837. The Cumberland Furnace in southern New Jersey was abandoned in 1840 after 15,000 acres had been stripped of their timber.[60]

Of course, the depredations of the charcoal cutters in the forests need never have been so great had coke been used for smelting. As early as 1709 Abraham Darby in England had proved that smelting with coke was possible, and because of the scarcity of fuel wood in Britain only 24 out of 77 pig iron furnaces were using charcoal by 1791, a proportion that had decreased to 11 out of 173 by 1806. In contrast, in the United States 439 out of 560 working furnaces were still using charcoal as late as 1856.[61] By making this conversion Britain became a low-cost producer of pig iron and castings, whereas the United States remained a high-cost producer.

The retention of the old technology in the United States was largely a question of the superabundance of wood fuel, although cheap timber did not necessarily mean cheap charcoal. It was also a question of the widespread preference for the sort of iron that charcoal made, a topic that will be discussed at greater length when we consider the antebellum iron industry.

Suffice it to say, there was still a total dependence on wood for fuel. The activities of the charcoal getters and burners left a deep imprint on the forest, destroying stands, thinning them, altering their composition if they regrew, and also altering the soil by stripping the turf and sterilizing the soil where the pits smoldered. Further afield, the forest was damaged by fires that got out of control.

5

The quickening pace: agricultural clearing, 1810–1860

> **With regard to the general aspect of America, the most correct idea you can form of it . . . is that of one immense forest, interspersed occasionally with patches, cleared for towns, or cities, or plantations and some times with natural meadows or prairies.**
>
> Adam Hodgson, *Letters from North America* (1824), 1: 173

> **Settlers in the wilderness have a natural antipathy to the sight of a tree and the axe levels all without distinction.**
>
> Henry Beaufoy, *Tour through Parts of the United States and Canada* (1828), 76

The fundamental change in the geography of the United States during the first half of the nineteenth century was the quickening pace of agricultural colonization that swept through the remainder of the eastern forests, around and over the Appalachians, eventually to reach the Lake States and the Gulf South. The removal of the Indian restrictions to expansion, and to a lesser extent of the French and Spanish restrictions, together with the rapid growth of population caused the dribble of people moving west in 1790 to become a flood after 1810. Whereas the population of the trans-Appalachian West was approximately 0.1 million in 1790, it had risen to 1.1 million in 1810, 3.7 million in 1830, and 9.9 million in 1850. The 200 years of gradual settlement that had produced an almost "motionless America" changed after the mid-1790s and the Revolution.[1] Thirteen new states were created during the years between Kentucky and Vermont (1792) and Wisconsin (1848); Texas had been annexed in 1845; and this all made for a very different America from the one that had hugged the plains and hills of the eastern seaboard for so many generations. The new settlements were basically agricultural in character, but urban centers blossomed with wonderful rapidity at strategic points in the emerging circulatory system of river transport along the Mississippi and Ohio rivers and their tributaries, on some canals established in the North, and on roads everywhere.

The nuclei of settlement established along the western fringe of the Allegheny Mountains, particularly in southwestern Pennsylvania at the forks of the Ohio and in the open country of the bluegrass region of western Kentucky, expanded after 1790 so that by 1820 the Mississippi had been crossed in several places, and tongues of settlement extended westward into Iowa, Missouri, and Arkansas. On the Gulf the long-established settlement in New Orleans and Louisiana expanded west and north; in the North the southern edges of Wisconsin and Michigan were being settled, and other, smaller frontiers of expansion lay in northern Maine and northern Florida. Everywhere in the western part of the country farmers were fronting onto the land between the forest and the prairie. Ultimately, of course, they were to encounter the almost treeless expanses of the prairie proper, an environment that demanded a new approach to farming and a reevaluation of the forest as a resource.

This was not the only new experience; several other novel changes were afoot. There was a gradual realization that maybe forest clearing had a deleterious effect on the wider environment by causing soil erosion, rapid runoff, and, perhaps, less rainfall. There was also the fact, though few would have recognized or been aware of it at the time, that the country was being transformed from a predominantly self-sufficient, rural, agricultural – almost peasant – economy to a highly commercial, urban, industrialized one. The expansion of settlement, the addition of over 26 million people between 1810 and 1860, and the quickened pace of economic activity had far-reaching implications for the forests, for in every way the demand for wood and wood-derived products rose phenomenally, and enormous areas were denuded of their forest cover. In the long view of the use and abuse of the forests of the United States, the years from about 1810 to the Civil War were pivotal and crucial.[2]

Clearing the forest and making a farm

In all this change one thing stayed basically the same. Farm making occurred in the familiar environment of the forest, and it was still hard, backbreaking work that used well-tried methods and technology. It is true that small prairie openings were being encountered by the settlers in western New York, and now increasingly in Illinois, Indiana, and southern Wisconsin, but they were regarded as peculiar oddities in the makeup of the forest, welcome, to be sure, because of the immediate pasturage and cultivable ground they provided but rarely settled upon exclusively until the late 1820s, especially if they were large.[3]

Paradoxically, the pioneers liked the forest because they were used to it; writing in 1836, James Hall said that traditionally "the idea of a wilderness was indissolubly connected with that of forest" in the minds of the pioneers. Their ancestors had settled in the woods, and it was there that "their ideas of a new country had been formed."[4] Also, they knew that raw, forested land could be converted to productive land by their own labor. In other words, the forest offered the opportunity to continue the application of the lessons of an old and well-known technology.

In emphasizing the importance of the unsettled forested lands of the West as the venue for colonization and farm making, it should not be overlooked that there were alternatives, such as the already established farms in the East that were up for sale and even the partially developed farms in the western forests. Alexis de Tocqueville said that it seldom happened that an American farmer settled for good on his own land; "especially in the districts of the Far West he brings land into tillage in order to sell it again, and not to farm it."[5] Therefore, there was a speculative element in forest clearance that first became evident in the later years of the eighteenth century as personal mobility increased. In 1806 Dr. Benjamin Rush had not been slow to point out that "the business of settling a new tract of land and that of improving a farm are of a very different nature," and some years later Timothy Dwight noted that there was a class of backwoodsmen who merely "prepared the way for those who came after them." Just as there were those who created half-formed farms, so there were those who were ready to take them up. "It is considered here a small affair for a man to sell, take his family and some provisions and go into the woods upon a new farm, erect a house, and begin anew," said the Reverend John Taylor about New York in 1802. The idea of moving from a half-cleared farm to start another was alien to most non-American observers: "They make light work of it here," said James Stuart in 1833, "and consider it to be merely a question of finance."[6] In addition to the half-created farms there were also pockets of land bypassed by earlier waves of settlement in northeastern Pennsylvania, southern New Jersey, and even Long Island.

Which location the would-be farmer chose depended ultimately upon a variety of factors, notably the availability of capital and credit and the possession of personal skills. The value he put upon social intercourse, religious homogeneity, or the prospect of immediate gains as opposed to long-term objectives also had its influence. Certainly, whatever he did, either buying and developing new land or buying an already established farm, it was costly.

Despite the fact that new settlers, particularly European immigrants, were advised to "pay a little for land *lately* cleared," the majority of settlers filed claims in untouched forest lands, convinced that their energy and enthusiasm to create a holding would compensate for their lack of capital and carry them through the vicissitudes of the early years.[7] That clearing was the most expensive item in the settler's budget was well known, as is apparent from what an innkeeper at Pontiac, Michigan, told Alexis de Tocqueville in 1831:

> The greatest expense is the clearing. If the pioneer comes into the wilds with a family able to help in the first work, his task is fairly easy. But that is generally not so. Usually the emigrant is young, and if he already has children they are in their infancy. Then he must either see to all his first needs of his family himself, or hire the services of his neighbours. It will cost him 4–5 dollars to clear one acre.[8]

In reality, the innkeeper had underestimated the costs of clearing, which could be much greater than that. One example of farm-making costs in the forest must suffice. In 1821 a New York farmer put forward a plan for making a small farm of 50 forested acres. Thirty acres were to be cleared and cultivated, five of which were to be occupied by the house, barns, orchards, and gardens, 10 acres for meadow and pasture, and 15 acres for crops. The remainder was left as a woodlot for fuel and fencing. It was calculated that the costs would be as follows, based on the purchase of good-quality stock:

Clearing 30 acres at $10 per acre	$300	
Fencing, 700 rods	70	
Log house and frame barn	200	
Outhouses, well, and orchard	150	$720
1 yoke of oxen	50	
1 horse	50	

2 cows	40	
2 hogs	10	
10 sheep	50	
Utensils and harness	50	250
		$970

This calculation did not include the cost of the land, which could be anywhere between one and five or six dollars per acre.[9]

Obviously, the first four items totaling $720 could be almost totally dispensed with for they could be the product of the farmer's labor, but he needed funds to tide him and his family over the farm-making period for a season or two at a rate of $50–$100 per annum. Thus, it is possible that up to $500 was the necessary minimum of capital needed in order to create a forest farm, together with good health, unremitting toil, and a great deal of frugality.

The chronology of making a farm and the rise of self-sufficiency are well illustrated by an account of a newly formed farm in northern Pennsylvania. Having acquired 100 acres, "with no other property than an axe or hoe," a healthy man might clear and sow 10 acres and erect a cabin during the first year. During the second year he would have enough crops to sell to be able to buy a cow, oxen, sheep, hogs, a plow and harrow, and some household utensils. In the third year he could clear 15 acres more and continue at this rate so that in 10 years his 100 acres would be cleared and fenced, and his stock would have been sold off from time to time to clear his debts and to provide money for subsistence.[10]

Not all pioneer farmers, however, were without capital, and the prospect of hiring labor to chop down trees, clear the land for the first crop, build a temporary dwelling and some fences was attractive and more common than is supposed. The "set-up" men were often common itinerant laborers wandering from place to place over great distances, armed only with their axes and grubbing hoes and usually asking about 50¢ a day, which was the prevailing wage in the early nineteenth century for unskilled labor. It was arduous work; some skilled axmen claimed that they could clear an acre in from three to seven days, but that was the work of the professional, and a strong one at that, and the rapidity of clearing also depended upon the density and the size of the trees.

Not all hired hands were menial itinerants. Many local farmers and townsfolk and young men and emigrants from the East out to make their way in the world cleared part-time until they accumulated enough money or their strength and health gave out. But such was the demand and such were the financial rewards that labor shortages on farms in the newly cleared frontier areas and in nearby towns were very great. Costs rose to 75¢ a day – or $5–$10 an acre – even if the men were available (see Table 3.1). Consequently, those who could afford to retained their set-up men as regular hands, who were invaluable in case the almost inevitable ague and illness attacked the farmer of the newly cleared block. Other set-up men who secured employment in a particular area often settled there and filed a claim for land. One special class of hired hand was the farmer who had bought a farm but could not clear enough in time to get sufficient cash returns to pay for food and other essentials for his family. He was forced to clear for others in order to save his own clearing.[11]

The problem of "how to subdue the land," as Jeremy Belknap put it,[12] remained, and methods of clearing during the nineteenth century did not change much from those employed in earlier centuries. The descriptions of Beaufoy, Belknap, Dwight, Hodgson, Stuart, and Tudor have much in common.[13] Generally, clear-cutting and girdling, both with burning, were the universal methods. The first was laborious and thorough but demanded capital to tide the family over, for no crop could be grown while the clearing was being done. The second was expeditious, deferred the real tasks of clearing, and allowed an immediate crop to be grown. Both left stumps. The only detailed account of costs is for an experiment in North Carolina for the late nineteenth century when both methods of clearing were compared. There was little to choose between them. With labor at 60¢ a day, girdling cost about $8 an acre and took 13.5 man-days, whereas clear-cutting cost between $10 and $12 an acre but took between 16.5 and 20 man-days. But many of the setting-up tasks, such as collecting fallen timber, were deferred in girdling, so the conclusion is that the expenditure of labor, and eventually of costs, in both methods was ultimately almost identical.[14]

Other variables entered into the clearing calculation, for which information becomes available for the first time. The season in which clearing was done affected the time taken and hence the cost. Although a spring chopping was quick, it meant that green wood made firing difficult. A summer chopping was marginally quicker because the wood was drier and took less effort to burn and therefore was cheaper. But a winter chopping was cheaper still and also had the advantage that it could be carried out during the off-peak period of farm work (Table 5.1). Therefore, whatever the number of days that chopping an acre took, the tidying up afterward would take from one day to three days, and costs varied by up to $3.25 an acre. It paid to cut in winter, and it paid even more to cut underbrush and fell timber, then leave it for a couple of years and burn it. To leave the splitting of rails to one's convenience also paid off.[15]

Time and hence costs do not appear to have decreased during the nineteenth century. In an analysis of 19 estimates made of the labor involved in clearing an acre of forest between 1800 and 1890 there is a range of between 13 and 36 man-days, the variation depending mainly upon the character of the forest, with a median figure of 20 days and a mean figure of 20.6 man-days. There is no evidence to suggest that the productivity of clearing by the farmer increased with time; admittedly, there were some advances in the technology of cutting in the lumber industry, which may have had some marginal effects, particularly in the North, but probably they did not filter down en masse to the individual farmer on his section.[16]

Both girdling and clear-cutting left stumps. It was hoped that the stumps would rot sufficiently in 5 to 10 years for them to be pulled out of the ground easily, and it was claimed that the stumps of girdled trees decayed more rapidly. The tools for stump clearing were the same as those that had always been used: axes, levers, spades, chains, and oxen. It is surprising, however, how little the task is commented about, possibly because it was a deferred task and rarely entered into the accounts of initial pioneering

Table 5.1. *Relative advantages and costs of clearing one acre at different times of the year (mid-nineteenth century)*

	Spring cutting	Summer cutting	Winter cutting	2-year lag, cutting to burning
Advantage	Quick	Moderately quick	Slack farm season	Convenience
Disadvantage	Difficult to burn	Easier burn	Easy burn	No immediate crop
Chopping cost	\$5.00	\$5.00	\$5.00	\$5.00
Brush burning				
Cost	\$1.00	\$0.31	—	Own labor
Labor days	1–2	½	Negligible	
Logging (3 men, 1 team)				
Cost	\$3.75	\$3.13	\$2.50	Own labor
Labor days	1½	1¼	1	
Burning				
Cost	\$1.25	\$1.25	\$1.25	Own labor
Labor days	2	2	2	
Fencing (drawing and splitting 300 rails)				
Cost	\$3.00	\$3.00	\$3.00	Own labor
Total cost	\$14.00	\$12.69	\$11.75	?
Total days of itemized labor	4½–5½	3¾	3	?

Source: Michigan Farmer 9 (1851), 70–1.

settlement. But all the indications from the few existing references are that it was a task that nearly equaled in effort the initial felling of the trees. In Massachusetts in 1841 it was estimated that two men and a yoke of oxen cleared three acres of stumps in 20 days, or about 13.3 man-days (to say nothing of the ox days!) per acre.[17]

From about 1850 onward various machines that worked on screw-and-lever principles were on the market, and one manufacturer claimed that two horses and two or three men could extract 100 stumps a day, but this seems an exaggeration, as it did not include any contingency for gathering the stumps and burning them. More realistic was the claim in 1851 that a winch pulled 800 stumps in 15 days with six men and three oxen, or 10 stumps per man-day of labor. Though some stump pullers made a marginal contribution to the lightening of the task, many were too light for the work they needed to do and broke; others were too heavy and cumbersome; and many were plain inefficient and did not work, wrenching out half the stump and leaving the rest in the ground. In 1889 it could still be claimed that no efficient stump puller had been invented and that the traditional tools of ax, lever, spade, and chain with the help of an ox still served best. Not until gunpowder for blasting out the stumps became the norm after the late 1880s, especially in the logged-off areas of the Lake States and the Pacific Northwest, was labor reduced, and then by about half.[18]

Clearly, the type of forest trees and the density of the stand were potent factors in the accuracy of calculations. Hardwood stumps like hickory and oak could rot in five years; northern and southern pines did not rot in 20 years in some cases. Redwood and Douglas-fir in the West and the white pine in the North left enormous stumps that were very difficult to move. No defensible estimate can be made for these variables, and one must take a reasonable norm, which suggests that stump pulling added about 12 days to clearing during the first half of the nineteenth century, making a total of 32 man-days to clear an acre completely. Costs seemed to stabilize at between $10 and $12 per acre, a figure considered to be the equivalent of one man's labor for one month.

With these calculations in mind, it can be appreciated that it was rare for the agricultural pioneer without help to clear more than 10 to 12 acres per annum during the first few years of settlement (e.g., 10 acres clear-cut at 32 man-days per acre = 320 days); a few people, in areas of lighter timber, could clear about 15 acres. There were too many alternative calls on the time and energy of the pioneer, such as home building, fencing, crop planting and cultivation, getting provisions, cordwood cutting, ash collecting, and so on. Indeed, the rate of clearing was likely to be within the range of 5 to 10 acres rather than more than 10 acres.[19]

Sometimes the task was speeded up when neighbors came to help in initial clearing and home building, but the bulk of the farm making had to be done by the pioneer himself, the only exception being in the West where the rapidly expanding population and the ever-present and profitable market for produce made it feasible for some farmers to hire labor to do some of the clearing at the going rate of between $10 and $12 per acre. But for the vast majority of farmers the task of clearing the woodland was a monotonous one of hacking, grubbing, and burning; it was a task not of a few years but of at least a generation.[20]

And so it went on until the farmer had as much land as he could handle. The remaining woodland would be used for providing fuel and as pasture for sheep for wool or for swine for pork. The originally cleared fields would, in time, be depleted of their inherent fertility and become "old fields" and converted to pasture or, occasionally, allowed to revert to woodland. Eventually the farmer, through his own efforts, got himself into a management problem; unless he could run more stock and sell them, the area of pasture became too great. In fact, continued clearing culminated not in the satisfying completion of the farm-making process but in a crisis. When this was apparent some farmers adjusted their rate of clearing in order to maintain a delicate balance between crops and stock and woodland lots.

The amount of land cleared

The amount of land cleared for agricultural purposes was rarely recorded and is impossible to estimate accurately except at a local level (e.g., the New York census).[21] Evidence of what happened is often slight and fragmentary because clearing was so widespread and commonplace as to be barely worth comment or record. The generalization of the map and the statistic is needed. Only when the forest was felled and the land rescued from its state of nature, or "improved," did it enter into the official statistical record.

With the introduction of the categories of "improved" and "unimproved" land in the federal census after 1850 a calculation can be made for the country as a whole. On the assumption that, in the heavily forested lands of the East, for land to be improved it had first to be cleared, there should be a close relationship between improved land and forest removal. Thus, if the counties of the United States are divided into those that are predominantly forested and those that are predominantly nonforested at the time of first European settlement, then the amount of improved land can be regarded as a measure of the amount of cleared land or of the amount of open land broken up. The division is a crude one, but it is assumed that the nonforested portions of the forested counties and the forested areas of the nonforest counties will tend to offset each other. Land withdrawn from farms can be subtracted from new, improved land to eliminate a potential source of error. The problems of calculating the cutover lands and of reclaimed, previously abandoned land are dealt with in Chapter 11 where they are more relevant to the discussion.

It is probable that some 113.74 million acres of land had been improved in the United States before 1850, the overwhelming bulk of which was carved out of the forests of the eastern half of the country, with only a small amount coming from either natural clearings or abandoned Indian clearings (Figure 5.1). At the very least, 100 million acres represent the culmination of two centuries of pioneer endeavor. Between 1850 and 1859 the amount cleared had risen to a colossal total of 39.705 million acres, equivalent to roughly one-third of all clearing carried out during the previous two centuries. It was a decade of maximum impact on the forest. At the same time, 9.139 million acres were improved in nonforest land. Figure 5.2 shows the distribution of clearing by states between 1850 and 1859. It was greatest in the areas of active colonization, such as Wisconsin (3.3 million acres), Ohio (2.8 million acres), Illinois (2.4 million acres), New York (1.9 million acres), Georgia (2.1 million acres), and Alabama (1.9 million acres).

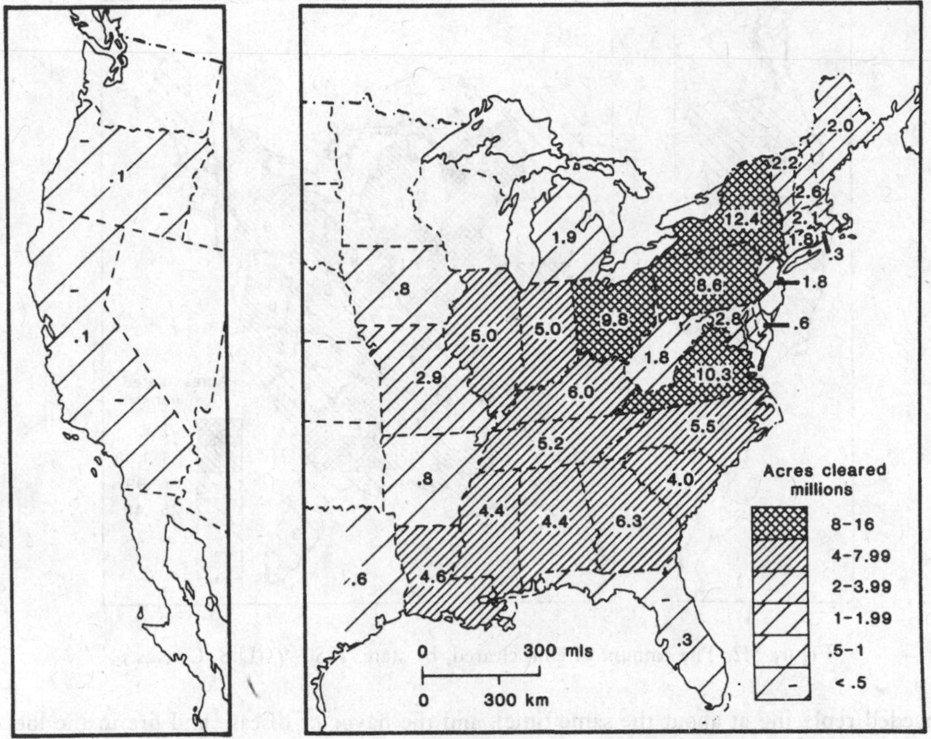

Figure 5.1. The amount of land cleared, by state, before 1850. (U.S. Census.)

Only slightly fewer acres were cleared on some of the older settled areas, such as Pennsylvania (1.9 million acres) and Texas and Kentucky (1.7 million acres each). Between them these nine states accounted for 58 percent of all land cleared during the decade.[22]

When clearing got beyond a certain point, however, the farmer faced a problem of achieving the correct balance between cleared and uncleared land. The previously unbroken forest became the farm woodlot, a source of essential warmth, shelter, and fencing material, and land values reversed. Ever the careful observer, James Stuart noted that in much of New York no more wood remained than was "necessary . . . for fuel and farm purposes." However, in some places, especially near Geneva, too much woodland had already been cleared: "Wherever a sufficient quantity of land has been cleared, the woodland of a farm bears as high a price per acre as the land actually cleared."[23]

The sizes of the residual woodlots varied, but they were rarely much below 20 acres because of the vast amounts of wood needed to feed the open fires. In 1818 President Madison thought 10 acres the minimum in the South and 17 acres the minimum in New England. When stoves became common the woodlot could be decreased a little. With a rate of consumption of 10 cords per annum for a couple of stoves and a forest yield of 30 cords per annum over 15 acres, a 45-year rotation of cutting was predicted as feasible in New England.[24] But such a simple calculation did not always work out in reality. Exceptionally cold winters, fencing (especially when all the set-up fences rotted and

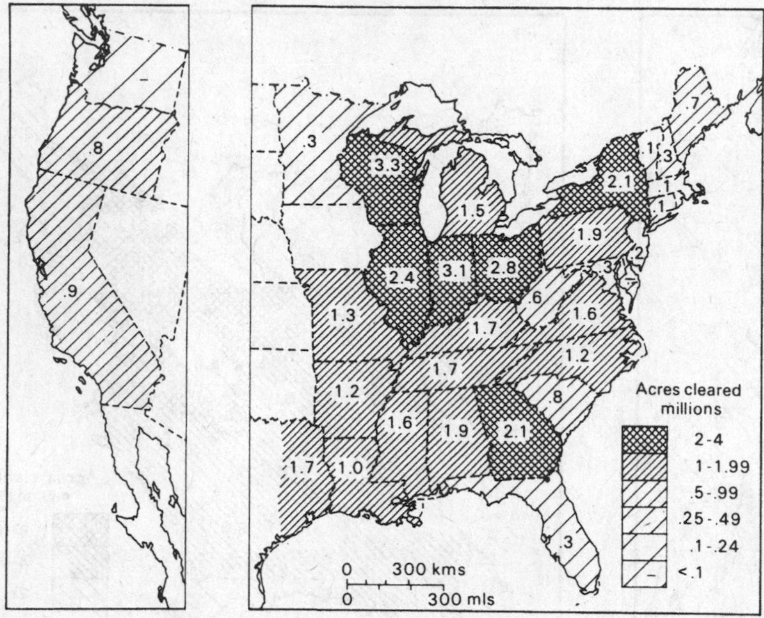

Figure 5.2. The amount of land cleared, by state, 1850–9. (U.S. Census.)

needed replacing at about the same time), and the havoc of disease and fire in the lot made such precision about the size of woodlots uncertain. Not until coal fires, wire fences, and brick houses replaced their wood counterparts did demand for wood decrease. What we do know, however, is that enormous amounts of woodland remained in even the most improved landscape and that the woodlot, that remnant of the great forest, was and still is a distinctive and characteristic feature of much of the American rural landscape. How the farmer managed his woodlot became a major issue in future years when fuel and timber supplies became scarce.

"A new creation": the landscape of clearing

The clearing of the forest fascinated travelers and topographers; a few commented on the processes of clearing but nearly all had something to say about the landscape of clearing in its various stages. A rich store of information and impression is left in the work of over a score of such people. Some of it is repetitious, some outright plagiarism of previous descriptions, but some of it adds new depth and insight to our understanding of the scene in the forests of early-nineteenth-century America and the attitude and action that helped make it.

One impression was dominant: America was still an overwhelmingly wooded country. There was forest everywhere, even in the towns where stumps still stood in the streets and the streets could be lengthened "only by cutting down fresh timber" (Plate 5.1).[25] When large stretches of cleared land were encountered, they attracted attention, as when A. Burnaby commented that nearly half of the 100-odd-mile journey he had

Plate 5.1. Embryo town of Columbus on the Chattahoochee. Hall remarked that this was "the most populous and least cleared part of the city." (Hall, *Forty Etchings*, plate XXVI.)

undertaken from Annapolis to Philadelphia in 1759 was so bereft of forest that it "had a different aspect from anything I have hitherto seen in America."[26] Yet America was so large that limitless forested wilderness could coexist with limitless development, and it was the progress of that civilization in development and the effort made to subdue the forest that attracted the most attention.

For foreign travelers, in particular, the American landscape presented an exciting and novel aspect. They had been used, and had adjusted aesthetically, to the "made" scenes of Europe, which were the product of centuries of honing the land to its shape and appearance, and they were also used to a situation where wood was scarce and needed to be conserved. Therefore, the prospect of a landscape in the process of being made and the prodigal waste of wood that accompanied that process stimulated a great outpouring of comment and opinion.

One such person who recorded his impressions was Basil Hall, an observant traveler from Britain who was also an artist. The scenes of clearing in western New York excited his literary and artistic imagination. To him the girdled forest was grotesque.

> Some of the fields were sown with wheat above which could be seen numerous ugly stumps of old trees; others allowed to lie in the grass guarded, as it were, by a set of gigantic black monsters, the girdled, scorched and withered remnants of the ancient woods. Many farms were still covered with an inextricable and confused mass of prostrate trunks, branches and trees, piles of split logs, and of squared timbers, planks, shingles, great stacks of fuel, and often in the midst of all this could be detected a half smothered log hut, without windows or furniture, but well stocked with people. At other places we came upon ploughs, always drawn by oxen making their sturdy way amongst the stumps like a ship navigating through coral reefs, a difficult and tiresome operation.

Plate 5.2. "Newly cleared land in America." The scene is Oak Orchard Creek, 40 miles west of Rochester, New York. (Hall, *Forty Etchings*, plate IX.)

In addition to this splendid description of cleared land, Hall later sketched the newly cutover land (Plate 5.2) and appended this caption to his etching:

> The trees are cut over at the height of three or four feet from the ground and the stumps are left for many years till the roots rot; – the edge of the forest, opened for the first time to the light of the sun looks cold and raw; – the ground rugged and ill-dressed . . . as if nothing could ever be made to spring from it. The houses which are made from logs, lie scattered about at long intervals; – while the snake fences, constructed of the split trees, placed in a zig-zag form, disfigure the landscape.

The scene, said Hall, had invariably a "bleak, hopeless" aspect, which had "no parallel in old countries."[27]

Others wrote in the same vein. Henry Tudor, on his way from Rochester to Canandaigua, saw large tracts of land all around "emerging into luxuriant cultivation from the boundless, and till now untrodden, forest." The whole scene, continued Tudor, was "literally a new creation," to which the countries of Europe presented no counterpart and which was "only to be witnessed in America." In the middle of the fields of grass and corn were "the stumps of a thousand giant trees" cut within two or three feet of the ground, and in other parts of the "freshly cleared wilderness was seen stalking through the blackened forest, the destructive element of fire, accomplishing with a more furious rapidity what the tardy axe had left unachieved." The clear-cutting "offered to the eye an aspect of desolation, that excited rather a painful and melancholy feeling," he continued. Even worse was the sight of the girdled trees, which produced "a moral sensation of pity." "It appeared to me in the excitement of my fantasy," said Tudor, "to be a wanton mode of destroying the tree by cutting its throat."[28] What is significant about Hall, Tudor, and the other British travelers is that they were less indignant about the destruction and waste of timber (they barely mentioned it) than they were about the aesthetics of the scene and the carelessness and untidiness it betokened.[29] Of course, they were looking at the forest through specifically British eyes. Nonetheless, despite the fact that their vision was colored by the experience and associations of other parts of the world, it was a scene that the native American had become almost unable to see and that in its obviousness was captured only by the foreign eye.

To the American the practical implications of the scene were more important than its aesthetic qualities. For William Nowlin, the "beautiful workmanship of nature" was subordinate to the joy of seeing his new clearing joining that of his neighbors; "Then it began to seem as if others were living in Michigan for we could see them." And Ulysses Hedrick thought his father's quest was "to transform prairies and forests into smiling fields of growing grain and grazing herds." Simply, trees and stumps meant toil; the cleared land meant production, food, and neighbors; and work now meant the sacrifice of current well-being for a more glorious future.[30] A few American observers did appreciate the beauties of the "arrangement" in the landscape, but the likelihood was that if they were struck by the scene it was by its rugged beauty, grandeur, and sheer magnitude.[31]

More typical were the observations about the densely settled Mohawk Valley in Tocqueville's travel notes. Admittedly, Tocqueville was not an American, but during his years of residence he developed a remarkable sensitivity to the ethos and reality of the country.

> In general, the whole country has the look of a wood in which clearings have been made. . . . Every sign of a new country . . . Man still making clearly ineffective efforts to master the forest. Tilled fields covered with the shoots of trees; trunks in the middle of corn. Nature vigorous and savage.

Tocqueville's description contained those elements of pride and awe that were often a feature of those who saw the clearings in the untouched forest as part of man's conquest of nature and as a battle of civilized man against an unfavorable environment. The experience of the "howling desert" and the need for "wilderness-work" described by Edward Johnson in early-seventeenth-century New England had never been eradicated from the American psyche. As Tocqueville traveled farther west the forest was disturbed less and less, and his description becomes full of phrases like "the vast dome of vegetation," "damp depths," and "sublime silence" as the endless miles of undisturbed vegetation left no variety of scene to describe. The oppressive solitude and silence depressed him. Then, to his relief, "We heard the echos of an axe-stroke. . . . In the centre of the rather restricted clearing made by axe and fire around it, rose the rough dwelling of the precursor of European civilization. It was like an oasis in the desert."[32] This scene of the clearing by the lone settler in the depths of the forest was the apotheosis of America until about the middle of the nineteenth century.

Similarly, Isaac Weld realized that he was unusual in being attracted to the wooded hill country to the northwest of Philadelphia, which was traversed by the Susquehanna and by many small streams and waterfalls, and from which distant views could be glimpsed of Chesapeake Bay.

> The generality of Americans stares with astonishment at a person who can feel any delight at passing through such a country as this. To them the sight of a wheat field

Plate 5.3. Contrasting landscapes in rural Pennsylvania. (Isaac Weld, *Travels through the States of North America.*)

or a cabbage garden would convey pleasure far greater than the most romantic woodland views. They have an unconquerable aversion to trees; and whenever a settlement is made they cut away all before them without mercy; not one is spared; all share the same fate and are involved in the same general havoc. . . . The man that can cut down the largest number, and have fields about his house most clear of them, is looked upon as the most industrious citizen, and the one that is making the greatest improvements in the country.[33]

As if to emphasize the point, Weld included an almost symbolic scene of dark, somber treescapes contrasted with light, open, cleared farmscapes (Plate 5.3).

The comparison of attitudes was brought out deliberately by Adam Hodgson, a British businessman and traveler, who attempted to highlight the "different impression which wood, as a feature in a landscape, conveys to an Englishman and an American." To the Englishman, he thought, trees conveyed impressions of "shade, or shelter, or rural beauty . . . [the] protection of game; – to add variety to park scenery, or to contrast with rich cultivation." On the other hand, in the imagination of the American:

if mere cultivation be not beauty, it is closely allied to it . . . and from its intimate connection with utility, which enters largely into his idea of beauty, it awakens many kindred associations.

Every acre, reclaimed from the wilderness, is a conquest of "civilized man over uncivilized nature;" an addition to those resources, which are to enable his country to

> stretch her moral empire to her geographical limits, and to diffuse over a vast continent the physical enjoyments, the social advantages, the political privileges, and the religious institutions, the extension of which is identified with all his visions of her future greatness.[34]

Of course, for some Americans the rediscovery of the beauty and virtue of the forest was about to begin. The Leatherstocking novels of James Fenimore Cooper, the poems of William Cullen Bryant, and the painting and writing of Thomas Cole and the Hudson River school were soon to be allied to sentiments of antiurbanism and patriotic feelings about the unique nature of the American wilderness. By midcentury the transcendentalist philosophy of Emerson and Thoreau had cast the forest in an entirely new light for Americans.[35]

Undoubtedly the most graphic impression of the landscape of clearing was given in Orsamus Turner's "Four sketches and descriptions of the pioneer in Western New York frontier lands," which was based on the actual experience of the sequence of events he had seen. The sketches are like four "stills" clipped out of a continually moving picture of the changing forest landscape. In Plate 5.4, the pioneer and his wife have been on the block for six months. He has cleared a patch "to open out" the wood and get logs for his cabin. The cows and sheep browse in whatever vegetation they can find. His nearest neighbor is miles away. Later (Plate 5.5), the pioneer has cleared a few more acres and enclosed them with a rail fence in front and a brush fence elsewhere. Corn, potatoes, and beans have been planted among the stumps, and in the background his scattered and distant neighbors have gathered together and are engaged in a "logging bee." Ten years later (Plate 5.6), the pioneer farmer has cleared 40 acres, and stumps in the background indicate that more clearing is underway; there are crops of corn and grass in the fields. The log cabin still stands but has been expanded and improved with a new clapboard barn. In the upper left we can discern a church and schoolhouse in a nearby growing town. Toward the end of his life (Plate 5.7), the surrounding area of forest has been cleared by the farmer and his neighbors so that only the ridge tops remain as woodlots. The house has been extended and "beautified," and a railway serves the village.[36]

Though we may vary the detail of the four sketches here and there, the scenes portray the essential truth of an experience that was repeated a few million times in the northern and eastern portions of the United States until the closing years of the nineteenth century.

At no time did Orsamus Turner dignify his four sketches as depicting "stages" in the settlement of the forest. They were merely arrested moments in the continuum of change in newly settled areas, and they happened at different times at different places, and none was inevitable. Yet in a sense they did depict something that had fascinated writers and observers of the forest scene such as Crèvecoeur, Benjamin Rush, and Timothy Dwight:[37] the "stages" of settlement and what they told of the likely progress of civilization and, let it be said, of the formation of typical American characteristics. Tocqueville said that in the areas of new settlement on the edge of the forest he "counted on finding the history of the whole of humanity framed within a few degrees of longitude." Hodgson also said that in his travels he had "traced man through every successive stage of civilization" and it was the rapidity of settlement that struck him. In a

Plate 5.4. The first six months. (Orsamus Turner, *A Pioneer History of the Holland Purchase of Western New York.*)

Plate 5.5. The second year. (Turner, *Pioneer History of the Holland Purchase.*)

Plate 5.6. Ten years later. (Turner, *Pioneer History of the Holland Purchase.*)

Plate 5.7. The work of a lifetime. (Turner, *Pioneer History of the Holland Purchase.*)

peculiarly perceptive passage he suggested that what was happening around him in America had a dreamlike quality. The sudden creation of towns and farms "in the centre of those forests, in whose solitudes, within a very few years, the Indian pursued his game, appears rather like enchantment than the slow result of those progressive efforts, with which, in the old world savage nature has been subdued."[38]

The edges of time and space were becoming blurred, so that what was "stage" and what was "place" were difficult to disentangle. "In successive intervals of *space*," said Hodgson,

> I have traced society through those various stages which in most countries, are exhibited only in successive periods of *time*. I have seen the roving hunter acquiring the habits of the herdsman; the pastoral state merging into agriculture; and the agricultural in the manufacturing and commercial.[39]

The sensation that time was being telescoped by distance was overwhelming, and central to that sensation was the rapidity with which the forest was being cleared and new landscapes and new geographies were being created. If forest clearing could be described as "perhaps the greatest single factor in the evolution of the European landscape" during historical time, the same could be said without any qualification of the North American landscape.[40]

The forest edge

While the bulk of the pioneer farmers during the early nineteenth century were hacking their way westward through the forests, there were some, like those from Kentucky and Tennessee, who had pressed northward via the tributaries of the Ohio and Mississippi into southern Illinois and, by 1815, had encountered the eastward-projecting edge of the grassland known as the "Prairie Peninsula." Others reached the openings and prairies of Ohio, Indiana, and southern Michigan by 1820 (Fig. 5.3). The story of the adaptation of the sylvan culture and ways to the new nonforested environment is not the main concern here; others have written about that in detail.[41] Nevertheless, the important point is that this was where the prairie began and the forest ended. A crucial and totally new experience was in store for the pioneer settlers in which the forest and its resources had to be reevaluated. The lack of timber for fuel, construction, and fences was made up for by massive lumber imports from other regions, particularly from the Lake States and later from the South. That led to the emergence of a continental system of timber transport and to the destruction of forests in other parts of the country. In addition, after an initial period of prejudice and technical inability to break the sod and farm the prairie, it became evident that far less labor was needed to create farmland than had been the case in the forest. That meant a massive westward shift in agricultural endeavor and later the slow and progressive abandonment of farms in many parts of the eastern forests for more favorable, "open" locations.

Many accounts of the new experience are dramatic and misleading. One has a vision of the pioneer agriculturalist emerging from the dark cell of the forest, where he had been imprisoned for centuries, into the harsh sunlight of the prairies where, half-blinded and

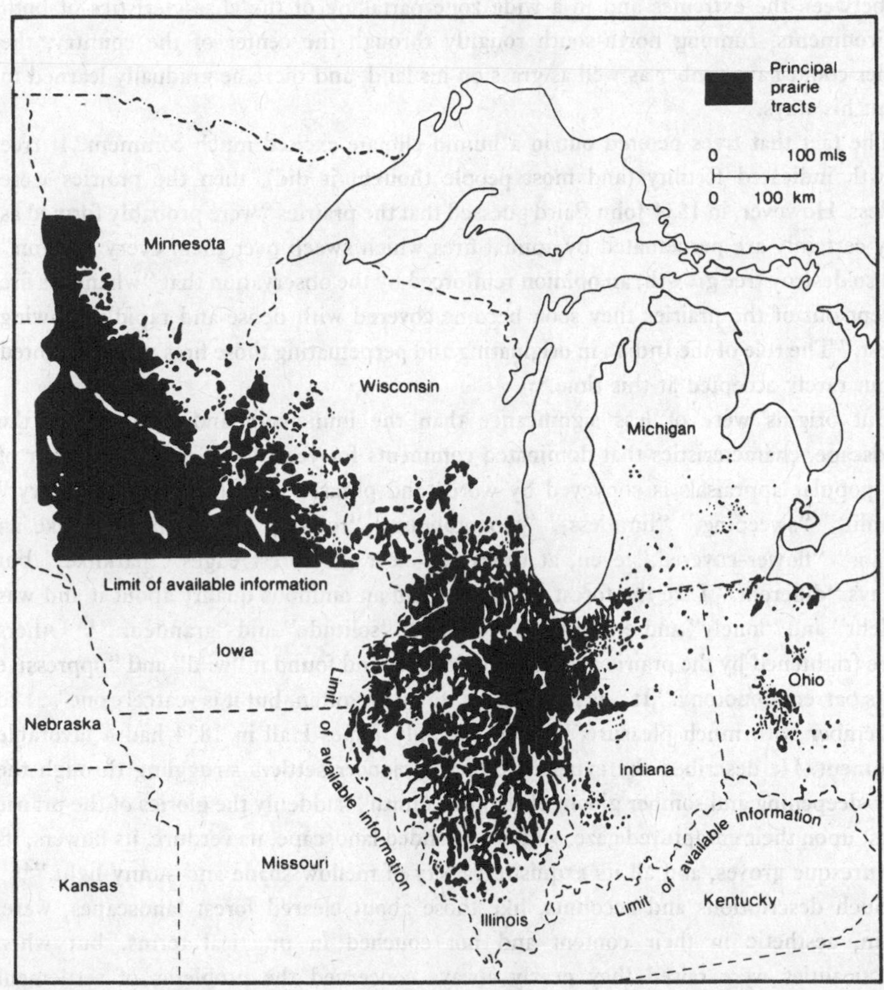

Figure 5.3. The prairie/forest edge in the northern United States. (After E. N. Transeau, "The Prairie Peninsula," fig. 1; and H. C. Prince, "The Real, Imagined, and Abstract Worlds of the Past," 8.)

bewildered, he stumbled about unable to adjust to his new freedom – even skulking back to the refuge of the familiar dark forest now and again. That impression is overdrawn and erroneous and arises from the fact that many scholars and observers have highlighted the differences between the forest and the prairie in order to accentuate the contrasts. In so doing, however, they have ignored two important points. First, even the woodland agriculturalist had encountered and had learned to value open ground, whether it was old Indian fields, the oak openings of western New York, Ohio, or Indiana, the barrens of Kentucky and Tennessee, or the savannas in Alabama and upper Louisiana. Second, the contrast between forest and prairie was not stark everywhere.

In between the extremes and in a wide zone partaking of the characteristics of both environments, running north-south roughly through the center of the country, the settler could have timber as well as grass on his land, and there he gradually learned to adapt his ways.

The fact that trees petered out in a humid climate excited much comment. If tree growth indicated fertility (and most people thought it did), then the prairies were useless. However, in 1834 John Baird guessed that the prairies "were probably formed as they certainly are perpetuated by annual fires which sweep over them every autumn" and so destroy tree growth, an opinion reinforced by the observation that "whenever fire is kept out of the prairies they soon become covered with dense and rapidly growing forest." The role of the Indian in originating and perpetuating those fires was also hinted at but rarely accepted at this time.[42]

But origins were of less significance than the immensity and openness of the landscape, characteristics that dominated comments for years. The particular flavor of the popular appraisals is conveyed by words and phrases like "brightness," "glory," "sunlit," "sweeping," "limitless," "monotonous," "rolling," "enthralling," "like an ocean," "flower-covered," even, at times and near the forest edges, "parklike," but always "different." Like the forest the prairie had an ominous quality about it and was "silent" and "lonely" and engendered feelings of "solitude" and "grandeur."[43] Others were frightened by the prairie and its frequent fires and found it "wild" and "oppressive in its barren monotony." It was not a scene "to be forgotten, but it is scarcely one . . . to remember with much pleasure."[44] Perhaps only James Hall in 1834 had a favorable comment. He described the experience of the pioneer settlers struggling through the ever-deepening and somber gloom of the forest until "suddenly the glories of the prairie burst upon their enraptured gaze, with its extended landscape, its verdure, its flowers, its picturesque groves, and all its exquisite variety of mellow shade and sunny light."[45]

Such descriptions and accounts, like those about cleared forest landscapes, were, again, aesthetic in their content and not couched in practical terms, but when practicalities were raised they nearly always concerned the problems of settlement without a supply of timber: "It will probably be some time before these vast prairies can be settled owing to the inconvenience attending the want of timber," wrote Harding in 1819, and this objection was repeated during later years; for example, Timothy Flint said that the Illinois prairie, "which strikes the eye so delightfully, and where millions of acres invite the plough, wants timber for building, fencing, and fuel."[46]

Nevertheless, many pioneers found the edges of the smaller prairies and the timbered edges of the larger ones very favorable places for settlement. For example, as early as 1812 John Bradbury wrote about southern Illinois:

> For the settler who is not habitually accustomed to the felling of trees and who has the courage to fix himself on wild land, this is much the best part of the United States, excepting Upper Louisiana. If he places his house at the edge of one of these prairies it furnishes him food for any number of cattle he may choose to keep. The woodland affords him the materials necessary for his house, his fire and his fences.

This advice was echoed by a few others, who advised the purchase of quarter-sections

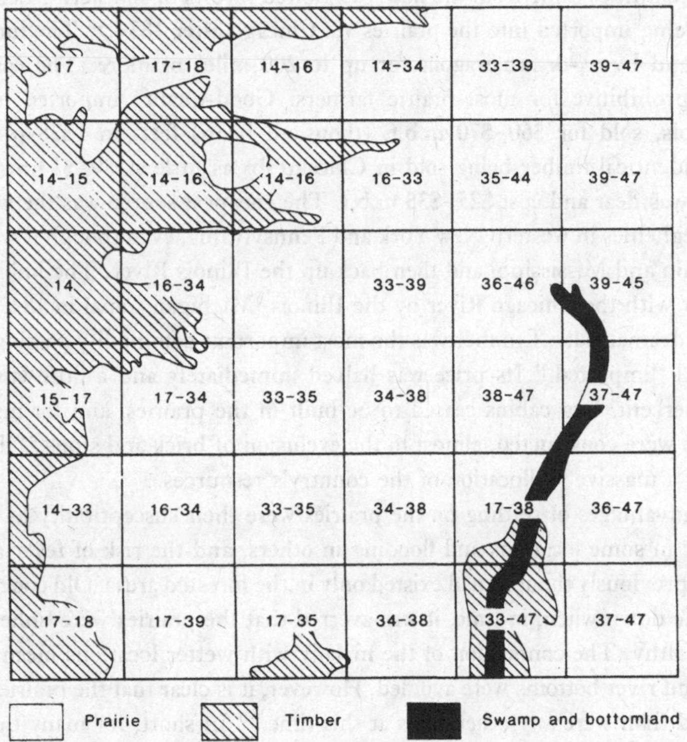

Figure 5.4. Entry dates and vegetation boundaries, Township TIN R6W, St. Clair County, Illinois. The figures refer to the year or years of entry; for example, top left-hand section was entered in 1817 and the adjacent section in 1817–18. (Based on D. R. McManis, The Initial Evaluation and Utilization of the Illinois Prairies, 1815–1840, 71.)

straddling forest and prairie land.[47] It became commonly accepted that the ownership of quite small woodland resources meant the control of large areas of the open range; in other words, prairie land without woodland was no good. Latecomers who could not find adjacent blocks of woodland and prairie tended to have fragmented farms with small timber lots located in the forest. All parts of the larger prairies were less accessible to woodland and therefore valued less for agriculture than were the small prairies. Consequently, the inland portions of land were settled up to two or three decades later than the edges. Terry Jordan, Douglas McManis, and Brian Birch have plotted these differences in sample areas (see n. 41 to this chapter), and one for a prairie township in St. Clair County, southwestern Illinois, is shown in Figure 5.4.

Of course, the lack of timber could be made up for by hauling supplies from more distant locations. Half a mile was said to be too much for some settlers who had been used to timber growing at their door, but in time between three and six miles seemed reasonable and acceptable. Therefore, hauling became increasingly common, until all local timber supplies were exhausted. After that the lumber came from much farther afield; huge quantities came from the newly exploited forests of southern Michigan and Wisconsin, being imported into the prairies via Chicago after 1834.[48] The lumber was hauled overland by ox-drawn wagons for up to 200 miles or more. The cost of this timber was prohibitive for most prairie farmers. Good-quality imported lumber at Peoria, Illinois, sold for $60–$70 m.b.f. (thousand board feet) in 1838 plus inland cartage, the identical timber being sold in Chicago for as little as $9–$14 m.b.f. Even local timber was dear and cost $25–$35 m.b.f. The cheapest source was the white pine from the Alleghenies in western New York and Pennsylvania, even though it was floated down the Ohio and Mississippi and then back up the Illinois River. The linking of the Illinois River with the Chicago River by the Illinois–Michigan Canal in 1847 changed the situation dramatically. Lumber was the most important commodity passing through the canal and "imported." Its price was halved immediately and a little later fell by another 20 percent. Log cabins ceased to be built in the prairies, and for many years frame houses were constructed, almost to the exclusion of brick and stone.[49] It was the beginning of a massive reallocation of the country's resources.

Other disadvantages of settling on the prairies were their susceptibility to fire, their lack of water in some locations and flooding in others, and the risk of fever and ague, which many previously thought had existed only in the forested areas. Old concepts were turned upside down when, in time, it was averred that the prairies were unhealthy and the forest healthy. The connection of the malady with wetter locations meant that the wet prairie and river bottoms were avoided. However, it is clear that the prairies were no more affected than were any other areas at this time.[50] In short, for many the prairies were too remote, too large, had too little timber and water, and were often too wet and unhealthy. But there were ways around these difficulties. Coal or imported wood or plantations could compensate for the lack of local wood; sod and dry bricks could be used for constructing huts; live hedges of Osage-orange, ditches, and eventually wire fences could take the place of wood fences; and wells could take the place of springs.

Even if all these problems were resolved, however, there remained one basic question that had to be answered: Was the prairie less, more, or just as fertile as the forest? If trees were regarded as an indicator of fertility, then the prairies had the worst soil the pioneers had ever encountered. This view, which was widely held, seems to have been dispelled quickly enough once the sod was broken up and the thick dark clay loam of the top horizon exposed, and soon report after report established the value of the prairie soils. The difficulties of breaking the tough, matted sod were overcome with the use of the common Yankee cast-iron breaking plow. With three to five oxen and two men, one to two acres could be broken in a day, and with the improved steel-tipped plow invented by John Deere in 1837 and in wide use after the Civil War, one man and a team of horses could break the same amount in an equivalent time. In other words it took about one-and-a-half man-days per acre to "clear" the prairie before 1850, and about one man-day per acre by the 1860s. And these figures must be compared to the 32 man-days per acre for the clearing and stump removal of one acre of forested land during the same period, to which would have to be added the time taken in plowing the soil. It was a mere fraction, between 5 and 10 percent, of the toil expended in the forests.[51]

After a time the differences between the forest and the prairie were not lost on the pioneer farmer who was spared the arduous and lengthy task of cutting and grubbing and who could improve his land through the agency of normal agricultural operations. The one disadvantage in prairie farming was the relatively short time during the year when sod breaking could be done and the cost of acquiring draft animals and plows compared with the cost of an ax, which was basically all that was needed in the forest. Consequently, much sod breaking was done by hired contractors. By 1837 James Hall was able to give the prospective farmer the advice that he had better

> settle in the midst of a prairie and haul his fuel and rails *five miles* than undertake to clear a farm in the forest. Transport was now less disadvantageous than clearing; the farmers of Illinois are beginning to be aware of this fact, and there are now many instances in which farmers, having purchased a small piece of land for timber, in the woodland, make their farm at a distance in the prairie.

It was not long before the idea gained ground that the soil quality improved the nearer to the center of a prairie one approached.[52]

In summary, then, it is clear that, rather than completely rejecting the prairies, settlers preferred the timbered margins and that they extended their occupation of the land out into the open ground in a ringlike series of waves, so that the very smallest prairies were covered with farms but the larger prairies remained empty, unsettled areas until after the 1850s. It was at this point, between forest and prairie, that the major problem of woodland settlement ended and where some of the problems of prairie settlement began. Until the prejudice against the grasslands was broken down and new techniques were available, the settler could not progress into these open lands. But by the time their obvious fertility, their natural pasturage, and the ease of breaking up the sod for cultivation were recognized, settlement in the Lake States to the north had progressed far enough for the development of a complex organization supplying timber to the prairies, the one necessity the grasslands lacked.

Domestic fuel supplies

The transformation of nearly 40 million acres of forest land into improved farmland between 1850 and 1859, to say nothing of the scores of millions of acres cleared during the first half of the nineteenth century, provided an enormous amount of wood for the energy requirements of the nation. Admittedly, some of the wood was used for fences and houses, some was burned up on the spot, and some was converted into pearl ash and potash or made into charcoal, but most of it must have been burned for heating in the home. Even a moderate estimate of yield, say, 20 cords per acre, from the forests felled between 1850 and 1859 gives a total of 794 million cords, which would be a large proportion of the estimated national consumption of 878.5 million cords of fuel wood during the same period.[53] Therefore, it is safe to say that new clearing supplied the bulk of the fuel needed and that the deficit was probably made up by recutting woodlots.

With such a large and constant demand, woodcutting remained a profitable sideline for the pioneer farmer and an indispensable aid in the making of his farm. It was worth spending a great deal of time and energy to cut and collect the wood and to take a load to a nearby market or merchant when on a journey to buy and collect essential commodities and utensils. The return of between one and two dollars, sometimes as much as three dollars, per cord might even pay for the hiring of set-up men to clear the land and certainly for the hire of corders at 25¢–75¢ per cord. At the very least it certainly covered the costs of food to tide the pioneer and his family over the initial period of clearing before a crop could be planted, so that comments such as the one made in Michigan that a settler could "chop out a home for himself . . . and his cord wood will purchase provisions" were common.[54] Cordwood was a bonus of extra income that he could not get otherwise.

A good indication of the time and trouble taken in cutting and carting wood is given in a number of diaries of farmers in Lewis, Montgomery, and Seneca counties in New York during the early and mid-nineteenth century. One of the farmers, Francis W. Squires, records that he or his father worked all or part of 44 days, or 14 percent of their total workdays, in chopping, splitting, cording, and drawing wood to the house or to the nearest town, Lowville, for sale. Another farmer, Richard Winne, spent all or part of 54 days, or 17 percent of his working days, in the same tasks, and another farmer, Henry K. Day, spent 39 days of his working year doing the same.[55] These are remarkable proportions, and they justify the oft-made assertion that the pioneer farmer was a part-time lumberman of sorts. Nonetheless, like other aspects of clearing the forest, such as pulling stumps and generally making the farm, journals and accounts are remarkably silent about the circumstances surrounding the gathering and sale of cordwood, despite the fact that they are full of details and advice about almost every other occupation in which a farmer could engage. Fuel wood was an unexpected by-product, an unasked for increment that was not enumerated. There is no other commodity of such importance about which we are so ignorant of its production and distributive institutions, and the complete story of America's first energy source awaits a great deal of research.[56]

The census of 1839 gives the first comprehensive view of the amount and location of the wood cut for sale. It shows that 5,341,445 cords were cut for sale, which must have been only a portion of the wood cut for fuel and which suggests that there was a large extracommercial market for consumption of fuel wood, perhaps 10 times the size of that reported, that the census never picked up. Of the total of 5.3 million cords sold, the greatest amount was cut in the states of the Northeast, where the greatest densities of population, particularly urban population, were concentrated (Fig. 5.5).[57] Obviously, ease of transport was the key to marketing this bulky commodity. The counties of coastal Maine were ideally situated to supply the trade to Boston and New York; for example, Hancock, Waldo, and Lincoln counties had census totals of 35,711, 37, 532, and 31,731 cords, respectively. Charlemagne Tower, who toured the coast and principal rivers of Maine in 1829, commented on the large stockpiles of wood on the banks of the Kennebec: "This is brought from the back country, where it is worth from a dollar to a dollar and a half, and is carried on sloops to the Boston Market, where it is sold for seven dollars a cord."[58] In New Jersey the coastal trade from the pine barrens northward to New York had been well established for almost a hundred years, and counties like Atlantic, Burlington, Cumberland, Gloucester, and Monmouth cut between 30,000 and 60,000 cords and were preeminent in the trade.[59] Some cordwood went to Philadelphia also, although that city was fairly easily supplied by rivers like the Susquehanna and

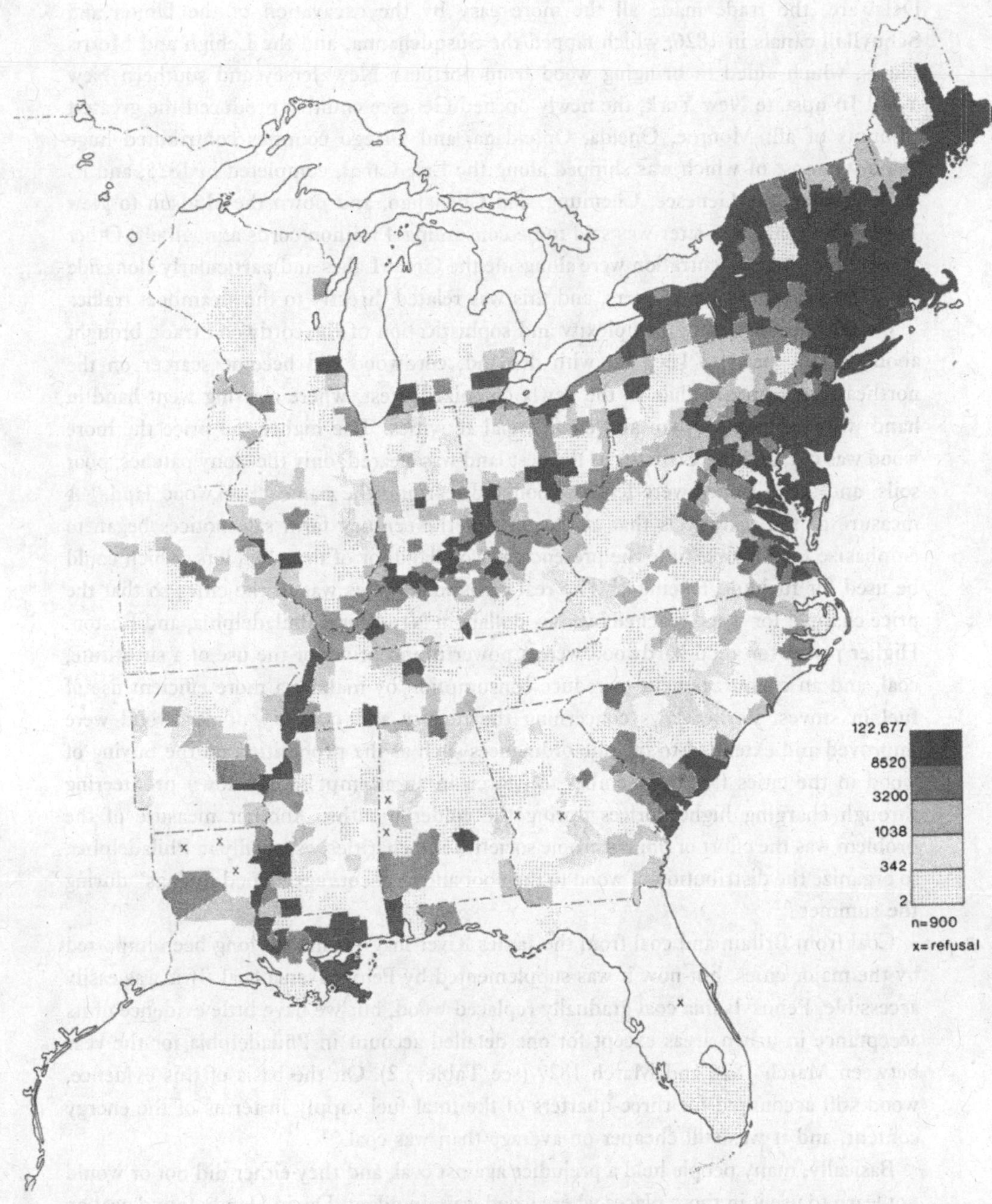

Figure 5.5. Cords of wood sold in the United States, 1840. (U.S. Census.)

Delaware, the trade made all the more easy by the excavation of the Union and Schuylkill canals in 1826, which tapped the Susquehanna, and the Lehigh and Morris canals, which aided in bringing wood from northern New Jersey and southern New York. In upstate New York, the newly opened Genesee country produced the greatest amounts of all; Monroe, Oneida, Onondaga, and Otsego counties contributed huge amounts, most of which was shipped along the Erie Canal, completed in 1825, and its feeder canals, the Genesee, Chemung, and Chenango, and down the Hudson to New York, which 20 years later was said to be consuming 4 million cords annually.[60] Other areas of cutting concentration were alongside the Great Lakes and particularly alongside the Mississippi and Ohio rivers, and this was related directly to the steamboat traffic.

Despite the increasing complexity and sophistication of the cordwood trade brought about by the need to keep up with demand, cordwood did become scarcer on the northeastern seaboard than in the newly colonized West, where clearing went hand in hand with profitable urban and commercial activities. The higher the price the more wood was cut, and increasingly, as the best land was cleared, only the stony patches, poor soils, and steep slopes were left as woodland, earning the name "backwood land." A measure of the scarcity is that at the turn of the century farm sale notices began to emphasize for the first time the presence of woodland, or of detached lots, which could be used for fuel and fencing.[61] The real crisis in supplies was in the cities so that the price charged for wood reached over six dollars in New York, Philadelphia, and Boston. Higher prices for good hardwood were a powerful incentive for the use of a substitute, coal, and an encouragement to reduce consumption by making a more efficient use of fuel in stoves. Earlier laws concerning the quality and quantity of cordwood were improved and extended to include ordinances such as the prohibition on the buying of wood in the cities from September to March in an attempt to cut down profiteering through charging higher prices during the colder months. Another measure of the problem was the effort of philanthropic societies and charities, especially in Philadelphia, to organize the distribution of wood to the poor and to encourage "wood savings" during the summer.[62]

Coal from Britain and coal from the James River in Virginia had long been imported by the major cities, but now it was supplemented by Pennsylvania coal. If it was easily accessible, Pennsylvania coal gradually replaced wood, but we have little evidence of its acceptance in urban areas except for one detailed account in Philadelphia for the year between March 1826 and March 1827 (see Table 5.2). On the basis of this evidence, wood still accounted for three-quarters of the total fuel supply in terms of the energy content, and it was still cheaper on average than was coal.[63]

Basically, many people held a prejudice against coal, and they either did not or would not learn to use it in those places where wood was abundant. David Handy found coal on the Warrior River, Alabama, above Mobile, and in 1840 he built boats to bring it down to the town. But nobody would buy it, and "He had to give it away and send a negro with a bucketful to show people how to light and burn it."[64] Conditions were slightly different in Pittsburgh, where from the very beginning coal outcropping from the sides of the Monongahela, Allegheny, and Young valleys was used for smelting and household

Table 5.2. *Fuel used in Philadelphia, March 1826–March 1827*

	Quantity	Average price ($)	Total cost ($)	Energy content (mill. b.t.u.)	% of b.t.u.	Cost per mill. b.t.u. (¢)
Wood	140,150 cords	4.50	630,675	2,937,544	75.8	21.5
Charcoal	3,200 tons	10.00	32,000	77,440	2.0	41.3
Coal						
Anthracite, Lehigh	25,545 tons	7.00	178,815	726,694	18.8	24.6
Bituminous						
Virginia	95,000 bush.	0.30	28,500	131,498	3.4	29.3
Liverpool	30,465 bush.	0.30	10,053			
Total			880,043	3,873,176	100.0	

Sources: Marcus Bull, *Experiments to Determine the Comparative Value of the Principal Varieties of Fuel*; and S. H. Schurr and B. C. Netschert, *Energy in the American Economy, 1850–1975*, 50–2.

heating. With temperature inversions, sulfurous smog accumulations over Pittsburgh were common: "a cloud of smoke hung over it in an exceedingly clear sky recalling . . . many choking recollections of London," said one traveler in 1800.[65] Later accounts make it clear that, as the use of coal increased in the Pittsburgh district, houses and buildings became black and "even the snow can scarcely be called white." Wood cost two dollars a cord delivered, but coal was six cents a bushel, which made it slightly cheaper than wood, and if the wood was of inferior quality there would have been an even greater difference in the cost per unit of energy purchased. François André Michaux said that coal was so plentiful that landholders would let anyone fell and cart wood at half the price of coal, merely to make a sale.[66]

From the middle of the eighteenth century onward people had been worrying about the dire consequences of the depletion of the sources of fuel and had been advocating an end to the wide, open, roaring fires that lost nine-tenths of their heat up the chimney and their replacement by more efficient enclosed stoves. An end to the "thoughtless prodigality" of timber destruction seemed desirable. Benjamin Franklin had long been a leading advocate of this, but basically he had little impact upon the preferences of the pioneer settler. To the pioneer the open fire meant more than necessary warmth; it was a cheerful thing to contemplate at the end of a hard day's work, and it was a focus for the family.

The ideas of Franklin and others were irrefutable – a well-designed, well-constructed stove could supply about four times as much heat as an open fireplace for the same amount of wood – but besides the sheer resistance to change there were some practical reasons for the early nonacceptance of stoves. In the rural areas the supply of wood seemed inexhaustible, and also there was the simple fact that to chop wood small enough to fit into a stove required a great deal of additional effort. It would have made little sense to waste more man-hours by cutting up the wood any smaller than the traditional cord in order to economize on the use of a seemingly abundant resource.

In the cities the real stumbling block to the early acceptance of the stove was its efficiency and reliability. Agricultural and scientific societies offered prizes and awards for the best designs, and between 1790 and 1845 over 800 patents were registered with the U.S. Patent Office, many no more than tinkerings and modifications of existing stoves but all adding up to more registrations than for any other commodity or object. In what was basically a preindustrial age in the United States the manufacture of stoves was an important technological advance and was a basis for further industrialization. It encouraged the manufacture of iron, and the industry improved after the 1820s and was able to cope with mass smelting and casting. By 1860 stove castings were made by 250 foundries, and manufacture had so expanded that it was estimated that New York and Philadelphia each produced 150,000 stoves annually, Albany 75,000, Cincinnati 50,000, Providence 30,000, and Pittsburgh 20,000. Clearly, at such a rate of production the manufacturers must have been supplying more than the urban markets, and sales must have penetrated the immediately surrounding rural areas.[67]

The expansion of stove manufacture demonstrated the stove's acceptance and use and was a measure of both its improved design and the rising cost of fuel. Franklin's crude box stoves made of six plates, which Schoepf had seen in houses in Philadelphia in 1783, were replaced by more elaborate and elegant models cast in one piece that served for cooking operations as well as for heating.[68] The exterior ornamentation was merely eye-catching, and the stove must have been one of the first manufactured and mass-produced items other than women's clothes and hats that attempted to attract sales by being visually rather than functionally "different." The stove was so generally accepted by the urban communities on the northeastern coast by the 1840s that it no longer was a topic for comment or concern during the 1850s.

Despite all the advances made in wood-burning stove design and efficiency, however, wood had the major disadvantage that it was weighty and bulky in relation to its content of energy when compared to anthracite, its main competitor. One cord of high-grade hardwood had only about 80 percent of the energy of one ton of anthracite coal but weighed about twice as much and certainly was more bulky. Poor-quality wood obviously had a lower energy content, which might approach only 50–75 percent of anthracite. It is true that the problem of the bulkiness and weight of wood was lessened considerably with the conversion of wood to charcoal; wood lost approximately half of its bulk and three-quarters of its weight. Nevertheless, the lengthy process of transformation meant that charcoal was the most costly fuel of all, and it gave only about the same heat per unit weight as did bituminous coal. These considerations of bulk, storage in urban cellars, and scarcity prompted city dwellers along the northeast coast to turn to coal. Coal production increased from a mere 881,000 tons in 1830 to 8.3 million tons in 1850. It then rose rapidly to 20 million tons in 1860 and was firmly established from then onward, production approximately doubling every decade from 1860 until 1890. Of the total amount of coal mined, anthracite accounted for approximately half. Nevertheless, despite this phenomenal increase in the production of coal, it consituted only 9.3 percent of the aggregate energy consumption of the nation in 1850 and only 16.4 percent in 1860 – the rest of the energy used was still derived from wood. The evidence is that the per capita consumption of wood fuel fell as coal penetrated the urban market,[69] but this is not to say that woodcutting declined, for the census of 1880 shows that woodcutters abounded in all states.[70] The drain on the eastern forest lessened a little, but the colonization and cutting west of the ranges produced enormous quantities of wood so that the farmers continued to enjoy their open fires for nearly another hundred years and to earn extra income by supplying fuel-deficient areas.

Minor pioneer products

For the pioneer settler the forest yielded a number of valuable products other than fuel wood. Foremost was the potash and pearl ash that commanded a ready market in the industrial East and overseas, and there were also skins and furs, ginseng, and other products such as maple syrup.

Potash and pearl ash

Potash and pearl ash were produced in small quantities wherever clearing took place in hardwood areas, but during the opening years of the nineteenth century production had become localized in areas of new settlement in upstate New York and adjacent areas of Ohio and New England (Fig. 5.6).[71] Here commercial production as a by-product of clearing reached its peak, and at the time of the 1840 census a total of 16,550.5 tons was being produced, 46 percent from the counties in Ohio on the Lake Erie frontage where individual values of 2,914 and 2,094 tons, respectively, were recorded for Geauga and Trumbull counties, then being cleared for agriculture.

The price of potash during the early years of the century fluctuated between $160 and $200 per ton, and its value as an export commodity exceeded that of naval stores in most years.[72] Its value is well attested by numerous examples. James Fenimore Cooper, in his novel *The Chainbearer*, set it at $200 a ton, and, indeed, it was known to have risen to $300 during the War of 1812. In 1822 Gov. De Witt Clinton of New York declared potash and wheat the two leading exports of the state. With such prices farmers produced as much as possible while they cleared and either purchased necessary supplies or bought their land with the income.[73]

By 1845 the center of production shifted even farther west into Ohio. Exports reached 24,219 tons in 1844 and then suddenly declined with the discovery of potash and soda salts in mineral waters and then later in mineral deposits. By 1864–5 exports had declined to a mere 2,633 tons, and although the price was still high, the once valued and traditional by-product of clearing dwindled into insignificance as the century progressed and cheaper and more easily obtained substitutes were found. The cleared forest was now burned on the spot without even the thought or prospect of saving its stored minerals.[74]

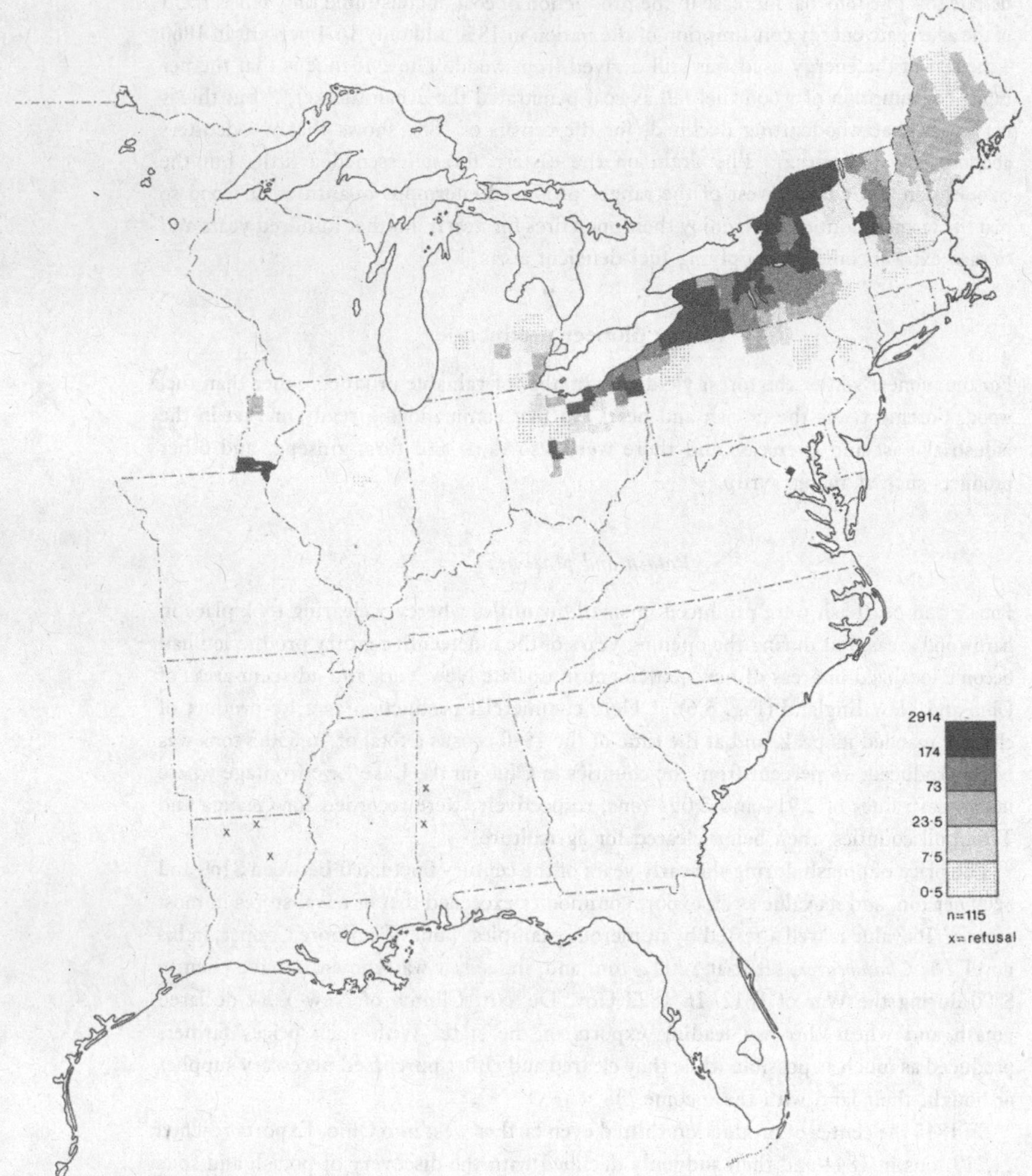

Figure 5.6. Tons of potash and pearl ash produced in the United States, 1840. (U.S. Census.)

Skins and furs

Although the trapping of animals for their skins and furs made little impact on the forest cover, it was an activity that was largely (though not wholly) carried out in the forest environment, especially during the early years of North American settlement. As such, skins and furs can be regarded as a product of the forest. Initially, of course, pioneer settlers had always regarded the animals of the forest as a sort of bonus for their hard work in clearing and making a farm. They were a bonus that in some instances provided essential clothing in the context of isolated pioneer life, epitomized, perhaps, by the factual and fictional characters of Davy Crockett, Daniel Boone, and Leatherstocking. Increasingly, however, mass-produced woolen and cotton clothing became available from the early nineteenth century onward for even the most remote pioneer communities, and trapping became a more specialized occupation supplying more fashion-conscious and discriminating markets in the East and overseas.

Information about hunting and trapping in farming communities in the East is tantalizingly slight, but at least the census of 1840 shows that the value of furs was a little over $1 million and that production was largely in the thinly settled areas west of the Mississippi and north of the Ohio, in the Adirondacks and northern Maine, and in the Appalachian uplands, all just out of the reach of agricultural settlement (Fig. 5.7). The greatest state total was for Missouri, of which 80 percent was for St. Louis, which suggests that it functioned as a collecting center for much of the western frontier.

By 1870 the value of furs had reached $4 million, but by that time Alaska was contributing the bulk of the national total through the fur seal business, most of the catch being exported via New York, followed by Boston, Charleston, Baltimore, and Lake Champlain.[75] Furs and skins as a product of the forest pass out of the record just after midcentury.

Ginseng and other products

Ginseng is the root of a small herbaceous plant, *Aralia quinquefolia*, which grew wild in the forest. It was collected and exported mainly to China for use as a medicine. The nature of "all other products" (see Fig. 5.8) was never specified, but they must have included maple syrup, wild fruits, and even wild honey. As in the case of potash, New York State was by far the greatest producer in total value, but the location of production was scattered there, as indeed it was elsewhere in the country. The only major regional concentration was in the northern Appalachian parts of West Virginia, eastern Kentucky, eastern North Carolina, southeastern Ohio, and parts of Tennessee, where pioneer hill farmers gleaned what they could from the forest to eke out their meager living. Most of the ginseng was exported via New York, but by the end of the 1860s San Francisco took an unbeatable lead with exports to Hong Kong and other ports in China. Exports fluctuated wildly from year to year but averaged $168,000 per annum during the 1840s and reached a peak of $6 million in 1875, after which they declined to insignificant proportions.[76]

In general terms, the minor products of the forest had proved valuable to pioneer

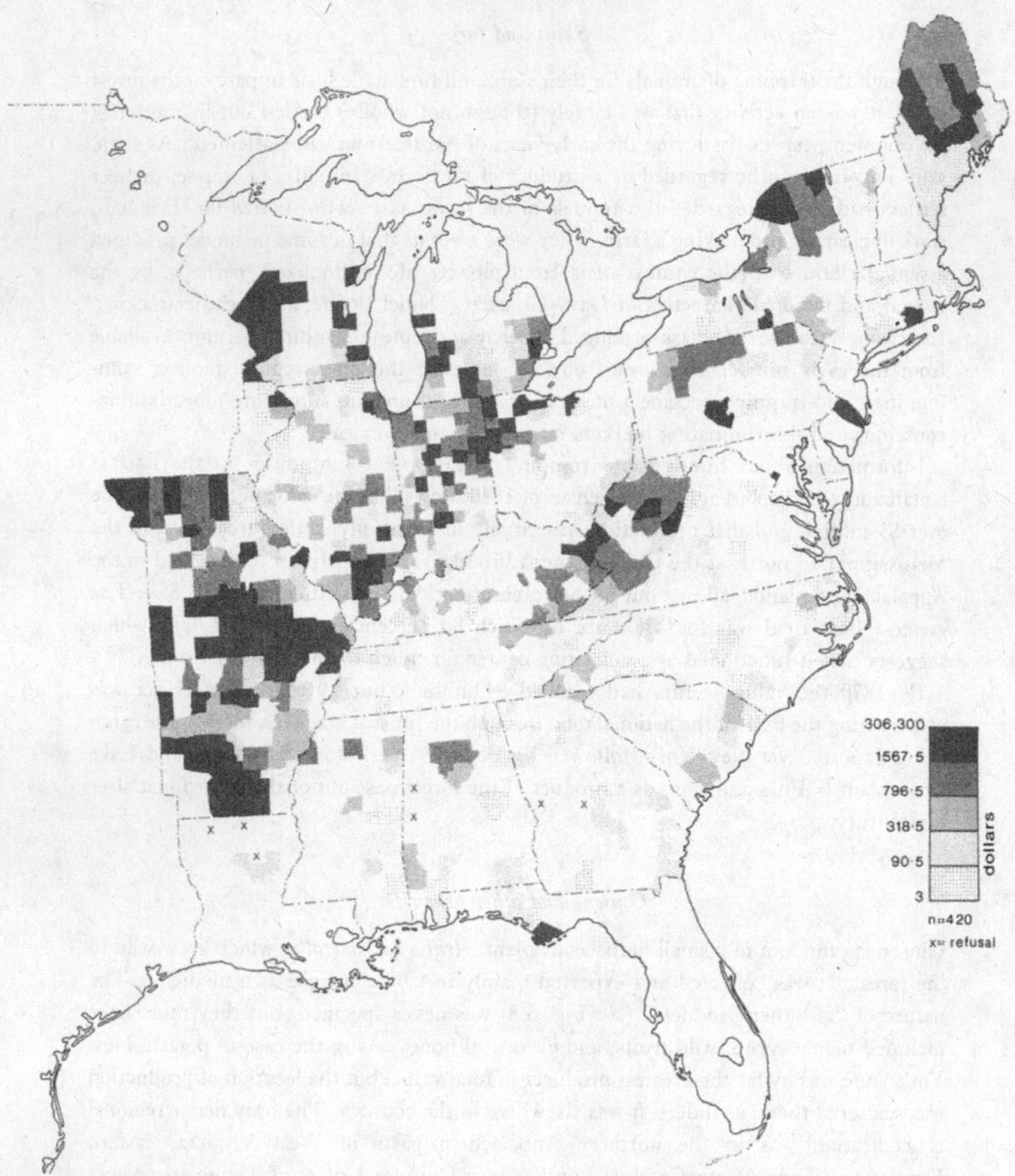

Figure 5.7. Value of skins and furs caught in the United States, 1840. (U.S. Census.)

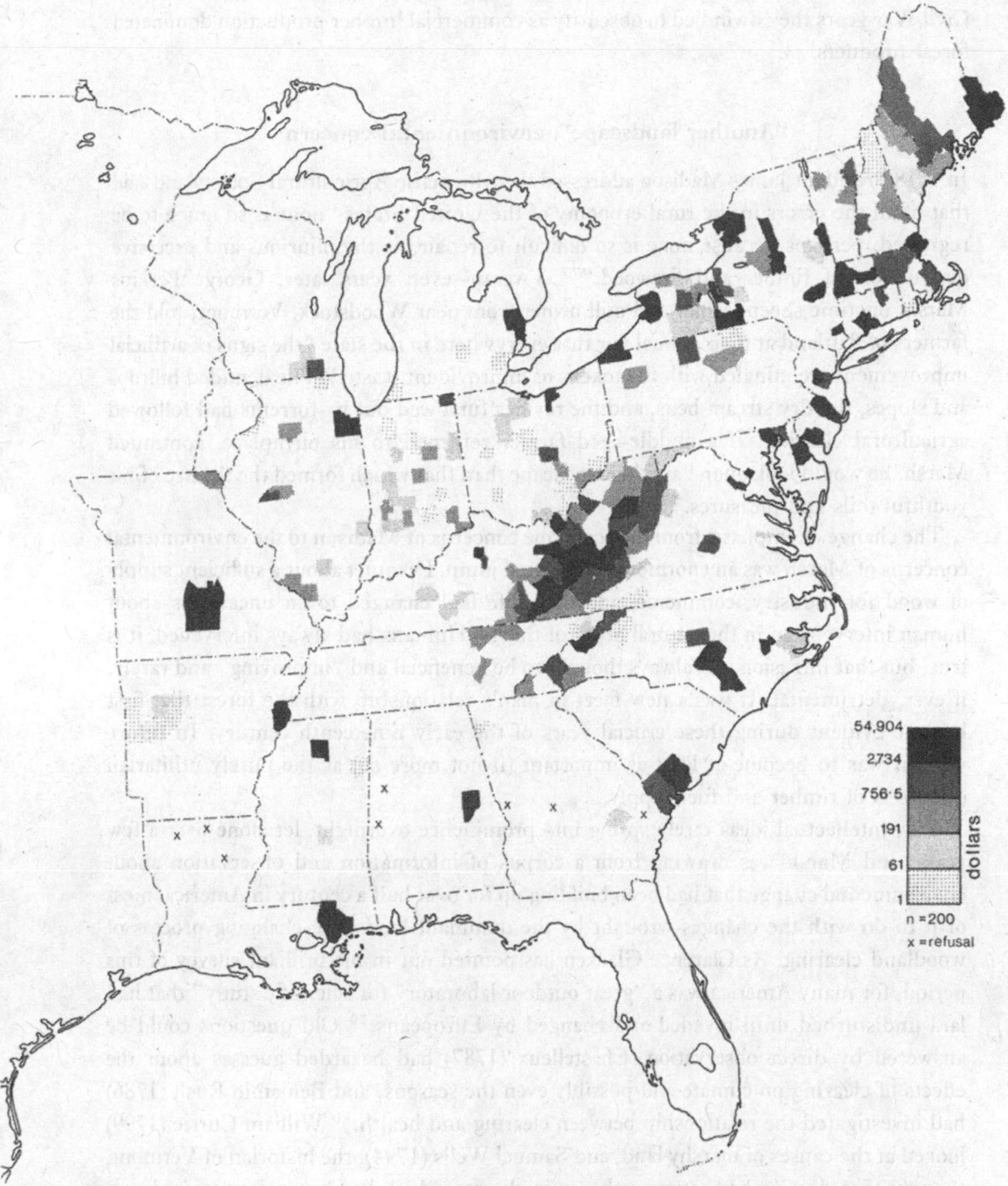

Figure 5.8. Value of ginseng and all other forest products in the United States, 1840. (U.S. Census.)

farmers during the early years of the nineteenth century, but after the upheaval of the Civil War years they dwindled to obscurity as commercial lumber production dominated forest products.

"Another landscape": environmental concern

In 1818 President James Madison addressed the Albemarle Agricultural Society and said that of all the errors in the rural economy of the United States "none is so much to be regretted, perhaps because none is so difficult to repair, as the injurious and excessive destruction of timber and firewood."[77] Twenty-seven years later, George Perkins Marsh, onetime sheep farmer and mill owner from near Woodstock, Vermont, told the farmers of Rutland at their annual fair that everywhere in the state "the signs of artificial improvement are mingled with the tokens of improvident waste." The denuded hilltops and slopes, the dry stream beds, and the ravines furrowed out by torrents had followed agricultural clearing. If a middle-aged farmer returned to his birthplace, continued Marsh, he would look upon "another landscape than that which formed the theatre of his youthful toils and pleasures."[78]

The change of emphasis from the economic concerns of Madison to the environmental concerns of Marsh was an enormous intellectual jump. Disquiet about a sufficient supply of wood for industry, commerce, and daily life had changed to an uneasiness about human intervention in the natural order of things. Humans had always intervened, it is true, but that intrusion was always thought to be beneficial and "improving" and rarely, if ever, detrimental. It was a new facet in man's relationship with the forest that first became evident during these crucial years of the early nineteenth century. In future years it was to become at least as important (if not more so) as the purely utilitarian questions of timber and fuel supply.

New intellectual ideas rarely spring into prominence overnight, let alone over a few years, and Marsh was drawing from a corpus of information and observation about environmental change that had been building up for over half a century in America, most of it to do with the changes wrought by the dominant landscape-changing process of woodland clearing. As Clarence Glacken has pointed out in his brilliant survey of this period, for many America was a "great outdoor laboratory for scientific study" that had laid undisturbed until invaded and changed by Europeans.[79] Old questions could be answered by direct observation. Chastelleux (1787) had hazarded guesses about the effects of clearing on climate and possibly even the seasons, and Benjamin Rush (1786) had investigated the relationship between clearing and health.[80] William Currie (1799) looked at the causes of marshy land, and Samuel Wells (1794), the historian of Vermont, thought that clearing had caused a change in climate, which had become more moderate and less predictable.

Count Volney's *View of the Climate and Soil of the United States of America* (1804) was a more sustained treatment of these themes. Through his reading and his discussions with knowledgeable Americans, Volney was convinced that the alteration of the climate was an "incontestable fact" and had occurred "in proportion as the land has been cleared." He continued:

> On the Ohio, at Gallipolis, at Washington in Kentucky, in Frankfort, at Lexington, at Cincinnati, at Louisville, at Niagara, Albany, everywhere the same changes have been mentioned and insisted on: Longer summers, later autumns, shorter winters, lighter and less lasting snows and colds less violent were talked of by everybody; and these changes have always been described in the newly settled districts, not as gradual and slow, but as quick and sudden, in proportion to the extent of cultivation.

Thus, there could be "no room to doubt the truth of a sensible change in the climate." Moreover, if stream flow was directly related to amount of forest cover and hence humus accumulation, then clearing, though necessary for agricultural expansion and improving health, could not but reduce stream flow and bring about eventual aridity. Man, therefore, was an agent of change who upset the harmonies of nature.[81]

More thoughtful and sustained still was John Lorain's *Nature and Reason Harmonized in the Practice of Husbandry*, published posthumously in 1825. Lorain saw nature as having a system or cycle of decomposition, growth, and change that returned humus to the soil, but continual plowing and cropping were the "hand of folly" that destroyed that cycle. Also, by destroying the protective vegetative cover man also interfered in the process of runoff from the uplands and hence deposition in the lowlands. In one blow, two fundamental processes in nature were affected. To illustrate his point he contrasted in some detail the clearing practices of New England and Pennsylvania farmers. Both were destructive in their clearing methods, the New Englander slightly less so because he tried to increase livestock numbers and hence grass cover and manure. However, his destructive use of burning ultimately matched the Pennsylvanian's continual cropping and lack of manuring. He concluded:

> Perhaps a better method could be devised for clearing woodlands, or a more profitable first course of crops be introduced, if it were not that by far the greater part of the animal and vegetable matter which nature had been accumulating for a great length of time, is destroyed in a day or two, by the destructive and truely inconsiderate and savage practice of burning.[82]

Thus, whereas nature had longtime processes, man had short ones. Rather than merely conquering and controlling nature, man was destroying it at a rate that was more rapid than his ability to adjust his activities. Permanent damage to the environment would therefore be the result, whether it was in the forest, on the prairie, or in the mountains.

In 1849 Marsh wrote to Asa Gray, the noted botanist, that having "spent my early life almost literally in the woods . . . I have had occasion both to observe and to feel the effects resulting from an injudicious system of managing woodlands and the products of the forests."[83] In succeeding years he was to combine his own experience and observation with the building blocks of writers in America and Europe to produce his brilliant synthesis, *Man and Nature*, which appeared in 1864. Here were the first stirrings of environmental awareness and the conservation movement, as it became known in the Western world. It started in America, and it started in the forest.

6

The quickening pace: the industrial impact, 1810–1860

> Well may ours be called a *wooden country*; not merely from the extent of its forests, but because in common use wood has been substituted for a number of most necessary and common articles – such as stone, iron, and even leather.
>
> James Hall, *Statistics of the West (1836)*, 100–1

The years from 1810 to the Civil War were a period of transition from a rural-agricultural society to an urban-industrial society in the United States. The forces at work in the economy during the early nineteenth century were stimulated by a period of intense technological innovation in many industries during the 1820s and also by a rapid increase in migration during the 1840s. It was the period typified by Walt Rostow as "takeoff" or the beginning of self-sustained growth. The debate about the exact date of takeoff and the reasons for it are not our concern, but it is clear that the quickening pace of industrial activity had far-reaching implications for the use of the woodland, for in every way the demand for wood, lumber, and lumber-derived products rose phenomenally.[1]

In some ways this increased demand for wood was strange, for a transition from a predominantly rural to an industrial economy, with its increased emphasis on the production of metal and the mining of fossil fuels, might have been expected to reduce the dependence on timber. But this did not occur in the United States; the use of wood not only lingered but increased rapidly in most industrial processes because of the ubiquity and abundance of supplies. No better illustration of this can be given than the widespread acceptance of the steam engine (the characteristic paleotechnic invention in Lewis Mumford's trilogy of ages of technology[2]) for riverboats and railways during the 1830s but the continuing use of wood fuel for firing boilers well into the 1880s. Similarly, the iron and steel industry was well established by 1880 and had long adopted labor-saving devices before it turned from charcoal as its principal fuel in the 1860s. Indeed, charcoal continued to be used until nearly the mid-twentieth century. As far as lumbering was concerned, steam engines were used in sawmills from 1830 onward but did not predominate over waterwheels until after 1880.

At the same time agricultural clearing continued unabated alongside the upsurge of industrialization. The steady stream of immigrants created unprecedented demands on the forests for farmland as the population swelled from 7.2 million in 1810 to nearly 12.8 million in 1830 and then to 31 million by 1860. Not only did the newcomers move into new areas of rural settlement, but they swelled the towns so that the proportion of people classed as urban rose from 7.3 percent in 1810 to nearly 20 percent in 1860. Wood and wooden products permeated American life. When James Hall said in 1836, "Well may ours be called a *wooden country*; not merely from the extent of its forests, but because in common use wood has been substituted for a number of most necessary and common articles – such as stone, iron, and even leather,"[3] he was thinking not only of houses, fences, furniture, and fuel but also of wooden roads, bridges, locks, nails, hinges, and machinery, all of which had not yet been replaced by iron. Wood was used for fuel. Wood was used for housing, some 84 percent of 54,000 houses constructed in 1839 being of wood. Every town and little settlement boasted of its furniture and wood fabrications works.

Charcoal and the iron industry

When, in 1810, no charcoal furnaces were in production in Britain and all were fired by coke or coal, not one coke furnace had been built in the United States. Not until 1835 were the first experiments made in America with coke, and not until as late as 1945 was the last charcoal furnace shut down. The American charcoal iron industry died a very slow death.[4]

The sheer abundance of wood in the United States was an obvious factor in the survival of this traditional technology, but the matter was more complicated than that. The demand for iron in preindustrial America was for a malleable, wrought material, one best produced by charcoal. The economy was still predominantly agricultural, and the product most needed was a semifinished bar or rod that could be worked into a great variety of items for forging by the local blacksmith or even by the farmer himself, ease of working being more important than cost.[5] A general-purpose, easily worked, nonbrittle iron was needed for horseshoes, pickaxes, or sledgehammers; an iron possessing a good cutting edge was needed for axes, scythes, sickles, and knives; and one strong enough for plates was needed for steam engine boilers. The demand for wrought iron, hand-rolled steel, and constructional material hardly made much headway until the railroads started to expand during the early 1850s, and even then they may not have consumed as much steel as is commonly supposed, most of it being imported. In 1849 domestic production of nails (used in wood, of course) may have "exceeded that of rails by over 100 percent."[6]

The iron industry between 1810 and 1860 was operating on a "preindustrial" (one could almost say medieval) technology, and its organization was almost "feudal." Charcoal was still used in preference to coke; the iron was still beaten with hammers to remove the carbon rather than being made by the puddling process in reverberatory furnaces, which allowed an impure fuel like coke to be used, and the iron was still beaten into shape rather than passed through grooved rollers to produce bars and rods. Most production was still coming from the semifeudal iron plantations where the ironmaster exercised a paternalistic, even feudal oversight, rather than from urban establishments and company towns.[7]

Primitive as the technology and organization were, however, they were ideal for producing the required multipurpose iron because charcoal contained almost no sulfur or phosphorus impurities, which combined to form brittle chemical compounds, as did most coals. In addition, because of its porosity, which allowed hot air to pass through it, charcoal produced an iron with a lower silicon content than did most coals, which meant that items made from it by traditional heat-treated tempering achieved a "fine grain" and were tougher and, in particular, kept a fine cutting edge. Also, as charcoal was easier to burn than coal, charcoal blast furnaces required only weak blast equipment, even as primitive as leather fireplace-style bellows driven by a waterwheel.[8] Under such conditions furnaces could not and need not get too large; therefore, not so much capital was needed in setting them up (they were nearly half the cost of a coke furnace). In every way the ease of working and producing charcoal iron gave it an advantage over coal-produced iron. Although it was possible to make reasonable "blacksmith's iron" from coke or anthracite, charcoal iron was of a known quality and therefore remained a premium metal that could compete despite its higher price. This was proved during the post–Civil War period when the known uniformity and freedom from impurities of charcoal iron meant that it could hold its own even when the demand changed to mass-produced industrial products. For example, the charcoal-produced iron railroad car wheel proved superior to its coal-produced counterpart because of the development of new variable casting techniques, and similarly, in this century, charcoal iron was specified for early automobile blocks before complex metallurgical and machining problems were overcome in the use of coke-produced irons.[9]

The technological simplicity, even crudity, of the charcoal furnace complemented the geographical conditions of antebellum America, where production and consumption were local and transport links were minimal. The limited production of the furnaces (on average, it was about one-fifth that of the contemporary anthracite furnaces during the late 1850s) meant a relatively limited impact on the forests. Were the furnace larger, the ironmaster would have to travel farther for his fuel, and his costs, already high, would have risen even further. Using our previous calculation that the average furnace produced 1,000 tons per annum and that it would consume at least 150 acres of wood per annum, then ideally four years' cutting would place a furnace in the center of roughly a square mile of cleared land, with the forest edge approximately half a mile away. However, in reality such ideal conditions were rarely met because agricultural clearing removed accessible woodland, and the ironmaster had to go much farther for his fuel supply. Therefore, there were sound economic reasons for keeping the furnaces small in order to obviate relocation and the loss of capital investment.[10]

In the older areas of smelting there appears to have been a fairly rapid depletion of the forests as the demand and production rose, from 165,000 tons in 1830 to 286,903 tons in 1840 to 563,755 tons in 1850 and 821,000 tons in 1860. Forest depletion was noticeable after 1840 when furnaces were being abandoned or relocated nearer to new wood supplies or even redesigned, either to make them more efficient or to adapt them to the cheaper anthracite coal or coke, while others were moved nearer to the coal or iron ore deposits (Fig. 6.1). This was just beginning to happen in eastern Pennsylvania where there were 12 anthracite furnaces producing about 15,000 tons of pig iron among 210 charcoal furnaces producing about 100,000 tons.

These changes in fuel technology and location can be given some precision from an analysis of the costs and profits of producing a ton of pig iron with various types of fuel in

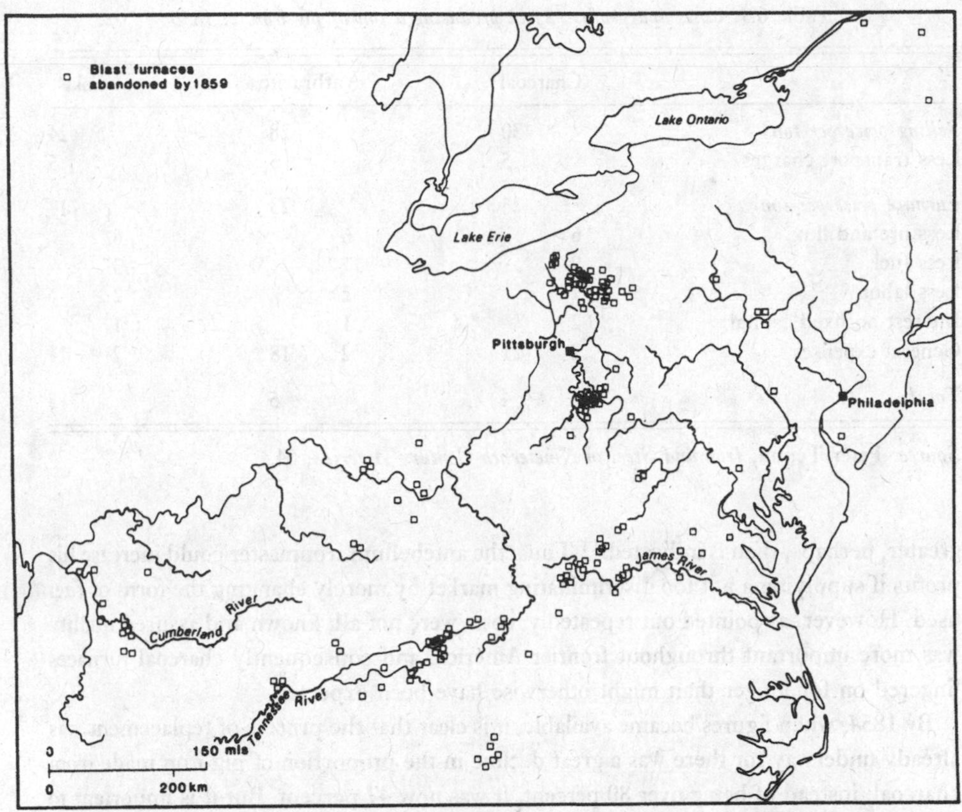

Figure 6.1. Abandoned charcoal-fired blast furnaces, 1859. (J. P. Lesley, *The Iron Manufacturer's Guide to the Furnaces, Forges, and Rolling Mills of the United States*, 30–7, 129–30, 249–50.)

about 1850. Although investment costs per ton of charcoal iron produced were about half those for anthracite or coke, production and transport costs were higher (see Table 6.1).

The procuring of fuel was obviously a major factor in the relative costs. It took eight tons of wood to produce two tons (200 bushels) of charcoal in order to smelt a little under one ton of pig iron, and the collection and intermediate process of charcoal making were more expensive than was mining coal. Wood was an extensive, scattered resource whereas mineral fuel was an intensively concentrated resource, considerations that affected locations and costs.

The difference in costs of iron produced by different fuels might have been even greater than is suggested in Table 6.1, where transport costs are assumed to be equal, whereas in reality they varied widely at the regional level. But it is known that whereas the smelting of one ton of pig iron required two to three tons of charcoal, it required only 1.7 tons of coke to do the same. The weight difference, though significant, was not all that great, but the difference in bulk was; charcoal was approximately 2.6 times more bulky than coke; therefore, it cost more to transport, making the charges for transport

Table 6.1. *Costs and profits ($) in producing a ton of pig iron, c. 1850*

	Charcoal		Anthracite		Coke	
Selling price per ton		30		28		24
Less transport charges		5		5		5
Furnace price per ton		25		23		19
Less ore and flux	6		6		6	
Less fuel	9		7		3	
Less labor	2		2		2	
Interest on fixed capital	2		1		1	
General expenses	2	21	2	18	2	14
Total		4		5		5

Source: Peter Temin, *Iron and Steel in Nineteenth Century America*, 64.

greater, perhaps, than is indicated.[11] Thus, the antebellum ironmaster could increase his profits if supplying a not too discriminating market by merely changing the form of fuel used. However, as pointed out repeatedly, costs were not all; known and assured quality was more important throughout frontier America, and consequently charcoal furnaces lingered on for longer than might otherwise have been expected.

By 1854, when figures became available, it is clear that the process of replacement was already underway for there was a great decline in the proportion of pig iron made from charcoal; instead of being over 80 percent, it was now 47 percent. But it is important to note that, though the proportion declined throughout the remainder of the century, the absolute amount of pig iron smelted by charcoal increased rapidly as new uses were found for this versatile material, the 1854 total of 306,000 tons of charcoal pig iron being doubled on two occasions as late as 1882 and 1890. Therefore, changes in fuel technology did not lessen demands on the forest.

The persistence of charcoal furnaces is highlighted by Lesley's survey on the iron industry as it was in 1859 on the eve of the great changeover to solid fuels in the East.[12] In that year there were 560 furnaces, of which 439 or 78 percent were still charcoal (Fig. 6.2). They were scattered throughout the eastern states with concentrations in the Hanging Rock district of southern Ohio, in the Allegheny Valley northeast of Pittsburgh, in the Juniata Valley in south-central Pennsylvania, and in the Berkshires on the New York/Massachusetts/Connecticut border. The bulk of the remainder of the furnaces were anthracite, and they were concentrated in eastern Pennsylvania, albeit interspersed with many charcoal ones. The location of the ore supplies determined the exact position of the furnaces as the balance of raw material consumption was heavily in favor of ore rather than fuel. Anthracite was brought downriver to the lower and middle reaches of the Delaware, Susquehanna, and Schuylkill; the charcoal furnaces remained in the forested interfluvial, intertransport locations where extensive fuel supplies remained, as well as along the Maryland tidewater where supplies could be brought in by sea.

In the Juniata Valley, scattered ore deposits were worked in the well-wooded hills, but

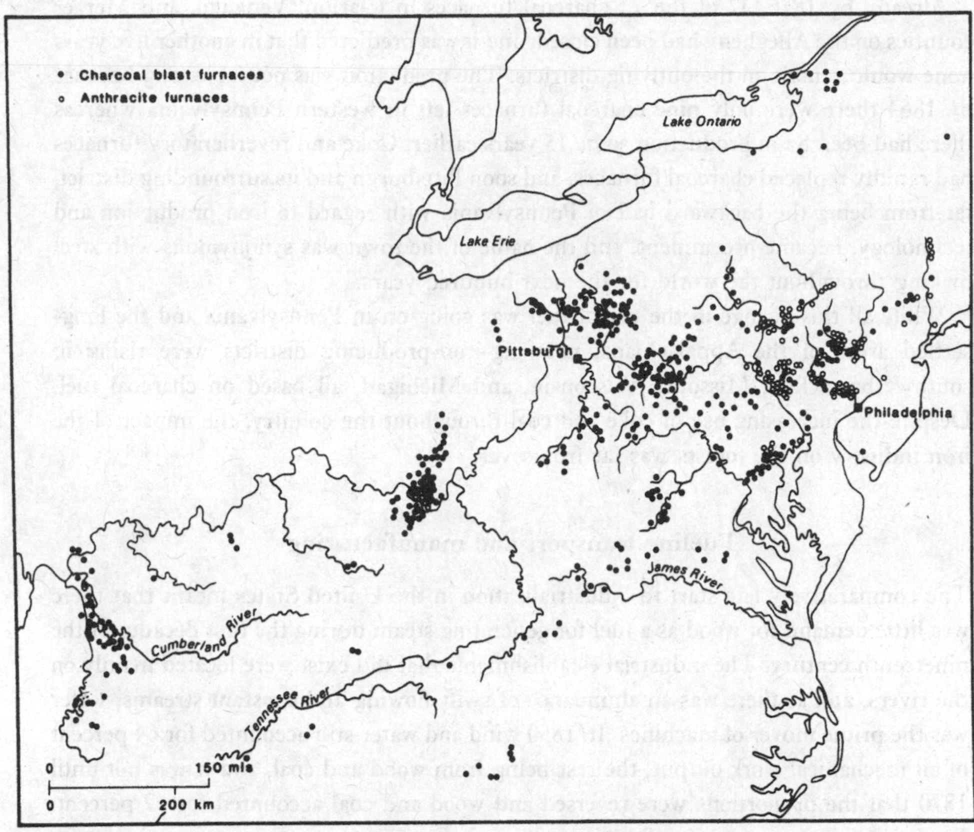

Figure 6.2. Charcoal and anthracite blast furnaces, 1859. (Lesley, *Iron Manufacturer's Guide*, 30–7, 129–30, 249–50.)

even here fuel supplies were at a critical level because of the longevity of smelting and clearing. As early as the mid-1830s some furnaces were hauling their charcoal 10 or 12 miles "with great expense and vexation," and by the late 1850s some were experimenting with alternative fuels such as anthracite but mainly coke, whereas others were idle, "blown out for want of charcoal" or "waiting for a second growth of timber."[13]

On the other hand, western Pennsylvania showed some interesting contrasts. Despite abundant deposits of bituminous coal and its early use for domestic purposes, it was hardly used for smelting. Only in the Shenango and Mahoning valleys north of Pittsburgh had raw coal taken over from charcoal by the time of Lesley's survey in the late 1850s, and the demand for charcoal iron in the pioneering West was still strong enough to keep the mass of furnaces in production on the Monongahela, Youghiogheny, and particularly the Allegheny. But declining fuel supplies due to agricultural clearing, the high cost of labor in this part of the state with the discovery of oil in 1859, and the discovery of the superb Connellsville coking coal in the same year were all going to promote rapid changes.

Already by 1856, 37 of the 73 charcoal furnaces in Clarion, Venango, and Mercer counties on the Allegheny had been closed, and it was predicted that in another five years none would remain in the outlying districts. The prediction was not far wrong because by 1864 there were only nine charcoal furnaces left in western Pennsylvania whereas there had been 85 in production some 15 years earlier. Coke and reverberatory furnaces had rapidly replaced charcoal furnaces, and soon Pittsburgh and its surrounding district, far from being the backward half of Pennsylvania with regard to iron production and technology, became preeminent, and the name of the town was synonymous with steel making throughout the world for the next hundred years.

While all this change in the use of fuel was going on in Pennsylvania and the long-settled areas of the Appalachians, new pig-iron-producing districts were rising in southwestern Ohio, Missouri, Wisconsin, and Michigan, all based on charcoal fuel. Despite the increasing use of coke and coal throughout the country, the impact of the iron industry on the forests was far from over.

Fueling transport and manufacturing

The comparatively late start to industrialization in the United States meant that there was little demand for wood as a fuel for generating steam during the first decades of the nineteenth century. The industrial establishments that did exist were located mainly on the rivers, and as there was an abundance of swift-flowing and constant streams, water was the prime mover of machines. In 1850 wind and water still accounted for 64 percent of all mechanical work output, the rest being from wood and coal, and it was not until 1870 that the proportions were reversed and wood and coal accounted for 67 percent. But mechanical energy has, of course, two major uses: to power stationary engines for industrial purposes (as the water power did) and to power moving engines for transportation as in riverboats and locomotives. In the latter, water power could not take part.

Although there are a few examples of steam engines imported for industrial manufacturing purposes from Britain as early as 1787 in New Jersey and 1791 in Pennsylvania, these were isolated cases, and it was not until the late 1820s that they became noticeable and not until the 1830s that they become numerous. By 1838 there were 1616 known and 238 estimated steam engines used in industry in the United States, and most were in Pennsylvania (383), Massachusetts (165), and Louisiana (274), the bulk being used for general manufacturing, except those in Louisiana, which were used for sugar processing. In addition to the stationary steam engines, there were steam engines for transport. Locomotives were a late introduction of the 1830s (there were only 337 by 1830) and steamboats were scarcely any earlier, despite Fulton's experiment with a steam paddleboat on the Hudson River in 1807, which was to be of worldwide significance. The acceptance of steam engines on boats was slow; only 243 existed in 1834, but the number then rose rapidly with the expansion of settlement across the Allegheny front into the interior river system and alongside the Great Lakes. By 1838 there were 700 steamboats (plus another 100 estimated), mainly in New York and Pennsylvania and in the states bordering the Ohio and Mississippi rivers.[14]

Steamboats

These early engines, standing or moving, were powered with wood fuel, except for a few in eastern Pennsylvania that used anthracite and some steamboats located at Pittsburgh, which used bituminous coal as early as 1818.[15] These were pinpricks in the total picture of the dominance of wood fuel, however. Basically there was a prejudice against coal that was not easily overcome, despite its greater energy content per unit of weight or volume. In 1837 experiments were carried out in steamboats on the Ohio River in Kentucky to determine the relative expense of coal and wood, and it was concluded that coal was about half the price but that it would not become a substitute until "the timber on the river bottom is cut down and consumed." Obviously, coal made some headway as a fuel for boats, for by 1844 it was said that large quantities of coal were mined at Hawesville and Cannelton on the Kentucky/Indiana border at seven cents a bushel, which was only about $1.75 per ton, but this price may not have been typical as the coal was mined from adits that sloped down to the river and there was no secondary handling of the coal into the boats. Bulky cordwood, on the other hand, cost anywhere between $1.25 and $6 per cord, the cost depending on the distance up the Mississippi away from the forests. It was found that "from 10 to 12 bushels of coal were fully equal to one cord of wood for generating steam and but few steamers pass without taking on some of this fuel."[16]

Nevertheless, there was still a remarkable prejudice against coal, and it was difficult to wean the steamboat operators away from the wood that grew in such abundance along all of the river lands and was such a clean-burning fuel. As the tonnage, the power, the speed, and the number of boats increased, so did the amount of cordwood used. For example, J. H. Morrison reported that in 1818 a 400-ton steamboat took 20 days to travel from New Orleans to Louisville and that it consumed about 360 cords of wood in the process, which is about one cord per 22 tons per 24 hours. By the end of the 1830s, however, 400-ton boats took only six days for the trip, and the rate of consumption had been reduced to about one cord per seven tons per 24 hours. From these and many other estimates, Erik Haites and his colleagues have calculated a decreasing tonnage per cord over the antebellum decades (Table 6.2) but an increase in the cords used as boats became faster and larger. To translate these figures precisely into a total consumption of cordwood used is difficult, but later estimates show that it must be measured in millions of cords.[17]

As steamboats rarely carried fuel supplies for more than 24 hours, it was common to take on fuel twice daily. Thousands of woodcutters or wood scavengers were available to supply any boat with wood, wherever and whenever it stopped. There were also many farmers who did the job on a part-time basis as they cleared their farms and earned a little extra cash. All these were known collectively as woodhawks,[18] and for the most part they were itinerant frontier wood choppers who wandered from place to place along the riverbanks supplying fuel to the boats. There were no chains of wooding stations. Each transaction must have been negotiated on the spot. It was a completely laissez-faire system of service that straddled the continent. In that sense it was like the fuel-wood business everywhere, even in the large cities where supply and distribution were only just beginning to be organized along modern business lines. Supplying the boats became

Table 6.2. *Cordwood consumption on steamboats, 1800–60*

	Tons/cords/24 hours	Total number of cords required per average round trip, Louisville to New Orleans
pre-1820	20	?
1820–9	12	483
1830–9	10	496
1840–9	8	504
1850–9	8	529

Sources: Erik F. Haites, James Mak, and Gary M. Walton, *Western River Transportation*, 144–6; and Erik F. Haites and James Mak, "Steam Boating on the Mississippi, 1810–1860, 56.

an important backwoods industry. Basil Hall, the intrepid English traveler and artist, described, in 1827, how his boat called at one of the "wooding stations" that lined the riverbanks only a few miles apart.

> About an acre of timber had been cut down to make firewood for the steam boats; and on the cleared space these little bits of rude huts had been perched on the tops of piles so that the flood just reached them. All the communication was by canoes, but how they get on when the waters subside or leave a stratum of six or eight inches of mud, guarded by forty thousand million mosquitoes, I do not know.

Typically, the wooding stations were operated by "squatters," people who had taken title to the land without grant and who cleared the land ahead of settlement but moved on when settlement impinged around them. Then they "take up their axes and retreat beyond the reach of those odious regulators of other people's affairs – judges and juries."[19] Working at all hours of the day and night, ready to load the boats whenever they called, the wood gatherers and sellers, or woodhawks, were a colorful, fiercely independent, if disreputable and uncouth, element of the river transport system, often debilitated by malaria and the alcohol taken to combat it.

James Silk Buckingham noted much the same scenes, and also the peculiar code of honesty that attended some of the transactions. Traveling up the Savannah River in 1842, he described how the steamer "halted several times at fixed stations to take in a supply of wood, as this is the only fuel used for the steam engines." The system was so customary that no one collected the $3 per cord; a note of the amount loaded by the crew was left at the fueling station, these notes were collected weekly, and the boat owner paid his bills in either Savannah or Augusta. In order to reduce the time lost in "wooding," the larger and faster boats that were developed in time began to tow flatboats upstream loaded with cordwood. Thus, with ensured supplies the journey was shortened and the empty boats cast loose to drift back to the wooding station.[20]

Confirmation of these descriptions of wood selling and fueling along the nation's major waterways comes from the census of 1840, which recorded the number of cords of wood entering into the commercial market.[21] In addition to the concentration of sales along the northeastern seaboard, noted before, the other very obvious feature was the concentration of sales along the major rivers and beside the Great Lakes (see Fig. 5.5). In 71 counties abutting either bank of the Mississippi, from the delta to the southern Wisconsin border, some 503,458 cords were sold. In 55 counties along the Ohio, from the Pennsylvania border to its confluence with the Mississippi, another 359,895 cords were sold, in all a total of 863,353 cords or 16 percent of all the cords sold in 1839. Some of the wood was consumed in the towns growing up alongside the rivers, but the bulk was used to fuel the hundreds of steamboats on the rivers. For 1829, James Hall had estimated that 525,000 cords were burned at an average cost of $2.25 per cord, to yield a total bill of nearly $1.2 million, and he also reported that the great flood of 1832 on the Ohio had swept away "vast quantities of firewood prepared for sale, at Louisville, Cincinnati and other towns, or accumulated at the woodyards and steam boat landings for the use of the steam boats."[22]

As late as 1856 coal still made little headway, and it was said that "steam boat engineers were of the opinion that such fuel could not make steam" and that only wood was feasible; indeed, after "eight years' exertion" the American Canal Coal Company at Cannelton, Indiana, was unable to sell more than 200,000 bushels per annum.[23] It was not until the lands along the riverbanks were cleared of easily accessible supplies that steamboat owners became interested in coal, and even then it was only on the Upper Ohio where the coal was relatively cheap. Before the advantages of coal could be fully realized, there were technical disadvantages to be overcome besides the inertia of established practices and custom. Coal required smaller fireboxes and grate openings and stronger-bottomed furnaces than wood, but to convert a furnace to handle coal only, when so much wood was available and coal was dear in places, was not practicable, especially as the lack of coal along the Lower Mississippi and Upper Missouri was a definite deterrent to its acceptance for long-distance travel.[24]

So, by the end of the antebellum period, wood still reigned supreme, with coal making significant inroads on the Ohio and Upper Mississippi, but even here few steamboats used coal exclusively. Steamboat boilers had been developed for the use of wood fuel, and the best results were still said to come from mixing equal amounts of coal and wood in the furnaces, not from using one or the other exclusively. The drain on the forests continued for as long as the steamboats ran.

Locomotives and industry

Perhaps the use of wood as a fuel for steamboats was less surprising than the continued firing of locomotives with wood well into the 1860s, and in some regions even into the 1880s. After all, locomotives were a later introduction, considerations of bulkiness and weight were more important than for boats (one ton of coal could replace up to four tons of wood at about half the cost), and railroads did emanate from the coal-producing, although in some places timber-depleted, states of the eastern seaboard. The Baltimore and Ohio tried anthracite coal in the 1830s, not because of considerations of weight but mainly because the British railways were using coal. However, this was an isolated instance at this time; the higher expenses of coal-burning engines per mile led to the almost universal use of cordwood until well after the Civil War, by which time over 30,000 miles of track had been laid.

As with the steamboats, a technical difficulty of conversion from wood to coal was the higher combustion point of coal, which meant less rapid ignition and lively combustion but a more intense heat once the fire was underway. This led to the rapid wearing out of the fireboxes (a point also brought up by the steamboat owners). It was not until structural and technical changes were perfected by the mid 1850s that coal could make any headway, and then only haltingly. Eventually, the substitution of coal for wood was a matter of costs and information. The railroads in the eastern part of the country and certain railroads in Ohio and Illinois faced rising cordwood consumption per mile as trains became longer and heavier, and prices increased as local supplies were depleted; for example, on the Boston and Providence the price rose to $7.40 per cord compared to about $4 in the Midwest. However, although wood was increasing in price, coal was not; it could be had for as little as $1.30 to $3 at the most. When the additional calculation is made that it was possible to get between 60 and 80 miles per ton of coal compared to 40 miles per ton of cordwood, it is obvious that considerable savings could be made by changing to coal. The recognition of the savings of costs and its acceptance did not occur simultaneously; they happened in different regions at different times and remained a function of the availability of wood either from controlled cutting or from the clearing of the land for agriculture. Albert Fishlow has calculated that by 1859 only about 350 or 400 coal-fired locomotives existed out of a total of about 4,000, but as the lines extended into woodless areas and coal became more plentiful the changeover was complete during the 1870s for most of the country. Only in the South, with its light trains and abundant pine wood, did many lines continue to burn wood into the 1880s and even later.[25]

Information is very scanty about the consumption of firewood by industry and transport during these years, but it was certainly slight compared with that consumed in the domestic hearths. The amount consumed in mechanical energy may have been between 6 and 9 million cords per annum during the 1860s and probably never exceeded 10 million cords per annum at what was possibly its peak of use around 1870. And that was only about 7 percent of the consumption of 138 million cords in the nation's homes during that year. It seems likely that locomotives were the heaviest users and that they may have used up to 6 million cords during the late 1860s, with an additional 2 to 3 million cords being used in steamboats and stationary engines. The only detailed record we have (Table 6.3) is for a later period in 1879, when the census suggests that wood for all mechanical purposes amounted to only 5.2 million cords or a mere 3.5 percent of the total of 147.2 million cords consumed in that year, or 4.5 percent if charcoal is included. It was negligible in the total,[26] and there is no reason to suppose that the proportions had altered materially from what they had been 20 or 30 years earlier, except that the proportions assigned to domestic use must have been even greater.

The annual consumption of wood for motive energy rose steadily from about 5 million cords in 1840 to 10 million cords in 1870, which would, at a conservative estimate of, say, 30 cords of wood per acre, require between 160,000 and one-third of a million acres of forest to be cut over annually. The indications are that consumption to keep the wheels turning dropped back and remained steady at the lower level of 5 million cords

Table 6.3. *Fuel wood consumption, 1879*

	Thousand cords		Percentage	
Domestic use		140,537		95.5
Industrial use				
Railways	1,972		1.3	
Steamboats	788		0.5	
Mineral operations	625		0.4	
Manufacturing	1,856	5,241	1.3	3.5
Subtotal		145,778		99.0
Charcoal for smelting		1,458		1.0
Total		147,236		100.0

Source: Charles S. Sargent, *Report on the Forests of North America*, 489.

(c. 160,000 acres cleared) for the remainder of the century and that domestic fuel consumption continued to dominate as the user of wood.

Naval stores

Of the minor products of the forest that entered into the local economies of the early nineteenth century, by far the most important was that of naval stores. The making of potash and pearl ash, the trapping of animals for furs and skins, and the collection of ginseng, maple syrup, and other forest products, already looked at in Chapter 5, came nowhere near in value, volume, or complexity to the production of resin, turpentine, pitch, and tar.

The collection, manufacture, and transport of naval stores remained primitive in character throughout the nineteenth century. The use of boxes on the trees, worked by intensive manual labor, did not (and could not) alter. The distillation process, although improved somewhat by the use of large copper vats, remained the same. Transport was still by river raft, road wagon, and sailing ship. In most cases and for most of the century the production of naval stores was an unorganized activity. The scale of production was limited; there was no competition. It was a by-product of the slave–plantation economy of the South, which individual owners indulged in if they wished. Only the marketing and distribution of the finished products overseas assumed typical nineteenth-century paleotechnic characteristics.

After the Revolution, production began to rise again and by the 1840s was running at a rate of about 200,000 barrels shipped out of the producing areas annually, an amount that jumped rapidly to a peak of 831,275 barrels on the eve of the Civil War, when it plummeted. This rise in production had been the outcome of a variety of factors: There was a general upswing in demand for goods throughout the country as population grew, more ships were being built throughout the world, and trade relations with Britain returned to some normalcy, particularly after the passing of the Free Trade Act of 1846, which facilitated the entry of goods into Britain. More important than all of these were the new uses found for naval stores. Continuing experiments culminated in 1842 with rectified spirits of turpentine (especially with the admixture of alcohol) being used for illumination. The mixture went under various names such as Camphine, Teveline, and Palmetto Oil, and it was the cheapest form of lighting known until kerosene was produced in large quantities in Pennsylvania after 1860.[27]

Although pitch, tar, and turpentine were produced at scattered locations throughout the nation, the commercial production of naval stores from the seventeenth century through to the mid-nineteenth century was heavily concentrated in a relatively small area of North Carolina about 50 to 100 miles from the coast between the northern and southern boundaries of the state with a slight extension into the adjoining portions of Virginia and South Carolina. In 1840, North Carolina produced 96.2 percent of the national total of 507,275 barrels, of which six-sevenths came from the 12 counties of Beaufort, Bertie, Bladen, Carteret, Craven, Duplin, Edgecombe, Greene, Jones, Martin, New Hanover, and Pitt (Fig. 6.3). The export outlets were New Bern, Washington, and, in particular, Wilmington.[28]

As demand rose, collecting became more reckless and trees were ruined permanently. Turpentiners moved south into the untouched forests of South Carolina and Georgia where even as late as 1880 a dense stand of pines could be bought for as little as $2 to $3 an acre. In their wake came the lumbermen. The prestige position of North Carolina as the leading producer was soon to pass away as the new stands farther south were more suitable for specialist mass-production "turpentine farms," which appeared just before the Civil War and became much more numerous afterward. The shift south is indicated clearly in Table 6.4, which shows the percentage of production of naval stores by value by states from 1840 to 1919, well beyond the period of immediate concern. Whereas production was almost wholly within North Carolina up to and including 1860, it shifted southward, first through South Carolina during the 1870s and 1880s, then through Georgia from 1890 to 1910, and then to Florida from 1900 through to 1919. There were lesser developments in Alabama, Mississippi, and later in Louisiana and Texas.

Because the center of production moved south, new ports sprung into prominence. Wilmington reigned supreme during the early years, exporting 600,000 barrels during its peak year in 1851, two-thirds going coastwise, mainly to the North, and one-third overseas. After the Civil War Wilmington lost its preeminence to Brunswick and then to Charleston, which reached its peak between 1875 and 1885 and in turn gave way to Savannah between 1885 and 1915, which had an even greater export volume, at times reaching 1.5 million barrels annually, three times that of Charleston. Jacksonville and Pensacola in Florida, Mobile in Alabama, and New Orleans in Louisiana became important locally during the early twentieth century.[29]

Increasingly, from 1890 onward, the naval stores industry became inextricably intertwined with the burgeoning southern lumber industry, and the large-scale corporations bought vast stands of timber and exploited them rapidly for stores in order that lumbering would not be delayed.

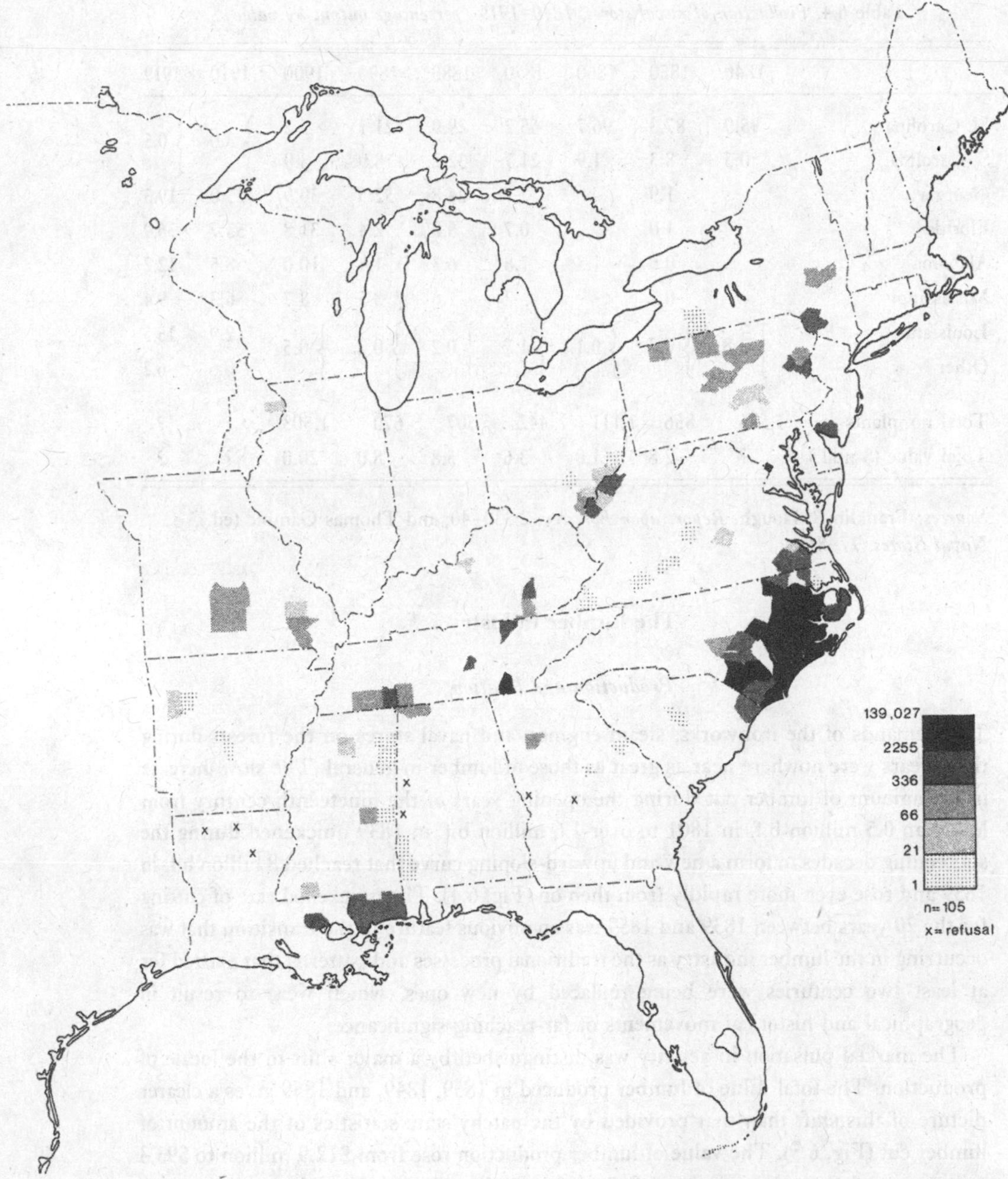

Figure 6.3. Production of barrels of pitch, resin, and turpentine in the United States, 1840. (U.S. Census.)

Table 6.4. *Production of naval stores, 1840–1919 (percentage output by value)*

<table>
<tr><th></th><th>1840</th><th>1850</th><th>1860</th><th>1870</th><th>1880</th><th>1890</th><th>1900</th><th>1910</th><th>1919</th></tr>
<tr><td>N. Carolina</td><td>95.9</td><td>87.3</td><td>96.7</td><td>65.2</td><td>29.9</td><td>21.1</td><td>5.2</td><td rowspan="2">3.0</td><td rowspan="2">0.5</td></tr>
<tr><td>S. Carolina</td><td>0.3</td><td>8.3</td><td>1.9</td><td>21.7</td><td>32.2</td><td>18.9</td><td>3.9</td></tr>
<tr><td>Georgia</td><td></td><td>1.9</td><td></td><td>2.7</td><td>24.8</td><td>52.5</td><td>39.9</td><td>25.0</td><td>19.3</td></tr>
<tr><td>Florida</td><td></td><td>1.0</td><td></td><td>0.7</td><td>5.0</td><td>2.4</td><td>31.8</td><td>53.7</td><td>36.9</td></tr>
<tr><td>Alabama</td><td></td><td>0.6</td><td>1.3</td><td>7.8</td><td>6.3</td><td>1.4</td><td>10.0</td><td>8.5</td><td>12.2</td></tr>
<tr><td>Mississippi</td><td></td><td>0.7</td><td></td><td>.2</td><td>1.6</td><td>3.5</td><td>8.7</td><td>6.3</td><td>9.4</td></tr>
<tr><td>Louisiana</td><td rowspan="2">3.8</td><td rowspan="2">0.2</td><td rowspan="2">0.1</td><td rowspan="2">1.7</td><td rowspan="2">0.2</td><td rowspan="2">0.2</td><td rowspan="2">0.5</td><td>2.9</td><td>15.5</td></tr>
<tr><td>Other</td><td>0.6</td><td>6.2</td></tr>
<tr><td>Total no. plants</td><td>1,526</td><td>856</td><td>1,111</td><td>442</td><td>507</td><td>670</td><td>1,503</td><td>?</td><td>?</td></tr>
<tr><td>Total value ($ mill.)</td><td>?</td><td>2.8</td><td>1.0</td><td>3.6</td><td>5.8</td><td>8.0</td><td>20.0</td><td>?</td><td>?</td></tr>
</table>

Sources: Franklin B. Hough, *Report upon Forestry*, 2:330–40; and Thomas Gamble (ed.), *Naval Stores*, 77–81.

The lumber industry

Production and location

The demands of the ironworks, steam engines, and naval stores on the forests during these years were nowhere near as great as those of lumber in general. The slow increase in the amount of lumber cut during the opening years of the nineteenth century from less than 0.5 million b.f. in 1801 to over 1.6 million b.f. in 1839 quickened during the succeeding decades to form a new and upward-sloping curve that reached 8 billion b.f. in 1859 and rose even more rapidly from then on (Fig. 6.4). The quickened rate of cutting for the 20 years between 1839 and 1859 was an obvious feature of the transition that was occurring in the lumber industry as the traditional processes and patterns that existed for at least two centuries were being replaced by new ones, which were to result in geographical and historical movements of far-reaching significance.

The marked pulsation in activity was distinguished by a major shift in the locale of production. The total value of lumber produced in 1839, 1849, and 1859 gives a clearer picture of this shift than that provided by the patchy state statistics of the amount of lumber cut (Fig. 6.5). The value of lumber production rose from $12.9 million to $93.3 million in each decade, a rise that reflected changes in real production during these years of relative financial stability.

In 1839 New York was preeminent and accounted for 30 percent of total lumber production by value. Maine followed with nearly 14 percent, and the other New England States together accounted for 10 percent. If Pennsylvania (8.8 percent) and New Jersey (1.9 percent) are included with New York and the New England States, then this northeast corner of the United States accounted for over two-thirds of the total

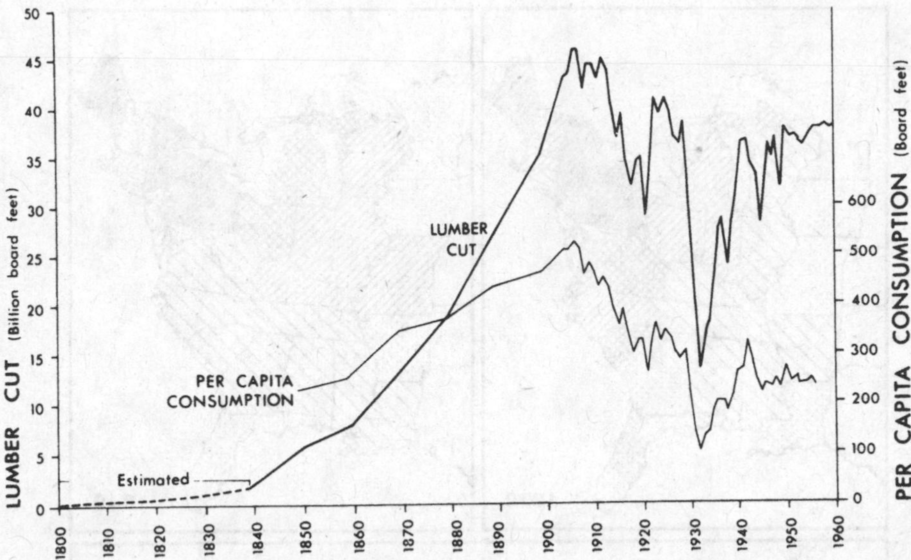

Figure 6.4. Lumber cut and per capita consumption, 1800–1960. (U.S. Bureau of the Census, *Historical Statistics of the United States from Colonial Times to 1957*, tables L87–97, L113–21.)

production, and it was the core area of experiment and change in the industry. Elsewhere, Michigan with 3 percent and Indiana with 2 percent were portents of what was to come, and in the South, Virginia and the Carolinas each accounted for about 4 percent of national production by value. In 1849 New York reached the peak of its production ($13.1 million), but, despite the fact that the northeast corner of the country was still the leading producing region it now accounted for only one-half of the national production as lumbering activity tended to even out across the continent as settlement spread west and mills sprang up to meet local demands and even to meet local deficiencies on the northeast seaboard. Indiana (3.7 percent), Ohio (6.6 percent), Michigan (4.2 percent), and Wisconsin (1.6 percent) began to show up in the national picture; Washington (2.3 percent) and California (1.6 percent) appeared for the first time. By 1859 the previous hints of a changing pattern of production became clearer. The Northeast now accounted for only one-third of total production, which was slightly less than the Lake and Central States combined. The Lake States of Michigan, Wisconsin, and Minnesota were emerging as a clearly defined high-production area, soon to be tied to the eastern consuming markets and to the prairies to the south and west by a complex continental transport network, linked by wholesaling nodes.

The two midcentury decades, therefore, can be looked upon as a significant pivotal period when the rate, magnitude, and location of production changed. These years, when personnel and technology diffused from New England, and from New York in particular, to the Lake States, were the prelude to the lumberman's massive assault on the forests of Michigan, Wisconsin, and Minnesota from about 1860 to 1890.[30] It was a time when new patterns, processes, and phenomena were being formulated that would

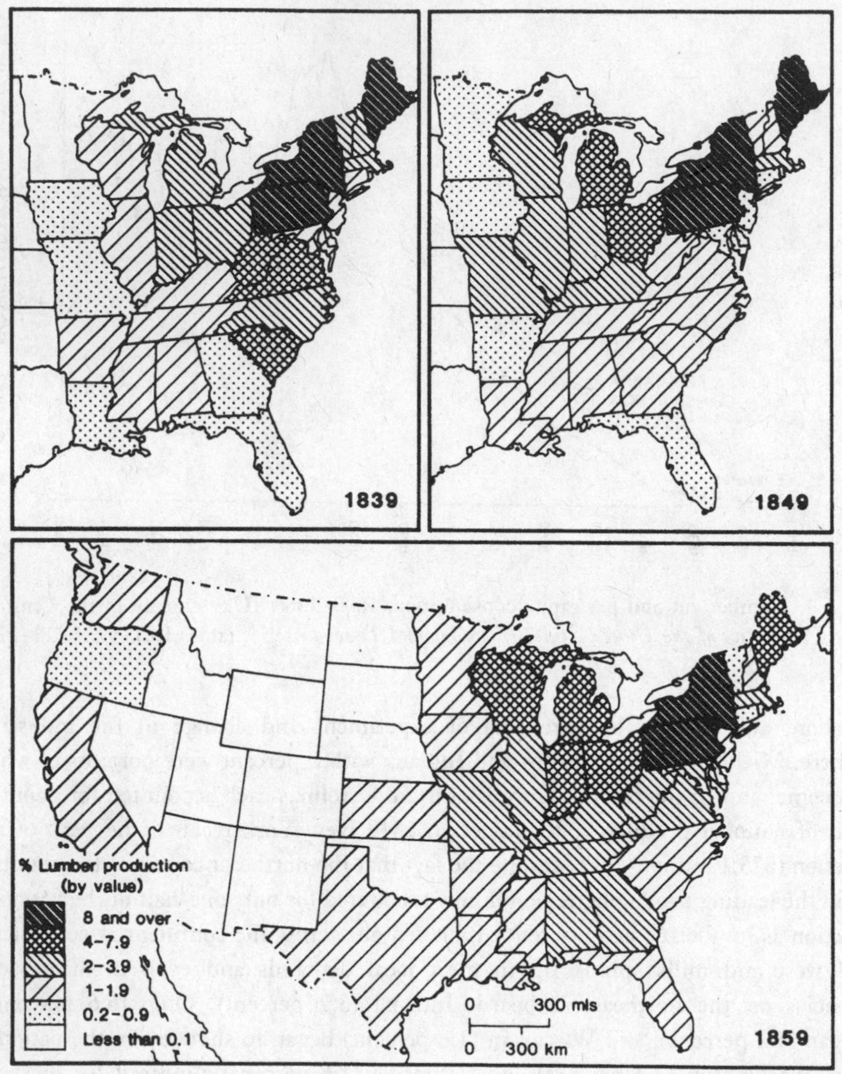

Figure 6.5. Lumber production in the United States, 1839, 1849, 1859; value shown as percentage of total United States production. (H. B. Steer, *Lumber Production in the United States, 1799–1946*, 11.)

ultimately affect forest exploitation during the latter half of the nineteenth and the early part of the twentieth centuries.

And yet, though there was much that was new in this era of transition, it is important to sort out the traditional from the novel and the innovative. In most cases the new was a slow evolution rather than an abrupt departure from the old, so that old and new existed side by side, overlapping and interpenetrating in any particular location. Evelyn Dinsdale has suggested that Mumford's trilogy of eras of technology – the eotechnic, the paleotechnic, and the neotechnic, already mentioned in relation to the iron industry – offers a ready and broad conceptual framework for understanding the changes in the lumber industry.[31] This is particularly so as each era was characterized not only by technical changes in power sources, tools, and methods but also by changing attitudes, social organization, scales of organization, and internal spatial arrangements.

Certainly, it is evident that the lumber industry in the United States up to 1810 was eotechnic in its characteristics. Either moving water or animals and men were the source of motive power. Wood and stone, with some metal parts, were the primary materials used, and the scale of operations was small. Tools were crude and hand-forged, production was restricted, distribution haphazard, and the market local. There was little competition and little incentive to initiate change. Simply, timber getting was still an adjunct to agricultural settlement, and the timber cut was mainly the by-product of the land clearing or the concern of a multitude of small mills that dotted the country to serve the agricultural population. Water was the only factor that limited the location of mills at the local level, but generally speaking they were evenly distributed. In 1840, for example, there were 31,649 mills in the United States, or an amazing average of 25 mills for every county proclaimed at the time. However, as pointed out earlier, because of the strong link between agricultural settlement and early sawmilling, the number of mills at any given location was more a reflection of the length of time an area had been settled than anything else. Therefore, most of the counties had a few mills, and a few in the older settled areas on the northwest seaboard had great concentrations (Fig. 6.6). The bulk of the counties, over three-quarters in fact, had fewer than 29 mills each; nearly half the counties had nine mills or less; and nearly one-tenth, in the largely untouched forests of the South, had none. On the other hand, of the 176 counties with 50 mills or more, 143 were in the states of New England, New York, Pennsylvania, New Jersey, and Delaware, and of the 76 counties with 100 mills or more, 74 were in these states.[32]

It is true that there were some areas where the timber produced was not consumed wholly by the farmers. The northeast seaboard from southern Maine to northern Virginia had a degree of specialization and a basis for commercial industry. A core of activity had existed there for centuries, supplying specialized products such as masts, ships' timbers, pipe and barrel staves, as well as general lumber, which had been exported along the coast overseas. A clue to the location of the commercial areas of production comes from looking at the number of men employed and the value of production in the 1840 census, both of which are indicators of specialized, commercial production (Figs. 6.7 and 6.8). The number of men employed is recorded as 22,042, well below the total number of mills. Far from being evidence that the census is inaccurate, the apparent discrepancy merely reflects the fact that the vast majority of mills were small-scale, one-man, even part-time affairs of active farmers who would not be employing any additional hired labor. In Figure 6.7 the pattern of men employed is largely what would be expected, with a major concentration in the New England States, New York, and adjacent areas of Pennsylvania, with scattered locations along the remainder of the East Coast.

By the same reasoning, the value of lumber produced by county shows nonlocal, that is, commercial, production. Slightly fewer than half (570) of the counties recording mills

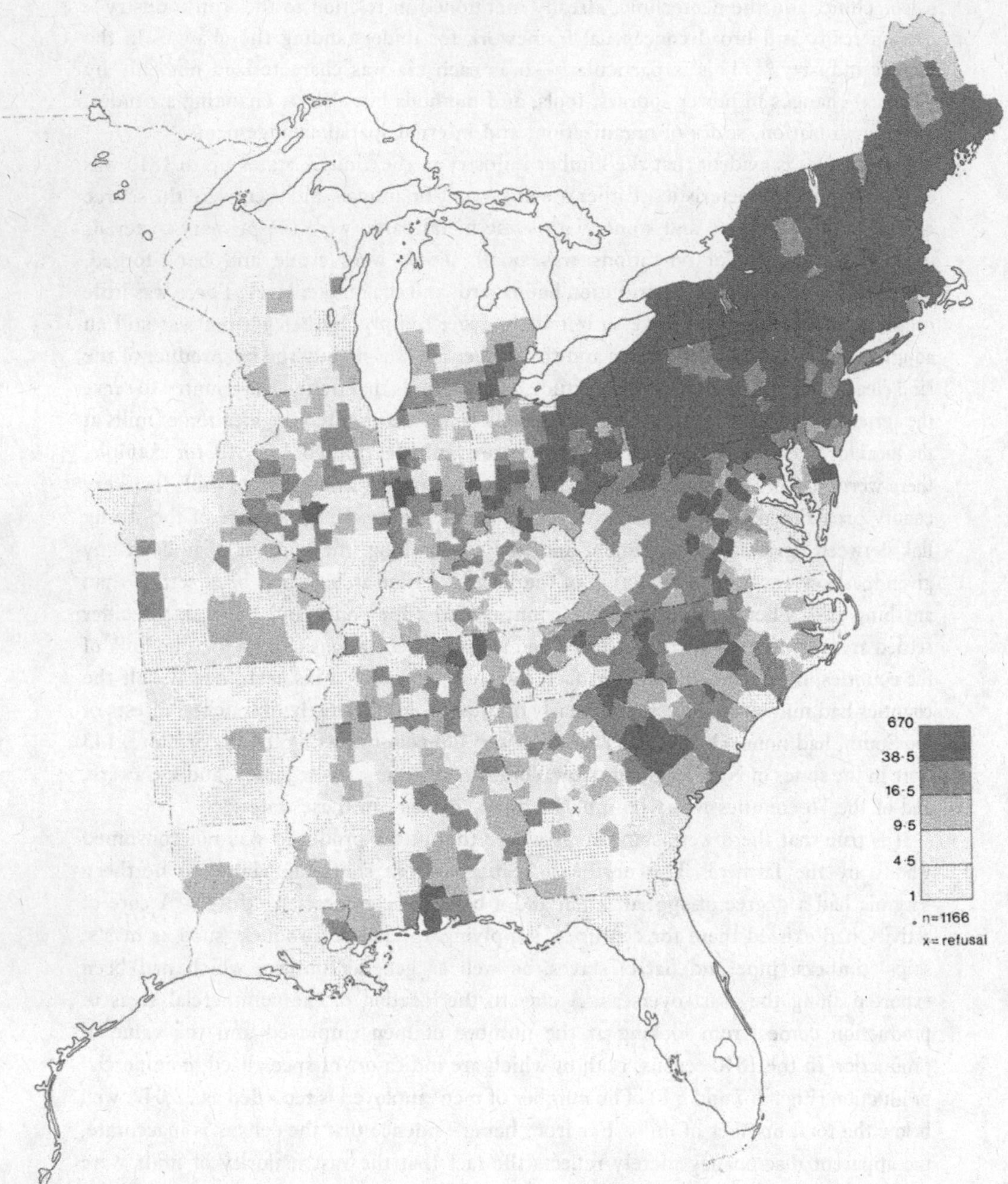

Figure 6.6. Number of sawmills in the United States, 1840. (U.S. Census.)

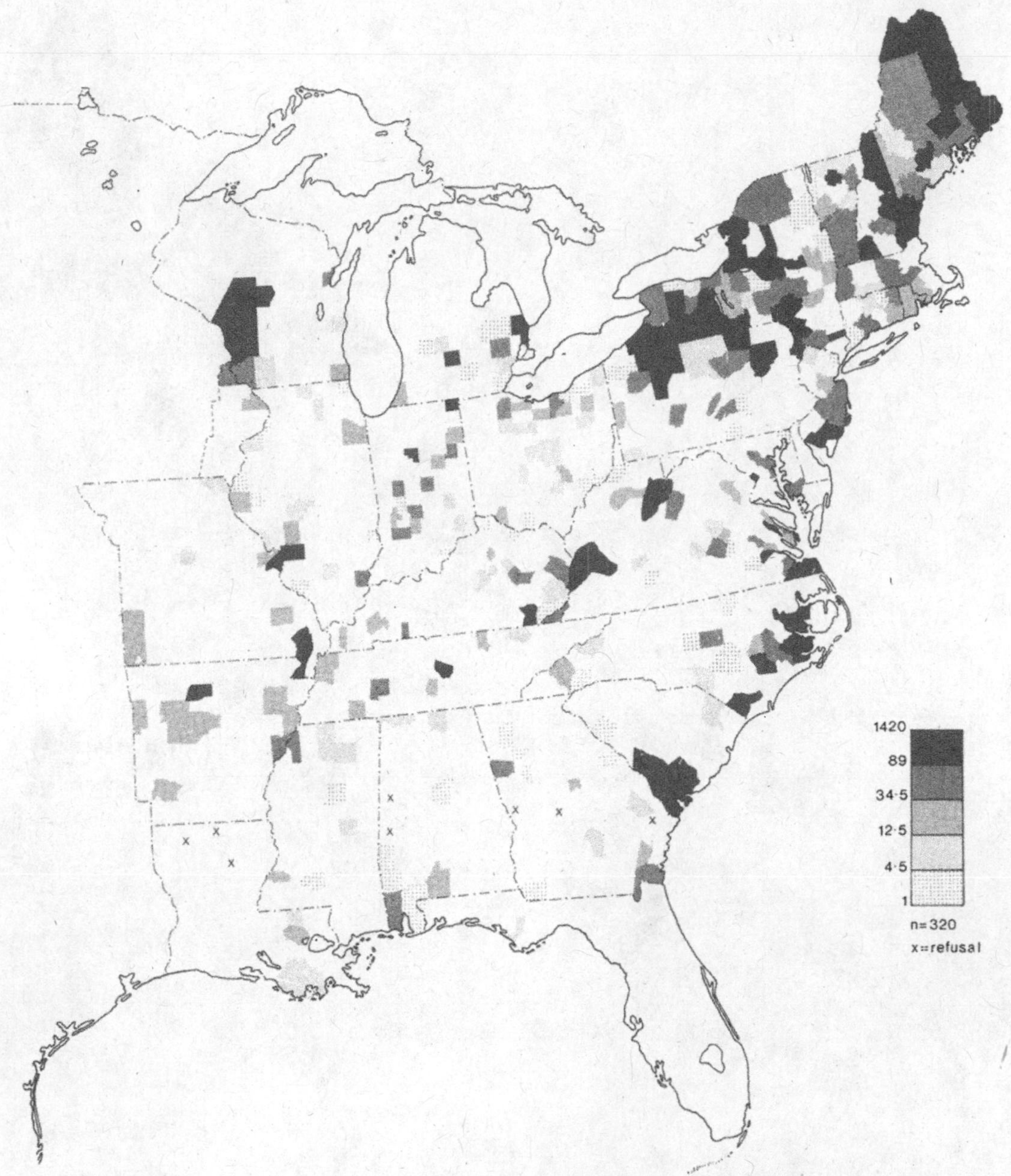

Figure 6.7. Number of men employed in sawmills in the United States, 1840. (U.S. Census.)

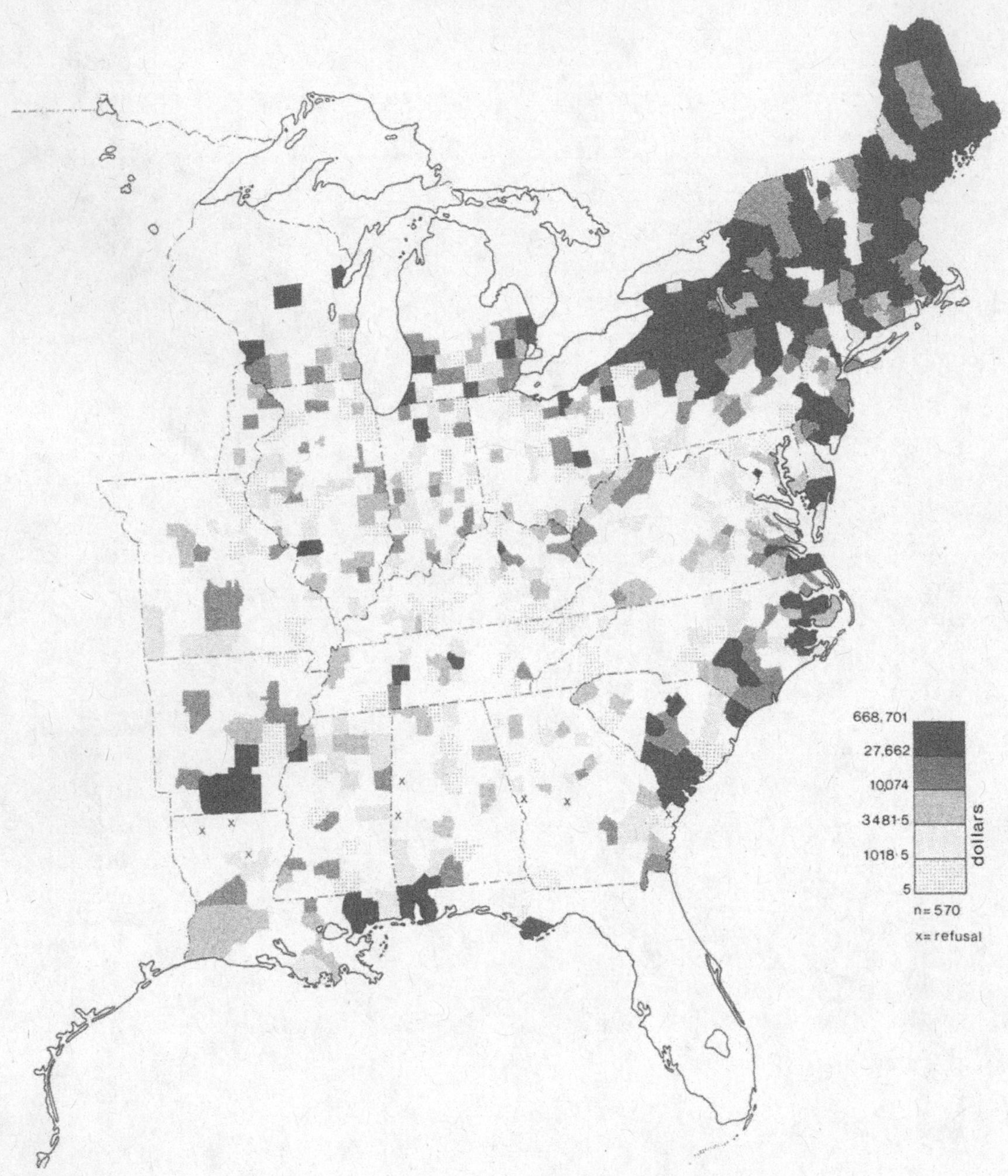

Figure 6.8. Value of lumber products, United States, 1840. (U.S. Census.)

Table 6.5. *Increasing productivity of saws, seventeenth to nineteenth centuries*

	Sawmill output per day (b.f.)
Hand-powered pit saw	100–200
Water-powered single blade (1621)	500–3,000
Water-powered single sash (1621)	2,000–3,000
Water-powered muley saw (c.1780)	5,000–8,000
Water-powered circular saw (1844)	500–1,200
Steam-powered gang saw (c.1850)	40,000+
Steam-powered circular saw (1863)	40,000+
Steam-powered band saw (1876)	10,000+

Sources: Alfred J. Van Tassel and David W. Bluestone, *Mechanization in the Lumber Industry*, 8; Ralph C. Bryant, *Lumber*, 3; James E. Defebaugh, *History of the Lumber Industry of America*, 2:8, 58; and Brian Latham, *Timber, Its Development and Distribution*, 209–18.

(1,166) recorded lumber values, which accords with the idea that most lumber produced did not have, for census purposes, a value recorded – not surprising in a pioneer economy in which small farmers bartered and exchanged commodities. Taken together, the number of workers and the value of lumber production are indexes of emerging commercial lumbering.

From the core areas of timber production, technology, and general trading expertise on the northeast seaboard came the impetus to change the scale and character of the industry during the two midcentury decades. The one- or two-man mill still existed, but it was giving way everywhere to bigger, more commercialized units, sometimes as large as 20 to 100 men, and lumbering became a specialized occupation with a great deal of division of labor as jobs became more specifically part of a total industrialized process. Basically, the technical advances that brought about the changes were four in number: first, improved saws, machinery, and the use of steam engines in mills; second, the improvement of the "local" transport with the development of the log drive; and third, the evolution of a continental transport system to link areas of surplus timber with areas of deficiency, which were becoming separated increasingly from each other. Finally, there was the development of wholesale centers at focal points in the transport system, which held together the multitude of movements between forests and markets.

Saws, machinery, and steam

The increase in the volume of logs moving downstream to supply the needs of the urban and industrial markets meant that there had to be a greater sawing capacity in the mills than before. In most places the number of sawmills increased, but, more importantly, the output of the mills was augmented by various inventions and improvements to existing machinery (Table 6.5).[33]

The most important of these inventions applied to the saws themselves. The old water-powered, single, up-and-down saw that cut 500–3,000 b.f. of timber a day was replaced by multiple-bladed gang saws and light and fast-moving muley saws, the efficiency of which was improved upon by new water-power transfers whereby the cumbersome wooden wheels were replaced by spiral vents, overshot wheels, and steel parts. Circular saws, invented, patented, and used in the English shipbuilding yards as long ago as 1777, did not appear in the United States until 1814, when Benjamin Cummings of Schenectady independently invented and made a crude hand-forged blade for his mill. Robert Eastman and John Jacquith of Brunswick, Maine, improved the saw in 1820, and one was used in Maine in 1821. From then on the circular saw came into more general use, especially after 1846, when Spaulding of Sacramento designed a curved ratchet, which held hard-tempered false teeth, with the result that the saw could be repaired easily by putting in a new tooth. Driven by steam, the saw could cut up to 40,000 b.f. a day, and even double that later. Its main disadvantage was that it tended to overheat and wobble at high speed. A similar story of invention and subsequent modification before acceptance occurred with the band saw, which was probably invented in France during the early years of the nineteenth century. It was used in Baltimore as early as 1819 but was not introduced into New York until about 1860 and made no significant headway until exhibited at the Philadelphia Centennial Exhibition in 1876, whereafter it became very popular.[34]

The adoption of various types of saws varied from one part of the country to the other. Some idea of the relative popularity of types is given by the situation in the mills of Massachusetts in 1875. Of the 2,148 saws, 83 percent were circular saws, 11 percent were still the old up-and-down type, approximately 1 percent were band, and 1 percent were muley saws. Gang saws barely entered into the picture; the remainder were jig and cylinder saws. At Williamsport and Long Haven, Pennsylvania, on the other hand, gang saws were far more numerous and accounted for 65 percent of all saws installed, muley saws for 26 percent, and circular saws for only 8 percent.[35] In the circumstances it is difficult to generalize. The reasons for the slow acceptance of the high-speed circular and band saws were related to the quality of the steel used in their manufacture and the welding procedures used in joining the parts, and these technological improvements were not widely available until the early 1870s, after which the saws were adopted more readily.

Associated with the improvement of saws was the improvement in the machinery for handling the logs in the mills and for preparing and finishing the timber. Edgers for squaring boards had been invented in 1825, and carriages for taking the logs past the saw in 1850. In 1828 William Wordsworth of Poughkeepsie, New York, invented and patented a planing machine that had an impact on the lumber industry second only to that of the saw, and by 1850 it had evolved into numerous specialized forms adapted to all sorts of purposes and final products.[36] A glance at the reports of the commissioner of patents gives some idea of the invention and modification that was going on in the lumber and associated industries during these years. For example, the first report of the commission showed that the number of patents issued during the previous 13 years and then expiring was 36. Three were for shingle-making machines; 10 for sawmills; 10 for mortising, joining wood and veneers; two for lathe cutting; four for stavemaking. This sort of thing was typical for all subsequent years.[37] All these inventions had the tendency to increase production by speeding up the flow of wood through the mills to produce a better result, and this was often achieved at no greater cost than before and certainly resulted in greater uniformity. In time these developments led to a greater consumption (Fig. 6.4).

One example must stand for many. In 1854 a British parliamentary committee of inquiry visited the United States and was astonished "at the wonderful energy that characterizes the wood manufacture of the United States."[38] The extensive specialization of machinery was particularly impressive:

> Many works in various towns are occupied exclusively in making doors, window frames, or staircases by means of self-acting machinery, such as planing, tenoning, mortising, and joining machines. They are able to supply builders with the various parts of the woodwork required in buildings at a cheaper rate than they can produce them in their own workshops without the aid of such machinery. In one of these manufactories twenty men were making panelled doors at the rate of 100 per day.[39]

The application of steam power to all these operations, whether sawing or woodworking, was, of course, essential to their widespread adoption and diffusion.

As far as sawing was concerned, the application of the steam engine was early but not so widespread as might have been expected. The first engine was said to have been applied to a mill in New Orleans in 1811, and there were others in Maine after 1822, but this singling out of "firsts" hides the fact that even as late as 1838, of 1,616 steam engines recorded as existing in America, only 208 or just under 13 percent were used in mills, 65 or 3.5 percent in wood planing, turning, or lathe making. The greatest number of steam mills then was in Massachusetts with 25, with a concentration around Boston; they were usually small affairs of less than eight horsepower (h.p.). There were other concentrations in Pittsburgh (12), Wheeling (11), and Charleston (13). In Michigan there were 26 engines with four new ones on order, and in general these engines were much more powerful than elsewhere, most being 12 h.p. and a few as great as 25 h.p.[40]

With steam power the mills were more reliable and worked with greater speed so that the output per mill increased rapidly and kept pace with the rising per capita consumption and the general increase in demand. After the 1860s many smaller neighborhood and cooperative family enterprises disappeared as mills became larger, lumber companies began to integrate vertically and horizontally, specialist retailers and wholesalers emerged, labor–management relations became strained, and small-scale production gave way to mass production and all that that entailed. However, just as there was an enormous overlap between the two ages in the iron industry and in the generation of mechanical energy, so there was in the lumber industry. To take one example, even as late as 1870 there were still 16,562 waterwheels in the lumber mills generating 327,000 h.p. compared with 11,204 steam engines generating 314,000 h.p., with the "old" regions of Maine, New York, and Pennsylvania having the greatest water-power capacity.[41] Steam power had still not begun to dominate, even at this late date.

The log drive

Wood is a heavy and bulky commodity of intrinsically low value. It is found in inaccessible, uninhabited places and often has to be transported many hundreds of miles to its market. During the early nineteenth century transport costs accounted for up to three-quarters of total production costs and probably never fell below half, the remainder being made up of the milling and sawing. Stumpage, the cost of the timber standing in trees, also entered into the calculation, but it was not great compared with transport.[42] In the United States, as elsewhere, not only was cheap transport needed, but flexibility of transport was also required because the cutting out of the resource meant that costly permanent links could not be afforded. Moreover, there was a need to eliminate the uncertainty and variability of supplies as a carefully geared work cycle of cutting in winter and sawing in summer needed to be maintained for a planned and well-managed industry. In response to the need to move more timber for greater distances and in larger quantities, the previous method of transportation by rafts and arks was extended but also augmented by new methods.

The major innovation of the early nineteenth century was the widespread use of the rivers and their tributaries for the log drive (see Figs. 4.2C and 7.10A). This innovation brought into play new methods of exploitation, organization, and equipment. It is commonly said that the log drive originated on the Schroon River, one of the tributaries of the Hudson, as late as 1813, as a means of supplying the great sawmills at Glens Falls, Sandy Hill, and Fort Edward and that it was pioneered by the Fox brothers of Warren County. It is true that the method of transport had long been known in north-central Europe on the Vistula and Neman rivers, on the Swedish rivers, and also in the mast-hauling trade of New England on the Piscataqua and Saco rivers, and even around the New Hampshire coast since the early eighteenth century. Nevertheless, the log drive differed from any previously known practice in that it considered the whole of the river catchment as a venue for exploitation and of the main stream and its tributaries as the medium for transport over distances never contemplated before. Moreover, much smaller streams could be used for log driving than for rafting, and that extended lumber operations considerably. It was a solution, therefore, not different in kind from all that had gone before, in that it was working within the limitations and conditions of the environment, but it was a solution that was certainly different in scale and organizational arrangement. The use of the rivers was a more feasible proposition than before because of the lesser selectivity that existed in the lumber industry. The masting days were all but over when only the longest and largest white pines were selected at many miles apart. Now all trees, pine or not, were cut in all directions up to a six-mile distance from the major rivers. Investment and operational costs in the log drive were low, and the ratio of men employed to board feet delivered at the mills varied enormously from one person/40,000 b.f. to one person/110,000 b.f., but generally the amount of lumber delivered per man by river logging was greater than by rafting.[43]

Cutting was done in the fall and in the early winter months, and then the logs were "skidded" or hauled to the water's edge by sled and stacked to await the spring flood and the drive.[44] (See Plates 6.1–6.5.)

Plate 6.1. From forest to market: felling and sawing logs. (*One Hundred Years of Progress.*)

Plate 6.2. Loading a sled. (*One Hundred Years of Progress.*)

Log driving needed a set of specialized persons. With the spring thaw the lumberjack set aside his ax and saw, donned his steel-spiked boots, took the peavey, a steel-spiked crook for moving and sorting out the logs, and became a log driver, which minimized the labor required in the forests. The job was hazardous; logjams were dangerous, and

Plate 6.3. Floating logs in the drive. (*One Hundred Years of Progress.*)

Plate 6.4. A logjam. (*One Hundred Years of Progress.*)

Plate 6.5. Loading the timber on a ship. (*One Hundred Years of Progress.*)

drownings and crushings were common. Accounts in the local newspapers of the breaking of logjams read like accounts of cattle stampedes. The analogy of a cattle stampede was more far-reaching than that, however, as the whole operation assumed the characteristics of a cattle drive along a stock route or grazing over a common or unfenced range. As one company or a few individuals were replaced by many, all using the same river system, there evolved an elaborate set of log markings (like brands) so that the logs could be identified, thefts stopped, and the drivers paid, usually on the basis of so many cents per m.b.f. delivered to the mill. By 1813 there were so many loggers on the Hudson that log marks had to be registered in the office of the town clerk of Queensbury, and similar regulations followed for the Ausable River (New York) in 1825. This cooperation of necessity was now compounded by state intervention and legislation as the lumber transport business became more complex. In Maine as early as 1821 the state intervened by enacting legislation to prevent and punish thefts, for "log lifting" was becoming as common as rustling was to become later. The possession of logs with their marks cut was prima facie evidence of guilt, and logs floating up or settling on the land of others could not be claimed until two years had elapsed. There were penalties for cutting out log marks, rewards were offered for information, and unmarked logs, or "prize" logs, were treated like stray cattle and auctioned off to pay the expenses of driving if unclaimed after some specified period.[45]

Much confusion and contention arose as the rivers became more and more crowded with operators, each one attempting to bring his logs to the sawmill and ultimately to the market. Marking helped, but there were two other developments in which both the state legislatures and individual lumbermen had a part to play: the opening up of the rivers to logging and their regulation and improvement through the formation of driving associations and a variety of private companies; and another cooperative exercise, the construction of booms for the sorting out of logs at the mills.

In the first case, the new methods of log collection and transport required freedom to use the streams unhindered by the claims of riparian owners and other stream users. During the early years of the century, rivers in New England, New York, and Pennsylvania were declared public highways for boats and rafts, and the clogging of the stream was forbidden. But, starting in the Raquette River in New York in 1810 and then with increasing frequency elsewhere, the definition of a public highway changed to include the "floating of sawn logs and timber," and by midcentury the log drive took precedence over all other forms of navigation. The organization of lumbermen to appoint a master driver was encouraged by state regulations in Maine. The lumbermen were assessed to pay the expenses of organization according to the number of logs they drove. This happened on the Kennebec as early as 1827, on the Androscoggin in 1830, on the St. Croix in 1836, and on nearly all the other main rivers and tributaries between 1845 and 1858. On the Kennebec the driving was organized as a stock company in which shares were purchased, and by 1837 its jurisdiction extended along the whole length of the river to Moosehead Lake. Sixty-four operators joined in 1835 and assessed themselves at $8,004 to have the logs driven. In this case the river was divided into sections and a different charge made for portions of the river depending upon the difficulty of the drive. The charge varied between 15¢ and 40¢/m.b.f., and this sort of division, according to the time spent and the difficulty of the drive, became common. All these regulatory procedures and voluntary restrictions and associations were repeated in one form or another in New Hampshire, New York, and Pennsylvania.[46]

Almost inevitably, so it seemed, the bringing together of the lumbermen and the rising demand for timber led many to a consideration of how the log-driving system could be made more effective. This was done by improving the river channels, building dams to increase the spring flood, excavating canals and slips around rapids, and even blasting rocks out of the riverbed. The splash dams, as they were known, were usually built on the minor tributary streams, which had little flow and which were secondary areas of exploitation. The dams accumulated a head of water that carried the logs down to the main river in the spring. The dams were not expensive and were considered expendable in this exploitative industry. Where there was less winter freeze and spring thaw, there was less need for splash dams, as in southern and western New York. On larger streams larger, more permanent dams with sluices were constructed, and a charge of a few cents per thousand (m.) logs that passed was made. One of the earliest known was erected by John Wood in 1828 on the Little Kazar River in Maine; he charged 4¢/m. and acquired the right to sell the logs if the owner did not pay him. Eight sluice companies were formed for negotiating rapids on moderately large streams between 1834 and 1861; 31 dam companies were formed with power to charge anything from 1½¢ to 25¢/m. Similarly, between 1821 and 1846, 29 canal companies were formed to excavate bypasses and canals, some schemes being such grandiose affairs they were never executed.[47]

The other major problem of log driving, that of how best to sort out the marked logs once they arrived at the sawmill, was overcome by the construction of booms by the major logging companies. Booms were rigid or flexible structures of chained logs floating across a river and sometimes anchored to intermediate piers. The booms could be used like pens for drafting branded cattle in a stockyard, and various gates could be opened and shut so that the marked logs could be sorted and assembled in bundles, counted, and sent to the appropriate mill.

The first well-documented boom was on the Androscoggin in 1789, and another was made at Brunswick in 1820. Others appeared on the Saco in 1824 and on the Kennebec in 1831. Subsequently, there were more, the largest being the boom at Bangor on the Penobscot built in 1825. The idea diffused south; other well-known large booms beyond Maine were the Williamsport boom on the Susquehanna built in 1846 (by John Leighton of Penobscot) and the Hudson River Boom Association of Glens Falls, organized in 1849. In Wisconsin and Michigan there were booms on all the major rivers by the end of the 1860s. Like all the works of improvement, sorting and grading was not a free service, and a toll or "boomage" was paid at rates varying from 25¢ to 40¢/m., which, when added onto the driving tolls, dam tolls, and canal tolls, made the movement of lumber more costly than the simple operation of log driving seems to suggest at first glance.[48]

All these improvements and developments were designed to cheapen transport and to increase efficiency of handling and certainty of supply, and even if there were boomage and driving costs and if some of the logs were damaged or lost, these disadvantages were more than offset by the increase in the quantity of lumber that passed through the mills. Transport, then, was the major pivot around which the entire process of converting the standing trees to lumber was forced to revolve. It controlled or influenced most other factors of production, such as capital investment, forested land (stumpage), and the availability of markets, finance, and labor.[49] Without the new system of the log drive as part of the production pipeline the new high-output mills could not have functioned efficiently and met demand. Later, logging railroads began to penetrate the forest and replace the log drive.

The continental transport system

While the log drive operated at the "local" level, there developed a transportation system at a continental level as areas of production and consumption became separated by greater and greater distances. Mills were usually located near areas of production rather than areas of consumption because milling reduced the bulk of this bulkiest of raw materials and at least doubled its value. The refined product, therefore, was more capable of being transported long distances by canal barge, river or lake boat, or railroad car.

Lumber was transported over distances that reached about 1,000 miles by the mid-nineteenth century. Admittedly, international trade covered similar or even greater distances by this time, but that was by oceangoing ship. It was not as if lumber was a high-value product; perhaps the nearest comparable commodity moved in bulk was wheat, but during the 1860s that was about four times as valuable as sawed lumber and 10 times as valuable as logs per unit weight – and even wheat was said not to be worth transporting when its price was low.[50] The continued internal movement of lumber for long distances irrespective of its bulkiness and low per unit value was a measure of its essential nature in the American economy and way of life, which had no counterpart in the world at this time.

As the distances became greater, so the hauling and handling processes became more complex. The lumberman knew that to be successful he had to eliminate the uncertainties in the production and supply of his raw material and replace the speculative element in his business by planning and good management – particularly in the field of transport. He had to become a specialist in hauling and marketing as well as in cutting. The intimate and often locationally close bond between timber getting, timber milling, and the lumber market was broken in most of the large population centers in the eastern states, and a specialized class of lumber merchants rose to prominence. Particularly important wholesale nodes in the new and emerging transport and distribution system were places such as Pittsburgh, St. Louis, Albany, Buffalo, Tonawanda, and Chicago, places never entirely without mills or processing plants, to be sure, but places principally concerned with the storage and distribution of the product along the new lines of communication that extended from the lake shores of Michigan, Wisconsin, and southern Canada to the northeastern seaboard and the interior of the continent in the Prairie States (Fig. 6.9). Other places like Philadelphia and New York were vast consumers. The character of lumbering was changing: Distribution was becoming as important as cutting, and they were both part of the same "pipeline" of production.

The magnitude and nature of the lumber deficiency are difficult to ascertain except by implication through either the rise of the new areas of production or the prices people in the prairies were prepared to pay for lumber. Certainly, by 1860 the amount of timber used in New England, on the northeast coast, and in the north-central and Prairie States had exceeded the amount of lumber cut in those regions, and the indications are that deficiencies were common on a local scale, particularly in New York after 1840 and in the western portions of the Prairie States after 1850. The deficiencies arose from a variety of causes. In the Northeast the rapid emergence of an urbanized and industrialized population was augmented by massive immigration. As some indication of this, New York and its suburbs had grown from about 63,000 people in 1800 to 700,000 in 1850, Boston from 38,000 to 212,000, and Philadelphia from 73,000 to 450,000 during the same period.

One measure of the deficiency was the amount of all forest products imported into New York State. From as early as 1835 to the end of that decade, the average amount per annum was 52,761 tons, by the early 1850s it had grown eightfold to 438,388 tons, and by the early 1870s it had reached an annual average of 1,020,104 tons.

To a certain extent the deficiency was artificial, as the universal prejudice in favor of eastern white pine, almost to the exclusion of all other timber except hardwoods for furniture, produced a shortage. These fixed habits and conservative and rigid customs

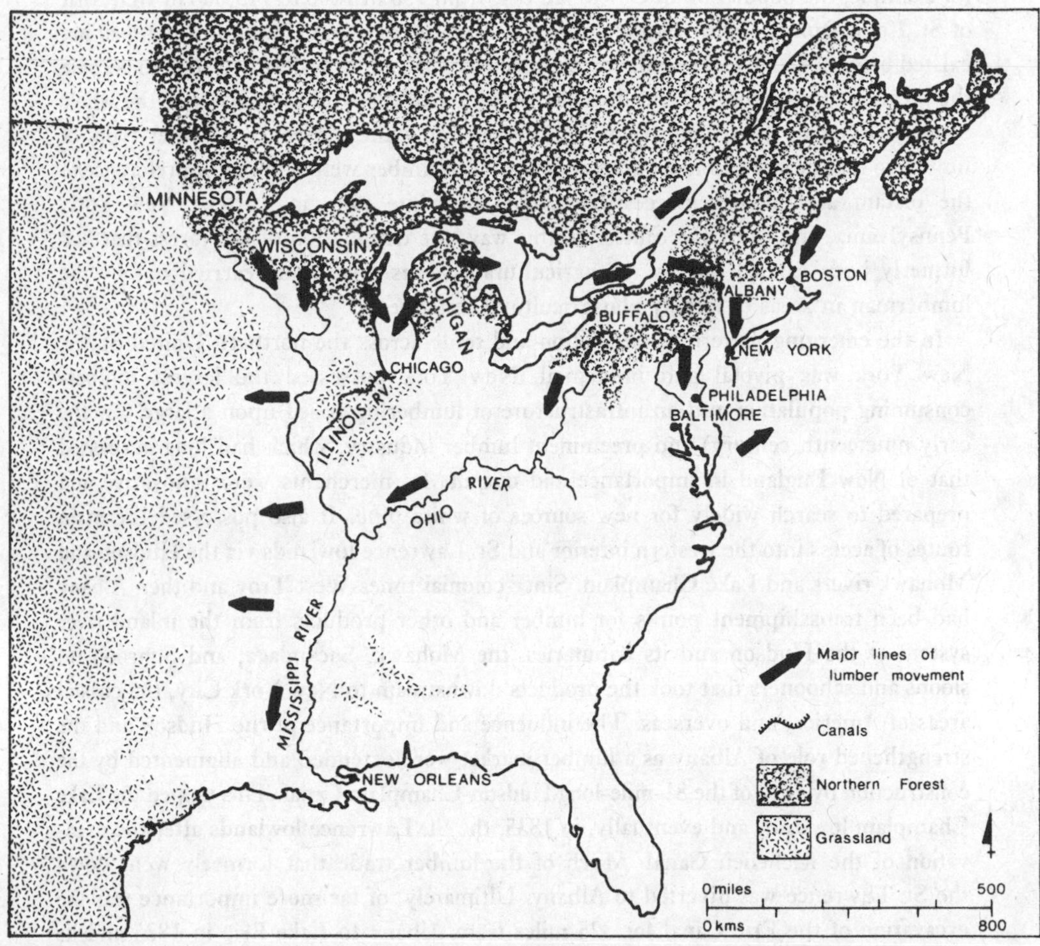

Figure 6.9. The continental log transport system.

were not broken down until about 1850 when the local stands of white pine were nearly cut out; then other species such as spruce and fir had to be accepted. Exploration for new stands of white pine was instrumental in fostering the rise of the lumbering industry in Michigan, Wisconsin, and, later, Minnesota. But the deficiency was real too, as increasing affluence, rising expectations concerning housing, furniture, and fittings, and the expansion of industry and the spread of railways all tended to generate a greater demand, which was reflected in the steadily rising per capita consumption of lumber (Fig. 6.4).

In the largely treeless portions of the western interior states settlement from about 1840 onward covered large areas rapidly because farmland could be created quickly without the backbreaking task of forest clearing. As settlement spread, the demand for timber for houses, barns, sheds, and fences was enormous. Urban settlement burgeoned;

for example, the population of Cincinnati rose from 750 in 1800 to 115,000 in 1851, that of St. Louis from 2,000 to 78,000 during the same period, and Chicago, which had not existed before 1836, had a population of 30,000 by 1850 and was destined soon to become the greatest lumber market and distribution center of the Midwest, if not of the whole country. Even in the plains the preference was for white pine over all other types of timber so that nearer and more accessible stands of timber were ignored in preference to the distant and almost inaccessible stands of white pine in western New York, Pennsylvania, and the Lake States. In this way, the denudation of the forest that had formerly been carried out by the agriculturalists was now being carried out by the lumberman in areas of largely nonagricultural settlement.

In the emerging pattern of circulation and trade across the northern United States, New York was pivotal and influential. New York possessed the largest lumber-consuming population and an infrastructure of lumbering based upon a large (for the early nineteenth century) and preeminent lumber industry, which had long surpassed that of New England in importance and output. Its merchants were innovative and prepared to search widely for new sources of white pine. It also possessed the main routes of access into the western interior and St. Lawrence lowlands via the Hudson and Mohawk rivers and Lake Champlain. Since colonial times West Troy and then Albany had been transshipment points for lumber and other products from the inland river system of the Hudson and its tributaries, the Mohawk, Sacandaga, and Schroon, to sloops and schooners that took the products downstream to New York City, the coastal areas of America, and overseas. The influence and importance of the Hudson and the strengthened role of Albany as a lumber market were extended and augmented by the construction by 1822 of the 81-mile-long Hudson-Champlain Canal. This tapped the Lake Champlain lowlands and eventually, in 1835, the St. Lawrence lowlands after the excavation of the Richelieu Canal. Much of the lumber trade that formerly went down the St. Lawrence was diverted to Albany. Ultimately, of far more importance was the excavation of the Erie Canal for 325 miles from Albany to Lake Erie in 1825 and its branch, the Oswego Canal, to Lake Ontario in 1828. These canals provided the main means of transport for bulky commodities, of which lumber was supreme, and the canals also stimulated lumber production in areas previously inaccessible to navigable rivers.

Initially, white pine cut in the northern part of the state was the basis of the lumber industry, and from 1770 logs were sawed at Glens Falls and the lumber then rafted to Albany for transport down the Hudson either to New York or to be bought by mills along the way. But during the 1850s supplies were not keeping up with demand, and the local lumbermen began to cut and buy up white pine from Allegheny and Chemung counties in the southwestern corner of the state and transport it to Albany by riverboat. In quick succession they moved to new sources of supply as the old ones were depleted. Next it was southern Ontario; by 1856 dealers even penetrated as far west as Port Huron and Saginaw in Michigan to acquire stands of white pine to meet the demands of an ever-increasing trade with the intention of bringing the timber along the Erie Canal. Already they were bringing hardwoods along the canal from Ohio ports like Toledo and Sandusky, and as early as 1847 dealers from Buffalo at the west of the canal had made trial shipments of white pine from Michigan. All these tentative moves were instrumental in paving the way for the vast expansion of lumbering in Michigan and Wisconsin, and later in Minnesota.[51]

These changing sources of supply were commented upon by a writer in a trade journal years later:

> To have spoken then [when white pine came from southwestern New York] of Michigan as a future source of supply would have subjected one to the audible smile of his contemporaries, and he might, with apparent propriety, have been told to go West. But, like the "march of civilization" our march has been westward for lumber, and today the pride of the Albany lumber market is her Michigan pine. Some may boast of "Ottawa stocks," some may boast of "Canadian clear," but away beyond these stands Michigan pine, the most admired, the choicest of all, and what everyone now wants when he wants the best. The more refined and advanced a people become, the greater the perfection they require in everything. To this lumber is no exception, and we only follow a law of our nature when we at this day raise our voices for Michigan lumber.[52]

The changing sources of supply are reflected, to a certain extent, in the graph in Figure 6.10, which shows the total tonnage of forest products carried in all canals in New York from 1837 to 1906 and the amount reaching the Hudson via the Erie and Champlain canals. There was a rapid growth of traffic during the 1840s that came from the lumbering in the southwestern portion of the state and the St. Lawrence lowlands. The peaks of the late 1860s and 1870s were made up largely of production from the Lake States, and the third peak during the 1880s was related to an increase in the demand for pulpwood. The decline after about 1890 was a result of railroad competition.

The wholesale centers

Although the continental system of transport pivoted around New York State, the three wholesale centers of Albany, Buffalo/Tonawanda, and Chicago, which extended for 800 miles across the northern part of the country, were crucial nodes in the organization of the movement of lumber from forest to farm and town. Each controlled vital waterways, which were the basis of the continental system.

Albany and the Hudson. Of all the nodes in the newly emerging system of lumber movement, the earliest and initially most important was Albany. It was situated at the junction of the Erie Canal and the major river system. The specially built timber terminus of the Erie Canal stretched about one-and-a-half miles alongside the Hudson, and on the thin strip of land between over 30 slips were constructed for the off-loading of the timber from the barges to the riverboats (Fig. 6.11). It was described as "an unrivalled facility for receiving and dispatching lumber." A peculiarity of the Albany yards was the fact that the task of unloading, sorting, and stocking was done by the barge hands for very little cost, a feature that kept Albany competitive long after the railways had undermined the supremacy of other retail lumber centers, so that even in the 1880s Albany lumber dealers were running lumber through their yards at 80¢/m.b.f. compared with $2.25 and $2.50/m.b.f. in Chicago.

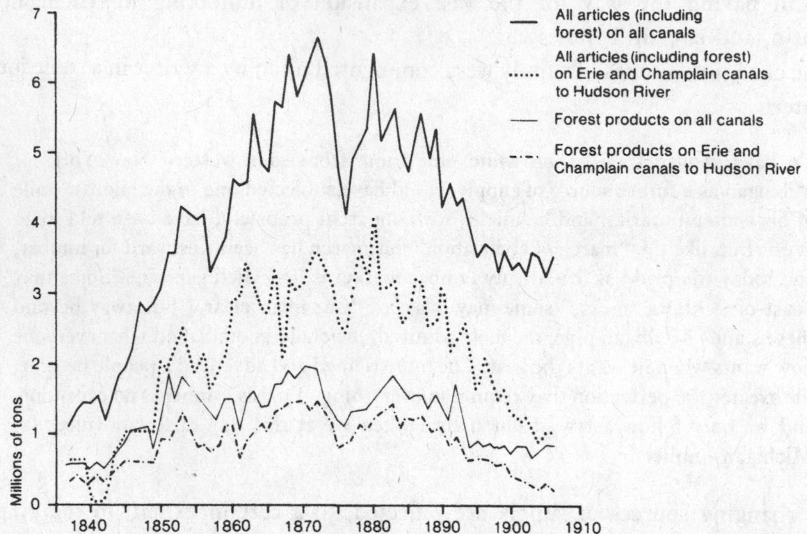

Figure 6.10. Tons of goods moved on New York canals, 1837–1906. (J. E. Defebaugh, *History of the Lumber Industry of America*, 2:419–20.)

The whole of the lumber wholesaling and retailing business became increasingly complex throughout the midcentury years, and this complexity was a measure of both the growth of the business of transferring lumber surpluses to areas of deficiency and the increasing specialization that characterized business during the transition from an eotechnic to a paleotechnic age. By the 1870s there were about 60 firms doing business in Albany (in addition to another 30 dealers in nearby West Troy), many vertically integrated, owning their mills, their boats for lake transport, and their barges. Some bought and sold on commission or on joint accounts, and many united into various forms of dealing according to the most advantageous combinations. There had been many changes in the scale and complexity of the organization of the lumber business since the early 1830s when "the direct patron . . . was the captain of a sloop or schooner who purchased his cargo upon thirty days credit and peddled it out in quantities to suit customers along the river, in New York City or wherever he could find a market. Now the boat captain merely acts as a paid agent for directing the transport of merchandise from buyer to seller."[53] Whereas the captain and his crew used to spend a week in loading 70,000 or 80,000 b.f., by the 1870s dock jobbers loaded large vessels with between 600,000 and 700,000 b.f. in two or three days. Rents to the Van Rensselaer family, who owned the strip of land between the river and the canal and leased it to companies to build warehouses, offices, and yards, rocketed. Whereas they totaled about $7,000 in the early 1830s, they were $80,000 in the early 1870s. Comparing the changes between the two dates, one local observer wrote:

> Then the gross sales yearly were about $1,500,000. Now one single house in the district has sales to that amount. Then, the rule was small stocks and a full

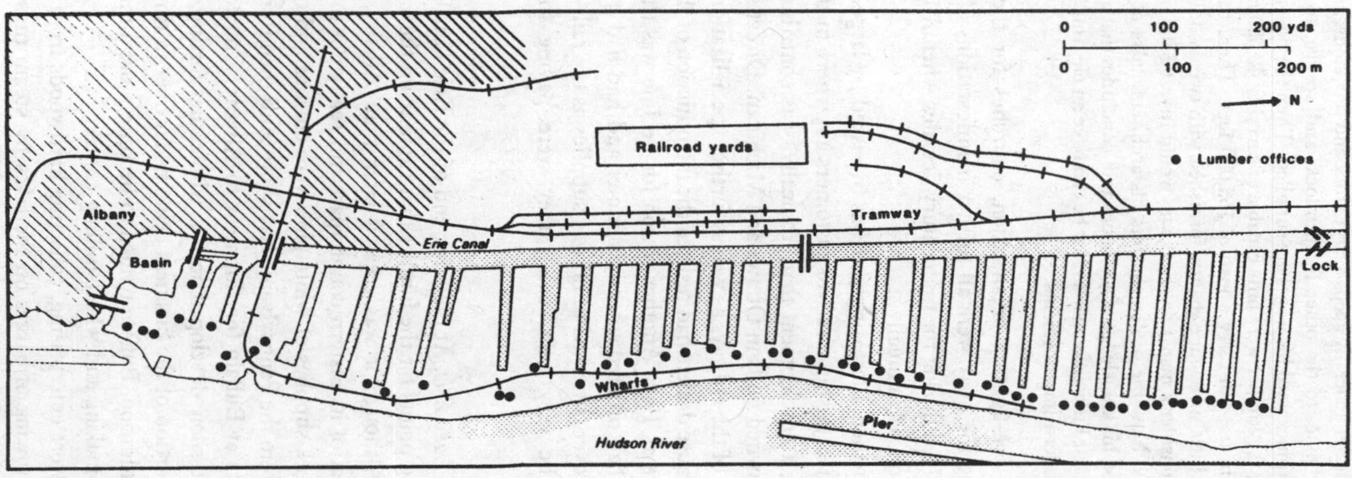

Figure 6.11. The Albany–Erie Canal waterfront, 1888. (Defebaugh, *History of the Lumber Industry*, 2:410–11.)

> assortment; now the dealer generally keeps a large stock and few kinds; and the buyer goes to one yard for pine, to the other for hemlock, and spruce, to another for hardwood and so continues until his wants are supplied. Then the boats used on the canal for transporting the lumber were only capable of carrying 40,000 feet. Some of the canal boats now in use can carry a load of 165,000 feet. Then, the dealer felt supremely happy in a little six by nine shanty, furnished with twenty dollars worth of fixtures, and would consider a man a prodigal who would invest $500 in a structure for business purposes. Now the dealer consults his architect, talks of Gothic and Corinthian, levies upon his knowledge of aesthetics and, concludes that a thousand or two either way makes but little difference so that he can have an elegant, commodious office, with all the modern improvements.[54]

A Board of Trade, a fire-fighting organization, churches for the workers, and a connecting tramway along the quay were all further manifestations of the new scale of organization that had evolved during the midcentury decades when Albany dealers were handling about 700 million b.f. annually.

But Albany grew no further. Increasingly, it was bypassed by large mills in the Lake States, which were prepared to ship direct to customers anywhere in the country in any quantity graded to order, a development that, ironically, was contributed to by Albany dealers themselves who owned mills in Ottawa and Michigan. Direct shipment was the outcome of the extension of the railways as well as further specialization in the trade, but it hit Albany severely because dealers had resisted the introduction of the railroad, which did not come until as late as 1906. Another reason for decline was that the New York jobbers, who previously had ordered stock in summer and had had it shipped to them before the Hudson froze over, switched to railway supplies as the railways were open all year. This eliminated the very high rentals they were paying for storage in the metropolitan area.[55]

Buffalo, Tonawanda, and Lake Erie. At the other end of the Erie Canal were Buffalo and Tonawanda, the receiving points for the lumber that came from the various ports along the lakes to the west. Although it was created later than Albany, Buffalo's rise in importance paralleled that of its eastern counterpart and really got underway by about 1850 when local New York supplies of white pine were cut out and cargoes, first from southern Ontario, then from the Pennsylvania lake shore, and then from Michigan, were transferred to canal boats at Buffalo for shipment to Albany. Buffalo was primarily a transshipment point but soon developed into a wholesale, sorting, and distribution center, the forwarding portion of the business being taken over in time by the growing port of Tonawanda. In addition, Buffalo tended to become more a manufacturing city, being a major furniture-making and wheat-milling center.

Buffalo's lumbermen were enterprising, collecting hardwoods from the southern Erie shore, inaugurating the towing of barges on the lakes by steam tugs in 1861, and the rafting of logs on the lakes. Initially, pine timber came from southern Ontario and walnut, oak, and other hardwoods from ports like Cleveland, Sandusky, Toledo, and Monroe, which forwarded timber from northern Ohio and Indiana so that by the early 1850s the lumber landed at Buffalo was approaching 100 million b.f. annually. Certainly, by 1851 about 38 percent of Buffalo's imports of white pine came from Detroit and ports to the west, some large boats coming from Grand Haven on Lake Michigan. However, for most of the 1850s ports like Erie, Conneaut, Ashtabula, and Fairview still siphoned timber from western Pennsylvania and northeastern Ohio to Buffalo.[56]

Initially, small sailing craft carrying 50,000–60,000 b.f. of lumber were the basis of trade on the lakes, but the high cost of handling induced some merchants to experiment in sending sawed timber in enormous rafts containing up to 3 million b.f. However, these were not successful; the lumber was damaged and much of it lost. Therefore, log rafts were used instead, and this meant the development of sawmills in Buffalo. But by the early 1860s even rafts gave way gradually, though not entirely, to schooners carrying up to 100,000 b.f. of sawed timber, which caused an ever greater reduction in freight rates to less than $1.25/m.b.f., by which time timber receipts at Buffalo were over 400 million b.f. per annum and those at Tonawanda had barely begun and were a mere 700,000 b.f.

Buffalo reigned supreme as the forwarding point of lake lumber so long as the trade was carried in sailing ships, and as a milling center so long as logs were rafted. It was the easternmost port of Lake Erie that the sailing ships could negotiate before the current increased in strength as the lake narrowed to become the Niagara River and plunged over the falls. But with the introduction of steam-propellered craft for carrying lumber and towing barges – introduced and promoted, ironically enough, by Buffalo traders – a new epoch opened up on the lakes. The steam-powered boats could travel against the current, and a location farther east at Tonawanda leaped into prominence as the major forwarding point for lumber to canal and railway. By 1873 the changeover in the mode of transport was so complete that all the schooners had been converted to barges to be towed across the lakes.

The growth of Tonawanda from "a swamp environed struggling hamlet" to the second-largest white pine market in the United States after Chicago was phenomenal. The Tonawanda Creek formed a quiet backwater on the Niagara River, and an island opposite its opening was so located that the main flow of the river went past it on the western side. The abundance of level, undeveloped cheap ground for stacking, the convergence of several railroads at this point as they left Buffalo for the West, and the proximity of the Erie Canal to the Tonawanda Creek made Tonawanda the cheapest location for the transshipment of bulky lumber from lake steamers to canal barges by a handling system similar to that of Albany.[57]

Figure 6.12 shows the rise of traffic at Tonawanda, where receipts of lumber surpassed those of Buffalo by the mid-1870s; from then on Tonawanda handled about half or one-third more volume of traffic than its neighboring parent and was second only to Chicago. Gradually, and then rapidly, from the mid-1880s onward the amount of lumber forwarded by canal fell as shipments in car-loaded lots direct to eastern retail yards increased. Tonawanda reigned supreme as the white pine trading center as long as the conservatism of the customers in the eastern states was maintained.

Chicago and the Mississippi. In the west another system of transport for distribution between areas of production and areas of consumption was emerging during the 1830s. It had two parts: that which focused the lake trade into Chicago and that which focused trade into the Mississippi River and its tributaries. They were never entirely separate from each other and were often in competition.

The settlement of the treeless areas of Iowa and Illinois and the extension of railways

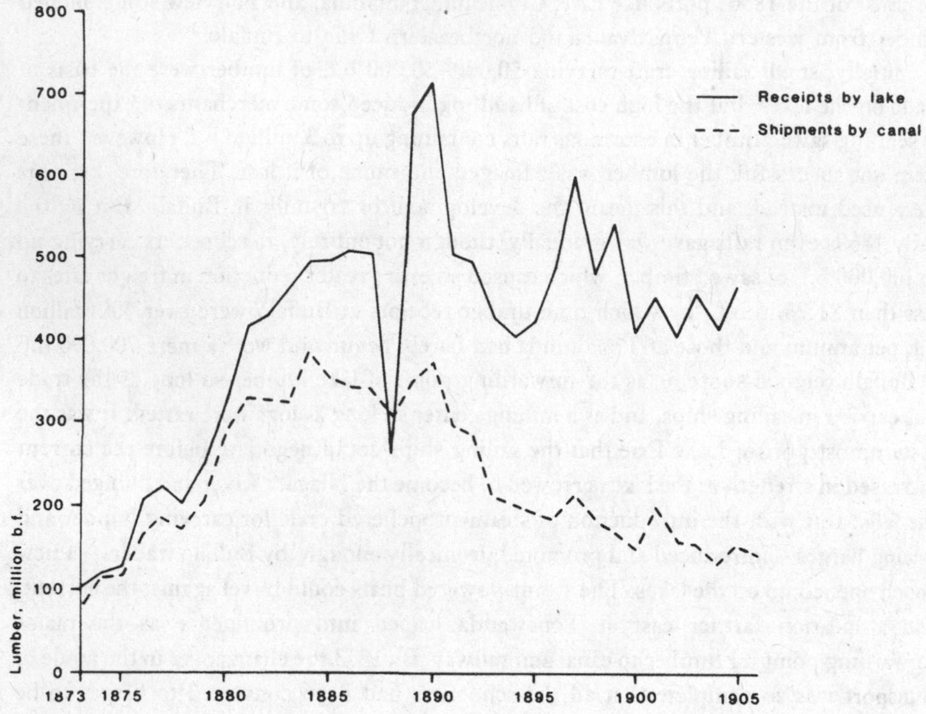

Figure 6.12. The Tonawandas: lumber receipts and shipments, 1873–1906. (Defebaugh, *History of the Lumber Industry*, 2:461.)

to the West provided a market for the lumbermen of the three Lake States, who competed vigorously among themselves but suffered little competition from any other lumber-producing area. In 1832 the first trial shipment of lumber was sent from Muskegon to Chicago, and the experiment was so successful that by 1856 Chicago had grown to surpass Albany as the primary wholesaling lumber center in the country; it contained about 150 firms employing over 10,000 people. Chicago's importance arose because it was the focal point of the trade emanating from rivers like the Menominee, Peshtigo, and Wolf in eastern Wisconsin and the Manistee, Muskegon, Grand, and Kalamazoo in western Michigan, all of which debouched into Lake Michigan and ran back a long way inland through the lumber-producing areas. Some lumber, but not a great deal, came from the Fond du Lac district in eastern Wisconsin via Rock River and was then unloaded at Rockford and overlanded to Chicago and elsewhere in southern Wisconsin and northern Illinois.[58] Some supplies were brought by log raft on the lake, which, of course, meant milling at Chicago, but by far the major part of the supplies were of cut lumber and came by sailing vessels. These were replaced by steam vessels, which were adopted readily during the 1860s because the long spells of prevailing southerly winds on Lake Michigan disrupted supplies and caused periodic lumber deficiencies at the yards in Chicago.

In 1847, when statistics of receipts became available, Chicago received 32 million b.f.,

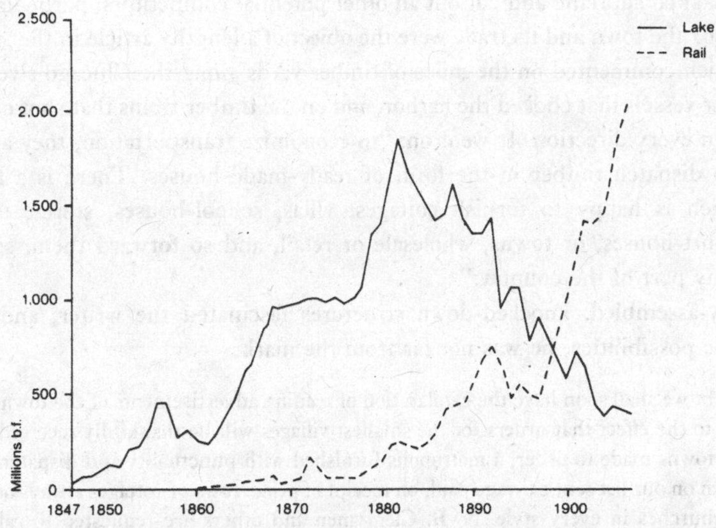

Figure 6.13. Chicago: receipts and shipments of lumber by lake and rail, 1847–1907. (F. E. Hough, *Report upon Forestry*, 2:49; and G. W. Hotchkiss, *History of the Lumber and Forest Industry of the Northwest*, 685.)

an amount that rose to 125 million b.f. by 1851 after the Illinois–Michigan Canal had been opened and topped 1 billion b.f. in 1869, a point from which receipts never dropped until the 1930s (Fig. 6.13). The trade in lumber accounted for about three-quarters of the cargo of all the vessels entering the port, and as the years progressed the ships came from farther and farther afield, from Lake Superior ports and even from the Michigan shore of Lake Huron. The bulk of the timber (between 60 and 85 percent) was shipped out of Chicago into canal barges on the Illinois–Michigan Canal and on the bullock wagons that trundled out across the prairies for up to 200 miles in all directions to supply the new farms and towns that were springing up everywhere. But, later, Chicago's own growth was so great that about half of the lumber was retained in the city for its own construction, especially after the disastrous fire of 1871.

Undoubtedly, it was the opening of the Illinois–Michigan Canal that cheapened transport into the prairies. When lumber was selling for between $9 and $14/m.b.f. in Chicago itself in the late 1830s and early 1840s, it was selling for between $60 and $70/m.b.f. in Peoria and even for $100/m.b.f. in less accessible spots away from the main trackways. With the opening of the canal in 1848, the Peoria price was halved overnight, and it fell by another 20 percent later on. Chicago merchants delivered lumber by canal boats up the Missouri as far west as Fort Leavenworth, Kansas. By 1858 inland transport had become so cheap that lumber cut and milled at Muskegon, Michigan, could be shipped by lake to Chicago, transferred to barge on the Illinois–Michigan Canal, reloaded on an Illinois River steamboat, and then resold to a St. Louis wholesaler at a price cheaper than that asked for Wisconsin lumber rafted down the Mississippi, supposedly the cheapest transport of all.[59]

Chicago reigned supreme and cut out all other potential competitors, particularly St. Louis. By 1867 the town and its trade were the object of a lengthy article in the *Atlantic Monthly*, which commented on the miles of timber yards along the Chicago riverfront, on the lumber vessels that choked the harbor, and on the timber trains that shot out over the prairies in every direction. It went on: "to economize transportation, they are now beginning to dispatch timber in the form of ready-made houses. There is a firm in Chicago which is happy to furnish cottages, villas, school-houses, stores, taverns, churches, court-houses, or towns, wholesale or retail, and so forward them, securely packed, to any part of the country."

The ready-assembled, knocked-down structures fascinated the writer, and as he mused on the possibilities, he was not far from the mark.

> No doubt we shall soon have the exhilaration of reading advertisements of the town-makers to the effect that orders for the smallest villages will be thankfully received; county towns made to order; a metropolis furnished with punctuality and dispatch; any town on our list sent carriage paid, on receipt of price: rows of cottages always on hand; churches in every style. N. B. Clergymen and others are requested to call before purchasing elsewhere.[60]

Rafting down the Mississippi and its tributaries was the other arm of the newly emergent circulatory system of the Continental West. As early as 1805 trade had developed down the Mississippi to New Orleans in lumber that came from the white pine stands of Pennsylvania via the Allegheny and Ohio. The tributary streams to these major rivers seemed "almost covered for miles with floating rafts" after the spring thaw. The Pennsylvania white pine trade reached its peak between 1832 and 1840 and then plummeted drastically as stands were cut out to almost nothing by 1870.[61] Ultimately, the lumber of Wisconsin origin was much more important and completely supplanted the Pennsylvania white pine. Shingles and squared timber sent down the Black River for sale along the Mississippi made up a thriving business by 1840. The Chippewa River and its tributary, the Flambeau, were the main source of white pine, which was milled at La Crosse and sent downstream. By 1853 supplies of lumber from the East were no longer trundled overland to the West, and the Wisconsin commissioner of emigration could say that "the course of the lumber trade may now be considered as permanently changed. The pines of Wisconsin now control and will soon hold exclusive possession of the markets of the valleys of the Mississippi and its great western affluents."[62]

The log drive was well established as the first stage in the transport of logs downstream to mills at, for example, Eau Claire and La Crosse, but other logs were sent downstream to milling centers on the Mississippi. Rafting began in 1843 when a group of men demonstrated the feasibility of rafting logs from the Black River to Nauvoo in Illinois, and the advantages of the practice were readily apparent. In the same year a boom on the St. Croix broke, and many logs went downstream; the lumber company tried to gather as many as it could at Stillwater, forming them into four huge rafts of 500,000 b.f. each, which sold at St. Louis for a good price. By 1850 there were rafts on all streams.

The log rafts were constructed in many ways – by ropes, chains, and clamped by poles but as trips got longer these construction methods were replaced by the "brail," a system by which logs were enclosed in a hollow frame held together by chains, ropes, or pegs. A standard brail was 600 feet long and 45 feet wide, and six of these together on the Mississippi contained at least 1 million b.f. and covered three acres. The lumber rafts consisted of 12–20 layers of planks or boards, which were formed into "cribs." A crib was 16 feet long and 12–16 feet wide, six or more making a "rapids piece" or "string," which could be taken apart easily when dangerous rapids or dams had to be passed. On the quieter waters of the Mississippi, up to five or more strings were coupled together side by side to make a single raft. As time progressed, rafts increased in size. By 1860 they contained 40–50 cribs and 300,000–500,000 b.f. of lumber, by 1864 they contained 1 million b.f., and by 1870 they had increased to a phenomenal 2.5 million b.f. and covered three to four acres, their tops stacked high with shingles, laths, and pickets; some became as extensive as 11–12 acres.[63]

The raft crews, like their Pennsylvanian predecessors on the Delaware and Susquehanna were basically farmers. Therefore, they returned upstream on foot or, when the towboats came into use during the early 1860s, returned upstream on them, there being about 100 or so steam towboats that plied up and down the Mississippi to guide the rafts. The towboats brought an end to some of the more picturesque aspects of the life of the rafting crews. A lumber journal in 1874 complained:

> The romance has gone out of lumbering to a great degree. The raftsmen who were handed down to posterity as living in the utmost contentment on the raft, and passing life in an unbroken round of cardplaying, fiddling, and dancing while they floated lazily down the stream have taken their places at the titanic oar, and a noisy raft-boat, powerful enough to haul an island out of the way, pushes the monstrous raft to its destination.[64]

Rafting from Wisconsin to ports along the Mississippi was cheaper than sending the lumber by land, but even so the expenses of the journey could consume about 50 percent of the receipts of sale; in 1857, when lumber was selling generally for about $20/m.b.f., the cost of running rafts from the Upper Wisconsin River to the Mississippi ranged from only $5 to $8/m.b.f., a rate which the scarcities of the Civil War maintained and even heightened but which fell in the late 1860s. The costs involved in rafting included the wages of the crews and the demands for "wharfage" made by the settlements alongside the river; for example, one village in 1868 demanded wharfage at the rate of 16.22¢ per m.b.f., and this was not altogether uncommon. In addition, river improvement companies in Wisconsin demanded the right to recoup their costs, and when towboats came into vogue the operator also needed paying, which rarely cost less than $1,000 per journey. All these costs could be forecast and calculated in advance, but the biggest variable of all was the level of water in the main streams. In 1872 it was estimated that the cost of moving a raft from Chippewa Falls to Eau Claire at high water was 7.74¢/m.b.f. but 11¢ in low water, which the presence of obstructions could raise to 14.6¢ and 23.25¢, respectively.

As the whole circulatory system grew and formalized during the midcentury years, key distribution points emerged, such as Winona, La Crosse, Dubuque, Clinton, Moline, Muscatine, Oquawka, Canton, Quincy, and Hannibal, to mention but a few. Of all points, however, St. Louis was outstanding, mainly because of its central location in relation to the new areas being settled to the west and its situation at the junction of the Missouri and Mississippi. Indeed, most of the timber was off-loaded to settlements on the west bank of the river, the river towns of Iowa receiving and distributing more than half of the lumber produced in the Wisconsin and Minnesota tributaries feeding into the Mississippi during the late 1860s, a proportion that reached the absolute figure of 450 million b.f. by 1873. In time settlements in Nebraska, Kansas, and Missouri became focal points, and Lake States lumber infiltrated as far west as Colorado and Texas. In all, central and northern Wisconsin shipped out 90 percent of its lumber via the Mississippi, some even going down to New Orleans and then to Europe and the West Indies.

It was a measure of the increasing demand for lumber and of the increasing complexity of the industry as it entered its paleotechnic era that the organization of distribution altered and had repercussions on the location of milling. Just as the old system of the schooner owners on the Hudson selling their load en route to New York City broke up with the emergence of a specialized system of distribution points, greater commercial organization, information dissemination, and rising specialization of demand, so the same happened on the Mississippi. Traditionally, the lumber was sold from the rafts to local independent wholesalers at each distributing point on the way downstream, the wholesalers taking what they wanted. If some of the lumber of the raft remained on reaching St. Louis, agents there undertook to break up the raft and dispose of its contents in time. By the opening years of the Civil War, wholesale yards on the Mississippi were sending orders to the mills direct before the rivers were opened after the winter freeze, and soon they were sending buyers to Wisconsin to contract for loads several months ahead of sawing. Inevitably, the wholesalers rose to positions of great prominence because of the cash they accumulated, something the mill owners rarely had because of the way in which their capital was tied up in the processing of the raw material. Mill owners borrowed from wholesalers and got in debt, so they began to set up their own wholesale and distribution yards at the major distribution points because it was here that profits could be made. Said one owner to his partners in 1860: "The retail dealers in my opinion are the ones who has [*sic*] made their money cheapest and easiest as well as has made the best investments.... I must still think Pileing and dressing our best lumber before its sale is the most profitable course."

With this incentive, places like Chicago, Milwaukee, and particularly St. Louis became the selling outposts of the Lake States mills, many mills forming partnerships with local dealers. Soon they were sending down logs rather than lumber to be cut at newly erected mills, thereby getting vertical integration of the industry and causing an extension of the manufacturing process as the mills came nearer to the market.[65]

Wisconsin towns complained of the drain of capital out of the region. Intense competition between Wisconsin firms led to price cutting and extended credit to buyers alongside the Mississippi, particularly during poor farming seasons when the supply of cash was short. When railways began to reach the Mississippi in 1861, Chicago dealers tried and partially succeeded in undercutting the river dealers, particularly in the period of high prices during the Civil War,[66] a two-cornered contest that was soon to have a third contestant when southern longleaf pine began to compete favorably with Lake States pine. But that is another story. By 1860 the twin circulatory systems of the

Midwest, those of Chicago and the Mississippi, were integrated enough to be in competition. Common markets and common sources of lumber meant that the 1860s were to be preeminently the era of the Lake States lumber industry, and it was to remain dominant for the next 20 years.

PART III

Regional and national impacts, 1860–1920

7

The lumberman's assault on the forests of the Lake States, 1860–1890

> Upon the rivers which are tributary to the Mississippi, and also those which empty themselves into Lake Michigan, there are interminable forests of pines, sufficient to supply all the wants of the citizens . . . for all time to come.
>
> Ben Eastman, *Congressional Globe*, 32nd Cong., 1st sess., 1851–2, App., 25: 851

Up to about 1860, lumbering dominated all other industrial and commercial uses of the forest, with the exception of wood cut for domestic fuel and the timber lost through agricultural clearing. Even by 1880, the relative weighting had not changed much (Table 7.1). What did change was the scale and thoroughness of lumber getting. A more concentrated and purposeful exploitation of the forest resource under new industrial techniques and organization became evident. Throughout the continent the total amount of lumber cut commercially increased phenomenally by nearly sixfold, from 8 billion b.f. in 1860 to over 45 billion b.f. in the peak year of production, 1906 (see Fig. 6.4).

The continuous and high demand stimulated innovation, exploitation, and migration from one region to another across the continent. By 1860 supremacy of production had already passed from New England to New York, and it was flowing to the Lake States. Although New York, Pennsylvania, and Maine were still important producers, Michigan, Wisconsin, and Minnesota cut the greatest volume and value of lumber. Lumbering in the South was beginning to achieve some prominence by 1890, and it carried on well into the twentieth century. And the Pacific Northwest was beginning to rank high in national production by 1900 when the Lake States were in decline (Figs. 7.1–7.4). The pattern of the spread of lumbering, then, was one of a continuously expanding wave of exploitation, which, despite local pauses and advances, moved with generally gathering momentum westward across the continent, with an important projection that swept down through the South. The wave seemed to have an ever-increasing height and volume as it reached each new region of exploitation in turn, first the Lake States, then the South, and then the Pacific Northwest (Fig. 7.5).[1]

However, although regions rose in prominence and supremacy of production passed on, the old regions were never completely eliminated, and they continued to thrive, albeit at a modified scale of production. For example, although Michigan's production rose between 1869 and 1889 to be roughly five times that of New York, New York's production remained relatively stable, and Albany continued as a major retail center, supported, it must be admitted, by substantial supplies from the Lake States.

Table 7.1. *Estimated value of forest products, 1880*

	$ million	%
Wood for domestic fuel	c.307	43.9
Sawed logs	140	20.0
Unsawed timber (poles, spars, etc.)	c.110	15.7
Wood for fencing	c.100	14.3
Fuel, railroads	5	0.7
Fuel, steamships	2	0.3
Fuel, manufacturing	8	1.1
Charcoal	5	0.7
Naval stores	6	0.9
Railroad ties	10	1.4
All other	7	1.0
Total	700	100.0

Source: Charles S. Sargent, *Report on the Forests of North America*, 9:485.

Supremacy of production may have moved elsewhere, but New York's position merely dwindled relatively on the national scale.

Each of the new regions of activity was different and distinctive, not only in the location and timing of its period of maximum exploitation but also in the technological phases and innovations that characterized that exploitation and their adaptation to the conditions of the local physical environment. But the years from about 1850 onward also saw the gradual decline of traditional methods of lumber cutting and getting and the emergence of a host of features on a national scale. The lumber and the forest products industries, in general, like industry everywhere in the United States, were entering a phase of vigorous expansion and what is often known as industrial capitalism. The change was partially a response to uncontrolled and uncontrollable overproduction and cut-throat competition in an unstable market, largely engendered by the development of new forms of transport, felling, mill technology, and new materials. Steam power meant the concentration of industrial activity and the beginnings of corporations and monopolies. Steel meant better and more efficient tools. The railroad meant reliable, faster, and more flexible transport. All meant increasing specialization of activity, concern for efficiency, greater competition, tight contractual agreements based on the time element, and the mass production of a standardized manufactured end product.

In the forest the systematic cutting of large areas replaced the cutting of individual trees. The large-scale ownership of standing timber was to become as important and fundamental an innovation as the technical advances in sawing and transport, and, similarly, the corporation was to become the means by which the abundance of production could be organized. The Bureau of Corporations report in 1911 recognized that the increasing centralization of control in the lumber industry was no longer a

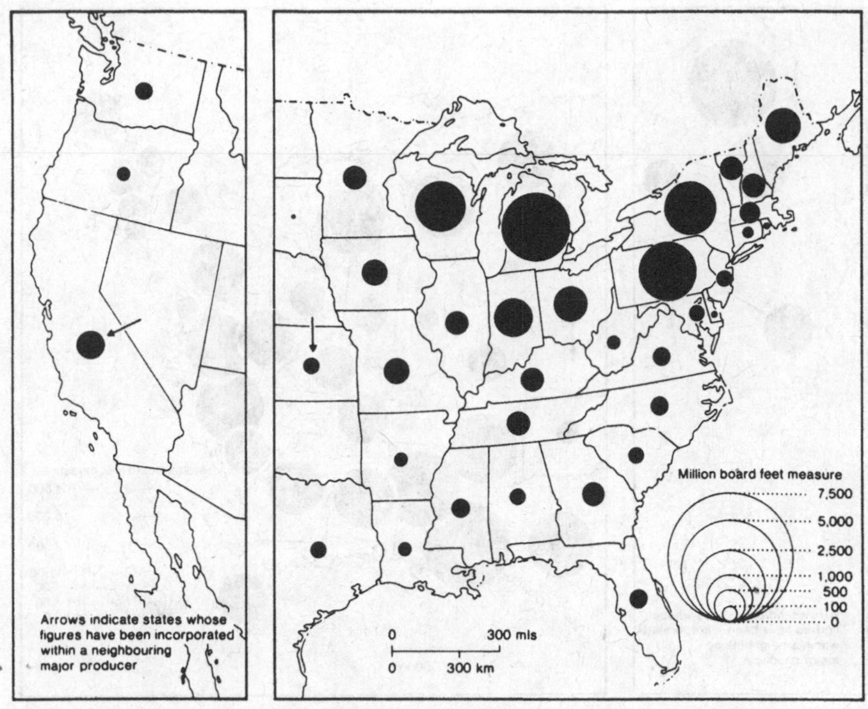

Figure 7.1. Lumber production, 1869, in m.b.f. by state. (U.S. Census.)

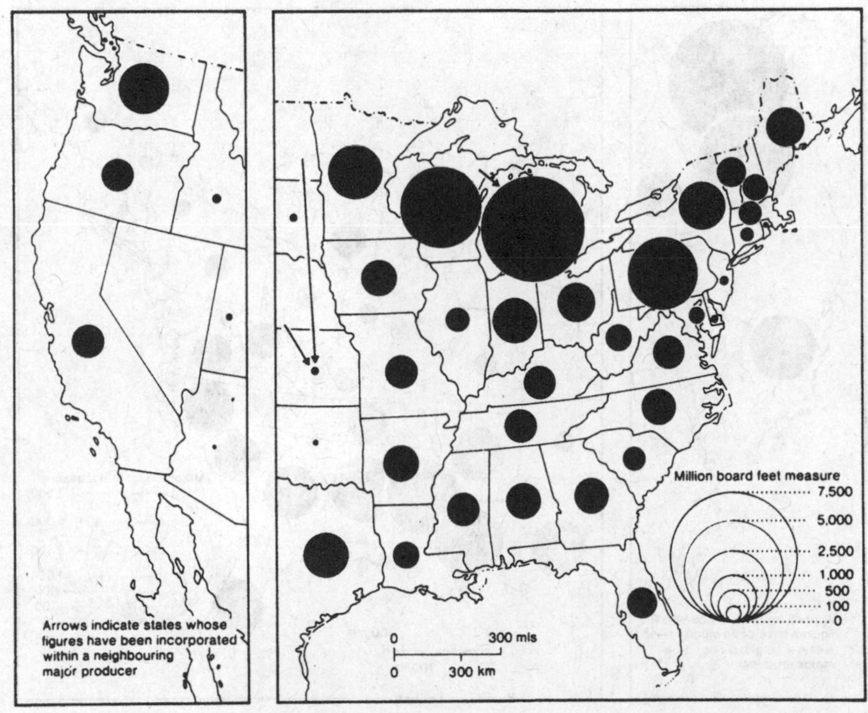

Figure 7.2. Lumber production, 1889, in m.b.f. by state. (U.S. Census.)

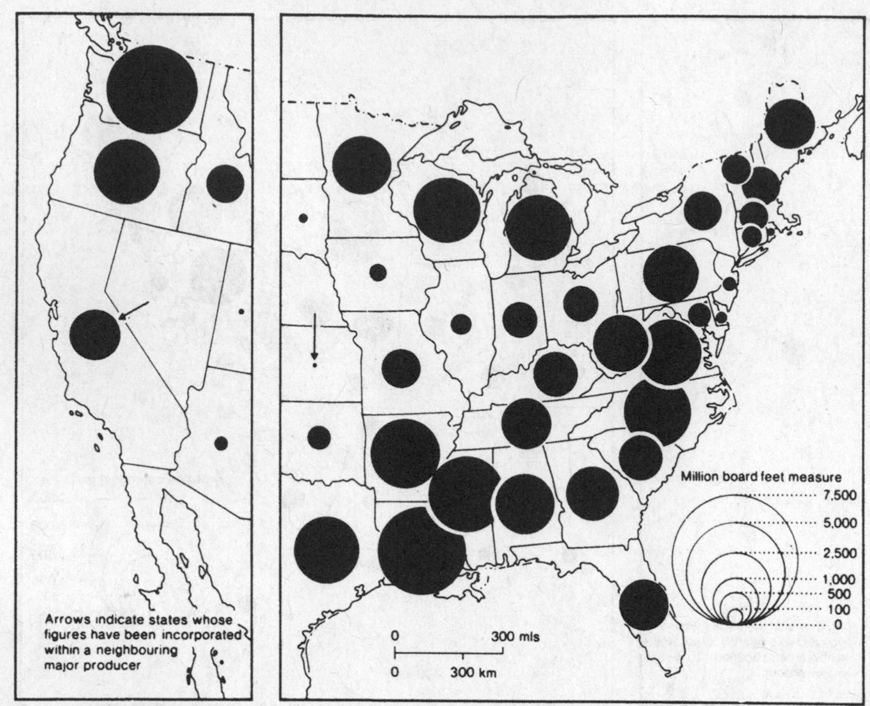

Figure 7.3. Lumber production, 1909, in m.b.f. by state. (U.S. Census.)

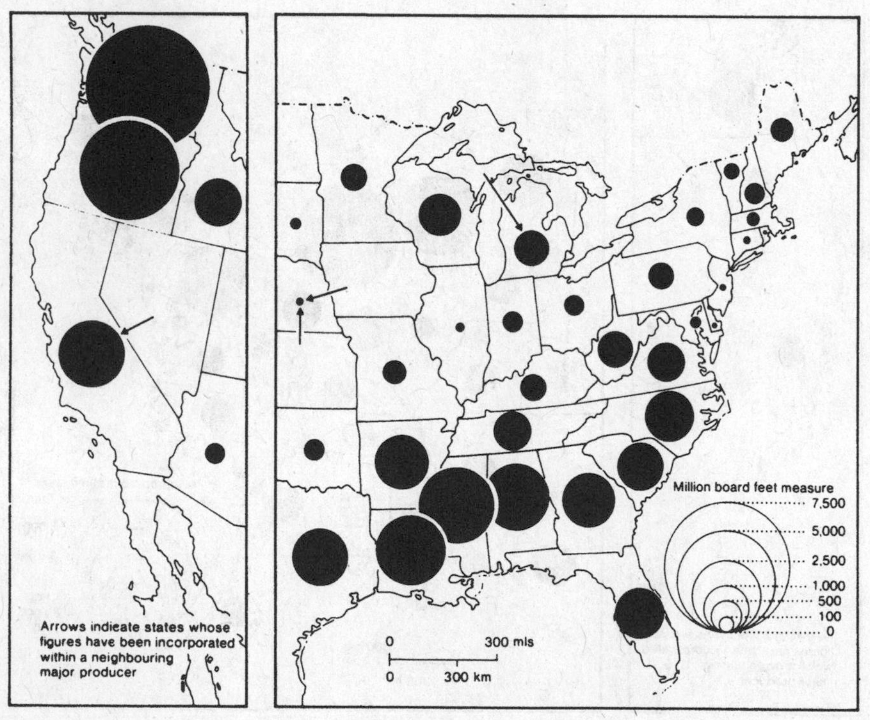

Figure 7.4. Lumber production, 1929, in m.b.f. by state. (U.S. Census.)

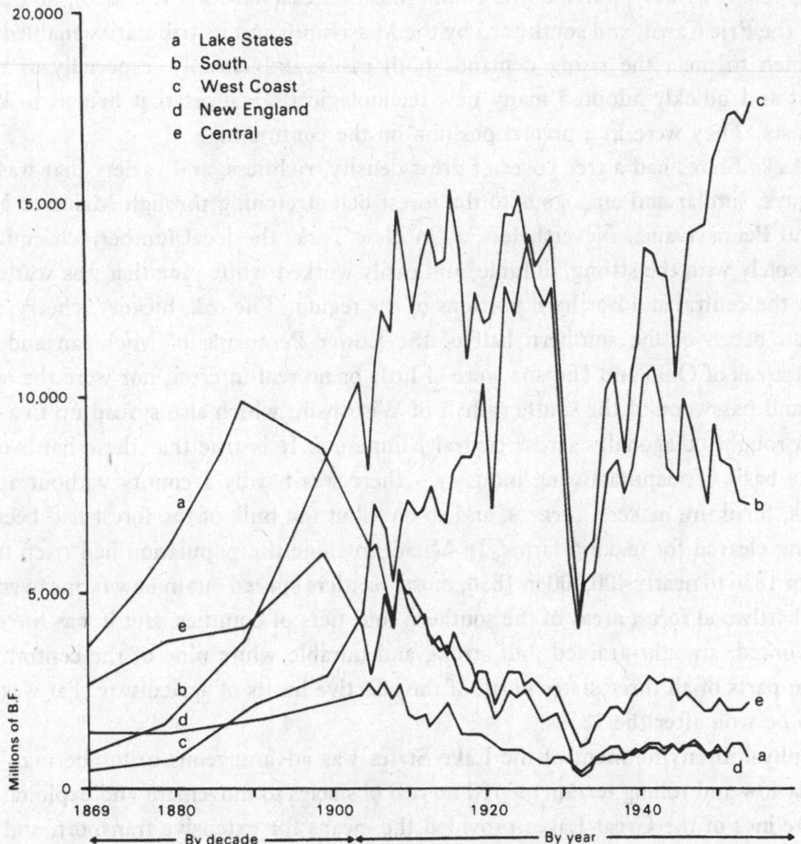

Figure 7.5. Production of lumber by major regions of the United States, 1869–1959. (U.S. Bureau of the Census, *Historical Statistics*, tables L113–21.)

matter of technical advance or "any economy of large manufacturing plants" but purely a matter of business organization. Monopoly and cooperation and devices such as trade associations minimized the worst effects of fluctuating production, and increasingly lumbermen sought institutional adjustments to lessen economic stresses. It was all evidence of what Bernard Weisberger has called the "organization of abundance" and the "new industrial society," and in the forests of the United States it first became obvious in the Great Lakes region.[2]

The Great Lakes setting

The assault of the lumbermen on the forests of the Lake States had already started in the 1840s as a response to the depletion of supplies in New York and the ingenuity of the Albany and Buffalo dealers in seeking out new sources of timber. In addition, the westward movement of the immigrant farmers into the relatively treeless Prairie States of the Midwest stimulated its own demand. Although the Lake States forests were located roughly in the middle of the continent and at great distances from the new markets, relatively easy water-borne communication eastward via the lakes, and after 1825 by the Erie Canal, and southward by the Mississippi and its tributaries enabled the lumbermen to meet the rising demands both easily and cheaply, especially as they invented and quickly adopted many new technological advances that helped to keep down costs. They were in a pivotal position on the continent.

The Lake States had a tree cover of great density, richness, and variety that was, in many ways, similar and analogous to the forest belt stretching through Maine to New York and Pennsylvania. Nevertheless, as in New York, the local lumbermen equated lumber solely with the strong, durable, and easily worked white pine that was scattered through the central and northern portions of the region. The oak, hickory, cherry, and American beech of the southern half of the Lower Peninsula of Michigan and the adjacent areas of Ohio and Indiana were of little or no real interest, nor were the oaks, maple, and basswood of the southern half of Wisconsin, which also spread up to a line that ran roughly diagonally across central Minnesota. It is true that these hardwoods were the basis of manufacturing industry – there was hardly a county without a few sawmills, furniture makers, turners, and so on – but the bulk of the forest had been or was being cleared for making farms. In Michigan alone the population had risen from 87,000 in 1836 to nearly 400,000 in 1850, mostly settlers spread out in new farms over the former hardwood forest areas of the southern four tiers of counties. But it was the soft, easily worked, straight-grained, but strong and durable white pine of the central and northern parts of all three states north of the effective limits of agriculture that was the prize to be won after the 1850s.

The physical environment of the Lake States was advantageous to lumbering. The relatively low and rolling terrain offered no real obstacles to movement and exploitation, the shorelines of the Great Lakes provided the means for extensive transport, and the multitude of substantial swift-flowing rivers that penetrated the very heart of each state could be used as log-driving streams and sources of power (Fig. 7.6).[3]

How much timber existed in pre–European days in the three Lake States is difficult to estimate because no accurate measurements were ever made of its extent or quality; timber cruising as an art or science was in its infancy. Putting aside some early vague descriptive estimates of stands of white pine in various localities, we find that the first comprehensive and detailed picture of remaining woodland was that given in 1880 by Henry C. Putnam of Eau Claire and George W. Hotchkiss of the Chicago Lumber Exchange, a figure then published by Charles Sargent in his 1884 census report: 84 billion b.f. (Fig. 7.6). These were, Sargent said, "estimates, and not facts," and therefore they were open to question and were undoubtedly understated to the point of being outright dishonest. Putnam and many of his informants were not disinterested investigators, for in creating the idea of scarcity and rapidly depleting supplies, they could expect the price of stumpage and lumber to rise, as it did. Further estimates varied widely but always upward; for example, in 1903 Filibert Roth suggested that the remaining pine stands in Michigan were about 190 billion b.f., much more than was thought previously. There was much confusion and uncertainty as to the true extent of the remaining forest.[4]

Whatever reliance is placed upon these figures, the point is that the forest reserves were enormous. They were not concentrated in one locality, however, but were scattered

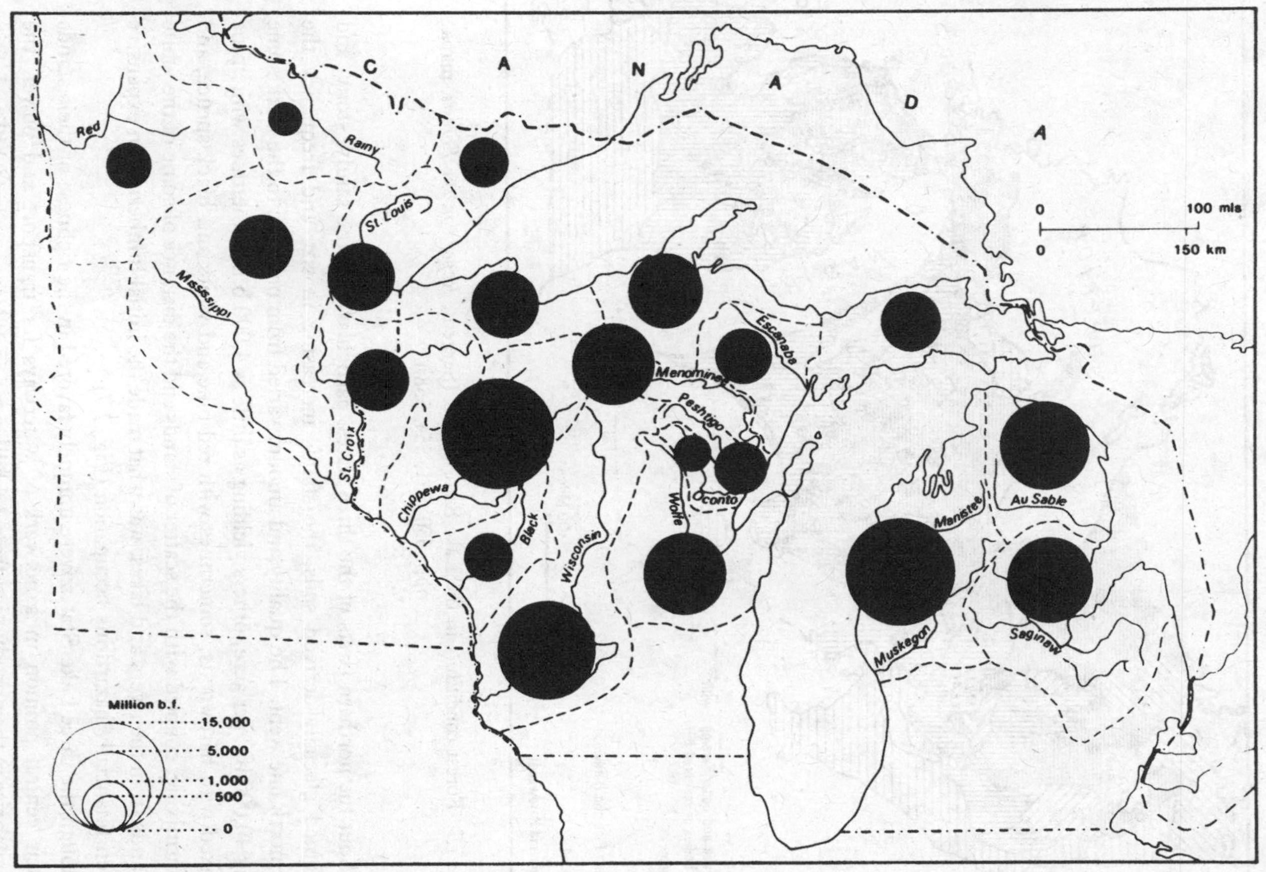

Figure 7.6. The rivers of the Lake States and estimates of standing white pine in major watersheds, 1880. (Sargent, *Report on the Forests*, 551, 554, 558.)

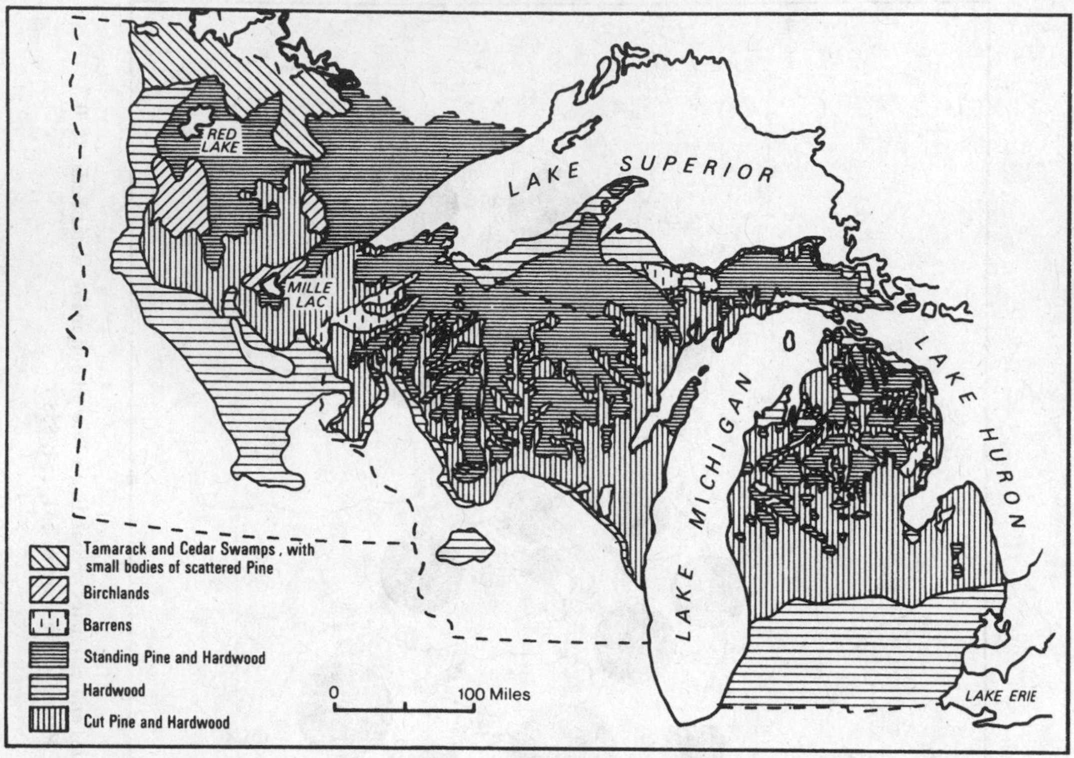

Figure 7.7. Forest conditions in the Lake States, 1881. (Sargent, *Report on the Forests*, maps opp. 550, 551, 554, 558.)

throughout the northern parts of the three states, particularly on the sandy, stony, and fine-grained glacially derived soils, the stands increasing in size and frequency the farther north one went. The quality and amount varied from one acre to the next, some yielding 40,000 b.f. per acre, others yielding as little as 4,000 b.f. Sometimes white pine was mixed with hardwoods, sometimes with red pine and white and black spruce, and this intermixing, coupled with the scatter of stands and the distance of some of the white pine stands from the rivers and lakes, was what made the initial lumbering an expensive and even financially hazardous occupation (Fig. 7.7).[5]

Although the three Lake States were uniquely favored in the richness of their stands and their central position in a network of waterways for transport and power, the spectacular rise in their production from 3.5 billion b.f. in 1869 to nearly 10 billion b.f. in 1889 can be understood only after a consideration of the technical and organizational changes that were being made during those years. The output of the saws and the mills increased enormously, the transport of the logs in the forests was speeded up and greater amounts moved, and the transport out of the region to the various markets was greatly enhanced, mainly through the expanding use of railroads. The organizational features of the whole business of lumbering also went through a radical transformation, which was both a cause and an effect of the increased production.

Technical changes

All the advances made in the application of steam to drive the mills, in improving the rivers for log driving, and in the splash dams that had either begun or were well underway in New York and Maine during the Civil War period were transplanted to the Lake States. This movement of ideas and techniques was not surprising because migrants from Maine, Vermont, and New York dominated the lumber business and became the key businessmen and entrepreneurs during the earliest years in Michigan and Wisconsin and only to a slightly lesser extent in Minnesota. In a study of 131 "leading lumbermen" in the three states, Frederick Kohlmeyer found that 48 came from New England (19 from Maine), 35 from New York, 14 from Pennsylvania, 5 from Ohio, 11 from other states, and 18 were migrants. It was little wonder that a congressman in Maine spoke plaintively about "the stalwart sons of Maine marching away by scores and hundreds to the piney woods of the Northwest." However, unlike the New England and New York producers, those in the Lake States had no inhibitions about the adoption of new methods and machinery. They were not faced with a static or declining production, which was detrimental to investment, nor were they encumbered to the same degree with the legacy of small, part-time operations, based on animal or water power. It is true that the small local sawmill typical of the colonial era existed, but it took a subsidiary role. From the start the system of the new producers was geared to a far greater volume of production; there appeared to be unlimited stands worth exploiting, and therefore it was worth investing and experimenting with new tools, modes of transport, and business organization. In this burgeoning age of technological and organizational innovation, only a few of the main changes that were to alter the whole concept and practice of lumbering can be pinpointed.[6]

Saws, mills, and steam

All those improvements already made to sawmills found a ready acceptance in the Lake States, as did many other innovations that speeded production, reduced waste, and made the use of labor more efficient. The water-powered muley and gang saws were being used extensively by the early 1830s in Michigan, but it was the application of steam to the saws that brought about the greatest increase in production. In 1854, 29 percent of Michigan's mills were steam-powered, the remainder being water-powered. By 1864 the proportion had risen to 49 percent, and by 1874 to 72 percent. By 1909, of the total horsepower generated for milling processes in the three Lake States, 90 percent was steam, nearly all the remainder was gas or electricity, and water constituted a negligible proportion.[7]

In the mill the circular saw, in particular, was favored. It allowed continuous, fast cutting. By 1876 nearly two-thirds of all saws in the Lake States (putting aside gang edgers) were circular saws. A serious disadvantage of the early models was the wide kerf that resulted from the saw plate's having to be thick enough to maintain its rigidity when in use and to counteract distortion by heat and centrifugal forces. For example, a circular saw, the bite of which averged five-sixteenths of an inch, could turn 312 feet into dust for every thousand feet of inch board sawn. If the saw could be reduced to one-twelfth of an inch, then only 83 feet would be lost in sawdust. Although certain improvements were made to circular saw blades that made them thinner, faster, and more dependable, the continued waste in sawdust and the limited thickness of the log they were capable of cutting, together with the increase in the price of lumber and dwindling supplies by the 1880s, meant that many mills relegated them to cutting rough timber only and began to adopt the band saw instead.[8]

The efficiency of the earliest band saws was impaired by the poor quality of American steel and the deficiencies of welding, a job that might have to be done from one to three times a day during normal operations. During the late 1860s, Jacob R. Hoffman of Fort Wayne, Indiana, introduced the band saw into the trade and began experiments with better steels, better welding, guiding wheels, and adjustable frames to maintain the correct tension. But Hoffman was more of a lumber manufacturer than a mechanic, and the thrust of experimentation moved to the Disston Saw Company. The company displayed a magnificent band saw at the Philadelphia Centennial Exhibition in 1876, and it caused considerable interest, not only among lumbermen but among the public because of its immense size. After 1880 factory-made saws appeared with iron frames and the means of adjusting the cutting edge, and large-scale manufacturers started production in Fort Wayne, Indianapolis, Cincinnati, Milwaukee, Menominee, Philadelphia, and Erie, so that by 1889 the band saw was beginning to replace the circular saw throughout the Lake States.[9]

Between about 1865 and 1875, a spate of new inventions and improvements in the mill itself helped to speed production and reduce labor requirements. Friction feeds, wire feeds, and direct steam-powered feeds increased the speed of the carriages that transported the logs past the saws. New labor-saving devices included the accurately adjusted setting works, head and side blocks, and a "steam nigger" for turning the logs for the saw. In all, the men hardly touched the logs. In 1873 a La Crosse lumberman patented the endless-chain method of bringing logs from the boom pond into the mills; another device was invented for washing sand from the logs on their way to the saw, as was a mechanical carrier for disposing of the finished lumber. In a mill on the Chippewa River the automatic carrying of the sawdust to the boilers of the steam engines was introduced in 1869. Double edgers and gang edgers expedited the finishing of the wood, and shingle-, stave-, lath-, and slab-making machines made it possible to utilize poor and awkwardly shaped material and thus reduce waste. With the advent of the railroad it became possible to ship millwork directly to the consuming areas, which encouraged the mill owners to install planing, flooring, matching, and molding machines, as well as artificial drying kilns so that the finished product could be taken as quickly as possible to the consumers.

The application of steam power to the mills was also important in centralizing operations and consolidating scattered interests into one major plant, thereby raising the level of production. The increase in the size of the mills can be gauged by a variety of measures. For example, the horsepower available to each mill increased; whereas in 1870 the mean horsepower available per mill in the three Lake States varied between 34 and 40 against a national average of 25 h.p., by 1909 it varied between 121 and 135 against a national average of 72. Similarly, the averge number of workers per mill rose from between 7 and 9 in 1860 to between 25 and 41 in 1890, dropping at the turn of the century with a decline in production. Again, the value of production per mill rose steadily but with a big jump between 1860 and 1870 as steam power and the new production methods were applied. In each case the comparison with the average for the United States, less the Lake States, was marked, workers per mill being consistently about three times the national average, production per mill being between three and four times the national average, the difference getting greater in the later decades of the century. Production per mill rose significantly; whereas the annual average of Michigan mills in 1850 was about 3 million b.f. with a range of production from 1.5 to 6 million b.f., some 30 years later annual outputs of between 10 and 20 million b.f. were common, some mills achieving an output of 50 to 60 million b.f. Mills capable of this scale of output were among the big businesses of the day.[10] In every way the application of steam power to milling had increased the size and production of the mills.

Transport: the modification of traditional means

River drives and booms. Of all the improvements of the productive process, transportation was the most important factor in the acquisition of profits. Even the poorest grades of lumber could be marketed if transport was cheap enough. It was little wonder that the multiplicity of waterways in the region were used to implement the well-tried methods of the log drive that had gradually evolved in New England and New York: the marking of logs, the building of splash dams and booms, and the use of rafts.

On the rivers, loggers had to secure the permission of the state legislature to construct dams and to levy tolls, and in 1851 the Michigan legislature gave county supervisors the right either to grant or to withhold permission, a right achieved in Minnesota in 1861 but not until 1917 in Wisconsin, by which time the railways had long superseded river transport. However, the many bills that went through the state legislatures leave a good record of what was done. There were, for example, 41 dams in the Menominee Valley, between 60 and 70 in the St. Croix Valley, 110 in the Chippewa Valley, and these were but a few of those constructed. Dams became more and more expensive to construct as they became larger, and as time went on more were needed because the best logging streams were cut out and less and less useful side creeks had to be utilized. Perhaps the largest dam (though not untypical in cost) was the Nevers Dam on the St. Croix River constructed in 1890, which was 614 feet long and 22 feet high and flooded the valley for 10 miles upstream. By the time the purchase of riparian rights and litigation costs were added to the cost of the construction, the Nevers Dam cost more than a quarter of a million dollars – and it was to last for one season only.[11] Everywhere snags and obstructions in the rivers were removed or dynamited and the system extended. For example, in the St. Croix River Valley alone, the miles of navigable river were increased from 338 in 1849 to 820 in 1869 – through careful clearing and the building of dams to maintain and control the flow of water.

Marking and cooperation became necessary because so many loggers were cutting in the same stream. In Michigan marks were regulated on the Muskegon River after 1859, and in all other counties a little afterward. Registration of marks was compulsory in Minnesota after 1854 and in Wisconsin after 1883. An indication of the scale and number of logging operations is the fact that by 1873 there were over 2,000 marks registered in the St. Croix Valley alone, and about 20,000 throughout Minnesota. However, the problem was even more complicated than the sheer numbers suggest, as loggers, landowners, sawmillers, and others might all wish to cut their marks in a single log.[12]

When lumbering really got underway during the 1860s, the mass of logs tipped into the streams soon clogged the lower reaches of the major rivers, sometimes stretching in unbroken jams for 3 to 10 miles. There was an obvious need for booms as a link and transition between the river drive and the sawmill operations, which included such jobs as sorting, scaling, and organized selling, and, if the logs were to be transported farther, as a link between the river drive and the raftsman. Methods of construction, boomage rates, the issuing of stocks and shares, the declaration of dividends, and all the accompaniments of booming operations that had evolved on the East Coast were replicated in the Lake States. The first boom was on the Black River in 1854, and the first really big boom on the St. Croix in 1856, initially at Stillwater and then at the head of Lake St. Croix. By 1870 every logging district had its booming and improvement company. Eleven large companies were founded in Michigan between 1864 and 1870 on each of the major rivers. Of these the Muskegon boom founded in 1864 was one of the largest, and it could eventually handle 300 million logs annually, and the Tittabawassee constructed in 1864, which controlled the Saginaw River system, was handling between 500 and 600 million b.f. annually from 1875 to 1883 (Fig. 7.8).[13]

The largest and probably most important booming company of all was in Wisconsin, at the mouth of the Chippewa River where it debouched into the Mississippi through a number of shallow channels, one of which was called Beef Slough (Fig. 7.9). The Chippewa and its tributaries flowed through some of the richest pine lands in the state. Most of the early mills were set up at Eau Claire and Chippewa Falls, but whereas these and other lumbering towns had developed alongside the drivable streams, after the Civil War other new milling communities were springing up along the Mississippi, and these were in sharp competition with the interior towns. Among the downstream towns were Winona and Wabasha in Minnesota; Prairie du Chien and La Crosse in Wisconsin; Dubuque, Clinton, Davenport, Muscatine, and Burlington in Iowa; Rock Island and Quincy in Illinois; and St. Louis and Hannibal in Missouri. Hostility between the two sets of interests developed, especially as the Wisconsin lumbermen had bought up much of the pine lands. Therefore, in order to ensure their supply of logs, various downstream firms combined in 1870 – largely under the leadership of Frederick Weyerhaeuser and his partner and brother-in-law, F. C. A. Denkmann of Rock Island, Illinois – to form the Mississippi River Logging Company to build a sheer boom at Beef Slough and turn the channel into an enormous reservoir to keep the downstream mills supplied. The boom acted as the linchpin in the whole of the new continental transportation system down the Mississippi, for without supplies for milling and making up rafts the downstream industry that served the rapidly populating prairie areas would have folded.

The Beef Slough boom was the epitome of the big business in lumbering, which had

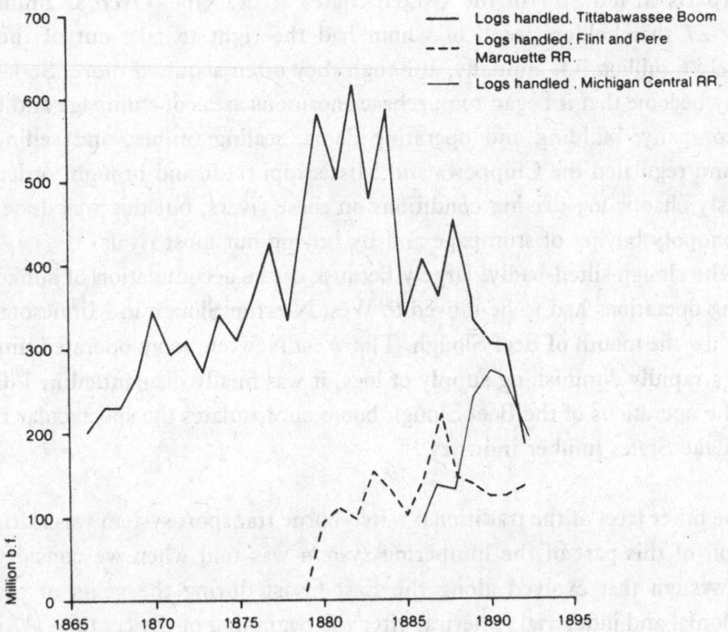

Figure 7.8. Logs moved, Tittabawassee River, Flint and Père Marquette Railroad, and Michigan Central Railroad, 1864–92. (Hough, *Report upon Forestry*, 1:516; and W. G. Rector, *Log Transportation in the Lake States Lumber Industry, 1840–1918*, 129.)

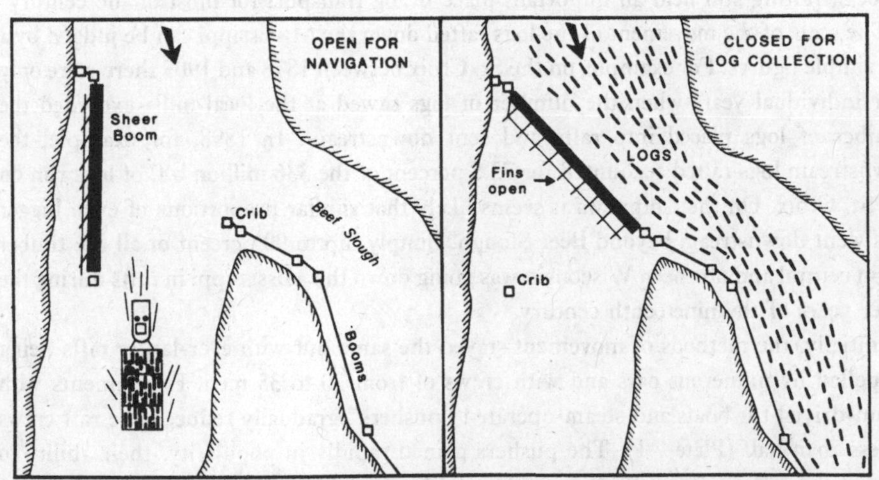

Figure 7.9. The Beef Slough boom, open and shut. (B. J. Kleven, "The Mississippi River Logging Company," 192–3.)

few counterparts in industry in the United States at the time. Over $1 million was invested by 27 shareholders, each of whom had the right to take out of the boom approximately 6 million b.f. annually, although they often acquired more. So large did the company become that it began to purchase enormous areas of stumpage and became a driving company, building and operating dams, scaling timber, and selling it. It controlled and regulated the Chippewa and Mississippi trade and brought order out of the previously chaotic log-driving conditions on these rivers, but this was done by the massive monopoly buying of stumpage and by buying out most rivals.

By 1880 the slough silted badly, largely because of the accumulation of sunken logs, and booming operations had to be moved to West Newton Slough in Minnesota, about six miles below the mouth of Beef Slough. The West Newton boom operated until 1904 but, due to a rapidly diminishing supply of logs, it was finally dismantled in 1906. The history of the operations of the Beef Slough boom encapsulates the spectacular rise and fall of the Lake States lumber industry.[14]

Rafting. The other facet of the traditional water-borne transport system was rafting, and the evolution of this part of the lumbering system was told when we considered the circulatory system that evolved along the East Coast during the years of transition between colonial and industrial America, after the beginning of the century. After 1860 rafting was operating on an even greater scale than before, aided by new inventions and improvements in technique. Some rafting was continental in scale, going down the Mississippi to St. Louis or across the Great Lakes to the northeastern seaboard; some was local, as for instance from the boom at Tittabawassee to the hundred-odd mills that dotted the banks of the Saginaw River between Saginaw and Bay City. But, continental or local, rafting still held an important place in log transport for most of the century.

The scale of the movement of the logs rafted down the Mississippi can be judged by a few simple figures. For example, on the St. Croix between 1878 and 1905 there were only four individual years when the number of logs sawed at the local mills exceeded the number of logs placed into rafts and sent downstream. In 1898, for example, the downstream logs rafted accounted for 72.5 percent of the 336 million b.f. of logs run on the St. Croix. On the Chippewa it seems likely that similar proportions of even bigger cuts went downstream beyond Beef Slough. Simply, up to 90 percent of all the timber cut in central and northern Wisconsin was going down the Mississippi in rafts during the latter years of the nineteenth century.

Initially, the methods of movement stayed the same but with ever-larger rafts being propelled by numerous oars and with crews of from 20 to 35 men. Experiments with steam-driven towboats and steam-operated "pushers" gradually reduced the raft crews to less than half (Plate 7.1). The pushers gained rapidly in popularity, their ability to control the rafts successfully being enhanced by guide ropes fixed to either side of the raft and pulled in by a newly invented double-spooled steam winch that could be operated by one man; as one drum pulled in the rope from one side, the other drum worked in a reverse direction and paid out more slack, and the steamboat thus acted like a huge rudder. Rafting soon became a matter of skilled piloting. By 1877, 95 percent of rafts were either towed or pushed. Another innovation was the use of towboats at right

Plate 7.1. Rafts of Lake States logs being pushed into position in readiness for their voyage downstream, upper Mississippi. Note the cordwood boat alongside the pusher, bringing fuel-wood supplies.

angles to the currents; these could be propelled forward or backward and therefore guide the raft. The total result of these innovations was that the number of boats actually decreased whereas the rafts increased rapidly in size to four or five acres by 1880, the only limitation on size being the width between the piers of the bridges, particularly the multiplicity of railroad bridges that were built after 1870.[15]

Ever since the early 1850s the New York and Buffalo entrepreneurs had experimented with bringing lumber from Michigan to the East by barge and schooner and logs by rafts to be milled at Buffalo. The problem of rafting the logs on the lakes was their susceptibility to storms and the all-too-frequent breakup due to the swell. As the white pine from eastern Michigan began to be cut out, Saginaw lumbermen began to log white pine in the Upper Peninsula and in Canada and raft it across Lake Huron and Georgian Bay. They quickly learned that a more flexible structure was needed than the Mississippi raft, and two Bay City lumbermen, Frank H. Durell and William Goldie, experimented with chaining large buoyant logs together and surrounding the free-floating mass with a floating fence, or "bag" or "balloon," as it was sometimes called. In 1885, 3 million logs were successfully towed to Bay City from the Lake Superior shore in Upper Michigan, and the "balloons" weathered the worst storm. A trade reaching 40 million logs per annum flourished during the rest of the 1880s with balloons as large as 8 million b.f. and 25 acres in extent coming across the lakes. In all, in the nine years from 1890 to 1898, 1.8 billion b.f. were rafted. Barges still moved a small quantity of logs, but they were of minimal significance.

The sudden increase and equally abrupt cessation in log transport from Canada arose because after 1890 the Canadian government removed the export duty on logs, but after 1898 the Dingley tariff restored the import duty on Canadian logs and the province of Ontario retaliated by stipulating that all logs cut on Crown lands had to be manufactured in Canada. The trade to Michigan stopped almost overnight.[16]

Despite these and many other improvements in rafting techniques, however, there were serious deficiencies in transport by water. There was the "shrinkage" of the cut; up to 10 percent of the cut was lost through sinking, stealing, and scattering through floods, the problem becoming worse as the lumbermen moved farther away from the rivers and utilized the less floatable trees like red pine, tamarack, and balsam fir. Constant litigation over damage to booms and bridges and the compensation to riparian owners added further to the costs of water logging. Little wonder that loggers were soon to turn to railways as a major means of transport.[17]

Transport: new methods

While the traditional means of transport continued, admittedly modified and improved, a technical revolution was going on in the forests with the development of ice roads, railroads, and power skidding. The need for innovation arose directly out of the desire to fulfill the rising demand that existed for timber and the profits that could be made, and also from the realization that the traditional means of moving logs had serious deficiencies – in particular, the reliance on an early freeze, heavy snow for skidding out the logs, and the spring thaw for moving the logs downstream to the mills (Fig. 7.10A).

If any one of these climatic conditions was not met, the loggers were hampered; prices rose in response to limited supply, milling operations were interrupted, and users in other parts of the country were encouraged to look to alternative sources. Short supply and high prices were usually accompanied by an unregulated response to cut more, and hence lumber prices would fall dramatically to less than half with overproduction. For example, dry winters in 1863 and 1864 sent the price of four-by-fours in Buffalo tumbling by 31 percent in 1865, and mixed lumber on the Chicago markets fell from $19.50 to $10 between March and June. The difficulties associated with the weather controlling cutting in the forest became more and more acute as logging operations were moving farther away from the rivers.[18]

Ice roads. In order to overcome these climatic problems and to even out the fluctuations of production, loggers experimented with various new devices for moving logs out of the forests. The initial "snaking out" of the logs by bullocks was replaced by the "go-devil," a rough sled that lifted the front end of the log to stop it from digging into the ground. This was soon replaced by the "bummer cart," a self-loading skidder in which linked wheels and a long tongue acted as a lever and one end of the log was hitched up under the wheels. More applicable to the Lake States, however, were the efforts to augment the climatic aids to logging. Michigan loggers widened snow roads to carry bigger sleigh loads, the ponderous oxen were replaced by horses, and more importantly, experiments were made to give sleigh roads an artificial covering of ice. The method of sprinkling the roads with water at night was probably invented by Elam Greeley of Wisconsin in 1872,

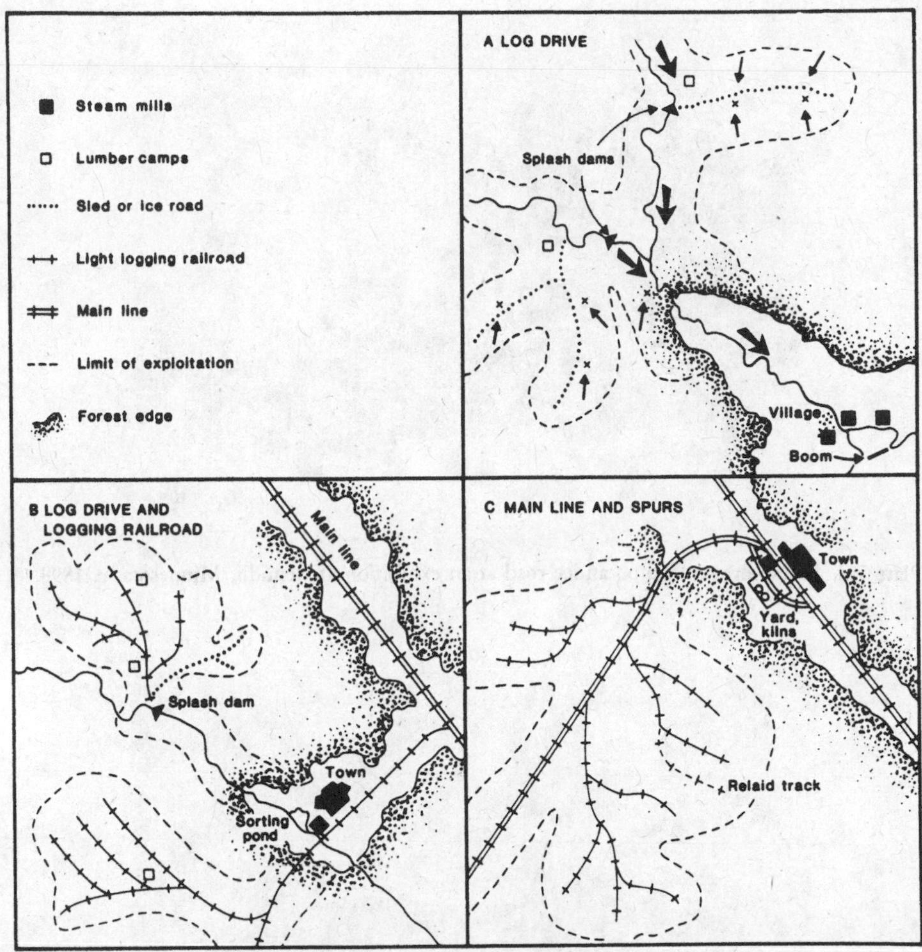

Figure 7.10. Paleotechnic forest exploitation

and the technique had begun to diffuse widely through all the Lake States during the late 1870s. In the following decade manufacturers developed snow plows to clear roads, rut cutters to make parallel grooves for the sleighs to follow, and sprinkling carts to fill the ruts with ice. By turning the snow roads into ice roads friction was reduced and larger loads than before could be moved. Sleighs grew wider and longer and loads grew higher, so that whereas during the 1870s a 3,000 b.f. load was considered usual, by the 1890s loads of over 30,000 b.f. and about 120 tons in weight were being carried, which was the equivalent of a logging railroad (Plate 7.2).[19]

Efforts were made to improve the haulage power of the sleighs by developing steam traction engines. The first was that of George T. Glover of Manistee, Michigan, who tested a machine in 1888. It had a caterpillar-like tread and was steered by a pair of front swivel bobs. It broke down continually and was not a success, but the idea was good, and it was the forerunner of the tractor developed by A. O. Lombard of Waterville, Maine.

Plate 7.2. Horse-drawn load on an ice road at an exhibition at Bemidji, Minnesota, c. 1890.

Plate 7.3. The Lombard log hauler on an ice road, Minnesota, c. 1908.

Lombard's tractor was basically a small steam locomotive on caterpillar tracks with a geared drive. He patented it in 1901, and it was highly successful for it could haul up to 100,000 b.f. on from 7 to 15 sleds (Plate 7.3). It required no animal feed, only refuse wood, and it replaced 60 men and 30 animals in the forests, thereby contributing to the stabilization of logging costs. The Phoenix Manufacturing Company of Eau Claire started producing a similar steam haulage machine in 1907, and this machine was successful and used widely. Nevertheless, however popular and economical these machines were, they never completely replaced animal power in the forests of the Lake States.[20]

Even with the use of animal motive power only, ice roads were not cheap to make because of the initial grading of the track and the spanning of swamps. Contemporary estimates of the 1880s suggest that they could cost as much as $1,000 per mile, a figure that was still too much for the small operator, who still prayed for a good fall of snow. When the loads carried increased with steam hauling, bridges had to be made stronger, downgrades banked, and "rolling stock" bought. Inevitably, many small operators could not compete, and ice roads became the new transport technique of the big operator.

Railroads and logging roads. The dependence on snow and ice could be decreased only if new methods of moving the logs could be found. The snow-deficient winter of 1877–8, in particular, encouraged "the desire for a more certain and controllable means for regulating supply," and the railroad seemed to be the answer. The fact that some railroads already passed through or near to the forests and were taking logs to the mills was important. For example, during the dry season of 1872 the Flint and Père Marquette Railroad started to carry logs and moved over 12 million b.f. to the mills, an amount that reached 87 million b.f. by 1880 (see Fig. 7.8). However, if the common carriers did not come close to or penetrate the forests, other means of carrying the logs had to be found. One such means was the tramroad, said to have been invented by Van Etten, Kaiser, and Company in response to poor snow conditions in 1872–3. The 13-mile tramroad was made of 9–12-inch diameter poles locked 5–6 feet apart, on which trucks with flanged wheels could run. Publicity in the trade journals about the "pole roads" led in four years to the building of about a dozen similar such roads averaging six to seven miles in length. Some of these were simply planks laid end to end on the ground.[21]

It was not long, however, before the lumbermen began to utilize iron rails. Although known since 1852 in Tioga, New York, the iron tramway made little impact on New York or Lake States lumbering until after the 1870s when white pine became scarce, there was a general decline in world prices, and all lumbermen were attempting to economize in order to sell more lumber. In 1876 W. Scott Garrish and associates were logging in Clare County, Lower Michigan, and built a seven-mile-long iron railroad from Lake George to the Muskegon River. It was successful during the winter of that year, and the Lake George and Muskegon River Railroad actually made a profit while other loggers went bankrupt. Quickly, other loggers followed suit, with important results. The *Muskegon Chronicle* reported that, since the "railroad fever" had taken hold of the lumbermen, "a large number of men and teams have been discharged and sent home. Railroad matters as it regards new roads for lumbering operators, is not all talk by any means. The parties interested are getting down to business and the work is being pushed vigorously."[22]

With the near doubling of lumber prices between 1879 and 1882 and the ever-retreating edge of the forest, the idea of logging railroads spread even more rapidly throughout Michigan, the rate of construction receiving a boost whenever there were

Plate 7.4. A Shay locomotive drawing 47 cars, with 393 logs totaling 61,000 b.f., through a logging settlement near Cadillac, Michigan, 1904.

warm winters. Between 1878 and 1880, 11 roads were built, then in 1881 alone a further 11 and in 1882, 32, to which another 14 were added the next year with about 500 miles of track. The number of logging railroads constructed fell as timber prices fell during succeeding years; nevertheless, by 1887 there were 126 railroads with an estimated capacity to move more than 1 billion b.f. of logs annually. The idea of logging railroads spread slowly to Wisconsin and later to Minnesota, so that by 1887 there were 89 in Michigan but only 11 in Wisconsin and one in Minnesota. There was a total of 850 miles of railroad in the three states with an annual capacity to haul 2 billion b.f. of logs. In actual fact, much more track was laid than these mileages suggest because after the timber was cut the track was quickly removed and used again. Most of the individual logging operations were highly flexible so that in some years more track was laid than the total length of the rail recorded. With remarkable ingenuity, Michigan lumbermen experimented and perfected new machinery for the railroads. For example, E. E. Shay designed a new locomotive in 1885 with gears so adjusted as to give it the greatest possible amount of drawing power (Plate 7.4).[23]

Most logging railroads were feeders to the main driving streams and the common carriers, but the linking of both was difficult. Logging roads were always of a different gauge than main railroads, and railroad gauges also varied. In Michigan alone only 14 of the 76 railroads had a standard gauge, and there were seven other gauges. The Bonita logging line alongside the Wolf River in northeastern Wisconsin was typical of the problem of multiple gauges and uncoordinated networks of logging tracks that converged on the Chicago and North Western Railroad (Fig. 7.11). But if the ice roads were expensive, the railroads were even more so. They cost between $1,900 and $7,600 per mile, which put them totally out of the reach of the small operator.

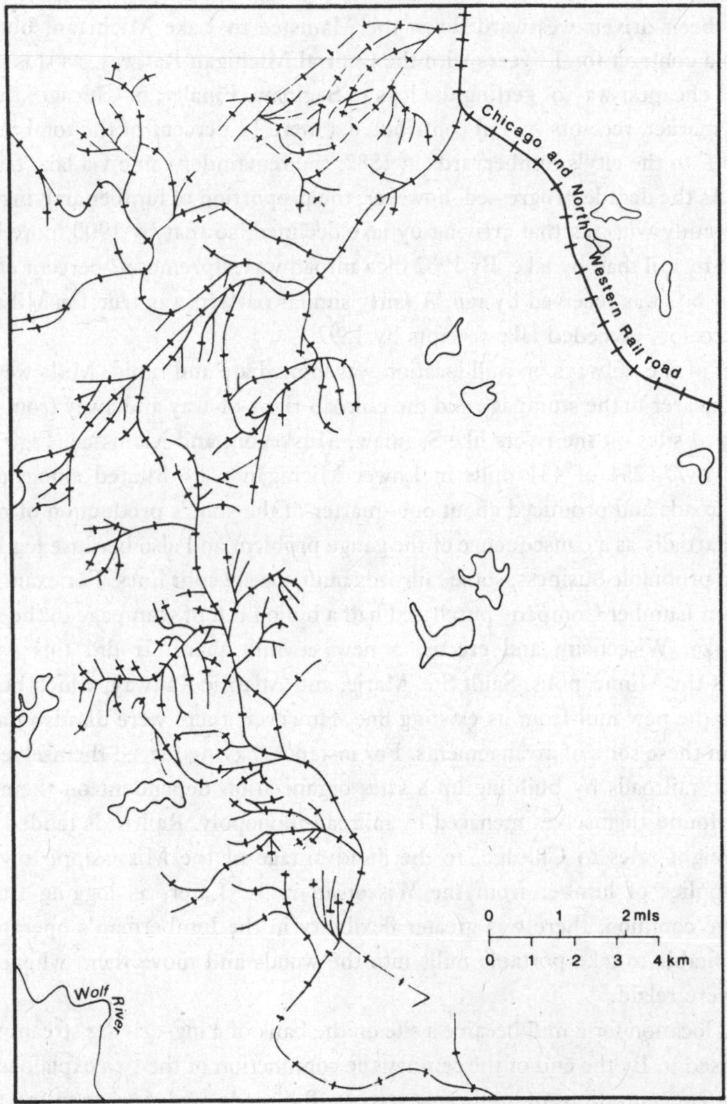

Figure 7.11. The Bonita line, Wisconsin, c. 1919. (After R. E. Rohe, "The Landscape and the Era of Lumbering in Northeastern Wisconsin," 19.)

In every way, the railroads and their versatile feeders caused a revolution in the movement of lumber away from the rivers. For example on the St. Croix River, formerly the epitome of the log-driving river, it is estimated that, by 1887, 270 million b.f. passed through the boom but 550 million b.f. went by railroads. In the Saginaw basin the Flint and Père Marquette handled 161 million b.f. of logs in 1888, 100 million of which were tipped into the Tittabawassee at Averill, but by 1892 it handled 192 million b.f., of which only 50 million was now tipped into the Tittabawassee, the remainder going directly to the mill by rail (Fig. 7.8). Similarly, in 1884 David Ward agreed to supply 400 million b.f. of logs from the interior of the Peninsula to the Saginaw mills. Normally, the logs would have been driven westward down the Manistee to Lake Michigan, but Ward entered into a contract for 12 years with the Central Michigan Railway, as this was the quickest and cheapest way of getting the logs to Saginaw. Finally, in Chicago, the main midwestern market, receipts by rail contributed a mere 13 percent of the total intake of 1.9 billion b.f. to the city's lumberyards in 1882; the remainder came via lake transport (Fig. 6.13). As the decade progressed, however, the proportion of lumber arriving by rail increased steadily whereas that arriving by lake declined, so that by 1900 more lumber was received by rail than by lake. By 1907 the railroad was supreme; 82 percent of a total of 2.4 billion b.f. was received by rail. A fairly similar pattern was true for Milwaukee, where rail receipts exceeded lake receipts by 1892.

The effect of the railways on mill location was immediate and rapid. Mills were built or relocated nearer to the stumpage and the railroad right-of-way and away from the old water-powered sites on the rivers like Saginaw, Muskegon, and Manistee (Fig. 7.10C). As early as 1877, 254 of 431 mills in Lower Michigan were located alongside nine different railroads and produced about one-quarter of the state's production of nearly 2 billion b.f. Partially as a consequence of the gauge problem and also because log hauling was clearly a profitable business, some railroads built special spur lines. For example, the F. W. Upham Lumber Company purchased half a billion b.f. of stumpage in the vicinity of Coon Lake, Wisconsin, and erected a new sawmill there. It did this with the agreement of the Minneapolis, Sault Ste. Marie, and Atlantic Railway, which built a 23 mile spur to the new mill from its existing line. However, there were disadvantages for the millers in these sorts of arrangements. For instance, having placed themselves at the mercy of the railroads by building up a sales organization dependent on them, many lumbermen found themselves menaced by railroad monopoly. Railroads tended to give favorable freight rates to Chicago, to the disadvantage of the Mississippi towns that received supplies of lumber from the Wisconsin mills. Later, as logging tramways became more common, there was greater flexibility in the lumberman's operations. It became profitable to take portable mills into the woods and move them whenever the tramways were relaid.[24]

The ideal location for a mill became a site on the bank of a log-driving stream where a railway crossed it. By the end of the century the conjunction of the two explained nearly all the larger concentrations of lumbering activity. Railroads made it economical to build sawmills as close as possible to the cutting site and for the mill owners to produce a wide variety of finished products, which could now be transported directly to the consumer. Prior to this, few mills bothered to season or finish their products, which could be damaged by a river journey, but with the introduction of the railway many mill owners found it profitable to install drying kilns to assist seasoning and planing mills to finish off products. This, incidentally, had a profound effect on the building and carpentry trades as the mill owners came to replace much of the work done formerly by those trades, and they even began to produce complete ready-to-assemble wooden houses.[25]

One example of the relocation of mills with the spread of railroads must stand for many (Fig. 7.12). In the Wolf River basin, northwest of Lake Winnebago in northern Wisconsin, mills were located exclusively on the river course in 1857. By 1885 the

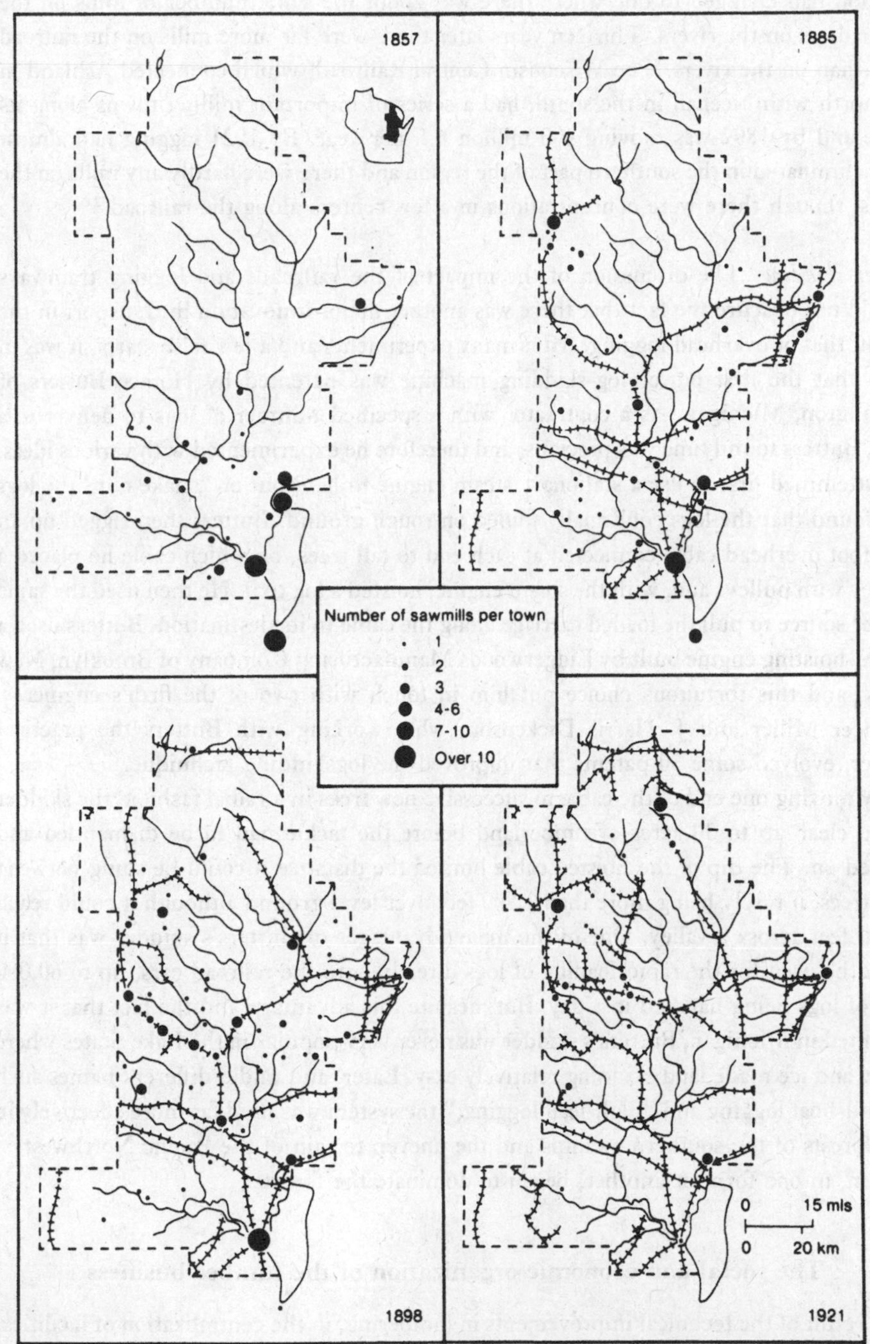

Figure 7.12. Logging railroads and sawmills, Wolf River basin, Wisconsin, 1857, 1885, 1898, and 1921. (After Rohe, "Landscape and Era of Lumbering," 17, 18.)

position had changed to one where there was about the same number of mills on the railroads as on the rivers. Thirteen years later there were far more mills on the railroad lines than on the rivers. The Wisconsin Central Railroad, which connected Ashland in the north with Neenah in the south, had a series of important milling towns along its route and by 1892 was moving 400 million b.f. per year. By 1921 logging had almost been eliminated in the southern part of the region and there were hardly any mills on the rivers, though there were concentrations in a few centers along the railroad.[26]

Power skidding. The discussion of the impact of the railroads and logging tramways should not obscure the fact that there was another major innovation in transport in the forest, that of overhead logging. After many experiments and a few false starts, it was in 1883 that the first power log-skidding machine was patented by Horace Butters of Ludington, Michigan. As a contractor with a specified number of logs to deliver to a mill, Butters found time was precious, and therefore he experimented with various ideas. He attempted first to use a stationary steam engine to haul out or "snake out" the logs but found that the logs continually fouled on rough ground. Butters then rigged up an 800-foot overhead cable connected at each end to tall trees, on which cable he placed a trolley with pulleys and, with the steam engine, hoisted a log to it. He then used the same power source to pull the loaded carriage along the cable to its destination. Butters used a steam-hoisting engine built by Lidgerwoods Manufacturing Company of Brooklyn, New York, and this fortuitous choice put him in touch with two of the firm's engineers, Spencer Miller and J. Harris Dickenson, who, working with Butters the practical logger, evolved some 50 patents that improved the logskidding technique.

By moving one end of the cable to successive new trees in a radial fashion, the skidder could clear up to 30 acres of timberland before the tackle had to be dismantled and moved on. The dip of the hoisted cable limited the distance it could be slung between two trees, it rarely being more than 1,800 feet over level ground although it could reach 5,000 feet across a valley. One of the main advantages of Butters's skidder was that it could be used for the rapid loading of logs directly onto the railroad cars, up to 60,000 feet of logs being handled in a day. But, despite this advantage and the fact that it was invented in Michigan, Butters's skidder was never very popular in the Lake States where snow and ice made land skidding relatively easy. Later, and under different names such as pull-boat logging and "high-lead logging," the system was used far more effectively in the forests of the southern swamps and the uneven terrain of the Pacific Northwest.[27] Steam, in one form or another, began to dominate the forests.

The social and economic organization of the lumber business

One result of the technical improvements in lumbering, in the centralization of facilities, and in the increasing use of railways was to bring about major changes in the social and economic organization of the lumber business on a scale hitherto unknown. The mode and place of work changed as mills grew larger and more complex. Lumber towns – special settlements devoted entirely to the manufacturing of lumber – appeared and then disappeared as the timber was cut out. There was the rapid evolution of corporations, and there was the establishment of monopolies and of lumber and trade associations. On the whole, society in the United States did not disapprove of these new forms of business organizaton, for the production of wealth and the acquisition of private property were laudable objectives which received, if not active, at least tacit government approval, assistance, and cooperation.

The development of the rapid and exploitative system of lumbering that accompanied the rise of these large companies and monopolies was not so welcome. Nevertheless, it was permitted because the forests were considered "boundless." Even if state legislatures had wanted to prevent depletion, there were a number of processes at work that would have made it difficult to stop – for example, population growth, technological advances, the inflow of capital from the East, the problem of dealing with the dispersed decision-making powers that flowed in the private property market, and the new financial techniques used to enlarge the scale of operations. An energetic minority existed with a clear vision of the short-run profit to be gained by rapid clear-cutting, a vision that could be translated into action without a restraining influence or regulatory agency to make that minority account for its actions. In any case, this minority perceived that its risks could be reduced by shortening the time span between capital invested and clear-cutting, followed by the speedy disposal of the land as cutover farms.[28]

Land ownership

Although technological advances and large-scale organizaton undoubtedly assisted monopolistic tendencies, the rise of the large mill and its production was not such a crucial factor in the power of combination and the exclusion of competition as was the ownership of land. It was quickly appreciated that technical changes leading to the enlargement of the plant and the increase in its output had their limits. The larger the mill the farther away it was likely to have to collect its logs. As transport accounted for up to three-quarters of the cost of lumber manufacturing, it was impractical to establish only a few great sawmills and bring timber a long distance. Transport costs would soon counterbalance any saving in manufacturing costs that might be attained through excessive size. The very largest mill could account for no more than 0.5 percent of the production of the United States. Therefore, important as the technological changes in the mills and in transport were in increasing output, the organizational changes in land ownership and the business structure were eventually going to be of far greater importance. Land ownership, therefore, was the key to monopoly.

Land was acquired fairly easily, if illegally at times, by a small group of speculators, and a scramble set in for the best pine lands. Land was acquired through cash sales at $1.25 an acre or through preemption homesteading or the acquisition of enormous land grants. During the 1860s, for example, Ezra Cornell, founder of Cornell University, acquired nearly 500,000 acres of pine lands in the Chippewa River Valley in many scattered locations and created what he called "a powerful pine land ring" that "exercised a fearful and terrible monopoly." Also, many of the millions of acres granted to the railroad, canal, and wagon-road companies and agricultural colleges, by one means or another, flowed back into the hands of the lumber companies. For example, of 760,000 acres granted by federal authorities to three canal companies in the Upper Peninsula of Michigan, over 88 percent ended up in the hands of the large timber companies.[29] Having acquired the land, the next job was to select the best pine lands. This was not a chance affair; there evolved a group of men known as "timber cruisers," highly paid and experienced foresters who spied out the best timber stands and established claims at the land offices. Finding the pine was one thing, but estimating the yield in thousands of board feet was another, as the estimate was the basis of investment in the lumber exploitation. Their decisions were crucial.[30]

In addition to reducing transport costs, the acquisition of as much land as possible ensured supplies of lumber and benefited companies by putting them in possession of a steadily appreciating asset, the value of which rose 20-fold to 50-fold in the 40 years after 1860. The simple fact was that the price of timber tended never to fall despite fluctuations in the price of lumber. This gave the timber owners an advantage over the mill owners, and with that advantage the timber companies could expand and diversify into the retail outlets, these being known as "line yard companies."[31]

By the time the Bureau of Corporations had issued its report in 1913, the monopolization of timber land ownership had gone so far that, of the 2.8 billion b.f. of timber estimated to be left standing in the coterminous United States, 2.2 billion were owned privately. Of the largest landholders, 1,802 owned nearly 80 percent of that total in the Pacific Northwest, the southern pine region, and the Lake States, of which a mere 195 landholders held half of the total. In the Lake States, concentration was extreme; in Minnesota, six companies owned 54 percent of the white and red pine; in Wisconsin, 10 landholders had 10 percent of all timber, and 96 landholders (each having 60 million b.f. or more) had three-quarters of the timber. The figures for Michigan were not much different. In the three Lake States as a whole, four landholders held 12 percent of the estimated total of 100 billion b.f. of timber, 17 held 22 percent, 44 held 37 percent, and a total of 215 landholders had a staggering 65 percent of all timber. In actual fact, the control of the timberland was greater than the figures suggest. The interweaving of interests, corporate and private, connected many holdings regarded as single, and the maps of ownership accompanying the bureau's report show how easy it was for companies to purchase land in a checkerboard fashion so that the intervening smallholders could be "blocked in." If, in addition, the intervening owners could not get rights-of-way for logging, the grip of the large companies was absolute, especially as some of the largest landholders also acquired a dominant position in common carrier railroad companies.[32]

Scale of operations

The increasing complexity and size of the machinery in the mills have been noted already, but the changes in the scale of operation can be seen in many other ways. In the forest the logging camps became larger. Before 1860 they were small affairs, often a mere collection of log huts and bark shanties. As time went on and operations became more thorough, however, the camps became more complex. One building was where the men slept, and if there were too many men another bunkhouse would be provided. The cookhouse had the kitchen and eating area. Other buildings might include an office, a storehouse, stables, blacksmith's shop, all, of course, made of logs, the cracks filled with a mixture of mud and moss. Logging camps did not generally become towns because the transient nature of logging demanded that the logging crews keep moving to new areas. Many shanties were built, in time, on wheels or skids so that they could be moved easily to new locations. Generally, there were no family camps but only temporary quarters for the logger, whose family, if he had one, would live in a nearby town or stay on the farm, which the logger had left during the winter in order to earn hard cash to tide him over.[33] The total complement of a camp could be between 40 and 60 men. Occupationally, the jack-of-all-trades in the camp was replaced by those with specialized jobs so that the old lumber jobs might be divided into a dozen categories, such as scalers, teamsters, log drivers, barn men, saw fitters, road foremen, and cooks, and when power skidding came into vogue, there was an even greater subdivision and variety.[34]

With the shift from water to steam power the mills became less and less constrained to riverside locations and moved to branch railways and even to common carrier lines. In many settlements special lumber districts arose devoted solely to the manufacture and storage of lumber and, increasingly, to pulp and paper making when the chemical and mechanical processes for making paper from softwoods were invented. Large areas were laid out as lumberyards, mills, and offices, as in the spectacular example of Albany, but this now occurred in a multitude of small towns where the sawmill and the grist mill dominated settlement. In time, settlements that were "company towns" were created by the lumber company to house and marshal its work force in one convenient spot, a development that eventually reached greater proportions in the South and the Pacific Northwest than it did in the Lake States.

The mills themselves became larger and more complex. For example, the mill of W. R. Burt and Company, constructed about seven miles below Saginaw City on the Saginaw River, was described as being "one of the most complete establishments on the river." It had a sawmill, stave and heading mill, barrel factory, shingle mill, saltworks, carpenter and blacksmith shops, gasworks, schoolhouse, and public library. Equipment in the mill included two gang saws and one circular saw, one upright saw with edging tables, and cutoff saws. During the season 150 men were employed, and the mill ran night and day, with an average cut of 100,000 b.f. every 12 hours. During the first half of 1874 it produced 14 million b.f. John McGraw's mill in Bay City was considered, locally at least, as one of the wonders of the world because of its size, 350 employees, and saw capacity. Working night and day, it could produce 40 million b.f. a year and double that if need be. Steam from the sawmill, which would otherwise be wasted, was used for drying the lumber in kilns. Although it was located beside the waterway, the mill did not rely on water transport as the Flint and Père Marquette Railroad had a branch line leading into the yard, and "cars from all the principal lines East and West may be seen loading here at all times."[35]

The trend toward an increase in size and scope of operations, the scales of economy achieved by the new machinery, the high capital investment, and the urge toward complete land ownership to ensure supplies meant the consolidation of small-scale family businesses into larger companies. Defebaugh mentions examples of small-scale family businesses in New York becoming large-scale speculative lumber operations (e.g., the Millard and Sharman families), and the same process was true of nearly all the companies in the Lake States. Enlargement also took place in an effort to maximize efficiency and minimize costs by attempting to integrate cutting and selling operations, but always over ever-extending lines of communicaton. It was the development at an early stage of what William Rector calls "the pipe-line theory of production." As a result, enormous business empires were set up – for example, Knapp Stout and Company, which had started in 1849 and by the late 1880s was operating mills at Menominee, Cedar Grove, Downsville, Rice Lake, and Prairie Farm, all near the sources of pine in Wisconsin. Most of the production of the Menominee mills went downstream to lumber yards at Dubuque and Fort Madison, Iowa, and to St. Louis, Missouri. The same firm also operated sawmills at Dubuque and St. Louis to cut logs brought down from Wisconsin. Similarly, D. M. Dulany of the Empire Lumber Company of Hannibal, Missouri, had mills at Eau Claire, Wisconsin, and Winona, Minnesota, which rafted much of their product to the wholesale yards at Hannibal. Other examples were the lumber empires of Frederick Weyerhaeuser and Orrin Ingram. Such arrangements were common and, in various combinations, dominated the Lake States lumber scene during the latter part of the century.[36]

One important point to remember in the growing complexity and scale of organization was that, between getting ready for logging in the fall, doing the cutting in the winter, and then driving or rafting in the spring for summer milling, the logger had enormous amounts of capital tied up in nonliquid assets for up to or even more than 12 months, while expenses were a constant drain on his resources. A man of limited capital had to resort to borrowing, and bankruptcy was a common occurrence if, say, weather delayed the log drive or there was a logjam. The whole system as it evolved with increasing production, long hauls to market, and more expensive stumpage favored the large-scale operator with a secure financial position, and it weeded out the little man.[37]

Trade associations

More complex still than the large-scale companies were the trade associations promoted by groups of individual companies in an effort to control production, rationalize transport between their far-flung operations, and stabilize the wildly fluctuating prices received for their products. They represented "a trend toward monopoly." An early example was the Lumber Manufacturers' Association of the North West, formed by lumbermen in the Mississippi basin portions of Wisconsin and Minnesota in 1882. The object of the association was to regulate production, but it found it difficult to obtain agreement among so many members, some of whom operated night and day in order to gain an advantage over those of their competitors who had agreed to shut down on agreed dates.

The largest and most successful trade association was the Mississippi Valley Lumberman's Association founded in 1891 (later to become the Northern Pine Association). By the turn of the century its members represented about 90 percent of all the pine producers in the areas of the Wisconsin and Minnesota tributaries to the Mississippi. Rather than attempt to curtail production, a move that had failed in previous associations, it set about direct price fixing, the task being made easier by the large syndicates and combinations that had already been set up by some companies for manufacturing and land ownership. Similar but smaller regional organizations were the Wisconsin Valley Lumberman's Association formed in 1895 and the Lake Superior

Lumber Manufacturer's Association in 1900, comprising 26 leading sawmill companies along the lake, and there were many more. Even manufacturers of windows and door frames formed their own association in 1880 – the Northern Wisconsin Manufacturing Association – and in 1910 loggers of particular types of timber formed their own associations, as, for example, the Northern Hemlock and Hardwoods Manufacturers Association, which included 75 companies. As most of these associations were price-fixing associations, some were subsequently proceeded against as a combination in restraint of trade under the Sherman Antitrust Law of 1895, but as they never achieved a complete monopoly, little could be done. The lumbermen always maintained that their activities merely produced orderly marketing, that they did not raise prices, and that they were inimical to monopoly as there were, after all, over 46,000 lumber manufacturers in the United States by 1900. No one association could possibly tie up the market. In any case, they argued, they were voluntary associations, and many members broke the rules and sold at lower than agreed prices.

In contrast, two other forms of organization did tend toward the creation of local and regional monopolies. First, there was the system of interlocking management. To take one example, Orrin H. Ingram of Eau Claire was at various times or was simultaneously director of the Chippewa Logging Company, vice-president of the Chippewa Lumber and Boom Company, president of the Rice Lake Lumber Company, president of the Empire Lumber Company of Minnesota, vice-president of the Standard Lumber Company of Dubuque, and a major stockholder in the Weyerhaeuser Timber Company, which was the holding company for the timberlands in the Pacific Northwest. Ingram also invested privately in timberlands in the South, in the Midwest, and on the Pacific Coast. To cap his management portfolio, he was president of the Eau Claire National Bank and the Eau Claire Water Works Company and treasurer of the Canadian Anthracite Coal Company, as well as having interests in many other enterprises.[38]

Second, there were the vertical integration and horizontal integration of operations and the syndicalism of people like Frederick Weyerhaeuser. Besides having a part in the management of 18 manufacturing concerns in three states, he was the leading organizer of the Mississippi River Logging Company and its affiliate corporation, the Beef Slough Manufacturing, Booming, Log Driving, and Transportation Company, which together controlled timber supplies and manufacturing, either directly or indirectly, throughout most of the Chippewa, St. Croix, and Upper Mississippi rivers. By owning timberland, mills, and marketing outlets, the syndicates acted as perhaps the most successful and powerful price-fixing agency and certainly the most economically integrated organization in the Lake States region and their dependent market territory in the Great Plains. By 1913 the Weyerhaeuser Timber Company had expanded well into the Pacific Northwest and, with its directly owned subsidiary concerns, owned 95 billion b.f. of timber, or 4.3 percent of the privately owned timber in the United States. In his way, Weyerhaeuser was the counterpart of Rockefeller in oil or Carnegie in steel. He was probably the largest of the "lumber barons," and though undoubtedly he was a millionaire many times over, most of his close associates in the Beef Slough venture became millionaires as well.[39]

In every way the business of lumbering in the Lake States had moved a long way from the simple situation of the one-man, even one-company, unit that it had started out as.

Besides getting larger and more complex, with milling companies, boom companies, and lumber companies, there was a centralization of organization and administration, there were mergers and consolidations, trusts and trade associations, high capital interest based on speculative buying and selling of shares, specialization of activity, horizontal and vertical integration, particularly with the control of stumpage, and everywhere an immense increase in production for each unit of time and effort expended. These were all part of the new, dynamic system of industrial capitalism that hoped to promote through these measures and innovations a rational, ordered, and stable environment for investment and production.

Production

Technology and business organization are interrelated in so many ways with production that it is both false and difficult to separate them. Nevertheless, the separation is a convenient one as it helps to isolate for consideration the underlying technical and organizational changes basic to an understanding of the scale and fluctuations of production in the Lake States. Here, as elsewhere in the country after 1860, there is a great increase in the amount of information available from the Census Bureau, the Department of Agriculture, and trade organizations about the magnitude of changes in production. Therefore, the scale of treatment changes. Whereas before it was possible to present a reasonably complete and comprehensive picture of events, the picture is now in danger of becoming an indigestible morass of detail, especially as rapid changes took place in the rate of production from year to year and from place to place in a relatively short time span. Therefore, only the main figures, trends, and locations are highlighted through summary statistics and distributional maps.

The graph of lumber cut in the Lake States (Fig. 7.13) shows the overriding characteristic of production: a rapid and spectacular rise between 1878 and 1883, a fluctuating peak to 1892, and then a steady and almost unrelieved drop to 1920. This almost parabolic curve, but on a much smaller scale, was the characteristic of nearly all the subregions of production in the Lake States as the wave of production flowed through them from east to west. Initially, most cutting was confined to Michigan, which at the time of the census of 1869 was producing nearly two-thirds of the total of 3.6 billion b.f. of the lumber cut in the Lake States. The total rose a little in 1879 but was only just over half of a record cut of 9.9 billion in 1889. During the same period Wisconsin was steadily increasing its proportion of the cut, and from 1889 onward Minnesota became an important producer (see Table 7.2). Of these totals white pine accounted for nearly all the production during the early years, but as supplies ran out other species began to figure more prominently. In all, about 320 billion b.f. of softwood timber was cut between 1830 and 1935, a quarter of that total being felled during the years of hectic activity between 1878 and 1883.

Undoubtedly, business depression and the gradual exhaustion of timber in the eastern pine-producing areas of New York and Pennsylvania had prompted many skilled workmen and entrepreneurs to move to the Lake States, which were receiving glowing and favorable publicity. The shipment of white pine to Albany in 1847, the completion

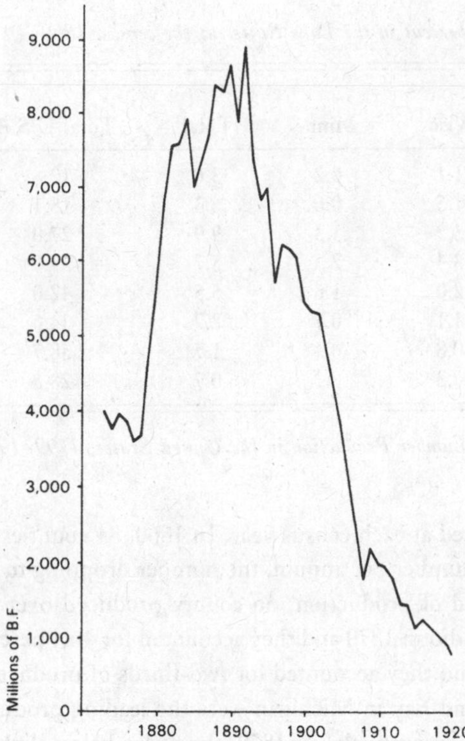

Figure 7.13. White pine lumber production, Lake States, 1873–1920. (Rector, *Log Transportation*, 42.)

of the railroad connection to Chicago in 1852, and the experimental rafting of logs down the Mississippi during the late 1840s all caused an enormous boost to production by linking distant markets to the sources of production.

Production in Michigan and Wisconsin prior to 1860 was mainly in the southern few tiers of counties, and by the time of the census in 1840, when both states were just being settled, there were 124 mills in Wisconsin producing a mere $202,239 worth of lumber products, and 491 mills in Michigan producing $392,325 worth of lumber. The low value of production per mill indicated that they were mainly local pioneer sawmills, small in scale and limited in their market, only a half a dozen or so in each state cutting white pine, the bulk cutting hardwoods for local farm use.[40]

From the censuses of 1860, 1870, and 1880, it is possible to get an overall picture of the trends and shifting locations of production in the three states. Unfortunately, however, the data were not collected on a county basis for the censuses after 1880, and therefore we cannot follow the distributional aspects for the peak census year of production in 1890 and the plummeting decline thereafter. Of the three indexes of production – the value of production, the number of mills, and the number of hands employed – only the value of production is mapped here (Figs. 7.14–7.16).

In general terms, two trends can be discerned: an increase in production per county and a northerly shift in that production. The value of production per county throughout

Table 7.2. *Lumber cut in the Lake States at the census, 1869–1939 (bill. b.f.)*

	Mich.	Wisc.	Minn.	Total	Total U.S.A.	Lake States as % of U.S.A.
1869	2.3	1.1	0.2	3.6	12.8	28.1
1879	4.2	1.5	0.6	6.3	18.1	34.8
1889	5.4	3.2	1.3	9.9	27.0	36.7
1899	3.0	3.4	2.3	8.7	35.0	24.8
1909	1.9	2.0	1.6	5.5	42.0	13.1
1919	0.9	1.1	0.7	2.7	34.6	7.8
1929	0.6	0.8	0.4	1.7	38.7	4.3
1939	0.3	0.3	0.1	0.7	28.8	2.4

Source: Henry B. Steer, *Lumber Production in the United States, 1799–1946.*

the Lake States increased at each census year. In 1860, 84 counties were producing less than $62,500 worth of lumber per annum, the number dropping to 41 in 1870 and 19 in 1880. At the upper end of production, no county produced over $1 million worth of lumber in 1869, but 10 did in 1870 and they accounted for 46.4 percent of production by value; 21 did in 1880 and they accounted for two-thirds of production. The counties of Saginaw, Muskegon, and Bay in Michigan were the leading producers, returning $5.1, $4.3, and $3.4 million, respectively, in 1870, the order being slightly reversed in 1880 when Muskegon, Bay, and Saginaw produced $7.7, $5.8, and $4.7 million worth, respectively.

The other trend was the northerly movement of timber exploitation and lumbering ahead of land settlement. The southernmost river basins were exploited first: the Grand and the southern section of the Saginaw system in Michigan; and the Wisconsin and, to a lesser extent, the Wolf in Wisconsin. Then, in succession, each river basin was penetrated from the south or from the lakeshores, the lumbermen going farther inland with each year of cutting.

In 1860, 70 percent of the 1,025 mills in Lower Michigan were located in the southern four tiers of counties in the hardwood forests. These mills were small, employing on average four or five men, which compared with an average of 15 or 20 men in the few larger mills that had been established and were operating in the Saginaw and Muskegon valleys and that also had a much higher level of production per mill. In Wisconsin, most of the 449 mills were located in counties along the Lake Michigan shore. Again, the southernmost mills were appreciably smaller than those to the north, particularly the couple of dozen inland mills along the Peshtigo, Menominee, Eau Claire, and Chippewa rivers.

By 1870 the number of mills had increased a little; there were now 1,157 in Lower Michigan, but the southern four tiers of counties had relatively fewer mills as production shifted northward into each of the major river basins, particularly the Saginaw, Muskegon, Grand, and Manistee. In Wisconsin the number of mills had risen to 536, high points of production showing up particularly in the Peshtigo/Menominee area and

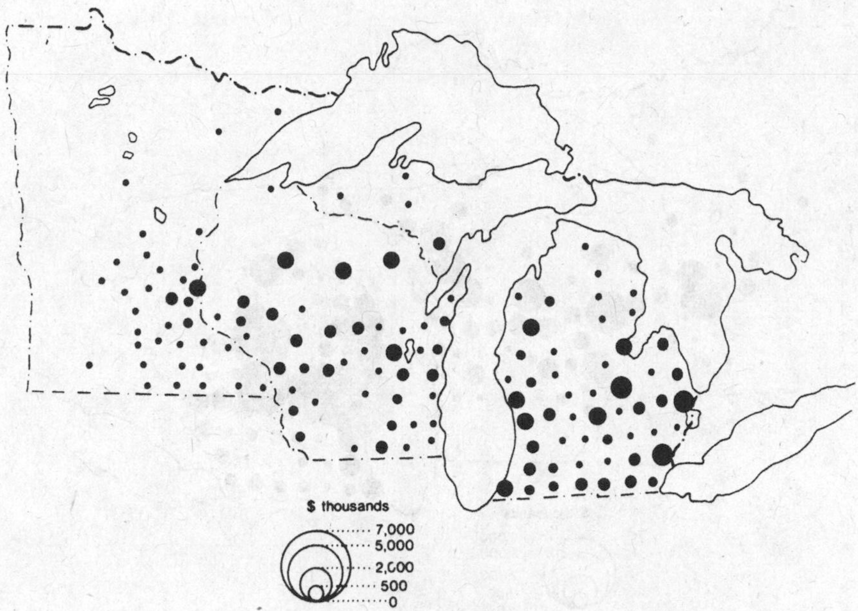

Figure 7.14. Production of lumber by value, Lake States, 1860. (U.S. Census.)

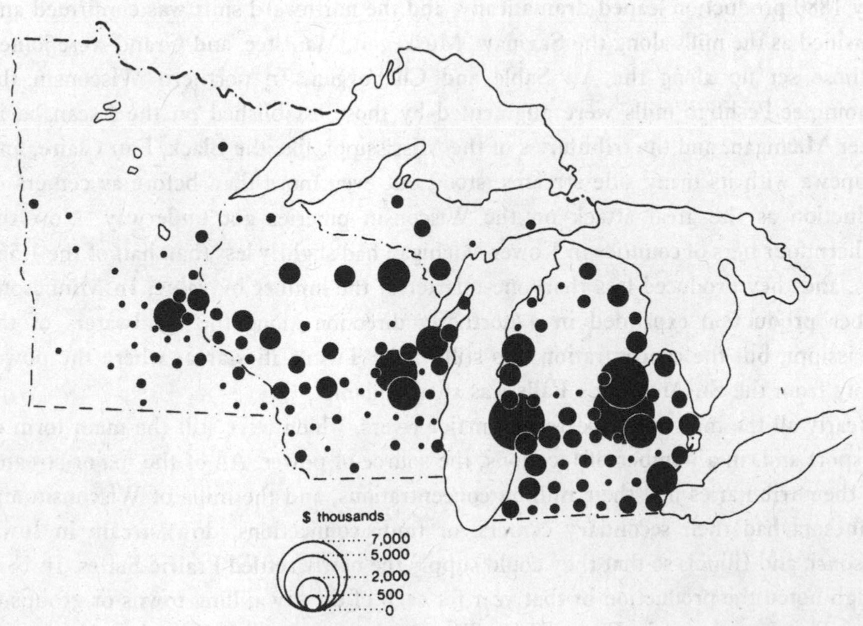

Figure 7.15. Production of lumber by value, Lake States, 1870. (U.S. Census.)

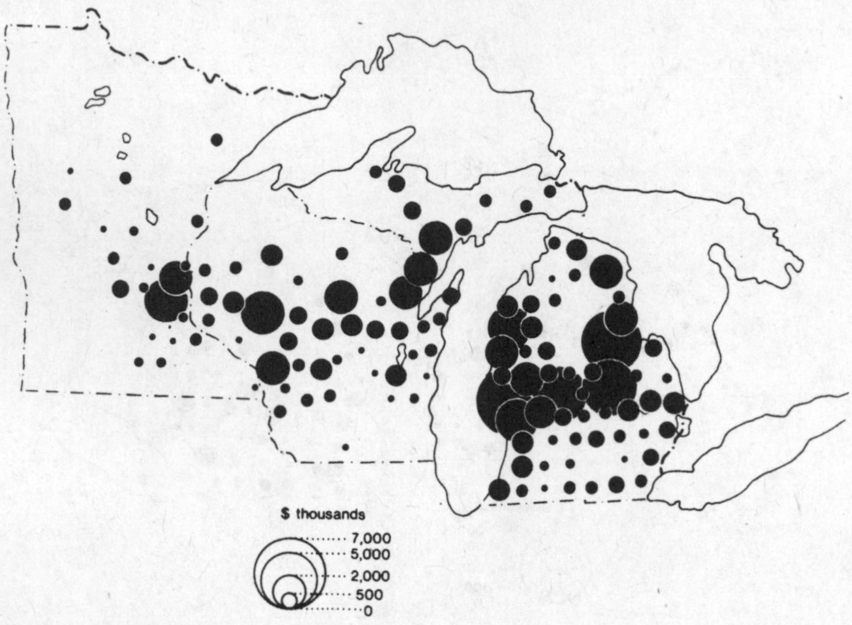

Figure 7.16. Production of lumber by value, Lake States, 1880. (U.S. Census.)

the Wolf River basin, while in Minnesota the St. Paul/Minneapolis milling center on the upper Mississippi collected logs from a wide area inland.

By 1880 production leaped dramatically, and the northward shift was confirmed and intensified as the mills along the Saginaw, Muskegon, Manistee, and Grand were joined by those set up along the Au Sable and Cheboygan. In northern Wisconsin the Menominee/Peshtigo mills were augmented by those established on the Escanaba in Upper Michigan, and the tributaries of the Mississippi, like the Black, Eau Claire, and Chippewa with its many side streams, stood out even more than before as centers of production as the great attack on the Wisconsin pineries got underway. Now the southern four tiers of counties in Lower Michigan had slightly less than half of the 1,563 mills, and they produced less than one-quarter of the lumber by value. In Minnesota, lumber production expanded in a northerly direction along the headwaters of the Mississippi, but the concentration was still in the Twin Cities area, where the power supply from the St. Anthony's Falls was so abundant.

Nearly all the mills were located on major rivers, which were still the main form of transport and, in a number of locations, the source of power. All of the major streams and their tributaries had their milling concentrations, and the mills of Wisconsin and Minnesota had their secondary centers, or trade connections, downstream in Iowa, Missouri, and Illinois so that they could supply the newly settled Prairie States. In 1876 Hough noted the production of that year for each of the sawmilling towns or groups of mills. They are shown in Figure 7.17. The absolute amounts in each location cannot concern us too much as they fluctuated from year to year – for example, the total

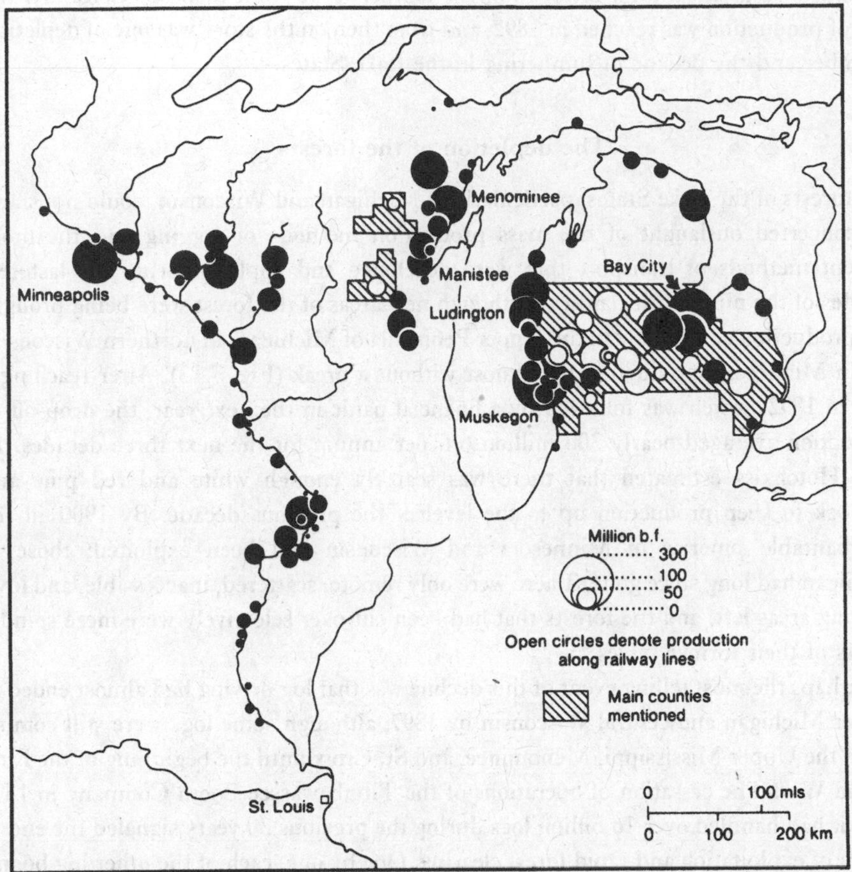

Figure 7.17. Great Lakes and Mississippi River: lumber production, 1876. (Hough, *Report upon Forestry*, 2:49–51.)

production represented in Figure 7.17 for 1876 was 3.66 billion b.f., a drop from 3.71 billion b.f. the previous year – but the relative amounts are probably correct. This map represents a "still" in an ever and rapidly moving picture of shifting patterns of production.[41]

One of the largest concentrations was on the 20-mile stretch of the Saginaw River Valley between Saginaw and Bay City. The 71 mills had a combined capacity of 718 million b.f. per annum and in 1876 had produced 537 million b.f. of lumber. Most of the mills had the added refinement of a saltworks on the premises, whereby "they were able to economise the waste steam from the engines in the evaporation of brine, and to use saw dust and slabs as fuel to great advantage." Yet by the time Sargent saw the same mills in 1884 their number had decreased through "the cut-out" of supplies, although the output of the remaining 66 mills had increased to be nearly 1 billion b.f. per annum.

What happened to production on a local scale during the peak years of exploitation is not known as the census does not record county statistics, and one must resort to the evidence of individual local histories of the lumber trade from then on. Certainly, the peak of production was reached in 1892, and from then on the story was one of depletion of timber and the decline of lumbering in the Lake States.

The depletion of the forests

The forests of the Lake States, particularly of Michigan and Wisconsin, could not stand the concerted onslaught of the mass production methods of logging and the more efficient methods of transport that were developed and applied during the last few decades of the nineteenth century. Although new areas of the forest were being brought into production continually in the Upper Peninsula of Michigan, in northern Wisconsin, and in Minnesota, production fell almost without a break (Fig. 7.13). After reaching a peak in 1892, which was followed by a financial panic in the next year, the drop-off of production averaged nearly 300 million b.f. per annum for the next three decades. In 1897 Hotchkiss estimated that there was scarcely enough white and red pine and hemlock to keep production up to the level of the previous decade. By 1900 all the merchantable pineries in Minnesota and Wisconsin had been exploited; those in Michigan had long since gone. There were only remote, scattered, inaccessible, and low-yielding areas left, and the forests that had been cut over selectively were mere spindly ghosts of their former selves.[42]

Perhaps the most telling event of this decline was that log driving had almost ended in Lower Michigan and central Wisconsin by 1897, although some logs were still coming down the Upper Mississippi, Menominee, and St. Croix until the beginning of the First World War. The cessation of operations of the Tittabawassee Boom Company in 1894 after it had handled over 16 billion logs during the previous 30 years signaled the end of an era of exploitation and rapid forest clearing. One by one, each of the other big booms on the Saginaw and Muskegon went out of operation so that by 1900 none remained and whatever logs were being cut were being transported by rail (Fig. 7.8). In Wisconsin, the last drive on the Chippewa River was in 1907, and the great West Newton boom (the successor to the Beef Slough boom) ceased operation in 1904 and was dismantled in 1906. With its dismantling went the disbanding of the great regional organization that had been set up so carefully by the Weyerhaeuser group. It is true, of course, that the new network of railroads was taking a bigger and bigger share of the log-moving trade, but even so, the log trade was diminishing rapidly because the timber had simply been cut out. There was a temporary return to hardwood production at the beginning of the century, but even this petered out, and by 1923 it was reported that "not a single saw mill on the Saginaw was cutting pine" and that a few were working hardwoods. As early as 1881 one mill had been reduced to milling "wormy pine" for bottle tops and coarse fencing.[43]

The rapid depletion of the forests had been foreseen. From the early 1870s onward many local newspapermen, always conscious of the support base of their local area, prophesied that lumbering must eventually come to an end if production continued at the same rate and forest fires consumed the remainder. Perhaps the most influential and certainly one of the earliest statements of all was that in 1867 by the three commissioners appointed by the Wisconsin legislature to report on facts and opinions "concerning the effects of clearing the land." Although the report was directed toward the supposed climatic effects of clearing the southern hardwoods, it did, at least, sound a warning note about the rapid depletion of the pineries, by taking into account the demands for fuel, railroad ties, and all other uses, which the commissioners calculated in 1867 to have equaled 1.5 billion b.f. from Wisconsin and Minnesota for that year alone. "How long will it last?" they asked. They could not calculate a life beyond 50 years ahead and advocated timber culture to maintain supplies. Filibert Roth was not prepared to calculate the life of the forest in 1898; already "half the mills of twenty years ago are no longer in existence, not because they failed to pay, but because their Pine supplies gave out."[44]

Long before the white pine lumbering actually ceased, changes were occurring in the lumber industry and the economic life of the region that were to signal the coming exhaustion of supplies. For example, lower grades of timber began to be used, and the lumbermen culled the forests to maintain supplies of white pine:

> Twenty year ago logs which would run 25 percent "uppers" [upper, or best-quality timber] were considered common; 40 percent was the rule, and as high as 75 percent "uppers" was sometimes obtained. Logs were then cut from the lower trunk of the trees below the tops, and only the largest trees were selected. Now land which has been cut over three times is gone over again, and lumbermen are satisfied if logs yield 20 percent "uppers."[45]

Consequently, loggers turned to other types of trees and lumber, particularly hemlock and hardwoods. It took time for the public prejudice against hemlock to be overcome; by 1897 it accounted for about 10 percent of all softwood cut in the Lake States, and a few years later it accounted for about one-fifth, eventually outstripping white pine. The hardwoods were used for railroad ties and the fabrication of wooden products, the enormous furniture industry of Grand Rapids growing during these years of the latter part of the century to consume over 50 million b.f. of lumber annually. One interesting and important sidelight of this trend toward the exploitation of hemlock and hardwoods was that they did not float well, and the "shrinkage" of the cut through losses increased and ate into profits so much that the movement of logs by sled and rail was emphasized.[46]

The impending decline was seen clearly by some lumbermen in the Lake States, who took steps to gain a holding interest in new lumber areas. In a sense, they always knew that they were operating on borrowed time. As early as the 1870s they began investing in southern pine, a trend that increased enormously during the 1890s. The Illinois Central Railroad cooperated by conducting tours for capitalists through the southern pine region, and people like Weyerhaeuser and his associates bought up hundreds of thousands of acres. By the turn of the century the same investment purchasing was happening in the Douglas-fir and redwood forests in the Pacific Northwest with the help of Pacific Coast railroads. Surplus capital was invested in mining and power development in the West. The Lake States lumbermen were clearly diversifying their operations. Some of the loggers simply died off, and their interests died off with them.[47]

Logging in the 1890s was becoming big business just when investment in equipment, in the longer hauls to market, and in stumpage were not paying. The price of prime stumpage rose phenomenally – whereas it was $2–$2.50 per acre in Michigan in the 1870s it had risen to between $8 and $12 by 1900, the biggest climb in the prices paid coming during the years from 1896 to 1900, if one could find the unbought lands. Some loggers were not prepared to pay these inflated prices and turned to papermaking with inferior timbers. There was a tendency for holders of timberland to consolidate their holdings and cooperate in order to economize in new logging methods, and as a consequence the little man was either squeezed into acquiescence or eliminated. These syndicates tended to pursue a policy of clear-cutting and to "cut out and get out," picking up and relaying tracks through the forests 20 times over in as many years in order to clear out the last substantial tree. Undoubtedly, clear-cutting was a response to high stumpage prices, timber scarcity, lack of capital or access to long-term credit, and also just plain greed, but the conception of the forest as a completely exploitable resource in which nothing was to be wasted did herald a new attitude. Clear-cutting was certainly the most effective method of cutting, and it even had Forest Service approval, for in its handbook of 1918 the Forest Service advised loggers to keep their "cutting area compact, and when making a skid take out all the timber tributary to it before you move on to the next one."[48] However, some individuals and groups of people were becoming concerned at these rapacious logging methods. Numerous pamphlets and tracts were written by Hough, Sargent, and Bernard Fernow, and by the governor of Wisconsin, and even some of the lumbermen themselves had come, by 1895, to favor some sort of state legislation in order to preserve the forests.[49]

The aftermath of depletion

Besides contributing to an awareness of the need to conserve the forest resource, the decline of lumbering resulted in two other changes: the decline and obliteration of the population and business activities of hundreds of communities, both large and small; and the effort of the lumbering companies and others to promote the agricultural settlement of the stump-ridden cutover lands that were once covered by the forests. Both of these changes had visible geographical implications for the landscape of the Lake States that had an enduring significance.

Declining towns

As the lumber industry declined, the population of the chief logging areas also began to decline. Wisconsin set up a Board of Immigration in 1895 to encourage settlement of the northern part of the state, where there were "dozens of cities and villages where the inhabitants have begun to wonder what will become of them when the timber is gone and the mills close down. Everybody has seen settlements very prosperous ten years ago which are now abandoned by almost all their former inhabitants."

Some lumber settlements managed to survive as retail and service centers for the surrounding rural communities, and a few more strategically located attempted to

Table 7.3. *Population of four western Michigan counties, 1880–1930*

	1880	1890	1900	1910	1920	1930
Lake	3,232	6,502	4,957	4,939	4,437	4,066
Manistee	12,532	24,230	27,856	26,688	20,899	17,409
Mason	10,065	16,385	18,885	21,832	19,831	18,756
Oceana	11,699	15,898	16,644	18,379	15,601	13,805

Source: After L. Alilunus, "Michigan's Cut-over 'Canaan,'" 192.

diversify their employment structure by attracting new industries. For example, after lumber milling declined by nearly half in La Crosse after 1890, the town encouraged the manufacture of agriculture implements, furniture, and confectionary and developed a foundry, a machine ship, as well as brewing, publishing, baking, plumbing, gas and steam fitting, tobacco, and planing mill products. Similarly, Oshkosh on Lake Winnebago developed an array of wood-using industries after 1870, such as door, sash, blind, and wagon and carriage manufacturing, and it became the country's largest producer of these items. Other industries based on the processing of agricultural products were also established. The town of Eau Claire followed much the same pattern of deliberate diversification. However, these were the successful centers, and many settlements were not so fortunate in attracting new forms of employment; ultimately, they disappeared from the map.[50]

Depending upon the location of the community within the general progress of what might be termed the lumber frontier, the decline in population became evident anywhere between 1890 and 1910, precision in these matters being conditioned by the decennial censuses. For example (Table 7.3), the four counties in western Michigan of Manistee, Mason, Lake, and Oceana, abutting Lake Michigan and straddling the pine-rich valleys of the Manistee and Père Marquette, experienced population increase through lumbering and some assisted agricultural colonization, and then decline. Lake County reached its peak in 1890 only to decline after that; Manistee peaked in 1900, and the other two by 1910.

The story of the life of the inhabitants of two typical declining sawmill centers in northern Michigan – Cheboygan and Alpena – was recorded in graphic detail.[51] In 1890 Cheboygan contained 6,235 people, and it was a thriving lumbering town, bustling with confidence about its future, having recently acquired a telegraph, telephone, railway, and paved main streets. Ninety percent of its income came from lumbering and lumber-derived industries, and there were 16 sawmills in the town with an annual production of 100 million b.f., two shingle mills, and numerous planing and processing mills.

> The smell of sawdust is in the air all over the town. It is impossible to escape it, for in those places where they are not sawing it they are burning it for fuel, or using it with slab wood to build docks, or spreading it on bar-room floors, or simply running it off into a huge pile to mold and rot.

After 1896 lumber production dropped rapidly because of dwindling local supplies and because the Dingley tariff cut off Canadian timber supplies, which were making up the shortfall. The mills were forced to turn to other timbers in order to provide the meagerest of farmers' supplies, like fence posts and barns and to acquire bark for tanning. By 1916 there were only two mills left in operation, and in 1928 the remaining mill burned down. The total number of manufacturing establishments of all kinds declined from a peak of 96 in 1899 to 26 in 1919 to 8 in 1939, and the number of employees in industry dropped from 1,282 to 225.

The contemporary account of the nearby lumbering settlement of Alpena was more graphic. Like Cheboygan, the town seemed to be made of sawdust:

> sawdust filled in the swamps, sawdust graded the street, sawdust extended the beach out into the lake; sawdust inclosed rows of piles or quays where the busy "dockwalloper" shoves the timber aboard the ship. But for the tall, fuming stack consuming the "pulverised plank" there would be a mountain like that of Cheboygan – sixty feet in height and ten acres in area. Until twelve years ago the rumble of a wheel or the beat of a hoof was never heard in Alpena. Now they have roadways on round cedar blocks.

As with Cheboygan, Au Sable, and many other milling towns, the timber supply of Alpena had already dwindled, and the economic future looked bleak. "Little is left of the elder order," wrote Rollin Lynde Hartt; "the whole land is rapidly being lumbered out." In addition, forest fires had "wrought a measureless havoc. Forests once dense with pine and hemlock, cedar and tamarack are left a sorry spectacle: beneath the underbush, above, the gaunt, infrequent skeletons of deadened, whitened, bark-torn trees. Only the Northern Peninsula lumbers as once Alpena lumbered." The lack of timber and the effects of the Canadian tariff were causing a reappraisal of what timber was left.

> Mills which formerly selected only the stoutest pine trunks now welcome the slender log, the crooked log, the rotten log, and the sunken log fished up from the river bottom. In place of beams for the western railway bridge or huge rafters for the Gothic church, Alpena busily turns out planks, shingles, spools, pail handles, veneering, and the wooden peg for furniture. It also makes manila paper out of hemlock pulp. It brings hemlock bark to its tannery. It combs its brains for inventions to utilize by-products, as does the Chicago pork-packer.

Undoubtedly, had logging and sawing been carried out more efficiently, many more hundreds of millions of b.f. of timber might have been won from the forests and thereby prolonged the life of the towns. The wasteful saws with their extravagant kerfs, the burning of refuse, the cutting of trees to leave high stumps, and the felling for roads and tracks all took their toll. Fire, in particular, was a curse in the forests, feeding on the slash that piled feet-deep on the ground after the timber was taken out, and it probably consumed about as much timber every year as reached the mill. The great Peshtigo fire of northeastern Wisconsin in 1871 devastated an area of 50 square miles and killed 1,500 people, and the Michigan fires of the same year consumed about 2.5 million acres. In 1885 nearly all the Wisconsin Valley was swept by fire, and in 1881 the eastern Michigan "thumb" was burned over, with the fire claiming 160 lives. In 1894 there was the great Hinckley fire in Minnesota that caused 418 deaths, and in 1894 and again in 1904 all the

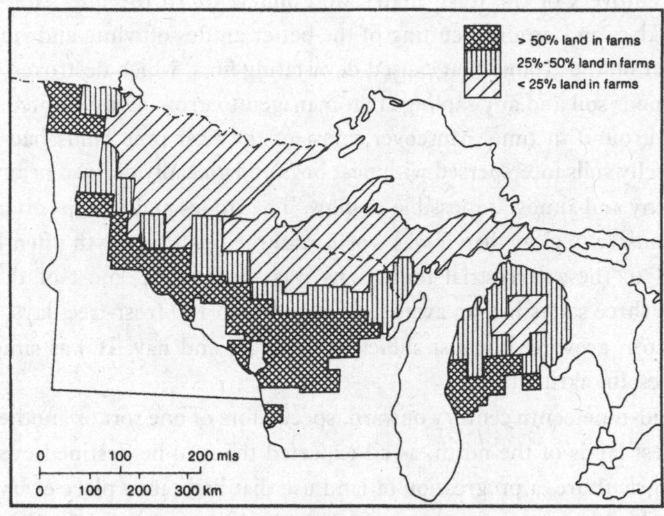

Figure 7.18. Lake States: cutover lands, 1920. (After J. D. Black and L. C. Gray, *Land Settlement and Colonization in the Great Lakes States*, 3.)

32 northern counties of Wisconsin were alight and several million acres of forest burned out. The majority of fires were caused by farmers trying to eliminate stumps and clear their partially cutover land, although locomotives also caused a great number of conflagrations.[52]

As the forest diminished and communities waned for lack of raw material supplies, the desire to "conserve" the resources that were left rose from the quiet murmur of the slightly eccentric and intellectual to the loud cry of practical people who saw their livelihood threatened.

The cutovers

Vast areas of the Lake States, probably totaling over 50 million acres and stretching from Lake Huron in the East to the Red River in the West in Minnesota, had been laid bare through the clear-cutting techniques of the highly mechanized and efficient lumber industry (see Fig. 7.18 for the extent of the cutovers in 1920). Unlike the hardwood forests of the eastern and central states, or even the hardwood forests of the southern portions of Wisconsin and Michigan, which had been taken up quickly for agriculture (which had been cleared primarily for that purpose, in fact), the stump-ridden land of the northern portions of the Lake States was a difficult legacy left behind by lumbering. The generally held solution to the problem was to sell this land for farming in order to bolster the local economy, now bereft of lumbering, and so expand the local tax base.

What had happened already in New England and particularly in western New York was scarcely any guide as to what to do with the cutovers because cutting there had been selective and destructive fire had not been prevalent, so that a certain amount of regrowth occurred, allowing the land to remain suitable for a mixture of forestry and farming. The cutovers of the Lake States were different. In the quest for immediate profits, the reckless and prodigal cutting of the better grades of white and jack pine had left a slash cover on the ground that caused devastating fires, which destroyed the humus in the already poor soil and any saplings that managed to grow. Only stunted bush grew to occupy the ground in time. Moreover, most of the best pine lands had been light sandy and gravelly soils interspersed with peat bogs and glacially derived heavy clays, the latter being stony and almost impossible to plow. The stones and stumps often made the land more expensive to clear than it was worth, and the bush regrowth often hid its true nature. Added to these terrestrial limitations was the climate; most of the northern portions of the three states had an average of only 100 to 130 frost-free days, which was too short for corn growing but just sufficient for grass and hay. It was simply too far north for successful agriculture.[53]

From the mid-nineteenth century onward, speculators of one sort or another had been eyeing the forest lands of the north, as all expected them to be destined eventually for dairying and agriculture, a progression of land use that had taken place everywhere else in the country.[54] There were, for example, large-scale speculators like Caleb Cushing with his great European-American Emigration Land Company, formed in 1868 with the intention of bringing Swedish farmers to the St. Croix district of Wisconsin. There were also land-grant railroads, such as the Wisconsin Central and the Chicago, St. Paul, Minneapolis, and Omaha, which held huge areas of land, and they pursued a policy of inducing American and European settlers to take up land alongside their lines. The Wisconsin Central and its agents, in particular, worked closely with the state Board of Immigration, advertising low-priced land and the advantages of living in the northern forests, in booklets printed in English, Swedish, and German. This was at a time in the early 1870s when home seekers were deterred by reports of drought and grasshopper plagues in the Great Plains and the approaching end of free land, and when the Lake States were equally determined not to be bypassed by the sweep of agricultural settlement. Agents worked in the large urban centers of Europe and America and sold land to the settlers on the usual company terms of approximately $5 per acre, with a $50 down payment and the remainder repayable in three annual installments at 7 percent interest. The railway companies hoped not only to sell their own land but to gain from the business that would arise as the land alongside their lines was settled, and they deducted fares and moving expenses if the homesteader actually bought land, even if it was not their own land. By 1890 the Wisconsin Central had disposed of 250,000 acres of its 838,628-acre grant, and the rate of disposal varied from 4,000 to 20,000 acres per annum during subsequent years. The companies were not concerned to "group" settlements, and consequently many settlers were located in out-of-the-way places destined to be the first to be abandoned in later years.

Undoubtedly, the movement to settle the cutovers was aided by the promotional activity of the Wisconsin State College of Agriculture and the publication of *Northern Wisconsin: A Handbook for the Homeseeker* by Dean William A. Henry, first published in 1896 at a time when there was a general upward thrust in enthusiasm about settling the cutovers. Its 192 pages and numerous illustrations of thriving farms played down the problems of the short growing season, the sandy and infertile soils, the toil of stump clearing, and the harsh struggle for existence that was the lot of the average settler in his tarpaper shed. The *Handbook* was peppered with testimonial letters of success from settlers, which presented the northern half of Wisconsin as a very attractive place in which to live. The towns were already there, Henry argued, supplementary employment existed in the lumber camps, and the land values would rise as settlement expanded and intensified. He was convinced that "northern Wisconsin will not revert to wilderness with the passing of the lumber industry, but will be occupied by a thrifty class of farmers whose well-directed intelligent efforts [will] bring substantial, satisfactory returns from fields, flocks, and herds."

Such official "booster" literature was common, and even the local newspapers urged the settlers to go north. State pride in seeing the northern areas well settled was a potent factor in the settlement of the cutovers and in their "rehabilitation." Pamphlets and literature were full of misleading statements and glowing case histories of success. The same boosterism was true of Minnesota, although to a lesser extent, and also at a slightly later date in Michigan, where the forest lands, it was said, merely awaited "the coming of the sturdy farming element to transform them into beautiful and productive farms and gardens." As one publicity bulletin for Michigan put it, "For a poor man to get a start on a farm, I would say, stay in Michigan, as land is cheap and you can raise almost any kind of crop on it, and if a man is not a hustler he will not make a success anywhere, if he can't in Michigan."[55]

The ruthless business methods of many of the Lake States lumbermen did not allow the investment of either the time or the money required to restore the land to its previous forested state. Rather, they preferred to move on to the South and then, later, to the Pacific Northwest. Naturally, some of the lumber companies, like the land speculators, saw the profit to be gained from turning the existing stump land into agricultural land, thereby lessening local county taxes on idle land. Around the turn of the century, some companies turned into settlement and colonization companies and lavished praise on their northern lands. For example, the Weyerhaeuser lands were partially conveyed to the American Immigration Company; the cutover land of Owen and Rust was conveyed to the Owen Lumber Company; there was a Rietbrock Land and Lumber Company; the land of H. W. Sage was sold by the Sage Land and Improvement Company; and there were many more.[56]

Frederick Rietbrock, for example, founded a commercially progressive settlement of German settlers in Athens in Marathon County, Wisconsin, based mainly on summer dairying and winter hardwood lumbering. Settlers came mainly from Milwaukee, but some came directly from Germany. The introduction of better-bred dairy herds was successful, and one of the settlement's Guernseys held the world's record for production of butterfat in 1905. Perhaps this was the most successful example of community farming in the cutovers. Another similar, though not so successful, venture was that of Heronymous Zeck, who founded a colony for Polish settlers from Chicago, Milwaukee, and Poland at Crivitz, in Marinette County in 1895; unfortunately, the colony went bankrupt by 1901. An alternative arrangement was for the lumber companies to sell their land to real estate speculators, who then sold to the settlers. There was usually a far lesser element of planning in these ventures, and sales were made without any forethought of what the land was like and how the settlers would farm it. One of the most flamboyant and prosperous of these operators was James ("Hustler") Agen of Superior, who is said to have sold $6 million worth of real estate in a decade. More conservative was R. Stannard Baker, whose firm had an annual turnover of about $1 million by 1900. But the biggest cutover colonist of all was the indefatigable promoter James Leslie Gates, or "Stump Land" Gates, as he became known, who at his peak controlled over 500,000 acres of stump land. Despite his rapid turnover he could not get enough land from the timber companies to remain solvent and went bankrupt in 1904. Later, Gates exercised his entrepreneurial expertise and tried to get J. P. Morgan and Frederick Weyerhaeuser to join him in a massive billion-dollar corporation to corner all the timber on the West Coast. These were the big operators, but in every village in the cutovers there was someone who sprang up to sell land to the settlers once the timber was cut down, mostly with exaggerated and misleading statements guaranteeing independence and financial success in two or three years, while neither the average rate of success nor the failures were mentioned.[57]

The other course of action open to the timber companies was to allow the land to lie idle after lumbering and let it be taken over by the state or county authorities in lieu of taxes. Tax-delinquent lands became a serious problem in time. Undoubtedly, too, resentment was felt by the local residents over the absentee ownership of such large blocks of pine land, and it became common for residents in the counties and towns in which these holdings were located to vote themselves extravagant local improvements, the costs of which were then levied on the absentees. This high taxation on the land then induced timber companies either to sell some of their lands or to clear-cut the timber prematurely, even if the prices were low. What resulted was a vicious circle in which rapidly rising taxes encouraged cutting and abandonment, but as abandoned lands paid no taxes a heavier burden was placed on the uncut land in order to compensate for the losses, which was yet another encouragement for the timber companies to migrate elsewhere. By 1907 there were 31.3 million acres of cutover land in the three states: 10.7 million in Michigan, 15.1 million in Minnesota, and 5.5 million in Wisconsin.[58]

Needless to say, in these northern lands the struggle to farm was harsh, and there was a high rate of failure among the settlers; in northern Michigan the rural population fell by 56,000 between 1910 and 1930. Of those that remained, a large number were located on either submarginal land or land irredeemably stripped of its timber, and they led a wretched life. Undoubtedly, land sales did reduce taxes by increasing the number of taxpayers in the district, but at a very great cost. The population of the 24 northernmost counties of Wisconsin had risen, from slightly over 120,000 in 1880 to 400,000 in 1900, but many were in lumbering companies and towns and not on the land. The hard sell on the cutovers was still to come, during the opening years of this century with the upsurge of immigration to the United States and to the Lake States in particular as the availability of free land declined in the West, but that is a story to be told elsewhere. The point is that the stage had been set and the cutover problem created, a problem, moreover, that was to be repeated in the South and to a much lesser degree in the Pacific Northwest lumbering areas. The northern portions of the three Lake States were dotted with many unpainted and sagging farmhouse structures, some mere tarpaper shacks,

Plate 7.5. The charred stumplands of the cutovers of northern Wisconsin. It was in such areas that settlers were encouraged to take up land and farm it.

derelict fences, with the occasional lilac bush and pile of stones heaped up in the fields that were virtually a desert, a mute and melancholy testimony of the abandoned homes and abandoned hopes of the ill-advised attempts to settle the cutovers (Plate 7.5).

Perhaps the best summry of the impact and results of the lumberman's assault on the Lake States forests is related in the events in the lifetime of one man, deeply committed to the cause of forestry. When a young man, Filibert Roth went to Fort Worth, Texas, in 1874 and was shown a Texas house built of Michigan white pine by a cattleman. About 30 years later Roth was in Manistee, formerly one of the premier lumber-producing settlements on Michigan's western coast. The city was in the process of building a new bathhouse, and the building contract specified that red pine from the Lake States was to be used for its construction. But the builder wanted a release from this restriction in his contract so that he could use long leaf pine from Georgia, which cost less than the locally grown product.[59] The South now dominated the lumber trade. Nevertheless – despite the emergence of new regions of production – the big mills, the continental transport system, the great fires, the cutovers, and the aggressive application of new technology and business organization assured to the Lake States an indelible image as the epitome of destructive lumber exploitation. In that the region still reigned supreme.

8

The lumberman's assault on the southern forest, 1880–1920

> Obvious to the native, enigmatic to the outsider, difficult of depiction by the human geographer, the piney woods possess a life and language of their own.
>
> Rupert B. Vance, *The Human Geography of the South* (1932), 109

The southern states were the next major region to participate in the continuously expanding wave of forest exploitation that swept across the United States. Between 1880 and 1920 production of lumber in the southern states rose dramatically from 1.6 billion b.f. to 15.4 billion b.f., the peak year of production coming in 1912, just at a time when output in the Lake States was declining so rapidly. The South was producing 37 percent of all the lumber of the United States in 1919, a position analogous to that of the Lake States some 30 years before, and for the opening decades of the twentieth century the South, together with the adjacent region of the South Atlantic seaboard, was producing only slightly less timber than all the other regions of the United States combined.

The onslaught across the southern forests reduced the original woodland area by nearly 40 percent, from nearly 300 million acres to 178 million acres by 1919, by which time a mere 39 million acres remained as virgin forest. Of course, agriculture also played a part in this reduction, but by far the greatest inroad was made by lumbering. In the pine forests alone, it was calculated that there was a staggering 90 million acres of cutover by 1920, of which only one-third was restocked with sawable timber, one-third with scrubby timber capable of being used for cordwood only, and one-third remained completely bare and barren.[1] Lumbering had left an indelible imprint on the landscape of the region.

Forest exploitation was not new to the South. The longleaf (*Pinus palustris*), shortleaf (*P. echinata*), loblolly (*P. taeda*), and slash (*P. elliottii*) pine forests had been felled for agriculture long before the Lake States had even been discovered, let alone settled. Naval stores had been gathered since the early seventeenth century, and lumbering had been carried out in a scattering of locations along the 4,000 miles of coast line from western Texas, around the Gulf and Florida, and as far north as Chesapeake Bay, as well as along the banks of the Mississippi and other large, navigable rivers that penetrated the pine forests.

Nevertheless, it was not until the 1880s that the acceleration in commercial lumbering came with the sudden and massive transfer of capital, technology, and know-how from the North, particularly from the area of declining resources in the Lake States. After that, the momentum of change was rapid, and eventually, by 1920, it even promoted the development of new uses of the forest through the medium of chemical processes, such as the development of paper, pulp, chip boards, and resin-backed boards – all in the space of 40 years.

Consequently, the story of the exploitation of the southern forests is lengthy and complex. The shift of capital and technology into the region during the 1870s and 1880s must be considered, as must the nature of that new technology and the way in which it promoted change in the forests, particularly at a critical time between the centuries when attitudes toward the environment and toward resources generally were undergoing serious examination.

The northern invasion

The steady rise of activity in lumbering in the South after the Civil War was largely the result of two quite different tendencies at work in the nation. There were many in the South who looked forward to greater industrialization. The war had demonstrated the basic weakness of an agricultural economy and society, and defeat had led to a determination on the part of many southern leaders not to seek solutions through politics but rather to adopt and emulate the methods of business, enterprise, and industry found in the North. However, despite the eagerness to acquire industry, the local resources for such change were not great. There was little capital in the region, and most of what there was had been tied up in land, slaves, and expensive houses. The one resource the South did possess in abundance, however, was land and its vegetation cover of timber, and that was thought to be the basis for future industrial growth.

Thus, the South during Reconstruction was a classic case of an underdeveloped region where the first industries to be developed were extractive and exploitative. The economy was lopsided, the land and its abundant resources were plundered in pursuit of quick gain, and little employment was generated through refining or manufacturing.[2] In a sense, it was the continuation of an old story that was encapsulated in the reminiscences of William Gregory of Graniteville on the South Carolina/Georgia state line. In 1860 he bemoaned the lack of domestic industry and the dependence on the North. He could not purchase a locally manufactured wooden bucket; his old homemade bucket had lasted for 11 years, but its Yankee replacement for only one year. While southern craftsmen starved from lack of work, northern workmen made quick cash profits. "Now," said Gregory, "such a sight [as a homemade bucket] is rarely seen, the homemade article being obliged to stand out of the way and make room for the better finished, finer looking wooden ware from the North." He continued:

> Nothing is more common now than to see at our store doors, hubs for log cart wheels which are often carried fifty miles into the country, often having travelled perhaps a thousand miles from the makers' hands, while it would be easier for the cart builder to get wood and turn them than go ten miles for them.[3]

What manufacturing of wooden articles did exist was confined to a few centers such as New Orleans, Savannah, and Mobile, and these did little more than service the local needs of those centers.[4]

At the same time as the search for industry was occurring in the Reconstruction South, the timber resources of the Lake States were being depleted. The vast areas of straight-grained southern pines – the shortleaf, the longleaf, the loblolly, and the slash –

looked like increasingly reasonable substitutes. It is true that there was a prejudice against the southern timber, as it was commonly thought that the tall, very fast-growing trees of a warm climate were not as strong as the slow-growing trees of a cold climate. Nevertheless, the prospect of a dire shortage by about 1880 quickly dispelled these notions.[5] Moreover, the Lake States lumbermen had learned that, together with the minimization of transport costs, the key to low cost and sustained production was the possession of cheap stumpage, and that, of course, could be obtained only if one got into the land market before the speculators. Undoubtedly the greatest profits in lumbering were made if the stumpage and the mill were owned by the same person. As one Michigan lumberman said, "Almost every man in Michigan who has made money out of pine has made it by the rise in stumpage rather than by manufacturing and selling it."[6] Therefore, the aspirations of the southern boosters and the needs of the northern lumbermen came together in the acquisition of land.

Initially, the southern lumber industry existed on the cash purchase of logs from a variety and scattering of independent landowners. The supply was adequate to satisfy the small-scale mill owners. What was not cut legally was cut illegally, either by purchasing a 40-acre section under the 1862 Homestead Act and then proceeding to clear the forest all about or by employing "dummy" homesteaders who, in time, relinquished their hold on their land to the timber company. Few people seemed to bother about the illegal taking of lumber; there was so much of it. This arrangement satisfied the small-scale operators, but larger mills were becoming established during the 1880s and the integration of an assured lumber supply with the production line became necessary for the continuous running of the mill. In any case, the cost of stumpage was so little that it was worth ensuring supplies through legal purchasing.

The previous system of unrestricted entry was upset when the victorious North replaced the 1862 Homestead Act with the 1866 Southern Homestead Act, which restricted land entries to 80 acres.[7] The object of the new act was to end cash sales and to promote small landholders, such as freed slaves, poor whites, and impoverished immigrants, on public land. In addition, an underlying aim was to break the hold that the old slaveowning aristocracy was perceived to have on the land and, certainly, to prevent further aggrandizement of estates. However, by the mid 1870s the southern land question ceased to be tangled up with Reconstruction issues and had become a problem of economic and business policy. Southern leaders wanted the act repealed, claiming that few entries had been made because the land was unsuitable for settlement, that few large agricultural estates had been created, and, in particular, that southern progress toward industrialization was being hindered by the restrictive operation of the act, which discriminated against lumbering, mining, and industrial purchases. They believed that the deplorable economic condition of the South would be relieved and its lumber industry stimulated by throwing open all the public lands to unrestricted purchasing.

Already there were signs that northern capital had entered the South in the form of massive land purchases, and this was a portent of future moves. With an almost pathetic eagerness, the southerners wanted more. Almost to a man they voted for the repeal of the act in 1876, and after an interval of five years, while the Public Land Office was reorganizing itself, all the land in the five southern states was opened for sale. In the meantime, Charles S. Sargent's report on the forest resources of America was published. It contained the estimate that there was a massive 333 billion cu. ft. of timber in the South compared to only 84 billion in the Lake States.

Speculators and lumbermen capitalists, their pockets bulging with the profits from their activities in the Lake States, came to the South and laid out enormous sums of money; the Illinois Central Railroad even ran special trains from Chicago to Mississippi and Louisiana, in particular, for their benefit. Companies sent "cruisers" to the South to assess the potentialities of the forests, and they returned with glowing reports of the unlimited resources at bargain prices, together with cheap labor and relatively flat and easy terrain for logging. One of the cruisers was James D. Lacey, who was also one of the largest land speculators. He later recounted his experience:

> In 1880 when I first went South ... we estimated what the value of Government land was. It was nearly all vacant then; and it was timberland. In 1889 it was offered at $1.25 per acre. We located several million acres for northern lumber companies. We estimated those lands would cut about 6,000 feet per acre, as they were then cutting timber. They were not going above the first limbs, the balance was left in the woods and burned up.[8]

If timberland could be purchased for as little as $1.25 an acre and it carried 6,000 or even, in some cases, 12,000 b.f. per acre, then lumber was being purchased at about 5–10¢/1,000 b.f., when average quality timber was being sold at $5–$10/1,000 b.f.

Generally, the speculators preceded the lumbermen, who were then obliged to buy back stumpage at inflated prices. But even those lumbermen who accompanied the first wave of purchasers seemed more concerned with speculation than lumbering, especially in Louisiana and Mississippi. For example, N. B. Bradley, a prominent Bay City lumberman, purchased 111,188 acres in Louisiana, all of which he subsequently sold off to other lumbermen. Similarly, a group of Chicago capitalists bought 195,804 acres, which they later sold to the Long Bell Lumber Company. The Brackenridge brothers of Oscoda, Michigan, bought and then resold 700,000 acres of southern pine and baldcypress lands in the same two states, and James D. Lacey of Michigan is known to have purchased and then sold 107,461 acres, although there is every indication that he must have been involved in many more transactions; for example, he is said to have purchased 160,000 acres in Calcasieu, Vernon, and Rapides parishes in southwest Louisiana.

Among the purchasers were some genuine lumbermen, such as Delos A. Blodgett of Grand Rapids, Michigan, who was reputed to have built the largest mill in Michigan in 1878. He acquired over 126,000 acres in Mississippi, an amount he later increased to a total of 721,000 acres by the purchase of state land in 1906, mainly in Jackson, Wayne, Greene, Perry, Marion, and Pearl River counties, only to resell much of it to other lumbermen between 1908 and 1912 for between $30 and $40 per acre. William C. Yawkey of Bay City took up at least 47,176 acres in Alabama, Florida, and Louisiana; Charles H. Hackley of Muskegon took up 89,000 acres in the Calcasieu basin, and Isaac Stephenson of Wisconsin purchased some 70,000 acres of federal land in Louisiana, which he pushed up to about 400,000 acres with purchases of state lands, later abandoning his mill in Menominee, Michigan, and relocating in Alexandria, Louisiana. Among the most perceptive of the purchasers were Henry J. Lutcher and G. Bedell

Table 8.1. *Purchases of 5,000 acres or more of federal lands by northerners and southerners in five southern states, 1880–8*

	Northerners		Southerners		Northern purchasers as % of all purchasers
	No. of purchasers	Acres purchased	No. of purchasers	Acres purchased	
Louisiana	41	1,370,332	9	261,932	83.9
Mississippi	32	889,259	11	134,270	86.8
Alabama	7	121,983	24	463,242	20.8
Arkansas	7	114,334	10	183,946	38.3
Florida	6	64,243	12	125,172	33.9
Total	93	2,560,151	66	1,168,562	68.6

Source: Paul W. Gates, "Federal Land Policy in the South, 1866–1888," 319–25.

Moore, who originated from the declining lumber center of Williamsport, Pennsylvania. Initially, they searched for timberland in Michigan and Wisconsin but finally purchased over 500,000 acres for as little as 50¢ an acre in eastern Texas and western Louisiana. They then moved their milling operations to Orange, Texas, where they were among the largest lumber producers in the South. Some local men saw and understood the significance of what was happening and also attempted to get into the land market. One such was John Henry Kirby, who slowly built up a holding of over 368,000 acres by 1907, initially while buying land for his employers primarily for its petroleum rights.

The participation of northern purchasers in the large-scale sales of federal lands was overwhelming in Louisiana and Mississippi, where the Illinois Central Railroad and the Mississippi River were key attractions (Table 8.1). Overall, nearly 69 percent of the 3.7 million acres of land sold in lots of over 5,000 acres were purchased by lumbermen from Chicago, Michigan, and Wisconsin, including Frederick Weyerhaeuser, "who was rapidly becoming the king of them all," and it is highly probable that a considerable portion of the remaining 31 percent acquired by southerners was made possible only by northern capital. Over and above these large-scale purchases by Americans from both North and South was the activity of such outsiders as Daniel F. Sullivan of England, who acquired about 150,000 acres in Alabama and 100,000 acres in Florida. He virtually controlled the lumber port of Pensacola through the ownership of railways, wharves, and timber yards, and by the time of his death in 1885 he was fast getting control of the larger and more important lumber center of Mobile.

In all, Paul Gates calculates that 5.7 million acres were sold by the federal authorities in the five southern states between 1877 and 1888. But large as this figure is, it comes nowhere near the amount disposed of by the states, which sold off tens of millions of acres of "swampland" around the Gulf for prices as low as 25¢ per acre. Undoubtedly the largest and most spectacular of these sales was that by Florida in 1881 of 4 million acres at 25¢/acre to a syndicate headed by Henry Disston, the Philadelphia saw manufacturer. At least 1.6 million acres of the Florida land was purchased by Sir Edward Reed of England, who had railway interests in the United States. This was not an isolated incident, as a Scottish company took up 500,000 acres of Florida timberland in 1883, and during the same year another British syndicate had purchased 1.3 million acres in the Yazoo delta lands of Mississippi and another 1.8 million acres in Texas.[9] Reuben McKittrick has estimated that by 1885 Texas had sold or given away about 32 million acres of its state land, much of it heavily forested.[10]

In the beginning the influential trade journal, *The Southern Lumberman*, had encouraged northerners to make these "lucrative purchases," but in time it too became alarmed at the extent and rapidity of the purchases, commenting that southern lumbermen had "stood by and looked complacently on, not seeming to realize the fact that the material for their future business was quietly but surely slipping away from them."[11] Though this trade journal was one of the first to realize that large-scale purchasing had brought little economic advantage to the region, it was not the only voice raised in alarm. Others became aware that much of the land was in absentee ownership, often by overseas residents, who were merely waiting for the rise in the price of the land before selling. It was clear too that many northern lumbermen had bought land merely in order to keep out competitors with mills in the Lake States, and others cut the lumber only to ship it to their mills in the North for finishing and fabrication. After nearly 20 years of endeavor, the South was nearly as bereft of industry as ever, and now its one resource, abundant land and its lumber cover, was being sucked out of the region. Perhaps, after all, a few thousand small-scale farmers might have been a better basis for prosperity for the region than its semicolonial dependency on the North, and local observers even began to express doubts as to the short-term availability of the timber resources left.

With these doubts and fears in mind, the southern legislators did a complete about-face in 1885 and began to cooperate with other reformist groups, such as antimonopolists, conservationists, and land reformers, who were all becoming concerned with the diminution of land for genuine farm settlement throughout the United States. With the change in mood, cash sales and homestead entries jumped from 200,000 acres in 1886 to 883,000 acres in the next year and to a dizzy 1.224 million acres in 1888 as the speculators and lumbermen made their last bids for southern forest lands. In desperation, a bill suspending large-scale purchasing in Alabama, Arkansas, and Mississippi was pushed through the Senate in April 1888, and within a month Louisiana and Florida also wanted to be associated with it. But it was too late; federal policies designed primarily for the disposal of agricultural lands were unworkable when applied to nonfarming regions. The best southern timberlands were owned by the North, and the product and the profits of the South were being creamed off for the benefit of people and organizations outside the region. Millions of acres of land for which the principal and best use was the growing of trees were being permanently damaged in the process.

Besides the very real tragedy of the decline of agriculture, society, and political life that the Civil War bequeathed to the South, perhaps the biggest and most permanent loss lay in the inability or failure of the southerners themselves to see that their vast timber resources offered a considerable degree of economic security and even prosperity. In every way, the story of William Gregory's wooden bucket was symbolic of what was happening throughout the whole of the southern economy.

Technological changes

The lumberman's attack on the southern forests had occurred before the massive land purchases of the 1870s and 1880s. Nevertheless, the purchases heralded a recognition of the existence of the extensive and untouched supplies of cheap pine and bald cypress and the commitment of the northern lumbermen to produce in the South. The technical knowledge and expertise, as well as the business organization that had grown up and flourished in the Lake States, flooded into the South as assuredly as had the capital that had initially bought up the land. Production, which had been stable but low, being geared to largely local needs, shot up sharply after 1880 as the new technology of exploitation was adopted widely; by late 1890 the curve of production had taken another upward turn, and by 1900 the burgeoning lumber industry of the South was producing more lumber than any other region of the United States. The enormous increase in production could not have been achieved without important technical changes in the cutting, moving, and marketing of timber, which in a chain of circular causation fed and even created the demand for greater production.

Transport

Water: rafting and log drives. It is commonly supposed that, because the development and spread of the lumbering industry in the South accelerated and first became evident during the later 1880s and 1890s, it was all based upon the railroad and that river and water transport was of minor significance. But in focusing upon the crucial role played by railroads we must not forget that the lumber industry had been in existence here for nearly 200 hundred years before the first railroads appeared in the United States. Before Independence portions of the Gulf South had had three national masters, the Spanish, the French, and the British. Each country attempted to ensure supplies of timber to its widely scattered colonial trading posts and agricultural settlements around the Gulf and Atlantic coasts, and each also promoted a complex export trade to other far-flung colonies in the Caribbean and South America, as well as to its homeland. Clearly, in this early trade, water-borne transport on river, swamp, and sea was the prime means of moving the bulky logs and lumber.

It is true, of course, that in the South as compared to the Lake States the forces of nature, in the form of swift-flowing water in a dense network of streams, minimal friction of snow and ice, and sudden spring thaws that filled the splash dams and sent the logs thundering downstream, were almost absent – except perhaps in Kentucky, Tennessee, and West Virginia.[12] Nevertheless, the South had rivers enough. Heavy rainfall during the summer ensured constant and even-flowing streams that crossed the pine forests for hundreds of miles. Moreover, most of the trees cut before about 1920 were easy to float as they contained a large proportion of heartwood, the product of 150 to 200 years of uninterrupted growth. The relative "flatness" of the South meant that rivers had a low gradient, that they were wide, slow-moving, and circuitous as they meandered through their flood plains, which did not make them good log-driving streams, but they were eminently suited to all other sorts of navigation. Hence, log drives were restricted to fast-flowing headwaters and small tributaries and to a few major rivers such as the Pascagoula and Pearl in Mississippi. Generally speaking, the rafting of logs and timber was much more common.

In the states facing the Atlantic coast rafting was fairly well established by the beginning of the nineteenth century. Cotton and rice plantation owners as well as small farmers had been moving farther inland for many years, and during the nonagricultural winter months they cut wood for the ready profits that were to be made by exporting lumber overseas. The logs were cut near the rivers, snaked out by oxen, and then sent downstream in rafts of either logs or roughly cut timber, which was sawed on the spot by small water-powered mills. The rafting was often done by independent operators, usually the small-scale farmers, who took time off during the slack season in winter to earn much needed cash, although sometimes they were satisfied to be repaid in kind, like the Mississippi farmers who were content if they got "plenty of pickled pork, cheap family flour, coffee, brogan shoes and a jug of joy."[13]

During the 1850s, for example, there were some 50 small mills around Aiken near the headwaters of the Savannah and Edisto rivers supplying cut lumber to Savannah and Charleston. Every 4,000 b.f. made a "bull," which was taken downstream by a crew of eight, who later returned upstream by steamboat or, in later years, by railroad. Compared to railroad freight charges at this time, water transport was much cheaper; if a lumberman lived, say, 40 miles from Charleston it would cost him $2.25/m.b.f. to move his timber by railroad or about a quarter of the price received for the product on the market, compared to about $1.50 if rafted. The price differential was so great farther south on the Savannah and Altamaha rivers that rafts of 50,000 b.f. were common.[14]

Such were the inroads of lumbering into the forests along the Atlantic Coast that by the time Sargent compiled his map of the forests in 1884 to show "areas from which the merchantable pine had been cut" there were narrow bands of cleared land extending from three to five miles from the sides of most of the rivers crossing the coastal plains (Fig. 8.1). In North Carolina the land alongside and between the Roanoke, Neuse, and other rivers leading into Albemarle and Pamlico sounds had been so affected over the centuries that it was said that "a larger proportion of pine forest on the coast had been destroyed . . . than in any other southern State." The Fear and its numerous tributaries, which focused on the exporting center of Wilmington, were likewise stripped, and loggers were already moving farther afield for logs. In South Carolina, Georgia, and Florida each river was a corridor of exploitation – the Little Pee Dee, Ogeechee, Altamaha, St. John, and many others. Only the Great Pee Dee and Santee were untouched because they were bounded by enormous, deep swamps, which were "ill-suited to lumber operations," although even here lumbermen hacked away at the trees from boats whenever summer floods allowed them to sail over the swamps.[15]

Despite these developments, it was on the rivers flowing into the Gulf waters that the

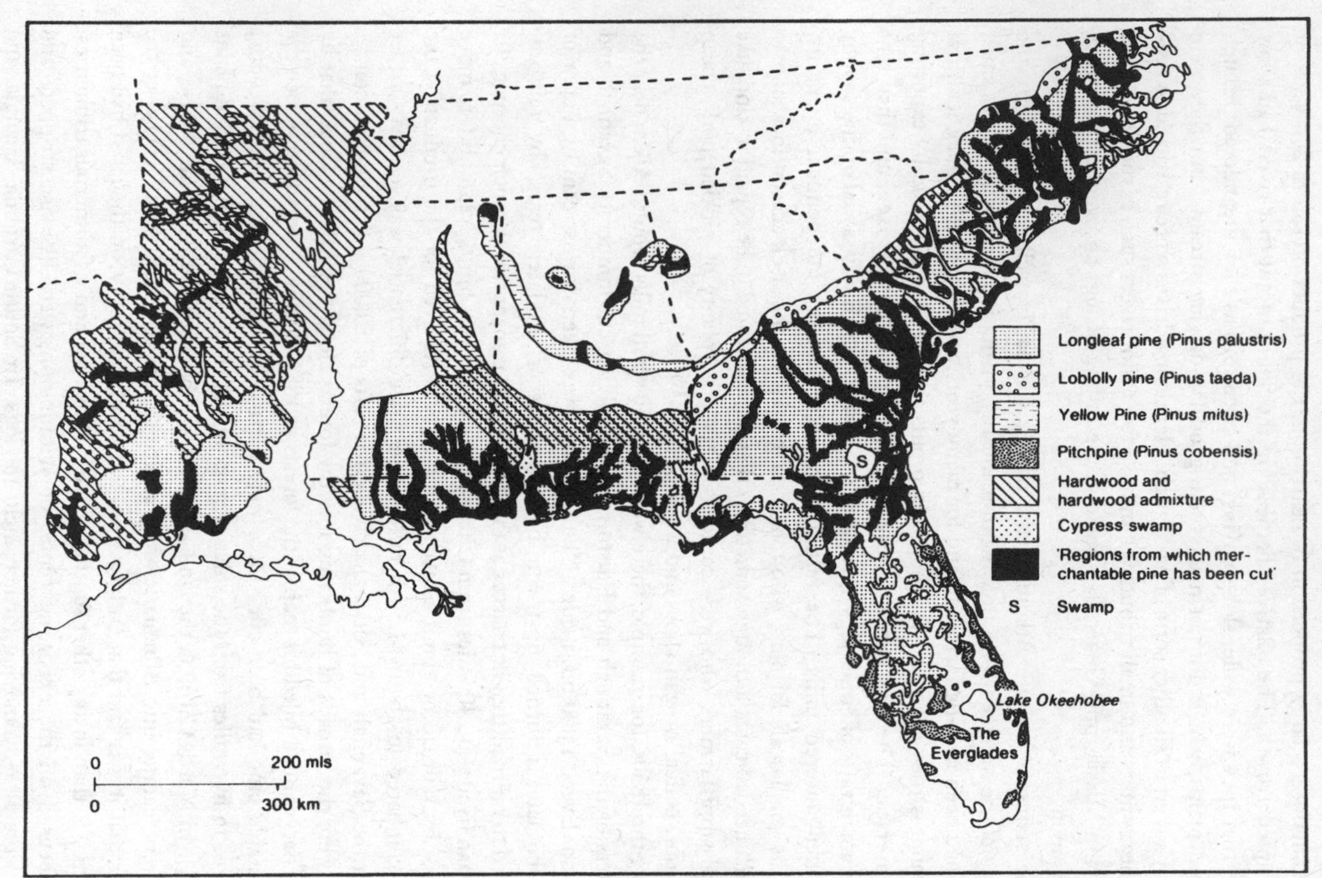

Figure 8.1. Forest conditions in the South, c. 1880. (After Sargent, *Report on the Forests*, maps opp. 518, 520, 522, 524, 530, 536, 541, 544.)

greatest and longest-lasting development of water transport took place. The banks of the Suwannee with its milling town of Cedar Keys, the Flint and Chattahoochee converging on Apalachicola, the Escambia and Conecuh and other streams converging on Pensacola, the Tombigbee and Alabama on Mobile Sound, had all been stripped of accessible pine timber a long time before 1880. The Mississippi system, the Red River, the Calcasieu, and the Sabine all handled untold quantities of lumber, much of it destined for the piling and building of New Orleans, the fastest-growing town in the South during the Reconstruction years, and also for a thriving export trade.[16]

Lumber on these streams came down by rafts because loose logs would have been a hazard to shipping. On some of the faster-flowing rivers, however, there is evidence of log driving. For example, on the Pearl and Pascagoula in Mississippi, over 5,000 different log marks were recorded, and logjams were common. There was a boom at Ferry Pass on the Escambia above Pensacola capable of holding 50,000 logs, a boom on the Choctawhatchee that held 10,000 logs, and one even on the Sabine.[17] Whether the whole of the river was used for log driving is difficult to say, but it is unlikely. The experience of the Pascagoula and other rivers like the Escatawpa, Biloxi, Wolf, and Jordan is instructive; on the Pascagoula, above the junction with the Red and Black rivers, log drives were normal in the fast-flowing tributaries, but on reaching the main river channel the logs were made up into rafts to be floated down to the mills at Moss Point on the Gulf Coast. However, during the early 1890s the great amount of lumber coming down the Pascagoula made a compromise system attractive. Loose logs were grouped together in a rapidly and easily constructed "bull pen" (the cattle-driving analogy being evident again), which consisted of a number of logs tied together to form an enclosure of about 100 logs' capacity; it was a curious local variation, somewhere between log driving and rafting. They were then sent downstream, and "So thick were the moving rafts, that they were seldom out of eye-sight of one another from the junction of the Leaf and Cickasawhay to Moss Point." To one oldtime raftsman, it seemed that "the reflection of the firelight [of the raft fires] on the water made the river appear in flames." When they arrived at Moss Point, they were broken up and sorted out inside an enormous boom that extended from the middle of the river and five miles up the eastern bank, the western portion of the estuary being reserved for navigation. The heyday of rafting was the period from 1890 to about 1910, during which time up to half a million logs came down the Pascagoula annually; though a large amount, this was only one-third of the logs rafted out from the St. Croix, Wisconsin, during 1883. By 1910 there was little timber left near streams in Mississippi, and by 1915 only the "deadheads," or sunken logs dredged out of the rivers, were left for milling.[18]

Thus, while rafting and log drives did form the basis of some transportation links on a regional scale, such links were of only minor importance and extent compared to those developed in the Lake States and the Northeast. Nor did water transport, except on the Atlantic Coast, aid the movement of timber to domestic markets. The Gulf Coast was oriented more toward the export markets overseas and to Latin America. The building of the north–south railroad system to link the forests with the markets was necessary before the lumber industry could develop fully.

Water: the baldcypress swamps. There was only one other part of the southern forest where water transport was supreme, and that was in the baldcypress swamps located in about 55,000 square miles of the Lower Mississippi Valley in the Yazoo lowlands, and particularly in the delta area of Louisiana below Baton Rouge, in most of the northern half of Florida, in the southeast corner of Georgia around the Okefinokee Swamp, and along the courses of individual rivers like the Little and Great Pee Dee and Santee in South Carolina or the Flint and Chattahoochee in Georgia.

The baldcypress was highly prized because of its durability and resistance to termites and to rotting under humid conditions (as well as for its moderate softness for working), and it soon acquired a reputation as a roofing material and even for the construction of tanks and cisterns. In its natural state it grew in a wide range of habitats and consequently produced a variety of color and form that was mistaken for various species, when in fact it was a single species, *Taxodium distichum.*[19]

Despite the wide range of habitats, however, most were distinctively wet – half-flooded swamps, abandoned river channels and bayous, and poorly drained bottomlands. Consequently, the tree was very difficult to harvest; the ground was usually far too boggy and soft for normal logging techniques with oxen and wagons, and the trunk almost invariably sank when felled. Sometime during the opening years of the eighteenth century, and certainly soon after the establishment and building of New Orleans in 1718, the French solved these problems. The trees were first deadened by girdling and thus lost sufficient sap to make them buoyant and float. The logs were felled from boats and towed out, or poled out through channels, or pulled out by men ("swampers," as they became known) standing waist-deep in water. With the realization that water was the key to exploitation, felling was usually carried out during the wettest season, and where land intervened, channels of considerable dimensions were dug.

These methods of exploitation did not change for over 150 years, and even as late as 1889 Dr. Charles Mohr described the scenes of activity in the baldcypress swamps of the Tensas River, Alabama, after three weeks of torrential rain.[20] The flood had made it possible to harvest the timber:

> No idle man was found on shore; everybody who could swing an ax, paddle a boat, or pilot a a log was in the swamp engaged in felling and floating cypress timber. All the mill hands worked in the swamps; fields and gardens were left untouched, and even the clerks from the stores were sent into the swamps as overseers.

In the swamps he came across a little boat in which two men were felling a large baldcypress; "it was astonishing," he commented, "to witness the steadiness and celerity with which they performed their work, considering the instability of their footholds in the narrow boats." Lopping and bucking was done in and even under the water, and then the trunks were poled out to a collecting pond where they were lashed together and sent downstream to the mill.

These intermittent and small-scale methods of cutting, milling, and shipping could not supply the rapidly growing demand for this timber during the late 1880s, and new techniques of extraction from the swamps were evolved. First, Horace Butters of Michigan experimented with his overhead logging cableway steam skidder, invented in

Plate 8.1. A two-drum pull boat in a set in a canal cut in the cypress swamps of Louisiana. Date unknown.

1883, by taking it to a North Carolina swamp and mounting it on a barge. Although the device proved reasonably successful, it had disadvantages: It had to be rerigged for each operation, the slack of the cable required five or six men to drag the hauling carriage back to the loading site, and the length of the operation was only about 700 to 800 feet, which necessitated expensive canal excavation in order to move the barge.

This machinery was improved in 1889 or 1890 by William Baptist of New Orleans, who developed the "pull-boat" system, which could skid logs for up to 3,000 feet. The steam engine was mounted on a shallow scow or pull boat anchored by cables to pilings, stumps, or trees and floating in an excavated enlargement of the canal called a "set." The engine operated two drums, the larger one carrying a heavy steel cable to pull the logs out of the swamp, at the same time unwinding a light wire rope from the smaller drum, which was used to return the main pulling line to the point from which the logs had been dragged, via a sheave block anchored just above ground level to a tree in the forest (Plate 8.1). Runs were cleared into the forests (often blasted out with dynamite) in a fantail pattern focused on the pull boat, each run being about 150 yards apart and costing anywhere between $3,000 and $5,000 per mile (Fig. 8.2). Branches splayed out from the runs. To prevent the logs from digging into the mud Baptist invented a steel cone, which slipped over the end of the log and acted as a buffer during dragging, but this was later abandoned in favor of the more rapid method of "sniping," that is, cutting the end of the log to a point. Since the pull boat did not utilize tall trees or towers like the overhead skidder, it was much faster to set up and could reach farther into the swamps. A problem was that with ground dragging so much soil and debris were dragged into the channels that often they became clogged, but this was overcome either by beginning

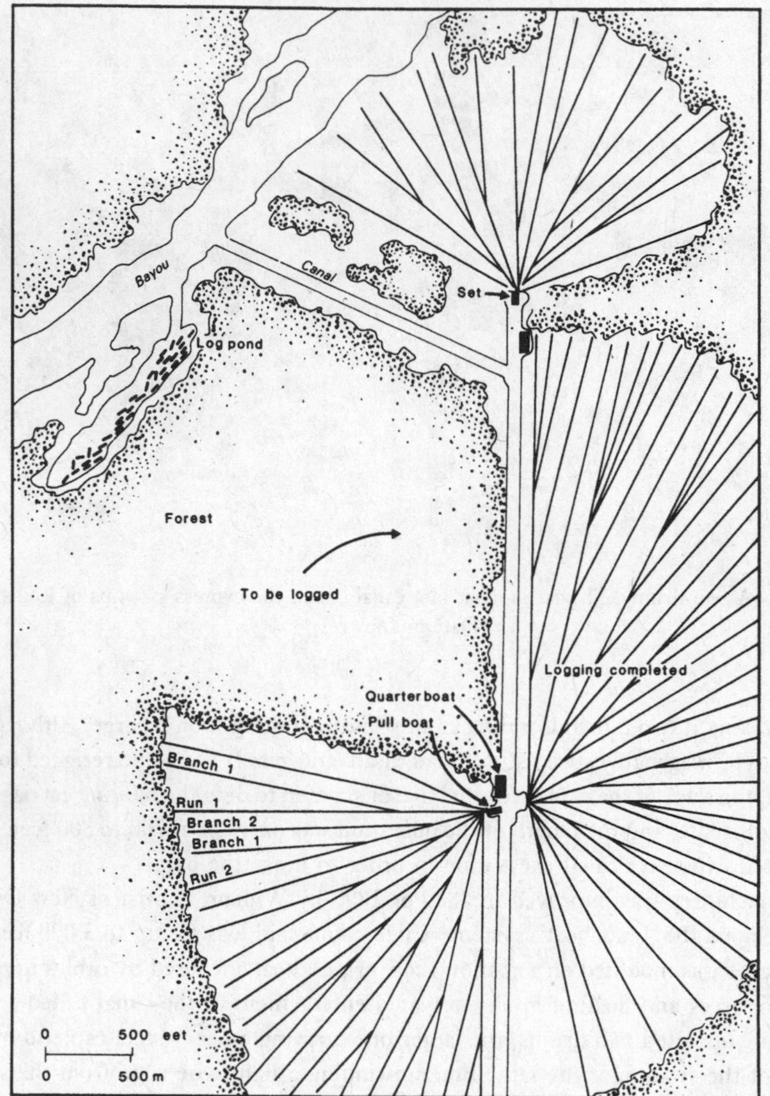

Figure 8.2. Pull-boat logging. (After R. C. Bryant, *Logging*, 234.)

work at the upper ends of the channel or by erecting a wooden barrier at the start of the run to catch the debris.[21]

The outcome of these inventions for the industrial exploitation of the baldcypress was enormous; production shot up as it and its unique qualities became more widely known and appreciated throughout the country. Production rose to about 1 billion b.f. per annum after 1905 through to 1913, but cutting was so ruthless and exploitation so complete that supplies were depleted to a point where production began to fall after that, and by the early 1950s there were supplies enough for only one mill to operate, and even that closed in 1956.

Plate 8.2. A slip-tongue cart in the forests of Louisiana, c. 1910.

Land: carts and loaders. As the great surge of activity got underway after about 1890 a whole spate of innovations came into being to replace the simple "snaking out" or dragging out of logs from the forest. In the southern pine forests the economic haulage by oxen to the rivers seemed to be limited to about three miles. Therefore, wheel carts, known as "bummer carts," probably invented in Michigan for summer use, became very popular in the South. A bummer cart was a self-loading skidder that was particularly useful in the level topography of the pine barrens. It consisted of two wheels on an axle, supporting a wooden haft that acted as a lever for lifting one end of the log onto it. A variation was called the "slip tongue," which had a movable shaft bolted to the lever. The forward movement of the horse pulled out the lever and raised the log off the ground (Plate 8.2). On a downhill run the load could be lowered to scrape the ground and act as a brake. Big-wheeled carts known as caralogs also became popular in the South. With wheels of up to 12 feet in diameter, they were driven parallel over one or more logs, the front ends of which were raised to the axle by a windlass.[22] After 1900 two-wheeled carts were outmoded by eight-wheeled logging wagons developed by John Lindsay of Laurel, Mississippi, in the very heart of the pine forests. These had wheels of three to five feet in diameter arranged in groups of four in a railroad bogie fashion to give greater flexibility and to reduce the risk of accidents. By 1904 there were said to be about 5,000 of these wagons in use in the South.

But long-distance haulage of logs over land by bullock power was prohibitively expensive because it was slow-moving; it cost 20¢ per ton mile or $1 per 100 pounds per 100 miles. This meant that haulage over a distance of, say, 400 to 500 miles raised the delivery price to between 10 and 20 times the mill price, and only a few prosperous settlers could afford that.[23]

However, as the pace of exploitation accelerated and the railroads proliferated through the pine forests, efforts were made to coordinate log-gathering and loading techniques with the railway, thereby speeding up the process of timber extraction and reducing its cost. Experiments with ground skidders with up to five cables powered by stationary steam engines began in Mississippi in 1894, Georgia in 1896, and Texas in about 1904. Up to 600 logs could be pulled in eight hours. Many companies then attempted to combine the power skidder with derricks in order to load the logs straight onto railroad cars. A later and more ingenious invention was that of the Bernhardt self-loader, first mentioned as being used in 1905 in Alabama. It was a self-contained unit that ran on tracks placed permanently on the tops of flatcars. The loader started at one end of the line of empty cars, next to the last car, backed up until all cars were loaded, and then returned to its own flatcar. Another important invention was the McGiffert loader patented in 1908 by Messrs. McGiffert and Decker of the Clyde Iron Works of Duluth, Minnesota.[24] Built to run on railroad tracks, it could be raised by hydraulic jacks to stand on four long legs, straddling the line. The cars could then be moved underneath and loaded. The self-loader could, of course, also be used as a skidder. Another and probably more important invention, which appeared during the same year, was the portable skidding tower placed on a railroad car. The tower replaced the tall tree in overhead cableway skidding and, with derricks, could act as skidder and self-loader combined. It greatly reduced the time involved in skidding because tackle could be rerigged in two hours compared to 24 hours as of old, thereby greatly increasing output.

The result of these developments was the coordination of railroad laying with skidding technology. The tracks were laid out in parallel lines from 1,200 to 1,400 feet apart and the timber logged halfway back from each side of the track. A crew of between 17 and 20 men working with a few horses or mules could haul in and load 35,000 b.f. of timber per line of the skidder, and as most had four lines, that meant 125,000 b.f. daily, and even greater amounts than that were recorded. The cost was a mere 75¢–100¢/1,000 b.f. moved.[25] Now the capital costs of logging equipment began to be greater than the capital costs of the mills and small independent operators in the forests were driven out of business. But the most lasting effect of the new technology of transport and log gathering in the woods, irrespective of which machinery was used, was to devastate the forest in a way that had never happened before. As logs were "yarded" into a central point, dozens of young trees were uprooted or snapped, and the natural regeneration of the cutover was prevented. W. Goodrich Jones, a pioneer in the crusade for reforestation in Texas, described the effects of the harvesting process during the early 1920s. The mobile snaking machine was like

> an octopus of steel with several grappling arms running out 300 or more feet. These grapple a tree of any size that has been felled, and drag it through the wood to the tram road. These become enormous battering rams and lay low everything in their way. Standing trees that are not pulled down are skinned so badly as to be worthless. The remains of the forest [are] like the shell torn area of France.

The method of logging was creating permanent wastelands in the the cutovers.[26]

The railroads. In the South the railroads were not just a useful adjunct to lumbering (as in the Lake States) but essential for any major exploitation of the forest. The rivers of the Gulf ran away from the main domestic markets, and those in western Texas were so variable in their flow as to be useless for transport. The Atlantic coast rivers were better located, for at least exports to the Northeast were possible. Without major north-south carrier lines, therefore, the raw material of the southern forests was almost totally closed to the markets of the industrial and urbanized North and Northeast, and completely so to the agricultural and largely treeless West. Similarly, exploitation in the forest – the haul from stump to saw – could not possibly achieve the volume required to feed the ever-growing economies of scale in machinery and organization that were the hallmark of late-nineteenth-century lumbering, unless accompanied by networks of main lines, spurs, and logging roads.

The building of the railroads had many effects. The most obvious was that, as well as being routes out of the region and to the market, they were also lines of exploitation. Mills would be set up at intervals along the line, and the devastation of millions of acres of pine land would follow, especially with the cruder methods of yarding that were adopted increasingly. But the revolutionary changes wrought by the railroads did not end with change to the vegetation cover and the visual scene. Because they were the spearheads of change, they introduced industry and industrial ways to remote and backward rural areas in what was a conservative society and culture that had not altered radically for nearly two centuries. Black and white alike entered the mill camps, where there were new sets of personal and commercial relationships. The backwoods, self-sufficient life of the white farmers was disrupted permanently, and the lingering paternalistic boss–worker relationship on the plantations and the drudgery of the tenanted cotton-farming system for the blacks were upset by the new opportunities for the use of large quantities of unskilled manual labor and fairly steady cash incomes.[27] Hundreds of new communities were created, but equally as many were obliterated as firms closed overnight; when the last tree was cut they moved on. Similarly, many existing towns expanded greatly if the railroad went through them.

When William Faulkner wrote about the coming of the logging railroad into his mythical Yoknapatawpha County, he encapsulated a whole gamut of experiences and emotions related to the changes that were occurring in southern life. The lumber camp and the railroad were the symbols of these changes; they were the intrusive machines in the pastoral Garden of Eden (Plate 8.3).[28] In Faulkner's story, one of the characters, Ike McCaslin, is shocked and amazed at the transformation of the pine forest from hunting ground to lumber mill, with its

> new planing-mill already half completed which would cover two or three acres and what looked like miles and miles of stacked steel rails red with the light bright rust of newness and of piled crossties sharp with creosote, and wire corrals and feeding-troughs for two hundred mules at least and the tents for the men who drove them.

He was so overcome with the sight that he hid in the caboose of the logging train, which lay still in the siding. Suddenly, it started and moved toward the "wall of wilderness" ahead.

> From the cupola he watched the train's head complete the first and only curve in the entire line's length and vanish into the wilderness, dragging its length of train behind it so that it resembled a small dingy harmless snake vanishing into weeds, drawing him with it too until soon it ran once more at its maximum clattering speed between

Plate 8.3. The bringer of change in the southern forests: a logging railroad near Oscilla, Irwin County, Georgia, 1903.

> the twin walls of unaxed wilderness as of old. It had been harmless then. . . .
>
> But it was different now . . . it was as though the train . . . had brought with it into the doomed wilderness even before the actual axe the shadow and portent of the new mill not even finished yet and the rails and the ties which were not even laid; and he knew now what he had not known as soon as he saw Hoke's this morning but had not yet thought into words; why Major de Spain had not come back, and that after this time he himself, who had to see it one time other would return no more.[29]

Ike's repugnance toward and urge to withdraw from the inroads of repressive civilization in the form of the mill and the railway were an echo of the deeply held feelings of other American heros, from Huckleberry Finn to Ishmael to Thoreau.

Railroads: the main-line carriers. It would be impossible to describe all the main lines constructed in the South during the late nineteenth and early twentieth centuries because of their number, their length, and their complexity. Between the end of the Civil War and 1910 the track mileage grew fourfold in the 13 southernmost states, from 9,183 miles to 38,893 miles, and they were crisscrossed by main lines. Also, it is difficult to separate the impact of the main-line roads from that of the logging roads because whereas in the beginning the logging roads were narrow-gauge tramroads, by the 1890s many were being replaced by standard-gauge lines and equipment and then incorporated into the main-line regional systems because of the advantages of reducing the breaks in gauge and hence breaks in bulk.

Certainly, it is safe to say that before 1884, when Sargent compiled his maps of the forest, the bulk of the forests in the Gulf South were untouched, and the impact of clearing was limited to the main lines converging on Savannah in Georgia, Charleston and Wilmington in South and North Carolina, the Baltimore and Ohio lines in northwestern Virginia, the Missouri–Pacific line between Texarkana and Little Rock, and the Illinois Central between Memphis and New Orleans via Jackson in Mississippi.

But all was to change after about 1890 when the number of lines increased and the full impact of the northern land purchases and expertise began to be felt, as shown by just a few examples in the adjacent states of Mississippi, Louisiana, and parts of Texas and Arkansas.[30] In Mississippi during the pre–Civil War period there were two main lines: the New Orleans, Jackson, and Great Northern (later the Illinois Central) and the Mobile and Ohio (later the Gulf, Mobile, and Ohio), both railroads a result of the rivalry of the ports of New Orleans and Mobile for the trade of the interior. The first of these new post–Civil War lines in these states was the New Orleans and Northeastern (later the Southern), constructed after 1884. It ran from New Orleans in a northeasterly direction through the forests of eastern Louisiana and southeastern Mississippi, largely in order to tap en route the vast resources of the longleaf pine forests. Between 1869 and 1903 the Gulf and Ship Island Railroad was built from Gulfport (on the Mississippi Gulf Coast) to Jackson to tap the untouched forests lying between the Pearl and Pascagoula rivers. With its branch logging lines to Laurel and Columbia, the railroad extended for over 300 miles, and there were said to be between 50 and 60 mills adjacent to the road. In 1902 the 74-mile stretch from Hattiesburg to Gulfport averaged one sawmill and one turpentine distillery for every three miles. The Mobile, Jackson, and Kansas City road, completed during the first years of the century, tapped all the forests of the state, running as it did from Mobile (Ala.) via Laurel (Miss.) and Jackson (Tex.), the small section between Newton and Shipman (Miss.) having 45 mills. A sixth major line was the New Orleans and Great Northern, which was a venture of the Goodyear brothers, and it ran 110 miles from the massive new plant of the Great Southern Lumber Company at Bogalusa (completed in 1907) via the Pearl River to New Orleans. On the Louisiana–Texas border similar lines were being built with an eye to penetrating and felling the forest. Among many was the Kansas City and Southern Railroad from Orange, Texas, to Shreveport, Louisiana, which opened up several of the heavily forested Louisiana parishes to large-scale lumbering. But the key railroad for exploiting the forests in this district was the Houston, East and West Texas Railroad, which ran for 232 miles from Houston to Shreveport via Logansport; it cut a swathe through the East Texas pine forests.[31]

The list of railroads built could be extended almost indefinitely; the simple fact was that as each line was built sawmills were erected alongside it at intervals of between three and five miles, sometimes even being constructed in anticipation of the road's being built. In 1884 Dr. Charles Wells wrote about a railroad that had just been completed through the central, unsettled portion of Alabama:

> Before even the survey of the Line had been completed, numerous sites for saw mills had been selected, and within a few months after, machinery had been hauled in wagons from Enterprise and Meridian, (its transport occupying days and often weeks,) . . . erected and set to work; until at the present date, upon the completion of the railway, there are millions of feet of accumulated sawed timber awaiting transportation to Northern and Eastern cities and to New Orleans for shipment abroad.[32]

When the immediately accessible timber was exhausted, logging spurs were run out 30 or 40 miles into the forests, creating small mill towns to serve as centers for the farms that sprang up on the cutovers. To take just one example: in southwestern Mississippi there were six major spur lines – the Natchez, Columbia, and Mobile (25 miles) belonging to the Butterfield Lumber Company; the Mississippi Central owned by the J. J. Newman Lumber Company, which accommodated 26 mills along its route by 1906, even before it was completed; the line of the J. J. White Company west from McComb: the Fernwood and Gulf Railroad of 33 miles; the Mississippi–Eastern Railroad owned by the Mississippi Lumber Company; and the Chickasawhay and Jackson road of the W. Denny Lumber Company, which became the Pascagoula and Northern Railroad.[33] The forests were being crossed by a network of trunk lines and spurs, each one a path of major exploitation. The pattern, when the lines were probably at their greatest extent in 1920, in one part of the longleaf forest, the Calcasieu basin in western Louisiana, is shown in Figure 8.3.

The impact of the main-line carriers varied from region to region. In Alabama in 1893 just over one-fifth of 45 million b.f. of timber was going out of the state by rail, the remainder by river and cart down to Mobile and the Florida coast. In Mississippi during 1891–2 the dense network of main carriers (Table 8.2) allowed over one-half of the lumber to be carried northward out of the state, the figure for the Illinois Central nearly doubling during the next year. Texas was a late starter in the lumber business, but even here great amounts of lumber were being moved by rail, nearly three-quarters of all lumber cut leaving via the railroads in the northeastern part of the state (Table 8.3). By 1900 nearly all of the lumber of Texas was carried by railroad, and of the 6 million tons of freight carried, lumber was by far the leading commodity in terms of both tonnage and revenue.[34]

The logging railroads. The main-line carriers and their spurs were the framework for forest exploitaton and marketing, but from these, and usually quite independent of them, extended the countless miles of small-gauge, easily movable logging tracks, which were put down, pulled up, and relaid time and time again as train, skidder, and loader were combined to strip the forest of all its trees. The logging railroad, however, was preceded by a number of local experiments to speed up and cheapen the laborious and costly movement of logs to rivers by oxen. Plank roads, pole roads, and stringer roads were tried. For example, the pole road first appeared in Alabama where the abundant straight and slender pines were peeled of their bark and laid in parallel lines like tracks along which trucks with huge flanged wheels were pulled by oxen and later by locomotives. At $25–$100 per mile, the pole roads were cheap to construct, and logs were moved at a price of between 16¢ and 30¢ per log for five miles compared to 65¢ per log for only two miles by oxen. Good as they were, however, they had their problems. They could not be constructed accurately and were not smooth-running. The pole tracks wore out quickly, and there were many derailments. But their greatest disadvantage was that they could not be connected to the main-line carriers.[35]

Disillusioned with pole roads, Emory F. Skinner, who operated mills at Pensacola, went to Lake Cadillac, Michigan, in 1883 to inspect the new narrow-gauge logging

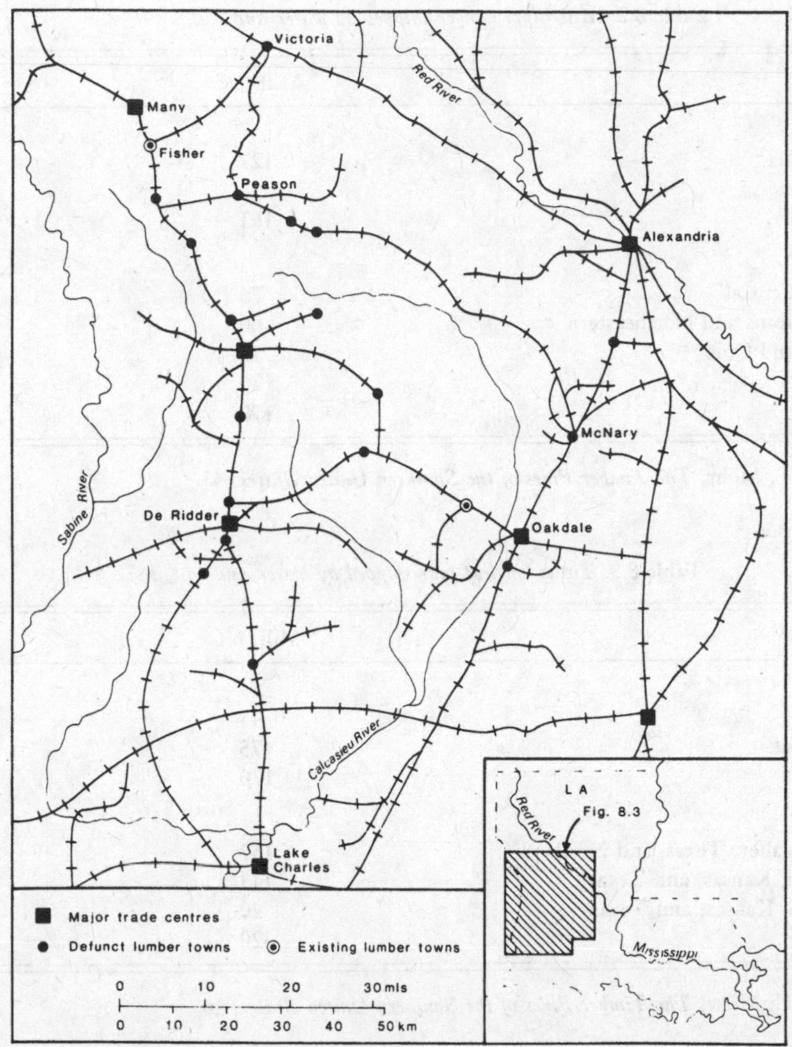

Figure 8.3. Main-line carriers and spurs, Calcasieu basin, Louisiana, c. 1920. (After G. A. Stokes, "Lumbering and Western Louisiana Cultural Landscapes," 251.)

roads, news of which had spread to the South. He was so impressed with what he saw that he bought and brought back to Florida six miles of track. The result was immediate and explosive as many hundreds of times that length of track was laid during the next few years. For example, the Southern Lumber Company of Alabama had over 100 miles of relayable track, and the Kaul Company of Tuscaloosa had about 75 miles of track, all laid out in a methodical pattern of parallel spurs through the forest in order to facilitate the maximum exploitation of the trees. In Alabama alone, about 750 miles of such lightweight private railroad were laid by 1913, and in Texas there were about 1,400 miles in 1910, the year of maximum mileage.[36] The complexity of the Fisher–Victoria tram

Table 8.2. *Mississippi lumber shipped by water and rail, 1891–2*

	Mill. b.f.	%
River		
Pascagoula	127	38.1
Pearl	36	10.8
Total	163	48.9
Railroad		
Illinois Central	78	23.4
New Orleans and Northeastern	60	18.0
Mobile and Ohio	12	3.6
Others	20	6.0
Total	170	51.0

Source: C. Mohr, *The Timber Pines of the Southern United States*, 43.

Table 8.3. *Texas longleaf pine shipped by water and rail, 1882*

	Mill. b.f.	%
River		
Orange	45	10.2
Beaumont	75	17.0
Total	120	27.2
Railroad		
Sabine Valley, Texas, and Northern	157	35.7
Missouri, Kansas, and Texas	143	32.5
Houston, Kansas, and Texas	20	4.5
Total	320	72.7

Source: C. Mohr, *The Timber Pines of the Southern United States*, 46.

system in the longleaf pine forests of Sabine and Natchitoches parishes in the Upper Calcasieu basin in about 1920 was typical of one stage in the kaleidoscopic pattern changing logging railroad networks (Fig. 8.4).

The construction of the logging railroads was both the cause and the effect of the large-scale lumbering operations that were becoming so common in the South. Although construction was relatively costly at $1,000 or more per mile, once established their operation was efficient and their running costs a mere quarter of those of ox haulage. Therefore, larger areas than ever could now be logged over. Conversely, however, without larger areas the expense of the tramways was not reasonable unless it was expected that the land would yield timber for at least 20 years. A rule of thumb used by the industry was that at least 1 million b.f. of timber had to be cut per annum per mile of railroad laid in order for the line to pay for itself, which also necessitated the clear-cutting of the area with all the small timber going for poles and paper pulp.[37]

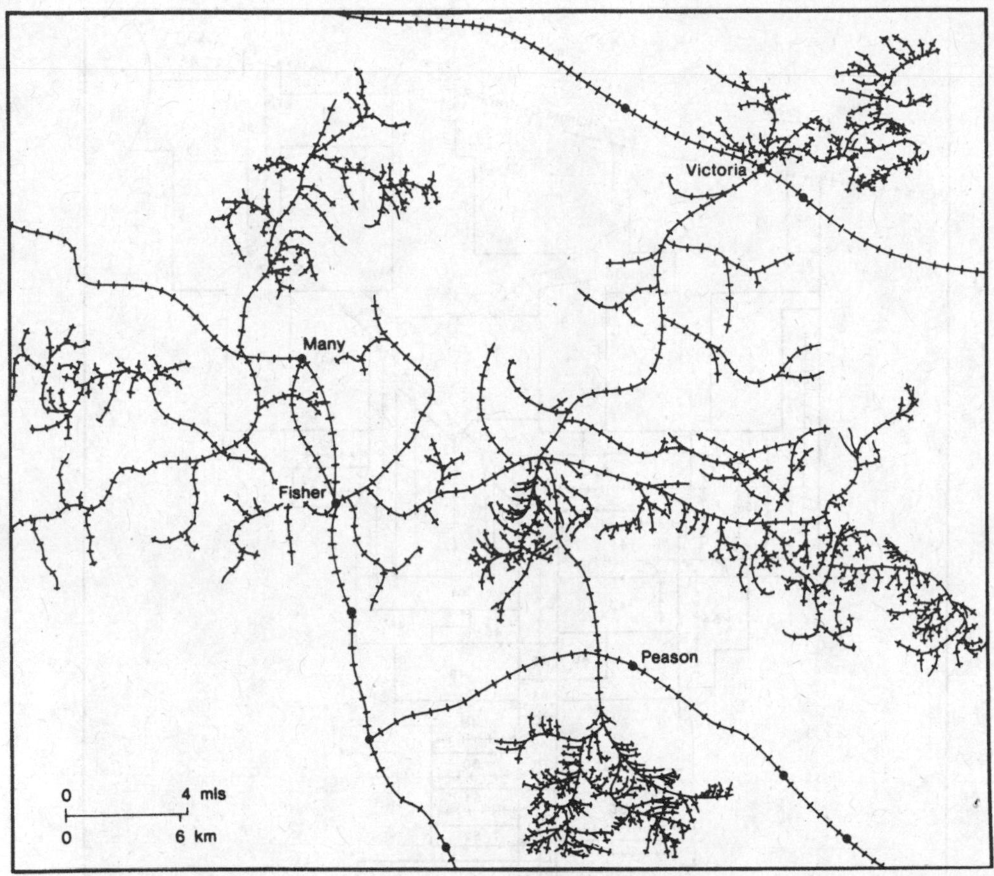

Figure 8.4. The Fisher–Victoria tramway system in Sabine and Natchitoches parishes, Louisiana, c. 1920. (Stokes, "Lumbering and Western Louisiana, 259.)

Of course, the process was cumulative, for once the logging roads became the usual means of transport and were accompanied increasingly with steam skidders and loaders, the small- to medium-sized operators began to drop out of production because of the capital costs involved, and the small landowners were powerless to market their timber because of the way in which they were "controlled" by the logging company and its tramroads and railroads.[38]

The logging railroads were not confined to the pine forests. In certain circumstances they were a reasonable alternative to pull boats in the Louisiana baldcypress swamps, especially when coupled with overhead skidders, when used in the original Butters way. If the trees were too far away from waterways to be logged by pull boats and the land too firm to be excavated into canals by dredging, then the logging railroad was preferable. It is true that sloppy terrain made construction difficult, but surfaces made of small timber cribwork or even like a corduroy road, with mill dunnage dumped as filling because it was lighter than crushed stone, made it possible to construct logging tramways at regular intervals to harvest the dense stand of high-quality timber, at a cost of between $9,000

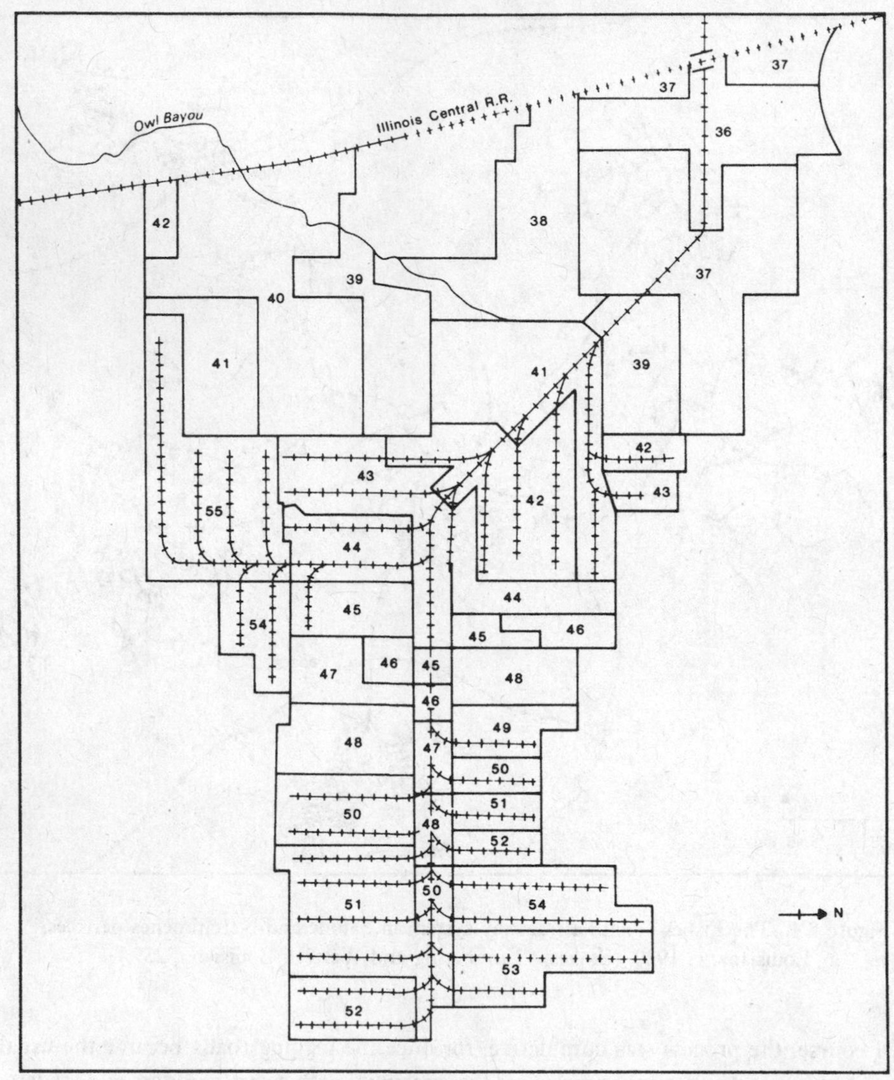

Figure 8.5. Railroad logging sequence, in Tangipahoa Parish, Louisiana, 1936–55. (Based on E. Mancil, "An Historical Geography of Industrial Cypress Lumbering in Louisiana," 120.)

and $15,000 per mile. With this system the pulling distance was usually 600 to 800 feet; therefore, a system of spur lines was built at right angles to the main line, usually 1,200 to 1,600 feet apart, and a car-mounted skidder could operate along each line and systematically clear the land in a predetermined sequence of exploitation, block by block, as in the example of Tangipahoa Parish, Louisiana (Fig. 8.5).[39]

Trucks. The logging railroads gave the large, integrated lumber companies a distinct advantage over the small, independent mill owners and landowners, all but eliminating them by the opening years of the twentieth century. But just as the technical advance of the gasoline engine brought about a reversal of trends in the sawmills by making small-scale logging possible again, so too did the complementary application of the gasoline engine to trucks and tractors. By the end of the First World War there was a greater flexibility of operation, which favored the portable "peckerwood" mills and also caused the breakdown of railroad "control" as small landholders used trucks to haul out their lumber. With the rising costs and declining markets of the late 1920s and early 1930s, trucks enabled an emphasis to be placed on operational efficiency (even in once cutover areas), a situation that could be contrasted to the previous emphasis on output by volume. The role of technology in influencing if not directing the pace, mode, and technique of exploiting the forest was displayed yet again.[40]

Sawmills

The delayed start to large-scale production in the South until after 1880 meant that all the advances in technology made in the North were applied immediately to the new mills in the South, which were, in many cases, no more than subsidiaries of or partnerships with northern firms. Almost from the beginning of expansion, the steam-powered circular saw was adopted in the mills to boost daily output, and it replaced the small-scale water-powered mill, which existed in hundreds of local communities, running a couple of sash saws and very often combined with a grist mill.[41]

Output per saw reached nearly 50,000 b.f. daily, which was often considered the level of "commercial" or "industrial" production by the end of the nineteenth century. Not until the introduction of the band saw into the South after 1890 could production of a single blade get up to 75,000 b.f. daily.[42] With these saws and with steam came all the adjuncts of commercialized and mechanized lumbering. Standard mill equipment came to include the mill pond for storage and continuous production, friction feeds into the mill, steam "niggers" for turning logs on the cutting deck, and edgers. Only a little later did planers, finishers, molders and matchers, hot-air kilns, and then more efficient steam driers become common. A lower and lower proportion of rough timber left the mill, and a finished, seasoned product was transported by rail out of the South for a nationwide market.[43] In addition, the more careful cutting of logs in the mill produced a greater proportion of marketable pieces, and waste was reduced by cutting edgings into small products like shingles and laths. Whatever remained was chipped for fuel and blown into the furnaces. In every way productivity was increased at the mill – by reducing waste, by speeding up the process, and by producing a more readily marketable product. As business grew and more feeder tramways were laid, whole batteries of saws were run in tandem, the largest mills accompanying the largest landholdings, although this relationship was not always clear in the longleaf pine zone.

Throughout the South after 1890, larger mills with at least two band saws and a daily capacity of between 100,000 and 150,000 b.f. and a cumulative annual capacity of 10 million b.f. began to increase rapidly. By 1900 there were 211 mills of the census group 5 type – that is, mills producing over 10 million b.f. annually – in the nine southern states stretching from Texas to North Carolina, and the percentage they were of all mills in those states was very nearly double that which similar-sized mills were of all mills in the rest of the United States. The total number of these very large mills grew to 352 in 1919, the greatest increases occurring in Louisiana (37 to 97), Mississippi (25 to 64), and Texas (38 to 50), all states except Georgia participating in this upward shift in group size during 20 years. A similar upward trend was evident in the group 4 and group 3 mills (i.e., 5–9 million b.f. and 1–5 million b.f. annually), and again the nine southern states had more of these mills as a percentage of all mills than was the case elsewhere in the United States. Inevitably, therefore, the South had a lesser proportion of very small mills, those cutting less than half a million b.f. annually accounting for only 57 percent of all mills compared with 67 percent in the remainder of the United States. In every way, the southern states, especially Louisiana, Texas, Mississippi, and to a slightly lesser extent Florida and Arkansas, stood head and shoulders above all other states in terms of "bigness." Only Washington and Oregon on the Pacific Northwest coast rivaled and surpassed them, and then not until after 1920.[44]

Like every other region of production, the South had its outstanding mills. For example, there was the mill of the Central Coal and Coke Company (4Cs) at Ratcliff in Houston County, Texas, which by 1900 had three band saws, a 52-set gang saw, and a daily capacity of 300,000 b.f. The mill of Henry J. Lutcher and G. Bedell Moore at Orange, founded in 1877 after they had migrated from Pennsylvania, was also capable of producing 300,000 b.f. daily by 1906, and with modernization in 1917 its output was pushed up another 25 percent. Probably the largest mill of all was that constructed by the Goodyear brothers at Bogalusa for their Great Southern Lumber Company. The Goodyear brothers and their mill were the epitome of change and big business in the South. They had started in New York, moved to Buffalo in 1872 and then to northern Pennsylvania in the early 1880s, where they established 15 small mills in the hemlock forests. But as supplies dwindled they migrated south in 1890, having first employed James D. Lacey, the famous cruiser, to assess the yellow pine stands, which they purchased for a ridiculously low price. They then selected the site for their mill on the Bogalusa Creek, Louisiana, just off the Pearl River and over the state border from Mississippi, which was just beginning to enact legislation to stem the flow of northern capital and control into that state. After an initial investment of $15 million, Bogalusa eventually became a town of 15,000, the center of 150 miles of main-line and 75 miles of side-track railroad that radiated out from the town to draw in lumber supplies to feed the gigantic mill, thus causing the clearing of between 40 and 100 acres a day.[45] In 1916 the company produced a record quarter of a billion b.f. of lumber in one year.

The trend to bigness was the hallmark of the paleotechnic age, but as forest exploitation in the South moved on into the neotechnic age of gasoline engine and electricity after the First World War a contrary trend was noticeable. The gasoline engine enabled small operators to set up portable mills in areas abandoned by the large mills as they had either gone out of business or moved on to the Pacific Northwest. These new mills were profitable, flexible, and mobile, moving to second-growth stands and scattered old-growth stands when the occasion arose and taking the logs out by truck instead of tying up capital in tracks and plant. The trend is difficult to discern easily, but the amount of timber cut by large (over 6 million b.f. per annum) mills and by very small mills suggests that during the late 1920s a noticeable shift in production was occurring toward smaller units. In 1919 two-thirds of the production came from large mills and one-third from small mills; but by 1929 less than half came from large mills and slightly more than half came from small mills, and the balance shifted further during later years.[46]

Business and social organization

By 1887 it was said that southern pine lumber was "rapidly winning its way into popular favour in nearly all parts of the United States."[47] The technical improvements in the mills and particularly in transport were making it possible to cut down and cart away untold thousands of acres of trees. But innovation and change were not limited to technics of lumbering, sawing, and carting; the scale and complexity of business and social life were entering a new commercial and industrial phase that was in clear contrast to all that had gone before. Much if not most of the timberland was now in massive holdings, settlements were created solely for company purposes and functions, labor was organized on a factory basis, and individual firms attempted to perpetuate the prevailing social system by banding together to act aggressively to combat common problems, whether those of overproduction, grading standards, low prices, or unfavorable freight rates. To be sure, all these manifestations of the new scale and complexity of business organization also became evident in the Great Lakes region in time, but they never developed there to the degree that they did from the very beginning in the South. The South became the epitome of industrial capitalism in the lumber industry.

Ownership of timberland

Examples have already been given of the massive purchases of land and of the growing scale of business operation. By 1920 there was a total of 352 plants capable of producing more than 10 million b.f. of lumber annually in the nine southernmost states. But far more spectacular than that was the unseen feature of the accumulation of huge areas of land in massive holdings. Adequate timberland facilitated the vertical integration of production, which was the key to the new-style large-volume, continuous output that resulted in low prices and reasonable, though not excessive, profits. The need for large amounts of timber is obvious when the economics of production are looked at in detail. It was estimated that a mill cutting 30 million b.f. annually needed approximately 20,000 acres of untouched forest in order to achieve a 10-year run, and most mill owners wanted a 15–20-year run.

It is difficult to come up with a definitive statement of the degree of concentration in the industry because of the problems of distinguishing between timber rights and fee-held land, between the holdings of speculators and those of genuine lumbermen, and between the interlocking directorships of various companies; however, we do know that concentration was great. The best overall view is that provided by the report of the Bureau of Corporations issued in 1914.[48]

The bureau's investigators concluded that the initial purchasing and subsequent

Table 8.4. *Concentration of ownership in the southern pine region, c. 1913, by size of holding*

Group	Group size (mill. b.f.)	No. owners	Bill. b.f. owned	% of total	% cumulatively
1	Over 25,000	—	—	—	
2	13,000–25,000	—	—	—	
3	5,000–13,000	3	22.1	3.5	3.5
4	3,500–5,000	8	32.7	5.2	8.7
5	2,000–3,500	18	47.9	7.5	16.2
6	1,000–2,000	38	50.2	7.9	24.1
7	500–1,000	92	58.7	9.3	33.4
8	250–500	148	51.3	8.1	41.5
9	125–250	251	42.0	6.6	48.1
10	60–125	367	31.1	4.9	53.0
11	Less than 60	?	297.7	47.0	100.0
Total		925	633.7	100.0	100.0

Source: Based on U.S. Bureau of Corporations, *The Lumber Industry*, 1, pt. 1: 30.

consolidation of land into holdings in the South had allowed a mere 925 persons or institutions to amass 336.3 billion b.f. of timber, or just over half of that which existed in the region (Table 8.4). Within that total, concentration was high. For example, of the 925, the largest 67 holders (classes 3–6) held 152.9 billion b.f., or nearly a quarter of all southern pine. In all, the 925 landholders held 46.6 million acres of timberland, which was more than 80 percent of the 58 million acres of privately owned pine in the South. The concentration of ownership was particularly great in the Gulf-facing states of Texas, Louisiana, Mississippi, and Florida, these four states alone accounting for 29.2 million acres of large holdings. In all cases, these holdings were, of course, land held in fee, and they did not include land over which timber rights were held, which covered many more millions of acres.

The size of holdings of individual owners was often difficult to ascertain. In addition to the genuine lumber operators there were many speculators of stumpage, and these usually had the largest holdings. There were 41 holdings above 300,000 acres in size in the United States, and 22 of those were in the South. They are listed in Table 8.5. Some were genuine lumber companies, and generally speaking they were far smaller than the land speculation companies, the majority of the lumber companies not even appearing on this list as they were usually less than 100,000 acres. Even in this list it is difficult to be sure of what the figures tell us as it is known that Kirby Lumber, in addition to its 365,000 acres in Texas and 15,000 in Louisiana, had timber rights over some 690,000 acres of fee-held land of the Houston Oil Company, which boosted Kirby's actual timber holdings enormously.

The only area in the South where the Bureau of Corporations attempted to unravel the complexities of land ownership and plot holdings was in the prodigiously rich longleaf pine forests of the Calcasieu basin in west-central Louisiana. Here, 74

Table 8.5. *Holders of over 300,000 acres of timberland in the South, c. 1914*

	Amount (thousands of acres)	Principal states
South Consolidated Land	1,625	Fla.
Southern States Land and Timber Co.	1,428	Fla.
Empire Land and National Timber Co.	1,172	Fla.
Florida Coast Line and Canal Transport Co.	610	Fla.
John Paul Interests	600	Fla.
Norfolk and Southern R.R. Co.	590	N.C.
Missouri Pacific Ry. Co.	571	Ark.
Missouri Lumber and Land Exchange	442	La., Mo.
Great Southern Lumber[a]	433	La., Mo.
Model Land Co.	427	Fla.
Long-Bell Lumber Co.[a]	399	La., Tex.
Texas Delta Land Co.	391	La.
Kirby Lumber Co.[a]	383	Tex., La.
Blodgett Co.[a]	368	Mo.
Southern States Lumber Co.[a]	364	Ark., Fla.
Wm. Buchanan	360	La., Ark.
Frost–Johnson Lumber Co.[a]	345	La., Tex.
Cummer-Diggins Interests	343	Fla.
Central Coke and Coal Co.[a]	343	Tex., La.
Crosset, Watzek and Gates	337	Ark., Ala., Fla.
Dowling Lumber[a]	317	Fla.
Camp Manufacturing and others	316	Fla.

[a]Known lumbering interests.
Source: U.S. Bureau of Corporations, *The Lumber Industry*, 2, pt. 3; 173–6.

landowners held just over 2.4 million acres of land, which contained an estimated 33.5 billion b.f. of timber, one of the richest stands in the South. The shadings on the map shown in Figure 8.6 correspond to the first three groups itemized in Table 8.6. The 14 largest landholders included Calcasieu Pine and Southern Timber Company (a company with Weyerhaeuser interests), Frost-Johnson, Long Bell, and Lutcher and Moore, which among them owned 802,189 acres and 12.3 billion b.f.

The full extent of their holdings is not fully represented on the map, however, because of the "control" they exercised over smaller, interspersed landholders and because of the interests that these companies had in other parts of Louisiana and the South generally. For example, the "outside" interests of the 14 largest landholders on the map included a further 1,468,912 acres of fee-held timberland, land with timber rights and cutovers in Louisiana, and another 2,103,874 in the other states in the South, a total of 5,025,933 acres or about three-and-a-half times the area on the map (see Table 8.7).

Few of these holdings were in contiguous blocks; rather, they were scattered around the country. They were created piecemeal, and on the whole one gets the impression that the holdings were the result not so much of the large-scale speculative merging of companies – as is said to have occurred in the Weyerhaeuser operations in the Lake

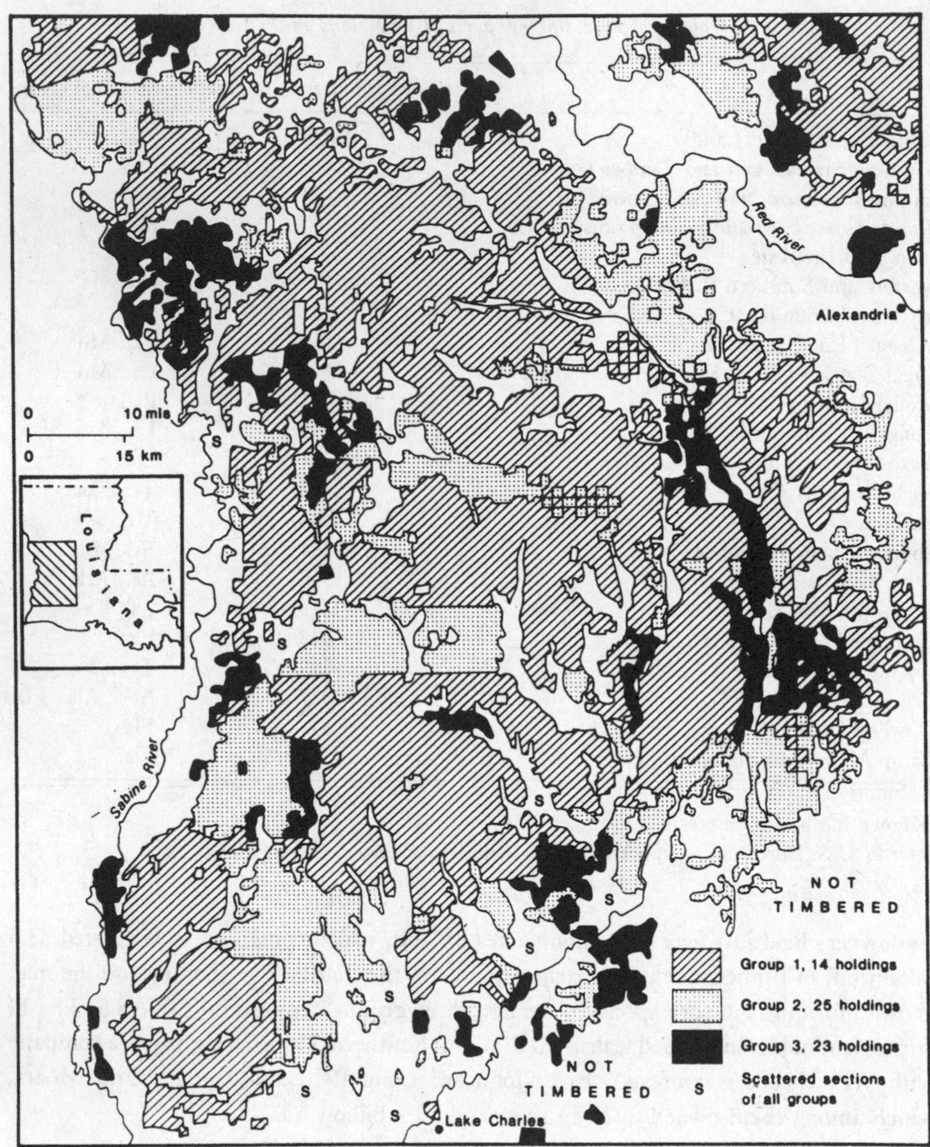

Figure 8.6. Large timber holdings in western Louisiana, c. 1914. (After U.S. Bureau of Corporations, *The Lumber Industry*, 2, pt. 2: opp. 18.)

States and the Pacific Northwest – as of a gradual buildup through the careful buying of land on the open market. However, the distinction is a fine and probably unrealistic one, for the result was the same: massive holdings of land on a scale never seen before.

Perhaps the outstanding example of large-scale ownership, and to a certain extent an exception to the general rule of northern purchases in the South, is the activity of John H. Kirby. As a young man he worked as a lawyer buying up land for an oil company. He

Table 8.6. *Timber acreage and timber stands of 74 owners in western Louisiana, c. 1913*

	No. of holders	Holding size (mill. b.f.)	Acres (thousands)	Total b.f. (billions)
Group 1 (on Fig. 8.6)	14	+ 60	1,453	21.8
Group 2 (on Fig. 8.6)	25	+60	652	8.4
Group 3 (on Fig. 8.6)	23	+60	238	2.6
Group 4 (not on Fig. 8.6)	12	Less than 60	21	0.2
Total plotted	74		2,364	
Not plotted	31	Less than 60	85	0.5
Overall total	105		2,469	33.5

Source: U.S. Bureau of Corporations, *The Lumber Industry*, 2, pt. 2: 137.

Table 8.7. *Acreage of timber (including timber rights and additional land) owned in Louisiana and in the southern pine region, by the 14 largest landholders plotted on Fig. 8.6*

	Fee	Timber rights	Additional land (cutover)	Total
Louisiana (on Fig 8.6)	1,453,147	—	—	1,453,147
Louisiana (not on Fig. 8.6)	687,999	407,875	373,038	1,468,912
Other parts of southern pine region	1,056,203	920,678	126,993	2,103,874
Total	3,197,349	1,328,553	500,031	5,025,933

Source: U.S. Bureau of Corporations, *The Lumber Industry*, 2, pt. 2: 137, 154.

quickly realized the enormous value of the timber on the land, especially as the adjacent areas of treeless prairie of north-central Texas were then being settled. He rapidly bought up vast quantities of stumpage and by 1890 was in a position to set up a mill in Houston to promote the Gulf, Beaumont, and Kansas City Railroad through his holdings, and by 1901 he had created 14 major lumber concerns throughout eastern Texas. By 1916 a new corporation, the Kirby-Bonner Company, was formed, with an output of half a billion b.f. annually, and it was probably the largest lumber company in the world at the time. The distribution of its branch offices gives some idea of the far-flung nature of its activities; there were branches in New York, Chicago, St. Louis, Kansas City, Oklahoma City, Indianapolis, New Orleans, Mobile, Galveston, Hattiesburg, and Havana in Cuba. Sales representatives worked throughout Mexico and Central and South America. The holding company for the whole operation was the Houston Oil Company, which also developed the lucrative oilfields. But, whatever the purpose of owning land, it did mean that Kirby could cut timber on well over 1 million acres in the longleaf pine lands of eastern Texas and western Louisiana.[49]

The company town

Just as the lumbermen and speculators owned the land, the trees, in fact whole counties, so many owned the towns and settlements around the mills, and by implication they "owned" the inhabitants too. There were hundreds of these little company towns, and they had much in common. Each one had a functional heart – the commissary – the department/general store owned and operated by the company, where the employees bought the bulk of the everyday items they needed. The common practice of paying wages not in cash but in checks redeemable in merchandise meant that prices in the stores were often "marked up" and wages were low. Surprisingly, this seems to have been tolerated to a large extent, although, admittedly, the lumber workers had little alternative but to acquiesce. Most were either ex-slaves or low-income whites coming off small farms, who were ready to work for long hours for low but at least assured wages. They had no possibility of alternative employment in secondary industry, and they had been used to trading by barter rather than by cash in the rural communities they had left. Their scope for protest was limited in the isolated lumber communities in the heart of the forest, and they were so dependent on the company for a living that it was not realistic to expect them to do anything to change their way of life. The only real challenge to the system came when employees began to own their own automobiles, particularly during the 1930s, and then they were able to bypass the commissary store and do their marketing in the cheaper stores in "normal" towns.[50]

Although black and white workers mingled in their jobs in the forests, their residences were usually segregated in the towns, as were those of the Mexican workers, if there were enough of them. Another form of segregation arose in Alabama and Georgia when mill owners used convict labor; then they had to build and man heavily guarded camps near the site of the mill but as far from the town as possible. But all was not bad; the towns did have another side. They probably had better facilities than the early "shack" towns that grew up around the smaller mills and also than the trade centers of the surrounding agricultural communities. Even the paternalistic, laissez-faire attitudes of the managers and owners – perhaps a hangover from the plantation heritage – often turned to the workers' advantage during times of illness, debt, and injury. To a certain extent, provided a man was skilled enough, there seemed little discrimination in jobs on account of color, and racial harmony prevailed so long as "no challenge to existing racial arrangements occurred." Nevertheless, it is also clear that the situation of labor in the forests of the South had its pernicious and unpleasant side. Many workers were exploited in the quest for profits. Attempts to organize labor, as in the case of the Brotherhood of Timber Workers in Louisiana in 1910, were a failure. Through the use of lockouts, dismissals, the physical breaking up of meetings, and the blacklisting of individuals involved in union activities, the employers stamped mercilessly and effectively on union activities up until the beginning of the Second World War in order to maintain their supremacy. The worst exploitation definitely occurred with the convict labor gangs, with the continental immigrant workers, and particularly with the black workers in the naval stores industry, a part of the timber industry where brutality was such that some employers exercised the power literally of life or death over their employees. It is easy to exaggerate these conditions, but equally it would be wrong to omit them; they were real events on the outer edge of the labor relations situation in the South, and they did give it a distinctive if at times somewhat disagreeable flavor.[51]

It is difficult to make accurate judgments about differences in ethos between lumbering regions, but the popular image of the fiercely independent, rugged lumberjack so prevalent in New England, New York, and the Lake States certainly did not appear in the South with its docile, captive labor, caught up in the sudden upsurge of the new and aggressive industrial system, which had been superimposed on the traditional, backward, and conservative society that had existed earlier. The southern forest laborer was different in other ways too; he was not a migratory worker living in a bunkhouse without women and children. More often than not, he was a settled family man, living in a company house in a company town, surrounded by a forest that could be wholly within a company county, encouraged to be a small part-time farmer, fire fighter, and warden as well as a lumberman. This difference in the work force had a lot to do with the lack of seasonality in lumbering in the South compared to the North, and with the different organization that business imposed right from the beginning in its quest to exploit the forests and impose order and stability on an industry characterized by rapid changes.

Associations and alliances

One result of the upsurge of activity in lumbering was that too many people got into the business, many with too little capital to cover the fixed and running costs. Lumbering had considerable overhead costs and long lead times, and much capital was tied up in stumpage and logs. Consequently, overproduction and continuous and intense competition between firms were the hallmarks of the trade in the South, and this resulted in depressed and variable prices, which was bad enough in the internal market but particularly injurious in the very substantial export market built up with Latin America and Europe, where consistent and reliable service by European suppliers provided intense competition.

"Everyman's hand is against his neighbor," wrote the *Southern Lumberman* in 1883, "and every man sells low because he is afraid his neighbor will undersell him. The result is that prices are ridiculously low, and good first class pine is going for a mere song."[52] The multiplicity of small units at this time had something to do with it, but many of the larger mills overproduced too. Some lumbermen were aware of the problem that was ruining the market, so that even by the early 1880s preliminary attempts were being made to form alliances and agreements based on some market research. Such was the growing enthusiasm for concerted action that the editor of the *Southern Lumberman* was inclined to believe that "the old style Southern lumberman" would be difficult to find in future years. It would be no loss:

> This generation does not read, never posts itself with regard to the state of the markets, runs its mills as long as a log can be got and sells for whatever price the dealer will give, knows nothing of the recovery of material, of fuels, of first class machinery and of modern methods. These mill men will down the market and prevent Southern timber and lumber from commanding their legitimate price in the market.[53]

A tangible result of the market chaos was the formation of various associations and marketing agencies in order to bring stability to the industry. On the whole, the lumber men were a very individualistic lot, and they found it difficult to cooperate. Many attempts were made to promote mergers, but these failed because of the loss of individual action. Many attempts were made to organize sales companies to handle the output of large groups of mills, but these, like many others designed to market the whole of output through a major wholesaler, all failed. Consequently, looser cooperative associations were favored. In 1890 the Southern Lumber Manufacturers' Association (later to become the Southern Pine Manufacturers' Association) grew out of the very much smaller Missouri and Arkansas Lumber Association founded in 1883. The association established a ratio between prices and various grades and lengths of timber in over 50 "official price" lists and thereby unified the great variety of marketing organizations that existed. "Recently we had a little meeting over at Memphis," said Mr. J. L. Thompson, president of the Association. "We had 70 billion feet in a room. . . . We all realized that there was only one evil. We took a vote on what was the cause of the troubles – 'over production'. Every solitary man agreed on that, and everybody was willing to suggest a remedy; everybody knew what the remedy was."[54]

The cutback of production engineered by the association had some effect and prices moved upward, which had the undesirable result of stimulating production so that once more there was a glut of lumber in the South. Then producers representing between 70 and 80 percent of southern output agreed to curtail production, which was successful temporarily, but again some firms saw their chance, jumped in, and produced as much as possible. It was chaos. Nevertheless, prices did rise generally during the early years of the twentieth century, a feature the public associated with the price fixing of the Southern Pine Manufacturers' Association and of the newly founded Mississippi Valley Lumbermans Association (later to become the Northern Pine Association). The Southern Pine Manufacturers published "Market Reports" and "Prevailing Price Information" in weekly trade journals, and although they purported to be retrospective reports on prices received, it seems clear that they were being used by the trade as favored minimum prices and enforced by agreement. This brought many firms under public scrutiny, and charges were made of restraint of trade.

The financial panic of 1907, and the general problems of uniformity of product, price, and overproduction, spurred mill owners to further efforts to control marketing, and they floated the idea of a gigantic $300 million lumber corporation, modeled on the United States Steel Corporation, to which all mill owners sold their plant and stumpage for stock holdings. The move received a surprising degree of support from many of the larger corporations, and a committee under the chairmanship of R. M. Weyerhaeuser was appointed to recommend plans for the formation of the corporation. But the new corporation would have created a monopoly situation on a scale never seen before in United States business, let alone in the forest industry, and on a scale never suggested since. The attorneys general of several southern states blocked its formation.

The continued rise in lumber prices and the mounting antitrust and conservation sentiment throughout the country culminated in the publication of the massive Bureau of Corporations report in 1913–14 on forest ownership and timber prices. Its whole tenor and conclusions were totally unsympathetic to the genuine problems of the majority of the lumber manufacturers, to controlled marketing and hence controlled production and the elimintion of fluctuating prices. Consequently, indidivual lumbermen attempted to get past these problems by means of greater vertical integration of their operations, which meant getting even bigger than they were already, and by forming an entirely new organization in 1914, the Southern Pine Association, which ran into all the old problems and accusations of fixing prices because it published its "Weekly Trade Barometer," which showed the production of the week and the value of sales and shipments. Elsewhere in the South, the baldcypress manufacturers formed a trade association in 1905, which also ran into all the same problems as had the yellow pine producers. On the whole, the exporters were more successful in getting their problems sorted out, and the Gulf Coast Yellow Pine Exporters' Association founded in 1892 persisted and did get uniform grading in the highly competitive overseas market.[55]

Everywhere in the industry, business methods were becoming more streamlined, more organized, and more cooperative, in line with the general trend in twentieth-century business methods to smooth out the fluctuations in the laissez-faire economy. The northerners were certainly in the van of these changes, and many southerners followed suit and "looked prosperous and were working shoulder to shoulder with the Yankees." Many, however, lamented the passing of an age, of "the old school of gentlemen, the midday mint juleps, and the easy-going business methods." The exploitation of the forests was transforming the society of the South.[56]

Production

The purchase of land and the advances in technology, transport, and business and social organization all culminated in a massive attack on the forests of the South, which reached its apogee by about 1920. The signs of such a massive inroad were already present in 1880 when Charles Sargent was writing his report. The Lake States forest had been logged over ruthlessly, and it was now the turn of the South. Sargent foresaw new centers of distribution emerging and undermining the preeminence of places like Buffalo and Albany, but he did not think that any one point would "attain the importance of Chicago." With the absence of an obvious focus for water communications in the center of the continent and with the growth of the railroad system, it was more likely that supplies from the South would go direct to the consumer. "In this way the Pine of Mississippi, Louisiana, and Arkansas will reach Kansas and Nebraska, and the whole country now tributary to Chicago, Western Texas and northern Mexico will be supplied by rail with the pine of eastern Texas."[57]

The supplies were certainly there; Sargent estimated that in toto there were 237 billion b.f. of pine timber of various sorts in eight Gulf and Atlantic-facing states plus Arkansas, of which nearly 60 percent was in the southwestern states of Texas, Mississippi, and Louisiana, and this estimate did not include baldcypress timber. Whatever credence one puts on this estimate of timber supplies by Sargent, the point is that relative to his contemporary estimate in the Lake States, the South had approximately three times as much standing timber as had the Lake States, Texas alone having nearly as much as Wisconsin, Minnesota, and Michigan combined. The southern estimates were not questioned in the next major report on the pines of the South by Dr. Charles Mohr, as he had made most of the Gulf States estimates for Sargent. If anything, Mohr was inclined to upgrade his previous estimates.[58]

The problem encountered whenever we look at production statistics in the Lake States or try to comprehend the astounding magnitude of the figures and their variation from year to year is repeated in the South. In fact, it is even worse. There are more state units with which to deal, and no one center ever reached the importance of either Chicago as a market or Saginaw as a production node, thereby enabling us to summarize the trends for the region. The detail of regional studies such as those of Easton and Collier on Texas; Norgress, Mancil, and Millet on Louisiana; Hickman on Mississippi; Massey on Alabama; and Boyd and Vance on the whole of the South need to be looked at for the finer-grained evidence.[59]

In general terms, the amount of lumber cut in the South showed an initial pattern of slow but steady growth until 1889, then a rapid and spectacular rise after 1890 (with the exception of the depression in 1904) reaching nearly 140 billion b.f. per annum in 1909. If the production of the Atlantic division of the Carolinas and Virginia is included in the southern division, then the South accounted for 46 percent of all timber cut in the United States (Fig. 7.5). Production fluctuated wildly at a high peak until the crash of 1929 but was not overtaken by the Pacific Northwest until the mid-1920s, thereafter to fall behind permanently in the mid-1930s.

Initially most of the cutting was in Georgia, which was the leading producer throughout the 1860s and 1870s, although all other states cut sizable amounts (Fig. 8.7). By the time of the big push of the 1890s Texas moved ahead of all other states and was confidently expected to "occupy in the Southern States the commanding position that the State of Michigan used to occupy amongst the white pine producing states."[60] It was closely followed by Alabama, Arkansas, and North Carolina. With the turn of the century, all the southern states, with the exception of South Carolina, Florida, and Virginia, were producing between 1.1 and 1.3 billion b.f. annually. The real differentiation in output among these states did not come until after 1900 when Louisiana surged forward and stood head and shoulders over all other states to produce a staggering 3.5 billion b.f, more than any other state in the country with the exception of Washington, a position it held as the leading or next-to-leading producer from 1906 through to 1919. It was only in 1929 that Louisiana's production fell back markedly and Mississippi was marginally the leader.

The broad temporal variation of production by states can be supplemented, at least in its early stages, by the spatial variation based on the census returns of the value of production for 1860, 1870, and 1880 (Figs. 8.8–8.10). To a certain extent the plantations were minimills supplying the needs of small communities of masters, tenants, and slaves, but over and above these thousands of unrecorded units a few hundred small sawmills show up in the census returns. In 1860 on the eve of the Civil War there were 375 counties with an annual production of less than $62,250 per annum, and another 44 with between $62,250 and $125,000. Above these were the main centers of production: the export port of Pensacola at the mouth of the Escambia with $780,000, followed by

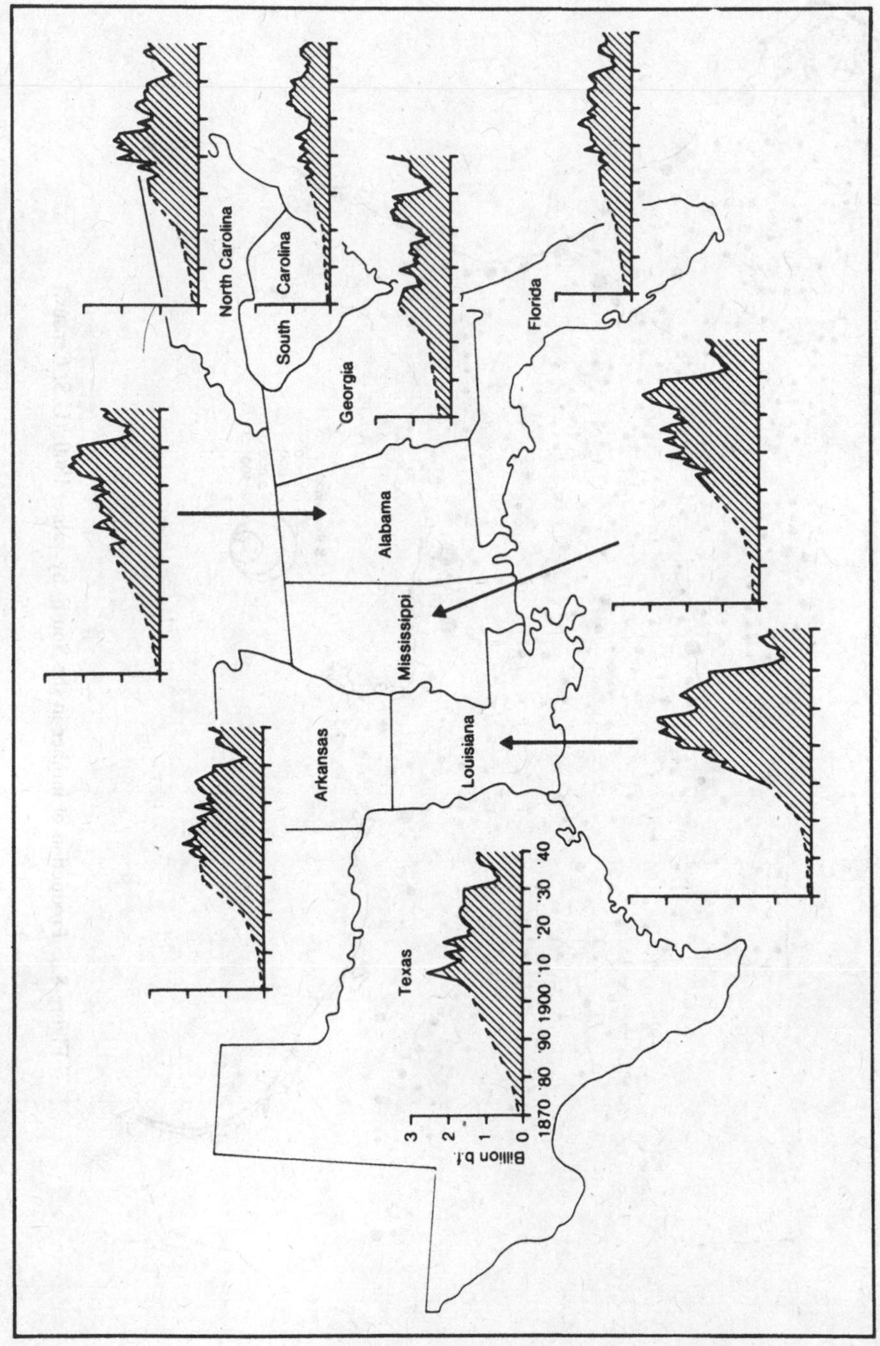

Figure 8.7. Production of lumber in the South, by state, 1869–1940. (Steer, *Lumber Production*.)

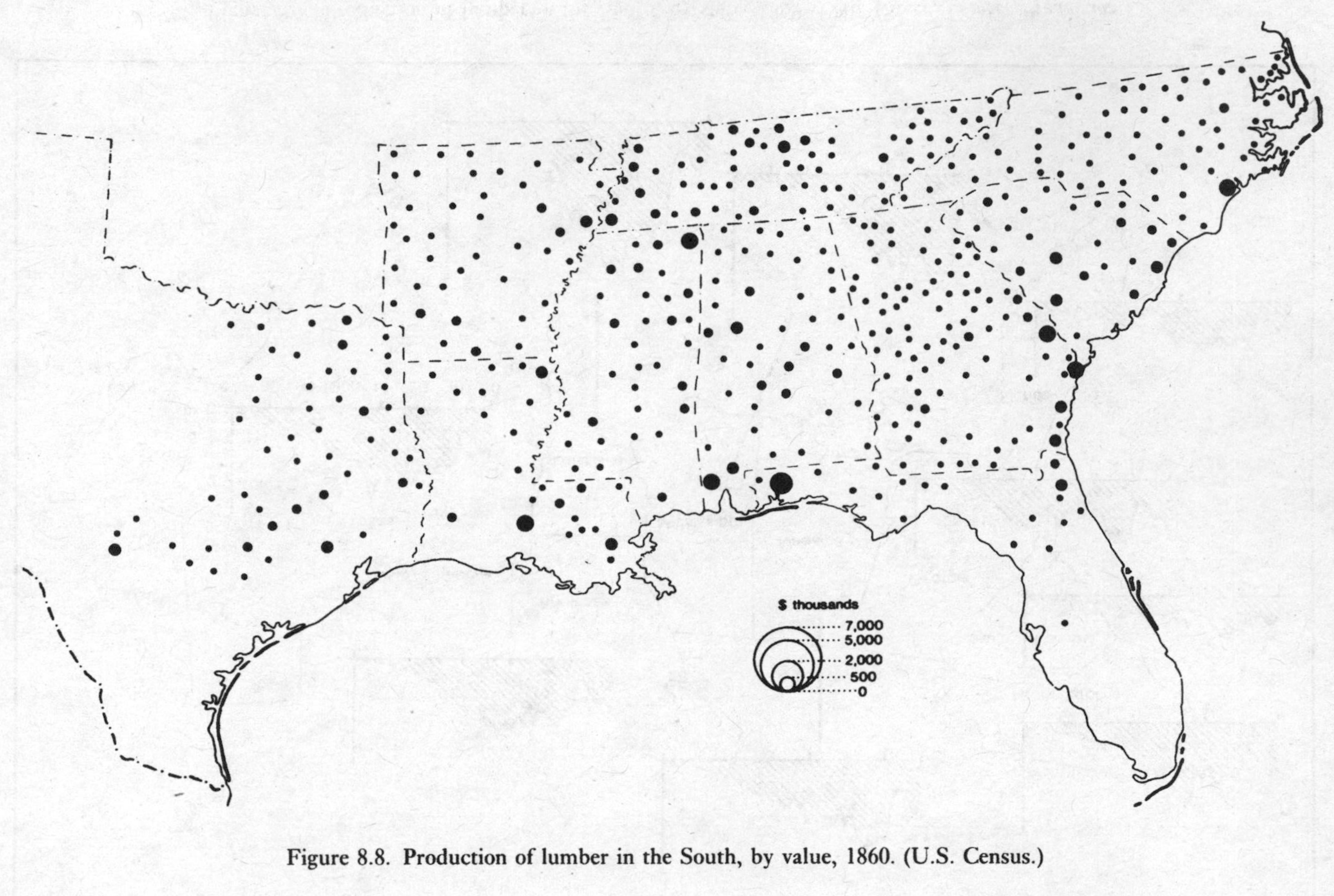

Figure 8.8. Production of lumber in the South, by value, 1860. (U.S. Census.)

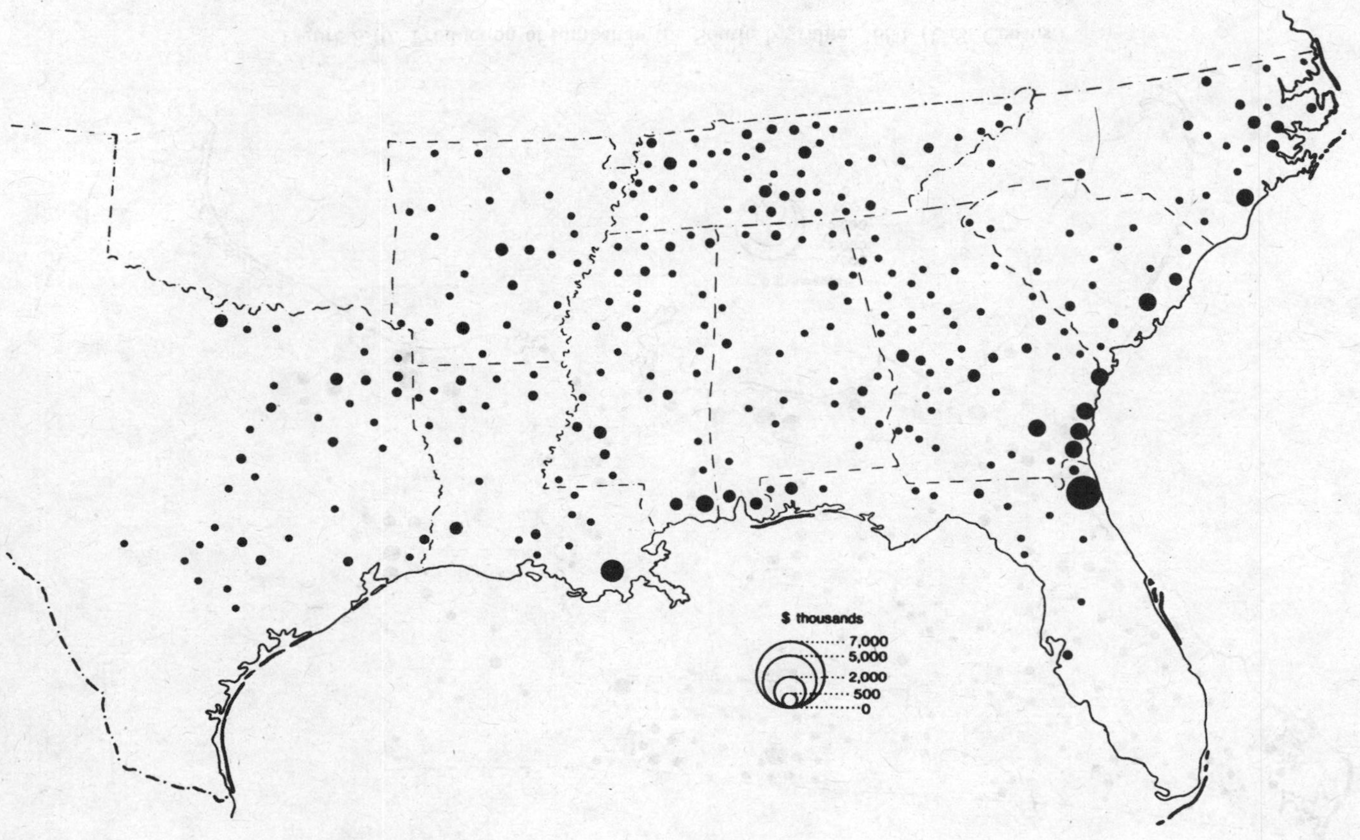

Figure 8.9. Production of lumber in the South, by value, 1870. (U.S. Census.)

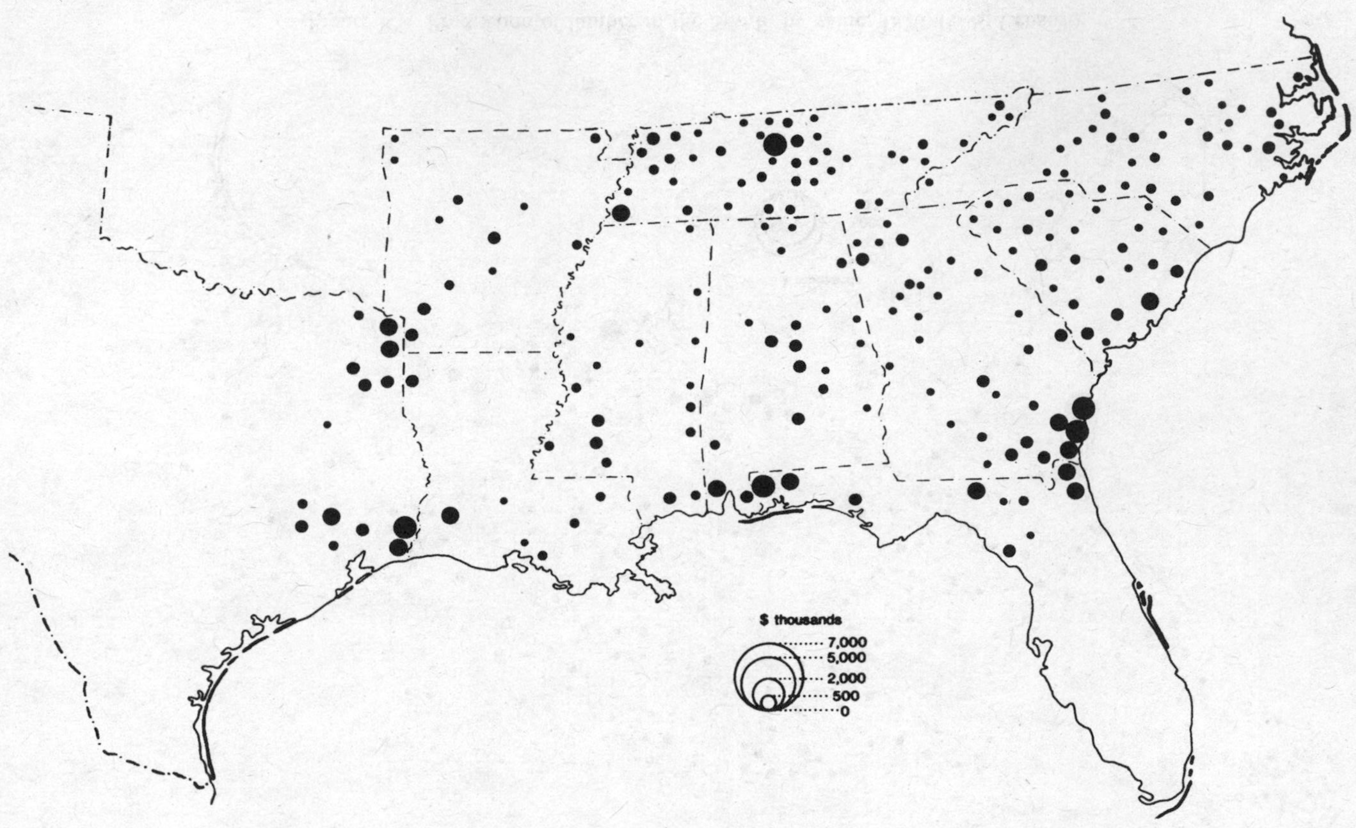

Figure 8.10. Production of lumber in the South, by value, 1880. (U.S. Census.)

Savannah with $381,000, Baton Rouge in the delta with $350,000, and Wilmington in North Carolina with $269,000. The map of 1870 (Fig. 8.9) shows that the events of the Civil War had caused a great sorting out of sites. The total number of counties reporting production had fallen from 436 to 289 in the same 10 states. The losses were almost exclusively in the smallest mills, which dropped in number from 375 to 226, but the larger centers grew larger. The export trade at Jacksonville reached well over $1 million in that year, followed by Bayou Parish in Louisiana ($955,000), and Memphis ($634,000), the latter in fact representing the production of the Lake States forests rafted south. There was a string of large centers of production along the Atlantic coast north of Jacksonville, Florida, through Georgia to North Carolina, such as St. Marys, Brunswick, Darien, and Savannah (Ga.), Charleston (S.C.), and Wilmington (N.C.) – all major lumber producing centers since the mid-eighteenth century – each exporting over 80 million b.f. of timber during these years. Moss Point on the Pascagoula now shows up as a major center of production. By 1880 (Fig. 8.10) the pattern shows further reduction in numbers and concentration in large units, with places such as Beaumont, Orange, and Texarkana being less prominent than before but others, such as Lake Charles (La.), Memphis, Nashville, Mobile, Pensacola, Jacksonville, St. Marys, Brunswick, Darien, and Charleston, still remaining as major production points. There is some shift in production toward the main railroad lines, but there are many unexplained blanks on the map.

Generally, the size of the production unit as measured by value increased at each census, as might be expected. In 1860 there were seven counties producing over $250,000 and representing 18.1 percent of the total value of production of $15.3 million, but by 1870 the comparable figures were 13 units representing 32.4 percent, and in 1880 19 units representing 39.8 percent of the total value.

The only other detailed view that we have of production in the South is from the map compiled by Henry Gannett, geographer to the U.S. Census, for the Twelfth Census in 1900 (Fig. 8.11). It shows the value of all lumber and timber products per square mile by county and is therefore not strictly comparable to the other three maps of production. Yet it does confirm some of the impressions of the previous distributions and, in addition, supports the idea of a rise of the inland centers dependent on railroad communications, particularly in Mississippi at Laurel and McComb, in Arkansas at Fordyce, and along the Texas border. The map also helps put the production of the South into the perspective of the United States in general.[61]

Depletion

"How long will the timber industry last in the South?" wrote A. E. Parkins in 1938.[62] It was a question that had been asked repeatedly and nervously for the past 60 years, and particularly when the depletion of the Great Lakes forests had become more and more apparent after 1880. The answer, of course, was not simple; it was a complicated equation, the solution to which depended upon the resources available, the rate of cutting, the rate of growth, and the rate of reforestation, if any.

Sargent's estimate of 237 billion b.f. of pine stumpage in the South was about three times that of his estimate for the Lake States, and it was confidently expected that it

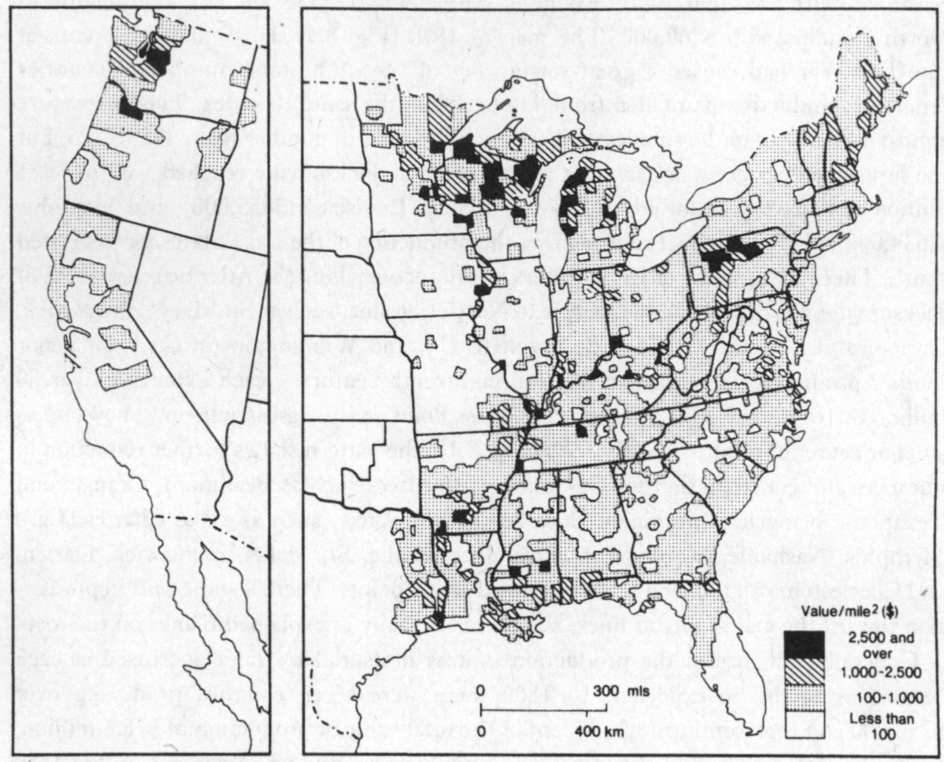

Figure 8.11. Value of lumber and timber products per square mile, 1900. (U.S. Bureau of the Census, Twelfth Census (1900), 9, pt. 3:803.)

would be enough to withstand the onslaught of mass production methods of logging that came with the railroads, power skidding, pull boats, and the massive mills. "While the North has lost (perhaps beyond repair) a great amount of its timber area," wrote F. P. Baker for Hough's survey in 1884, "the South has still a vast amount of forest which may be husbanded and saved." At that time so much of it had "scarcely been touched" that the "land of forests" could always offer new areas of production. On the basis of the rate of cutting in 1880, the pine forest would last far into the foreseeable future: "the pine supply in Texas will last 250 years," he forecast, "in Louisiana 100 years; in Mississippi 150 years; in Alabama 90 years; in Georgia 80 years; in Florida 30 years; in Arkansas 300 years; in North Carolina 50 years; and in South Carolina 50 years."

Most local agents agreed with these optimistic assessments: "Not more than 20 percent of its [Louisiana] woodland has been cleared of timber," wrote the reporter from New Orleans, "and it has been done mostly for agricultural purposes. . . . More timber must yet be removed by the farmers for 25 percent left remaining is thought to be sufficient. There is plenty of room for the lumberman and the saw-mill."[63]

As the 1890s dawned, the situation seemed much as predicted; the census of that year gave an upward revision of the standing pine timber at 300 billion b.f., and this reassured those who looked at the upward-sloping curves of production in Texas, Louisiana, and Mississippi, in particular. Hereafter the estimates of stumpage varied; the next two, in 1892 by B. E. Fernow and in 1898 by G. W. Hotchkiss, were downward at 250 and 187 billion b.f., respectively. The Yellow Pine Manufacturers' Association pushed the estimate up to 300 billion in 1905, and the Bureau of Corporations to 384 billion b.f. in 1914. However accurately carried out, the estimates were bound to vary according to the size of the tree considered suitable for logging, the height above the stump at which the tree was cut, the scale of the measurement used, among many variables, and all these varied over time. Yet, after the high point of 1914, which seemed a watershed year in concern about the depletion of the forests, estimates took a marked turn downward and the forests were not now predicted to last for more than 30 years by any calculation possible.[64]

Of course, not everyone was so complacent. The local agent in Mississippi reporting to F. P. Baker in 1884 noted that two-thirds of the timber in Copiah, Hinds, Madison, Rankin, and Yazoo counties had already been cleared and that those counties around Meridian in the east-central part of the state were all cleared within five to eight miles of the railroads. He sounded a warning almost unique at the time: "English and Northern capitalists are fast purchasing our magnificent pine forests. The avarice of capitalists and the great number of saw-mill men, if not in some way checked, will ere long destroy the grand pine forests of this section."

Baker also noted two disturbing things. First, the present rates of production "were not a criterion for future rates," particularly if the railroads kept spreading as they had been doing. Second, cutting was already leaving large areas bereft of timber, "a little over 200 square miles in a single year" in Mississippi, Alabama, and Florida. In addition, turpentining was already wiping out another 150 square miles of forests annually in Mississippi and Alabama, and all this destruction did not take into account the ravages of disease and fire. Charles Mohr issued similar warnings in 1889: The exploitation was being carried on with the sole object of the "speediest returns without heed for the future," and he reminded the skeptical that the forces that led to the depletion in the North were being exported to the South so that the depletion of the southern forests "will scarcely take a much greater length of time."[65]

But, on the whole, these were lone voices; the cutting went on at an ever-increasing pace. Few lumbermen cared how much they cut so long as they cut. Many were so heavily in debt because of their policies of expansion (particularly in the laying of railroads and tramroads) that anything that reduced logging costs and maintained prices was pursued. Steam skidding may have scourged the surface bare, but at least it had wrung the last penny out of the timber stand, and there was always the cutover to dispose of afterward as agricultural land.

For many lumbermen, the stumpage was security for their enormous debts, and therefore continuous operation to meet payments of interest and principal were necessary. Moreover, they reasoned, tree growth was so slow that stumpage was bound to decline in value every year due to blowdowns, fire, and disease; therefore, the sooner it was cut the better. State taxation policies were also perceived to make matters worse. In Mississippi, for example, timberland and farmland were taxed similarly on either an assessed value or an annual *ad valorem* tax basis. Lumbermen objected; unlike farmland, forests had no annual return, and therefore they wanted to cut out and get out as quickly as possible in order to stop paying taxes. The tax situation appeared to get worse as the stumpage prices rose; land valued at $1.25 an acre in the late 1890s rose to $5 in 1905 and $15 in 1909. In addition, as the tax burden in most southern counties was laid on land or timber, the valuations increased as the demand for services increased. A few examples must suffice to illustrate the point. The Blodgett holding in Pearl River County, Mississippi, was revalued from $808,000 to $1,334,665 in a year, and this raised the tax on the company by nearly $16,000. Other lumbermen pointed out the great differences in the taxes on stumpage in the South and the low taxes on stumpage in Canada. Edward Hines, another lumberman with large concerns in Pearl River County, logged off his 241,000 acres of land quickly in order to sell it as cutover, but because he left a few small trees on it the land was not classified by the county assessors as cutover for tax purposes. Hines then promptly laid them waste, thus eliminating all natural regeneration.[66] Not until the system of taxation was changed to a severance tax, where the tax was paid on the resource once it was removed or severed from the soil, could popular attitudes toward the exploitation of the timber lands ever begin to alter.

"The nameless towns"

The growth of the southern lumber industry after the widespread introduction of the railroad and of steam power to drive machinery meant that most mills were locationally footloose and were not tied through inertia to settlements with existing functions, such as ports, or to sources of power on rivers. There were some exceptions, of course, like the long-established export centers of Orange, Mobile, Pensacola, Jacksonville, Darien, Savannah, Wilmington, and Charleston, and there were a number of towns at strategic points on the railroads inland that retained their function, as, for example, Laurel, Meridian, Jackson, and Hattiesburg in Mississippi, or Alexandria, Shreveport, and Baton Rouge in Louisiana, or Brewton and Montgomery in Alabama. But the majority of mills were located as near as possible to the supplies of lumber.

Most of the settlements grew up along or at the ends of spur lines of the logging railroad, which snaked out into the forests from the main lines. Sometimes only temporary lumber camps were established, but more usually more substantial settlements were created with churches, schools, and a modicum of facilities. As one writer commented in 1887, "in much of the pine woods of the South the sawmill is the forerunner and foundation of the town."[67] For example, Bogalusa in Louisiana was created around a mill. A site on the Bague Lusa Creek on the Pearl River was chosen by James Lacey for the Goodyear brothers to set up the mill for their Great Southern Lumber Company. On February 7, 1906, the first pines were felled, and over 14 million b.f. went into construction of the mill and houses of the new town, which was ready for occupation in November 1907. Immediately, 3,000 men were employed in the mill and its associated works, and the city grew to a population of over 15,000 within a few years.[68]

Generally, the lumber towns grew with a wonderful rapidity as the attack on the forest began; however, unlike Bogalusa, which remained as a permanent settlement, they vanished with equal quickness once the forest was logged out. Their story was captured in the evocative but quite accurate account written by R. D. Forbes in 1923, entitled "The Passing of the Piney Woods."[69] The lumber town he visited presented an air of desolation and decay:

> No wonder the hotel was empty, the bank closed, the stores out of business: for on the other side to the railroad, down by the wide pond that once held beautiful, fine-grained logs of Louisiana longleaf pine, the big sawmill that for twenty years had been the pulsing heart of this town, was already sagging on its foundations, its boilers dead, its deck stripped of all removable machinery. A few ragged piles of graying lumber were huddled here and there along the dolly ways in the yards where for years lumber had been stacked by the million feet. . . . The mill had "sawed out" – had cut its last log six months before. Within the town grass was beginning to grow in the middle of every street and broken window lights bespoke deserted houses.

In county after county across the South the pinewoods had "passed away." "Their villages are Nameless Towns, their monuments huge piles of saw dust, their epitaph: 'The mill cut out.'" There was really nothing fanciful in the description by Forbes, who as a former director of the Southern Forest Experiment Station had seen it happening everywhere across the South. The Georgia Board of Forestry reported in 1922 that "Dismantled mill plants and deserted communities throughout the lumber regions of the state are forbidding reminders of the migration of an industry which, under wise and proper management of our forestlands, should be a permanent and leading industry in the state."[70]

The human and social cost of this constant uprooting and moving was never calculated. The general picture of the temporary nature of the lumber towns can be particularized by the example of the exploitation and settlement of the Calcasieu basin in western Louisiana, east of the Sabine and south of the Red River, an area of some of the greatest company concentration of ownership (Fig. 8.12). In about 1892 the longleaf pine covered about 2.5 million acres of the basin, and although some cutting was going on at Lake Charles the bulk of the timber was untouched. During the next few years this inaccessible forest area was crisscrossed with railroads, and many company towns sprang into existence at strategic points. By the mid-1920s most of the forest was logged out and the towns began to disappear. Along one railroad 23 mills, each cutting more than 100,000 b.f. daily, went out of existence in the space of five years, and 16 large mills in Vernon and Beauregard parishes, employing at least 400 men each and with dependent town populations of about 1,000, were abandoned by 1933. For example, Fullerton in Vernon Parish was established in 1906. By 1910 it had a population of 1,238; during the next few years it grew to 2,412. The mill stopped operation in 1926, and the population was a mere 148 inhabitants by 1930. Logging ended in Rapides Parish in 1924 when the entire population of McNary Town moved to a new lumbering community called McNary in eastern Arizona. The move was witnessed by H. H. Chapman:

> Two months ago, Louisiana had this thriving town of 3,000 persons. As the forests became denuded of pines the employers of the villagers began looking for a new site . . . . In two long special trains half the town was started westward to build a new village. Today the last of the inhabitants left. In 52 hours they will be at a point 80 miles from the new activity . . . and they will be back home in McNary.[71]

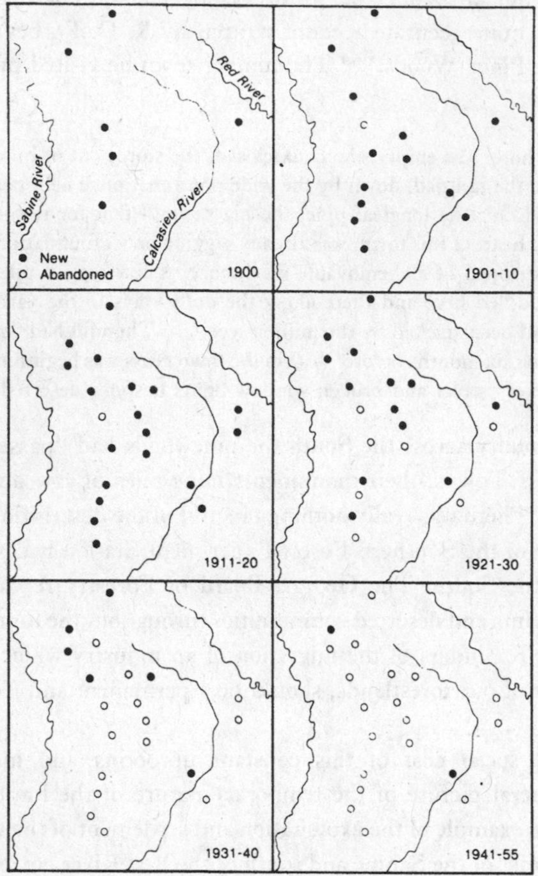

Figure 8.12. The birth and death of mill towns in the Calcasieu basin, Louisiana. (After Stokes, "Lumbering and Western Louisiana," 252.)

This spectacular wholesale movement was the exception, however, as it was far more usual for the companies to ensure their future by buying new stumpage in the West and taking only a few key personnel with them.

The same story of rapid change could be told for the baldcypress-milling towns of the lower delta. There were 150 mills in 1915, but as the supplies ran out the number dropped to 90 in 1920 and to 4 in 1940 until the last one ceased operation in 1956. These were the foundations of towns such as McElroy, Garyville, Lutcher, Garden City, Paterson, and many more; others, such as Palquemine and Iberia, have been completely obliterated.[72]

Wherever the large mills moved out or stopped, whether in the pinewoods or the baldcypress swamps, their place was taken by a multitude of smaller and often portable gasoline-powered mills, which either made the most of the timber left or cut the meager second growth on the cutover. But these were even more temporary than the lumber towns, and virtually no permanent settlement resulted from their operation.

The cutovers

When the land was stripped of its timber and the thin soil scraped off by the power-skidding operations, the timber companies moved their mills and let the towns die. But they were left with an expensive liability; the land yielded nothing but was still taxable. Either they abandoned it completely as tax-delinquent land or they sold it off as farmland, a process encouraged by the prevailing sentiments about the promotion of small-scale, owner-occupied farms. The cutovers were a complex problem with economic, social, ecological, and aesthetic implications. In one particular parish in Louisiana, they

> stretched wearily away from the rusting rails of the mainline track. Those nearest the mill, which had been cut last or had been gone over a second time during the recent era of high prices, were desolate indeed. Not a living pine tree remained on acre after acre. The extraordinary demand for every available stick of timber . . . the pitiless system of taxing annually every board foot in standing trees, and the sweep of slash fires had done their work.

This description was not "sensational"; it was "a commonplace" by the early 1920s, and even before the First World War the demand for timber had resulted in such ruthless cutting, not only to maintain production but to wring the last penny out of the land.[73]

As in the Lake States, the final profits from the cutovers came from selling them to would-be farmers. Therefore, lumber companies and their subsidiary land corporations used high-pressure advertising techniques to dispose quickly of the cutover land.[74] Articles were written in the lumber journals extolling the superior merits of the newly cleared land. Lumber companies in Mississippi cooperated in the hard sell by forming the Mississippi Land and Development Association, which established demonstration farms to aid in the promotion. Special trains were put on to bring northern farmers to the South.[75] The lumber companies were not alone in hoping that the northern farmers would come; the states were "vitally interested" because the cutting of the timber had caused land values to slump and taxes to decrease, and the railroads were apprehensive that freight would disappear.

Although the climate was adequate for plant growth (unlike the cutovers on the margins of cultivation in the Lake States), not all the southern cutover lands were capable of being farmed successfully. In Louisiana, much of the shortleaf pine forest was roughly coincident with poor Eocene sands and gravels, and, similarly, large areas of the longleaf forest were associated with Oligocene and Miocene sediments, all easily eroded and relatively infertile (hence the name "pine barrens," once the areas had been cleared). In the baldcypress forests, the problem of the cutover was of a different nature. Here, nearly 1.6 million acres of land were flooded or were floodable swamp that needed to be drained, but at least they were fertile enough to be farmed if that was done.[76] In Mississippi and Alabama the soils of the denuded forests were slightly better, but even so the cutover land was a liability that was worth getting rid of at any price.

All in all, the farming of the cutovers was not profitable. Stump clearing was expensive and the work a backbreaking, thankless task. Most of the sandy soils were infertile, and the swamps needed expensive draining. The southern climate was enervating for those

Table 8.8. *Total cutover land, restocking and not restocking, 1907 and 1920 (thousands of acres)*

	Restocking		Not restocking	
	1907	1920	1907	1920
Ala.	11,858	7,500	1,845	6,500
Ark.	11,416	7,500	100	850
Fla.	12,058	1,700	3,015	5,300
Ga.	12,209	9,800	1,429	5,000
La.	6,667	5,700	1,292	3,530
Miss.	3,985	6,000	2,657	3,000
N.C.	10,115	9,000	1,950	1,200
S.C.	7,362	5,500	1,400	1,900
Tex.	3,439	2,700	3,000	2,724
Total	79,104	55,400	16,688	30,004

Sources: For 1907, U.S. Congress, Senate, *Report of the National Conservation Commission* "Reduction of Timber Supply through Abandonment or Clearing of Forest Lands" (1909), 2:639. For 1920, USFS, *Timber Depletion, Lumber Prices, Lumber Exports, and Concentration of Timber Ownership* (Capper Report) (1928), 20.

not used to it. As in the case of the Lake States, more settlers gave up than stayed. After a few scant crops, which barely helped them pay their way, and with their savings gone for improvements, they just walked off the land, the little clearing in the regrowth and the abandoned shack being the only evidence of their years of hard work. Nevertheless, a few did make a success of farming. They usually had land on very favorable soils near to urban markets, and with the aid of legume crops like peas, beans, peanuts, clover, and particularly the Japanese legume *Lespedeza*, they built up the soil fertility through the addition of nitrogen. Truck crops, the grazing of sheep, and improved herds of cattle and swine (both particularly successful in the mild climate with the open-air maintenance of stock) led to mixed farming. Others succeeded by allowing herds to graze indiscriminately over the unfenced cutover, and quantity compensated for quality.[77]

The definition of what constituted cutover varied, and many estimates of its area were bandied about, a complication being the clearing of the forest by cotton cultivators and its subsequent abandonment. The only consistently reliable figures are those of the Forest Service. It compiled tables for 1908 and 1920 that suggested that although some of the cutover was being restocked with timber a larger amount was being left idle. As shown in Table 8.8, the total amount of land affected by lumbering in the nine southernmost states (where less land was converted successfully to farms) was in excess of 95 million acres in 1907 and 85 million acres in 1920. But whereas only 16.6 million acres or 17.4 percent of the total was left idle in 1907, the figure had risen to 30 million acres or 35.1 percent of all cutover land in 1920. The increase in idle land was marked in

Alabama, Florida, Georgia, and Louisiana. Even then, the restocking land was not all going back to sawtimber; about half of the 55.4 million acres were growing only stunted cordwood. In 1907 it was estimated that of the 9.5 million acres felled nationally, 11 percent went into farms, 60 percent was restocked in some form or another, and 29 percent remained bereft of all purposeful land use, the high percentage of idle land being affected mainly by the situation in the Lake States. But by 1920 it was the South that was presenting the worst problem.[78]

As the dimension of the devastation became evident in the South and the ineffectiveness of selling the land to would-be farmers became more obvious, the lumber industry, in the form of the Southern Pine Association, cosponsored the first Cut-Over Conference of the South in New Orleans in 1917 with the Southern Settlement and Development Organization, an agency formed by a group of railroads to promote settlement and truck and better livestock farming in the southern states.[79] More conferences followed, but they foundered as they were motivated by a desire to promote settlement, which was never questioned, despite mounting contradictory evidence. The idea of using the lands for reforestation never really surfaced as a viable alternative, and grazing and further farming were recommended. The problem was perceived to be taxation, for just as taxes on forested land had led to its rapid denudation so taxes on cutovers inhibited reforestation with its long time scale of returns. Taxation was an issue that was going to consume a lot of time, energy, and legislative effort during the forthcoming years, and not until a severance tax could be introduced, that is, a tax on the timber once cut, would timber operators view their activities differently. Although it was not fully acknowledged at the time, repeated and devastating fire in the cutovers was another major reason for not planting.

Reforestation

The devastation of the forests, the disruption, cost, and economic and social upheaval of the movement of the mills to new stands, and the problem of what to do with the cutovers all pointed to one solution, reforestation, but the acceptance of that solution was going to be a protracted affair.

As early as 1890 Charles Mohr, the botanist, had directed attention to the rapidity with which the shortleaf pine seeded naturally on the "tens of thousands of acres" of old fields and cutovers regarded by most as "unfit or unprofitable for agriculture" in Mississippi. The lesson seemed obvious; it could be possible to replenish the timber resources, which were "fast falling through ever progressing destruction of the original forest without other outlay than simply assisting nature in her efforts to recover from injuries sustained in the wholesale destruction of the forests."[80]

But the implications of Mohr's observations were not appreciated. People were not aware of the extent of destruction or of the rapidity with which the southern pine could regenerate compared with the northern pine. The rapid invasion of scrub oak was considered to be the "natural" and only consequence of pine felling. Above all, the attractive economic advantages of clear-cutting and the use of steam skidders to reduce costs left a land surface completely bare of the mature trees capable of bringing about the natural reseeding Mohr had so admired. Reforestation as an alternative strategy to natural reseeding was not even thought of as the taxation system on the cutovers militated against all long-term planning for a slowly maturing natural resource like trees. Potent and real as all these reasons were for not doing anything, the lack of action did seem to have a deeper-seated and more complex cause. Whereas firms in the later-maturing lumbering region of the Pacific Northwest, like Crown Zellerbach and Weyerhaeuser, were experimenting with tree planting and fire control as early as 1901, and people like Austin Cary in New England were studying selective cutting and growth rates, nothing similar was happening in the South.[81] It was as if the original invitation issued during the immediate antebellum period to "come in and get our resources" was being taken to its logical conclusion by northern and native southern lumbermen alike, with little thought for their own future or for the future of the communities dependent upon their activity. In that and in many other ways, the South was still an under-developed section of the larger American community.

There were some important exceptions to all this. The example and action of Henry Hardtner of the Urania Lumber Company, Urania, Louisiana, who since 1904 had experimented reasonably successfully with allowing pines to reseed on his cutovers in order to keep his mills running, proved conclusively that southern pine could triumph over scrub oak if razorback hogs were kept out and fire was carefully controlled in its natural cycle, an experiment he later extended to actual reforestation. His enthusiasm was indirectly instrumental in getting the Louisiana legislature to pass measures in 1904 to establish a department of forestry, but this measure like many other "firsts" for Hardtner – including the creation and chairing of a Committee for the Conservation of Natural Resources (1908–10); the creation of a conservation fund, derived from a severance tax, to be used for fire protection; and the reduction of taxes on cutovers replanted to forests (the Timber Conservation Contract Act, 1910) – passed on to the statute book but was never operated upon because of the lack of funds and/or because it simply was ignored. Hardtner was alone in planting about 27,000 acres of his land to various southern pines under a timber conservation contract in 1913, but the fact that no one followed his lead for nearly nine years highlighted the complacency of mill owners about their supplies and their short view of their operations.[82]

The emphasis on Hardtner should not obscure the fact that there were other individual pioneer conservationists in the South who had an equally important though perhaps less well-known effect upon their local areas. Such people were John Barber White of Missouri[83] and particularly W. Goodrich Jones of Texas, who recommended a sustained-yield program and fire control in cooperation with the U.S. Bureau of Forestry as early as 1900. Such a program was needed, he thought, before the East Texas forest was cut bare: "what escapes the big mill is caught by the little mill and what the little mill does not get the tie-cutter and rail-splitter soon has chopped down." Influenced by the example of Hardtner, Jones fought for a state forestry department (which came into being by 1915) and changes in land taxation based on a severance tax, the proceeds to be used for reforestation.[84]

Soon, even the most "boom-minded" companies realized that reforestation was simply good business. The Goodyears in Bogalusa feared that their investment of over

$15 million (before even a board came out of the mill) was in danger of being lost. The cutovers were increasing at the startling rate of nearly 20,000 acres per annum so that the mill was, in effect, getting farther and farther away from its supplies and was in danger of being abandoned, Bogalusa facing "the dismal prospect of becoming a ghost town with a mountain of saw dust as a monument to its passing." After Goodyear visited the sandy Landes district near Bordeaux in southwestern France and saw the enormous area planted with pines to stabilize the dunes, he came back to the South convinced of the practicability of large-scale reforestation. He visited Hardtner's mills and forests in 1920 and was so impressed with what he saw that he immediately started a forestry program at Bogalusa under the supervision of his chief ranger, F. O. Bateman, fencing the lands to keep out the marauding hogs, cutting breaks to control fires, and planting a trial area of 800 acres in the first year and a further 15,000 acres during the next few years, with plans for over 10 times that amount. The small area planted initially was a consequence of the company's emerging policy of cutting the remaining forest selectively, thus encouraging continuous growth and production. During the depression years the company was taken over by the Gaylord Container Corporation, which in turn was taken over by the Crown Zellerbach Corporation in 1955, under which company the reforestation program was extended greatly so that by 1963 the planted forest exceeded 200,000 acres and was said to be the largest continuous area of man-planted forest in the world.[85] The patchy nature of the diffusion of ideas and action depended initially upon the success that states had in revising their taxation legislation. Louisiana's severance tax did not become fully operative until 1922, and it was not adopted in neighboring Missouri until 1940, by which time land had been either abandoned as cutover or purchased (1.5 million acres) by state and federal governments in order to create state forests. Later the tax issue assumed far less importance.[86]

Eventually, the impetus for widespread reforestation came when it became clear that the waste of the cutovers need not be tolerated. By the late 1930s lumber companies were beginning to appreciate fully the advantages of selective cutting and fire protection in order to achieve an even yield. Even more significant, perhaps, in the long run was the gradual realization that southern pines made excellent newsprint. This came mainly through the campaigning efforts of a Georgia chemist, Charles H. Herty. It was a popular belief that pines were an unsatisfactory raw material of newsprint as they contained too much resin and made yellow paper. After many experiments Herty proved that young pines were basically resin-free and could provide white paper. Southern pulp had the advantage of being cheaper than the northern black and white spruce product and was obviously not dependent on Canadian supplies, which had a stranglehold on the market. Additionally, growing young trees provided a lucrative and useful cash crop for southern farmers devastated by the depredations of the boll weevil. The land was producing what it grew best – pines – and when that was realized a major land-use battle was won. Despite much opposition from Canadian and northern producers, the Southland Papermill near Lufkin, Texas, was opened in 1940, marking a milestone in the shift of pulp production from North to South that has continued ever since (Plate 8.4). When the pulpwood revolution is coupled with the even longer-established manufacture of wrapping and packaging papers and paperboard, together with the

Plate 8.4. A fairly large modern mill in the South, Huttig, Arkansas.

invention by William H. Mason of Mississippi of the process for turning mill waste and poor timber into wallboard (Masonite) and the suitability of pines for manufacturing rayon (which Herty predicted), then it becomes clear that the southern forest was destined to move into a new era of prosperity.[87] The cutovers created during the first couple of decades of the century were after 1940 becoming a thing of the past.

9

The last lumber frontier: the rise of the Pacific Northwest, 1880–1940

. . . the lumbermen's westward migration ended. The blue Pacific Ocean, not another ridge of green forests, now met their gaze. Thoughts shifted toward permanence.

Harold K. Steen, *The U.S. Forest Service* (1977), 174

To the West only, of all our heritage of magnificent softwood forests, can the country look to an increasing cut; but even here there are already local evidences of depletion, warnings that the conclusion of the story will be the same as that of other regions and in far less time than has been estimated.

USFS, *Timber Depletion, Lumber Prices* (1920), 13–14

Experiment and change

The ever-expanding and generally westward-moving wave of lumber exploitation that had begun in New England during the seventeenth century, swept through New York and Pennsylvania during the mid-nineteenth century, carried on to the Lake States during the 1880s, and then swept down through the southern states during the 1890s had leaped the Great Plains and reached the Pacific Coast by the beginning of this century. Lumber production in the Pacific Northwest rose rapidly after 1900, and 10 years later it surpassed that of the Lake States. By 1920 production nearly equaled that of the South and was some 30 percent of the national total. Thereafter, production in other regions declined, and that of the Pacific Northwest stabilized so that it exceeded that of all the other regions, a position it maintained and strengthened. Since the mid-1950s the region has contributed consistently nearly one-half of the nation's timber.

But the story of the Pacific Northwest was different from that of the other regions, and it was certainly more than an account of forest destruction and a set of production statistics. First, because the soils were poor, slopes steep, rainfall heavy, and the level land too high and too cold, agriculture flourished in only a few isolated lowland places, the only sizable developments occurring in the largely forest-free Willamette Valley, the central valley of California, and the great Columbia basin east of the ranges.[1] In addition, the conventional wisdom that big trees meant fertility had to be modified because the sheer size of the trees in the West resulted in the almost physical impossibility of individual pioneer clearing for agriculture. In any case, the massive cost of clearing the big timber and pulling the stumps could not be covered by the sale of the by-products of the process, such as potash, pearl ash and timber, as had happened in the earlier areas of settlement in the Northeast. Thus costs often exceeded the combined value of the timber and the new farmland created, and the result was that the "natural" sequence of land use, where forest became cutover farmland, rarely happened. The stump-ridden cutovers were not easily converted into agricultural land.[2]

Second, it was no coincidence that in this, the last lumber frontier, where the green westward horizon of trees gave way to the blue horizon of the Pacific Ocean, other considerations began to hold sway.[3] This was the last of the public lands, the remaining untouched, unalienated stock of forest left in the country. There were many people who thought that this sizable section should not fall into private hands to be exploited for individual gain but should be preserved and managed by federal authorities for the good of the nation as a whole, especially at a time when timber prices seemed high and available resources were becoming scarce. Thus, the Pacific Northwest, California, and portions of adjacent states in the Rocky Mountain region were the scene of the first federal intervention in forest exploitation, in the form of the national forests created.

Not only was this the last of the public lands and the remnant of the West, it was in many ways the most spectacular of lands, a treasure house of natural wonders and scenic beauties that captured the imagination and concern of the increasingly environment-conscious society emerging in the United States at the end of the nineteenth century. That foremost among these wonders were the very trees of the redwood and sequoia forests themselves only added to the concern felt by many that this unique forest should be preserved for its own sake and that other forests should be preserved as a part of the general setting for the scenic beauties of the region. Thus, although the Pacific Northwest became the epitome of big business and big-scale lumber organization after 1900 because it was the last of the major forested regions to undergo exploitation, equally it was the scene of numerous experiments in preservation, ownership, and management.

Curiously, these experiments were not confined to federal bodies alone but also extended to the lumber companies, which pioneered methods of combating the waste of fire, sustaining the yield, as part of an overall strategy to achieve order and stability in the industry, which was probably easier in the Northwest than in any other major region because of less entrenched interests and fewer operators, thus prolonging the life of the forest. The capital investment in stumpage, plant, and associated services was now so great that the lumber companies could not afford, even if they had been able, to pull up roots and move on to new stands. They had seen what had happened in the other major lumbering regions, and they could see what was happening in the Pacific Northwest. If the destruction continued at the rate at which it was moving during the early decades of this century, then no amount of abundant supplies would free them from an uncertain future. Thus the last lumber frontier was in many ways the first of a new kind of frontier. It was a region that linked the previous three centuries of destruction and desolation of the forest to the present century of inquiry and concern about the forest. The story of the West Coast's timberland is not primarily one of deforestation and destruction as was that of the Lake States and the South, if for no other reason than that little land was cleared permanently for agriculture. Rather, the story was one of experiment and change.[4]

Tidewater lumbering, 1850–83

Although the burgeoning exploitation of the forest after about 1900 and the subsequent reassessment of where it was all leading are the subject of this chapter, it must not be overlooked that a thriving lumber trade had been in existence on the West Coast since Hispanic times and well before the California gold rush of 1849. That trade had laid the foundations for many of the patterns and processes that were to follow. Despite the intense local activity, however, it was limited and small-scale. Just as the physical attributes of the Great Lakes aided the exploitation of their magnificent forests, so the physical aspects of the West Coast states hindered the exploitation of their equally magnificent forests. Along the Oregon and Washington coasts there was a narrow belt of Sitka spruce (*Picea sitchensis*) with admixtures of western juniper (*Juniperus occidentalis*) and western hemlock (*Tsuga heterophylla*); and the western slopes of the Cascades from Puget Sound to northern California were clothed in a uniformly thick mantle of Douglas-fir (*Pseudotsuga menziesii*) and some Jeffrey pine (*Pinus jeffreyi*) in all but the highest locations, where forest gave way to open pastures and alpine vegetation. Farther south in California the redwood (*Sequoia sempervirens*) forests spread along much of the fog-drenched coast. In the Sierra Nevada, ponderosa, Jeffrey, and sugar pine (*P. lambertiana*) predominated, with patches of giant sequoia (*Sequoiadendron*). Because of their density and extent, however, the early exploitation of these forests was limited to a scattering of points along the coast possessing a combination of inland access and safe anchorages, such as at Santa Cruz, Monterey, Santa Barbara, and Puget Sound.[5]

Distances were immense. Not only did 2,000 to 3,000 miles separate the lumbermen from their main markets in the East, but nearly as great a distance separated the Canadian and Mexican borders. Consequently, problems of isolation dominated the nineteenth-century exploitation of the western timber stands. The lack of external and internal intercourse slowed down the growth of the region. Even when the transcontinental railway linked the coasts in the early 1880s, the high freight rates and the ready availability of the lumber in the Lake States and the South dampened the potential for local production. Whatever stimulus there was came more from the development of international markets around the Pacific basin, in Australia, New Zealand, Hawaii, Chile, China, and Peru. These could be served relatively quickly and cheaply by sail and thereby made large-scale lumbering possible far earlier than would otherwise have been the case.[6]

In addition to the problems of isolation caused by distance, internal movement within the region was severely restricted by the nature of the terrain. The high coastal ranges of the Cascades and the Sierra Nevada, the rugged terrain, and the steeply cliffed coasts hindered inland and coastal movements. There were only a few safe harbors along the stormy and fog-bound coast – Puget Sound and San Francisco Bay – and less-safe harbors such as Grays Harbor and Tillamook, Coos, and Humboldt bays were developed later. Therefore, along most of this forbidding coast any rock, promontory, or inlet that offered a shelter from the surf and the storms was used. These "dogholes," as the sailors called them, were of two kinds, both equally unsatisfactory. Along the rocky northern

California coast, near the redwood forests, the loading of ships alongside jutting headlands were accomplished only by sending the timber thundering down long apron chutes that jutted out over the cliff edge and into the hold of the vessel waiting below. Loading ships in the bar harbors that were established at the mouths of the numerous short streams that ran off the coastal ranges in Oregon and Washington was equally unsatisfactory. Rarely were the harbors deep enough to offer safe anchorages, their mouths usually being clogged by sand deposits or bars. They were made usable only by having steam tugs haul the lumber vessels over the bar, and so essential was the tug that it was the key to profitability in many north coast lumber operations.

The short and fast-flowing streams that plunged to the coast off the mountain ranges offered little access to the inland forest stands, unlike those on the eastern side of the Cascade Range. The turbulent streams were of limited use for driving logs, and often the canyonlike nature of the valleys (particularly in the Sierra) inhibited access by rail or road. The lack of large, natural drainage networks meant that exploitation proceeded in a series of isolated river basins along the coast, each one separated from the other, each one being operated at a less than economical level for the scale of production and organization that was being developed nationally during the later part of the nineteenth century. Only around Puget Sound was there a favorable combination of a safe harbor, nearby timber, and access inland at many points. The other safe harbor of San Francisco achieved a similar success as a lumber center by being the focus of the sawed timber that was transported from the numerous small, submarginal mills along the northern California coast. It became the major market and "nerve center" from which manufacture was increasingly controlled "as well as the locus from which the maritime commerce that linked the sawmills and their Pacific markets was directed."[7]

The gold rush of 1849 upset that relatively simple local system; the resultant demand for building and mining material far outstripped the few meager shipments that came from the East Coast and the South of the United States. In addition, the subsequent development in the post–gold rush years of the trans-Pacific shipments of lumber to Australia, Hawaii, Chile, and northern Latin America by the West Coast traders in the search for markets led to a thriving export demand long before western timber was railroaded to national markets in the central and eastern parts of the continent. The overall result was an enormous boost to local production. San Francisco was the core and trading center of the new enterprises with mills scattered around in the redwood forests at Bodega Head, San Jose, Redwood City, and Santa Cruz, all within a relatively short haulage distance by land or sea to the central transshipment point. By 1853 demand rose significantly, lumbering spread, and the merchant millers looked farther afield in order to establish mills in the untapped forests of northern California, Oregon, and Washington. The ever-extending maritime-lumbering empire of Asa Mead Simpson was illustrative of this. Soon, other recognizable centers of milling began to emerge all along the West Coast, particularly along the northern California coast in Mendocino County, around Humboldt Bay, and farther north in Puget Sound – all the new regions of production transporting their cut to distant markets, all relying on a complex system of ocean transport focused on San Francisco.[8]

Like all other initial lumbering concerns anywhere in America, the West Coast industry had begun with a host of small-scale units that flourished or were extinguished according to the vagaries of local demand. But the alternative, though sometimes coincident, stimuli of the Californian and a variety of Pacific markets had cushioned this nascent industry from the worst effects of fluctuating demand that usually plagued such embryonic enterprises. The financial collapse in 1854 sorted out the weak firms from the rest, but the overseas trade saved many. Those that prospered were those that firmly dominated what came to be known as the cargo trade, and these were located mainly around Puget Sound and along Humboldt Bay.

Outstanding among these was the firm of Pope and Talbot, which still flourishes today. In 1852 its first mill was established at Port Gamble on Puget Sound, and since then the firm has "enjoyed a history of remarkably steady growth and expansion."[9] Its history is a vignette of the early West Coast lumber trade, though few concerns were ultimately as successful. Andrew John Pope and Frederick Talbot of East Machias, Maine, landed in San Francisco on December 1, 1853, in the wake of the gold rush. Like most of the people who made money from the Gold Rush, they did not go to the diggings but supplied those who did. With two compatriots from Maine they purchased a boat to lighter ship cargoes to the shore in the bay. The venture was so profitable that they bought bigger ships and, foreseeing the trend in docking techniques as the port grew, rented a beach lot and opened a timber yard. The business prospered; relatives in the lumber industry in Maine shipped cargoes to the yard, cargoes of lumber were purchased from New Zealand, and cargoes were sold upriver from Sacramento. The pair began to realize the profits that were to be made by combining the ownership of oceangoing craft with the selling of the cargo they contained, and by careful management and some good fortune they prospered.

Subsequently they moved to Puget Sound to be nearer the source of the lumber supply and established the mill at Port Gamble, a focal point of the many channels of the inlet. By 1862 they owned a fleet of 10 vessels and were shipping about 19 million b.f. of lumber annually to foreign and domestic markets. They duplicated the original mill and then in 1864 rebuilt the original to double capacity. By 1875 they were shipping 43 million b.f. annually. Seven years later the company owned a fleet of 16 vessels and four tugs, had established outpost plants at Utsalady and Port Ludlow nearer to the timber stands at the mouth of the sound, all producing a combined capacity of 350,000 b.f. daily, and they owned upward of 150,000 acres of stumpage.

The success of the Puget Mill Company was due to a combination of cautious financial management, by which the company never allowed itself to get into debt, and the knowledge that the funds of the Pope and Talbot enterprises in Maine were always available to back them if needed. As Thomas Cox points out, however, the Maine connection was of little significance as Pacific Coast funds were "transferred eastwards to buttress the faltering family business in Maine and in an ill-fated milling operation that relatives purchased in Quebec." In addition, Pope and Talbot pursued a policy of market diversity with up to 40 percent of the cut going to overseas markets annually, and aggressively pursued new markets whether in San Diego or Hawaii.[10]

The success story of Pope and Talbot was repeated by a few other mills, notably the Port Blakely Mill Company under Charles Hanson. These mills shared a number of features in common. They were all located around Puget Sound. None originated in the Northwest; they were spawned by the capital earned originally in trading and shipping from San Francisco. The fact that all owed their success in large part to the ownership of fleets of lumber carriers and tugs had various implications. They all had offices and yards in San Francisco, the main transshipment point and only harbor capable of accommodating overseas trade ships; their large capital investment meant that they all were intent on achieving market stability and therefore participated in diversified foreign markets; they all moved from trading and cutting other people's lumber to owning their own timber stands; all had the most modern and up-to-date steam milling machinery and had their labor forces concentrated in milling towns; all cut and dealt with Douglas-fir, or "Oregon-pine," as it was known.

In highlighting the trend toward vertical integration backward into the source of supply in the Douglas-fir forests surrounding Puget Sound, one must not suppose that it was the only part of the West Coast where entrepreneurs were engaged in the quest to get big. For example, after a number of starts, Asa Mead Simpson set up a mill at Gardiner on the Umpqua River near Coos Bay. Soon he owned another five mills scattered in different forest areas in order to get all types of lumber for his yards. He eventually owned 24 vessels and even established his own shipbuilding and repair yard in his quest for total horizontal and vertical integration and independence from all others.[11] Other large mills were established later at Grays Harbor, mainly by Puget Sound interests.

Along the redwood coast most mills were small. Because of the terrain of the area and the sheer size of trees, production was always more expensive than on the sound. Once the stands immediately surrounding the mills on the coast had been cut, the trees from inland had to be hauled out by ox team or logging railroad or floated out on freshets in the steep streams. For most of the winter it rained so heavily that logs could not be moved. Consequently, the uncertain supply of logs closed the mills from time to time, and production per mill was never so great as around the sound.[12]

Nevertheless, the successful mills were those that followed the same pattern of development as those on the sound, which meant they had to do more than just produce lumber. They had to own vessels, and offices and yards in San Francisco, and they had to pioneer market diversification around the Pacific basin as well as along the West Coast, particularly in southern California, in Los Angeles and San Diego. Above all, they had to exercise managerial and financial caution. Hence, by 1865 there were eight mills in the redwood forests of Humboldt County producing 12.6 million b.f. annually. By 1875 Kentfield, Buhure, and Jones cut 16 million b.f., Dolbeer and Carson cut 12.7 million b.f., and the Occidental Mill at Eureka cut a prodigious 75 million b.f. in an unusual year in 1875.[13] But prominent and important as these few large mills were in the Douglas-fir and redwood coast regions, they were not typical. The fact was that the bulk of the timber was harvested by time-honored methods of axmen and ox teams. It went out by sail via a host of small-scale dogholes along the coast, the lumbermen being attracted by the obvious profits to be made from the cutting of the abundant timber close at hand. All these ports were small, marginal, and underfinanced operations beset by uncertain prices and undependable transport (Plate 9.1).

Plate 9.1. A doghole on the rocky north California coast. The logs were tipped into the waiting ship from the high-level track.

Some idea of the regional pattern of lumbering is given in Figure 9.1, which shows the value of sawed lumber by county for the census years 1860, 1870, and 1880 for the three states of Washington, Oregon, and California. In 1860 production was concentrated in the redwood forests in the counties around San Francisco Bay: Sonoma, Mendocino, and Humboldt to the north and Santa Ana, Santa Clara, and Santa Cruz counties to the south. Other local concentrations in lumbering were in the goldfield areas of the western slopes of the Sierra Nevada, such as Placer, Butte, Nevada, Sierra, and El Dorado counties. In the Puget Sound area only Kitsap was prominent at this time.

By 1870 the value of production along the California coast had quadrupled, and lumbering activity was pushing farther north up the coast into Del Norte County, emphasizing again the importance of seaborne communication. Unfortunately, the data for Washington are missing. By 1880 the intensity of production was thickening and spreading everywhere, but particularly in the Douglas-fir region in Oregon and along the margins of the Willamette Valley, where agricultural settlement was pushing into the lower slopes of the ranges. Around Puget Sound the pattern of activity in King, Snohomish, and Jefferson counties was a forerunner of the intensive lumbering that was to come.[14]

San Francisco persisted as a consuming center and transshipment point. It is difficult to know how much of the cut of the coast went to the bay, but the records of Kentfield, Buhure, and Jones show that 534 out of 555 lumber cargoes that left Humboldt Bay in 1887 were destined for San Francisco, either to be consumed there or to be transshipped elsewhere. Nevertheless, Figure 9.2 is an indication of the trend of events; the economic influence of San Francisco was felt everywhere at this time.[15]

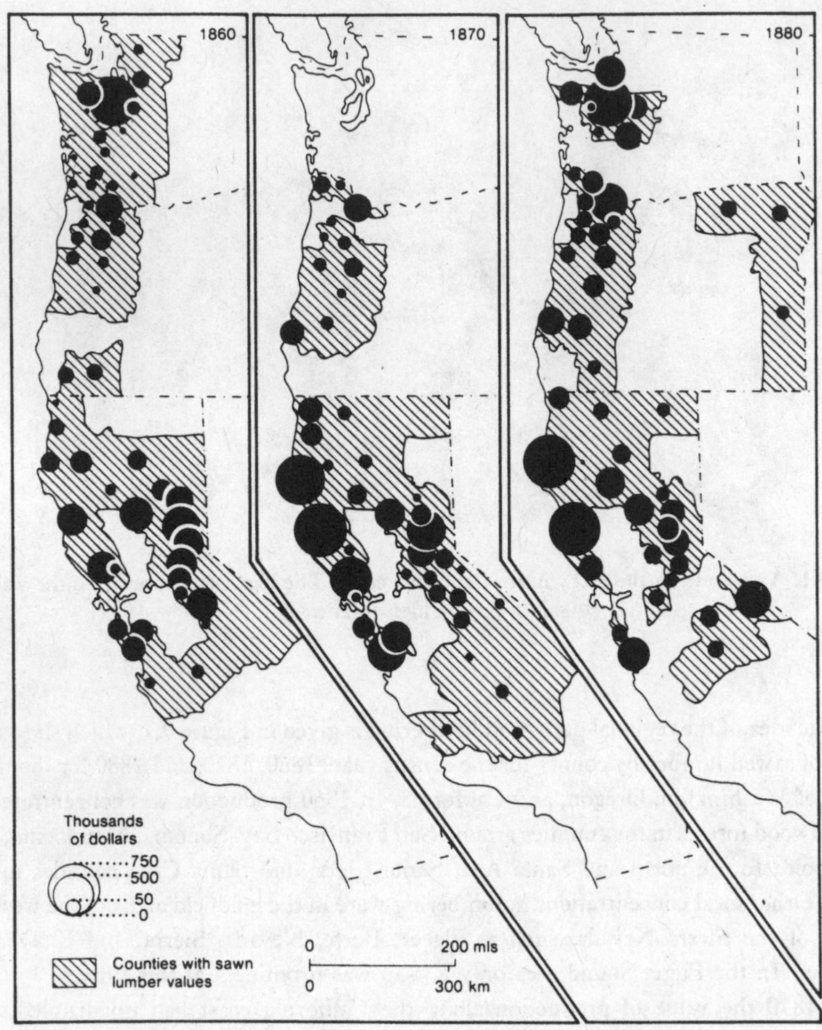

Figure 9.1. Pacific Coast: production of lumber by value, 1860, 1870, and 1880. (U.S. Census.)

Transition, 1883–99

The rugged physical nature of the West Coast and its isolation from the rest of the country had more or less determined the broad outline of exploitation of the forests up till about 1880. During the next couple of decades, many features of the earlier age lingered on, for example, the use of the ax, of ox teams, of water power for mills, and of sailing ships for transport, but a number of other developments were going to have their effect. First, there were changes in the relationship of the region with other timber-consuming and timber-producing sections of the country brought about by the development of intercontinental railways and changes in overseas markets. Second,

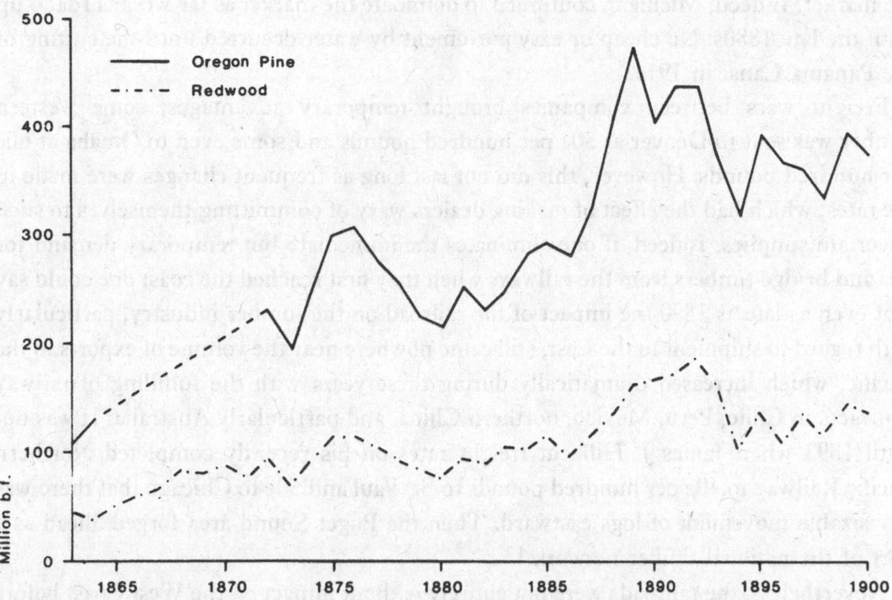

Figure 9.2. Lumber cargoes received in San Francisco, by species, 1863–1900 (mill. b.f.). (Based on T. Cox, *Mills and Markets*, app. 3, table 2.)

many changes of a technical nature were rapidly evolving in the forest and the mills, which collectively were going to have far-reaching effects on production and which were a prelude to the great boom in production that was to follow in the twentieth century. Third, other impacts on the forest, such as agriculture and grazing, were going to have ramifications.

Changes in the region

Despite the high hopes of the West Coast producers for increased trade resulting from the completion of the transnational Northern Pacific Railway to Tacoma in 1883, the Union Pacific Railroad to Portland in 1884, and an additional link eastward from Portland by the Southern Pacific in 1887, the results were disappointing. The new links did not bring about the creation of vast new markets that were thought to lie in the Great Plains east of the Cascades and the Rockies. It was partly a matter of perception of the geography of the continent; in their desire to see distance minimized, West Coast people had shrunk in their minds the vast mountain and intermontane expanse of nearly 1,000 miles that separated them from the "treeless" plains being settled and cultivated so actively during the 1870s and 1880s. In addition, the forested lands of the intermontane states such as Idaho and Montana proved to be good suppliers for the railroads, which in any case reached them first. Consequently, the mills of competitors east of the Cascades lessened the market for the coastal mills initially.[16] Perhaps more important, however, was the fact that high railroad freights on lumber to the East prevented penetration of the market. Indeed, Michigan continued to dominate the market as far west as Idaho up until the late 1880s. No cheap or easy movement by water occurred until the cutting of the Panama Canal in 1914.

Freight wars between companies brought temporary advantages; some western lumber was sent to Denver at 50¢ per hundred pounds and some even to Omaha at 60¢ per hundred pounds. However, this did not last long as frequent changes were made in the rates, which had the effect of making dealers wary of committing themselves to such uncertain supplies. Indeed, if one eliminates the immediate but temporary demand for ties and bridge timbers from the railways when they first reached the coast one could say that even as late as 1890 the impact of the railroad on the lumber industry, particularly with regard to shipment to the East, still came nowhere near the volume of exports to the Pacific, which increased dramatically during these years with the fulfilling of railway contracts in Chile, Peru, Mexico, northern China, and particularly Australia. It was not until 1893 when James J. Hill cut freight rates on his recently completed Northern Pacific Railway to 40¢ per hundred pounds to St. Paul and 50¢ to Chicago that there was any sizable movement of logs eastward. Then the Puget Sound area forged ahead as a part of the national timber network.[17]

Nevertheless, the railroads were not entirely without impact on the West Coast before 1893. First, population numbers grew rapidly as a result of the breakdown in isolation and the promotion by the railroads of real estate speculation and trade generally. This was reflected in building booms during the 1880s in southern California with the completion of the Santa Fe line, which, of course, also increased the local demand for timber.[18] Second, the attraction to building lines for the railway companies had been the enormous blocks of land they were granted by the federal government. When these blocks were used by the companies for logging or subsequently sold off to logging companies together with offers of advantageous sawmill sites, spatial monopolies were created in the Pacific Northwest by the turn of the century that were to rival anything seen in the South. The land grants were not essential for the expansion of the eighties, but they became an important part of the boom after 1900.

Just as population came with the railroad, so did capital for investments and a general business interest in the region. Given the fine forest stands, it is not surprising that experienced lumbermen from the Lake States should cast their eyes covetously on the region, just as they had in the South. As early as 1869 a local writer, A. J. Dufur, had compared Oregon with the Lake States and New England: "The numerous water-powers, the vast forests of excellent timber, much of which is easy of access from tide-water, require only capital and men to render Oregon equal, if not superior to either of these localities [New England and the Lake States] for its lumbering wealth."[19]

In 1884, Sargent, in his voluminous *Report on the Forests of North America*, had added a more up-to-date and quantified opinion: "Fire and ax have hardly made a perceptible impression upon this magnificent accumulation of timber," he wrote, and his map of timber densities across the continent seemed to prove the point (see Fig. 2.4). Some of the stands in western Washington could yield up to 200,000 b.f. per acre, and over 20,000 square miles could yield more than 25,000 b.f. per acre. He was not prepared to even hazard a guess at the total reserves, with the exception of the redwoods, which amounted, in his opinion, to nearly 26 billion b.f.[20] Of the latter, he said they were "the only real substitute for white pine," a comment that was bound to strike at the heart of every Lake States lumberman, many of whom were in the throes of moving their operations from the declining white pine stands of the Lake States to the southern pineries or to Canada or were even thinking vaguely about the inexpensive stumpage of the Pacific Northwest. Thus, in 1889 Chauncey and Everett G. Griggs, father and son, purchased 80,000 acres of timberland from the Northern Pacific Railway on behalf of a syndicate and founded the St. Paul and Tacoma Lumber Company of Tacoma. The Griggses were precursors of others who came later during the 1890s, including Weyerhaeuser, Dubois, Brookings, Woodard, Walworth, Weatherwax, C. A. Smith, Bradley, and Carlisle, as well as John D. Rockefeller and English syndicates. When the southern stands began to look as if they too were becoming cut out during the 1920s, Keith, Crosset, Ingham, and particularly R. A. Long also moved to complete the two-pronged movement of eastern interests.[21]

All these newcomers tended to ignore the Pacific cargo trade with its scattered and uncertain foreign markets and its centralization of operations on San Francisco, neither of which conformed to their concepts of vertical integration from stump to market, which they had pioneered in the Lake States and even in the South. They had their eyes very firmly on the market in the interior of North America in which they had contacts and experience. It was a market in which high-quality produce was in demand, which could be fulfilled by the certainties of rail rather than by the uncertainties of sail, and it was in contrast to the local market for green, rough, and ungraded timber such as ties and planks that predominated on the West Coast and particularly overseas. Their operations were aggressive and well financed; those of the coastal traders were disorganized and lacking in capital. Despite the efforts of the coastal traders to increase their efficiency by improving the capacity and reliability of the cargo ship with deeper hulls, retractable keels, and steam motors to get them over the bars of the shallow estuarine harbors, few of the cargo mills could compete with the new scale and mode of organization. They were distant from the railroads, rarely had lumberyards for sorting and storing in advance of orders, and merely cut as required. Their plants were outmoded and of small capacity, and few had finishing or planing equipment.[22] What competition and different market orientation did not eliminate during the 1880s the depression of the 1890s finished off, and one by one the small cargo mills disappeared, leaving only a few large survivors such as Pope and Talbot. Only where mills were located where they could service both rail (the intercontinental and the expanding local network) and cargo markets, such as around Puget Sound and on the Lower Columbia, was the market base broad enough to allow them to survive.[23]

Thus the embryonic timber industry of the West Coast – dominated by California capital, operated by New England lumbermen, and largely dependent on overseas and coastal trade – was changed. Eastern capital and technology, Great Lakes lumbermen, and a national American trade replaced them. It was the prelude to a great boom and the transformation of the forest scene.

Changes in the forests: technology

Logging in the nineteenth-century Northwest was wasteful and backward. The plentiful nature of the trees, the contract system of paying for the daily total of board feet produced, and the enormous girth and defects of the tree butts were all reasons for the local tradition of cutting the tree 10 to 12 feet above the ground level from a series of working platforms or springboards inserted into notches on the side of the tree.[24] Sargent commented on the general wastefulness of the practice when he said:

> loggers cut only timber growing within a mile or a mile and a half of shores accessible to good booming or shipping points or which will yield not less than 30,000 feet of lumber to the acre. Only trees are cut which will produce at least three logs 24 feet long, with a minimum diameter of 30 inches. Trees are cut not less than 12 feet and often 20 feet from the ground, in order that the labor of cutting through the thick bark and enlarged base may be avoided while 40 or 50 feet of the top of the tree was entirely wasted.[25]

The difficulty of cutting the large trees arose partially from the persistent use of the single-bitted ax, and it was not until 1878 that the superior long-handled, double-bitted ax began to be stocked and used. It was cheaper and held its steel edge longer. In addition, it was a fact that West Coast lumbermen felled trees exclusively by means of an ax; the crosscut saw was used only when the tree was on the ground. When it was realized that ax and saw (the latter lengthened and improved by raker teeth that prevented binding) could be used in combination, the time needed in felling was reduced by some four-fifths, and the direction of the fall was controlled with a consequent lessening of damage to the tree. Production in the redwood and Douglas-fir forests increased accordingly to keep pace with the sawmill demand.[26]

River transport had figured little in the movement of the massive logs of the region, unless the trees were growing near the edge of water, because of the steepness and turbulence of the streams and the periodicity of the flow. Ox teams yarded or skidded the bulky logs over steep terrain without the benefit of frozen ground in winter as in the Lake States and New England. Therefore, the specifically West Coast invention of the skid road came into being, where small-diameter trees were trimmed and laid crosswise at close intervals in a corduroy pattern across the skidding patch, the logs often being daubed with grease or animal fat to reduce friction (Plate 9.2). Skid roads extended up to two miles into the forests, and they increased the area capable of being exploited, but by the early 1880s this was not enough to feed the capacity and output of the new-style mills, and new methods of getting the logs out of the forest had to be found. Modifications to yarding and skidding were introduced, horses replaced oxen because they were more agile and had greater speed, and skid roads were integrated more closely with new logging railroads. Splash dams were built on unlogged streams in order to increase the flow.[27] The Columbia River and Puget Sound were two favorable locations where rafting (eastern-style) could be undertaken in order to bring logs to the mills from farther afield. Later, large cigar-shaped "rafts" were towed from the Lower Columbia around the coast, particularly to San Diego.[28]

However, all these innovations and local peculiarities had a marginal effect on log

Plate 9.2. The B. F. Brook logging camp, Cowlitz County, Washington, c. 1892. The teams and skid roads were central to operations of the camp.

removal, and animal-powered skidding dominated forest exploitation until it suddenly went into abrupt demise with the advent of mechanized yarding.

The steam donkey engine was the invention of John Dolbeer, a West Coast logger who tried out his machine in the redwood forests near Eureka in 1881. His earliest machine offered little real improvement in speed or distance over animal power. Hemp ropes were bulky to handle and they stretched, power was limited, and the oxen had to haul the rope back to the cutting area. The engine remained suitable for yarding only. But during the mid-1880s, with the advent of steel cables, double drums that allowed a light cable to carry a main yarding cable back to the cutting area as another was drawn in, and the introduction of more powerful machines, the system allowed logs to be dragged from a greater distance. One by one, the redwood loggers adopted the Dolbeer donkey for yarding, as did the Douglas-fir loggers in Washington and Oregon during the late 1880s. Bigger logs than ever before were hauled out of the woods, and costs were nearly halved. Animal power disappeared very quickly, and production rose phenomenally (Plate 9.3).[29]

The problem remained of getting the logs from the donkey engine and out of the forest. Some could be steam-yarded, as was happening after 1890, but the logging railroad, the favored means of transport of the Lake States lumbermen, grew in

Plate 9.3. A Dolbeer donkey engine and skid road in the redwood country near Fort Bragg, Mendocino County. Possibly about 1890.

popularity as more of them moved their operations to the Pacific Northwest. The first logging lines were laid in the Douglas-fir forests around Puget Sound and west toward the Olympic Mountains. The big, high-capacity mills located on the sound needed to draw on a wide circle of supplies in order to keep running. Haulage distances increased from the one to two miles of the ox teams and skid roads to 10 to 15 miles, with little or no increase in costs. So successful were the logging railroads that they spread rapidly into the Lower Columbia basin and then into the redwood forests of Humboldt County in California. As the more difficult terrain of the ranges was encountered, the railroad builders displayed remarkable ingenuity, building immense trestle bridges over canyons and ravines, taking lines up steep slopes by inclines and switchbacks, and around tight curves by the means of geared locomotives.[30]

Sometimes the terrain was too difficult, in the Cascades and particularly in the Sierra Nevada, for a railroad to ascend, and therefore massive and complex flume systems were built. Characteristically, they linked the sawmills in the inaccessible high ground (3,000–5,000 ft.) to the lowland planing mills located on the common carrier lines that skirted the uplands.[31] Many of the logging railroads were independent of the common carrier lines – of which there were, in any case, few at this time – and ended in tidewater log sumps or remote landings and sawmills. But later they were carefully integrated with the main lines into a complex system of exploitation of the remotest portions of the forest.

While transport was becoming more efficient, and thus handling larger quantities of logs, the same was happening in the mills – in fact, the developments in the mills were often the impetus to changes in transport. All the developments that have been seen already in the Lake States and the South were adopted in the Pacific Northwest. Steam power rapidly replaced water power in the mills, rising from 51 percent in 1869 to 90 percent in 1904, as the manufacture of lumber became steadily heavier, faster, and more reliable.

Circular saws replaced the old-fashioned muley and up-and-down saws in the 1860s, and the invention of the means to replace worn or broken teeth by Nathaniel W. Spaulding of Sacramento, in 1859, greatly extended the life and usefulness of circular saws by many years. The disadvantages of the circular saw were its wobble, its extravagant kerf, and its inability to cut more than half its diameter, which became significant as the logs got bigger, and these problems led to the introduction of double and even triple saws during the 1870s and finally to the band saw in the mid-1880s. Daily output capacities rose from a mere 40,000 b.f. to 250,000 b.f.[32]

Kilns for drying and planers and molders turned rough, green lumber into high-quality finished goods as the local market became more sophisticated and demanded various forms of dressed timber. The value of the finished product rose accordingly, it being estimated that planing raised the value of a given quantity of timber by 100 percent, manufacturing doors by 200 percent, manufacturing sashes by 300 percent, and making moldings by 500 percent.[33]

The new railroad in the forest and the equipment in the mill required vast outlays of capital, and a circle of cumulative causation ensued as more capital investment required the maximization of production and more production required greater capital investment. Local operators tended to rely on their own capital resources or on those of the San Francisco timber merchants, but they were little match for the capital of the Great Lakes timber men, with their expertise and the backing of stock companies. It was as if the eotechnic were pitched against the paleotechnic. Everything pointed to the need for larger operating units over bigger areas. For example, the investment in logging railroads was wasted unless one had access to all the timber alongside the line, because logging railroads operated efficiently only if they had access to a large volume of logs. This meant owning large blocks of timberland so that there could be no obstruction to the right-of-way or hold-up of supplies from small, independent landholders. Thus there had to be a heavy investment in stumpage and also in logging camps, because the independent loggers were sidestepped. The Lake States lumbermen, in particular, were aware of this relationship, which they had pioneered in the Lake States and had experimented with in the South, and they started to buy up large tracts of timberland in advance of opening sawmills at a later date. Of about 21 timber holding companies acquiring tracts in the redwoods during the period from 1880 to 1902, 10 were from the Lake States. Toward the end of the period, the giant Weyerhaeuser Timber Company began to acquire tracts of timberland in anticipation of setting up mills. It had happened in the South, and it was going to happen all over again, but with even greater effect, in the Pacific Northwest.[34]

For those small cargo mills that still relied on the overseas markets, technical changes at sea were going to underline their marginality. Ships became larger, and the output of a single mill was too small to make up a paying load on a long run across the Pacific. In any case, the dogholes along the coast were too small to accommodate the new, larger vessels, even though tugs were employed to take them over the bars and ships were built with retractable keels. In time only the safer, deeper openings of Puget Sound, Grays Harbor, and the Lower Columbia with their existing large mills could profit from the technical changes in shipping. Additionally, the trend in overseas markets was beginning to resemble that in the internal market; that is, the market was for a more finished timber, and low-grade products were not so easily off-loaded as formerly. Thus, the small cargo-trade mills were just as vulnerable as were the inland mills to the trend toward diversification of products and to the capital investment in machinery that entailed.[35]

Overall, the cumulative impact of these changes in the forest was to boost production. The total amount of lumber cut in the three states rose fourfold during the boom of the 1880s from 0.6 billion b.f. in 1879 to 2.6 billion b.f. in 1889, particularly in Washington, only to slow down during the depression years of the 1890s to reach 2.9 billion b.f. in 1899. Lumbering was moving from being simple, small-scale, and scattered to becoming complex, large, and concentrated.

The impacts of agriculture and grazing

Many people thought that the same progression of land use would take place in the West as had occurred in the eastern United States. Forests would be felled and homesteaders encouraged to create farms under the Homestead and Timber and Stone acts. However, the poor soils, steep slopes, heavy timber, and heavy rainfall made land clearing difficult. Despite the active promotion of "family forest farms" by people like John Minto of Oregon, the reality of the environment and the heavy buying of the homestead plots by lumber companies and speculators meant that few such farms were created. In Washington alone, just under 3 million acres were alienated between 1860 and 1870 without restriction. Much of it was acquired by Puget Sound lumber companies, and, as some of it was also in the Northern Pacific Railway grant, much also went to that company. Under the Timber and Stone Act just over 2 million acres were sold between 1879 and 1905, again much of it bought by lumber companies.[36]

Of the forested lands, only the lightly timbered and scrub-covered country, such as that in the Willamette Valley and the edge of the northern Sierra Nevada, was settled. The maps that accompanied the various reports of Henry Gannett's Geographical Division of the United States Geological Survey show that relatively little forest land had been changed permanently, something the manuscript maps of the Geographical Division of the Forest Service, compiled between 1913 and 1915, confirm. West of the Cascades (see Table 9.1 and Fig. 9.3), only 2,771 square miles in Washington, particularly around Puget Sound, and 799 square miles in Oregon (a total of 2.3 million acres) were classified as "cut" by 1903 and therefore, by implication, ready for occupation. In addition, however, there were 582 square miles in Washington and 7,102 in Oregon (a total of 4.9 million acres) that were "timberless" or "open country" – some cleared long before the survey and in farming use, some naturally open as in the Willamette Valley.[37] But how much of the total of approximately 7.2 million acres was

Table 9.1. *Classification of land in the Pacific Northwest, 1903*

	West of Cascades		East of Cascades		State	
	Sq. mi.	%	Sq. mi.	%	Sq. mi.	%
Washington						
Merchantable timber	16,554	67	17,691	42	34,245	51
Timberless	582	3	22,620	54	23,202	35
Cut area	2,771	11	1,458	3	4,229	6
Burned	4,069	17	557	1	4,626	7
Rocky and barrens	315	1	2	—	317	1
Mountain meadows	261	1	—	—	261	
Total	24,552		42,328		66,880	
Oregon						
Merchantable timber	15,089	53	13,754	22	28,843	32
Open country	7,102	25	44,874	73	51,980	58
Cut area	799	3	280	—	1,079	1
Burned	5,159	18	936	2	6,095	7
Barren	359	1	1,983	3	2,352	3
Total	28,508		61,827		90,349	

Sources: Henry Gannett, *The Forests of Washington: A Revision of Estimates*, 12–13; and Henry Gannett, *The Forests of Oregon*, 11.

settled and used for farming and how much was cutover and/or regrowth are difficult to assess. Probably no more than 250,000 acres were cultivated at that time. The problem was that even if the forest was logged over the massive stumps remained, and sometimes the lower quarter of the trunk was also left. Their removal was almost beyond the means of the average pioneer settler. Only the introduction of dynamite during the early years of the twentieth century could clear these lands, but at a massive cost.[38] In any case, the urge to create new farms in the western United States was beginning to abate, particularly as the ease of creating them on the treeless Great Plains was becoming more obvious.

The problem was twofold, however. Farms could not be made, and logging companies could not sell their cutover to would-be farmers and thereby relieve themselves of the burden of taxes on their logged-over land. There were two possible solutions to the problem of these tax-delinquent lands: Either pay the state taxes, which few companies did, or let the land revert to the state, which many allowed to happen. Thus the history of land use in the lumbering areas of the Pacific Northwest took a different turn from what had happened elsewhere. Instead of the cutover stump land being given over to marginal farms, most of it remained uncultivated, and, provided there were no devastating fires, another crop of timber grew in time. Sargent got it just about right in 1890 when he said:

> It seems reasonably certain . . . that, whatever may be the fate of the forests which now cover western Washington territory and Oregon, they will be succeeded by

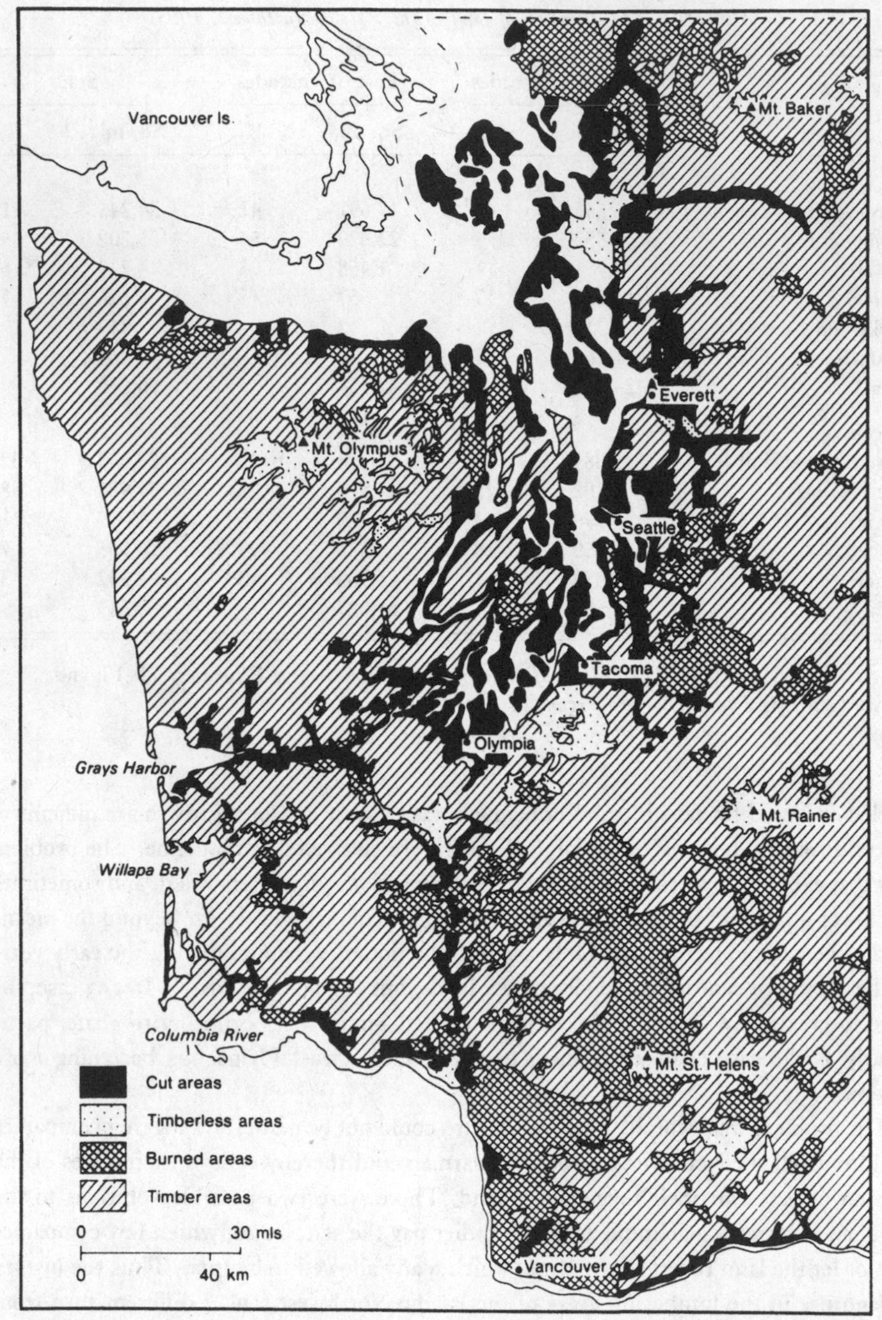

Figure 9.3. Classification of land in western Washington, 1902. (Simplified from G. Plummer, F. G. Plummer, and J. H. Raynor, *Map of Washington Showing Classification of Land.*)

> forests of similar composition, and that this whole region, ill adapted in soil and topography to agriculture, will retain a permanent forest cover long after other great forests of the continent have disappeared.[39]

Fuel gathering for domestic use and for mining must have made significant inroads into the forests, but it is largely unrecorded. The woodcutters worked in the foothills of the Sierra Nevada and Cascades in proximity to settled areas, culling the forest for easily won timber.[40] The gold-mining areas southeast of San Francisco produced a heavy demand, not for ore reduction, as in the prodigiously wasteful smelters in the copper/silver mines such as Tombstone and Bisbee in Arizona where millions of cords of juniper were consumed annually, but more for shoring up, housing, and general warmth.[41]

Though the agricultural impact on the forest remained minimal, the grazing of stock in the upland ranges did not. Grazing expanded rapidly both before and during these years, and this had an effect on the upland forests. From the time of the earliest settlement in the West sheep and cattle were driven up from the dry and hot lowlands to summer pastures in the lightly timbered and open, high valley, upland pastures of the Sierra Nevada, the Cascades, and the Coast Ranges. Inevitably, these grazing lands were sometimes damaged by fire as pastoralists "greened up" the range and sometimes by overgrazing. With the development of irrigation works in the lowlands and the need to provide supplies of water to the new towns in the region (and later for hydroelectric projects, particularly in the Sierra Nevada), questions of water supply and quality became important. Soon moves to protect sources by limiting grazing were advocated by a variety of interested bodies such as recreational groups in Oregon, those concerned with water supply to Portland, and scenic preservationists everywhere.[42] The issue came to a head in the 8 million acres of forested land in the first group of reserves created by the federal government in 1893 – the Cascade, Sierra, and Pacific (later part of the Mount Rainier National Park). When a committee set up by the National Academy of Science in 1896 toured the West to investigate the creation of even more reserves, it was joined by John Muir, and subsequently Muir wrote the part of the committee's report concerned with grazing, which roundly condemned the practice. Over 90 percent of fires in the ranges were started "by shepherds who make fires in the autumn to clear the ground and improve the growth of forage plants for the following year," he claimed, and in Oregon and California "great bands of sheep, often owned by foreigners . . . driven in the spring into the high Sierra and Cascade ranges . . . [ate] every blade of grass, the tender, growing shoots of shrubs and seedling trees," leaving a swath of destruction behind them, contributing to erosion and flooding, and hastening the evaporation of water in summer and the melting of snow in spring. The sheep should be excluded, he said.[43]

Most recreationalists in the West and the Department of the Interior, which at this time controlled the forest reserves, agreed with this judgment. Naturally, the western stockmen did not. John Minto, a practical sheep farmer and nature lover, disagreed strongly with Muir's assessment, pointing out that the sheep-grazing industry was not as migratory as was made out, that it was certainly not the preserve of foreigners, and that it was not grazing that caused floods but the chinook wind that melted the snow rapidly.

Additionally, trees did not necessarily help to keep an even stream flow because snow actually lay longer in timberlands than in the open.[44]

Faced with conflicting local interests and contrary assertions, the Department of the Interior felt that it needed an independent investigation, and in 1897 Frederick V. Coville of the Department of Agriculture was asked to report on the effects of grazing. Coville's report was not overly antagonistic to the sheepmen; nevertheless, he thought that some regulation was necessary and that certain areas such as water collection catchments, scenic sites, and huckleberry fields should be permanently excluded. His solution was to grant five-year permits, the number of sheep allowed being based on the average grazed during the previous five years. Regulations based on this report were set up in Oregon and Washington, and, although Muir stomped through the states and uttered his celebrated condemnation of the "hoofed locusts," the regulations remained intact for many years.[45]

The investigations and decisions concerned with grazing in the timberlands of the Northwest had a number of far-reaching and important results that were to affect the pattern of events there and elsewhere throughout America. First, the solution of the grazing controversy by accommodation and compromise firmly established the Forest Service as the overriding manager of the national forests and promoted forestry interests. As Will C. Barnes put it, "say what you will the grazing men of the Forest Service were the shock troops who won the West for forestry."[46] Second, investigations into local conditions on the spot and with local people became the accepted procedure for solving disputes, rather than having recourse to Washington, D.C., and a distant bureaucracy, and this approach brought a high rate of success. For example, Gifford Pinchot, Albert Potter, and Frederick Coville investigated other disputes between sheepmen and irrigators in Arizona in 1900 and again in California in 1902 and came up with acceptable solutions. Third, the disputes over the Cascades grazing drove a wedge between the utilitarian preservationists such as Pinchot, Potter, and Minto and the purist preservationists epitomized by Muir and the Sierra Club. It was a clash between moderates, who accepted the role of government to regulate and placed aesthetic considerations among a hierarchy of uses, and militants, who thought that wilderness alone was the highest value that could be bestowed on an area. It was a dispute that was never to be healed, and one that attained national significance in the celebrated inquiry in 1913 to investigate the damming of Yosemite National Park's wild Hetch Hetchy Valley.[47] Fourth, the grazing disputes highlighted issues of watershed management, which were particularly important in California and the semiarid Southwest but which were later to have national significance with the passing and implementation of the Weeks Act of 1911. The Weeks Act authorized the purchase of public lands in order to protect watersheds of navigable streams, predominantly but not exclusively in the East of the country, and the law was used in effect to create new national forests on purchased land.

Although the subsequent history of grazing regulation is not our immediate concern, one can note that after 1900 a free permit system to control animal grazing in the national forests was established on the recommendation of Pinchot and Coville, and animals were to be excluded if it was thought that they would do permanent harm to water catchments. This practice was adopted by the Department of the Interior and later by the Forest Service when it took over the reserves under Pinchot in 1905. Later still, fees

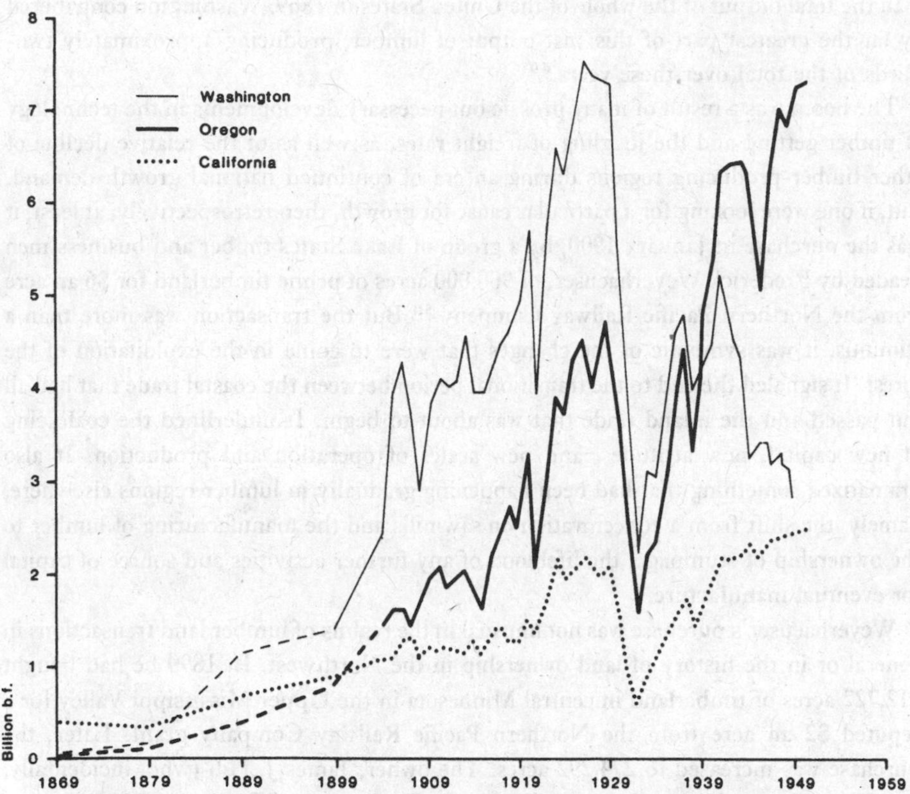

Figure 9.4. Lumber production on the Pacific Coast, by state, 1869–1949. (Steer, *Lumber Production*.)

were charged for grazing. Whatever the form, however, there was a gradual move toward regulation based on a carrying capacity acceptable to range user and forester. It was, says one commentator, a form of sustained-yield concept.[48]

The long boom, 1900–40

Ownership

The census return of production for 1899 showed an increase of 750 million b.f. in the output of the mills of the West Coast over the previous decade, which was fairly evenly distributed between the three states. The modest increase was an occasion for sober self-congratulation that the industry had held its own, despite the depression of the early 1890s and the problems of volatile overseas markets and high freight charges to internal markets in the East. Few people realized, however, that it was also the beginning of a long boom in the Pacific Northwest, fueled by the Klondike gold rush of 1897 and the building boom in Seattle, which was to continue with minor fluctuations until the world economic depression of the early 1930s (Fig. 9.4). The boom was going to take West Coast production to a staggering annual total of 14.1 billion b.f. by 1929, which was more than the total output of the whole of the United States in 1869. Washington contributed by far the greatest part of this vast output of lumber, producing approximately two-thirds of the total over these years.[49]

The boom was a result of many prosaic but necessary developments in the technology of timber getting and the juggling of freight rates, as well as of the relative decline of other timber-producing regions during an era of continued national growth demand. But, if one were looking for a particular cause for growth, then retrospectively, at least, it was the purchase in January 1900, by a group of Lake States timber and business men headed by Frederick Weyerhaeuser, of 900,000 acres of prime timberland for $6 an acre from the Northern Pacific Railway Company.[50] But the transaction was more than a stimulus, it was symbolic of the changes that were to come in the exploitation of the forest. It signaled the end to the transitional period between the coastal trade that had all but passed and the inland trade that was about to begin. It underlined the coalescing of new capital, new attitudes, and new scales of operation and production. It also dramatized something that had been happening gradually in lumber regions elsewhere, namely, the shift from a concentration on sawmills and the manufacturing of lumber to the ownership of stumpage, the lifeblood of any further activities and source of capital for eventual manufacture.

Weyerhaeuser's purchase was not unusual in the realms of lumber land transactions in general or in the history of land ownership in the Northwest. In 1890 he had bought 212,722 acres of timberland in central Minnesota in the Upper Mississippi Valley for a reputed $2 an acre from the Northern Pacific Railway Company grant. Later, the purchase was increased to 274,292 acres. The owner, James J. Hill (who, incidentally, was Weyerhaeuser's next-door neighbor in St. Paul), was willing to let Weyerhaeuser control the lumbering provided that he (Hill) controlled the transport of the lumber to the Lower Mississippi region, for as Hill was to say later in 1905, an acre of timberland was worth more than 40 acres of agricultural land to a railroad because of its capital appreciation and the inevitability that the end product would have to be transported a long distance to a consuming region.[51]

Weyerhaeuser's motives in purchasing are clear enough – a large supply of superb raw material that would last for decades – but it is probable that he was also prompted to move into the Pacific Northwest during the closing years of the century because of the rapidity and even arbitrariness of presidential decisions about proclaiming forest reserves (13 in 1897). If things continued as they had, there might be little forest left for private purchase.

Shortly after completing the Washington purchase from the Northern Pacific, Weyerhaeuser, through his West Coast manager George S. Long, set about consolidating his patchwork of odd-numbered sections into more solid blocks by purchasing nearly 100,000 acres from many small-scale owners and a further 221,000 acres of even-numbered sections from the Northern Pacific (Fig. 9.5). Prior to Weyerhaeuser's transactions, lieu scrip was issued to the Northern Pacific to compensate it for its lands becoming incorporated in the newly created Mount Rainier National Park and the Pacific National Forest Reserve. Through skillful manipulation of the Mount Rainier legislation, the railroad company had been able to exchange worthless mountainous

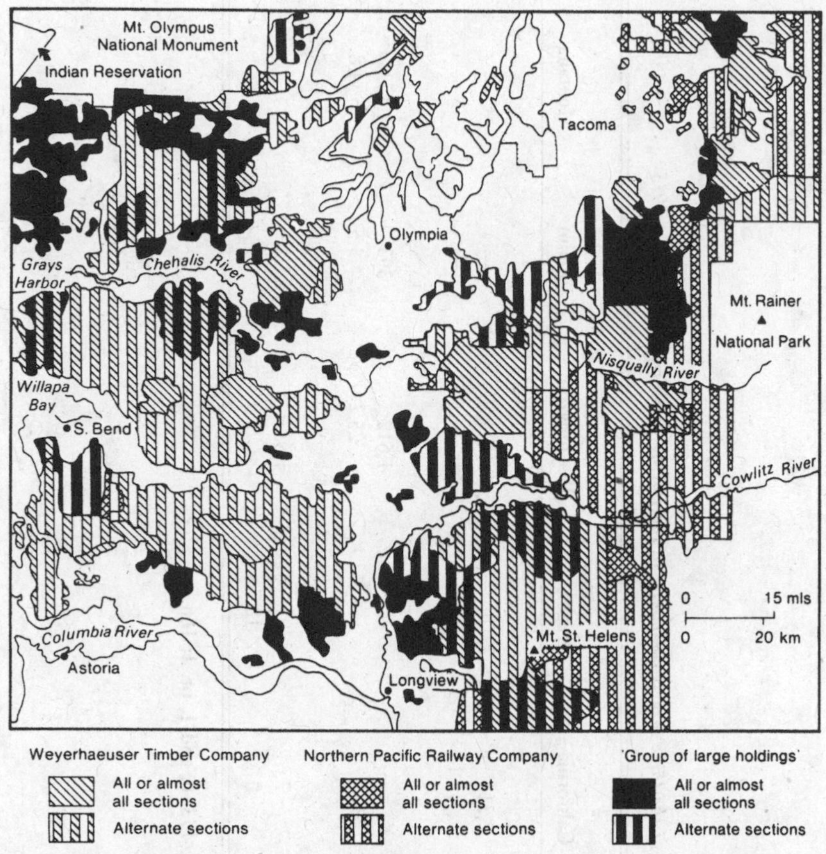

Figure 9.5. Land held by F. K. Weyerhaeuser and other large landholders, northwestern Washington, c. 1913. (Simplified from U.S. Bureau of Corporations, *The Lumber Industry*, vol. 2, pt. 1.)

land for the best timberland in the Northwest, selected by its own cruisers. Almost immediately the forest-lieu land was sold by the Northern Pacific to the Weyerhaeuser Timber Company. Yet more purchases followed, including one in 1902 for 200,000 acres of Douglas-fir from the Northern Pacific in Oregon for $5 an acre.[52] The remarkable readiness of the Northern Pacific Railway Company to sell so much timberland so frequently to Weyerhaeuser lay in the fact that the railroad badly needed working capital and also wanted to stimulate lumbering and general economic expansion and thereby gain financial benefits from the movement of lumber eastward and general trade and passenger traffic westward. Lines without loads were useless.

The Weyerhaeuser purchases and acquisitions were not unique in the Northwest. Although he held 1.9 million acres of timberland, two other companies, the Southern Pacific Railroad and the Northern Pacific Railway Company, still retained control of even larger areas in their original land grants (see Table 9.2 and Fig. 9.5). Vast as these areas were, however, when the amount of timber growing on them is considered, the

Table 9.2. *Size and standing timber of the holdings of the three largest companies in the Pacific Northwest (PNW), c. 1913*

	Investigating area	PNW	California	Oregon	Washington	Idaho	Montana
Southern Pacific Co.							
Acres (thousands)	4,522	4,393	2,314	2,079	—	—	—
B.f. (billions)		105.6	35.1	70.5	—	—	—
Northern Pacific Co.							
Acres (thousands)	3,196	3,158	—	—	1,612	393	1,153
B.f. (billions)		34.5	—	—	25.2	4.5	4.8
Weyerhaeuser Co.							
Acres (thousands)	1,936	1,936	7	393	1,533	3	—
B.f. (billions)		95.7	0.1	18.7	76.9	?	

Source: U.S. Bureau of Corporations, *The Lumber Industry*, 3, pt. 3: 173–6, and 1, pt. 1: 166.

Table 9.3. *Concentration of timber ownership in the Pacific Northwest c. 1913, by size of holding*

Group	Group size (mill. b.f.)	No. owners	Bill. b.f. owned	% of total	% cumulatively
1	Over 25,000	3	237.4	23.5	23.5
2	13,000–25,000	5	101.4	10.4	33.5
3	5,000–13,000	15	97.4	9.6	43.1
4	3,500–5,000	18	71.0	7.0	50.1
5	2,000–3,500	26	64.6	6.4	56.5
6	1,000–2,000	67	91.9	9.1	65.6
7	500–1,000	86	59.2	5.8	71.4
8	250–500	96	34.3	3.4	74.8
9	125–250	176	31.9	3.1	77.9
10	60–125	222	18.3	1.8	79.7
11	Less than 60	?	205.6	20.3	100.0
Total		714	1,013.0		100.0

Source: U.S. Bureau of Corporations, *The Lumber Industry*, 1, pt. 1: 29.

Weyerhaeuser company, together with its direct subsidiaries, was the second-largest owner of standing timber (95.7 billion b.f.) after the Southern Pacific Railroad (105.6 billion b.f.), and when all of the interlinked interests of the Weyerhaeuser family and associated interests were grouped together throughout the country they controlled 291.1 billion b.f. of timber (see Table 13.2).

Vastly smaller than the three giants in the Northwest were another 61 owners who made up "a group of large holdings" and who held 334.4 billion b.f. of timber, or 33 percent of all the stumpage of the western part of the states of Oregon and Washington (Table 9.3). In total, then, 64 owners held 56.7 percent of all the timber in the region.[53] Most of the large ownership blocks in western Washington were on the accessible, lower slopes of the Cascades, penetrated by the Cowlitz and Nisqually rivers and their tributaries, and on the southern and western sides of the Olympic Mountains and the upland nearer the coast between the Chehalis and Columbia rivers (Fig. 9.5). Together, these areas probably had the greatest concentration of softwood trees in the world. The higher and less accessible timberlands toward Mount Rainier had been set aside as a reserve in 1893, and much of the surrounding and unreserved upland forest was still held by the Northern Pacific Railway Company under the terms of its land grant or was in national forests. Weyerhaeuser's holdings, however, stretched to an elevation of about 5,000 feet, to the very foot of Mount St. Helens. At a much later date, the establishment of mills at Willapa Bay on the coast, at Longview near the Cowlitz–Columbia junction to the south, and at Everett on Puget Sound, north of Tacoma and just off the northern edge of Figure 9.5, completed the strategy of timber exploitation by the company.

It is impossible to establish with any accuracy any purchases by the largest corporation or the "groups" as defined by the Bureau of Corporations, after its detailed ground surveys were completed between 1907 and 1910. However, it is known that in

1914 the Weyerhaeuser company purchased a further 225,000 acres, this time in the ponderosa and sugar pine timberland to the south, in the Klamath River plateau area lying astride the Oregon–California state boundary.

The Weyerhaeuser strategy of acquiring and holding large blocks of stumpage was carried on by judicious purchasing at a reasonable price wherever possible, and inevitably it led to criticism. Whereas small local lumbermen feared a crippling monopoly control on their operations and hurried to get out – the situation hastened by the prevailing uncertainties caused by overproduction and low prices – the larger operators welcomed the purchases as a move toward the control of cutting and the stabilization of production and therefore of prices. For its part, the Weyerhaeuser company, through the wise direction of its local manager, George S. Long, thought that it was important to dispel fears of monopoly control and "wished to foster an image of co-operativeness" with other timber producers. To that end it had judiciously sold off 36,000 acres of timberland to small operators by 1904, in order to keep them in business (and their own West Coast operations solvent), and a total of 250,000 acres by 1920 and thus showed that even though it was a monopoly it was, at least, a flexible one.[54]

Mindful of its experiences of cutting out and moving on in the Lake States and later in the South, the company aimed to acquire stumpage supplies for the future and not to manufacture immediately. It rejected overtures from Pope and Talbot, who wished to sell their extensive Puget Mill Company for $6 million, even though the company owned 186,000 acres of timberland. At that time the mill was an encumbrance in the long-term strategy.[55]

Generally speaking, the years from 1900 to 1905 were difficult ones for the lumber industry, with uncertain prices, and the Weyerhaeuser company was not prepared to risk capital on the large-scale plant it would have built inevitably. Conversely, it did not want to injure small local mills. There were, as well, intense differences of opinion in the company about future policy on the West Coast. Weyerhaeuser wanted to establish a mill; other board members and the local manager, Long, thought that lumber prices did not yet warrant entry into manufacturing. The sudden demand fueled by the San Francisco earthquake and fire was encouraging but short-lived, and in 1907 the lumber industry was plunged into worse depression than before. The one purchase of plant was the small and dilapidated Bell Nelson mill at Everett in 1902. It was a modest manufacturing venture, which was managed carefully and economically. It acted as a pilot project in which much was learned about the sawing and marketing of Douglas-fir, and other floundering mills were purchased. The Bureau of Corporations report published in 1914 was severely critical of the Weyerhaeuser company for hoarding timber and neglecting manufacturing, which in a sense was a valid criticism. Ever sensitive to its public image, the company began to consider seriously the desirability of opening another plant. With the peaking of timber prices during 1912–13 and the imminent opening of the Panama Canal (1914), which promised to bring the eastern markets within easy reach, the company embarked on the building of the massive Mill B at Everett and thus began its serious entry into Pacific Northwest manufacturing, which it has more or less dominated ever since.[56]

The fortunes of the Weyerhaeuser Timber Company could be followed further, but it was exceptional. Its size, stumpage, and solvency allowed it to withstand the onslaught of the depression, to buy even more mills, land, and cutover, and to start a widespread, planned sustained-yield program on its lands. But the smaller firms cut frantically during the uncertain early years of the century just to stay in business, and the forest suffered from rapacious clearing.

The new technology

The improvements in tree felling, log getting, and lumber sawing that were developed during the 1880s and the 1890s laid a firm foundation for an expanded lumber industry on the Pacific Coast after 1900, but they were not sufficient in themselves to bring about the very rapid rise in production that occurred during the early years of the century. That demanded a new technology and, in particular, the development of means of overcoming the steep slopes, the inaccessibility, the lack of driving streams, and the presence of large trees, which were common to most of the western forests. In Mumford's terms the new technology that evolved was decidedly neotechnic. Electricity and gasoline were the prime movers and power sources. Strengthened steels, capable of running at high speeds and light in weight, were the materials of the age. Eventually, chemicals in the form of fertilizers and pesticides, together with careful genetic breeding, were to become important. Many of these developments had their origins on the West Coast and/or were developed there.[57]

The greatest labor in the forest was the felling of the trees. Nothing had replaced muscle power, although the double-bitted ax and the crosscut saw introduced during the 1880s did make the task easier and faster. The search for a mechanically powered saw was being carried out all the time. Early experiments with steam and pneumatic saws did not prove very successful because of the large amount of heavy backup equipment needed in what was a very mobile operation. Nevertheless, the invention in 1927, attributed to Andreas Stihl of Germany, of a portable gasoline-powered chain saw altered this in time. The adoption of the chain saw was slow, and not until Joe Cox, an Oregon logger, improved parts of the design in 1947 did the new invention really take off. Immediately, the chain saw replaced the crosscut saw as the main felling tool, and it eliminated most of the ax work done in the forest. In addition, with the big timber trees of the West, significant reduction of waste came about through the adoption of the saw because it enabled fellers to use what became known as the Humboldt undercut. A horizontal cut was made first, followed by an upward-sloping cut from below. In this way the wedge of the undercut was taken from the stump and not from the butt of the log, which was left with a squared end, and so between 3 and 7 percent of the best timber was saved.[58]

During the early decades of the century, the movement of logs in the forest was accomplished by cable yarding with stationary Dolbeer donkey engines, whereby the logs were drawn in along radial paths to the engine. But this method of collection had the grave disadvantage that stumps, rocks, and debris hindered the progress of the log to the landing, which restricted output. Additionally, cable hauling was inseparably connected with clear-cutting. It was destructive of new growth, young trees being broken and

Plate 9.4. High-lead yarding somewhere in the Pacific Northwest.

seedlings torn up, which led to logging "deserts." High-lead yarding, which overcame some of these problems, appeared perhaps as early as 1896 and certainly by 1904 (Plate 9.4). The invention of high-lead yarding was not specific to the Northwest; Horace Butters had experimented with it in Michigan during the early 1880s, and it had become important in the southern baldcypress swamps, where it was known as "pull-boat logging."[59] Nevertheless, wherever it was pioneered, the point is that it was particularly well suited to the conditions of the West with its broken terrain, steep slopes, and densely forested large holdings, which owners hoped to liquidate quickly. A large, straight tree was topped and trimmed of its limbs to form a spar pole. A block was rigged up near the top and a main cable slip slung through it, the other end tethered to a stump in the general area of logging. To the main cable were attached cable nooses or chokers. When the main cable sagged, the chokers could be attached to one end of the felled and trimmed logs and then, with the donkey engine turned on, the main cable tensed and the logs were lifted and dragged swinging and crashing into the spar tree. With other tackle the logs could then be loaded onto railcars.

A further development of high-lead logging was skyline logging, which was introduced in 1906 and was common by about 1915. A second tall tree was selected in the general area of logging as a tail spar onto which was attached the cable and carriage. Then the logs could be hauled in well clear of the ground, an advantage in the rough terrain of the western logging region. In time steel derricks mounted on railcars took the place of spar trees. These gave greater flexibility and speed to skyline logging and increased output by up to tenfold. By 1919 there were nearly 3,000 cable yarders at work on the West Coast: 1,700 in Washington, 825 in Oregon, and 462 in California. By the early 1920s diesel-powered cable-winding units were appearing, and the powerful but costly, cumbersome, and difficult-to-move steam-powered Dolbeer engines were abandoned. But hardly had the change to diesel engines gotten underway than the tractor appeared as an alternative yarding device, and there were few additions to stationary yarding devices after 1935.[60]

Undoubtedly, the introduction of the crawler or caterpillar tractor was the greatest change in the western forests since the invention of the Dolbeer donkey engine. In conjunction with the trucks for hauling logs, it effected a revolution in logging practice that spread to most other lumbering regions after 1940. Steam tractors with large-diameter wheels were used first in California and Oregon as early as 1893, but they were top-heavy and did not have great tractive or drawing power. It was the discovery of the efficiency and versatility of caterpillar track by Charles H. Holt in 1904 (probably building on the experiments of Alvin O. Lombard of Maine in 1900 with his snow haulers) and the adoption of the gasoline engine to the tractor in the next year that brought about the first big change in the accepted means of timber haulage in the forest. At first the tractor was used almost exclusively as a substitute for animals and the logging locomotive because of its ability to haul trailers of logs out of the woods, particularly as it did not require an expensive permanent way. Skidding was still done by oxen and donkey engines. But refinements of the machine and the track mechanism in 1911 made it more maneuverable, and therefore it became possible for tractors to ground-skid logs directly from the stump. With new forms of "bummer carts" (first developed in the Potlatch forest, Idaho), large loads could be handled, particularly as tractors could work in softer ground than even animals could deal with. Longer lengths of log were cut, thus reducing limbing and bucking costs in the forest. The impetus of armored tank construction during World War I led to the perfection of tractor design, and with the development during the early 1920s of a steel arc or A-frame hoist attached to the rear, the tractor was transformed overnight into a versatile lead-line skidder, cable yarder, and loader because logs could now be lifted at one end and snaked out to the landing or out of the forest.

Each invention and refinement improved the tractor's mobility, versatility, and load-carrying capacity. Soon tractors were proliferating throughout the West, but particularly among the smaller-girth trees of the ponderosa and sugar pine forests in California. Compared to the new steel cable yarders they had many advantages. The yarding engines could weigh up to 200 tons, sometimes took days to move and erect, and could

Plate 9.5. Revolution in harvesting the timber of the forest: trucks, tractors, and hoists.

cost up to $100,000 to install. The tractor, on the other hand, was a relatively cheap, mobile towing unit that could be worked by a single man. Moreover, scarce and scattered stands could be logged, and individual trees in denser stands could be logged selectively. After 1923 sales of heavy cable-yarding equipment began to decline, and by the mid-1930s tractors were a practical alternative to high-lead and skyline logging. They replaced the old donkey engines and also the new diesel winding engines that had started to appear in the forest. During and after the Second World War diesel-powered tractors had become powerful and durable enough to handle the biggest redwood and Douglas-fir logs in the roughest terrain, and they virtually replaced all other means of haulage. Their use lowered the damage done to seedling regeneration (Plate 9.5).[61]

Just as the tractor opened up the logging areas of the Douglas-fir forests in the difficult terrain of the Cascades and Coast Ranges, so the complementary invention of the logging truck allowed the logs to be taken out of the forest. Few stands were too small or too remote to escape the impact of these versatile machines.

Although logging trucks had been used for hauling before World War I, they were not sturdy or powerful enough to become a viable substitute to logging railroads for another two decades. The logging railroads maintained a dominant position in log hauling. They had increased steadily in number and mileage throughout the early 1900s until by 1929 there were some 270 lines in California, 450 in Oregon, and over 1,000 in Washington, extending for a total of over 7,000 miles. Even as late as 1929, when the Weyerhaeuser mill at Longview was started, 194 miles of logging railroad were laid to serve it. The changeover from logging railroad to truck was sudden and decisive. In 1930 a mere 6 percent of logs were moved by truck; a decade later the pecentage was over half. The supremacy of the railroad was over, first on the spur lines and later even on the main routes.

The reasons for the rapid changeover were varied and interdependent and occurred mainly during the years of the depression. First, trucks were designed with detachable trailers that permitted flexibility of movement and the carrying of any length of log.

Advances in diesel engine performance and pneumatic tire design also helped. Second, the cost of building logging railroads soared to anywhere between $50,000 and $1 million a mile, depending upon the difficulty of the terrain, an enormous investment that was warranted only if the density and volume of timber logged were great enough. At the same time, previous cutting over of large areas left remnant blocks in inaccessible areas that could be reached not by rail but only by truck. In addition, the sharp drop in production during the depression years meant that the lines could not operate at full capacity, which they had to do in order to repay costs. Conversely, trucks could cope with small, scattered stands of old-growth timber that could be reached only via steep inclines and that were harvested on an intermittent basis. In short, the truck was the key factor in the rise of the Pacific Northwest lumber industry after 1920, particularly in Oregon where vast stands of Douglas-fir on the slopes of the southern Cascades could not have been tapped economically by logging railroads.[62]

Just as these developments were becoming apparent during the early 1930s, the blade-mounted tractor, or bulldozer, made its appearance. This was decisive, as tracks could be made cheaply (c. $2,000 per mile) and quickly, often as a part of fire access and firebreak measures.[63] Almost overnight the number of tractors and trucks shot up dramatically. The expanding public road system also played a part in the conversion. Between 1925 and 1940 thousands of miles of fast all-weather highway were constructed at no cost to the truck operators, with 10 different crossings of the Coast Ranges linking the interior to the Pacific Coast, 8 across the Cascades, and 10 more up into the high valleys. In a rough calculation of truck logging operations in Oregon in 1933, Alfred Van Tassel found that 86 percent of the total length of haul was made over public roads. Simply, although truck hauling was dearer than logging railroad hauling per unit mile, the public road system allowed truck operators the opportunity to disregard amortization costs on routes, which railroad operators could not ignore on their tracks. As the road system expanded, the rail system contracted. Thousands of miles of track and hundreds of locomotives were abandoned and sold, if possible, for scrap. A totally new concept of forest exploitation had evolved and now prevailed, and the western forest of the 1940s was totally different from the forest of 1900 or even that of 1920.[64]

Despite the general trend to largeness, the technological innovations of the depression years, such as the crawler tractor, the logging truck, the chain saw, and flexible electric and gasoline power, spawned the very opposite trend with the emergence of small-scale, individual, independent operators, or "gyppo" loggers, as they are called. Lower man-power requirements for these tools, their versatility, and low capital input and risk compared to, say, logging railroads meant that the gyppos could handle all operations from stump to mill nearly single-handed. They purchased and harvested small and isolated parcels of trees and contracted at piece rates to supply the mills. They handled jobs that the large operators could no longer cope or be bothered with.[65] It was another new pattern in the forest, somewhat akin to the rise of the peckerwood mills in the South.

While all was change in the forest during the 1920s and 1930s, there were few fundamentally new developments in the sawmills but rather refinements of existing modes of production. There were more automatic handling and sorting machines and quicker and more efficient planers and surfacers, all of which helped the production of timber. Perhaps more important was the shift from steam power to electric power. Whereas electric horsepower in the industry was 7 percent in 1914, it had risen to 44.6 percent in 1929 with a marked concentration in the Pacific Northwest and in the large mills. The shift would have been greater had there not been a contemporaneous move to smaller, portable gasoline- and diesel-powered mills in the forests. One result of this was a vast reduction in the number of firemen in boiler houses and maintenance men in general, but particularly those engaged in the inspection of the complex shafts, belts, and pulleys associated with the direct drive of steam. An incidental development of the elimination of belting and shafting was that mills could be laid out in a more flexible fashion than before; the steam engine was no longer the starting point of a layout, and individual operations could have more direct power from almost custom-made motors. The cumulative effect of many of these changes was that production rose by between 15 and 20 percent. Costs also dropped because electricity meant a reduction in fire hazards and hence insurance costs. In the timber yards trucks, tractors, straddle trucks, and lumber buggies replaced horse-drawn carts and manpower and promoted flexibility, mobility, and greater efficiency. At the dockside, gantries and monorail crane systems reduced the loading times in ships to a fraction of what they had been before, so that up to 80,000 b.f. could be loaded in an hour.[66]

In general, then, the story of the new technology in the Pacific Northwest was partially a response to the increasing difficulty of getting accessible, good-quality timber, which put up costs in an area remote from its main markets and growing at a time of decreasing national demand and increasing economic stress. There was an improvement in machines and the means of transport that allowed greater flexibility in the size, location, and mode of operation, an improvement that was partially responsible for the ability of the industry to meet changing demands placed upon it in the succeeding decades.

From Yacolt to Tillamook

At first glance the huge timber holdings, the invention and adoption of the fastest and most versatile technology for felling, transporting, and producing timber, and the soaring regional production and expanding cutovers that became evident in the Pacific Northwest after 1900 seemed to be the perfect ingredients for the unleashing of an assault on the forest that would rival anything seen in the Lake States or the South. But ultimately it was not like that. The magnificent stands were able to withstand the felling. The boom only got underway at a time when fears of a timber famine, the movement to reserve areas of forest and conserve supplies, and a declining national demand for timber were becoming matters of concern (Fig. 6.4). Recession was a jolt that caused a massive reassessment of current practice. Almost from the beginning of industrial logging at the turn of the century, elements of caution and concern were evident among the lumbermen, and these found their practical expression mainly in cooperative fire control and protection, and also in other collaborative endeavors. The Northwest led the nation in its technical and organizational experiments, aided, to be sure, by a very sympathetic and cooperative Forestry Service, and provided a paradigm for all other regions of how to manage the forest.

In one sense the problem of fire control in the Northwest was less complex than in other forested regions. Unlike the South and the Lake States, agriculture had not made massive inroads into the forest land, and logging had not left such a great legacy of cutovers with all their problems of divided responsibility, neglect, and fire hazard. In another sense the problem of fire was more easily tackled. Alarm at the large size of individual timber holdings obscured the fact that private ownership accounted for only about one-third of the timber, the other two-thirds being held roughly equally by state and federal authorities. Mixed ownership, relatively few owners, and shared and mutual concerns promoted cooperation. Undoubtedly, however, it was the fires of early September 1902 in Washington and Oregon that triggered events. The fires consumed about 600,000 to 700,000 acres of forests, destroyed over 7 billion b.f. of timber, caused $13 million worth of damage, and claimed 35 lives.

The conflagration started from a number of small autumn slash and debris fires and the fires of end-of-season picnickers and campers. These were fanned by strong and persistent northwesterly winds into a vast blaze, which by the middle of the month had devastated large areas in Marion, Clackamas, and Tillamook counties in northern Oregon, and Skamania and Clark counties in southern Washington. The fires were particularly devastating in the Lewis Valley around a then little-known place, Yacolt.[67] It was this "conflagration" that was the "catalyst" to cooperative action. Between the Yacolt burn in 1902 and the great Tillamook "blowup" in Oregon in 1933, fire control grew, so that it was probably "the Northwest's most enduring legacy."[68]

Thick, choking smoke and "dark days" that hindered navigation along the coast were no new phenomenon. Travelers had witnessed extensive grazing fires in Oregon during the late 1840s, and in 1868 over a million acres were burned along the West Coast from Washington to northern California. Also, it was common for the graziers to fire pastures behind them in the ranges as they drove the sheep herds down into the plains during the fall. Toward the end of the century the situation, especially in the lumbering areas, was serious (see Fig. 9.3). Said one commentator: "Personal observation warrants the assertion that destruction of timber by fire has never been greater anywhere than in Washington Territory, especially along the Sound. It is simply fearful, criminal!" It could not go on. The country was no longer unsettled as in the past, and the potential loss of life and property rose accordingly. In particular, Frederick Weyerhaeuser and the other recent arrivals from the Lake States had laid out millions of dollars on long-term investments by purchasing stumpage. The recurrent and characteristic nuisances of the past were now the menaces of the future.[69]

It was not surprising that the manager of the biggest timber holding in the Northwest – George S. Long of the Weyerhaeuser Timber Company (which lost 20,000 acres in the Yacolt burn) – emerged as the forceful leader of a campaign to promote forest protection. The company assumed its role of leader as quickly and readily as the smaller companies seemed willing to bestow it. Discreetly, Long lobbied the state legislatures of first Washington and then Oregon. In March 1903, Washington became the first state to pass a Forest Fire Law. Under it a state fire warden was appointed, with deputy commissioners in each county who could appoint up to five rangers and could issue burning permits. In 1905 Oregon enacted similar legislation but with a closed season on burning (June 1–October 1), and Washington adopted the same closed-season approach in the same year under a new Forest Protection Law that allowed the appointment of a Board of Forest Commissioners with wide antifire powers.[70]

Though few people questioned the desirability of fire protection, many objected to public money being spent to safeguard private forest holdings. Long, ever mindful of the public distrust of the big company and Weyerhaeuser's monopoly image, labored quietly to achieve his ends by subtle and discreet means. When the legislation of 1903 and 1905 did not fulfill all the needs, he began to encourage the moves already underway by individual companies to patrol their own property because the intricacies of ownership caused problems that could be overcome only by cooperation. The earliest formal patrol agreement came in 1905 when the Booth Kelly Lumber Company joined Weyerhaeuser and two other lumber companies in a joint patrol of the heavily forested lands of Lane County in Oregon. The lessons learned from this venture were not lost on Long. Such privately financed ventures gave the protection that the inadequately financed state system lacked, and they illustrated to the public at large the willingness of the lumbermen to put their own house in order. Similar organizations spread over the West so that there were eight major fire protection associations in Washington, Oregon, and particularly Idaho by 1908. "Within 10 years 40 formal organizations existed" throughout the country, in "California, Oregon, Washington, Idaho, Montana, Michigan, Wisconsin, New Hampshire, Maine, Pennsylvania, West Virginia and Kentucky. Dozens more were set up in the South during the 1920s; by 1933, Georgia alone boasted of 87."[71]

The spread and success of the movement were as much a response to the lack of enforcement in some states as they were a result of compulsion in others, such as Oregon and California. But the movement's mushrooming after 1910 was a direct response to the disastrous Idaho/Montana fire of 1910, which wiped out 3.3 million acres of prime timberland and claimed 85 lives. William Greeley, who was the Forest Service district forester in the region, reacted immediately by successfully organizing cooperative patrol districts, a system that spread rapidly. In the following year a further impetus was given when the Weeks Act, which provided funds for federal state cooperative measures, went through Congress.

Though many of the fire protection associations passed into oblivion as the state and federal organizations assumed their role, those in the Pacific Northwest endured. In 1909 several of the associations had met, under the chairmanship and encouragement of Long, who was trying to push the idea of cooperation to control fires and chaotic markets and prices even further. Out of these discussions was formed the Pacific Northwest Forest Protection and Conservation Association, and when California and British Columbia joined in the next year its name was changed to the Western Forestry and Conservation Association. Under the able leadership of E. T. Allen, the patronage of Long, and the benevolent eye of Weyerhaeuser and Gifford Pinchot during its early years, this influential consortium of private timber men achieved many things, including further cooperative fire control measures embracing all forest landowners, education of the public on forest fire hazards, and, later, drawing attention to other forest problems such as pest and disease control, a fire weather service, and taxation anomalies.[72]

When William Greeley became head of the Forest Service in 1920 he faced a protracted battle with the embittered but still very influential ex-head of the service Gifford Pinchot, who wanted federal coercion and regulation of private timber companies. Greeley had seen at first hand how the cooperative fire protection and other collaborative measures had worked in the West, and he was convinced that that experience could be repeated elsewhere as the first step toward a genuine self-regulation on the part of the whole United States timber industry. To that end he skillfully steered the Senate Select Committee on Reforestation, which was considering future timber supplies under the chairmanship of Sen. Charles L. McNary of Oregon, to a position where it recommended cooperative state/federal action on many matters but particularly on fire control, something the lumber industry wanted badly, if only out of sheer self-interest. The outcome was the Clarke–McNary Act of 1924, which was a landmark in the administration and preservation of the forests and which embodied the essence of Forest Service policy from then on – that cooperation inspired voluntary action. In 1928 Greeley resigned from the Forest Service, and perhaps it was not coincidental that he became the executive secretary of the West Coast Lumbermen's Association, from which position he pursued his cooperative aims even further until he retired in 1946.[73]

The contribution of the Northwest to cooperative action did not end with the passing of the Clarke–McNary Act. In August 1933 a fatal combination of natural forces of drought, high wind, and low relative humidity led to another major conflagration: the Tillamook burn or "blowup" in Oregon, which rivaled the Yacolt burn of 21 years before. Nearly 250,000 acres of Douglas-fir went up in what was one of the most spectacular fires ever known, a fire that, in the public imagination, rivaled symbolically the Dust Bowl in its manifestation of waste. But out of devastation came positive results. For the first time manpower to fight the fires was organized efficiently with the use of the Civilian Conservation Corps (CCC) and the backup of the army, all in the spirit of the New Deal, and a cooperative salvage operation was mounted to reclaim the burned timber. The repetition of fires in the Tillamook area in 1939, 1945, and 1951 continued the devastation and pointed to the need for vigilance, but eventually some 70 percent of the 10 billion b.f. of timber originally dismissed as lost was reclaimed from the successive burns and harvesting.

Unfortunately, the extent of burning and the emergency salvaging left little vegetation for natural restocking. The severity of the burn became a challenge that ultimately led to a new field of cooperation. In 1948 in an unprecedented move in American forest history Oregon voters approved the issue of bonds to finance the replanting of 225,000 acres of Tillamook, a task completed by 1973. Thus, by the time of the first of the successive Tillamook burns in 1933, the Northwest was already providing a model of how to manage the forest (Plate 9.6). Just as the leadership in production had passed to the Northwest, so leadership in cooperative programs of management went that way too, and more was to follow. The last lumber frontier was indeed a new sort of lumber frontier.[74]

Production and supply

The timber resources of the Pacific Coast were enormous. Although they accounted for a mere eighth of the total area of the forests of the United States, the forests of the Pacific Coast were estimated to contain between one-half and nearly two-thirds of the timber

Plate 9.6. Part of the area burned in Tillamook, Oregon, 1932. The loss of all the trees prevented natural reseeding. Salvage operations were still under way when this photograph was taken 20 years later, before the start of the program of replanting the area.

resources of the country in 1920 and 1933, respectively. They were probably the last great remnant of temperate forest on earth. But there were problems. First, large as the forest was, cutting was calculated during the 1920s to be running at a greater rate than restocking. Second, the easily accessible timber was being cut out rapidly, leaving only the difficult-to-get stands.

By 1922 Greeley estimated that cutting during the earlier decades had been running at about 10 billion b.f. per annum, with additional losses of 1.3 billion b.f. through fire, disease, and other causes, all of which was said to be about 3.5 times the rate of growth. This estimate was supported in general terms by George Peavey in his *Oregon's Commercial Forests* (1922). R. W. Vinnedge, a prominent local lumberman, voiced concern about the future supplies: "To what extent will the Pacific northwest, the last great timber storehouse, be able to take up the stock caused by the depletion of the mature forests of the other producing regions?"[75] The concept of the region as the "storehouse" or "reservoir" that would make good the deficiencies of the rest of the country was becoming increasingly common, and it underlined the contemporary view of timber as a nonrenewable resource in an expanding timber-consuming market.

How much was actually cut in toto is difficult to calculate as the cut was recorded for decennial census years only before 1904. However, between 1904 and 1939 a total of 320.5 billion b.f. came out of the forest of the three westernmost states (plus Neveda).

When waste, the depredations of fire and disease, and so on, are taken into account, that amount would have to be multiplied at least 1.5 times to give an overall total of 480 billion b.f. As most of this was the product of mature forests, which must have yielded 50,000 b.f. per acre at a minimum, then 9.6 million acres, or 150,000 square miles, of forest land would have been cleared, which does not include clearing before 1904. This is probably a conservative estimate, as in 1920 the Capper Report calculated that only 55.5 million acres remained of the original 77.2 million acres of forest in the West.

Whatever the exact amount cut during these years, the results were local timber shortages and local forest devastation. The presence of large and ever-expanding areas of cutover stump lands provoked much debate about the future of the forest lands as well as about the desirability of the Northwest for farming. In 1909 Harry Thompson calculated that, of the 8.7 million acres of land in Washington west of the Cascades, 27 percent was abandoned as cutover and a further 5 percent was agricultural or pastoral land created largely out of the former cutover. The cutting was confined almost entirely to the non–national forest land and was more widespread and serious in the accessible lowlands surrounding Puget Sound, where over half the land had been cleared, and in counties abutting the Columbia River (Fig. 9.6).

By the time of the Capper Report in 1920, Greeley was prepared to look back at the rate of devastation in the past and project the life expectancy of the forest in the future. For example, in Grays Harbor County there were 750,000 acres of timber in 1900 and only about 75,000 acres of cleared land; by 1922 there were 355,000 acres of stumps, and the remaining timber might last for only another 25 years. In King and Snohomish counties on the west side of the sound, the life expectancy of the timber was 35 years. Originally, forest edged the very suburbs of Bend, Oregon, but by 1922 cutting had left "a practically unbroken waste for 6 or 7 miles to the west and south" and lumbering operations were extending up to 30 miles away. At the then current rates of exploitation the forest would last for only another 5 to 30 years. Away from these areas of active cutting, the forest remained largely untouched. With these examples in mind, Greeley warned that it would be "very unwise to over-estimate the resources of the coast as much less than the total stand is readily available." Logging operations were being "pushed back to the less accessible timber in the rougher mountain regions," and as truck and tractor transport had not made their impact yet, the forest of the backcountry seemed beyond the reach of commercial exploitation.[76]

There was, of course, a basic fallacy in the reasoning about depletion that was not appreciated in the almost emotional concern about the destruction of the "virgin" forest. Much of the Pacific Coast forest was a mature forest in which fire, disease, and decay were just about keeping pace with growth. In this state of near equilibrium it was not producing any more timber. However, when the old timber was removed, the way was cleared for the growth of a younger and more vigorous forest, the annual rate of growth of which would well exceed that of the original forest. Inevitably, there would be a lag between cutting and new growth, but in the long run the destruction of the old forest would bring about a greater yield. By the time of the 1933 Copeland Report, the point was still not appreciated fully. The virgin stands of Douglas-fir and ponderosa pine remained the final hope of the lumbermen; they were "The one great remaining

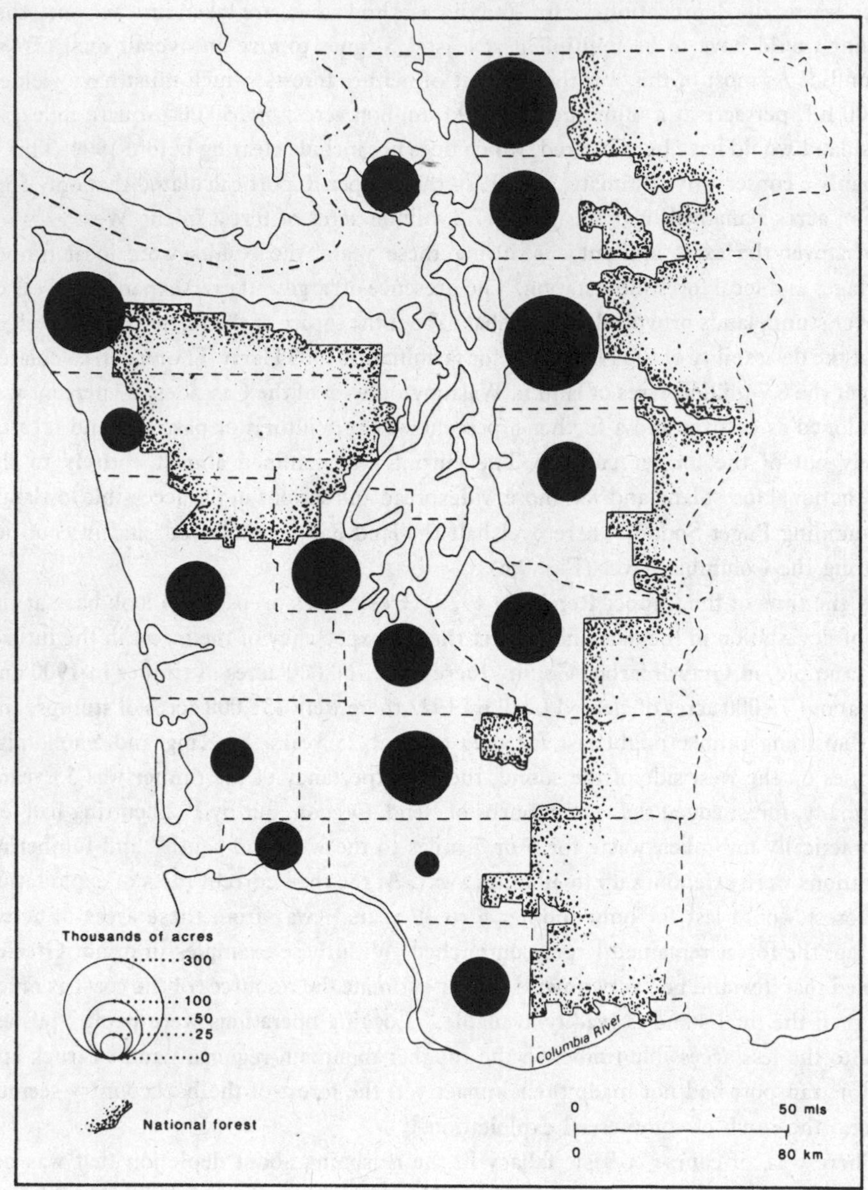

Figure 9.6. Western Washington: forest reserves and cutover land, 1909. (Based on H. Thompson, *'The Cost of Clearing Logged-off Land' for Farming in the Pacific Northwest*, 5.)

reservoir of saw timber, and that soft wood" which was an important asset in "helping to tide over the interval which apparently must elapse before the East can be organized on a more satisfactory forest producing basis."[77]

The exhaustion of timber on easily accessible land and the changing technology after 1920 were instrumental in causing a shift in the location of timber getting. Basically, the

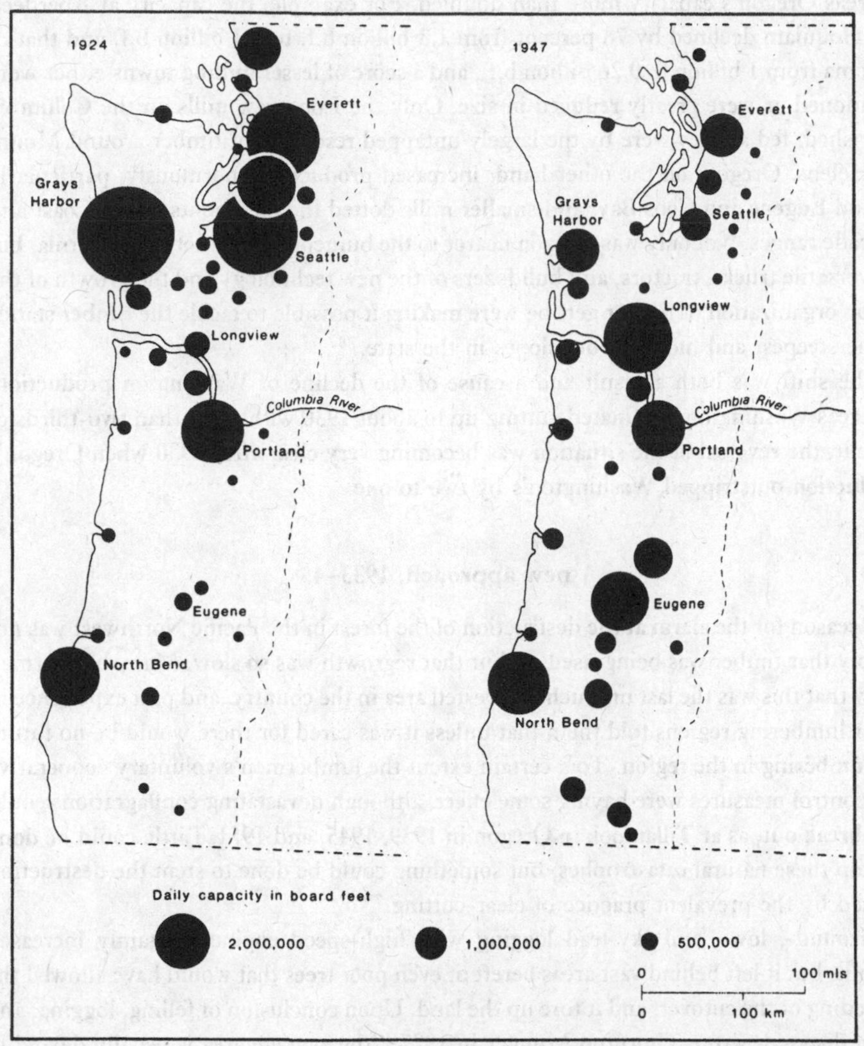

Figure 9.7. Installed daily capacity of lumber centers in Washington and Oregon, 1924 and 1947. (Based on W. R. Clevinger, "Locational Change in the Douglas Fir Lumber Industry," 24.)

shift was twofold: from the tidewater to the inland, and from Washington in the north to Oregon in the south. The shift is shown very simply in Figure 9.7, which compares the daily installed capacities of mills in 1924 and 1947. In 1924 there were major concentrations of milling on the tidewater (with rail facilities) on the eastern side of Puget Sound and in Grays Harbor County at Aberdeen and Hoquiam. In contrast, there was little installed capacity in Oregon except in Portland on the Columbia River and Coos Bay on the coast.

By 1947 half of Washington's capacity had disappeared as private acreage of old-growth timber declined and the cost of procuring supplies from farther afield rose, whereas Oregon's capacity more than doubled. For example, the capacity at Aberdeen and Hoquiam declined by 78 percent from 1.3 billion b.f. to 0.3 billion b.f. and that of Tacoma from 1 billion to 0.26 billion b.f., and a score of lesser milling towns either were abandoned or were greatly reduced in size. Only the Longview mills on the Columbia flourished, fed as they were by the largely untapped resources of timber around Mount St. Helens. Oregon, on the other hand, increased production enormously, particularly around Eugene and Coos Bay, and smaller mills dotted the piedmonts of the Coast and Cascade ranges. Not only was Oregon nearer to the burgeoning market in California, but the versatile trucks, tractors, and bulldozers of the new technology and the growth of the gyppo organization of timber getting were making it possible to tackle the timber stands of the steepest and most remote slopes in the state.[78]

The shift was both a result and a cause of the decline of Washington production. Whereas Washington dominated cutting up to about 1930 with more than two-thirds of the cut, the reversal of the situation was becoming very clear after 1930 when Oregon's production outstripped Washington's by two to one.

A new approach, 1933–45

The reason for the alarm at the destruction of the forest in the Pacific Northwest was not simply that timber was being used up but that regrowth was so slow. Many lumbermen knew that this was the last untouched forested area in the country, and past experience in other lumbering regions told them that unless it was cared for there would be no future for lumbering in the region. To a certain extent the lumbermen's voluntary cooperative fire control measures were having some effect, although devastating conflagrations could still break out, as at Tillamook in Oregon in 1939, 1945, and 1951. Little could be done to stop these natural catastrophes, but something could be done to stem the destruction caused by the prevalent practice of clear-cutting.[79]

Ground-, low-, and sky-lead logging with high-speed engines certainly increased output, but it left behind vast areas bereft of even poor trees that would have allowed the reseeding of the cutover, and it tore up the land. Upon conclusion of felling, logging, and slash disposal, wrote Thornton Munger in 1927, "the average area is usually devoid of living trees. The smaller trees left by the 'fallers' have for the most part been knocked down by the logging lines or burned in the slash fire, and [of] the few larger trees which were culled and left standing some if not all have been killed in the broadcast burning." The presence of steam engines in great accumulations of debris and slash meant more fires than ever before. The result was the creation of stump deserts; they were common in the South and now they were appearing in the West.[80]

Rapacious cutting reached its apogee during the 1920s and early 1930s. The forest was viewed as a resource that could be exploited rapidly in order to pay for the heavy capital investment needed to deal with the large trees of the West. In addition, in the past many companies had gone into debt in order to buy stumpage, which they thought would appreciate in value rapidly, but when the full force of the depression struck after 1929 these firms tried frantically to liquidate their assets of standing timber. Thus overproduction and overcapitalization coalesced to promote the rapid and ruthless cutting of timber and led to vast destruction and little regeneration.

It was against this background of change in what Copeland called "the one great remaining reservoir" of timber in the country that the thoughts of some West Coast producers turned to managing the forest in order to achieve continuous tree production by ensuring that the amount cut annually corresponded to the amount grown annually. The concept of sustained yield had many origins. It came from western Europe – especially Germany – but by the time it reached and was applied to North America it had to be adapted to meet local conditions. In particular, unlike the balanced age distribution of managed European forests, the forests of the United States were predominantly stagnant old-growth forests and had reached a point of equilibrium. It was inevitable, therefore, that annual cut would exceed annual growth at some point in their exploitation and before a younger and more dynamic forest replaced them. This was a real and unavoidable problem. Gradually, therefore, the concept changed. Sustained yield of forest products was modified to the idea of a long-term even flow of forest products, a notion that was bolstered during the early 1930s by concern for the future level of timber supplies and the stability of lumbering communities and local economies. The problem remained, however, of how to achieve an even flow and "harvest timber at its optimum silvicultural maturity when the initial age distribution of the trees deviated so far from that of a fully regulated forest." Planting new trees – reforestation – seemed the obvious answer, although no one could wait the 500 years necessary on the West Coast for full value of the planting.[81]

To a certain extent some of the urgency in achieving a more conservative approach to logging practice and its effects on regeneration was minimized by rapid changes in the technology of production, particularly of transport. Ground-lead logging ended abruptly during the mid-1920s, and the slightly less destructive high- and sky-lead logging replaced it. More significantly, the crawler tractor increasingly replaced all forms of lead logging and so lessened destruction, and the abrupt and rapid introduction of the truck for log transport achieved the same result. Other than in the even-aged Douglas-fir stands, logging could now be selective, and young, growing trees could be avoided in lumbering. At the same time, the advent of the new and versatile machinery changed the scale of operations in many parts of the forests. Small-scale loggers – gyppos – subcontracted to large producers, and exploitation of the forest was no longer dependent on large-scale, expensive, and semipermanent plants, the cost of which could be repaid only if the forest surrounding them was stripped bare. Nevertheless, despite these technological advances, the need of a philosophical commitment to replanting and "farming" the forest remained.

No one person or organization can be credited with providing either the theoretical underpinning or the practical example for programs of sustained yield, or even-rate production. There were many influences. For example, as early as 1895 Fernow had asserted that forestry "is exactly the same as agriculture, it is the application of superior knowledge and skill to produce a wood crop." Later he said that one could look forward to the day when the forest reserves would be "treated as a crop rather than as mine or quarry from which we take what is useful and then abandon it as waste." In 1909 George S. Long, manager of the Weyerhaeuser Timber Company, proclaimed in an address that "Timber is a crop," and in 1922 William Greeley wrote his highly influential article, "Timber: Mine or Crop?" for the USDA *Yearbook of Agriculture*. In the South during the teens and twenties, Henry Hardtner in Louisiana and W. Goodrich Jones in Texas (appalled at the destruction caused by steam skidders and mechanical harvesters) were vocal advocates of practical experiments in reforesting the cutovers.[82]

Then in the 1930s the focus swung back to the Pacific Northwest. David T. Mason, who became a consultant forester in Portland, was an ardent advocate of the sustained-yield approach, which he defined as "the policy of managing . . . lands for permanent timber production." He influenced subsequent events markedly when he was appointed to the Federal Timber Conservation Board in 1931 and then, with William Greeley, had a large part to play in the writing of Article X of the short-lived National Industrial Recovery Act (NIRA), which required lumbermen to adopt conservation techniques. They were required to take practicable measures to

> safeguard timber and young growing stock from injury by fire and other destructive forces, to prevent damage to young trees during logging operations, to provide for restocking the land after logging if sufficient advanced growth is not already present, and where feasible to leave some portion of merchantable timber (usually the less mature trees) as a basis for growth for the next timber crop.

Subsequently, Mason became chairman of the committee advising on the management of the federally owned Oregon and California Railroad Revested Lands (O&C Lands), 2.5 million acres of prime Douglas-fir timber lands, where he was able to put his sustained-yield ideas into practice.[83]

At about the same time the Weyerhaeuser Timber Company had to decide what to do with the many thousands of acres of cutover land on which it had not paid taxes since 1928 in the belief that the land was not suitable for agriculture or capable of sale. By 1936 the lands were due to revert to the state of Washington. The company surveyed the lands and its own stocks of standing timber. It was faced with the sobering fact that, although about two-thirds of its original timber stand remained, much of it was mature or overmature timber susceptible to disease, fire, and decay and was actually a declining resource where growth was not exceeding losses. On the other hand, although the derelict land was clearly unsuitable for conversion to agriculture, it was eminently suited to growing trees. Nevertheless, the cost of converting cutover to productive forest through replanting was from four to six times that of buying up 50-year stumpage in a depressed market. Seen in that perspective, the company's decision to rehabilitate, with the cooperation of the Forest Service, some 120,000 acres of the old Clemons Logging Company lands around Grays Harbor can only be called an act of faith. Designated a tree farm in 1914, the name encapsulated simply and comprehensively the idea that "Timber is a crop." The Clemons Tree Farm became the first in a nationwide program to encourage good management in privately owned timber lands, which by 1981 encompassed 40,713 farms and covered 80.8 million acres.[84] The Clemons Tree Farm was symbolic of a new era in the history of man's use of the forest. Perhaps for the first time, the land on which the forest grew best and continuously was considered more important than the stumpage owned. The productive capacity of the forest overshadowed the sheer ownership of stumpage.

10

Industrial impacts on the forest, 1860–1920

> ... there is no other interest of the country so great as that of the forests, considered from the commercial or pecuniary point of view, while in other aspects its value is altogether beyond computation.
>
> Nathaniel Egleston, in USDA, *Annual Report of Commissioner of Agriculture 1884*, 450

The destruction of the forest for lumber in each of the major regions of exploitation in turn did not constitute the only commercial and industrial inroads into the timber resources of the country during the latter half of the nineteenth and the early years of the twentieth centuries. In addition, fuel, charcoal, and railroads were important consumers of wood.

The total amount of wood consumed by all these inroads is difficult to ascertain, if only because one has little idea of the contribution made by the domestic farmers/clearers to production. Nevertheless, a reasonable and probably conservative estimate would be that the total consumption of all sorts of wood (fuel, lumber, pulp, veneer, poles, etc.) stood at 3.76 billion cu. ft. per annum in 1859. It nearly doubled to 6.84 billion cu. ft. in 1879, and very nearly doubled again during the next 27 years to reach 13.38 billion cu. ft. in 1907, a figure it has never reached again, hovering as it has between 10 and 11 billion cu. ft. per annum for the remainder of this century.[1]

The breakdown of these uses as a percentage of the whole is depicted in Figure 10.1, which shows that despite the emphasis on lumbering (which has been the subject of the last three chapters) the gross volume of lumber did not equal that of fuel until after 1890 and that fuel cut for domestic and industrial uses rarely fell below one-third of all timber cut until after 1922. Additionally, miscellaneous demands on the forests, such as poles and piling, continued to be important, and even such a seemingly trivial item as the wood for lead pencils was consuming 110,000 tons of redcedar per annum. New industrial uses for wood, such as in the making of pulp, began to appear from the turn of the century onward. The forest as a source of raw materials for industry and for fuel was of overwhelming importance. In 1865 Frederick Starr calculated that wood and its derivatives paid "more than one-half of the entire internal revenue of the United States," and a mere 18 years later Nathaniel Egleston made further elaborate computations, which led him broadly to the same conclusion:

> Our cars and ships are the products of the forests. The thousand tools of our various handicrafts, the machineries of our factories, the conveniences of our warehouse, and

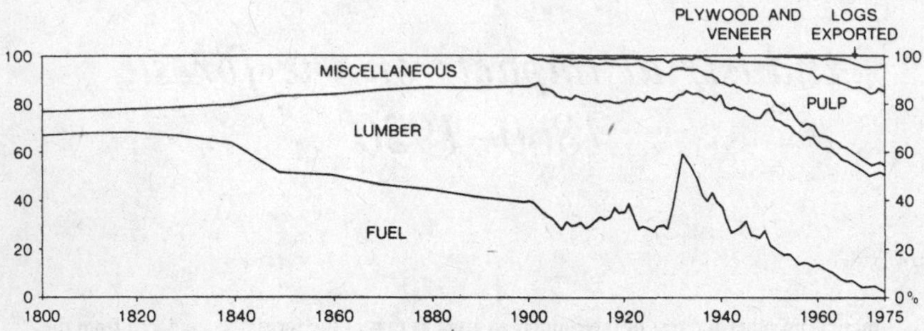

Figure 10.1. Utilization of wood grown in the United States, by roundwood equivalent, 1800–1975; major uses are expressed as percentages of the total. (M. Clawson, personal communication.)

> the comfort and adornments of our dwellings are largely the product of our forests. Behind all the varied industries and conveniences of life stand the forests as their chief source and support.[2]

The American dependence on wood was overwhelming and ubiquitous. Wood entered into every walk of life. Of its many uses in industry, three are examined here: wood as fuel for warming the home; wood converted to charcoal for use in making iron; and wood consumed by railroads, both as fuel for steam locomotives and as ties in the network of tracks that splayed across the nation and bound its economy together.

Changes in fuel use

Fuel for domestic hearths, charcoal furnaces, and mechanical and locomotive energy was still an outstanding product of the forest during the closing years of the nineteenth century, lumber gaining an ascendancy over fuel in percentage terms only sometime between 1880 and 1890 when the flow of cheap supplies from the South gathered momentum. Of course, in a situation of ever-expanding utilization of wood the absolute amount of fuel cut did not decline, and in every year until 1940 (with only a handful of exceptions during the later 1920s) more fuel was being cut in absolute terms than at any previous time, consumption actually reaching a peak in 1933. The availability and cheapness of wood in rural areas made it the ideal fuel, and it was not until electrification became generally available in rural areas in the mid-1930s that wood consumption dropped off.

The contribution of wood to the total energy consumption of the country was impressive during the late nineteenth century, but it became less in both absolute and relative terms as other sources such as coal, oil, and gas took a larger share of a total consumption that experienced a sustained growth and expansion, increasing 17-fold between 1850 and 1955. The choice of fuels at various times depended upon their availability, comparative price, advances in technology of production and utilization, and shift in consumer preference. Fuel wood still accounted for four-fifths of energy used as late as 1860, but by the mid-1880s coal overtook it as the principal source. By 1900 fuel

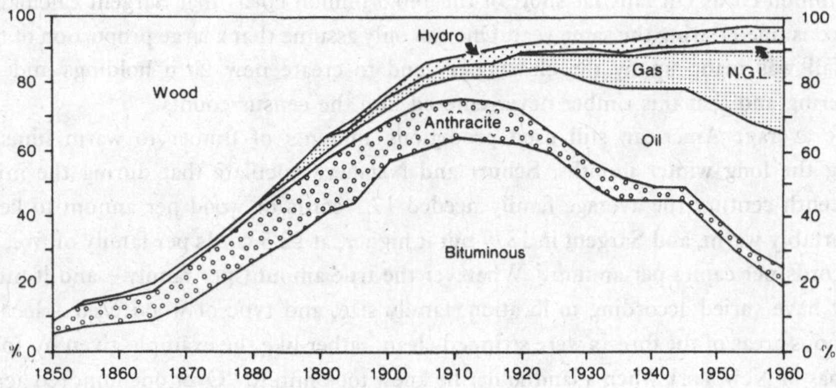

Figure 10.2. Energy sources in the United States as a percentage of total energy consumption, by five-year periods, 1850–1960. (Based on S. H. Schurr, and B. C. Netschert, *Energy in the American Economy, 1850–1975*, 36–7.)

wood was probably contributing only 21 percent of the nation's energy needs, and by 1920 a mere 7.5 percent, by which time coal had passed its peak, and oil and natural gas were making their impact felt, so that by 1955 their share in the total energy supply had become nearly two-thirds of the total (Fig. 10.2).[3] But, once again, it is both important and salutary to stress that although the relative position of wood changed in the picture of aggregate energy consumption its contribution was large in absolute terms. Even a share of 5.4 percent in 1940 was probably equivalent to nearly 4 billion cu. ft. of timber, which was more than in any year in the nineteenth century with the exception of 1899.

Domestic hearths

All the points made in Chapter 5 about the domestic fuel consumption and production between 1810 and 1860 can be made again in varying degrees for the subsequent years. Between 1860 and 1910 about another 151.6 million acres of forest land were cleared, which at the very modest estimate used before of 20 cords per acre would give the astronomical figure of 3.032 billion cords of wood, an amount that would have supplied at least half of the total national consumption of fuel wood during the same years, which was estimated to be over 6 billion cords.[4] The shortfall was made up, presumably by the recutting of woodlots over the 50-year period and, to a certain extent, by the waste coming from lumber production, although as time went on the mills became increasingly imaginative in the way they utilized every scrap of wood, and pulp production began to consume more and more small timber from lumbering.

Cutting remained a profitable sideline for the farmers, who continued to spend up to one-third of their time cutting, hauling, and splitting timber, particularly during the winter months when they could do little else. The location of this activity is difficult to ascertain with any certainty, but the Tenth Census does itemize the numbers of cords cut on farms only, and also the value of the cut, for the year 1879.[5] However, the total of

51.4 million cords cut falls far short of the 140.5 million cords that Sargent calculated were consumed during the same year. One can only assume that a large proportion of the shortfall was made up by the clearing of land to create new farm holdings and by lumbering and that this timber never entered into the census counts.[6]

The average American still used prodigious amounts of timber to warm himself during the long winter months. Schurr and Netshert calculate that during the mid-nineteenth century the average family needed 17.5 cords of wood per annum to keep comfortably warm, and Sargent in 1879 put it higher, at 22.5 cords per family of five, or 4.56 cords per capita per annum.[7] Whatever the true amount per family – and it must surely have varied according to location, family size, and type of wood used – locally enormous areas of the forests were stripped clean, rather like the example given by John Thomas of New York when a landholder he knew took him to "Over one hundred acres of land, once densely covered with timber, but now entirely cleared for the sole purpose of supplying his family with firewood during the forty years he had resided there."[8]

Despite the relative abundance and inexpensive nature of wood, coal (particularly anthracite) began making great inroads into the wood market in the urban areas and near the coalfields. It was true that wood was scarce near the urban areas because of the centuries of clearing around them, but a more important factor in the penetration of coal was that it weighed less and was less bulky in relation to its energy content than was wood, by a factor of as much as one-third or a half again. The physical storage of wood caused difficulties in the urban areas, especially as it was still being delivered in the traditional measure of the four-foot cord, which, in any case, did not fit easily into the new stoves that were now common in most houses.[9] But it was the invention during the late 1850s of household furnaces in the cellars of urban dwellings for the central heating of all or part of the house or apartment that really demanded large and reliable quantities of fuel. As early as 1830 or 1840 shipments of coal from Pennsylvania went up and down the East Coast, and increasingly urban fuel dealers changed from "wood" dealers to "coal and wood" dealers, and they cut the traditional cords into lengths of 15 inches that could be delivered to cellars and used in stoves. The increasing construction of apartment houses in the cities during the last decades of the century brought further changes. Few, if any, had open fireplaces because of the fire risk and because space was too valuable for the storing of wood, which would be used only in the cold season. By the late 1930s dealers along the Atlantic Coast had ceased to handle firewood.[10]

The diffusion of coal as a fuel into the predominantly fuel wood setting of nineteenth-century America is shown in Figure 10.3. It represents a stage beyond that of initial penetration. By 1880 coal predominated as the main fuel in four main areas: New York and its immediate environs, the Hudson Valley, Massachusetts and eastern Pennsylvania; western Pennsylvania and eastern areas of Ohio; northern Illinois and western Iowa; and two adjacent areas in the West, along the Platte and Arkansas rivers in Nebraska and Kansas. It addition, coal was the domestic fuel in nearly every city with a population of over 15,000. One example of local change must stand for many. In Fairfield County in southern Ohio it was said in 1880 that "wood is not used for fuel more than half as much as it was eight or ten years ago. Many farmers having timber on their lands find it cheaper and more convenient to buy and burn coal."[11] Around the core areas of coal use lay a broad belt of land stretching from eastern Nebraska and

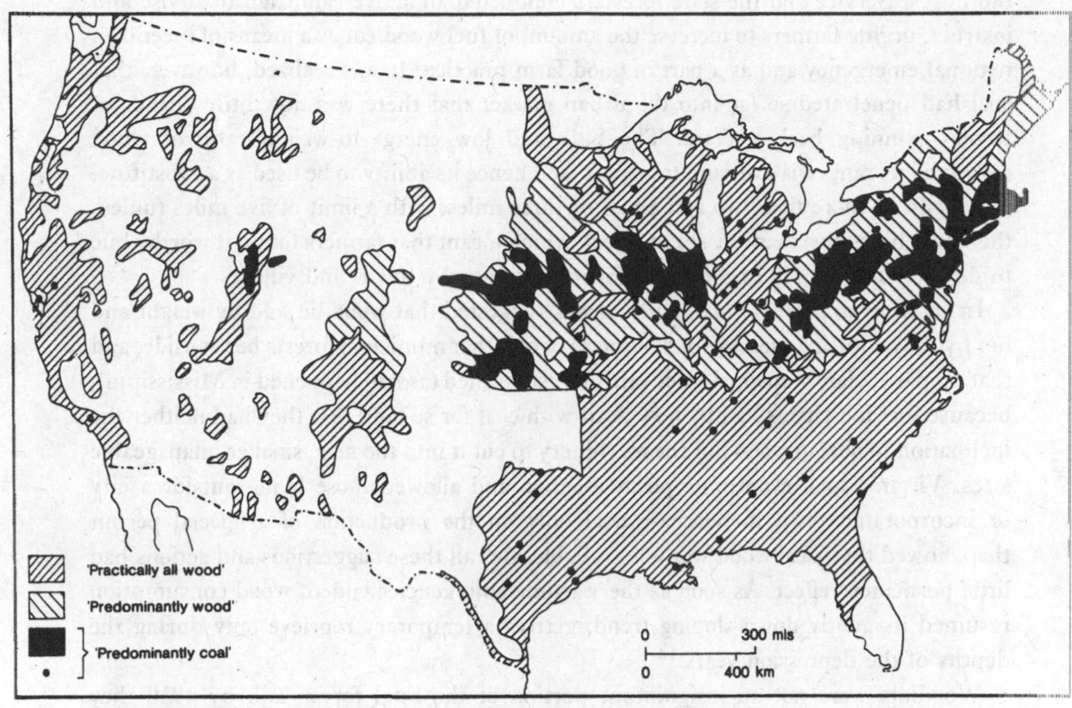

Figure 10.3. Type of fuel used in the United States, c. 1880. (Based on Sargent, *Report on the Forests*, 488–9.)

Kansas across the continent to New York and Pennsylvania in which coal had made some inroads as a fuel but in which wood still predominated. The rest of the settled portion of the country constituted a third fuel use zone: across the entire South and West and throughout Wisconsin, Minnesota, Michigan, Indiana, and inland Maine, where the fuel was "practically all wood."[12]

There is little to suggest that the picture presented in Figure 10.3 altered very much during the next few decades except that coal consumption went up and wood consumption dropped slightly, more coal penetrating the intermediate zone across the Northeast of the country and all the growing urban areas, wherever they were located. By 1908 consumption of wood may well have dropped to 84.1 million cords, with four-fifths of the consumption being on farms, a proportion that reached nearly 95 percent in the agricultural states of the South and the West but declined to as low as 60 percent in the vicinity of the intensely industrial portions of the Northeast.[13]

However, there was a renewed though short-lived interest in the subject of fuel wood and the management of woodlots by farmers. The advent of the First World War and the consequent diversion of manpower, an actual shortage of coal during the winter of 1917–18, and measures to conserve fuel and minimize transport movements focused attention on the remnants of the forest after clearing as a source of the nation's fuel supply. Consumption of fuel wood went up by at least a quarter from 82.7 million cords to 102.9 million cords in 1917, and wood was seen to be "coming into its own again," as the Forest Service and the state foresters conducted an active campaign to advise and instruct, urging farmers to increase the amount of fuel wood cut as a means of meeting a national emergency and as a part of good farm practice. It was realized, however, that coal had penetrated so far into the urban market that there was now little chance of wood's winning back a share. The bulk and low energy-to-weight ratio of wood restricted its range via overland transport, and hence its ability to be used as a substitute, to usually no more than two and a half or three miles, with a limit of five miles (unless there was bulk movement by sea or river), which meant that farmers found it worthwhile to do the hauling only to their own farms or to nearby towns and villages.

In an effort to promote marketing, it was suggested that wood be sold by weight and not by the confusing measure of volume as of old, that municipal forests be set aside, and that, in particular, municipal woodyards be established (as had happened in Mississippi), because fuel merchants had by now dealt with coal for so long that they had neither the inclination to deal in wood nor the machinery to cut it into the new, smaller, manageable sizes. Virginia took even more positive action and allowed those living outside a city or incorporated town to purchase coal only on the production of a special permit that showed that local wood was not available. But all these suggestions and actions had little permanent effect. As soon as the wartime emergencies ended, wood consumption resumed its gently down-sloping trend, getting a temporary reprieve only during the depths of the depression years.[14]

Woodlots were not an insignificant portion of the total forest, and by 1910 they covered a staggering total of 143.3 million acres on 4.3 million farms in the eastern United States. Early views about their productivity and likely life were pessimistic,[15] but these were replaced during later years by more balanced assessments, which were based on more exact appraisals of their extent and yield.[16] Their average size was a little less than 30 acres, although extreme sizes varied from less than five acres or small farms in New England to 150 acres on partially cleared holdings in northern Minnesota and South Carolina. All in all, the woodlot was a conspicuous feature of the landscape of the eastern United States. Because the woodlots were often a remnant of the original forest left after clearing for agriculture, their area had decreased gradually by 15 percent between 1880 and 1900, although regionally there were great variations, as in New England where woodlots had increased with farm abandonment and reversion to forest, and in the Lake States where the inclusion of stump land and uncleared forest in farms with the extension of settlement northward had boosted woodlot amounts (Fig. 10.4).

The woodlot had, of course, other uses than fuel, some of which were of very direct benefit to the farmers. They were shelters for stock during extreme climatic conditions, especially if planted as shelterbelts. They contained up to 15 percent of the nation's timber supply, although the full value of this lumber was rarely realized by the farmers because of the variable quality of the stand due to poor management, their ignorance of timber grading, and consequently the low price received from the dealers. Woodlots were also the source of rough constructional timber and poles and fencing, but above all they were a source of fuel worth some $170 million in 1910, double that of the total crop of tobacco or rye and barley combined. Indeed, in Vermont and New Hampshire the woodlot products (mainly fuel) were the second-ranking crop, and in Maine, South

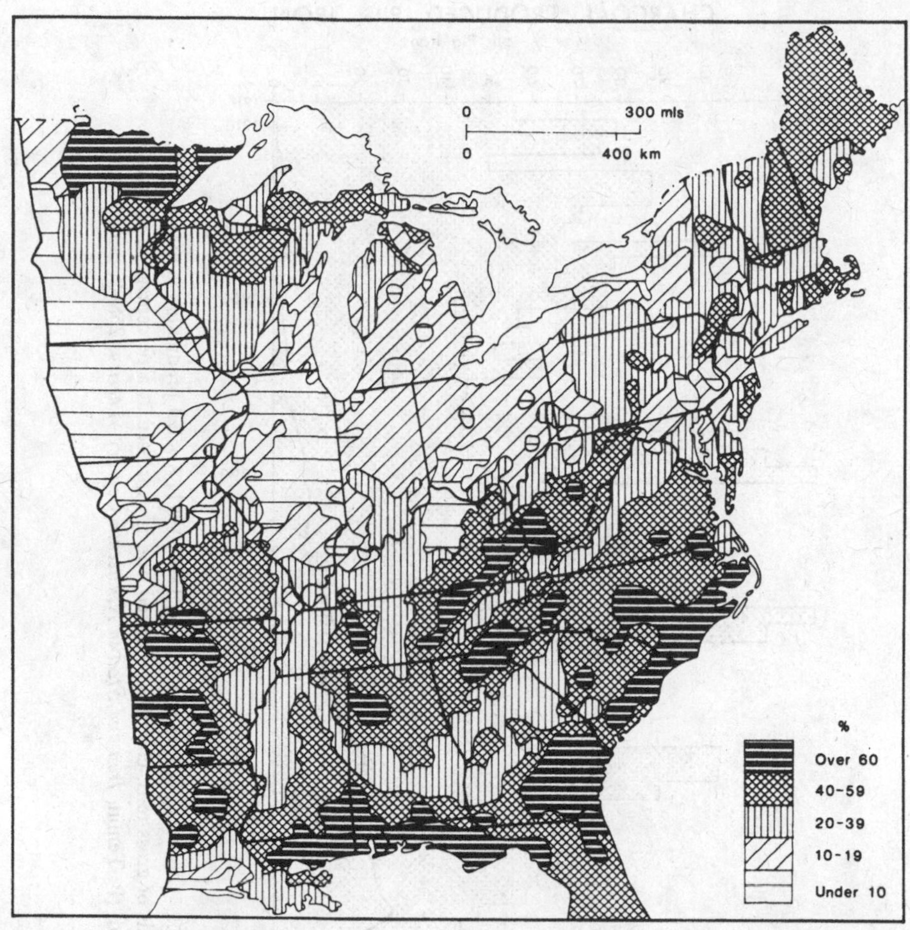

Figure 10.4. Proportion of farms in the eastern United States covered by woodlot, 1910. (Based on E. L. Frothingham, *The Status and Value of Farm Woodlots in the Eastern United States*, 4–5.)

Carolina, Georgia, Alabama, Mississippi, and Arkansas they were the third-ranking crops.

Even as late as 1940, some 7.7 million homes still used wood for heating, a figure that had doubled to 14 million homes in the energy-conscious seventies and early eighties, underlining the nature of substitution in fuel supplies as a reaction to price differentials and the abundant nature of the wood supplies of the country.[17]

Charcoal and the iron industry

The changes in location, fuel use, and technology that were beginning to become evident in J. P. Lesley's survey of the iron industry in 1856 gathered momentum as America moved slowly into the truly industrial age after the Civil War.[18] Charcoal was replaced

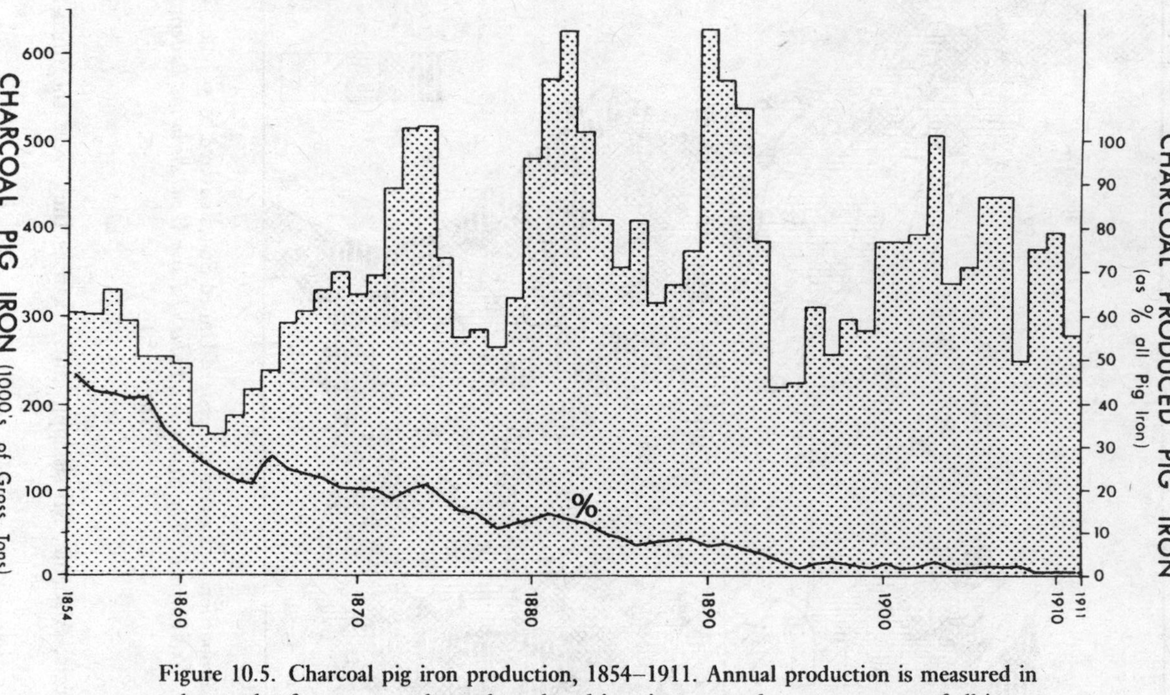

Figure 10.5. Charcoal pig iron production, 1854–1911. Annual production is measured in thousands of gross tons; charcoal-produced iron is expressed as a percentage of all iron produced. (P. Temin, *Iron and Steel in Nineteenth Century America*, 268–9, table C-3.)

by mineral fuels. The proportion of iron smelted by charcoal dropped from 45 percent of all iron in the mid-1850s to 25 percent at the close of the Civil War. However, its declining percentage was within a rapidly rising total output, which meant that charcoal iron production was not eliminated and fuel gathering in the forests continued. On the contrary, the demand for the versatile metal increased as the economy expanded so that more was being produced in nearly every year after 1860 than before, with production reaching over half a million tons on 11 occasions during the last couple of decades of the century, only to decline after about 1909 (Fig. 10.5).[19]

It is true that the number of furnaces fell rapidly from 385 in 1859 to a mere 33 in 1920, but much of this decline was because the small preindustrial furnaces with a capacity of less than 2,000 tons per annum were eliminated in the older areas of production in Pennsylvania and Ohio and increasingly large (20,000 tons capacity or more) furnaces were being built in the new areas of production farther west. The new locations were in Michigan, northwestern Wisconsin, Missouri, western Tennessee, northern Alabama, and northwestern Georgia, all near to easily extracted sources of iron ore and in proximity to extensive forests (Fig. 10.6 and Table 10.1). With the change in the number of furnaces came a change in their appearance and efficiency. The typical, squat (30-foot), truncated masonry pyramid of the preindustrial furnace was slowly replaced by the new, tall (55–65-foot), narrow chimneylike furnace. High-pressure hot blasts could be passed more easily through these furnaces, which were much more economical in their use of wood fuel. The hot air was collected from the top of the furnace rather than being allowed to escape as in the old furnaces, and it was now pumped at high pressure by a steam engine rather than by a waterwheel.[20] Consequently, the amount of charcoal needed to smelt a ton of pig iron decreased. Whereas early antebellum furnaces used anywhere from 150 to 250 bushels of charcoal per ton, the late-nineteenth-century furnace used between 72 and 120 bushels per ton, the median value of 18 furnaces known to be operating between 1881 and 1926 being 91 bushels per ton.[21] One problem was that the weight of the ore in the more efficient tall furnaces caused the brittle, traditionally made charcoal to crumble. That was overcome by distilling the wood in brick kilns and iron retort systems, which produced less ash, stronger charcoal, and also valuable and salable wood chemical by-products from the condensed gases.[22]

Indeed, the charcoal industry did not stand still in the face of competition from solid fuels, and it adopted innovations fairly quickly. One could go so far as to say that because charcoal was an inherently more expensive fuel than coal (despite the abundance of timber) the charcoal industry had to become supercompetitive; hence its very slow death. In fact, charcoal furnaces became more like coke furnaces and were operated accordingly, some even changing from charcoal to coke and back again, depending upon the availability of fuel supplies and upon the price and demand for different types of iron.

With these changes in technique came changes in the organization of production; the old-fashioned iron plantation disappeared with the old-fashioned furnace. The charcoal ironmaster became more efficient and aggressively competitive, so that after about 1860 he reserved his premium product for fine cutting tools, small arms, steam boiler tubing, crank shafts, axles and gears, and, above all, railroad wheels, which consumed the bulk of

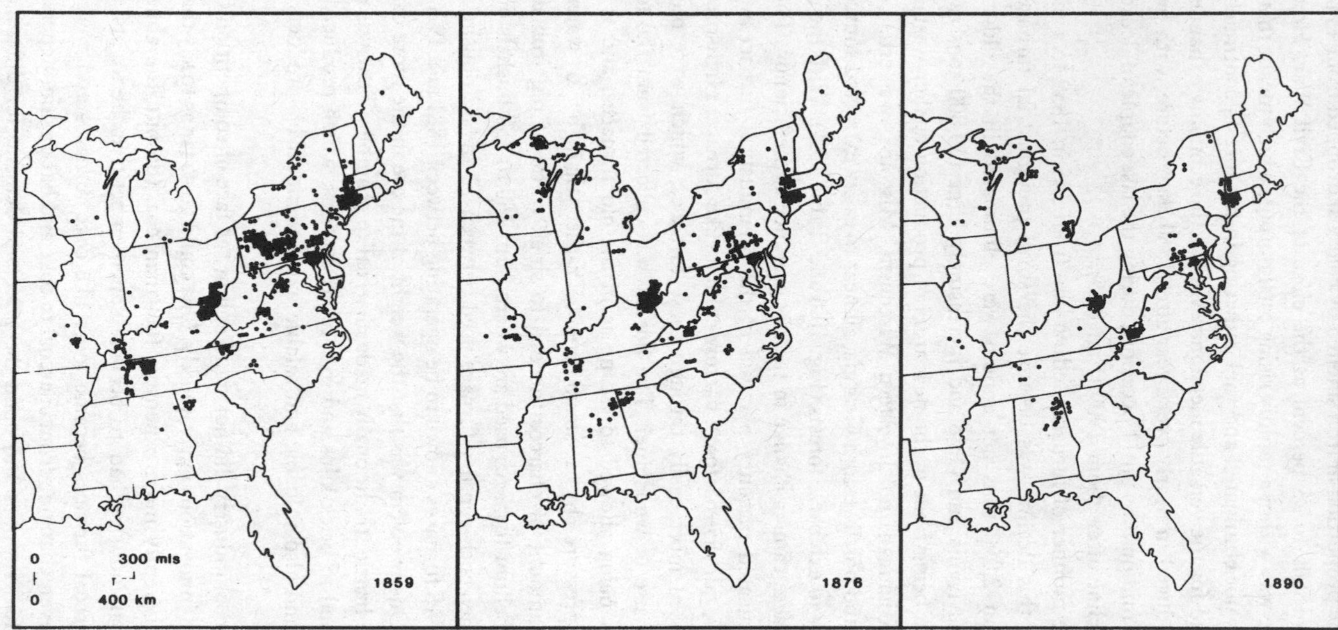

Figure 10.6. Location of charcoal blast furnaces, 1859, 1876, and 1890. (After R. H. Schallenberg and D. A. Ault, "Raw Material Supply and Technological Change in the American Charcoal Iron Industry," 447–9.)

Table 10.1. *Charcoal pig iron production, by state, 1866–1910 (to nearest 1,000 tons)*

	Mass. Conn. N.Y. Me.	Pa.	Ohio	Md. Va.	Ky. Tenn. Ga.	Ala.	Mich.	Mo.	Wis.	Total
1866	—	58	88	—	—	—	35	—	—	181
1872	55	45	96	50	77	13	87	46	28	497
1880	64	43	69	47	45	38	154	16	43	519
1885	34	12	18	23	42	78	143	22	20	392
1890	47	18	26	24	64	110	259	34	95	677
1895	17	5	12	—	15	21	102	2	51	225
1900	24	4	9	7	29	65	176	38		352
1910	19	5	11	2	11	40	292	73		453

Source: R. H. Schallenberg and D. A. Ault, "Raw Material Supply and Technological Change in the American Charcoal Iron Industry," 461–2.

the output, and he left the less demanding market for rails, horseshoes, and cruder constructional bars and rods to the coke iron manufacturer. After about 1890, however, advances in ferrous metallurgy generally began to catch up with charcoal iron and erode its few remaining markets. Bessemer converters and, particularly, open-hearth steel furnaces produced large quantities of high-grade steel, and the advent of chrome nickel alloys to strengthen coke steel made it a serious competitor for the specialized products. In particular, the railroad wheel succumbed to the new, strengthened steel because of the demand for wheels capable of withstanding greater weights and speeds. With its main market drying up, the charcoal iron industry began to falter.[23]

Although the impact of the charcoal furnaces on their immediate surrounding forested areas was great, nowhere did the devastation become so widespread during the pre–Civil War years that the industry died, nor did charcoal become so costly that it was uneconomical to use, which are the conventional explanations for the shifts of the industry to the West.[24] Rather, as Schallenberg and Ault suggest, it was either the exhaustion of the local iron ore or the steady progress of the technology of steel production and manufacture that eventually squeezed the charcoal industry out of existence. For example, there is no clear evidence that the price of charcoal in the older, settled areas of the East, where competition for supplies would have been greatest, was higher than in the newly settled areas of the West where supplies were abundant. In fact, prices may have decreased as superior techniques of production resulted in salable by-products. Despite its brittleness, charcoal was transported fairly long distances to the furnaces, which were not necessarily in the middle of the forests but were often in urban or near urban locations. River transport and even railroad transport were used by many furnace owners, like those near Baltimore or in Virginia.[25] Again, the number and capacity of furnaces in Michigan during the 1860s and 1870s were nearly as great in the largely nonforested Lower Peninsula as in the densely forested Upper Peninsula. The important and common factor was not proximity to fuel, which could be transported, but accessibility to the cheap high-grade ores of the Superior ranges, which could be moved easily by lake transport. Charcoal was supplied over distances of 300 miles in New England, New York, Pennsylvania, Alabama, and Michigan in specially built railroad wagons.[26] Indeed, transport of fuel was an inevitable accompaniment of increasing furnace capacity because, with furnaces of 30,000–40,000 tons capacity per annum, fuel had to be collected over a greater radius than ever before.

To emphasize that wood exhaustion was not the primary reason for either the locational shift westward or the decline of the charcoal iron industry during the late nineteenth century does not, however, detract from the fact that large areas of the forests were permanently affected by cutting for charcoal. Depending upon the density of the trees in the forests, the amount of land cleared varied from place to place and from time to time. Using a modest estimate of 150 acres of woodland to produce 1,000 tons of pig iron, it is possible to calculate that the number of acres cleared in order to fire the furnaces could have ranged from 25,000 acres (39 square miles) per annum in a low-production year like 1862 to 94,000 acres (147 square miles) in a high-production year like 1890, although it should also be borne in mind that many forests near the furnaces were cut over at 25–30-year intervals, or sometimes oftener. For example, a detailed survey of the 837 square miles of Vinton and Jackson counties in the Hanging Rock district of southeastern Ohio shows that 60 percent of the forest was cut clear between 1850 and 1860, down to four-inch diameter trees, and that the forests regenerated sufficiently for recutting to be carried out during the early part of this century. The same was partially true of neighboring Scioto County, which was "originally well wooded . . . but a large proportion was turned into the charcoal to be used for smelting purposes in the large furnaces." However, here the forests did not regenerate; "after the forests had been cut, nothing was done to ensure their rejuvenescence; cattle were permitted to eat the young trees that came up and to trample over them; in some instances the cleared districts were burned over to secure a better pasture," and consequently some furnaces were abandoned for want of charcoal.[27]

A clue to the rotational cutting practices is contained in the estimates of the amount of woodland reserved by the smelting companies in order to keep the furnaces going permanently, an estimate that was, of course, linked to the calculation of the amount of regrowth that occurred after cutting. The ratio of acres of woodland reserved to tons of iron produced varied enormously, from 2.67:1 in Kentucky to 37.3:1 in New York, the median value of eight observations being five acres of woodland for every ton produced. For example, at the Woodstock Iron Company, Anniston, Alabama, which produced 6,100 tons, it was said that "20,000 acres are sufficient for a permanent supply. Lands cut over are reserved for growing another crop, and the furnace now has 4,000 acres of young trees." At the Center Furnace, Kentucky, which produced 3,900 tons, it was calculated that "about 10,440 acres would yield a permanent supply if no accident, such as fires, happened to the growing timber. Many furnaces in Kentucky have large tracts of woodland for growing new supplies. They are cut off once in 28–30 years." At the Hamilton Iron Works, Missouri, the ratio of woodland to a ton produced was about 8.4:1 because, "the timber being small and of slow growth, 30,000 acres would probably be

Table 10.2. *Charcoal iron: bushels of wood used per ton produced, 1880–1900*

	Bushels, charcoal	Tons, pig iron	Bushels per ton
1888	53,910,000	480,000	112.31
1890	67,672,000	628,000	107.75
1900	31,422,000	385,000	81.61

Source: U.S. Bureau of the Census, Twelfth Census (1900), 10, pt. 4: 18.

required to keep a furnace permanently supplied. More than half the land in the region is good for nothing except for growing timber. No second growth has been cut and it would probably require 30 years for it to become large enough for profitable cutting." At the Champlain Ore and Iron Company, Elizabethtown, Essex County, New York, the production of a mere 1,600 tons of iron seemed to require 60,000 acres for permanent supply, although of that "30,000 acres are reserved, and a new growth may be cut in from 15 to 20 years."[28] These examples could be multiplied endlessly; suffice it to say, the area considered necessary for a reserve stock depended as much on the management of the timber as on the rate of regrowth.

A further and final complication in the calculations is that the amount of charcoal needed to smelt a ton of pig iron decreased over time with the advent of better smelting and charcoal-making techniques, thereby lessening the impact of charcoal making on the woodlands as the years progressed. Whereas furnaces during the early nineteenth century consumed well over 200 bushels per ton smelted, the amount of charcoal needed fell rapidly toward the end of the century, and between 1880 and 1900 it was reduced further from an average of 112.3 bushels to 81.6 bushels per ton (Table 10.2).

The only cross-check we have for the amount of forest destroyed is for particular census years. For example, in 1880 the production of charcoal pig iron reached 480,000 tons, and 52.91 million bushels of charcoal were used to smelt it. If the charcoal from 150 acres of woodland is considered sufficient to smelt 1,000 tons, then some 72,000 acres would have been cleared in that year. However, if, as was commonly suggested, between 1,000 and 1,200 bushels of charcoal were the product of one acre of woodland, then between 54,000 and 45,000 acres must have been cleared. There are clearly some discrepancies, depending on the statistical transformation used. An additional complication is that 15.6 million bushels of charcoal went into rolling and steel mills and forges and bloomeries, which must have meant the clearing of another 13,000 to 15,600 acres.[29]

Because of the many variables and uncertainties involved, one cannot be too dogmatic about the amount of woodland cleared in any one year, or indeed for the whole of the post–Civil War years of the nineteenth century. At best we can make a rough summary calculation that the 20.4 million tons of charcoal iron produced between 1855 and 1910 may have required, in toto, the forest growth of 4,800 square miles of well-stocked woodland. If we assume a 25-year regrowth and cutting over the previously harvested ground, then that figure drops to about 3,000 square miles. Impressive as the area is, however, it is salutary to compare it with the amount of land calculated in Chapter 11 as having been cleared for agriculture during the same period. Clearing for iron production is only 1.3 percent of the land cleared for agriculture or, if regrowth is taken into account, a mere 0.8 percent. With this comparative perspective it can be seen that charcoal iron production had relatively little impact on the forests of the country as a whole. However, it had an impact out of all proportion to its area because it was a concentrated impact that could be seen, understood, and calculated, and that, unlike agricultural clearing, was almost an alien intrusion into post–Civil War landscapes and society, which were still overwhelmingly agricultural and rural in their appearance and consciousness. As such, clearing for charcoal commanded special attention and comment, and it contributed, perhaps disproportionately, like the railroads but with less justification, to the heightening awareness of the destruction of the forests during this critical period of forest use. Whatever the exact truth of the situation, however, most people living in iron-producing districts would have agreed with N. W. Lord in 1884 in his assessment of what was happening in the Hanging Rock district of Ohio: "The disappearance of the forests under the demands of the furnaces, which is now so apparent throughout the region, increases every day the difficulty of obtaining the necessary fuel, and marks very plainly the fate of the charcoal iron industry."[30]

The railroads

Of the many industries and activities that depended on the products of the forest, those of railroad construction and operation made perhaps the greatest impact. The impact was real in terms of the cubic feet of lumber consumed in railroad ties, trestle bridges, station buildings, telegraph poles, snow fences, and fuel, to mention but the main uses to which lumber was put (Plate 10.1). But the impact was also symbolic. For some people the railroad was to be deplored; it was an intrusion and an alien element in the pastoral scene, the harbinger of change and destruction – in short, the "machine in the garden."[31] On the other hand, for others it was the technological culmination of the paleotechnic age. It caused urban commercial and industrial functions to flourish, and it even seemed to promote the formation of the nation by linking together the various parts of the continent. As an early bulletin of the Forestry Division pointed out in 1888, "the pioneering days are rapidly disappearing before the energetic push and advancement of railroad building and settlements." Earlier, in 1868, E. F. Palmer had given this symbolism a pictorial representation in his picture entitled "Westward the Course of Empire Takes Its Way," in which a railroad crossed the scene and disappeared into the distance, splitting the picture diagonally into two worlds, "civilization" and "wilderness."[32]

Whichever group one belonged to, however, there was little doubt that the railroad meant change, and in the context of the forest that could mean only one thing: the diminution of the timber stand. In 1866 Andrew Fuller had noted that "even where railroads have penetrated regions abundantly supplied, we soon find all along its track timber soon becomes scarce. For every railroad in the country requires a continued forest from one end to the other of its lines to supply it with ties, fuel, and lumber for building cars."[33] Certainly, during the 1860s supplies of fuel wood near railways in the

Plate 10.1. Baird Creek, near Longview, southern Washington: the impact of the railroad on the forest – for ties, trestle bridges, and fuel, and as a means of transporting the bulky timber to mill and market – was enormous.

Northeast had become scarce. In southern New England, New York, Pennsylvania, and eastern Ohio, and also around Chicago, the demand of the urban population pushed up prices, and the demand for firewood, building timber, and charcoal, together with the high value of land in agricultural use, pushed railroad fuel to between $7 and $8 a cord. The New York Central Railroad ran irregular services during the winter of 1864 because of shortages and problems caused by burning coal in wood-burning locomotives, and "energetic agents were sent back into the country, and by offering high prices and making great exertions to supply the road, in mid-winter the trains began to resume their regularity." Starr calculated that the daily consumption of fuel wood by all railroads at that time was 21,555 cords, which would have meant an annual drain on the forests of at least 6.5 million cords, which in value terms was two and one-half times that of coal mined.[34]

Nevertheless, all the evidence points to the fact that the demand for fuel from wood-burning locomotives was falling throughout the 1870s and 1880s. In 1878 Hough had conducted a survey of 38 railroad companies that operated nearly 11,000 miles of track, and only 254 out of 2,424 locomotives, or a mere 10.5 percent, still used cordwood. Nearly half the wood burners in Hough's survey were on one line only, the Lake Shore, Michigan Southern Railway (Buffalo to Chicago), but even this line used a mere 87,236 cords of wood compared to 261,719 tons of coal.[35]

Although the demand for wood fuel was becoming almost insignificant, it was gradually realized that the amount of timber needed for general railway purposes, such as buildings, stations, telegraph poles, fencing, and crossties in particular, would reach astronomical proportions if the length of track constructed continued to expand at the rate it had in the past. Track length more than doubled from 32,000 miles in 1864 to 90,000 miles in 1875, and it was set to nearly double again during the next decade, and again during the next.[36]

The resulting destruction of the forest was enormous and widespread. To take one example: Ohio, which lay in the pathway of many east–west railroads, had over 6,000 miles of track by 1870 enclosed by over 10,000 miles of wooden fencing, running on more than 10 million ties, over 16 miles (in aggregate) of wooden bridges, and 10 miles of trestles, with locomotives consuming about 700,000 cords of wood fuel. The average replacement rate for ties was between six and seven years, for bridges five and a half years, and for trestles seven years. The felling required to keep pace with this rate of construction but even greater rate of decay and replacement made it difficult, concluded Daniel Millikin, to "begin to conceive the demands which this new invention will make upon the woods." By 1876 the railroads were deemed to be making greater inroads than agriculture into the forests, "which were being removed entirely too rapidly," the whole balance of agriculture itself being put in jeopardy by "this denuding process."[37] By 1884 residents in Trumbull County in eastern Ohio noted that their "majestic forests" were "fast disappearing, to aid the rapid strides of public improvements"; in Belmont County, three new railroads had furnished "a special market for lumber to be used for tunnels, ties, bridges, etc.," which caused considerable destruction; in Fairfield County, "a vast amount of lumber" had been used in railroad construction; in Tuscarawas County, the "railroads had culled the forests . . . to the extent that there is *no first class* or 'heavily wooded lands' left." In Muskingum County, the construction of the new Baltimore and Ohio line had unleashed "a mania for buying woodland, stripping it of its timber, and then selling it for agricultural purposes." Everywhere, the advent of the railroad meant felling.[38] But it was the building of the land-grant railroads across the prairies, such as the Union Pacific, the Central Pacific, the Burlington, and the Santa Fe, which used such enormous amounts of wood in largely treeless regions, that raised timber-depletion consciousness. On the prairies speed of construction (at a rate of up to eight miles a day), in order to beat competitors, and the need to get a rapid return on investment prompted the companies to find quick and cheap answers to engineering problems. Steel, earth, and stone were long-term and costly solutions to bridges, banks, tunnels, and the like; moreover, they required some skilled labor. Wood, on the other hand, was a familiar, well-tried, and relatively low-cost material, which could be improvised, fabricated, and adapted in numerous ways by unskilled labor. It could be brought onto the site of construction by established means of communication via rivers, or it could come behind the railroads as they splayed across the prairies.

Although it is true that in many areas of the country the railroad cut a swath through the forest – an image of destruction enhanced greatly by the publication in 1884 of Sargent's maps of forest destruction, which showed broad zones of clearing of at least five miles on either side of any railroad – in many areas the laying of a railroad passed almost unnoticed. However, the simple equation that railroad equaled forest destruction became so firmly fixed in people's minds as the decade wore on that, like the charcoal industry, the railroad assumed an importance out of all proportion to its actual impact, great as that was in places.

The vague notions of how much wood was being used were pinpointed and given some precision by Hough's inquiry of 1878 in which he noted that lines required between 2,200 and 3,500 ties per mile, the most common number being 2,640. But, because the timber rotted and decayed as it lay embedded in ballast, the ties had to be replaced about every five to six years so that between 15 and 20 percent of the track mileage had to be renewed every year.[39] As was often the case, Hough did not take his observations to their logical conclusion and calculate the amount of forest destroyed as a result of construction and maintenance. It remained for others to do that. In 1883 Egleston, in his annual Forestry Division report, drew attention to this matter and hazarded a guess that about 3 million acres must have been destroyed to supply the existing track mileage with ties and that 472,400 acres were needed annually (or 12.6 million acres in all) just to keep up maintenance, figures he put at 567,714 and 17 million acres a few years later.[40] There the matter seemed to rest until the publication in 1887 of the first technical and professional paper of the new Forestry Division, which was devoted to "The relation of railroads to forest supplies and forestry." This bulletin thrust the railroad tie issue into the forefront of the forest depletion debate, which was gaining momentum during the closing years of the century. M. G. Kern of St. Louis, agent for the division and author of the bulletin, posed the question in blunt and uncompromising terms:

> Considering the stupendous amounts of timber already withdrawn from native forests, the annual demands of railways now in operation, and the increase in mileage from year to year, it becomes necessary to take a more accurate survey of the fields of demand and supply, unbiased by the popular delusion of the inexhaustible forest wealth of America. The necessity is no longer either to be ignored or lightly treated as in the past.[41]

By making reasonable and conservative computations, Kern was able to calculate that existing tracks and poles consumed 3.1 billion cu. ft. and that maintenance and new construction consumed another 0.5 billion cu. ft. annually. When due allowance was made for types of timber needed (poles or dimensional timber) and for the density and yield of the forest, the annual drain was nearly 300,000 acres: 249,214 acres for the maintenance of tracks and poles and 47,673 acres for the construction of new track. This figure for the annual cut was less than Egleston's sheer guesses of a few years earlier but no less startling. It had been calculated in a reasonably precise and quantitative manner, which lent it an authenticity and stature that Egleston's estimates never had.

Whatever the precise figures, however, the concern over the impact of the railroad industry on the forests was genuine and it was great. The concern was heightened by the manner of cutting. The railroad owners preferred hardwoods: white oak (*Quercus alba*) and, to a lesser extent, other hardwoods like chestnut oak (*Q. prinus*) and black locust (*Robinia pseudoacacia*), which had the requisite qualities of strength, elasticity, and resistance to rotting (Fig. 10.7). But they were convinced that the ties lasted longer if they were taken from young, "second-growth" timber; if they were hewn rather than cut; if the outer sapwood was discarded and only the inner heartwood used; and only if they were felled in the winter rather than in the spring or summer months. The result was the total destruction of substantial stands. As one lumber journal put it:

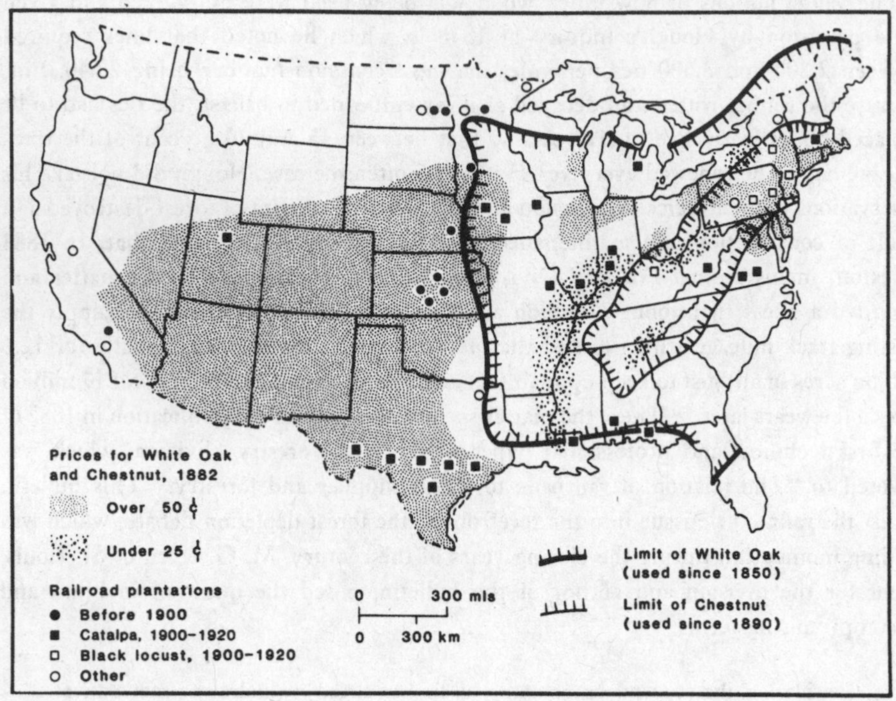

Figure 10.7. The range of oaks and chestnuts, the hardwood tie market (1882), and railroad plantations. (Based on S. Olson, *The Depletion Myth*, 15, 18, 81.)

> There is no branch of the lumber industry where there is more waste of raw material. . . . Each tie is split from clear wood, and it takes about 35 feet of clear lumber to make a merchantable tie . . . . When to this is added the percentage of "culls" that are arbitrarily rejected by the inspectors on behalf of the railroads at the owner's expense, it will be found that each tie represents about 75 feet of good merchantable lumber in the standing timber destroyed for it.[42]

This wasteful mode of exploitation had little scientific basis. A spring or summer felling did not hasten decay because of the presence of sap in the outer wood although a winter felling was incidentally advantageous in that it meant the wood stood for several months when fungi growth was at a minimum and the wood was seasoned to a certain extent. This wasteful, even uneconomical mode of exploitation was not held in check by market forces, as might be expected. Farmers anywhere near stands of the favored species either cut trees on their own land during their winter off-months or worked for the railroads, and tie supply made an important contribution to farm capital and purchasing power in the least-developed parts of the country. The farmers cut readily and abundantly and oversupplied the railroad companies, who then controlled the selection of the quantity, quality, and price paid by merely adjusting the freight rate. This had the effect of keeping the price down. In addition, the ever-increasing spread and integration of long-distance and branch lines throughout the 1870s and 1880s allowed new sources of timber to be tapped so that the railroads, as consumer, never felt the pinch of higher prices, the specter of scarcity, or the need to economize by using substitute materials. They were capable of creating and supplying a massive demand with no change in price.[43]

The result was that immense quantities of tie timber were cut down and delivered to the railroads at below their real value. "When one locality is exhausted," wrote Kern, "the scene of slaughter of the valuable young timber is simply shifted to another." The preference for oak hardwoods meant that areas that had, on the whole, escaped the worst ravages of commercial lumbering, such as the Appalachians, the Ozarks, southern Wisconsin and Michigan, Illinois, and Indiana, now became major crosstie-producing areas. The tributaries of the Ohio that flowed off the Appalachians, such as the Tennessee, Cumberland, Green, Kentucky, Licking, and Kanawha, were main lines of production and movement. In Missouri, the oaks of the Ozark plateau fed the railroad network that radiated out from St. Louis in all directions, and indeed, St. Louis became the main crosstie market in the country at about the time of World War I. Although well over 60 percent of the ties were some variety of oak and another 5 percent chestnut oak, the shortfall had to come from other species elsewhere in the country, and the price of oak and chestnut oak ties was correspondingly high (Fig. 10.7). For example, hemlock and tamarack came from the northern areas around the Lake States (3.4 percent) together with redcedar and white cedar (6.8 percent), all of which flowed south through the major lumber market of Chicago to be distributed across the prairies. Individual Chicago merchants like Edward Ayer, who controlled two-fifths of the Chicago redcedar tie market, could fill individual orders ranging from 1 million to 2.6 million ties for companies like the Burlington and the Santa Fe. For the first time the eastern slopes of the Rockies became a major source of commercial timber as crossties were floated down the east-flowing rivers to meet the lines advancing westward across the prairies.[44]

For the railroad companies the supply of timber never seemed in doubt throughout the 1880s and 1890s. The tapping of new sources, the extension of long-distance and branch lines, and the abundant oversupply of ties by the farmers ensured a stock of timber that was more than equal to their needs. But the government foresters and the conservation publicists thought otherwise; from Fuller to Hough, and through the succession of personalities such as Egleston, Kern, and particularly Fernow, there developed a concern about timber depletion in which the crosstie industry was a central focus. The railroads were identified as a major cause of the depletion because they had spread into all regions of the country, and, moreover, their effects on the forests could be calculated precisely from the length of the line and the rate of replacement. In addition, the railroad was a popular target because of the latent dislike of "big business" among some and the apprehension of its disrupting effect on society by others. Bernhard Fernow, who became chief of the Forestry Division in succession to Egleston in 1886, saw the railroad tie question as a new and positive means of promoting forestry interests and the fortunes of his fledgling division. In addition, the crosstie question was one that fit well into Fernow's concept of attempting to resolve forestry depletion by propagating among large users the idea of the rational and economical use of timber. From here on, nearly every annual report of the division drew attention to the impending dearth of supplies and the role of the railroad in bringing about that depletion.[45]

According to Kern, the "reckless system of forest clearing" for crossties could not go on indefinitely; the forest supplies would run out. He suggested two remedies: the preservation of timber to prevent rotting, and the planting of new trees to make good the deficiencies. The impregnation of wood with chemicals had made little headway in the United States, although the use of zinc chloride and creosote had been known in Europe since the beginning of the century. The cost of creosoting was high, and whereas it had a good effect in maritime conditions its efficacy did not seem proven for crossties. One problem with treatment of all kinds was the mistaken idea that the fungal growth was a symptom rather than a cause of decay and that treating the outside of a crosstie could therefore arrest or prevent decay even if fungal growth had already taken hold in the heart of the unseasoned wood. Creosoting worked, as did most impregnating methods, only under intense pressure, which was such an expensive process that a treated softwood tie of eastern hemlock or redcedar might ultimately cost as much as an untreated oak tie but last no longer. Consequently, although the information on treatment was available from about 1885 onward, few railroads favored treating ties. Supplies of ties continued to come forward throughout the 1890s, and whatever economies the railroad companies needed to make were effected by other means, such as better purchasing procedures, better stock keeping of supplies, mergers and consolidations. It is true that a few railroads did preserve their ties. From as early as 1881, the Santa Fe had established a treatment works at Las Vegas, New Mexico, and the Union Pacific at Omaha, Nebraska, but these were isolated instances, and preservation plants made little headway until the price of timber rose in response to dwindling supplies after 1900. Then the 10 plants in the country increased rapidly to 70 in the space of a few years and reached 102 at the outbreak of the First World War. From then on a very substantial proportion of ties of all kinds were treated as methods became more reliable and costs of treatment declined both absolutely and relatively.[46]

Kern's second solution, planting trees, was far less effective than preserving in combating railroad timber depletion but far more attractive and eye-catching. Tree planting was an emotive issue that had already captured the imagination of the settler on the plains west of the ninety-sixth meridian. Arbor Day had been proclaimed in 1872, the timber culture bill went through Congress a year later, and the rainmaking debate was still in full swing during the late 1880s and early 1890s, at the very time that the railway-induced timber depletion debate was getting underway.[47] Andrew Fuller had articulated the concept of plantations back in 1866 when he said, "How simple it would be for the railroad companies to have a few acres of forest trees every few miles all along and contiguous to the line." By 1872 the Burlington and Missouri River Railroad had invested moderately in plantations, as had the St. Paul and Pacific in southern Minnesota, not for timber, it must be stressed, but for snow fences and windbreaks and as a part of their settlement encouragement and promotion programs to expand traffic along their lines.[48]

Perhaps the first serious experiment to grow trees for timber was by H. H. Hunnewell in 1877 at Farlington in Crawford County, Kansas. Hunnewell, who was president of the Kansas City, Fort Scott, and Gulf Railway, had planted about 640 acres of catalpa (*Catalpa bignoniodies* and *C. speciosa*), which did well and were first thinned out in 1894–5. This, and many similar small projects, attracted enormous attention from all people interested in forestry, and, indeed, if one looks at the Forestry Division bulletins and the general literature on arboriculture, such as the proceedings of the two-part American Forestry Congress held in Cincinnati and Montreal in 1882 with papers on catalpa and black locust by eminent figures like Hough and Warder, one might be excused for thinking that these quite insignificant experiments were the main and even the only concern of forestry during these decades. The reports were glowing: "these stately blocks of young and thrifty trees, sprung up as if by magic on a treeless plain," was common enough, and under the pressure of price rises between 1895 and 1907 and the Forest Service interpretation that prices would continue to rise because of timber depletion, more companies planted trees along their tracks.[49] Yet, in reality, the plantations were few and insignificant. In all, about four dozen locations can be identified, and perhaps, at most, about 15,000 acres were affected during the space of about 30 years, which, if the trees had all come to maturity and been good, merchantable timber, would have supplied ties for less than 10 days at the 1910 rate of consumption (Fig. 10.7). The enthusiasm over catalpa, black locust, cottonwood, willow, and eucalyptus introduced from Australia was largely optimistic. The plantations did not make the money their promoters forecast, and many were located outside their natural limits. However justified the plantations were as snow fences, windbreaks, or aesthetic settings for stations, they came nowhere near to filling the timber needs of the railways. By 1915 or thereabouts, the experiment in planting was largely over.[50]

The railway timber depletion debate had one other immediate effect. When Fernow took over as chief of the Division of Forestry in 1886 he was convinced that besides planting and preservation savings could also be made by substitution and timber science. Substitution was simple enough and merely consisted of promoting the use of steel, stone, and live hedges where wood had been used before. The role of timber science was more complex, for Fernow believed that savings of anywhere between 20 and 25 percent could be made if the proper selection of timber for bridges and ties could be made. He realized that the railroads were becoming efficiency-conscious and that they had access to distant supplies and could, therefore, select the best timber for the job, unlike the domestic user or the burner of cordwood. Under his direction, thousands of tests were carried out on the strength and durability of timbers, on air seasoning, on tie preservation, and on every aspect of "timber physics" and "material research." For example, tests proved that the formerly despised chestnut oak was perfectly interchangeable with the favored white oak as a tie timber and that southern pines were all of very similar quality and were equal competitors with northern white pine as a bridge timber, both of which findings appealed to the economy-conscious railroads capable of bringing in exotic timber from new regions of supply. Yet, Fernow's experiments were stopped in 1896 when Congress diverted funds from material research to tree planting and forest management. Nevertheless, the important test centers that arose out of his program became in 1910 the nucleus of the Forest Products Laboratory in Madison, Wisconsin.[51]

What, then, had been the impact of the railroads on the forest resources of the country? Kern's figure of nearly 300,000 acres of forest destroyed every year had been lost sight of, to a certain extent, in the tangled debates and conflicting claims of preservation and planting, substitution and scientific testing, none of which could

Table 10.3. *Estimates of crossties used and acres of forest cleared, 1870–1910*

<table>
<tr><th></th><th>Miles of track</th><th>Ties renewed annually (millions)</th><th>Ties used on new construction (millions)</th><th>Total ties annually (millions)</th><th>Acres of forest cleared (thousands)</th></tr>
<tr><td>1870</td><td>60,000</td><td>21</td><td>18</td><td>39</td><td>195</td></tr>
<tr><td>1880</td><td>107,000</td><td>37</td><td>21</td><td>58</td><td>290</td></tr>
<tr><td>1890</td><td>200,000</td><td>70</td><td>19</td><td>89</td><td>445</td></tr>
<tr><td>1900</td><td>259,000</td><td colspan="2">91</td><td>91</td><td>455</td></tr>
<tr><td>1910</td><td>357,000</td><td colspan="2">124</td><td>124</td><td>620</td></tr>
</table>

Source: Sherry Olson, *The Depletion Myth*, 12; and U.S. Bureau of the Census, *Historical Statistics of the United States from Colonial Times to 1957*, vol. 2, table 392.

diminish significantly the overall impact of the railroad on the forest. In general terms, Kern's calculations corresponded with the 1911 estimates of the Forestry Service, which calculated that, with an annual average renewal rate of 350 ties per mile of track and an average cut of ties of 200 per forested acre, some 290,000 acres were being destroyed in 1880 and some 445,000 acres in 1890 (Table 10.3).

As Sherry Olson has pointed out in her penetrating study of the railroad tie industry, these figures of nearly half a million acres cut annually by the turn of the century must also be augmented to take into account wood used for the construction and replacement of buildings, rolling stock, bridges, and so on. What she does not point out, however, is that the mileage of urban electric railways was also increasing during the years from 1890 onward, adding about 30,000 miles to track mileage by 1910.[52] Thus the railroads were using one-fourth to one-fifth of the nation's annual timber production during the latter part of the nineteenth century, and the contemporary view that the railroad was "the insatiable juggernaut of the vegetable world," though a little exaggerated, was essentially correct.[53]

Each of the impacts on the forest examined here declined with time. First coal, then electricity and gas were substituted for firewood in heating homes; charcoal gave way to coke for smelting iron; fuelwood was replaced by coal on the steam locomotives. Although nothing has really been a successful substitute for wooden ties, the number needed has declined as railroads have decreased the miles of track maintained. None of these uses have important impacts in the late twentieth century, but from our contemporary viewpoint it is all too easy to forget that the abundance and widespread availability of wood were probably the mainsprings of the country's industrialization during the second half of the nineteenth century. Without its wood America would not be the country it is today.

11

Agricultural impacts on the forest, 1860–1920

The original necessity of sweeping off the forest has imparted to many a propensity for cutting and slashing which has not stopped within reasonable boundaries.

John J. Thomas, "Culture and Management of Forest Trees" (1864), 43

It does not follow, because any portion of the country is now or has been for a long time treeless, that it must remain so, or that this is its natural condition.... It is impossible to say as yet where a tree cannot be made to grow.

Nathaniel Egleston, in USDA, *Annual Report 1885*, 184

The complex and many-faceted relationship of mankind with trees was never so well illustrated as in the interplay of agricultural themes that unfolded during the latter half of the nineteenth century. On the one hand, trees were destroyed by the millions of acres as forest clearing for agriculture reached its peak during the decade of the 1870s. On the other hand, the trees were planted by the millions as the movement of farmers into the Plains moved toward its peak during the same decade and later.

But the relationship was more complex than these connected but opposing trends suggest. Perhaps the first real concern about the extent and severity of clearing came from agriculturalists rather than lumbermen, because it was the agriculturalists who saw and appreciated what clearing could do to the land itself, to the soil, the stream flow, the microclimate, and the fish and animal life. Their concern, coupled with the ever-increasing store of scientific knowledge about the environment, was to lay a basis for thinking about the environmental health of the forest and the integrity of forest ecosystems.

Thus, for the first time, the forest was viewed as beneficial, rather than solely as an impediment to get rid of or as a source for the supply of timber. Trees might affect climate by increasing rainfall or decreasing wind velocities – one could not be sure – but trees certainly affected stream flow and stopped the land from coming apart. It was all a part of a very complicated and interwoven story of events and concerns at the end of the nineteenth century, which led ultimately to the move to preserve and manage the forests that were left.

Clearing and making a farm

Although the destruction caused by commercial, industrialized lumbering tends to dominate the story of the forests during the 60 years between 1860 and 1920 (see Chapters 7–9), the inroads made by the agriculturalists were immense. "Every field was

Table 11.1. *Amount of land cleared and man-years expended in forested and nonforested areas, 1850–1909*

	Forested		Nonforested	
	Acres (millions)	Man-years (thousands)	Acres (millions)	Man-years (thousands)
1850–9	39.7	4,268	9.1	48
1860–9	19.5	1,973	19.4	68
1870–9	49.3	4,243	48.7	122
1880–9	28.6	2,471	57.7	139
1890–9	31.0	2,486	41.1	68
1900–9	22.4	1,705	51.6	86
Total	190.5	17,146	227.6	531

Source: M. L. Primack, "Farm Formed Capital in American Agriculture, 1850–1910," tables 3, 4, 7, 8.

won by axe and fire," said one Ohio pioneer, and it was not far from the truth. Even near the prairie edge where the forest petered out and trees became scarce, the habits of generations were not easily shrugged off. Early settlers were still "tree destroyers, not culturists," and they did not cease their destruction until the paucity of trees became obvious.[1]

If the earlier calculations of the amount of land classified as improved in predominantly nonforested counties during the 1850s are continued for the next five decades, the conclusion must be that a total of approximately 150 million acres of former forest land was settled and cleared during these years. This compares with a total of approximately 218 million acres of former prairie settled during the same period (Table 11.1).[2] Even if we bear in mind all the qualifications that must attend these calculations and make allowance for local variations here and there, we cannot deny the magnitude of the impact of agriculture on the forests. Clearing during the 50 years between 1860 and 1910 resulted in slightly more land succumbing to the ax for farming than during the whole of the 250 preceding years.

The fact that during these years slightly more than a quarter again as much land was settled in the open prairies or Plains has obscured the importance and extent of farm making in the forests. The Plains have caught the imagination: They were a new environment that required new tools and techniques of settlement, and they were the locale of new legislation, the Homestead Act of 1862 and its subsequent amendments. They are fixed in the popular mind as being perhaps the *only* area of new settlement at this time. But this is simply not true. The glamour of the open spaces – the range, the cowboy, as well as the grim pioneering of the agriculturalist in his sod house, in fact everything that made up the amorphous concept of "the West" – needs to be put into perspective by considering carefully the long-drawn-out process during the second

Plate 11.1. "The war on the woods." Pioneering in the forest was still commonplace little more than a hundred years ago. Lycoming County, western Pennsylvania, c. 1880.

half of the nineteenth century, what our Ohio pioneer called "the war on the woods" (Plate 11.1).

The making of a farm in the forest – building a shelter, laying out fields, and clearing some trees – had become such a commonplace by midcentury that even less attention was paid to recording its progress than before. In the predominantly wooded states of the East and South, the number of farm holdings increased over threefold, from, in round figures, 1.4 million in 1850 to 4.8 million in 1920, and although we cannot be sure of the exact number created in the forest we can be fairly certain that, given the value placed on wooded land by farmers and on wood as a constructional and marketable commodity, the vast majority must have been in or near the forest.

The effort needed to clear the forests remained immense. At the beginning of the second half of the century the time expended on clearing an acre of the forest still stood at about 32 man-days.[3] There is little evidence to suggest that the component of 20 days for actual tree felling altered during this period, although it may have varied locally. For example, the trees in the South and the Southeast were of smaller girth than those in the Lake States and the Pacific Northwest; nevertheless, there is evidence to suggest that it often took much more than 20 days to fell an acre of the northern forests, and therefore the errors may well cancel each other out if we take a national view. Another problem is that productivity in the lumber industry increased rapidly in the period from 1890 to 1910 with new ways of handling and transporting lumber, and although these improvements may well have affected the rate at which the cutover lands were created, it is unlikely that this gain in productivity was one that was passed on to the individual farmer clearing his own land. For him, the making of his farm was still a combination of unremitting toil and a great deal of frugality, made possible only by the attainment and maintenance of good health. Drawing on his own and his father's experience of making a farm in the forests of New England, Horace Greeley said:

> I would advise no one over forty years of age to undertake, with means, to dig a farm out of the dense forest, where great trees must be cut down and cut up, rolled into log-heaps and burned to ashes where they grew. Where half the timber can be sold for enough to pay the cost of cutting, the case is different; but I know right well that digging a farm out of the high woods is, to any but a man of wealth, a slow, hard, task.[4]

In contrast to the effort of tree felling, the component for clearing the stumps probably fell quite dramatically, from about 12 man-days in 1860 to 10 man-days in 1880 to 6 man-days in 1900 and perhaps to as low as 4 man-days in 1910, which, when combined with the time taken to fell the trees, gave a minimum of 25 man-days for clearing an acre of land.

The rise in productivity was a result of the search for and adoption of new methods, particularly in the difficult-to-clear cutover lands in the Lake States and the Pacific Northwest, where the stumps were larger than in the hardwood areas and where they rotted slowly, if at all. Between 1909 and 1924, article after article was written about the techniques of clearing these northern lands, each with precise instructions on removal methods and costs.[5] One major innovation was the use of dynamite for blasting out the stumps. Mention of its use in England was made in American agricultural journals in the mid-1870s,[6] and dynamite seems to have become popular in America after 1890 when the problem of the large pine stumps in the Lake States intruded upon the activities of the sellers of the cutovers and they had to find a reasonably cheap solution. Stump pullers based on horse-powered capstan winches, tripod-mounted stump lifters, and, later, donkey engines, using a variation of overhead steam-skidding devices over an area of up to 1,500 yards around, came into prominence, with or without blasting. The burning of the large stumps by boring holes into the base and inserting inflammable materials; char pitting, which was the packing of the base of the stump with kindling and covering it with clay, as in charcoal making; and even the use of patented stump-burning machines whereby a gasoline-powered blower fanned flames at the base of the stump, were all alternative or supplementary methods of stump removal.[7]

From a large range of evidence available, it can be estimated that by 1910 an acre cleared by blasting required four man-days, by stump pulling five days, and by burning three days, and as stump pulling was the most common method of removal, five days was the usual time taken. Despite the decline in man-days involved, however, the insignificance of the energy expended in preparing the open prairie land compared to the forests remained. In fact, in the prairies the improved breaking plow had reduced the time involved to a mere half a man-day. The result of this increase in productivity was that the total amount of land cleared in the forest decreased whereas that in the prairie accelerated (see Table 11.1). The relative ease of settling the prairies was indicative of the radically shifting center of agriculture in the country; whereas the forest had the advantage of offering the continuation of an old and well-known technology in farm making, the prairie gave quick returns, provided capital was available.

Table 11.2. *Number of acres (millions) cleared from previously abandoned land*

1850–9	0.0
1860–9	0.5
1870–9	8.3
1880–9	3.5
1890–9	2.9
1900–9	2.4

Source: M. L. Primack, "Farm Formed Capital in American Agriculture, 1850–1910," table 5.

The apparent simplicity of the above calculations is upset by two considerations. First, around 1890 the cutovers were being brought into production in the Lake States, a little later in the South, and around 1910 and after in the Pacific Northwest. Obviously, the labor requirement for bringing these lands into production was less than that for bringing the hardwoods into production, because only the stump and brush remained to be cleared. Therefore, estimates of the labor required in the forest after 1890 may well be exaggerated. Second, some forest land was brought into production that was previously abandoned, cleared land, which would obviously require much less effort in clearing. One can only assume that the land abandoned in one decade would become a part of an increase in a subsequent decade. The acres so cleared from previously abandoned forest are therefore calculated on the assumption that "any county that had experienced a decrease in acres and that subsequently experienced an increase, the increase up to the amount previously abandoned was reclaimed abandoned land."[8] Again, the amount of labor expended in bringing these lands into production would have been considerably less than in virgin forest stands and could, on average, have been as little as five man-days per acre. The rate of clearing, per decade, of previously abandoned land is shown in Table 11.2. This represents a component of the total area of forest land cleared, and it was particularly important in the cotton-producing and Appalachian states of the South and East, namely, the Carolinas, Georgia, Alabama, Louisiana, and Mississippi, and Virginia, West Virginia, Kentucky, and Tennessee.

The locale and extent of clearing

The locality of clearing shifted as pioneer settlement and farming moved across the United States. A general view of the amount of clearing can be gained from looking at Figures 11.1–11.5, which depict the number of improved acres in predominantly forested and nonforested lands at each decade between 1860 and 1910, data that are also tabulated by major region in Table 11.3.

During the decade 1860–9, nearly 20 million acres were cleared, nearly half in the Lake States of Minnesota, Michigan, and Wisconsin and the three immediately

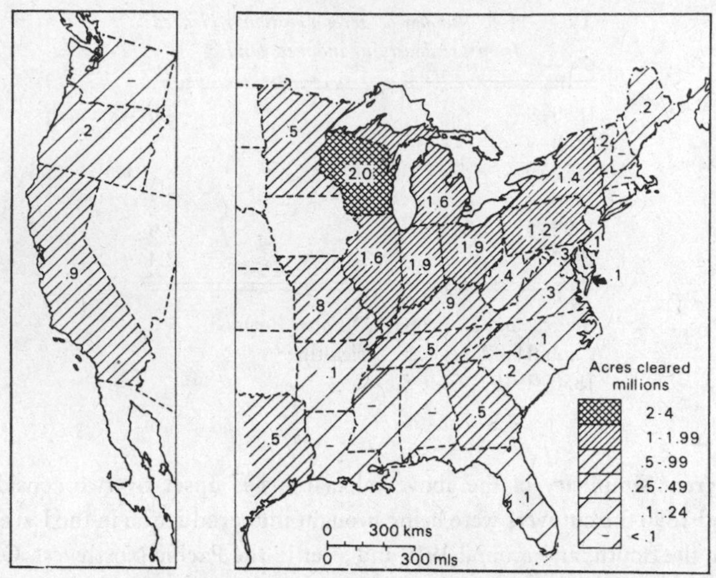

Figure 11.1. The amount cleared, by state, 1860–9 (millions of acres.) (Based on M. Primack, "Farm Formed Capital," and U.S. Census.)

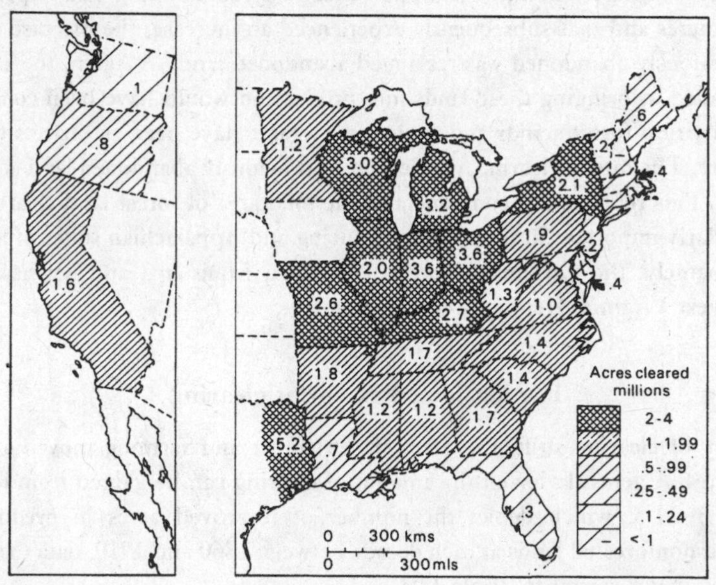

Figure 11.2. The amount cleared, by state, 1870–9 (millions of acres.) (Based on M. Primack, "Farm Formed Capital," and U.S. Census.)

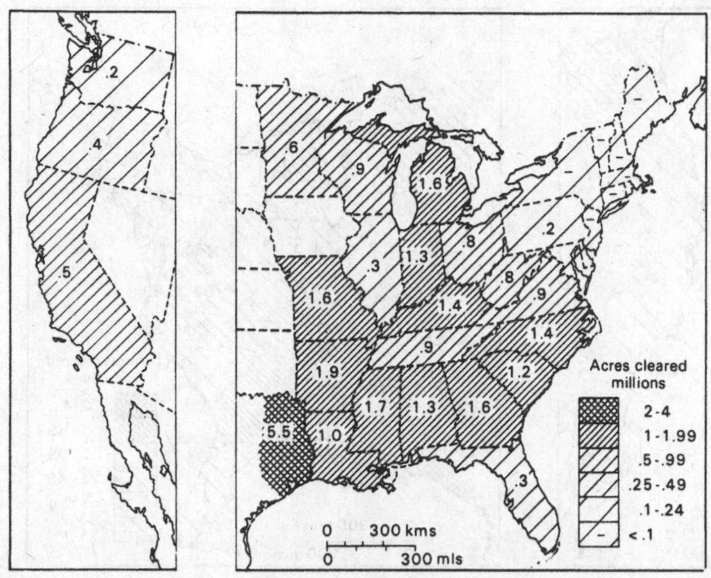

Figure 11.3. The amount cleared, by state, 1880–9 (millions of acres.) (Based on M. Primack, "Farm Formed Capital," and U.S. Census.)

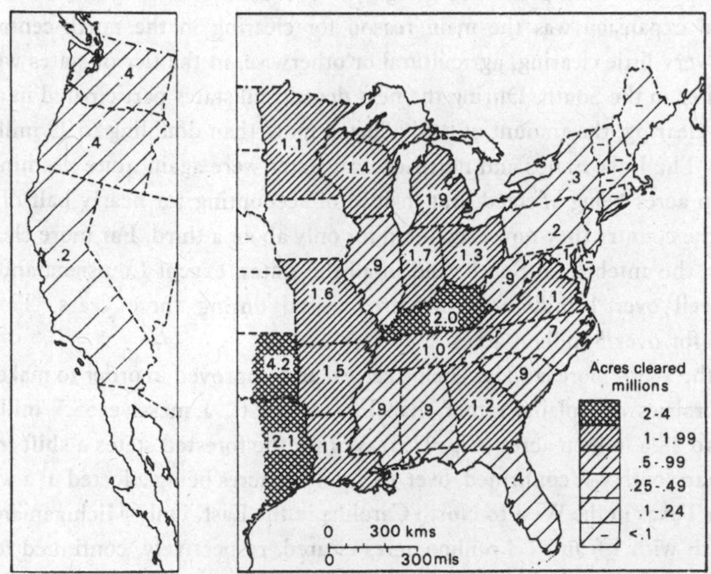

Figure 11.4. The amount cleared, by state, 1890–9 (millions of acres.) (Based on M. Primack, "Farm Formed Capital," and U.S. Census.)

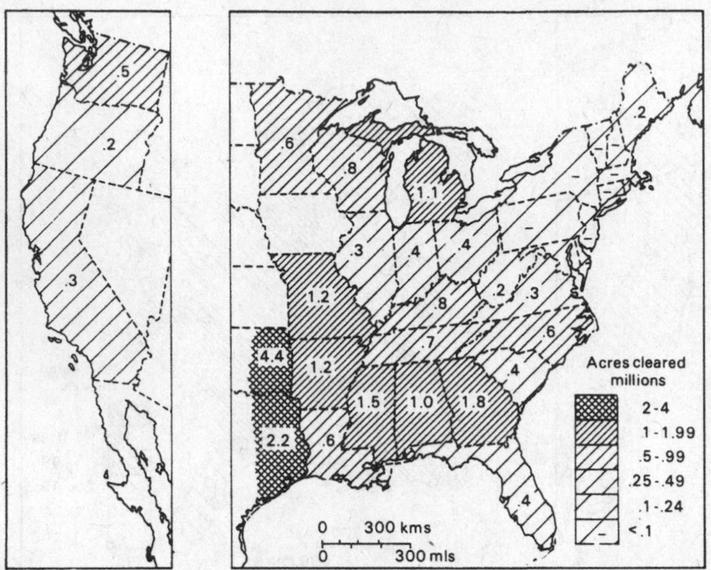

Figure 11.5. The amount cleared, by state, 1900–9 (millions of acres.) (Based on M. Primack, "Farm Formed Capital," and U.S. Census.)

adjoining states to the south in the north-central region, namely, Illinois, Indiana, and Ohio (Fig. 11.1). At this time, only a small proportion of clearing in the Lake States would have been cutover land. Most would have been in the highly desirable deciduous forest, interpersed with prairie openings, in the southern fringes of these states. Agricultural expansion was the main reason for clearing in the north-central states. There was very little clearing, agricultural or otherwise, in the tier of states west of the Mississippi or in the South. During the next decade, all states participated in a massive increase in clearing, the amount of land affected more than doubling to 49 million acres (Fig. 11.2). The Lake States and north-central states were again quite prominent, with 16.5 million acres being affected, but instead of accounting for nearly half of the land cleared in the country they now accounted for only about a third. Far more clearing was going on in the antebellum South, and in all the states, except Louisiana and Florida, totals of well over 1 million acres were cleared during these years, Texas alone accounting for over 5 million acres.

During the 1880s more land was being settled and improved in order to make farms in the open prairies and plains rather than in the forests, a massive 55.7 million acres compared to 28.6 million acres (Fig. 11.3). Within the forested states a shift in clearing from north to south was confirmed, over 16.7 million acres being affected in a wide arc of states from Texas in the West to North Carolina in the East. Only Michigan and Illinois in the North with 1.6 and 1.3 million acres cleared, respectively, continued to have an amount of land affected that in any way rivaled the amounts being cleared in the southern states. The 1890s represented a very similar picture, but with slightly more clearing going on, again, in the north-central states. Settlers were looking with new favor

Table 11.3. *Amount of forest clearing each decade, by major region, 1850–1909 (in thousands of acres)*

	1850–9	1860–9	1870–9	1880–9	1890–9	1900–9	Total
Northeast	1,417	619	1,339	62	80	228	3,745
Mid-Atlantic	4,283	2,811	4,278	295	646	87	12,400
Southeast	6,822	1,869	7,612	6,213	5,410	3,712	31,638
North-central	8,245	5,566	9,199	2,420	3,804	1,200	30,434
South-central	9,025	1,803	9,156	8,168	7,489	5,767	41,408
Lake	5,093	4,188	7,347	3,131	4,575	2,497	26,831
Southwest	3,018	1,422	7,796	7,148	8,020	7,832	35,236
Pacific	1,774	1,188	2,590	1,166	983	1,063	8,764
Total	39,677	19,466	49,317	28,603	31,007	22,386	190,456

Source: M. L. Primack, "Farm Formed Capital in American Agriculture, 1850–1910," table 2.

on the remaining forested lands in the East because now the more easily taken up and climatically favorable prairie lands had been preempted (Fig. 11.4).

The final decade from 1900 to 1909 saw the slowing down of agricultural clearing in nearly all states with the exception of some in the South, and for the first time there was clear evidence of some reversion of once cleared land to forest (Fig. 11.5). Nevertheless, the amount of land affected – 21 million acres – was still sizable and significant.

Ohio: microcosm of forest clearing

This necessarily generalized view of clearing based on the crude statistics of the state unit leaves many questions unanswered as to how and why clearing either accelerated or slowed down in various states at various times. Work on a local scale has to be undertaken in order to interpret these patterns successfully. Not all states can be examined, but one example on a local scale is Ohio, which stands at a locational and temporal crossroads in the utilization of the forest in the United States. Early settlement between about 1790 and 1850 was largely a repetition of the pioneer experience of the eighteenth-century eastern seaboard, but as the century progressed increasingly commercial attitudes toward farming and settlement prevailed, which were more characteristic of the newly emerging, agriculturally rich midwestern region. Ohio was also the scene of some of the earliest attempts to record and assess accurately the impact of clearing on the forest, which makes this example particularly valuable.

The state contained approximately 24.3 million acres of land, the overwhelming bulk of which was covered by deciduous broadleaf forests of elm and ash on the Erie shores and inland, beech and maple through most of the center, and oak and hickory in the southern portions and southeastern part of the state.[9] It seems likely that only about 930,000 acres (slightly over 4 percent of the land surface area) was not forest originally, and this open country consisted of prairies, oak openings, and barrens, particularly in the central portions of the state in, for example, Logan, Champaign, Clark, Madison, Delaware, and Licking counties. These nonforest vegetation types were often associated with moraines and preglacial drainage lines. Other open country consisted of wet prairies, bogs, and swamps that developed in areas of inadequate drainage or ponding on glacial outwash deposits or the remnants of old lake sites, such as the great Black Marsh, which stretched across about 1,500 square miles of the northwest corner of the state.

The settlement of Ohio proceeded westward from the Pennsylvania border during the 1770s and 1780s and was mainly to the south of the Greenville Treaty line that divided Indian lands from lands open to European settlement.[10] By 1800 Ohio's population was 45,345, and it rose rapidly to nearly 2 million by 1850 as land was converted into farms and small-scale local industries were established. In the census returns of 1850, 9,851,493 acres or 40.5 percent of the state was listed as "improved" land, largely if not wholly cleared from the forest. All parts of the state were affected, but a core area in which over one-half of the land was cleared lay in a broad belt extending across the state, from Cincinnati in the southwest and widening out to the Pennsylvania border in the east (Fig. 11.6). The highest percentage of land cleared was in Lake and Summit counties near Lake Erie where 64 percent was improved. On either side of this central belt the proportions of cleared land fell off, sometimes rapidly as in the direction of the Black Swamp in the northwest and in the uplands in the southeast. With 143,807 rural holdings in existence in 1850, there were approximately 68.5 acres of improved land for every holding.[11]

The processes and landscape of clearing during these midcentury years were exactly the same as they had been on the eastern seaboard. One pioneer in Bloomfield, Ohio, wrote "home" to Connecticut about his new, raw surroundings in the forest from

> the Portico of a log house fronting the south. The Portico is supported by six columns, I am not a conoisseur [*sic*] in architecture, and cannot tell you what order they are, but I can tell you they have rec'd no polish or ornament from art other than being divested of their bark. The mansion house on your canvas must be placed nearly in the center of a small opening or clearing like an island in the midst of a vast ocean of dark green forest. Your imagination must supply the rest as I am but poor in description.[12]

Very little imagination was needed, in fact, as so many of the first wave of immigrants to Ohio were from New England. The landscape of isolated farms in the middle of a clearing, just like the place names, bore a familiar resemblance to those "back East."

It is, of course, possible to trace the development of clearing during each subsequent decade from the evidence of the census returns. But these distributions are not presented, as they merely show a universal decrease in the amount of woodland in all areas, albeit with a great variation in local rates of change. Suffice it to say that another 2.7 million acres of woodland were felled in the next 10 years, to 1860, to give 12,625,344 acres of cleared land (or 52 percent of the state), most of the clearing going on in the same central belt of counties as before. By 1870 a further 1.8 million acres were cleared to give 14,469,133 acres of cleared land, or 60 percent of the state, the greatest amounts of clearing going on in the peripheral outer counties that had low values before. The really

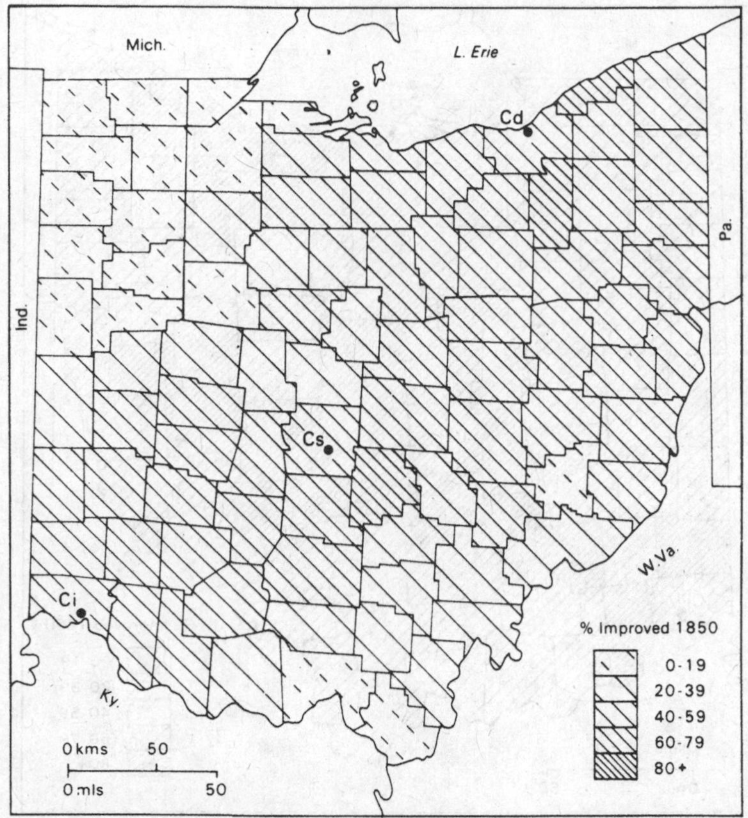

Figure 11.6. Ohio: percentage improved, by county, 1850. (U.S. Census.)

big change, however, came in the years between 1870 and 1879 when 3.6 million acres were cleared to produce the distribution shown in Figure 11.7. Much of this new surge of clearing was in the peripheral counties of the state, particularly in the northeast where the Black Swamp was drained and crossed by mail roads and highways and the surrounding forest lands were cleared.[13] By 1880 nearly three-quarters of the state had been cleared of its forests.

The degree of devastation revealed by the census of 1880 caused an uproar as thinking farmers and officials alike expressed grave concern over the destruction of the natural timber stand. It is true that warnings about the rate and degree of clearing had already been sounded by Millikin, Klippart, and Hough in 1871, 1876, and 1884, respectively,[14] but the simple extrapolation forward of the trend in clearing witnessed during the previous couple of decades was alarming and added a new urgency to the debate. In 1885 a newly formed Ohio State Forestry Bureau decided to conduct a comprehensive survey of the state's forests. It sent out a questionnaire to "township officials and intelligent farmers," who were asked, among other things, "What causes are in operation to produce the waste and decay of the forests in your township?"[15] The order in which the

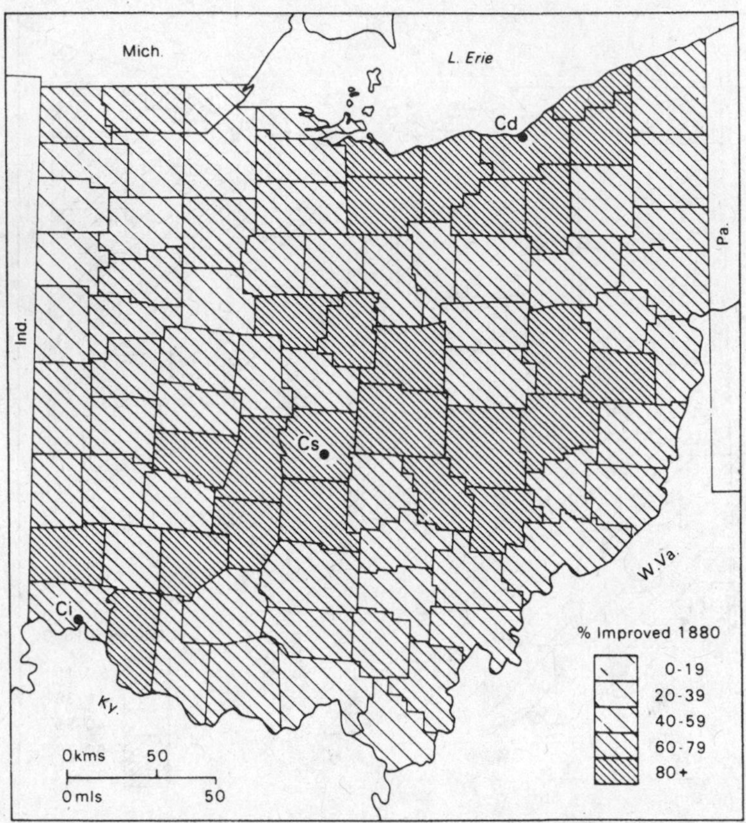

Figure 11.7. Ohio: percentage improved, by county, 1880. (U.S. Census.)

answers are listed is assumed, with good evidence, to indicate the degree of importance that they are given, and they have been ranked accordingly in Table 11.4 and plotted in Figure 11.8. Some respondents gave only one cause, some five, but most two or three; hence the total number of replies does not add up to any theoretical total.

Opinion was fairly unanimous that clearing for agricultural purposes, whether to create new farms, to enlarge the farm holding, or to enlarge individual fields, was the main cause of forest destruction. It was listed as the first-ranking cause in 90 percent of the 88 counties and was a reason in nearly all the others, for quite simply, as Mr. George Adams of Perrysburg in Wood County said, it was "every farmer's ambition to have his woodland turned into a great corn-field." Timber getting for constructional purposes or for fuel was the main cause of destruction in eight counties, five of them in a contiguous block in the iron-producing area of Hanging Rock in the southern part of the state, namely, Vinton, Jackson, Pike, Scioto, and Lawrence. Charcoal getting was also the first-ranking cause of destruction in Belmont and Columbiana counties, which had numerous local furnaces along the Ohio. After Pennsylvania, Ohio was the most important iron-producing district in the United States at this time, well above its nearest rivals, and its blast furnaces consumed prodigious amounts of charcoal.[16] Timber getting was also a

Table 11.4. *Reasons given, by county, for forest destruction in Ohio, c. 1884*

Reason by rank	Agriculture	Timber and fuel	Cattle	Fire	Insects	Draining	Gold	Drought	Storm/ wind	Total
1	79	8	1	—	—	—	—	—	—	88
2	2	36	15	6	3	3	—	—	—	65
3	2	6	26	9	6	1	1	1	—	52
4	—	1	7	4	9	2	4	2	1	30
5	—	—	1	2	4	1	4	—	1	13
Total	83	51	50	21	22	7	9	3	2	248

Source: Adolph Leue (ed.), *First Annual Report of the Ohio State Forestry Bureau* (1885).

secondary cause of forest destruction in 37 counties and a tertiary cause in five others, and this conforms with all that we know about the interrelationship that existed between pioneer farming and part-time timber getting.

The grazing of sheep and particularly cattle was another important secondary and tertiary cause, often coupled with, and a consequence of, both agricultural clearing and timber getting. In one case only – Hamilton County – was it a primary cause of destruction, and it was said that 96 percent of the forested area in that county had been felled in order to provide grazing for sheep. Damage by grazing was particularly important in the group of adjacent counties of Miami, Logan, Champaign, Clark, Greene, and Clinton, and also Pickaway and Ross in the west-central part of the state. Here Kentucky graziers had established themselves in the succession of small prairies, barrens, and oak openings that ran through these counties, first as transhumant pastoralists and then as permanent graziers.[17] It was suspected by some county correspondents that the cattle owners "fired the woods" surrounding the plains in order to obtain more and better pasturage, as in Scioto County. They were not the only incendiarists – hunters smoked out game, and tobacco farmers cleared an acre here and there – but in all cases "reckless hands burn the bush without caring whether the fire is kept within bounds."[18]

The depredations of fire and insects and the drying out of the soil due to draining were of relatively little importance in number or rank. Similarly, there were only 14 instances of climatic causes of destruction, such as cold weather, storms, and drought, and none of these ever got above third rank.

There is ample descriptive evidence to bear out these assessments of the causes of forest destruction. To take but a few: the local correspondent in Crawford County in north-central Ohio commented on the totality of the original woodland cover and then went on:

> its adaptedness [*sic*] to agriculture attracted attention at an early period. The forest, then the great barrier to progress, had to be cut without regard to value; a portion of the trees felled were split for rail fences, another cut up into fuel, and the rest was burned, for it had no market value then.

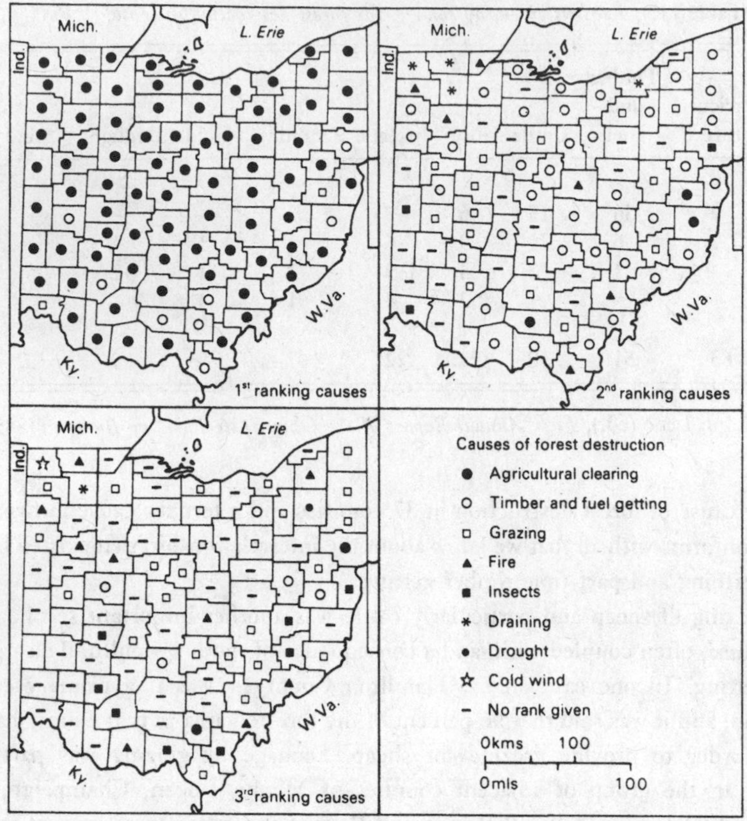

Figure 11.8. Ohio: the causes of forest destruction, 1885. (Adolph Leue [ed,], *First Annual Report of the Ohio State Forestry Bureau.*)

Although over 80 percent of the country was cleared, and although "the woodman's ax is at rest, the forest area is decreasing from year to year" because pasturing of stock prevented second growth and foraging caused the tree roots to be exposed, which were then subject to extremes of heat and cold, causing the trees to die. The process of destruction had gone so far and timber had become so scarce that another correspondent, James Robinson of Polk Township in the same county, said, "woodland is worth more than cleared land," and similar evaluations were made by many other correspondents.

In Belmont County, fronting the Ohio River near the border with Pennsylvania, the local spokesman bemoaned the "wholesale slaughter" of trees for agriculture and grazing: "Hundreds of acres of them are annually cut down and sawed into lumber, large quantities being left on the ground to rot or piled up and burned."

The construction of railroads nearby was providing a ready market for the farmers. For example, in Muskingum County the new Baltimore and Ohio Railroad had seemed to unleash a "mania for buying woodland, stripping it of its timber and then selling it for what it will bring for agricultural purposes." But, even putting aside these special local demands, cutting continued: "the trees seem to be looked upon as cumberers of the ground, to be cut down and got rid of as soon as possible." Many farmers argued, speciously, that it was best to cut the oaks now as they were old and dying anyway and were not being replaced by younger trees because the hogs were destroying young shoots and acorns. "Whether these arguments are founded on fact or not," commented the correspondent, "they are urged by the land-owners as an excuse for cutting off the timber, and the process is going on at a rate that will extirpate the forests in a few years." Mr. John English of Union Township, Auglaize County, had no doubts about the cause of destruction: "owners want to cut and sell every thing that will sell at all," he said. Some necessity, but mainly plain greed, was the cause of forest destruction.

After 1880 the rate of clearing in Ohio decreased, possibly because of an increased awareness of the detrimental effects of denudation and runoff. Only 257,733 acres were added during the decade to the total of 18,081,091 acres. From 1890 to 1899 just over another 900,000 acres were cleared to bring the amount of improved, cleared land to a record 19,244,472 acres, a total amount that was never equaled. It fell to 19,227,669 acres in 1910 with the reversion of many fields to forests. Overall, the pattern of change between 1880 and 1910 was stark. Nearly every county east of a line drawn from Lorain on the Erie lakeshore to the south of the state experienced some decline in the amount of cleared land; that is to say, large areas reverted to forest, especially in the environs of large urban areas. In contrast, nearly every county to the west of the line had more land cleared. A broad belt across the center of the state had up to 10 percent less forest; 13 counties a little farther to the northwest had up to 20 percent less forest; and in a triangular area across the extreme northwest corner of the state in what was once the badly drained Black Swamp, a mixture of marshes and scrubby beech–maple forests, 9 counties had up to 62 percent less forest, making them among the most open areas in the state. By 1910, after which new agricultural clearing was minimal, Ohio could be divided in half by a north–south line running very nearly through the center of the state (Fig. 11.9). To the west of the line over four-fifths of the land was cleared, most of it in crops; this was part of the corn belt.[19] To the east between half and four-fifths of the land was cleared, but generally the forests and the woodland accounted for a greater proportion of the land than in the west. Only two counties – Scioto and Lawrence – in the very southernmost tip of the state had less than half of the land cleared. A century of agricultural advance and clearing had left its mark.

Another and more detailed view of clearing than on the scale of the state can be gained by focusing on the smaller scale of the township. Figure 11.10 shows the decreasing cover of forest in Cadiz Township in Green County, Wisconsin, near the Wisconsin/Illinois border in 1831, 1882, 1902, and 1950.[20] The township was originally almost totally covered by upland deciduous forests, dominated by elm, basswood, and sugar maple, except for an area of oak–hickory forest in the northwest corner and a small portion of prairie and surrounding oak savanna in the southwest corner. Taking the township as a whole, the woodland was reduced from nearly a total coverage in 1831 to 29.6 percent in 1882, to 9.6 percent in 1902, and to a mere 3.6 percent in 1954. The remaining forests are now owned by the farmers and are a source of firewood and occasional sawtimber. The experience of Cadiz Township was common, to a greater or lesser degree, to most wooded counties in the hardwood forests of the United States.

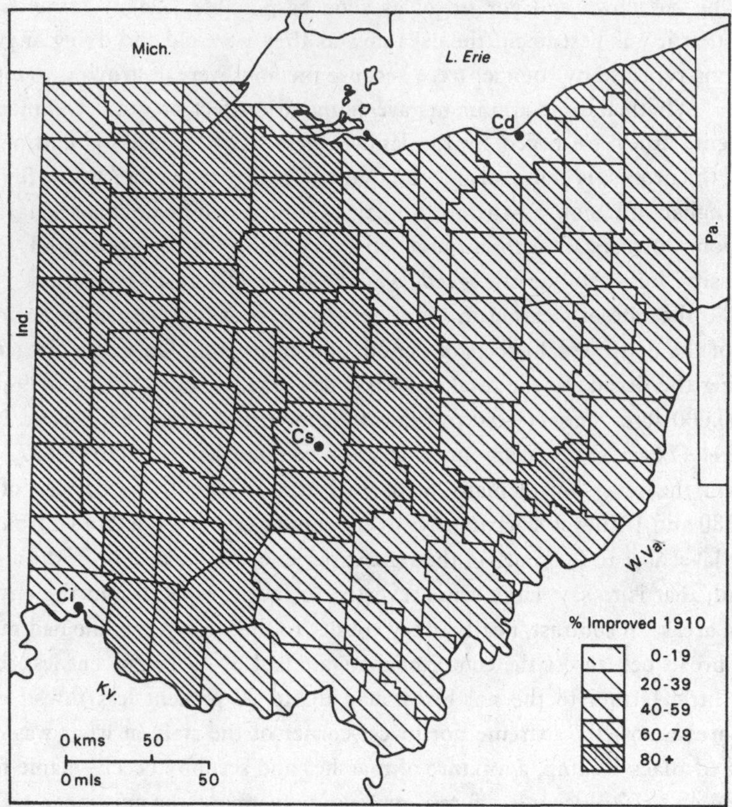

Figure 11.9. Ohio: percentage improved, by county, 1890. (U.S. Census.)

The stripping of the forest by the farmers continued throughout the century, throughout the continent; but, here and there, there were indications that the farmer had given up waging "war on the woods" and that the woods were advancing again. There were hints of that in individual counties in Ohio after 1880, and it was very clearly so for the whole state after 1900. But after 1920 the reversion became a general trend throughout the country, if only because the advent of the tractor meant that farmers needed between 20 million and 30 million acres less to feed their horses and mules, which acres could now be devoted to food crops. Though a new chapter was about to open in the farmers' relationship with the forests, the story of the nineteenth century had basically been one of cut and burn and cut and burn again.

Early warnings

If we consider the immense amount of forested land affected by the expansion of agriculture in the eastern United States, it should not be surprising that some of the earliest warnings about the deleterious effects of clearing should come from those with some close connection with the land. By 1860 about 140 million acres of once forested

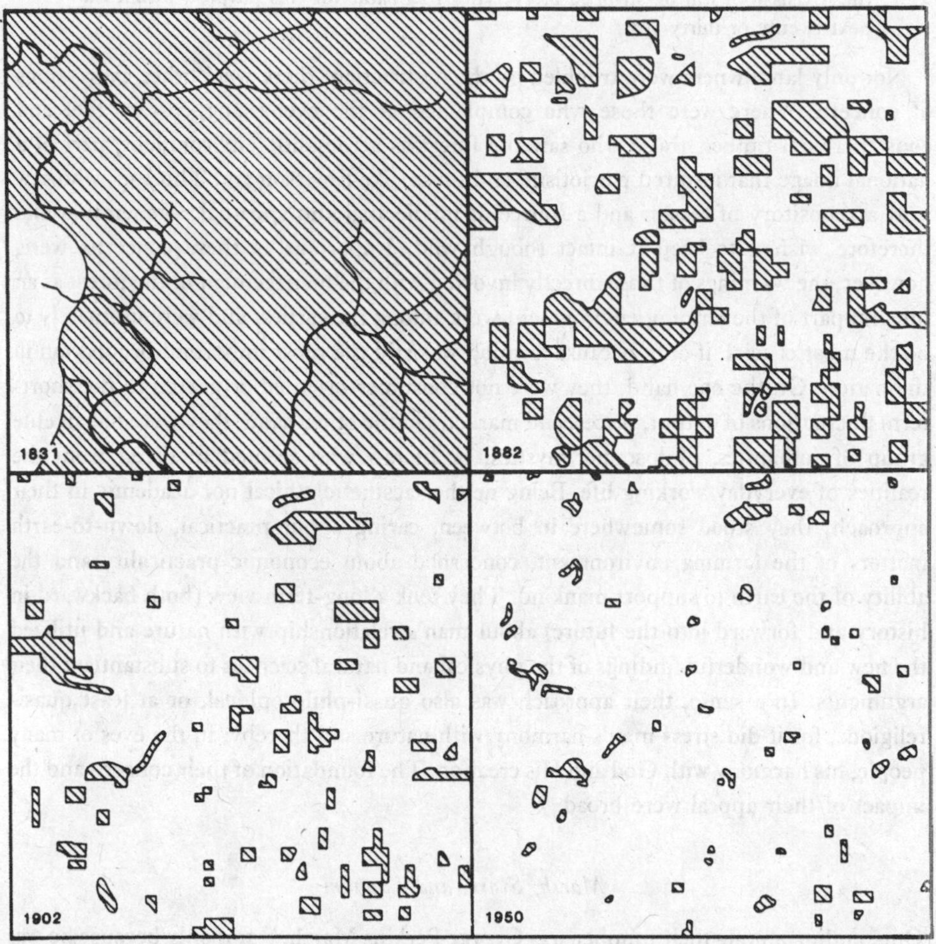

Figure 11.10. Woodland cover, Cadiz Township, Green County, Wisconsin, 1831, 1882, 1902, and 1950. (J. T. Curtis, "The Modification of Mid-latitude Grasslands and Forests by Man," 726–7.)

land had been cleared, to which another 150 million acres was to be added during the next 50 years.

In 1864 John J. Thomas of Union Springs, New York, commented on the national situation, and summed it up pretty well: "No people has ever equalled the American in boldness of enterprise and active industry. They have, within a comparatively brief period, . . . formed a powerful and prosperous nation." Unfortunately, he continued, their enterprise had not always been sufficiently discriminating:

> The original necessity of sweeping off the forest has imparted to many a propensity for cutting and slashing which has not stopped within reasonable boundaries. . . . The rapid disappearance of our forests should excite the serious attention of land owners generally. . . . at the present rate of consumption . . . the whole region east of

the Mississippi will be stripped of everything valuable for this purpose within the next twenty or thirty years.[21]

Not only landowners were troubled by the destruction; there were other expressions of concern. There were those who complained at the ever-rising cost of fuel and constructional timber, those who saw the forest as a romantic environment, part of a national image that inspired patriotism, and those, like Emerson and Thoreau, who saw it as a repository of reason and a source of intellectual and spiritual regeneration and therefore wished to keep it intact though wild. Important as these concerns were, however, the warnings of those directly involved with the management of the forest as an integral part of the farming environment were among the earliest and were ultimately to be the most crucial, if only because farming was still the most important occupation in the nation. On the one hand, they were not consumer-oriented or geared to the short-term fluctuations of output, prices, and market. On the other hand, they were not an elite group of romantics, philosopher/mystics, or literati who seemed divorced from the realities of everyday working life. Being neither aesthetic/ethical nor academic in their approach, they stood somewhere in between, caring about practical, down-to-earth matters of the farming environment, concerned about economic practicality and the ability of the earth to support mankind. They took a long-term view (both backward in history and forward into the future) about man's relationship with nature and utilized the new and wonderful findings of the physical and natural sciences to substantiate their arguments. In a sense, their approach was also quasi-philosophical, or at least quasi-religious, for it did stress man's harmony with nature and thereby, in the eyes of many people, his harmony with God and His creation. The foundation of their concern and the impact of their appeal were broad.

Marsh, Starr, and Lapham

Outstanding among their number was George Perkins Marsh,[22] not only because he was probably the first American to enunciate clearly the deleterious consequences of prolonged forest clearing but also because of his sheer intellectual range and erudition. In Chapter 5 we have examined how, by 1849, Marsh had built solidly on the early works of Chastellux, Lorain, Rush, and Volney to produce his talk to the Rutland County Agricultural Society on clearing in his native Vermont. Then, in 1861, Marsh was appointed American ambassador to Italy, and it was there that his greatest work, *Man and Nature*, was written. It was published in 1864. Feeding on his early Vermont experience, stimulated by the examples of the devastated landscapes of Mediterranean Europe, and drawing on a vast historical, philosophical, and scientific literature now available to him (and he could read 20 languages), he compiled his highly readable account of man's impact on his environment through time. The theme of the injury to earth by man dominates the book, and whereas earlier writers had emphasized the beneficial effects of clearing in America, Marsh emphasized the deleterious effects, pointing out "The dangers of imprudence and the necessity of caution in all operations which, on a large scale, interfere with the spontaneous arrangements of the organic and inorganic world."[23] More than a third of the book dealt with the forests, although the American forests were featured only briefly. Much of the section on forests was devoted to a painstaking review of the literature on forest influences, such as the relationships of trees to temperature, precipitation, soil formation, flora and fauna, and human health and disease, and to the role of forests in maintaining the flow of springs and in retarding erosion and flooding. Marsh's conception of the forests went far beyond their value for timber production, and he adumbrated a concern for the total forest ecosystem. He ended his book with a question that was crucial then in relation to the forests and is crucial now in a global sense; he asked "whether man is of nature or above her." He was sure himself that nature was in a harmonious balance except for the disturbances made by man and that whatever man could disrupt in the natural environment he could eventually restore and control with the wise and intelligent use of his knowledge and the findings of science. That was an important conclusion for those concerned with forest depletion and its mitigation. In a sense, *Man and Nature* was a philosophical treatise documented with technical details. The findings of emerging science and technology were converging with the seventeenth-century conception of nature as a divinely designed balance and harmony.[24]

Man and Nature had an immediate, although short-lived, impact. Its eloquent style and the evidence it displayed of broad reading and the exercise of a new kind of synthesis opened up vistas of hope for those who saw the forests fall before the ax at a frightening speed. But they were a minority, and in a society "dominated by an almost unlimited faith in possibilities of material progress" the book dropped into obscurity until revived in the mid-twentieth century.[25]

One early visionary who was influenced by Marsh was the Reverend Frederick Starr of St. Louis, who wrote a comprehensive and persuasive article in the *Annual Report* of the commissioner of agriculture one year later, entitled "American Forests: Their Destruction and Preservation." In the opening sentence he set out the scope and depth of the concern of his inquiry.

> There are few subjects so closely connected with the wants of society, the general health of the people, the salubrity of the climate, the production of our soils, and the increase of our national wealth, as our forests; and yet no considerable interest of our country has received so little attention at the hands of the people, and enjoyed so little fostering protection from the government.[26]

Starr's concern was frankly utilitarian and practical, and in a sentence that was to be quoted and paraphrased many times over in the future he conjectured that the things that were most fundamental to a nation's growth and prosperity were "cheap bread, cheap houses, cheap fuel, and cheap transportation for passengers and freight," and wood and the forests entered largely and consistently into each of these "four great departments of industry and living." In a "simple array of important facts," he listed in detail the causes of the decrease in the forests: the consumption of fuel, the ravages of the Civil War, the demands of manufacturing industry, and above all, the impact of agricultural expansion. With regard to the latter, he drew on the information in the census on "improved land," and he calculated that in New York State alone over 500 acres of forest had been destroyed daily between 1850 and 1859 to make way for farming, and up to 10,000 acres daily for the country as a whole during the same period. In the light of our calculations about the extent of destruction, his figures are interesting. He maintained that about 50 million acres had been cultivated between 1850 and 1859, of which 30 million acres "were lands either previously or during those years heavily timbered." This can be compared to our calculation of 39.7 million acres of woodland cleared and 9.1 million acres of nonwoodland cleared (Table 11.1).

Starr's solutions to the overall destruction were threefold: first, to educate landowners about the relative merits of different types of timber for fuel, construction, and so on, and the resultant virtues of more scientific and selective culling of the forests; second, to promote national scientific experiments and encourage the promotion of a specifically "American" literature on the subject, as most books on forestry were written in either French or German; third, to encourage government promotion of forestry. However "expensive and extensive" were private efforts, nothing would "so much impress the great mass of the people with the immediate and pressing importance of the action as to have Congress make some movement worthy of a subject so grand and an interest so vast."[27] Only a government-endowed corporation would avoid the evils of "the spoils system," the frequent changes in personnel, and achieve a national view to a truly national problem. Starr noted that efforts were already being made to promote American and some foreign trees in botanical gardens and nurseries. Somewhere in the world there must be trees capable of growing quickly in a variety of American situations that would answer the American need for cheap timber. Thus the scene was being set during the early years of concern about the forest for a concentration on tree culture – arboriculture – rather than on forestry, the latter word barely entering the language with its present meaning until the mid-nineteenth century.[28] Drawing on Marsh's *Man and Nature*, Starr reiterated the "Warnings from History," which showed that other great nations in the past had been reduced to a mere wilderness either by excessive forest clearance for agriculture or by overgrazing. The time for action was now, because "growth is slow and restoration tedious, while destruction is rapid and injury instantaneous."[29]

Whereas Marsh had emphasized the environmental consequences of continued clearing, Starr had laid much more emphasis on the economic consequences. It was left to Increase Allen Lapham of Wisconsin to combine both in his *Report on the Disastrous Effects of the Destruction of Forest Trees Now Going On So Rapidly in the State of Wisconsin*, published in 1867.[30] From now on the two strands of the argument against clearing were coupled intimately, each being invoked to support the other in the quest for a tamed and managed nature.

Lapham was no new hand at considering the consequences of forest clearing; he had issued a warning as far back as 1854 about the ruthless cutting by agriculturalists then going on in the hardwood forests of southwest Wisconsin, but his advocacy of plantations to make good what he saw as an impending deficiency fell on deaf ears. Lapham and others agitated for action through the state horticultural and agricultural societies, but it was not until after the publication of Marsh's *Man and Nature* in 1864 and Starr's article the following year that Lapham was asked by the state legislature to write a report. Both Marsh and Starr were totally absorbed by Lapham, the cover of the

Wisconsin report carrying a quotation from Marsh: "Man has too long forgotten that the earth was given to him for usufruct alone, not for consumption, still less for profligate waste." The report's early pages paraphrase Marsh's strictures on the need to preserve trees and repeat all the warnings of the environmental consequences of too much forest destruction. Lapham was commissioned particularly to inquire into the effects of agricultural forest clearance on the incidence and effects of the southerly and westerly winds in Wisconsin – cold in winter and dry in summer – and this represented new work. But he copied, almost word for word, long sections of Starr's assertions and statistics that wood was fundamental to a nation's growth and prosperity.

Lapham foresaw a time when the demands of agriculture for land, and of industry and the home for fuel, would strip Wisconsin of its timber. The state would move gradually through a progression of landscapes, from forest to farmland, from farmland to cattle range, and from cattle range ultimately to desert:

> The small farm with its neat house, orchard and garden, its fields of yellow grain and tall corn, – the home of the happy family, will become part of a cattle range, for these alone can retain a foothold, until that other more distant day shall come when the winds and droughts shall reduce the plains of Wisconsin to the condition of Asia Minor. Trees alone can save us from such a fate.[31]

Yet, for all the concern about the destruction of the forests expressed in the Lapham report, it was totally ineffective. It was agriculturally oriented and certainly had no effect in stemming the cutting of the great northern pineries, because it and the public generally accepted the dictum that cheap wood was fundamental to national prosperity. The report recommended tax incentives to encourage settlers in the more populous southern portions of the state to plant and preserve belts of timber between fields; and it provided an exhaustive list of the best trees to plant. The Wisconsin legislature did enact tax exemptions and bounties to encourage farmers to plant, but it seems that no one availed himself of these concessions, except perhaps for one farmer, Walter Macon Ware of Hancock County, who planted 1,876 white pines to commemorate the Declaration of Independence. As a correspondent, L. B. Longworthy of Rochester, New York, told Lapham, drawing on his experience of over 80 years: "you may speak and reason like an angel and you will never be able to effect any great or radical change in the habits of our people nor induce them to expend time and money that they can't realize the almighty dollar [from] in about the period of the wheat crop or fattening pigs." Posterity, commented Longworthy, was not in the vocabulary of his fellow citizens – only prosperity.[32] Longworthy's judgment was harsh but, as experience had shown, essentially correct. It was left to those with higher ideals and longer views to do something about stopping the clearing of the forest.

The correct balance

Associated with the concern about clearing was the notion of the "correct balance" between woodland and nonwoodland, a concept of environmental balance pursued briefly by Marsh with origins going as far back as John Evelyn's *Sylva* of 1664. The Wisconsin report had stated categorically that "a country destitute of forests as well as one entirely covered with them is only suited to the condition of a barbarous or semi-barbarous people. . . . It is only where a due proportion between the cultivated land and the forests is maintained that man can attain and enjoy his highest civilization.[33] This question of balance was one that troubled Americans, because if the beneficial environmental properties of the forests were proven (and many thought that they were) and if they were coupled with the concept that the forest was a part of a distinctive national heritage that was a part of "wilderness," then the forests were of supreme importance in American life and livelihood. These notions of forest influences, both moral/aesthetic and ecological, were ultimately to become major factors in the acceptance by the public of forestry in the United States.[34] How much of the U.S. forests had been cleared? How much forest did the United States have compared to other countries? Was America going the way of Spain or Italy, or was it more akin to, say, Germany, which was increasingly being looked upon as the ideal with its carefully balanced and expanding economy and well-tended forests? The rise and fall of nations fascinated Americans, who saw their nation at the start of a glorious ascendancy.

The Bureau of the Census took the lead. Whereas land had been divided simply into improved and unimproved in 1850 and 1860, the bureau decided in 1870 to separate "woodland" from "other unimproved" land, which could, of course, include prairie, marshland, or even abandoned fields. The result was that in 1870 it was calculated that there were 188.9 million acres of improved farmland, 159.3 million acres of woodland, and 59.5 million acres of other unimproved lands. The information was of limited value, however, as it excluded all land not in farms and represented only about a third of the territory of the coterminous United States. Yet it was the only information available, and the statistician of the Agriculture Department used it to make calculations on a state basis. Of the 2.3115 billion acres that made up the United States, 562 million acres were woodland, which meant that 24.3 percent of the country was wooded, a proportion that was clearly less than that for Norway, Sweden, and Russia with 66 percent, 60 percent, and 30.9 percent, respectively, and only a fraction less than Germany with 26.6 percent.[35] Whatever the "correct" balance was in absolute terms, those concerned about agricultural clearing heaved a sigh of relief; the United States was well up in the International League table, a position that was improved upon with a reworking of the figures in the 1875 report, when it was decided that there were 583 million acres of woodland, or 25.2 percent of the total land area, which placed the United States close to Germany.[36]

The same statistics on the amount of woodland in farms, together with a vast amount of information from government departments, individuals, and the Land Office, were used by William H. Brewer, a Yale botanist, in a map of "The Woodland and Forest Systems of the United States" for Francis Walker's *Statistical Atlas of the United States*, which came out as a companion volume to the census reports.[37] It showed that total, or even near total, woodland cover was rare and that most places had a partial cover only. Whether this was due to natural factors or to agricultural clearing was not evident, but Brewer was fairly certain that it was the result of "prodigal use and needless waste."

The pendulum of concern swung back and forth between local (state) and national forums. In 1871 Daniel Millikin of Hamilton, Ohio, wrote a prize essay for the Ohio

Agricultural Society on "The Best Practical Means of Preserving and Restoring the Forest of Ohio."[38] Drawing on the 1869 agricultural returns for the state, he drew attention to the fact that 14.4 million acres had been "improved" out of a total of some 24 million acres that were originally almost totally forested. The destruction had arisen from the "small regard, or indeed contempt," with which the pioneers had looked on the forest, an attitude that was the result of a long-drawn-out episode of farm making that went back not only to early Ohio pioneer colonizers but to the abodes of their ancestors on the eastern seaboard. The pioneer knew that, "as he made war on the woods, so he prospered." However, Millikin was convinced that it was time to make peace; timber depletion was affecting lumber and fuel supplies and prices, the imported supplies from Michigan and Wisconsin would soon, in his opinion, either run out or stop as they were diverted to the treeless West, then being settled, where people were willing to pay higher prices. In Ohio itself the local climate was being affected by excessive clearing, and the expanding telegraphic and railroad systems were consuming vast quantities of woodland, as were the implement and container makers and the makers of charcoal.

Writers in nearly every timber-growing state in the North were beginning to voice concern, not only about the agricultural inroads into the forest and the massive depredations made by the lumbering companies, which were becoming evident in the Lake States in particular,[39] but also about the attitude of people in general toward the country's resources. Maine had already appointed a commission on forest policy in 1869; Charles Sprague Sargent sounded warnings in Massachusetts a few years later; and in New York a commission was set up to inquire into the effects of lumbering on the watersheds in the Adirondacks, which fed the Hudson and other rivers, and particularly the Erie Canal. No action was taken immediately, but in 1883 no more forested state land could be sold, and the nucleus of the Adirondack State Park was in the making.[40] In Ohio, J. H. Klippart summed up the prevailing attitude as a case of "After the deluge." That, he continued, "is so unmistakeably written upon all developments of these resources as to cause a feeling of regret and sadness rather than a joyous anticipation for the condition in which we are transmitting nature's Bounties to future generations."[41]

The duty of governments

It was perhaps fortuitous that this growing mood of unease about the country's resources in the future, as well as the problems of settling the subhumid plains west of the ninety-sixth meridian, should have prevailed when Hough delivered an address on "The Duty of Governments in the Preservation of Forests" to the American Association for the Advancement of Science in Portland, Maine, in August 1873. Dr. Franklin B. Hough was a physician and an amateur botanist, geologist, and historian in his native New York. But his passion for statistics and the gathering of data found him engaged in census taking and compilation for the state of New York, and later, in 1870, he was made superintendent of the United States Census.[42] There was nothing particularly new in Hough's address; it drew heavily on Marsh and all the other writers who had gone before, and it reiterated the warnings about the multifaceted effects of clearing, particularly the threat of a decline of civilization and a balanced society. Yet Hough had responded to the sense of urgency that was abroad, and his address caught the right people at the right time.[43] So moved was the convention by the argument that it appointed a committee, which included Hough, George B. Emerson of Boston, and the botanists Asa Gray and William Brewer, to advise it on further action. The committee issued a report, which contained the warning on "the injuries that may result from the destruction of the forests, and the exhaustion of our supplies of timber," and which evolved into a recommendation to Congress that it create a federal commission of forestry to inquire into the amount and distribution of woodland, the rate of consumption, the influence of forests on climate, and the methods of forestry practiced in Europe.

But nothing was done until August 1876 when Congress authorized the establishment of a Division of Forestry within the Department of Agriculture and Hough was appointed the first agent of the division.[44] Despite the concern, however, interest in the impact of agriculture, which had been the initial impetus to the early warnings on depletion, began to wane as the 1870s progressed. Economy and society were still traditional enough for most people to feel that agricultural expansion was the natural – almost "God-given" – order of things and that it should be encouraged in order to promote the ideal of the self-sufficient family farm. Consequently, the majority of people did not care about, even if they were aware of, the inroads that agriculture had made on the forest cover. On the other hand, lumbering and industrial activities on the scale that they were assuming during the 1870s were intrusive and alien, went against the ethos of the nation, and now became a matter of concern. When statistics became available about the amount of wood cut, shipped, or received, as they did when the commercial transactions of the lumber companies, railroads, and ports were collected painstakingly by Hough in his *Report upon Forestry* for 1878, 1880, and 1882, the evidence of the "new" attack on the forests by lumbering seemed overwhelming.[45] The figures were almost incomprehensible in their millions and billions, and they lacked summary analysis and long time series; nevertheless, they were alarming enough in their magnitude and widespread nature to suggest that the forest was being cut down in large quantities.[46]

Carl Schurz, then secretary of the interior, recognized that Hough's work was not based on original field investigations, and as his department handled the census he was anxious to get some original documentation into the 1880 returns to provide a firmer basis for legislation. Through the good offices of the Smithsonian Institution, he had Charles Sargent appointed to prepare a report for the Tenth Census on the distribution and value of the forests based upon a systematic, geographical, and economic investigation. Sargent traveled throughout the continent over the next three years, and finally his massive *Report on the Forests of North America* was ready for publication in July 1883 and appeared the next year.[47] Nothing quite so extensive or comprehensive had appeared before, for it not only considered the distribution of various forest trees and forest density but also presented an analysis of various economic uses of the forest and its products, almost country by country. It was comprehensive, up-to-date, and authoritative, totally apolitical and devoid of practical recommendations; those Sargent was prepared to leave the legislators. The maps of clearing in various states were totally new, and the accounts of the inroads in various counties added up to a valuable picture of the destruction of the forest.[48]

Yet neither Hough's first three reports nor Sargent's report of 1884 made much reference to the impact of agriculture. In his first volume Hough tabulated the census returns of improved and unimproved land for 1850, 1860, and 1870, but because large areas of nonfarmed land were omitted he dismissed them because "calculation based upon them would . . . invoke much uncertainty." Sargent's emphasis was on industrial inroads; maps were produced only for the obvious lumbering states and showed areas of "cut pine" or "regions from which merchantable pine has been cut." The largely denuded and now predominantly agricultural states such as Ohio, Indiana, New York, and Illinois were ignored.[49]

Perhaps there was nothing unusual in all this; the census figures were difficult to manipulate and interpret correctly, agricultural clearing was a commonplace, "natural" occurrence, and lumbering and mining were the new menaces, with their "reckless, ruinous treatment of the forests." The official view was summed up neatly in a Forestry Division report of some years later:

> The farmer, having a portion of his farm only covered with trees, will almost naturally be prudent in the consumption of them and can easily be led to see that sweeping them off at once for cord wood or lumber, although it might put a desirable sum of money in his pocket, would be to lessen the amount and injure the quality of all the crops of his cultivated fields, and that in the end he would be the loser. But the lumberman is open to no such considerations. . . . His aim and interest are to level the trees and convert them into lumber as speedily as possible. He sees in the tree, or thinks he does, so much money, and he aims to secure it by the most rapid means.[50]

The faith in the agriculturalist was touching but almost totally misplaced; the analysis of the lumberman was largely correct but a trifle hypocritical, as it was the population at large that demanded the enormous quantities of cheap wood the lumberman obligingly supplied. But time had made the past enchanting. America was moving into an urbanized era, and sentimental glances backward to the golden age of rural living when all seemed simple and all seemed good were becoming increasingly common.

Ohio again

And there the matter of agricultural clearing lay until the publication of Egleston's *Report upon Forestry* in 1884, which contained an article by Hough on "The Decrease of Woodlands in Ohio."[51] This article is interesting because it not only relates to Ohio, which has been looked at in detail already, but also marks the revival of interest in agricultural impacts, which had lain dormant for over a decade. The article became widely quoted and was used as a case study, it being, in Hough's words, "a convenient means of studying with some care and in detail by counties the progress of clearings, and in an inverse proportion the exhaustion of its woodlands." It demands some attention.

The figures Hough used were culled from the county assessors' returns for Ohio, which recorded "improved," "unimproved woodland," and "unimproved other" lands annually. The assessors' returns were first tabulated by Klippart in his article, "The Condition of Agriculture in Ohio," published in 1876, in which he compared the first census return of 1853 with that of 1869. Almost without comment, Klippart presented lists of county figures for the two dates and calculated (often inaccurately) the reduction in the unimproved woodland as a percentage of all land returned, which of course was only land in farms. These figures were copied by Hough in 1884, and he also added the return for 1881. Then, despite his strictures of many years before about the difficulty of interpreting the census returns because they did not include all land in a county, and actually noting that the assessors' returns did not include all the land in the county, he proceeded to treat them as if they did. His county percentages and his elaborate statistical diagrams were hopelessly incorrect. Nevertheless, despite these shortcomings in computation and interpretations, his conclusions about the decrease of the woodland cover from about one-half in 1853 to about one-quarter in 1881 were roughly valid. The case against agriculture seemed proven, and the example of Ohio was an eye-opener for those who had never considered the impact of agriculture on the forests, and that included most people.

Concern about the quantity of land cleared always assumed that too much clearing was qualitatively bad for the land. Again, the case of Ohio and the evidence of the 1885 report on Ohio forests are informative. The indiscriminate clearing of steep slopes and poorly developed soils had led inevitably to erosion and permanent destruction of the land. This was particularly true in the dissected and hilly country of the Allegheny plateau in the southern portion of the state. Thus, when the hillsides of shale in Pike County were stripped of forest and cultivated they quickly eroded and had "soon become a [*sic*] nearly desert as any lands in the State ever become." Lowland and valley bottoms were cropped continuously for 50 years in Greene County, and total exhaustion of the soil was the result. In Belmont County the correspondent ventured to suggest that the losses were beginning to outweigh the gains in new farmland and that the adoption of a more scientific form of farming would bring far more profit than extending the area to be farmed by new clearing. In Columbiana County, a Mr. J. Jenkins of Winona spelled out the arguments more forcibly:

> We have gained pasture and farming lands on the hills. We have gained in the aggregate productions of the region and in population by extending the cleared land.
>
> On the other hand, we have subjected ourselves to cloud-bursts on the naked hills, which once drank in the descending floods in the porous woodland soil, but now the solid compact hill-sides throw off the floods into the valley, at times an irresistible torrent, carrying destruction and death in its course. We have lost in the destruction of untold millions of feet of timber, sold at prices low compared with present rates. . . . When we consider also the fact that, as a result of our interference with the restorative operations of nature, we have stamped out all prospect of the renewal of the forest-growths by close pasturing and continuous ploughing; it is, indeed, a question whether we have not lost instead of having gained by the change.

The idea that losses might be outstripping the gains, that the farming environment might be irreversibly damaged, was not new, but it was an argument that had now filtered down to the farmers. When the township correspondents for the 1885 Ohio report were asked if clearing had any "noticeable effect upon the climate, soil and water supply," and if so what, 381 people replied. The replies were in varying degrees of detail from a typical reply like one from Mr. E. B. Smith of Colton in Henry County – "Yes; warmer, less rainfall, and more productive soil" – to the extended consideration of John S. Patton of Hill Fork, Adams County:

> Yes, Very perceptible, especially on soil and water supply. Streams that twenty-five or thirty years ago were fair mill streams are now worthless for that purpose, rise rapidly and run out in a few hours. Thirty-five years ago this township had four grist and saw mills run by water, while none exist at present. The soil washes a great deal more than it did a number of years ago. On rolling or hilly land the farmer must be vigilant or his farm will be cut to pieces by washes. The climate is different, which I attribute to the clearing away of the forests.[52]

Of the 381 respondents, only 25 of them or a mere 6.7 percent denied that clearing had had an effect. The practical farmers were convinced that a deterioration had set in and that it was largely due to their own forest felling. Whether this realization was the result of long and hallowed folklore or independent observation or the extended propaganda campaign of nearly 20 years' duration is difficult to say. But it is ironical that, at the very time when it seemed as though the message of excessive clearing had finally reached and been accepted by the farming population, agricultural clearing more or less died as an issue. The focus of agricultural expansion shifted from the forested East to the Midwest and the largely treeless Plains, and there was even some reversion of cleared land to woodland in New England and New York. At the same time, the tree experts' and propagandists' attention was being diverted from agricultural clearing to the activities of mining and lumbering and, most importantly, the campaign to plant trees on the Plains in order to increase rainfall.

Planting in the Plains

By the end of the Civil War, Americans were confronting the problems of settling the Great Plains. The sporadic forays on the forest/prairie edge in Wisconsin, Illinois, and Kentucky in the 1820s became a total advance during the next few decades as hundreds of thousands of would-be farmers moved westward through the oak–savanna woodlands of southern Minnesota and the complex mosaic of bluestem prairie and oak–hickory forest that stretched through southern Iowa, most of Missouri, northern and western Arkansas, and into the easternmost edges of Kansas and Nebraska. In 1865 the frontier of agricultural settlement extended to about the ninety-sixth meridian, at a point where the woodland ran out, except along the main river courses.

Westward settlement had reached a critical point. It was no longer a question of deficient supplies of wood for housing, fencing, and fuel; they could always be supplied (admittedly at a price) from the expanding lumber mills of the pineries in the Lake States or the South. It was more basic than that; there were *no trees at all.* American farmers were accustomed to judging the value of land by the kind and density of trees it supported, and land devoid of trees was perceived to be unfit to grow crops. Moreover, in the collective imagination, the interior of the United States was destined to be the Garden of the World, but in order to establish itself in the vast open plains, "the myth of the garden had to confront and overcome another myth of exactly opposed meaning, although inferior in strength – the myth of the Great American Desert."[53]

Ever since the reports of the expeditions of Zebulon Pike in 1810, Stephen Long in 1823, and Lt. Gov. K. Warren in 1850, the emphasis on the aridity of the western plains and their uselessness for agriculture had become a common talking point and a warning for the westward-moving farmer. The plains were likened to the Sahara, Arabian and central Asian deserts, and they were clearly fit not for agricultural settlement but only perhaps for migratory pastoralists who followed their flocks and herds, like the Bedouins of Arabia and the Tartars of the Asian steppes.[54] But these people were uncivilized and perpetual outlaws from stable society. The inescapable conclusion seemed to be that if the Plains were not settled by agriculturalists then it was likely that life in the interior would degenerate into a semibarbarous state, with warlike, lawless brigands roaming the region and cutting off the East from the coastal West, which clearly did not fit into a concept of a unified country. In many ways the supposed degenerative influences of the Plains were akin to the supposed degenerative influences of the uncleared forest, as both were wild environments that reduced man to savagery. Only the "middle landscape" of cultivated, well-tended land with the correct proportions of arable land, pasture, and woodland could produce the civilizing influence that was necessary for law, order, and Christian morality, as well as the promotion of the traditional American frontier type, the independent yeoman farmer. The question of "balance" posed by Marsh, Lapham, and many others came to the fore again.

As settlement moved slowly up the valleys of the Platte and Kansas rivers, it seemed as though the myth of the desert and the specter of the menacing, Asiatic-like horsemen were being dispelled. The farther west the settlers moved the more they were convinced that rainfall and the associated vegetation types were all that distinguished the bleak upland of the western parts of Kansas and Nebraska from the fertile mixed prairie/forest of the eastern parts of those states. But how did one increase rainfall? The answer to that question lay in past forest destruction. If the clearing of the forest caused a decrease in rainfall and a drying out of land (and to many that relationship seemed accepted and proved), then would not the planting of trees increase the rainfall? It was a nice twist of the coin that Marsh had so carefully minted for the woodsman in the forested areas. Thus tree planting became, perhaps, the major answer out of many put forward for increasing the rainfall of the Plains.[55] The tree was both problem and solution; its absence was judged by many to be responsible for the limited precipitation, and its introduction was a remedy for this problem. The evidence in many states, particularly Illinois, Iowa, and Missouri, that the control and elimination of grassland fires resulted in trees rapidly encroaching on and placing grassland on the forest edge seemed added confirmation that trees were the natural inhabitants of the Plains.[56]

The literature developing the theme that trees and moisture supply were related was of long standing and was voluminous, and one can do no more than highlight parts of it. The idea went back a long way in western European literature. One early and clear statement was that of the comte de Buffon during the late eighteenth century, who, after talking about the "burning sands" of Arabia, assured his readers that "A single forest, however, in the midst of these parched deserts, would be sufficient to render them more temperate, to attract the waters from the atmosphere, to restore all the principles of fertility to the earth, and, of course, to make man, in those barren regions enjoy all the sweets of a temperate climate."[57]

Buffon's audacious assertion of forest influences had little impact at the time, but it did touch upon a rich vein of folk belief that began to gain prominence as the nineteenth century wore on, bolstered by the widespread assertion that the rainfall was already increasing. A notable example of this was the writing of the Missouri trader, Josiah Gregg, who published his *Commerce of the Prairies* in 1844. It was based on his experiences in the Plains during the previous decade. He wrote:

> The high plains seem too dry and lifeless to produce timber; yet might not the vicissitudes of nature operate a change likewise upon the seasons? Why may we not suppose that the genial influences of civilization – that extensive cultivation of the earth – might contribute to the multiplication of showers, as it certainly does of fountains? Or that the shady groves, as they advance upon the prairies, may have some effect upon the seasons? At least many old settlers maintain that the droughts are becoming less oppressive in the West. The people of New Mexico also assure us that the rains have much increased of latter years.[58]

In 1849, in the same year that Marsh corresponded with Asa Gray about the need for forest management, Daniel Lee, a medical doctor and amateur meteorologist, wrote a report for the United States Patent Office on "Agricultural Meteorology." Basing much of his information on the work of previous writers, particularly Alexander von Humboldt, Lee talked about the adverse effects of cutting down the trees, thus reducing evaporation potential and hence rainfall. But he did, at least by inference, point out that the planting of timber in regions where it had been removed or not even planted could only result in rainfall.[59]

Interest in rainmaking began to increase as settlers pushed further westward to the true prairie environment and with the construction of the Union Pacific Railroad through Nebraska to the West Coast. Pronouncements by Joseph Henry, the director of the Smithsonian Institution, that man could alter climate added considerable professional weight to the changing view of the desert based on folklore alone,[60] but, perversely, it was Marsh's *Man and Nature* (1864) that proved to be the real catalyst to the upsurge of writing by the rainmakers after the Civil War. *Man and Nature* was pillaged and distorted in order to provide proof that rainfall followed reforestation, whereas Marsh's emphasis had been, more often than not, that desiccation followed deforestation. Whereas Marsh had emphasized the negative, deleterious effects of forest clearing, his work was scoured for anything that would emphasize the positive, beneficial effects, and his hints that reforestation *might* restore the land to its former fertility and productivity in time were seized on.[61] Thus Marsh's remarkable synthesis of European literature, experience, and observation was bent to give a scientific respectability to the writings of the rainmakers.

One of the first to follow this up was Joseph Wilson, commissioner of the General Land Office, who urged that the Homestead Acts be amended so as to require every homesteader to plant a few hundred trees in order to encourage greater rainfall and the expansion of settlement "beyond the 98th meridian."[62] No official action was to be taken for a long time, but year after year Wilson thundered on in his annual reports about America's destiny in the Plains. In 1868 he wrote a lengthy article, "Observations . . . on Forest Culture." In general, he castigated his fellow countrymen for their "habits of indifference as to the value of trees" and stressed the need for forest conservation in the East and the establishment of forestry schools, which was sensible. But his argument for the establishment of "a thriving forest at some point west of the 100th meridian" to increase rainfall repeated all the old, garbled information, much of it based on a selective use of Marsh. With assurance he predicted that "If one-third the surface of the great plains were covered with forests there is every reason to believe that the climate would be greatly improved, the value of the whole area as a grazing country wonderfully enhanced, and the greater part of the soil would be susceptible of a high state of cultivation."[63]

Before the Land Office could formulate a land policy for the Plains that would incorporate these suggestions, Ferdinand V. Hayden of the Federal Geological and Geographical Survey lent his support to the Wilson thesis. In 1867 Hayden published his report on the geology of Nebraska, and in this, and in a number of other geological reports by him and his associates on other parts of the West, he was at pains to emphasize that the Great Desert was in retreat before the advance of the trees planted by the settlers. He wrote:

> It is believed . . . that the planting of ten or fifteen acres of forest-trees on each quarter-section will have the most important effect on the climate, equalizing and increasing the moisture and adding greatly to the fertility of the soil. The settlement of the country and the increase of timber has [*sic*] already changed for the better the climate of that portion of Nebraska lying along the Missouri [River], so that within the last twelve or fourteen years the rain has gradually increased in quantity and is more equally distributed throughout the year. I am confident that this change will continue to extend across the dry belt to the foot of the Rocky Mountains as the settlements extend and the forest trees are planted in proper quantities.[64]

This was reassuring and inspiring news, seemingly based on scientific observation and presumably carrying the stamp of federal government approval. Although Hayden claimed that his observations rested on scientific evidence, he was emotionally involved in the promotion of settlement on the Plains, and he acted more like an official "booster" than a scientist. He edited the reports of the surveys "more or less as one might edit a journal forming the mouthpiece of a literary coterie."[65] But, biased as his reports were, they found general favor. The government wanted to extend settlement westward; the railroad companies, like the Kansas Pacific, had already encouraged tree planting and wanted to promote settlement too; and most Americans wanted to see the desert removed from the map of the continent.

The years 1872 and 1873 seemed to be formative and critical, both locally and nationally, in the development of forest matters in the Plains. Booster literature continued, potential farmers were actively recruited in the East and in Europe, and belief in the rainfall/forest syndrome was mounting in popularity. In Nebraska, in particular, a group of academics, administrators, and politicians were in the rainmaking business in a big way. Among these was J. Sterling Morton, a pioneer farmer, newspaper editor, and later president of the American Forestry Association and secretary of agriculture in the second Cleveland administration (1893–7). Morton originated and successfully established Arbor Day, an idea that came from his almost obsessive desire to encourage the planting of fruit trees in Nebraska. In early January 1872 he elaborated his ideas:

> There is a beauty in a well ordered orchard which is a joy for ever. It is a blessing to him who plants it, and it perpetuates his name and memory, keeping it fresh as the fruit it bears long after he has ceased to live. There is comfort in a good orchard, in that it makes the new home more like the old home in the east, and with its thrifty growth and large luscious fruit sows contentment in the mind of a family as the clouds scatter the rain. Orchards are missionaries of culture and refinement.... Children reared among trees and flowers growing up with them will be better in mind and in heart, than children reared among hogs and cattle.[66]

Morton's thinking was eclectic and confused; orchards were the epitome of refinement and of the "cultured landcape," and they were also the bringers of rain, but whatever their exact purpose they were one more element in the strategy of what Walter Kollmorgen has called "The Woodman's Assault on the Domain of the Cattleman." "America [had] long embraced the symbol of the man leaning on the hoe while rejecting that of the man sitting on the horse," and the cattleman and all that he represented had to be kept out of "The Garden."[67]

Arbor Day was proclaimed on April 10 (later changed to April 22), immediately achieved great popularity, and was adopted in all states (except Delaware) as well as in many foreign countries, particularly Australia. The ceremony was soon construed to mean planting of any sort at all. Its observation spread to schools (first in Cincinnati in 1882) and became associated with a public holiday. It was accompanied by a formal ceremony, the singing of songs, and the solemn recitation of appropriate parts of the Bible. Arbor Day began to take on a quasi-religious tone that suggested the cleansing of the soul and the promotion of Christian values; it had an instant and widespread impact, which certainly focused attention on tree planting, although not, perhaps, on forestry. For some, the whole ethos of mysticism was anathema. "I have always considered 'Woodman, spare that tree,'" wrote Horace Greeley in his inimitable style, "just about the most mawkish bit of badly versified prose in our language, and will never guess how it should touch the sensibilities of anyone." However, he did not oppose the planting of trees, "mainly because I believe it will pay."[68]

Meanwhile, the earlier suggestions of Wilson and Hayden that homesteaders be required to plant trees on their sections, or even be rewarded with extra land if they complied with tree-planting regulations, now began to find wide public favor. Debates in Congress in 1872 had shown that body to be receptive to the idea, and it also had the backing of the commissioner of agriculture as well as President Grant.[69] The leading congressional advocate of tree planting was yet another Nebraskan, Phineas Hitchcock, who introduced the timber culture bill to the Senate. The object of his bill was quite clear: "to encourage the growth of timber, not merely for the benefit of the soil, not merely for the value of the timber itself, but for its influence upon the climate."[70] The act, which offered 160 acres to homesteaders who planted a designated percentage of their land to trees, was passed on March 13, 1873, and remained in operation for 18 years.

At a practical level the act promoted very little tree planting and soon became discredited because it was used by speculators on the eastern side of the Plains as a means of falsely acquiring land to sell back to pioneers and on the western side of the Plains by cattlemen to secure or enlarge their operations by buying up water frontages, hemming in other landholders, and generally enlarging their holdings. By 1882 the commissioner of the General Land Office called attention to the abuses of the act and said: "My information is that no trees are to be seen over vast regions of the country where the timber culture entries have been most numerous."[71]

He called for a total repeal of the law. It was not until 1891 that it was finally repealed, by which time 40 million acres were entered but only 9.8 million acres had been patented, over four-fifths in the five adjacent Plains States of Nebraska (2.5 million acres), Kansas (2 million acres), South Dakota (2.1 million acres), North Dakota (1.2 million acres), and Colorado (0.6 million acres). And it is doubtful if even a quarter of a million acres were ever planted to trees. On a more emotional level, however, the act was a legal and administrative expression of the prevailing sentiment of the country, whereby useless (for agriculture, that is, but not for grazing) land was converted to productive land by the planting of trees; "it was a beautiful dream," wrote Benjamin Hibbard, in his classic study of public land policies, "but the substance of the dream was for the most part as unreal as such visions usually are."[72]

Nevertheless, the rainmakers had been encouraged, and they continued to be, throughout the 1870s and 1880s, by the pronouncements of these and other Nebraskans. Outstanding among these was Samuel Aughey, professor of botany at the University of Nebraska, and Charles Dana Wilber. Aughey had worked with Hayden on the Nebraska survey and was thoroughly convinced of the rainfall/forest relationship. But he added a new idea. The geological investigation of fossil remains revealed that Nebraska might well have possessed a near tropical climate during the Tertiary epoch; therefore, he mused, could not Nebraska be made to revert to its former state?[73] Later he formulated another rainmaking theory: that it was cultivation that increased the absorptive power of the soil and "that has caused and continues to cause an increase in the rainfall of the state."[74] Charles Dana Wilber, a close friend and admirer of Aughey, seized on this pseudoscientific idea and gave it imaginative overtones, especially when he coined the terse epigram, "Rain follows the Plow." For Wilber, the plow became "the avant courier – the unerring prophet – the procuring cause" of the new march of settlement into the Plains.[75] The very act of cultivation ensured the success of cultivation. It was heady stuff.

Separately and together, Aughey and Wilber (with the assistance of many others of similar standing, such as Charles E. Bessey, the botanist Frank H. Snow, and Harvey Culberton) dedicated themselves to the cause of rolling back the Great American Desert and extinguishing it from the map and the landscape of North America.[76]

Later, other theories of rainfall making accompanied tree planting and plowing, such as, for example, the use of electrical currents, the firing of guns into clouds, and the influence of standing water and irrigated land on the atmosphere. But tree planting and plowing found widespread acceptance and prevailed the longest, and of these two, tree planting seemed to have the most basic and rational, worldwide acceptance, particularly in Europe. Therefore, the tree planters became the chief defenders of the faith and "the most stubborn to yield to critics."[77] Whenever people questioned whether rainfall could in any way be manipulated, as they did increasingly during the 1890s, the tree planters jumped in first with critical replies on the nature of the observations and the techniques of analysis.

It is against this background of the promotion of cultivation on the Plains of the West during the early 1870s, as well as the background of diminishing timber supplies and the deleterious effects of clearing in the East, that we must view Hough's address to the American Association for the Advancement of Science in August 1873 and the memorial of the association to Congress during the following year.[78] Both representations approved of the Timber Culture Act as a means of modifying the climate and saw it as a basic step in the campaign to heighten public awareness of and involvement of the government in forest issues.

But Hough was not merely a spokesman on forestry issues in general; he did contribute personally to thinking on the subject of rainfall and the forests. He echoed Marsh about the deleterious effects of forest destruction and the lessons of history and was convinced that the Plains had once been forested. Therefore, he felt justified in fully subscribing to the idea that to reforest the Plains would restore them and make them fertile for agriculture. For example, Hough's first *Report upon Forestry*, written in 1878 after he was appointed the agent to the newly formed Forestry Division, contained over 100 pages on the "Connection between Forests and Climate," and he later revealed his convictions in his *Elements of Forestry* (1882)[79] and in a sketch he drew for a Treasury Department report. By evaporation, shading, and cooling, forests increased the humidity and reduced the temperature, and if the dew point was reached, precipitation would result. He went on, in explanation of his diagram:

> I have often seen in a "dry spell" a passing cloud send down filaments of rain which vanished and dried up before they reached the ground. I have also seen, where this cloud passed a forest of considerable area, that these filaments of rain become a copious shower, which dried up when they came over the heated and naked fields beyond.[80]

The tradition of defending the rainfall/forest syndrome died hard in the Division of Forestry – whether through genuine conviction or a nagging sense that it was politically and popularly useful to stress this positive, beneficial side of forestry, which affected a large part of the continent, is not known. Hough's successor in the division, Nathaniel H. Egleston, was doubtful that the forests could increase rainfall but nevertheless insisted that no one could deny that they had a direct influence on the distribution of the rain that fell. From that, he concluded that one-fourth of the Plains should be planted to trees.[81] Bernhard Edward Fernow, the Prussian-born, Prussian-trained forester who came to America in 1876 and who succeeded Egleston, became a vocal defender of the concept even when it was being questioned more and more. Fernow answered his critics in his annual reports and elsewhere; for example, in 1888 he entered into an extended defense after Henry Gannett, geographer to the United States Geological Survey, presented a lengthy paper before the Washington Philosophical Society in which he said that he knew of no theoretical grounds on which the belief that trees increased precipitation could be based. In fact, Gannett's statistical analysis of the existing Plains rainfall records suggested to him that the presence of forest may even have decreased rainfall! Fernow replied, characteristically, that much more research was needed before the relationship could be disproved.[82] Yet, even that cautious approach was abandoned when he addressed the Nebraska State Board of Agriculture in January of 1891, when he went so far as to assert that the real problem of the Plains was that not enough trees had been planted. The settlers had failed the forests, not the forests the settlers.[83]

But in reality the argument was at an end. Climatic occurrences were going to triumph over popular opinion and official propaganda. The years 1889 and 1890 were dry years for large parts of the Plains, and although the following two years brought some relief from the drought, the widespread crop failure of the 1893 season, which was the worst on record, devastated the western Plains. Farmers either appealed for aid or simply abandoned their farms, up to 90 percent becoming derelict in parts of South Dakota and the population of western Kansas alone dropping from 96,000 to 76,000.[84] The Forestry Division had one more chance to put the case for forest influences in 1893 when it published a lengthy report on the topic, but the evidence was now mounting against the relationship. Analysis by meteorologists, such as Mark W. Harrington, chief of the Weather Bureau, and George E. Curtis, suggested that no evidence existed for a positive response, although Fernow still demurred and wanted further research.[85] In the meantime, the true character of the climatic variability of the western Plains had been demonstrated. They might not be a part of the Great American Desert, but they certainly could not be turned into the Garden of the World by the planting of trees.

Shelterbelts

By the end of the century, the argument that the forests influenced climate was almost dead. Its scientific basis was being questioned increasingly, and the inability of trees to modify the climate during the droughts of the 1890s had effectively dashed all hopes of success. But trees had another effect on the climate. It was a more modest and more local effect than producing rainfall, to be sure; nevertheless, it was one that was generally accepted, namely, shelter – protecting houses and crops from the force of the wind, stopping snowdrifts on highways and railroads, and providing shade for houses and stock during the summer.

The beneficial effects of shelterbelts were appreciated widely. In the once wooded northern and eastern parts of the country excessive clearing had left many areas exposed to "the biting and blighting effects" of the northerly winds. The reports of Thomas in New York (1860), Lapham in Wisconsin (1867), and Millikin in Ohio (1871) had all castigated the farmers for their thoughtless action and urged them either to retain trees as shelterbelts or to plant new ones if none existed in order to exclude whatever summer and winter winds affected those areas.[86]

In the more open Prairie States, the need for belts of trees was even more obvious, and the fact that shelterbelts, like woodlots, could be culled for fuel and fencing material made them doubly attractive. Consequently, trees were planted around most midwestern and western farms. Samuel Edwards of Bureau County, Illinois, pointed out that once the settlers had moved a few miles away from the trees "one of the first improvements was the planting of a grove, usually of locust."[87] In addition to planting groves, the prairie farmers were enthusiastic about the planting of "live" fences in this timberless region. The Osage-orange (*Maclura pomifera*) was very popular because of its speedy growth, sharp thorns, which made it an effective barrier to cattle, and thickness, which protected the crops from winds. It was cheap to establish and did not require expensive timber. Before 1880 the live hedges exceeded by a factor of two or three all other types of fences on the Plains, but when barbed wire and woven wire became readily available, with a labor requirement per running measure of only about one-sixth that of the hedge, very few hedges were planted.[88]

Shelterbelt planting was an individual, local, and small-scale operation and was not so susceptible to the grandiose plan and the continental result that tree planting for rainfall was supposed to give. One exception, however, was the idea of Joseph Konvalinka of New York, who in 1889 proposed that successive shelterbelts be planted across the northern boundary of the United States from the Rockies to the Atlantic in order to reduce the velocity of the cold winter winds, which dried out the soil and withered crops at the extremities of the seasons. Such a system of shelterbelts would "improve the climate of this country so as to excel that of Italy, since the United States enjoyed the advantage of a more southerly and favourable location."[89]

In his enthusiasm for his idea Konvalinka had conveniently forgotten that Italy had the barrier of the Alps to the north, which shielded it from much of the continental influences of the European winter. And there the matter lay for many years. The extravagant claims of the rainmakers had caused a reaction to the assertion that deliberate tree planting could purposefully change the climate, and skepticism reigned supreme.

The erosion of the soil by wind, however, was another matter, and it was, moreover, one that was slowly being recognized visually and appreciated theoretically. In Canada, the government launched an antierosion shelterbelt program in Manitoba in 1901 and issued seeds to farmers free of charge so that some 60 million trees were planted during the next 20 years.[90] In the United States similar concerted action came much later. Early experiments by Charles A. Keffer and William C. Hill of the Division of Forestry on suitable tree types and localities for planting came to nothing, and it took the droughts of 1910 and 1911 and massive dust storms to induce the government to take action. The Department of Agriculture was authorized to set up experimental stations – one at Ardmore in Fall River County, South Dakota, to experiment with drought-resistant grasses to stabilize the soil, and one at Mandan, North Dakota, to experiment specifically with tree planting in the Dakotas and the eastern parts of Wyoming and Montana. The numerous and quite extensive plantings from 1916 on showed that the establishment of shelterbelts was feasible and that they were generally beneficial to crop yields.[91]

Though the droughts of 1910–11 brought about a greater appreciation of the value of planting trees for windbreaks, almost inevitably, so it seems, they also brought about a slight resurgence of interest in the tree rainfall syndrome. People in the forestry profession flirted with this will-o'-the-wisp again. If only it could be proved what a boost to forestry it would be! Raphael Zon, "the architect of research in the U.S. Forest

Service" who had been investigating the effect of forests on flood control, had not completely disavowed belief in the forest/rainfall relationship. In his lengthy report of 1912, "*Forests and Water in the Light of Scientific Investigation*," he concluded that though forests did not prevent floods they could ameliorate their destructiveness and then went on to say that "Forests increase both the abundance and frequency of local precipitation over the areas they occupy as compared with that over adjoining unforested areas, amounting in some cases to more than 25 percent."[92] In the same year he addressed the Society of American Foresters on the relationship between the forests on the Atlantic plain and the humidity of the interior. But that was the end of it; the idea was now so discredited that Zon pursued it no further.[93]

However, Zon's influence in agriculturally related forestry matters was not at an end; he was always vitally concerned that research in forest matters should be with "immediate and practical problems." Therefore, when he was appointed director of the Great Lakes Forest Experimental Station in 1923 he had the opportunity to pursue many programs of research that seemed relevant to the surrounding areas, particularly to agriculture, such as the value of different sorts of land use in protecting watersheds, erosion rates, forest/water flow studies, and the effects of forest windbreaks.[94] It was the latter for which Zon will be the most remembered and which is most relevant to our story of man's relationship with trees. The droughts and dust storms of the 1930s brought an intense human discomfort and suffering in the Plains that became symbolic of the hardship of the depression years in the United States. Zon saw, rightly, that the answer to wind erosion was a coordinated approach involving land-use planning, grazing control, diversification of agriculture, water conservation, strip cropping and terracing, new cereals and soil-binding grasses, and the planting of shelterbelts. Whether Zon's stature and influence as a forester were such that he convinced the Roosevelt administration of the desirability of the shelterbelt program, as some have suggested, is not clear, but he definitely advocated that agriculture on the Plains, "although beset by many hardships, holds out promise of becoming thoroughly established if protective tree planting and other ameliorating measures are given their proper function on the land."[95] He was careful not to be too extravagant in his claims about the benefits the shelterbelt would bring:

> While the planting of trees will not change climatic conditions as a whole, it will alleviate or modify many unfavourable features of existing conditions, principally through the diminution of surface velocity of the wind by the successive forested strips. Also the trees undoubtedly would help in some situations to conserve soil moisture to some extent, at least by retarding run-off and evaporation, and lessening cooling.

The general effect would be not "the creation of more rainfall . . . but the more economic use and conservation of the available rainfall."[96]

By 1934 Congress had appropriated $528 million for drought relief, and a portion of it was used to establish the shelterbelt, which was to be about 100 miles wide and stretch for about 1,150 miles from the Canadian border to the Texas Panhandle, roughly astride the ninety-ninth meridian, covering approximately 1.28 million acres (Fig. 11.11, Plate

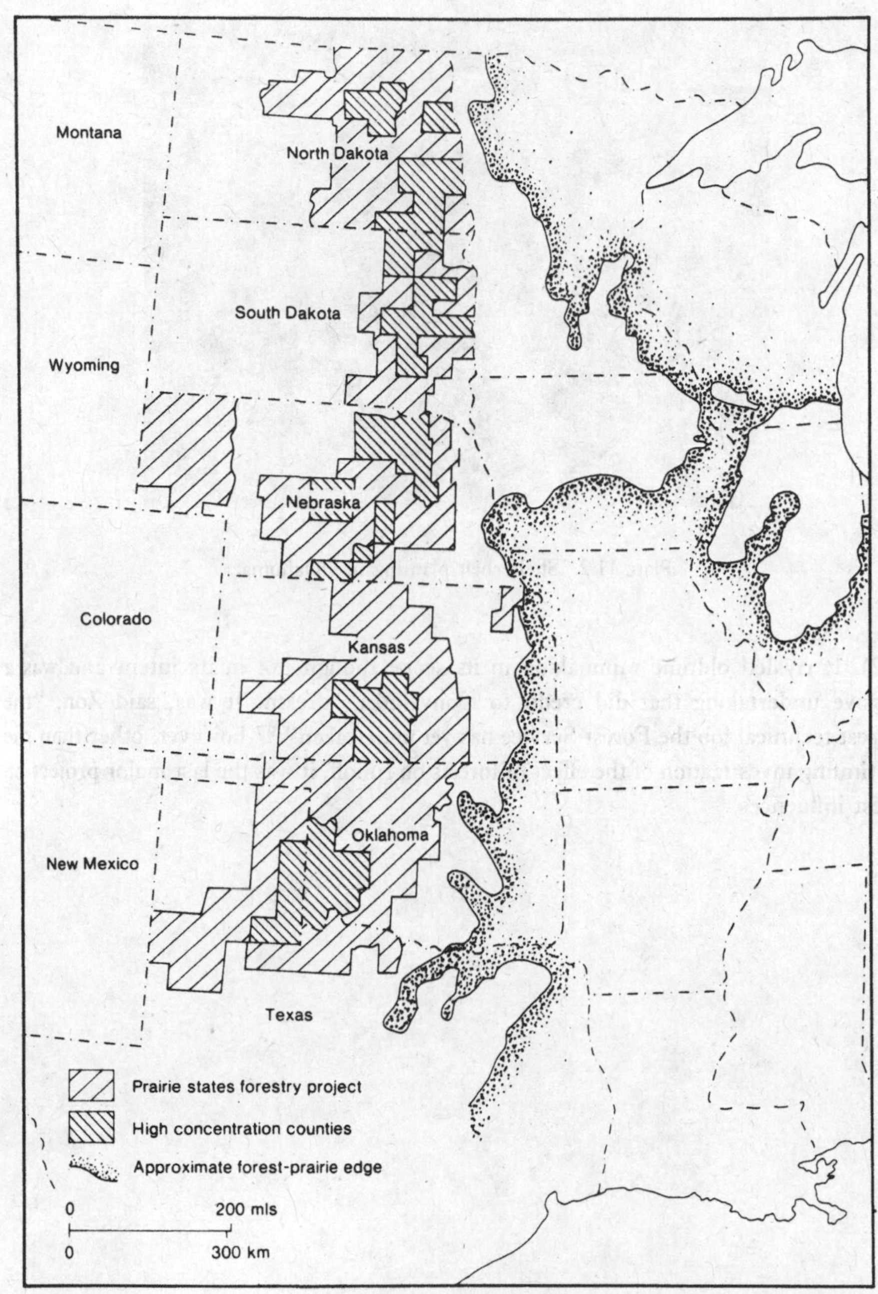

Figure 11.11. "Area of shelterbelt planting to date, Jan. 1941." (National Archives GRG 95, Map Series 94, A-65.)

Plate 11.2. Shelterbelt planting in Oklahoma.

11.2). It rivaled oldtime rainmaking in its scope though not in its intent and was a massive undertaking that did credit to Konvalinka's dream. It was, said Zon, "the biggest technical job the Forest Service has yet undertaken";[97] however, other than the continuing investigation of the effect of forests on runoff, it was the last major project on forest influences.

PART IV

Inquiry and concern: questions about the forest, 1870–1933

12

Preservation and management, 1870–1910

The subject of Forestry in our country opens a very wide field of enquiry embracing many particulars. . . . in fact, there is hardly any limit to the enquiries which arise in connection with this subject.

Nathaniel G. Egleston, in USDA, *Annual Report* (1883), 444

The rapidity with which this country is being stripped of its forests must alarm every thinking man.

Carl Schurz, U.S. Department of the Interior, *Annual Report* (1877), xvi

"The day is coming"

Most Americans knew firsthand, or could at least appreciate, the experience of John Thomas of Union Springs, New York, who wrote in 1864: "During a ride through a portion of the country that I had well known twenty years ago, but had not seen in this interval of time, I was struck with the havoc that had been made in the woods."[1] Everywhere the forest was being reduced in size due to the combined impacts of agricultural clearing and industrial felling and lumbering. The steady demand of the settler for new land and the insatiable appetite of the ever-expanding population for greater amounts of timber and timber-derived products created an immense drain on the forest resource. The conspicuous consumption of lumber and the even more conspicuous waste through milling and fire caused many thinking people to question the wisdom of continued felling and also the conventional assumption of the inexhaustibility of the resource that had always assumed such an important place in the life, livelihood, and landscape of the country.

When the Civil War was over people began to take stock of the situation. "The nation has slept because the gnawing of want has not awakened her," wrote Frederick Starr in 1865. "She has had plenty and to spare; but within thirty years she will be conscious that not only individual want is present, but that it comes to each from permanent national *famine* of wood." In the next year Andrew Fuller warned that "Every civilized nation feels more or less the need for an abundant supply of forest trees [and] America has felt this need the least." But "the day is coming, if not already here, when her people will look back with regret to the time when forests were wantonly destroyed."[2]

By the early 1870s the day had come. The concern was no longer academic – if it had ever been – and a new urgency entered into the discussion. Whereas in the past supplies of lumber and land had always seemed obtainable from the opening up of some new region, there now seemed to be a definite limit to the land available. The public domain was being closed. Admittedly, settlers were moving into the forested areas of the Rocky Mountains and the Pacific Coast, which contained about one-third of the known forest stand, but this was the "last timber frontier," and there was no more after that.[3] In the East the forest seemed in danger of being reduced to a pathetic remnant of its former glory, if not eliminated entirely. In 1872 the commissioner of agriculture commented that there was "much apprehension of the ultimate destruction of our forests and of a great scarcity of timber at no distant day," and five years later there seemed to him to be no way in which the country could avert "a timber famine."[4] Few people would have been prepared to dispute that forecast with much conviction.

The commissioner was merely voicing the concern that had been expressed for some time. During the late 1860s Marsh, Starr, Fuller, and Lapham had all raised the specter of a timber famine, and the debate about the "correct balance" between farming and forest land had arisen largely from the realization of the extent and the rate of forest clearing for agriculture. Admittedly, agricultural clearing was not the sole concern; there were anxieties over industrial inroads and also about the "interconnectedness" of different phenomena in the environment, which had been the very special concern of Marsh. For example, did the absence or presence of trees affect the climate? Were river flow and water supply, whether in the navigable rivers of the humid East or the irrigable rivers of the arid West, dependent on the degree of forest cover in the catchment? Was soil degradation a function of the vegetational cover? The worries were neither academic nor ethical but arose from a genuine concern about the ability of the nation to support itself and led directly to profound questions about the greatness of the nation.

The nature of the inquiry into forest depletion became more complicated as the years progressed, if only because it slid easily and naturally into a debate about forest preservation and conservation. In all its wonderful variety, richness, and size, the forest had many different users and admirers, each with his or her own view about its utilization and future. There were the lumbermen who wanted to run a profitable business, and there were the industries that depended on the lumber; there were the farmers who wanted the land the forest covered, and there were the domestic and industrial users who wanted the cordwood for fuel; there were also the arboriculturalists who wanted to make rain and the incipient foresters who wanted to protect streamflow and encourage rational management; and there was a federal government, which was becoming conscious that it owned and even had a responsibility for land that was timber-covered, and state governments and their settlers, who wanted to use those forests. Above all, there were a new set of users and admirers who emerged strongly for the first time during these years. There were those who found solace and peace in the vast solitude of the forests, and there were those who thought that the forests were a unique part of the American heritage that should be protected for future generations. Both the aesthetic and the patriotic preservationists inclined toward the idea that mankind needed the forest in order to "re-create" himself, and this concept of recreation in natural surroundings grew stronger as the century progressed and ultimately was to become perhaps the major mover in national forest policies.

Needless to say, the voices and interests and the attitudes and perceptions were not only numerous but often conflicting. Moreover, they often proceeded simultaneously along parallel and nonconverging tracks with little cooperation and were using completely different vocabularies and assumptions. Occasionally, some people like Hough, Fernow, and Greeley and some lumbermen moved from one sphere to another, but on the whole there was little coordination.

It is little wonder, therefore, that it is difficult to unravel successfully the events that occurred during the last three decades of the nineteenth century which led to the formulation of a national policy on the forests. The task is particularly difficult because the articulation of that evolving policy passed through various levels and various types of mouthpiece, ranging from stubbornly opinionated individuals like Pinchot, Muir, and Theodore Roosevelt, to professional societies and organizations, to governmental departments and bureaus, and to sectional interests in Congress. By about 1870, however, it seemed as though concern was crystallizing into six questions:

1. Are the forests worth preserving?
2. How should the forests be managed, and who should do it?
3. Who owns the forest?
4. How much forest is left?
5. How should the forest be protected?
6. How should the forest be used?

Perhaps with the exception of the all-important fourth question – How much forest is left? – these questions were rarely formulated in such clear terms because of the numerous and conflicting interests already mentioned and because the questions were so interrelated that the issues could not be separated so neatly. However, to sort out events into these six aspects of inquiry and concern does help us to understand the tangle of events during these years.

Are the forests worth preserving? 1870–90

Preservation was simply the reverse of destruction; therefore, if destruction was bad preservation must be good. The simplicity of the argument was overwhelming, but what to do beyond regretting the passing of the once great forest was a problem, especially in the light of the prevailing ethos of pioneer agricultural settlement and the role of private enterprise in industrial development. However, as shortages, higher prices, and timber thefts on public lands became more marked, and as romantic and patriotic notions of the value of the forests grew, a few Americans began to toy with the idea of deliberate preservation. But the ultimate goal was not that simple. Even if a portion of the forest were cordoned off from destructive exploitation, did one aim to preserve it from all subsequent use, totally untouched and uninjured, or did one aim to conserve it, that is, use it wisely? After all, the forest was a living and renewable resource; it could not stay the same one year after the other. This conflict of purpose always lurked just below the surface of the debate, and although it was rarely acknowledged in the early years, it became an issue of great importance by the end of the century.

The issue of preservation probably had five main sources of inspiration and action – the federal government in the western public lands, the newly created Forestry Division, the professional organizations, the nature lovers, and, paradoxically, the lumbermen – and all had different aims and emphases.

The governmental response: the public lands

Throughout its history the government of the United States had seemed content to dispose of the vast domain of public land that lay under its control as quickly as possible. There was a general and deeply rooted belief that the only thing that could be done with forested land was to clear and farm it. But, with the realization during the 1870s that there was a foreseeable limit to the amount of public land left and that much of it, particularly in the Lake States and the South, had already fallen into the hands of "the railroad magnate, the cattle king, the mining baron and the timber monarch"[5] rather than into the hands of the small family freeholder, steps began to be taken, however falteringly, to protect the land and its resources, particularly the timber and water.

The causes of concern were many. They arose from the haphazard and fraudulent application of old land laws as well as from the implementation of new but carelessly framed laws enacted during the late 1870s. In all cases, the inequities of the system of disposal seemed to emphasize the seriousness and ruthlessness of timber felling, particularly by the independent loggers, and to point the way toward the preservation of the forests. America's "last lumber frontier" in the Rocky Mountains and on the Pacific Coast seemed in danger of disappearing entirely.

The abuses were of many kinds but arose mainly from the rapid disposal of land under swampland grants, grants for education, and "the whole hydra-headed system of grants and concessions to railroads."[6] In addition, the Preemption, Commutation, Homestead, Desert Land, Timber Culture, Public Sales, and Private Entry laws were all available to lumbermen and speculators, who could achieve their greatest and quickest profits by stripping the land of its timber. In turn, easygoing and understaffed agencies often failed to notice the abuses or just turned a blind eye to them. A few examples must suffice. A determined speculator could manipulate the Preemption and Homestead acts, which were designed to aid the family farmer to establish himself. The Preemption Act did not require permanent residence, and although the Homestead Act did, its provisions could be sidestepped by commuting the claim to a preemptive entry after six months and then buying the land for $1.25 an acre. Both were later described as "mere pretensions for getting the timber." Similarly, the Timber Culture Act of 1873, which was designed to encourage planting, was actually used as a means of stripping timber. The claimant did not have to plant trees for four years, which gave him plenty of time to strip the land of any existing trees and graze it bare.[7]

Some of the most notorious abuses came under the grant systems to the railroads. The permission to cut timber for constructional purposes along the right-of-way went well beyond the immediate vicinity of the railroad and beyond the needs of actual construction, and timber was felled for sale wherever possible. The situation became worse in time because in 1880, Secretary of the Interior Henry Teller ruled that the phrase "adjacent to the line of road" applied to timber growing anywhere "within fifty miles of the track, and even beyond the terminus of the road." In Washington, some railroads were built in thickly timbered areas with no pretense of carrying passengers and only to strip off the trees for miles around, and the Northern Pacific, completed in July 1887, made surveys of its own designating the land for sale and cutting the timber growing there. What actually belonged to the railroad and what belonged to the government were open to considerable dispute.[8]

Examples of abuse of the homesteading laws and abuse by the railroads could be multiplied endlessly in locations as far apart as Wisconsin, Washington, and Alabama. Suffice it to say, the Department of the Interior was virtually powerless to control events in a practical way. All it could do was to deprecate what was happening, warn of the consequences, and campaign for regulation and preservation. In the first of many similar statements, the secretary of the interior warned in 1874 that the rapid destruction of the timber in the country, "especially that which is found on the public lands, is a source of great solicitude to all persons who have given the subject any consideration. If the destruction progresses in the future as rapidly as in the past, the timber lands of the Government will soon be denuded of everything that is valuable."[9]

The warnings about the ultimate effects of unregulated and illegal cutting in the West and about the need to preserve the forests came increasingly from two men, J. A. Williamson, commissioner of the General Land Office, and Carl Schurz, secretary of the interior.[10] Both realized that a real problem existed in that the timber on the public lands was held by the government but without protection, and yet, paradoxically, it was virtually given away to speculators who bent the law. But timber was a necessity in the West, just as it was elsewhere in the country, and no legal or honest way existed for the sale of timber at a reasonable price, a situation that almost inevitably promoted theft, graft, and dishonesty. The dilemma was made worse because most of the timberland was not of agricultural value and therefore could not be cleared legitimately as a prelude to farming. Sensing that there was little that they could do to change the law, Williamson and Schurz attempted to give the existing law some teeth. In an attempt to eliminate graft and dishonesty and to enforce the existing legislation, they issued a circular in 1877 that made agents appointed by the General Land Office in Washington, D.C., responsible for the public lands, rather than local land agents. Both of them urged the executive branch of government to enforce the existing laws; otherwise, said Schurz, "if we go on at the present rate, the supply of timber in the United States will, in less than twenty years, fall considerably short of our home necessities." In his lengthy and comprehensive report, Schurz urged the government to take effective measures not only to stop timber theft but also to "Preserve the forests still in its possession by keeping them under its control, and by so regulating the cutting and sale of timber . . . as to secure the renewal of the forest by natural growth and the careful preservation of the young timber."[11]

One immediate result of the circular was the introduction of a forestry bill by Sen. Preston B. Plumb of Kansas. It aimed to reserve valuable timberland from sale and concessionary grants but to dispose of the less valuable lands. Controlled sale of timber from reserved forests and the creation of a forestry corps were other proposals.[12] It was the first of a line of abortive bills for forest preservation that would lie strewn across the tables of both houses during the course of the next 20 years. The public at large and particularly their representatives were not ready for such measures.

Paradoxically, Schurz's circular made matters worse. With the more stringent enforcement of the existing legislation, demands for access to timber supplies became more vociferous. In 1878 the Free Timber Act was introduced by western representatives, and though it met genuine and reasonable local demands it provided scope for the excessive abuse of the forests. The act allowed residents of the eight Rocky Mountain States of Colorado, Nevada, New Mexico, Arizona, Utah, Wyoming, Idaho, and Montana the right to cut timber on "mineral lands, for building, agricultural, mining or other domestic purposes."[13] The act was badly drafted and open to various interpretations, so that the government and its agents were incapable of finding those people who felled timber in forests beyond the "mineral lands," even if the "mineral lands" could be defined satisfactorily. The situation was made worse in 1882 when Henry Teller of Colorado, now secretary of the interior and ever the friend of western interests, managed to broaden the scope of the act to allow "lumber dealers, mill owners and railroad contractors to cut timber even for commercial purposes, and for sale as well as for use."[14] Everywhere, individuals and companies stripped the timber without making payment to the government and with no regard for the consequences to the environment. For example, around the Comstock mines in Nevada thousands of acres were felled between 1870 and 1893, and here, as in all cases, the government was incapable of finding the culprits who had cut illegally or of sustaining a case that would lead to their prosecution. With only 15 agents (gradually increased to 55) throughout the West, the government could not even detect abuses, especially as the western territories (many soon to become states) were sympathetic to anyone residing within their boundaries and basically antagonistic to the federal government and its agents. Agitation mounted during the 1880s, as it was claimed government interference deprived local citizens of access to timber, which they now had to import from hundreds of miles away when it grew on their doorstep. A campaign was mounted for free timber for domestic purposes without regard for the mineral character of the land, and these western desires were translated into legislative forces in Congress when Montana, North and South Dakota, Washington, Idaho, and Wyoming achieved statehood between 1889 and 1890. Eventually, legislation (the Permit Act) to provide for the concession of free timber was passed early in 1891, with trifling opposition in the House.[15]

But the Free Timber Act was not the only means whereby the western settler got access to timber and the speculator wrought havoc in the forests. The Timber and Stone Act of 1878, which applied to the three western coastal states and Nevada, was judged to be "even more vicious in its provisions." It allowed the miner and the farmer to take timber by clearing the land in order to make good a claim or start a farm – "taking the timber necessary to support his improvements."

Whether such clearing was going to be incidental and subordinate to cultivation as the prime cause of entry was going to be difficult to determine, and many individuals and lumber companies abused the law. In the redwood district of Humboldt County in northern California, lumber companies hired, for a payment of $50, anyone they could muster in the streets of Eureka or from the crews of vessels that were tied up in port to apply for land and then pass it over to the company.[16]

The result of the Free Timber and the Timber and Stone acts was, in effect, to donate the timberlands to the inhabitants of the West. Secretary of the Interior Carl Schurz forecast that they would "stimulate a wasteful consumption beyond actual need and lead to wanton destruction. . . . in a few years the mountains of those states and territories will be stripped bare." He called for the repeal of the acts because they were more likely "to hasten the destruction of the forests . . . than to secure the preservation of them." By 1885, when Henry M. Teller, secretary of the interior at that time, could look back, he judged that the Timber and Stone Act in particular had operated "simply to promote the premature destruction of the forests."[17]

The story of timber abuse in the public lands and the early attempts to protect and preserve the forests had one further and final episode. Over the years it was claimed that the enforcement of the 1877 circular had been "cruel and harsh." Therefore, another act to liberalize timber getting was passed in 1879, which excused retrospectively those who had stolen timber from the public lands on payment of $1.25 per acre. It was, said Mr. Bragg, a congressman, "an act to license timber thieves on the public domain," and it cut the ground from under the feet of those who had striven to protect the forests.[18]

John Ise is probably correct when he says that it would be wrong to take a high moral attitude toward timber stealing during the early settlement days. Timber was a pioneer necessity, yet there was no general law for the purchase of timber on public land, and Schurz had the difficult task of enforcing the law while caring about the consequences of forest destruction. The congressionally inspired Public Lands Commission of 1879, consisting of John Wesley Powell, Thomas Donaldson, and A. T. Britton, was critical of the land laws and, in a general way, upheld Schurz's concern. But Schurz got nowhere; he was accused of overreacting and of attempting to transplant a German silviculturalist's aims and prohibitions to the different physical and social environment of the United States. In reality, the criticism was less that he was concerned with forests, he reminisced, than that "an officer of the government dared to interfere with the legitimate business of the country. Members of Congress came down on me." Schurz was one of the first public servants to attempt to change public opinion, something that was to be increasingly common in later years.[19]

During the next decade, further attempts were made by Commissioner of the General Land Office Sparks to recover fraudulently acquired land. He recovered nearly 3 million acres but lost his job in the process, and the system of getting timber on the public lands remained confused and arbitrary until the Forest Reserve Act of 1891, section 24 of which contained provision for the proclamation of the forest reserves. Only then did public opinion and governmental action coincide in the common pursuit of preserving the forests.[20]

The governmental response: the Division of Forestry

While the government and its agencies struggled to resolve the conflicting aims of use and preservation in the forests on the public land of the West, the problems of forest depletion on private land in the East gathered momentum. The issue of forest preservation was first posed squarely by Franklin B. Hough in his address 'on the Duty of Governments in the Preservation of Forests," which he delivered to the American Association for the Advancement of Science at its meeting in Portland, Maine, in the summer of 1872. The destruction he had witnessed and recorded in New York during the 1850s and 1860s and the subsequent shortages of constructional timber and fuel, together with the experience of the rest of the country since then, convinced him that either Congress or the state legislatures had to consider the subject of the "protection of forests and their cultivation, regulation and encouragement." Hough's argument for preservation was based largely on simple economic grounds: Timber entered into nearly every aspect of life, and demand for it was rising. In addition, the forest might affect rainfall, a consideration that was assuming significance because of the advance of settlement into the western Plains. He did not propose an elaborate plan of protection but was confident that the American people would "work out a practical system, adapted to our social organization and our general theory of laws."[21]

After a number of years, and an abortive attempt in the House to set up a commission of inquiry into forest destruction, Hough was appointed as the country's first forestry agent in 1876, with the principal task of collecting and disseminating information on American forestry. With a meager budget he was literally a one-man show. He collected data assiduously and compiled four source books of information: *Report upon Forestry,* which appeared in 1877, 1880, 1882, and 1884. The practical importance of his work was realized by Congress in May 1880 when it passed a bill to make forestry investigations a regular branch of the Department of Agriculture. It is significant that forestry was seen then and forever after as a logical accompaniment, result, and adjunct to agriculture, not as an adjunct to the public lands administration, which was a recognition of its national rather than its more purely western importance. Hough received a new commission to travel to Europe to investigate the sophisticated forest management systems there, and while he was away an elderly clergyman, Nathaniel H. Egleston, took over as chief. On his return, Hough prepared most of the fourth *Report upon Forestry* under the supervision of Egleston.[22] Like its predecessors, it contained copious statistics on production but also Hough's special report on the deforestation of Ohio, which struck Egleston with particular force as evidence of the need to preserve the forests. "The appearance of our forest supply is deceptive," he wrote; "we have the most urgent motives for the careful and economical use of what forests remain."[23]

It has been almost customary to deride the amateur and limited efforts of Hough and Egleston and their lack of action and success in forest preservation. Even Fernow, who succeeded them, commented somewhat disdainfully that "they lacked technical knowledge of the subject, the first a most assiduous worker being a writer of local histories and gatherer of statistics, the second a preacher," both constantly quarreling with each other as to where the credit for work should go.[24] Though this was true, the reality needs careful qualification. They were severely limited by their meager budgets and facilities. Hough had a salary of $2,000 and a desk; Egleston had a budget of $10,000 and employed three field agents. Their meager financial backing, and hence limited action, was only a reflection of the lack of interest at large in stopping forest destruction. Yet they performed adequately their informational and advisory functions, the collection of hard facts being the first step in raising public awareness and promoting action. Moreover, Hough's stubborn enthusiasm for topics of forest destruction and preservation had kept the issues alive during the 1870s when interest was at a low ebb. Perhaps more important still was the fact that they appreciated very quickly some important truths about forest preservation in America. For example, Egleston foresaw that a change in attitude toward the forests was an essential prerequisite for legislation. The time had come to regard them not as an obstruction in "the way of agriculture, . . . but as a great national treasure, to be cherished as such and made the most of. Our spendthrift treatment should be stopped, and a more thrifty course adopted."[25]

Having questioned the myth of the inexhaustibility of timber supplies and the view that timber was a cumbrance on the ground, Hough and Egleston argued that the government both could and should take positive action in the forested lands of the remaining public domain, withholding and reserving land for sale and stopping illegal cutting, thus showing that it "allots a value to its timber as well as to its land." Both realized, however, that the bulk of the nation's timberlands and the supply of timber were in the private hands of farmers and lumber companies, mainly in the East, and that given the prevailing sentiment toward property the government could do little to prevent the "reckless and wasteful" destruction of timber on these lands. With a fine sense of American pragmatism, Hough perceived that if that was so, then the government could merely "adopt measures for ascertaining whatever experience has taught or that experiment can ascertain, and it should make known to its citizens whatever is worth knowing in this line of enquiry."[26] He still looked, albeit in vain, for that "practical system adapted to our social organization and our general theory of law" – in other words, a specifically American solution to the problem of forest preservation. Nevertheless, unknown to him, it was emerging imperceptibly as public officials and lumbermen began to share common views about the difficulties of the lumber industry and the need for some sort of cooperative action.

In March 1886, just before Forestry became a recognized division of the Department of Agriculture, Egleston handed over the administration to Bernhard Fernow, a German-born, German-trained forester who had immigrated to the United States only 10 years before and who, since arriving, had worked assiduously, if ponderously, in the fledgling American Forest Congress to promote forestry interests.[27] Fernow judiciously rejected the idea of imposing federal regulatory powers over private landholders because he understood their difficulties. Instead, he wanted to emphasize "simply the example of a systematic and successful management" and the dispensation of technical information and "advice and guidance" to private landholders who sought it. He wanted to prove "the usefulness of the Forestry Division by actually making it useful."[28] Fernow, therefore, sought to influence major users of wood like the railroads, the naval stores industry, the tannin industry, as well as the lumbermen in general to cooperate with the federal government and to use wood economically and rationally in order to reduce consumption. In addition, extensive studies of the life history of individual trees were carried out by Dr. Charles Mohr on the eastern and southern pines and by Prof. V. M. Spaulding on the white pines of the Lake States, followed by an extensive program of laboratory testing on the technological quality and character (e.g., strength, durability, elasticity, etc.) of American forest trees, a comprehensive science that Fernow called "timber physics." With these moves he gained the confidence of the lumbermen.[29]

Despite Fernow's brilliant record and almost missionary zeal in promoting professional forestry in a private capacity before becoming chief of the division, his career never quite fulfilled the promise that might have been expected. Unlike his predecessors he was a highly technically trained forester but had the disadvantage, as he admitted, of being "a foreigner who had first to learn the limitations of democratic government," which was in such contrast to the authoritarian administration of his native country. Consequently, he was cautious in his actions and did not want to enter the political arena or provoke political argument. Perhaps it was with the criticism leveled at his compatriot, Schurz, in mind that he delayed taking up his appointment as chief "because a politician and not a foreigner was needed in the position." He retreated into the world he knew best, the world of scientific knowledge and experiment concerning trees, and the promotion of professional forestry. Forest preservation was a political issue, and though Fernow was quite prepared to articulate his views in a commission or a committee he had no liking for engaging in interbureau rivalry or for the wheeling and dealing of the Washington, D.C., political scene.[30]

Just as Fernow had belittled his predecessors, so Fernow was belittled by Pinchot, his successor, but much more seriously and with much less justice. Fernow's work on timber physics was likened to Hough's on statistics, and Pinchot wrote that Fernow's "bent was to lecture rather than to lead," thus by implication boosting Pinchot's claim to be in the van of forest conservation; he even claimed that he initiated conservation and that he invented the term as applied to natural resources in 1902.[31] But in justice to Fernow, while recognizing his limitations as a political activist and noting the prevailing political and social sentiments of the 1880s, one must conclude that he was no laggard in seeing the way ahead.

His thinking covered most of the practical and philosophical points. In his first annual report he had stressed that the remaining forested public lands should remain in the hands of the government, and he accurately forecast that, although these lands would not provide enough timber for future needs, the government could use them to demonstrate to private concerns the proper methods of forestry and at the same time "preserve intact the headwaters of mountain streams and regulate their flow." Again, Fernow realized that there were difficult practical and administrative problems in relation to forestry in the public lands. The Department of the Interior, which adminstered the public lands, needed a forestry bureau to manage the forests, but, conversely, his own Forestry Division needed land on which to carry out experiments and display its expertise; therefore, its contribution to management could only be limited.[32] He did not, however, have the political influence to merge the two. Finally, although Fernow emphasized "timber physics" in his attempt to influence major timber users, it must be stressed that he was not against nor did he neglect positive forest management, as his detractors have suggested.[33] "The expertise already gained by the Division," commented Fernow, "proved that preservation was not merely planting trees but also cutting them"; in short, "preservation . . . does not consist of leaving forests unused, but in securing their reproduction." At other times Fernow likened forestry to agriculture: "It is the application of superior knowledge and skill to produce wood crops, managing them in such a way that the largest amount of the best timber is produced in the shortest time [for] the largest income . . . without leaving out of consideration the other benefits accruing from . . . forest cover." Or, as he said another time, the forest should be treated "as a crop rather than a mine or quarry from which we take what is useful and then abandon it as waste."[34] These were all as clear statements of the conservationist's philosophy of positive management, as opposed to the preservationist's philosophy, as one could wish to find. Undoubtedly, Fernow would have "used" the forest if he had had the opportunity.

Professional responses and pressure groups

While the forces of preservation struggled to assert their view through governmental agencies during the 1870s and 1880s, two other groups added weight to the campaign: the professional foresters and the wilderness conservationists, the latter group ultimately becoming the driving force in forest preservation.

Naturally enough, professional forestry was concerned with many more issues than preservation alone. Indeed, the initial concern of Dr. John A. Warder, often regarded as the founder of professional forestry in the United States, was horticulture and tree planting. Warder retired from medical practice in 1855 to establish a 300-acre farm at North Bend, Ohio, which was, in fact, an experimental station for forestry. In 1873 he was appointed as a commissioner to the International Exhibition in Vienna, and that experience transformed his limited concept of forestry into a wider one entailing, among other things, the relationships between forest cover, rainfall, and river flow. In 1875 Warder convened a forestry congress in Chicago of lumbermen and foresters that led to the creation of the American Forestry Association, which merged in 1882 with the more virile and aggressive American Forest Congress, whose members included activists like Hough and Fernow.[35]

Partially in honor of Warder, the first congress was held in Cincinnati in April 1882. Baron Richard von Steuben, an influential German forester who was in the United States to participate in the centennial celebrations of the victory of Yorktown in which his ancestor had taken such an important part, fired the participants with concern about their country's forests. In a follow-up congress in Montreal during the next year, a draft constitution of the aims of the professional foresters was drawn up by Fernow, which included in its objectives "the conservation, management and renewal of the forest." During the following years the congress (changing its name to the American Forestry Association after 1889) met annually, and there is no doubt that Fernow was the driving force behind its initial success. He was secretary from the beginning until 1889, the chairman of the Executive Committee from 1888 to 1898, and editor of its *Proceedings* from 1885 to 1893, and thus he was able to mold opinion on forest policy in both a private and an official capacity.[36]

The association acted as a lobbying group on forest matters but did not pursue preservation very actively. Rather it concentrated on bread-and-butter issues to professional foresters, promoting forest education in state agricultural colleges and universities and eventually succeeding in establishing the first department of professional forestry at Cornell in 1898, of which Fernow became the head in the same year. At the same time, but quite unrelated, Dr. A. Schenck opened a private forestry school at the Vanderbilt estate, Biltmore, in North Carolina, which subsequently trained a number of influential people in forestry.[37]

While this branch of professional organization grew during the 1880s, the public at large was served by an increasing number of journals that diffused information on forestry matters. G. Bird Grinnell's *Forest and Stream* campaigned for forests as well as for wildlife protection, and Charles Sprague Sargent's *Garden and Forest*, which ran from 1888 to 1897, pursued forest matters generally, although with an arboricultural emphasis. These were not technical professional journals, however, and a number of those were floated during the 1880s. The *Journal of Forestry* edited by Hough lasted a year; the *Forestry Bulletin* published by Fernow lasted for three issues. Gradually, however, the obscure *New Jersey Forester* established itself as the main professional journal, and going through a succession of names, each one indicative of some contemporary concern and emphasis in forest policy and management, such as *The Forester*, *Forestry and Irrigation*, *Conservation*, *American Forestry*, *American Forests and Forest Life*, has ended up as the well-known *American Forests* of today.[38]

Because forestry needed to be made respectable and to establish itself as a serious subject, the early professionalism had leaned toward technical, economic, and educational matters, and it was therefore left largely to the amateurs to fight for the preservation of the forests. Since at least the 1850s, transcendentalist thinkers like Emerson and Thoreau had skillfully laid the philosophical groundwork for the movement to preserve wild country. Wild country, be it mountain, plain, or forest, afforded the freedom and solitude to explore one's untapped qualities and to experience ultimate truth and reality, stripped of and unsullied by the trappings of civilization. But the appreciation of the wilderness to which they contributed "led easily to sadness at its disappearance from the American scene." Like John Thomas of Union Springs, Thoreau deplored the steady and ruthless clearing of the forests of Maine, which seemed to him like a "war on the wilderness," and he was prompted to ask "why should not we . . . have our national preserves . . . our forests . . . not for idle sport or food, but for inspiration and our own true recreation?"[39]

This novel and even revolutionary intellectual stance, which regarded the forest as beneficial and worth preserving, was given a romantic visual expression by artists like George Catlin, Thomas Cole, Asher Durand, and Frederick Church, a poetic expression by William Cullen Bryant, and a literary expression in, for example, the novels of James Fenimore Cooper and the writing of Horace Greeley. In time, the American landscape was invested with a distinctiveness that was a source of pride for the nation and that created a sense of mysticism for the individual. Visits to Europe by Cole and Horace

Greeley and their reactions as Americans to the age-old domestication of the landscape convinced them of the need for America to preserve some of its wild space. Greeley urged his fellow Americans "to spare, preserve and cherish some portion of your primitive forests."[40] Though the American scenery had a wildness and sublimity that the European landscape did not possess, most travelers had to admit that the accretion of antiquities did give the European landscape a dimension and depth that the American landscape did not have. But, even here, the slowly changing appreciation of the forest was providing an answer to that deficiency, and Americans could ask, increasingly, "What monument erected by human skill could be compared to forests of gigantic trees, like those in Maine, Mississippi, and California, which have outlived the empire of Rome with all its grandeur and architecture, and may yet live after all modern nations have become lost in the history of the past?"[41]

There can be little doubt that an important reason for the changed attitude of the public to forests and the wilderness, as the century progressed, in general lay in the writings of John Muir. Muir echoed the thoughts of Emerson and Thoreau but "articulated them with an intensity and enthusiasm that commanded widespread attention." His books were "minor best sellers, and the nation's foremost periodicals competed for his essays." Born in Scotland and raised in Wisconsin, Muir settled in San Francisco in 1868 but moved quickly to the Sierra Nevada where he found an environment conducive to the development of his philosophy. He took an almost phenomenological approach to nature, in which the total immersion of self in one's surroundings produced a mystical experience and an appreciation of the rest of Creation. "The clearest way into the Universe," Muir wrote, "is through a forest wilderness," because there one could experience the totality of creation in an undisturbed way.[42]

The formative years of Muir's life coincided with the rising concern over forest depletion, and although he made some concessions to the concept of the wise use of undeveloped lands in his early writing his attitude hardened later into one of purist preservation, which fit in well with the ideas of the early agitators. In his article "God's First Temples: How Shall We Preserve Our Forests?" he first advocated public ownership of forests and their control by government. Muir's uncompromising championship of the wild was all-pervasive, and besides his two obvious achievements in the creation of an enlarged Yosemite National Park in 1890 and the formation of the Sierra Club in 1892 (of which he was president for 22 years), he also succeeded in permanently altering national attitudes in a way the transcendentalists never had.[43]

Before Muir's influence became widespread, the breathtaking national showpiece sites of Yosemite, Yellowstone, Grand Canyon, and, of course, Niagara had been described, painted, and talked about enough for most people to know about them and to feel that they should be preserved from spoliation. What was needed, wrote one commentator, was "A Great Park where the great natural elements of our national strength should be collected and preserved for the admiration and benefit of posterity."[44] Of course, the "great natural elements" were scattered across the continent, and they could not be gathered together in one spot. The answer lay in the creation of a number of parks. Sentiment was transformed into action when the federal government granted money to California in 1864 in order that the Yosemite Valley could be preserved as a park "for public use, resort, and recreation." This was followed in 1872 by the establishment, again by the federal government, of the Yellowstone National Park of over 2 million acres in northwest Wyoming. In both cases the impetus was preservation based on the desire to protect the natural curiosities and wonders – the waterfalls, geysers, and hot springs. In other words, it was largely a museum function. But by the mid-1880s there was a great change in public opinion. When the proposal to lay a mineral railway through Yellowstone Park came before Congress it was defeated convincingly. The majority of congressmen thought that not merely the hot springs but the total aesthetic qualities of the park, particularly its "sublime solitude," were worth preserving and should not be sacrificed for economic gain. Such an attitude was a practical manifestation of Muir's sort of thinking, which had crept up on people unawares.[45] By September 1890 an act to establish Yosemite National Park was passed with little discussion in either house.

Whatever forest had been preserved so far had been purely incidental to the preservation of the wonders of the newly discovered West. The true preservation of trees, however, first occurred in the older, settled areas of the East. In New York State the westward surge of settlement during the first couple of decades of the nineteenth century had swirled around and left relatively untouched the forests, lakes, and mountains of the Adirondacks. Only lumber companies had penetrated the area to any degree, locating their mills at the power-generation sites on the rivers that flowed in a radial pattern from the upland core.[46] However, as the urban population increased on the eastern seaboard, the Adirondacks attracted a different sort of attention as an area in which scenery, solitude, and sports opportunities abounded, ideas very much influenced by the popular local writing of S. H. Hammond (1857) and H. H. May (1869), so that by the late 1870s the Adirondacks were described by Verplanck Colvin as "the summer home of thousands, a public pleasure ground – a wilderness park for all intents and purposes."[47]

The recreational uses increased as the timber getting decreased, but then, from the mid-1860s onward, alarm was expressed by the recreationalists at the new wave of timber exploitation triggered by the demand for pulp and paper, which made even small trees and second-growth timber useful. In 1872 a New York State Park Commission was set up (one of the commissioners was Hough) and instructed to investigate the establishment of a forest reserve and public park. It recommended that no more state land be sold and that tax-delinquent land that became available as farming was abandoned and timber companies cut out the forests be retained by the state forever.[48] The argument was not purely aesthetic, dealing with what a later report called the "sentimental, pleasure seeking [and] sanitary character" of the Adirondacks; it also had a strong utilitarian thread running through it. Concern over the decreasing water levels in the Erie Canal and even the Hudson River and the consequent decrease in supply of water to industry, even the possible creation of a potential railroad monopoly to the West if the canal dried up, made the investigation essential, and they were potent arguments for protecting the forests on the watersheds. Even the sporting lobby, as evidenced by the magazine *Forest and Stream*, was prepared to play up the utilitarian argument during the 1870s and 1880s in order to achieve forest preservation. The active interest of the

New York Chamber of Commerce and Board of Trade and Transport ensured that further commissions and inquiries followed, and public opinion was further stimulated at mass meetings addressed by prominent citizens, including Carl Schurz.[49] By 1885 an act to create a forest preserve of 715,000 acres was passed, with the object of preserving the land in its wild state but also allowing some controlled logging. For the growing body of purist preservationists, however, this was just not good enough. In 1892 the "blue line" of the larger Adirondack State Park of 3 million acres was drawn. The wording of the act, as Roderick Nash pointed out, heralded a great change in motivation; for the first time a recreational rationale for forest preservation was elevated to "equal legal recognition with more practical arguments." In 1894 the formation of the park advanced yet a step further when an amendment to the state constitution stipulated that the preserved forest should be "forever kept as wild forest land" and that not only could the land not be sold but the timber on it could not be "sold, removed, or destroyed."[50]

It was during these years of the Adirondack Forest Preserve movement that the idea of an eastern national park, later changed to a national forest, was conceived. The two eastern areas upon which interest centered were the White Mountains of New Hampshire and the southern Appalachian region of North Carolina, both areas of "mountain scenery of the most picturesque and sublime nature." The problem was that the eastern lands, unlike the western lands, were in the hands of private individuals, and if the government wanted to set aside any of the land it first had to purchase it. Charles Sprague Sargent, Francis Parkman, and others campaigned vigorously from about 1892 onward, with the backing of local bodies like the North Carolina legislature and the West Virginia Academy of Sciences. However, the American Forestry Association under Fernow felt less able to back the proposals to preserve these eastern lands for scenic purposes, even with the added bait of a program of selected cutting, while vast areas of commercial forests on the public lands in the West were crying out for attention. Ultimately, the agitation for the eastern national forest foundered on the constitutionality of the federal purchase of land, and not until the Weeks Act of 1911, which legislated cooperation between state and federal governments for the purchase of private land for national forests, did anything happen.[51]

However, the lack of success with the eastern national forest was more than balanced by the success in the Adirondacks. Even more important, the idea was now firmly fixed in the political, public, and individual mind that wild forest land had its own value, which was a totally nonutilitarian argument of great portent for future action in the preservation of the forests (see Fig. 12.1).

The lumberman's lobby

By about 1890, the debates about preserving the forests on the public lands, the agitation of the Forestry Division and professional forestry, and the groundswell of opinion about the forests' recreational and wilderness potential were coming to a head and even coalescing. They had been played out against a background of real and imagined timber shortages, the migration of lumber companies across the continent to new areas of

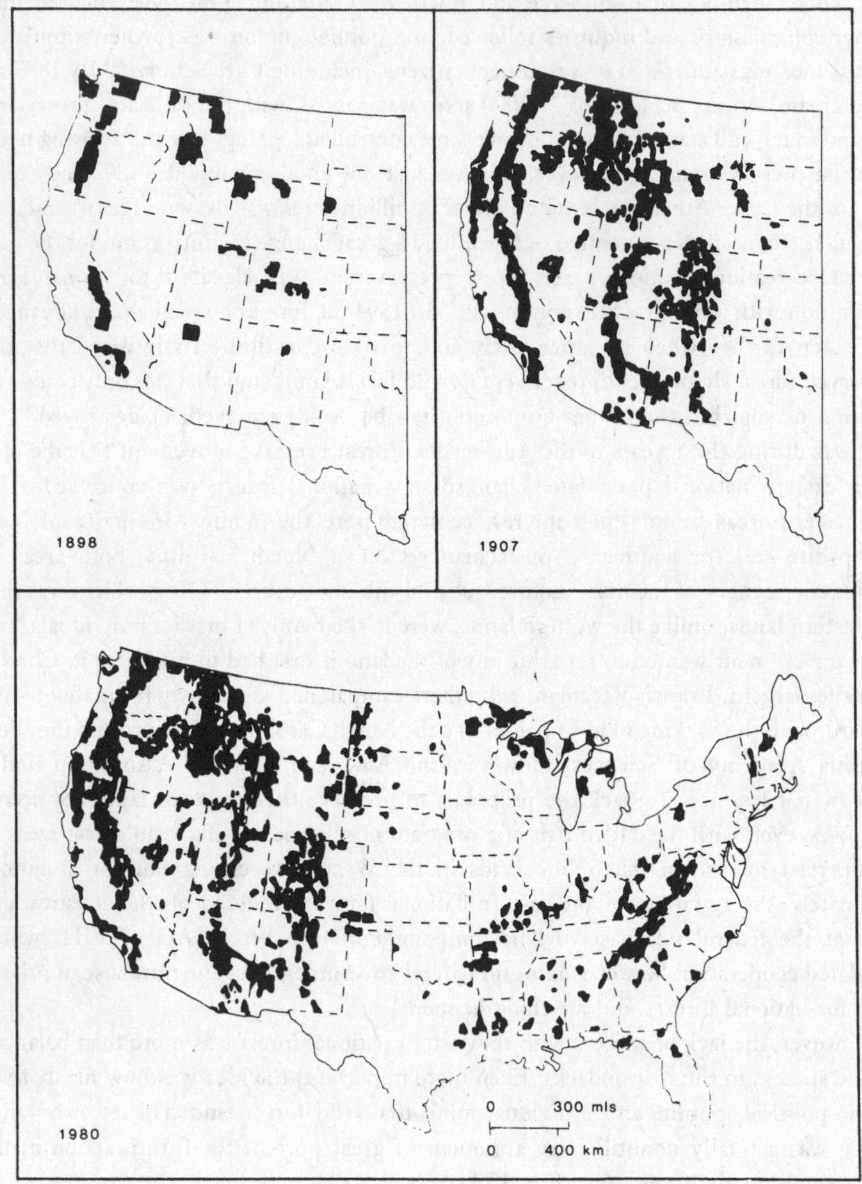

Figure 12.1. The national forests of the United States, 1898, 1907, and 1980. (National Archives GRG 95, container 108 and official maps.)

exploitation, and also a mounting concern over the management of watersheds, whether for the purposes of irrigation, navigation, water supply, or flood control. Increasingly, people looked to government to take the lead in solving these difficult and disputable issues because only it could transcend the problems of individual ownership and local interest and therefore be prepared to take the broad and long-term view so necessary with slow-growing trees. Hough, Schurz, Egleston, and Fernow had all urged greater government involvement and industry cooperation, but the real problem was how to break down the national pioneer mentality and the rationale of "progress" and also how to overcome sectional interests in Congress, particularly those of West versus East, in order to produce a majority to allow the passage of any conservation measure.

Rudyard Kipling summed up the situation well after a tour through the states in 1889:

> The great American nation, which individually never shuts the door behind its noble-self, very seldom attempts to put back anything that it has taken from Nature's shelves. It grabs all it can and moves on. But the moving on is nearly finished and the grabbing must stop; and then the Federal Government will have to establish a Woods and Forest Department the like of which was never seen in the world before.

All those people who had been accustomed to "hack, mangle and burn" as they pleased would inevitably object to the infringement of their rights, he thought. But what Kipling had not realized, and indeed few others had, was that the lumbermen were also seriously concerned during the 1880s about the chaotic conditions of their industry, where overproduction, glut, low prices, and close-downs plagued them, conditions they were unable to control, so that, paradoxically, the lumbermen, the object of the conservationists' wrath, were willing agents of conservation. William Robbins argues persuasively that in order to achieve their aims of order and stability in the "grabbing" they "capitalized upon public appeals for conservation and 'forestry'" by insisting that remedial measures were geared to the industry's needs and the specific requirements for stabilizing the market. Thus they "used conservation propaganda as a tool to achieve stability" and enmeshed the legislative and regulatory measures into their long-term aims. In this they were aided by a willing bureaucracy, eager to show that forestry both was practical and "paid off." For example, lumbermen, certainly the larger ones, welcomed the reservation of forest areas in the West because this eliminated or limited competition in that region. This is not to say that all lumbermen moved in "regimented concert or conspiratorial fashion," but they did tend to move in one direction in order to protect themselves. Only the small operators and individual settlers were to be adversely affected by future legislation and regulation.[52]

The forest reserves

Between 1870 and the end of the century well over 200 bills relating in some way to forestry and tree planting had been introduced into Congress, and though it would be possible to look at some of these in detail the exercise would be largely without meaning as none were successful in creating forest reserves or controlling cutting.[53] Nevertheless, the constant barrage of measures and proposals must have helped to raise the consciousness of Congress in forestry matters. Two important reports toward the end of the decade, however, cannot go without comment.

First, in 1887, the second Bulletin of the Division of Forestry, on "Forest Conditions of the Rocky Mountains," with a large-scale map of the forest areas of the ranges, was published.[54] Never before had such a detailed and accurate account of forest conditions over such a large area or such an accurate map of forested areas been available as a source for informed discussion and debate. Second, in the same year, Edward A. Bowers, lawyer, special agent of the Land Office, and later secretary of the American Forestry Association (1889–91) and assistant commissioner to the General Land Office, wrote a report entitled "Plan for the Management and Disposition of the Public Timberlands," which was transmitted to Congress the following year. It was a reasoned plea from an acknowledged expert in the field for the withdrawal from disposal of all timberland in the public domain, the setting up of a forestry bureau in the Department of the Interior, and the general tightening up of the timber-theft laws and penalties.[55] Later that year a comprehensive forestry bill was drafted by Bowers and Fernow and introduced by Senator Hale of Maine. The Hale bill incorporated many of Bowers's proposals, including the creation of "permanent forest reserves" (but by presidental decree) and the administration of the public land forests by a special forestry bureau, not in Bowers's own Department of Agriculture but in the Department of the Interior, a move that might minimize friction. The bill did not pass, but parts were incorporated into later successful bills, so that even if it did not become law it became at least "an instrument of education."[56]

Meanwhile, the American Forest Congress kept up a campaign of petitioning and memorializing presidents and Congresses almost annually, but its position was strengthened late in 1889 by Fernow's suggestion that the group join forces with the American Association for the Advacement of Science. A delegation of the most influential members of both organizations met in the Department of the Interior, apprised the secretary of the interior, John W. Noble, of their concern about "reservation and protection" of forested lands, and then took the somewhat unusual step of transmitting their resolutions directly to the president, not Congress, a move that would be significant in later years.[57]

The pressure seemed to be paying off. Rep. David Dunnell of Minnesota introduced a new bill (based on the Hale bill) in March 1890, but consideration was deferred to the next session of Congress until a subcommittee had worked on an overall land reform bill. By September 1890 Fernow could say confidently that "signs of an approaching change and fundamental improvement in the treatment of our forests are clearer than ever" and that there was "a very general awakening" of the need for forest preservation among diverse groups of people.[58] But when the change came it was in a most unexpected fashion, without fanfare and, one suspects, more by sheer chance than by a change of conviction. Earlier in 1890 the long-awaited bill to revise the land laws was introduced, entitled "A Bill to Repeal the Timber Culture Laws," which by now everyone recognized to be ineffective and open to grave abuse. The bill was passed and then went through a most complicated series of procedures until it landed in the Senate, which then passed it on to a conference committee. It was there that an obscure rider (section 24, subsequently known as the Forest Reserve Act) was attached (illegally, so it transpired), which stipulated that the president "May from time to time set apart and reserve [land] in any State or territory having public land bearing forests . . . wholly or in part covered with timber or undergrowth . . . as public reservations."[59] What Pinchot was later to call (quite correctly) "the most important legislation in the history of Forestry in America," and which had taken a generation of agitation, planning, and pressure to bring about, was hurried through Congress almost unnoticed.[60] It is true that a number of events had occurred which might have had some bearing on public opinion. For example, the Yosemite National Park bill had been passed a few months earlier, and the joint memorial from the American Forest Congress and the American Association for the Advancement of Science requesting that public lands be withdrawn from sale, "Until it can be determined what portion of them is situated within the natural watersheds of streams," had been read to Congress during the early stages of the "Bill to Repeal the Timber Culture Laws," along with a message from President Harrison urging adequate legislation "to the end that the rapid and needless destruction of our great forest areas may be prevented." But, given the previous record of Congress, it is unlikely that these events alone would have had much effect. From a close reading of the *Congressional Record* one gets the firm impression that Representative Payson, who was in charge of the bill, had steam-rollered it through a distracted and inattentive House, as repeated calls for it to be read, printed, and seen were denied, sidestepped, and evaded.[61] A peculiar combination of chance circumstances favored the passage of the bill and its rider, rather than a change in the attitude of Congress or the lobbying of any one of the prominent forest preservationists at the time. Whatever the reason, few people realized the full implications of the measure, which at last conferred power on the president to remove land from the public domain, thereby enabling the forests to be preserved. The aim of preservation had been achieved.

How should the forests be managed? 1890–1910

The 1891 act and the evidence of the subsequent proclamation of over 47 million acres of forest reserves during the next 10 years seemed to demonstrate conclusively that the forests were worth preserving (Fig. 12.2). But the battle was only half-won. Preservation alone was not enough. The forest was a living entity, and even if fences had been put around all the timber stands in the West they would not have remained unaltered. In attaining the 1891 act the forest reformers had given little thought to the management and use of the forests, let alone to the damage caused by fire and theft. Management, said Special Agent Baker to the American Forest Congress in Washington in 1884, was "a new field," and the focus of concern shifted gradually from the question "Are the forests worth preserving?" to "How should the forests be managed?" That, of course, was a deeply philosophical, as well as practical, question because of the often opposing views of the two basic schools of thought, preservation and conservation. There was the additional problem of who should administer the managing and protecting. Despite all these unsolved questions, the years from 1890 to 1910 were formative ones during which many questions were posed and few solved. Nevertheless, there was a sharpening of the focus on planned, long-range management or some form of sustained-yield concept, which contrasted markedly with the "grabbing and moving on" of the commercial lumbermen and pioneers described by Kipling.

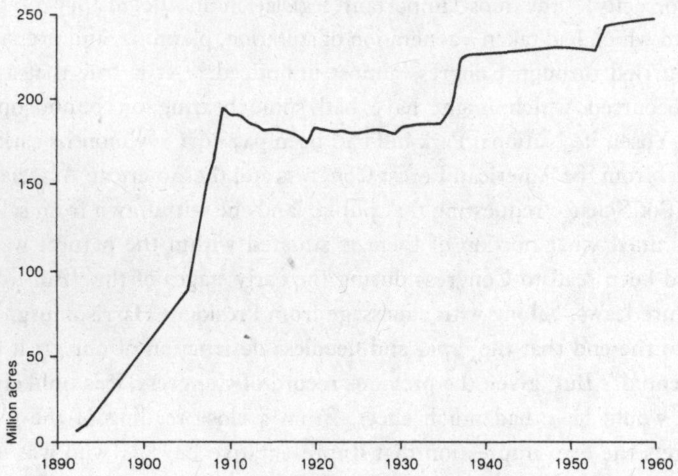

Figure 12.2. The area of national forest and other lands, 1890–1960. (U.S. Bureau of the Census, *Historical Statistics*, table L10.)

Enter Gifford Pinchot

Almost immediately after the passage of the 1891 act, President Harrison created 13 million acres of reserves (Fig. 12.2). His successor, Grover Cleveland, added another 4.5 million acres about a year later,[62] but the mounting concern expressed in many quarters about the lack of administration and management of the existing forests and reserves convinced Cleveland that it would be foolish to create any more reserves before Congress decided what to do.

The impasse coincided broadly with the return to America in 1891 of Gifford Pinchot, a young Yale graduate with a passion for forests. He had recently completed his training in Germany and France and had seen at first hand the managed forests of the Continent. Pinchot was a man of considerable financial means, considerable ability, and even greater enthusiasm and self-confidence; he was also a young man "in a hurry."[63] He was quietly contemptuous of Fernow, who had offered him a position as assistant chief of the division in 1890 (which he rejected), and even, in time, of Sargent, whose patronage he had accepted. In 1892 he became forester on the Vanderbilt estate at Biltmore, near Asheville, North Carolina, where he claimed to have initiated the first practical experiment in the United States in sustained-yield management, fire control, and selective logging practice and from where he undertook extensive work as a professional consultant to private forest owners.[64]

Pinchot saw the mounting debate on how to manage the forest reserves as an ideal opportunity to put into practice his knowledge and experience, and undoubtedly, too, as an opportunity to further his career. In fact, the initiative for forest matters shifted from Congress where it had matured during the 1880s to the president during the 1890s and eventually to Pinchot himself and his associates, who by sheer chance included in time President Theodore Roosevelt.

The elements and the personalities in the management debate were many and complex, but, to simplify matters, one could say that on the one hand there were the recreation and wilderness lovers, like John Muir and Robert Underwood Johnson, who saw the forest reserves mainly as extensions of the existing national parks, and on the other hand there were the people in the Department of the Interior, like Secretary Hoke Smith, who pressed for "legislation for a wide comprehensive system."[65] The Forestry Division, under Fernow, broadly endorsed the position of the Interior Department, stressing that the "multiplication of national parks . . . was not the intent of the law" of 1891, that it was "to prevent useless destruction of public property to provide benefit and revenue from the sale of forest products . . . and altogether to administer this valuable and much endangered resource for present and future benefit." But Fernow was conscious that because the division owned no forests it could do little in a practical sense to further management policies and could merely "suggest plans for a more rational treatment of our forest reserves."[66] Consequently, Fernow spent much of his time promoting, advising, and commenting on the many bills to manage the forests (particularly the Paddock bill of 1892) that went in and out of Congress with confusing speed and change. The division between the two approaches was not wholly clear-cut, however, for even as late as 1895 John Muir, the personification of the wilderness philosophy, conceded that it was "impossible . . . to stop at preservation" and that the forests must "be made to yield a sure harvest of timber while at the same time all their far-reaching uses may be maintained unimpaired."[67]

Another personality in the debate was Charles Sprague Sargent, the now influential and respected curator of the Arnold Arboretum, Harvard, and the author of the great 1884 *Report on the Forests of North America*. His philosophical stance in the debate is not altogether clear, but the fact that he resurrected his earlier proposal that the military should be responsible for the protection and administration of the forest reserves suggests that he wanted positive action and orderly administration of the reserves, however used.[68]

The conflicting ideas and lack of progress and the fact that the various versions of the bill proposed by Congressman Thomas R. McRae in 1893, "To Protect Forest Reservations," dealt only with the reserves and not with all the public land forests infuriated Pinchot, who held firmly to the "wise-use," conservationist approach. He realized that he had to bypass Fernow and the division and, to an extent, the Interior department, which held the forests, and this could only be done by creating a completely new and independent body that transcended the interbureau rivalry. Although he thought that Sargent was totally out of touch with modern forestry methods and concerned only with forest preservation, he recognized Sargent's stature in forestry, scientific, and governmental circles and therefore through Sargent agitated for a national forestry commission, a long-held ambition of those active in forestry circles. Together with Sargent, William A. Stiles, editor of Sargent's *Garden and Forest*, and Robert Underwood Johnson, editor of *Century Magazine*, he met to propose a forestry commission, which was to make a thorough study of the nation's forests and their management. Along the way they enlisted and gained the support of the American Forestry Association but failed to get the backing of Fernow, its most active and prestigious member. Perhaps true to his cautious nature in matters political, Fernow pinned his hopes on the current revision of the McRae bill then going through Congress and opposed the formation of the committee, which he thought would jeopardize the bill's success.[69]

With unflagging energy and subtle maneuvering, the pressure group prospered. Johnson wrote and printed an editorial for his *Century Magazine* entitled "Hope for Forests," which was a popular statement about the desirability of positive management. He sent a copy to the secretary of the interior, Hoke Smith, and asked him for his opinion, and Smith in turn approached Dr. Walcott Gibbs, head of the National Academy of Sciences, asking him to appoint an advisory committee to present views on the practicality and necessity of preserving timberlands, on the effect of forests on climate, soils, and water runoff, and on the legislation needed to prevent present malpractices. When the chairman of the National Academy of Science did appoint such a committee it was chaired by Sargent and included Gen. Henry L. Abbot, army engineer; Prof. William H. Brewer, the noted Harvard botanist and author of one of the first reports on forests in the country; Arnold Hague of the United States Geological Society; Alexander Agassiz of Yale; and Pinchot as secretary.[70]

Almost at once, Pinchot was critical of Sargent who he felt was dragging his feet, did not understand forestry, and was interested only in arboricultural ideas and concepts. During the summer of 1896, ever impatient and in a hurry, he went off about six weeks ahead of the committee to inspect the reserves of the West, where Muir met him. Given their different philosophies they got on surprisingly well and became close friends, perhaps because both had an all-consuming love of forests. Yet, as Nash points out, they were soon going to fall out completely because Pinchot's "ultimate loyalty was to civilization and forestry; Muir's to wilderness and preservation."[71]

When the committee returned from its western inspection internal disagreements arose between Sargent on the one hand and Pinchot and Hague on the other about the purpose and strategy to be pursued in the forthcoming report. Sargent wanted to move immediately to ask a sympathetic but soon outgoing President Cleveland to create more reserves. Pinchot and Hague wanted the committee to decide on a management strategy and to approach Congress to enact it before more forests were reserved. To Pinchot's dismay, Sargent's views prevailed, and Sargent wrote (probably unilaterally) to the secretary of the interior. On February 22, 1897, just about two weeks before leaving the White House, President Grover Cleveland established 13 new reserves in South Dakota, Montana, Wyoming, Idaho, Washington, Oregon, Utah, and California, containing over 21.2 million acres, with no mention of how they were to be administered, managed, or used.[72]

The outcry from the western states was as loud as it was unanimous. Not only were whole communities engulfed by some of the reserves, as were some 20,000 people in the Black Hills of South Dakota, and citizens everywhere debarred from getting timber, but the proclamation was regarded as further evidence of the ignorance and arrogance of the East concerning the legitimate interests of the West. Congressmen were now goaded into doing something quickly, and the result was the Forest Management Act, which was passed on June 4, 1897.[73] To Pinchot it was an "irony of fate" that "Sargent's greatest and most vital contribution to the forest movement in America came through his inexcusable handling of the National Forestry Commission,"[74] but there is plenty of evidence to suggest that, although Sargent was startled by the ferocity of the campaign from the West against the reserves, he had really played a clever game to force Congress into action after it had dithered for so long over various editions of the McRae bill. Fernow certainly saw it that way, and so did Frederick H. Newell, chief hydrographer to the U.S. Geological Survey, who was able to say very shortly after the bill was passed that the new reserves meant that "a greater number of persons would be induced by self-interest to urge upon Congress the enacting of laws which public interests alone have not been sufficient to bring about. The wisdom of this argument," he continued, "was seen in the demand from the West for immediate action on the part of Congress." On May 1, the National Forestry Commission presented its report to Congress, which, after concluding that the "segregations of these great bodies" of forest could not be withheld from the people of the West, recommended that they must be made to perform "their part in the economy of the nation."[75]

The battles in Congress over the exact wording and provisions of the bill for managing and preserving the existing forest reserves were complex and numerous, but almost inevitably, after the uproar in the West, the emphasis was on a distinctly utilitarian view of the forests, an emphasis probably lobbied for by the lumbermen. Basically, the secretary of the interior was given power to regulate "the occupancy and use" of the reserves, which could not be established except to "improve and protect the forests" or to "secure favourable conditions of water flow" or to "furnish a continuous supply of timber for the use and necessities of the citizens of the United States."[76] Therefore, although the wording of the bill failed to specifically grant full commercial use, it certainly did not deny it or exclude it. For example, although grazing was not explicitly recognized as a legitimate use of the forests it was not forbidden, nor was lumbering or the generation of hydroelectric power, now becoming a major consideration in the Western watersheds. In addition, mining and agriculture were allowed in lands considered "more valuable than forest lands."[77]

The act worked fairly well, with the exception of the fraud associated with the forest-lieu provisions. These were won by the West at the last minute in order to allow "settlers and owners" to exchange land within the newly reserved areas for equivalent land outside, but they were also used by large lumber companies and railroads (who were also owners) in order to rapidly fell timber on reserved lands and then trade in the cutover for timbered land elsewhere. Over 3 million acres were affected. Needless to say, the lumber companies supported these provisions, as they did the reserve system generally because it eliminated competing private ownership from public timberlands and boosted the prices of existing privately held land and its timber.[78] Taking the long view, however, one could say that with the 1891 Forest Reserve Act and the 1897 Forest Management Act the foundations had been laid for all subsequent action by establishing principles of land withdrawal and management in the forests of the West and, later, in some of the forests of the East.

Pinchot in control

Early in 1898 Fernow resigned as the chief of the Division of Forestry to become the first professor at the new School of Forestry opened at Cornell, and Pinchot was appointed in his place. The division was poorly staffed, poorly funded, and with the 1891 Reserve and 1897 Management acts the impetus for change had passed to the Department of the Interior. Fernow was "neither a money raiser nor a politician," wrote his biographer, Andrew Denny Rodgers, and "government work had been difficult for him."[79] In addition, although Fernow had had the satisfaction of seeing his endeavors in drafting and redrafting the many versions of the McRae bill incorporated into the Forest Management Act, he must have been disappointed at not having carried the American Forestry Association (almost his own creation) along with him in his opposition to the National Forestry Committee and, paradoxically, in not having been asked to be on the committee itself. Perhaps a more fundamental disappointment was the fact that, despite his constant statements that the forest needed to be managed and used and that the forester was "a sower as well as a reaper, a planter as well as a logger," he became stereotyped as a "timber physics" and preservationist man, an impression undoubtedly cultivated by Pinchot in the furtherance of his career.[80] In any case, Pinchot was impatient with advancing knowledge through research and wanted the resolution of policy problems achieved through his conservation campaign.

In all fairness, however, Pinchot achieved his reputation and success not merely by belittling others but rather by his own energy, initiative, insight, and adroitness at manipulating public and congressional opinion. He sensed brilliantly the mood of the times: that the country was concerned over the rapacious laissez-faire economy so well exemplified in the forest by the "cut out and get out" mode of destructive exploitation; that the western states were becoming anxious about the integrated development of watersheds to provide water for irrigation, domestic use, and even hydroelectric power; and that forests were an emotive issue, not only to the treelover, the rainmaker, the sportsman, and the wilderness preservationist and aesthete but also to the man-in-the-street who was concerned about increasing timber scarcity and rising prices as consumption of lumber climbed dizzily at the turn of the century to reach 46 billion b.f. per annum in 1906, an amount never attained before or since.[81] Perhaps most important of all, Pinchot realized that the loggers themselves were frightened by the dwindling supply of timber, the frenetic shifts of price and output, and the immense costs of moving increasingly expensive machinery and plant to the forest stands rather than managing the forests in such a way that the lumber could be brought continuously to the plant. Even they were aware by 1899 that moving from forested region to forested region across the continent had almost come to an end – there was no more land – and that the only answer from now on was the wise use and management of the forest that remained.

Initially, Pinchot concentrated all his energies on the western forests, where he thought he had a chance of successfully establishing his new management program prior to influencing private landholders in the East. He set himself three distinct yet obviously interrelated aims: to transfer the forest reserves from the Land Office to his own Division of Forestry (renamed the Forestry Bureau in 1901); to raise the status of the Forestry

Plate 12.1. Forest Service Chief Gifford Pinchot with President Theodore Roosevelt during an Inland Waterways Commission excursion on the Mississippi River, 1907.

Division; and to add even more forests to the reserves. Pinchot was fortunate that the period during which he embarked on these ambitious aims coincided with the presidency of Theodore Roosevelt, who fortuitously came to power after McKinley was assassinated late in 1901. Roosevelt was young, flamboyant, headstrong, and keenly interested in the outdoors – that "damned cowboy," as he was called later – and his personality and interests coincided perfectly with Pinchot's. They seemed to understand each other intuitively; both were imbued with the same crusading zeal to change the current way of looking at land and society in turn-of-the-century America, and both had gifts of personal leadership and "a dramatic instinct for using situations and setting the stage." It was claimed that Roosevelt consulted Pinchot more than any other person in Washington, and so Pinchot was able to rule like a benevolent autocrat (Plate 12.1).[82]

On taking office, Pinchot saw immediately that the Land Office consisted mainly of lawyers, had no trained foresters, and was, he argued, "hopelessly involved in a maze of political appointments, legalistic routines and personal favoritism." Moreover, if forestry was a biological science, then the management of the forests fell logically to the Department of Agriculture.[83] While assiduously cultivating and convincing congressmen and government officials, lumber and railroad executives, and even presidents of the justice of his case for the transfer, and of his wise-use policy of management, Pinchot attacked the Land Office for its incompetence and even abuse of the law over the handling of the forest-lieu provisions of the 1897 act. It had been, he claimed, planning for additional reserves in areas in which the railroad companies already had extensive holdings laid out in checkerboard fashion, and he also attacked the Land Office over its mishandling and the consequent fraud in Oregon and California where worthless school lands in the forest reserves were swapped for valuable timberland outside the designated forest areas.[84]

These charges were serious enough for the secretary of the interior, Ethan Hitchcock, to take the unprecedented step of recommending an investigation into his own department. Hitchcock was sympathetic to Pinchot and wanted forestry affairs transferred to the Department of Agriculture. The upshot was that the commissioner of the General Land Office, Binger Hermann, was dismissed, and then Roosevelt appointed a Public Lands Committee in 1903, consisting of William A. Richards, Hermann's successor; Frederick H. Newell, hydrographer of the U.S. Geological Survey, conservation advocate, and close friend of Pinchot; and Pinchot himself, to inquire into these charges of land abuse and related matters.[85]

Of the committee's many recommendations affecting the public lands, those relating to the forests were confined to rescinding the forest-lieu provisions and the segregation of agricultural lands in forested areas capable of being homesteaded. But the main effect of the report was that it made a strong factual and, at the same time, emotional plea for conservation that was in tune with public sentiment,[86] thus paving the way for the transfer of the forest reserves to the ever-growing Forestry Bureau, which Pinchot was building up carefully. It now consisted of 150 foresters with an annual budget of nearly $400,000, which compared to Fernow's two foresters, nine other employees, and $20,000 a few years before. By December 1904 events moved fast and far enough for Roosevelt to propose the transfer of the forests to Pinchot's bureau, and then a bill was introduced into the Senate on January 13, 1905.[87] With consummate timing and wide national publicity, Pinchot staged a grand session of the American Forestry Congress in Washington, D.C., from January 2 to January 6, with the expressed purpose of influencing opinion. It was a grand affair with nearly 400 delegates whose names read like a rollcall of the prominent and influential, not only in forestry but also in the lumber industry, railroads, grazing, irrigation, and government. Its *Proceedings* had little substance but were an exhortation to promote the wise use of the forests, with lumber barons like Weyerhaeuser and James Hill preaching the need for lumberman and forester to unite. Roosevelt opened the congress with a stirring address on "The Forest in the Life of a Nation," in which he first thanked Pinchot for his work in bringing about "a business view" of the forests and then posed the crucial question of "how best to combine use with preservation."[88] If there were any doubting congressmen, they would have been swept along by the occasion, and sure enough, on February 1 the forest reserves passed to the jurisdiction of Pinchot's Forestry Bureau. In the same year the bureau became the Forest Service, and a little later, in 1907, the forest reserves became the national forests.

Pinchot's ability to achieve his aim of transferring the forest reserves from the Land Office to the new bureau by 1905 was due to a combination of his personal qualities, which enabled him to forge friendships and alliances even with the lumbermen, and his foresight in pursuing policies during the previous seven years that boosted the value of the Forestry Bureau.

On taking over from Fernow, Pinchot realized that without any land his division could flourish only if foresters, highly trained in management, could convince private lumbermen that they had something of value to offer. The management program Pinchot wished to offer was that of the wise use of the forest, or the planned development of the reserves; if one took the long-term view, it was possible that the forests could be made to sustain a high yield (even increasing it) and provide a continuous supply of timber. Naturally, this appealed to the lumbermen, some of whom could move no longer but many of whom were sensible enough to realize that clear-cutting and the new technology of cable yarding and dragging was destructive and sensitive enough to realize that the image they presented of big business and exploitation made them unacceptable personally in forestry, public, and government circles during this progressive era. Many grasped the opportunity to alter the practice and alter the image, and by 1903 timber owners in the Pacific Northwest were said to have reached "a harmonious understanding" with Pinchot,[89] a state of affairs that eventually began to permeate the trade journals and was clearly manifest in the utterances at the 1905 Forestry Congress in Washington, D.C. Thus, as the lumbermen found themselves partners rather than opponents of government forestry under Pinchot, their traditional suspicion changed to support. In addition, in order to gain the lumbermen's support Pinchot actively approved of higher lumber prices because he thought that this would encourage sustained-yield forestry and lead to less wasteful milling and logging. Likewise, he was not opposed to tariffs on imported forest products or to forest monopolies, as proposed in the southern pine industry, because these might lead to a stabilized and innovative industry, always provided, of course, that the industry accepted his management program.

So successful was Pinchot's policy of providing management advice that very early on he found that the division could not meet the demands placed upon it, and it was because of this demonstrable need that the division was able to grow in size so rapidly.

The idea that the nation could have its forests and use them appealed initially to the preservationists like John Muir, but Muir broke away from Pinchot when the latter made it clear (as indeed Fernow had before him) that the management policy should have little to do with "beauty or pleasure" but should stress "economics." "Every other consideration was secondary," said Pinchot in later years. Conservation was simply "good business."[90]

Pinchot's efforts to bring the large and powerful corporations and interests as well as the Western congressmen behind him in the furtherance of his aims did not end with the lumbermen. As early as 1897 it was his insistence that the forest reserves be used for sheep grazing that led to Muir's break with him, and soon after he severely criticized the Land Office for restricting grazing. He actively discouraged sportsmen's associations, like the Boone and Crockett Club, from attempts to promote bills to reserve areas of the forests for game breeding, as this would conflict with getting the livestock owners on his side. After meeting western livestock owners at their annual convention in 1904 and convincing them that he would certainly allow grazing in the reserves if the reserves were transferred to him, Pinchot returned to Washington, where he could write, "Everything goes well and the transfer looks promising."[91]

Almost from the beginning of the campaign to reserve the forests a stock argument was that not only would the timber be preserved but flood control could be achieved and water conserved for irrigation. Roosevelt connected the two in his first annual message to Congress on December 3, 1901, when he fused the lack of water and dwindling timber supplies into one issue, saying that "The forest and water problems are perhaps the most vital internal questions of the United States." Pinchot and Newell, who were by now confidants of Roosevelt and the nucleus of a group of federal scientists and technicians dedicated to the search for greater efficiency and growth, had probably written this section of his speech. Pinchot realized that in one sense water was the most valuable natural resource to the westerners and that, whereas forest conservation was essentially an eastern movement resented in the West, irrigation was uniquely western. The policy of reservation of the forests pursued by the Land Office since 1897 did not appeal to the westerners, who complained that the forests were "locked up" and that they could not construct dams for irrigation, water supply, and hydroelectric power generation. Consequently, Pinchot strove to promote his multiple-use concept of natural resources, and although he never entirely convinced all westerners of his intentions they were more favorably disposed to seeing the forest reserves under the Forestry Division than under the Land Office of the Department of the Interior.[92]

In essence, then, the program of management that the Forestry Division offered under Pinchot was one of rational planning to promote efficient development of the natural resources of forest regions. In offering this comprehensive policy, Pinchot not only drew on the ideas of his predecessor, Fernow, but was not all that different from other federal scientists and administrators who saw the possibility of expansion, of opportunity and optimism for the future, with the application of science and planning to the environment. It was an approach that Hays has accurately and enduringly dubbed *Conservation and the Gospel of Efficiency*. Implicit in the policy were other concepts of "the most productive use for the permanent good of the whole people" and of "the greatest good to the greatest number for the longest time," both of these phrases being contained in a letter from the secretary of agriculture, James Wilson, to Pinchot, in which were set out the guiding principles of the new forest act, which were to become the guidelines for all subsequent forest legislation.[93]

Bit by bit Pinchot's policy of management emerged as a utilitarian, highest-use, multiple-use policy, which evolved as much, one suspects, from the practicalities of gaining support from various groups for the transfer of the reserves to his control as from any deeply held philosophy. What was his thinking and what was Roosevelt's are difficult to unravel, but almost inevitably the policy was going to create problems because no user could be wholly satisfied.

First, he had completely alienated the preservationists and aesthetes by his frankly utilitarian approach, and what they saw as collusion with the lumbermen. Second, the western congressmen were incensed yet again by the seemingly arbitrary proclamation of reserves by Roosevelt in 1907 on behalf of a federal government that behaved like "an alien landlord." As Pinchot was perceived, quite correctly, as Roosevelt's alter ego in forestry and in conservation matters generally, it was inevitable that he received as much criticism as did Roosevelt. But the resentment of the West went much further than that. Under the Forest Reserve Act the government was to sell its timber to the highest bidder. This discouraged the smaller logger, who either owned no land abutting the reserves or did not have transport and therefore did not feel that it was worthwhile to bid against the large lumber corporations and syndicates. It was estimated that Weyerhaeuser and other lumber companies in the Pacific Northwest acquired virtual control of about 4 million acres of prime timber that had been included in the 1907 forest reserves.

The misgivings of sections of the public about the partnership between the lumbermen and the Forest Service and about management practices were now compounded by suspicion about the distinct economic advantages the lumber companies had gained, particularly as they had welcomed the new reserves so enthusiastically. Increasingly, the fear crystallized that wise-use policies were creating new monopolies of lumbermen and other big-business interests that were excluding the little man and that the total conservation program was not the common man's program it was proclaimed to be. The image of the small-scale yeoman farmer was still strong enough and desirable enough in the Pacific Northwest during the opening years of this century to cause even figures like Pinchot and Roosevelt to stumble a bit. Third, the graziers who had previously enjoyed free grazing in the reserves objected to the new charges made on them, especially as it seemed the ideas had been Pinchot's originally. "Pinchot's ideas became Roosevelt's as far as the public domain was concerned," one commentator on the West has written, and Roosevelt confessed to Robert Underwood Johnson that "in all forestry matters I have put my conscience in the keeping of Gifford Pinchot."[94] Pinchot's ability to create a network of active, working relationships with foresters, lumbermen, statesmen, hydrographers, and other administrators was a unique gift that he alone possessed, and consequently he seemed to display an extraordinary talent for overcoming practical obstacles in order to achieve his objectives. Thus, given his close personal relationship with Theodore Roosevelt, it is little wonder that his final aim of adding to the forest reserves was not difficult. In 1901, when Roosevelt came to office, there were 54 forest reserves covering 60.1 million acres. After the 1905 act these were increased to 83 reserves covering 85.6 million acres, and in 1906, only two days before signing an act prohibiting the creation by the executive of any more forest reserves in the six northwestern states of Washington, Oregon, Idaho, Montana, Wyoming, and Colorado, Roosevelt boldly proclaimed 21 more reserves in these states, bringing the area affected to over 150 million acres (Figs. 12.1 and 12.2).[95]

Exit Gifford Pinchot

With the firm control of an enlarged national forest of about 150 million acres under a revamped and revitalized Forest Service, Pinchot had achieved his immediate aims. However, these aims had not been achieved without opposition and the creation of hard feelings, so that the promising beginning to a conservation program in the western forests, which was to be a forerunner of programs throughout the country, was not to be sustained. Pinchot's concept of the multiple use of the forest resource, coupled with the charging of fees to generate income for the service and to ensure that use would be "brought about in a thorough and businesslike manner," made it seem that the government was doling out timberland free of charge to the lumber companies, which was not far from the truth. Many heads of bureaus in Washington, D.C., and members of Congress were resentful of Pinchot's position of influence with Roosevelt and his seemingly cavalier conduct in bypassing accepted channels of communication.[96]

To the West it seemed as though Pinchot and Roosevelt saw the timber reserves as the panacea for all the ills of the forested lands. The governor of Washington declared that "Gifford Pinchot . . . has done more to retard the growth and development of the Northwest than any other man." Western anticonservationists (for that is what they had become) called a meeting in Denver in June 1907 to protest federal policies in the West. Pinchot, Newell, and Garfield, the new secretary of the interior, attended and tried to parry the blows by denying the creation of monopolies and declaring themselves in favor of the little man, but the graziers stacked the meeting and passed resolutions asking for virtual abolition of government control of the forest and grazing lands.[97]

As Roosevelt's term began to draw to a close, he and Pinchot looked for new ways of keeping the conservation movement alive and in the public eye. As much of the opposition had come from the West, often in the form of criticism of the East, it seemed reasonable and politic to extend the arguments to the eastern half of the country and so make conservation issues national issues. In a bid for popular support, Roosevelt had already called the first Inland Waterways Commission in March 1907 (which included Pinchot, WJ. McGee, and Frederick Newell, among others) to make a thorough study of the multiple use of the nation's waterways for the improvment of navigation, development of power, irrigation, flood and lowland protection, as well as the prevention of soil erosion in the uplands, thus demonstrating that multiple and wise use was applicable to all resources in all places. This was a popular move that caught the mood of the times, but it also demonstrated what was becoming abundantly clear, that Pinchot and his colleagues had appropriated the term "conservation" to mean wise and multiple use and had imbued it with social and even political overtones. The argument gradually but powerfully emerged that the conservation of resources was a moral duty that paralleled the conservation of the life of the nation, of national attitudes, and even of patriotism. It became an all-embracing concept for political action that, together with the "Square Deal" and "equalocracy," was a major plank in the platform of the Bull Moose campaign.[98]

It was while Pinchot was on this commission that he hit upon the idea of a Governors' Conference, a spectacular gathering of governors and experts at which the philosophy of conservation could be proclaimed before the people, bypassing an unwilling and even hostile Congress that could yet be bludgeoned with the "big stick" of opinion. Financing the conference largely out of his own pocket, Pinchot managed it adroitly, he and McGee writing most of the papers, steering the discussion along the lines they wanted, and keeping out such discordant and disruptive elements as John Muir and Robert Underwood Johnson, who might throw doubt on the whole wise-use policy. At the same time they welcomed the support of the representatives "of the public" like Andrew Carnegie and James J. Hill. It was difficult for Pinchot to get recommendations out of the conference, but he steered it toward the compilation of a National Resource Inventory, compiled by a National Resources Commission, which was harmless enough and could offend no one, being merely a national accounting of stock, just what any ordinary businessman would do. Conservation, after all, was merely sound business.[99]

During the latter half of 1908 the new National Conservation Commission began its meetings. It consisted of four sections, Water, Forests, Land, and Minerals, headed by four sound, trustworthy, and conservation-minded section leaders: McGee, Overton Price of the Forest Service, George Woodruff of the Department of the Interior, and Joseph A. Holmes of the Geological Survey. Pinchot was the overall chairman, and Henry Gannett, geographer to the U.S. Geological Survey, supervised the compilation of the commission's enormous three-volume report, which appeared in January 1909.[100]

But what might be called the "big hurrah" was over. Congress strangled the National Conservation Commission by failing to meet its expenses, other than those of the publication of its report. And with the passing of an amendment to the sundry civil bill proposed by Rep. James A. Tawney of Minnesota, it prohibited all bureaus from doing work for a commission appointed by a president without legislative permission. In future, the enormous amount of work done by various government bureaus for the Pinchots, McGees, and Newells of the Washington world would not be repeated.[101]

Roosevelt's successor, Taft, was not inclined to put himself out to defend the sometimes rash acts of his predecessor, especially as he even doubted their legality at times, or to weather the storm of animosity that Roosevelt had stirred up. As public sentiment changed, he distanced himself increasingly from those people in the administration intimately associated with the Roosevelt brand of conservation, with its alliances with business interest groups. Garfield was replaced by Richard Ballinger, a westerner with anticonservation sympathies, as secretary of the interior; Newell and Pinchot were too difficult to dislodge as no one was prepared to accept the positions knowing full well that the majority of the staff in the Reclamation Bureau and Forest Service owed such allegiance to their chiefs that the prosecution of the jobs would be impossible.[102]

Because forest policy demanded close cooperation and liaison between the Departments of Agriculture and the Interior, Pinchot continued his old practices of dealing directly with Interior subordinates, without reference to Ballinger. Ballinger objected and became increasingly protective about Interior affairs, so that he and Pinchot were in constant conflict during 1909, as, indeed, were he and Newell in the Bureau of Reclamation. Pinchot tried to influence Taft as he had Roosevelt, but Taft was lukewarm to Pinchot's approaches and lost patience with him, even instructing the Department of Agriculture to clamp down on Forest Service publicity, which seemed excessive. Eventually, Pinchot forced the issue with Taft by attempting to discredit Ballinger over the Alaska coal holdings case. The intricacies of the Ballinger–Pinchot affair are complex and still hotly debated, but the upshot was that Taft dismissed Pinchot for insurbordination early in 1910.[103]

Although Pinchot was now out of federal government, never to return, he continued to act as a goad, spur, and agitator for another 35 years in his capacities as leading spokesman for the Progressive party from 1910 to 1917 and as governor of Pennsylvania from 1923 to 1927 and from 1931 to 1935. He crusaded not only against the privatization plans of Ballinger but also for the transfer of the national parks to the Forest Service (the National Parks Service was eventually created as a special section of the Interior Department in 1916). But Pinchot was walking an intellectual and political tightrope. In order to whip up support for his conservation policies during 1907 and 1908 he gladly accepted the support of a new group who saw conservation as a moral crusade against materialism as well as for the preservation of resources. But Pinchot could not have it both ways. His wise-use policy was rank materialism, and he advocated the development of resources. Even as late as 1913, as star witness in the celebrated Hetch Hetchy inquiry, he argued for the "highest-use" and "wisest-use" concepts, meaning the building of the dam and the flooding of the valley, which showed how he was becoming increasingly out of touch with the trend, personified by John Muir, that would ultimately win the day. In the debates over forest management during the early 1920s his conservation proposals were becoming authoritarian, misdirected, and tainted by association with big business.

Pinchot's influence faded gradually after 1923 although it was never entirely dead, his utterances conditioned increasingly, Louis Peffer suggests, by "thwarted ambition, bitterness, and determination for revenge" that stemmed from his sense of martyred dismissal. He certainly rewrote the history of conservation in the self-seeking *Breaking New Ground* by deliberately claiming too much for himself and omitting or belittling the sterling work on the forests of predecessors like Hough, Schurz, and particularly Fernow and the contributions of his contemporaries in formulating irrigation and water use policy, particularly those of Frederick H. Newell and Marshall O. Leighton of the U.S. Geological Survey, Representative Francis G. Newlands of Nevada, and George H. Maxwell, a California water law specialist. His warped judgment of issues and his actions after his dismissal probably retarded the wider movement for forest conservation through the succeeding couple of decades and up until the time of the New Deal.[104] Nevertheless, for all his obvious faults and prejudices, the story of the United States forest would never be the same again. The story of the forest reserves, their management and enlargement, and of the creation of an efficient Forest Service was the story of Gifford Pinchot. He bridged the gap between the amateurism of the end of the nineteenth century and the professionalism of the twentieth century, a bridging that saw the forests move from the concern of a handful of thinking individuals to the concern of a nation.

13

Ownership, supply, protection, and use, 1900–1933

It has always seemed to me that the turn in our forest economy from the timber *mine* to the timber *crop* came in the decade ending with 1930.

William B. Greeley, *Forests and Men* (1951), 124

The passionate debates and the protracted deliberations about the preservation and the management of the forests during the years straddling the turn of the century had been played out against a backdrop of concern about the increasing concentration of ownership, the diminution of supply, and the devastation of fire – all the outcome of 40 years of uncontrolled and accelerating lumbering, land disposal, and agricultural clearing. The questions of who owned the forests, of how much forest was left, and of how the forests would be protected and used were rarely articulated as such, but they were never far from the minds and were sometimes on the lips of foresters and politicians as they struggled with the problems of preservation and management. Essentially, people took a pessimistic view, fueled by the gathering and publication for the first time of a mass of estimates of land still under forest and of the amount of standing timber that remained. These estimates varied wildly, but whether high or low, they were usually accepted as a part of some future scenario in which demand outstripped supply.

Nevertheless, all was not gloom. The years from 1900 to 1933 were years of intense debate, experiment, and adjustment concerning the issues of ownership, supply, and management. They were also years that bridged the transition between the the non-planning and non–state intervention of the previous 300 years of American settlement and the planning and direction of the last 50 years. They were in many ways the prelude to the modern forest.

Who owns the forest?

Despite the conversion by 1910 of some 150 million acres of public lands into national forests and other reserves, the government held slightly less than one-fifth of the standing timber of the United States. The bulk of the commercial forest still lay in private hands, overwhelmingly in the Pacific Northwest, the South, and the Lake States, and decisions taken in these areas were going to have more far-reaching effects on the timber cover and supply of the country than anything the federal government did in the national forests. In particular, there was a widely held suspicion that lumbermen were fixing prices and even influencing forest policy. Sen. Alfred B. Kittredge of South Dakota introduced a resolution in the Senate calling for an investigation of high timber prices, asserting that "the lumber Trust is king of combinations in restraint of trade."[1] This suspicion, together with the certainty that ownership was becoming concentrated in too few hands, prompted Congress to investigate the lumber industry to see if its practices were in the public interest, a popular move during the antitrust years of the Roosevelt administration, that was probably instigated directly by Roosevelt and Pinchot.[2]

The Bureau of Corporations began its investigations in 1906,[3] but it was not until 1913 and 1914 that it published its massive three-volume report. Its findings were not wholly unexpected, but they were nonetheless startling. Lumber companies had been able to accumulate large holdings in the Pacific Northwest, Idaho, California, Louisiana, Florida, and to a lesser extent in the Lake States by the adroit use and manipulation of the liberal land laws enacted during the previous 40 years, particularly the railroad and wagonroad grants and the swampland grants. Whereas about three-quarters of the standing timber was publicly owned in 1870, now four-fifths was privately owned. Of the estimated 2,200 billion b.f. of standing timber in private hands, the Bureau of Corporations surveyed and investigated 1,747 billion b.f., or four-fifths, primarily in the three major producing regions of the Pacific Northwest (including California and Idaho), the pine regions of the South, and the Lake States. In all, the investigation covered 88.6 million acres, 60 percent of which was owned by 1,851 companies.[4]

Of the total of standing timber investigated, 840 billion b.f. were owned by a mere 215 companies, the three largest of these carrying 238 billion b.f., or 11 percent of the country's timber (Table 13.1). The three largest companies were located in the Pacific Northwest. The Weyerhaeuser Company owned 96 billion b.f. (on about 2 million acres); the Northern Pacific Railway, 36 billion b.f. (on about 3 million acres); and the Southern Pacific Railroad, some 106 billion b.f. (on about 4.5 million acres, 2 million of which it subsequently forfeited in 1916). These massive holdings were acquired through federal railroad grants, 80 percent of Weyerhaeuser's land having been bought from the Northern Pacific Railway in 1900 and 219,000 acres having been secured from special forest-lieu selections created in the formation of the Mount Rainier National Park.[5] The staggering magnitude of these three holdings alone is revealed when one realizes that in the aggregate they represent more than double the standing timber growing in the three Lake States.

The concentration was greater than the figures suggest because of the ability of the large landholders to control blocks of smaller, scattered holdings, especially the checkerboard of odd sections bequeathed by government to the Northern Pacific Railway in Washington and the New Orleans Pacific grant in Louisiana (see Figs. 9.5 and 8.6).[6] Similarly, the interweaving ownership, interlocking directorships, and ownership of subsidiary companies, as well as the control of the means of transport in and out of the forests, meant that the parent companies exercised an invisible control over large areas of the forest. For example, if all Weyerhaeuser's interests and those of his stockholders and directors in other timber companies were added to those of the main company, his control of the forest resources would rise from 96 billion b.f. to 292 billion b.f. (see Table 13.2), somewhat more than half the timber to be found in the national forests.

Table 13.1. *Concentration of ownership of standing timber, Pacific Northwest, South, and Lake States, c. 1913*

Grouping (bill. b.f.)	No. of holders PNW	South	Lake	Total	Timber (bill. b.f.)
Over 25	3	—	—	3	238
13–25	5	—	—	5	102
5–13	12	3		15	120
3.5–5	18	8	4	30	115
2–3.5	26	18	6	50	116
1–2	67	38	7	112	149
0.5–1	86	92	27	205	132
0.25–0.5	96	148	34	278	96
0.125–0.25	176	251	69	496	85
0.060–0.125	222	367	68	657	56
Total	711	925	215	1,851	1,209
Below 0.060	n.a.	n.a.	n.a.	n.a.	538
Total					1,747

Source: U.S. Bureau of Corporations, *The Lumber Industry*, 1, pt 1: 21, 29, 30, 31. (These figures on the concentration of ownership differ slightly from those in the text of the bureau's report.)

Table 13.2. *The wider Weyerhaeuser interests, c. 1913 (billion b.f.)*

	PNW	South	Lake	Total
A. Weyerhaeuser Timber Co.	95.7	—	—	95.7
B. Companies in which Mr. W. has an interest	35.3	7.3	6.9	49.5
C. Companies (other than B) in which 1 or 2 stockholders in A have interests	48.3	14.7	2.1	65.1
D. Companies (other than B or C) with which one or more associates of Mr. W. are associated	49.2	26.7	5.7	81.6
Total	228.5	48.7	14.7	291.9

Source: U.S. Bureau of Corporations, *The Lumber Industry*, 1, pt. 1: 100–2.

In addition to the Big Three in the Pacific Northwest, three other companies owned over 1.5 million acres, nine owned between 0.5 and 1.5 million acres, and 40 owned between 300,000 and 500,000 acres. But with the exception of the Big Three, large landholdings did not necessarily mean large timber stumpages, as differential growth rates favored the Pacific Northwest over all other regions. For example, the next five biggest owners of standing timber were Charles A. Smith and Thomas B. Walker (both of Minneapolis), who had holdings in Oregon and California; N. P. and W. E. Wheeler of Endurance, Pennsylvania, with holdings in California and Oregon; A. B. Hammond of New Jersey, with holdings throughout the West; and the Chicago, Milwaukee and St. Paul Railroad, with holdings in Washington and Idaho. The holdings of these five were 344, 771, 396, 300, and 395 thousand acres respectively, but in the stumpage-size league tables they were only thirtieth, tenth, twenty-second, twenty-fifth, and twenty-first.

At the other end of the scale, however, there were a large number of smaller holdings, over 20,000 of less than 1 billion b.f. in Washington and Oregon alone. In the hardwood areas of the East anywhere between 143 million and 153 million acres of farm woodlots were distributed among 4.6 million owners and contained 360 billion b.f.

The bureau was convinced that the policy of the great companies was to cut as little as possible in order to reserve the stumpage and get a better price in the future. "Many of the men who are protesting against conservation and the national forest system because of the 'tying up' of natural resources," it concluded, "are themselves deliberately tying them up far more effectively for private gain." The purchases were not made because of any particular economy accruing to the large manufacturing plants, reasoned the bureau, because even the largest sawmills of the early twentieth century cut less than 0.5 percent of the total annual output of lumber or less than 0.05 percent of the annual cut.[7] In any case, the economics of manufacturing effected by enlarging the capacity of the plant was offset or counterbalanced by the increased cost of transporting the bulky raw material of lumber from an ever-increasing distance. At a certain point, enlargement of the plant merely entailed the duplication of machinery. There was only one conclusion: Large-scale purchases were speculative, and the "formidable process of concentration" of land ownership was aimed at monopolizing raw material supply and thereby extracting excessive prices from the public in the future, as well as exercising considerable political pressure on tax policy.

In a sense, however, the Bureau of Corporations' conclusions were neither fair nor timely, and it was probably prejudging the situation in order to produce a politically "acceptable" result.[8] The heady rhetoric of Progressive politics and thinking condemned every large organization as being inevitably against the public good. From the lumber companies' point of view, however, large purchases of raw material had been a reasonable and sensible solution to an old problem. Given the technology and knowledge (or lack of it) of forest management at the time and the volatile nature of the lumber market, lumbermen could maintain production only by cutting out the big timber and getting out. They had been caught a number of times by diminishing supplies and had had to move from one region of production to another across the continent, Weyerhaeuser being the classic and spectacular case. But there were no new regions left after the Pacific Northwest, and so now they bought to ensure their future. In addition, large holdings were proportionately cheaper to protect from fire and trespass and cheaper to log than small holdings. Moreover, a large holding justified the building of a mill and feeder logging lines and therefore gave an operater some independence. For example, a mill with a capacity of 20 million b.f. a year needed a supply of 400 million b.f., or 15–20 years' supply, which in a region of light stumpage may have been 100,000 acres. Therefore, large holdings were worth more in proportion to their acreage and stumpage than small holdings.

The bureau's argument that the public was being, or was going to be, held ransom by the monopoly situation never seemed to come about. Per capita consumption of lumber fell rapidly from 539 b.f. in 1900 to 325 b.f. in 1920 and kept on falling steadily afterward. The lumber companies were supplying a diminishing and competitive lumber market and were in no position to force up prices, although the consumption of other lumber products, particularly pulp, partially offset the decline during these years. In general, prices rose no more than they had before and actually dropped significantly from 1911 to 1917. They rose steeply after the end of World War I during 1919 to 1920 due to shortages, poor weather, transport difficulties, and panic buying by dealers at panic prices, but then they tended to level off with minor oscillations until 1932, when they rose again.

In fact, big changes were going on in the large holdings even as the bureau was deliberating its findings. Most companies were no longer buying stumpage, and the tendency toward acquiring speculative holdings beyond operating requirements was lessening. As Wilson Compton pointed out in his influential – and, from the lumber industry point of view, sympathetic – monograph, *The Organization of the Lumber Industry*, the lumbermen were caught in a vicious circle of falling prices and increased charges.[9] The lumber companies realized that they had considerable capital tied up in the mature, long-term timber investments that were, in effect, a carrying charge or rent that could double every seven or eight years. This acted as a brake to the movement to build up enormous speculative properties that had been in full swing before 1910. Some companies, it is true, cut regardless of price in order to cover charges and steal a lead on competitors, but that actually depressed lumber prices.

By 1920 the Capper Report initiated by the Senate could not come up with any definite conclusions about ownership; it was believed that the tendency was to put the timber holdings on a rational basis, adjusting their size to a practicable scheme for sustaining investment in a sawmill or logging improvement rather than carrying large surpluses beyond operating requirements. In the Northwest, timber holdings were passing from being long-term speculative holdings to being blocks of raw material connected to particular manufacturing plants; the largest holdings were being reduced, not increased, as the big companies realized that unused stumpage was a liability and not an asset.[10] The value of stumpage was not equal everywhere, of course; it depended upon location, proximity to lines of transport, and transport costs from forest to mill. Some stumpage was so inaccessible as to have a zero value; other tracts were of high-grade timber, accessible and even strategically located at the outlet of valleys that controlled the backcountry. The evidence is that the price for the Douglas-fir stumpage, the most important species in the Northwest, might have varied slightly from year to year after 1900, but it did not exhibit a clear upward trend until the early 1940s, despite fears during the First World War that prices would "take off." Companies were concentrating instead on large manufacturing groups, consolidating units and acquiring stumpage only to supply their plants for 20–25 years ahead, during which time it was hoped that improved management and sustained-yield practices would allow regrowth of stock. They also economized by forming better marketing arrangements and trade associations.

Details about size of holdings tail off dramatically after 1920, largely, one suspects, because that sort of inquiry no longer yielded the same political returns as it did during the prewar years. Certainly, the massive Copeland Report of 1933 did not bother with a detailed analysis of the size of holdings anywhere in its 1,600-odd pages. It was less concerned with the iniquities of large forest holders than with the iniquities of private holdings of any size versus the supposed virtues of public ownership, and it advocated the massive purchase of private lands by federal authorities. Forest holdings were categorized simply as public or private, and this primary division was analyzed by regions only. One thing is clear, however: Despite the creation of national forests and the purchase of land by states and counties to create local public forests, private ownership still dominated the forest scene. Of the 495 million acres of commercial forest in 1933, 88 million acres were held federally, 10.6 million acres were state-held, and 396 million acres were in private hands – 270 million acres in industrial forest units and 126 million acres in farm woodland.[11] In other words, 80 percent of the forest land was still held privately, and even more significantly, it contained over 90 percent of the growth capacity of timber. The original Copeland program of massive purchases to increase the federal share barely got off the ground, although 8 million acres were purchased. A testy Congress, the continuing economic difficulties of the depression, and World War II altered priorities and concerns.[12] The question of who owned the forest was becoming far less important than the question of how to promote the maximum growth of timber commensurate with the conservation of the stand in any forest, private or public. The amount of forest, its depletion and its growth, and how to promote the latter were now the important concerns.

How much forest is left?

Much of the concern about the monopoly of the forest resources presupposed a finite and diminishing resource basis. It was a common notion that a timber famine was on the country and that the big companies were going to eke out the remaining stands at ever-increasing prices. In view of the destruction of the forest cover that most people had witnessed during their lifetimes, the idea of timber depletion was perhaps not totally unwarranted. The warning had gone out repeatedly since the early 1860s, and in 1872 the commissioner of agriculture had declared that "a timber famine" was inevitable.[13] The word "famine" was particularly appropriate, for wood ranked next to food as a basic ingredient in American life, so that a dearth of wood was not merely a shortage but the equivalent of the withdrawal of a major life-support commodity, the absence of which could adversely affect the very nature of life in the country as it had been known up to then. In a more sober but equally somber mood, James E. Defebaugh pointed out that up until the beginning of the twentieth century the country had been "drawing on the surplus," but by the time he was writing in 1905 it was starting to "draw upon the capital fund" of timber and therefore could no longer afford to be lavish in its use of timber but had to adopt conservative practices.[14] But the issue was not entirely closed. A few perceptive people did not agree. Henry Gannett, geoprapher to the Geological Survey, was scornful of the lack of knowledge displayed by the alarmists about the extent and stand of the forests and, in a remark attributed to him in 1891, said:

> It's all bosh – this talk about the destruction of our forests. There is more wood growing in the United States now than there was one hundred years ago, more than we want and can use; . . . Don't you know that all the abandoned farms in New England grew up in wood? Don't you know that in the South more than half the land is covered with forest? And all along the Pacific Coast the timber is simply inexhaustible?[15]

Gannett's optimism was unique among people seriously interested in the forest at the beginning of the century, a period when one official after another was issuing dire warnings about the coming dearth.

But who was right? Was the forest being skinned mercilessly, or was there sufficient timber for the next hundred years? Was the by now conventional wisdom of an imminent timber famine in accord with the facts of forest growth and replacement? The analogy of a financial ledger was a good one because politicians, professional foresters, and public alike could then see the debits and credits at any given time. The debits, or drain, were reasonably well known; the credits, or growth, were a mystery. In order to produce a balance sheet, two major investigations had to be carried out. First, the location and extent of the forest had to be determined; second, the stock of timber in the forested areas had to be calculated as did its rate of growth and depletion, the calculation of growing stock being possible only after the forest extent and content had been determined. These were questions that obsessed writers and researchers on the forest throughout the first third of the century, and they are still crucial questions that are posed continuously in our more resource-conscious society of today.

The location and extent of the forest

Of the two basic questions about forest resources, the one about its location and extent was probably the first to be answered. Despite the lack of precise records on the topic, people began to piece together the evidence of surveys, plat books, personal accounts, census material on agriculture, all in conjunction with fieldwork, to come up with some approximation of where the forest was. As early as 1858 Prof. Joseph Henry had published a crude map of the eastern forest, the prairie, and the western forests, but the big advance came with William H. Brewer's map of "The Woodland and Forest Systems of the United States" (1873), which accompanied the Atlas based on the Ninth Census and which was followed a decade later by Sargent's map of the "Relative Average Density of Existing Forest" (1883) (see Figs. 2.3 and 2.4)[16] Not only were the edge between the forest and the prairie and the various "openings" shown more accurately than ever before, but in both maps an attempt was made to estimate the density of the timber stand. Nevertheless, from comments by Brewer it is obvious that the distribution, to say nothing of the density, was little understood.

There was no census of the forests, nor was there any follow-up of the pioneer work of Brewer and Sargent for the Ninth and Tenth censuses. In fairness, perhaps, there would not have been any improvement of the national picture until the local picture was filled in in more detail. A big advance came with studies like that of Col. Edgar T. Ensign, special agent for the Forestry Division, whose report on the forests of the Rocky Mountains in 1887 was accompanied by a map of the distribution of the forests in the region. His work was refined by the activities of the Division of Geography and Forestry of the Geological Survey under Henry Gannett, which mapped over 70 million acres of the western forest reserves when the forest-lieu provisions were introduced.[17] Another outstanding example of detailed mapping was that of Thompson, Matthes, and Cudlipp for the forest/prairie edge in Oklahoma in 1899.[18] Not until the West had been thoroughly explored and the forests mapped could vegetation maps of the continent be drawn. After 1905 Schantz and Zon were producing crude maps of forest (Fig. 2.5), differentiated according to dominant tree associations, which were later amended in detail a number of times up until the early 1950s as more accurate information came to hand.[19]

Because so many of the maps of the extent of vegetation were based upon the evidence of the plat books and on field surveys of the surviving remnants of the forest, nearly all attempted to reconstruct the distribution of the "original" vegetation, that is, the vegetation as it was before European settlement in the early seventeenth century (the impact of the Indian was conveniently forgotten or ignored). Consequently, the estimates of the area of the "original" forest before European man interfered with it are affected by the accuracy of the maps. Many of the estimates made of the forest area (and forest volume and growth) between 1895 and 1977 are shown in Table 13.3. Broadly, the estimates fall between c. 494 and 1,000 million acres, the most commonly accepted figure being 850 million acres of "commercial" forest and 100 million acres of noncommercial forest, the "roundedness" of the figures indicating the uncertainty.[20] Therefore, of the total land area of 1,904 billion acres of the coterminous United States, about half was forested originally and about two-thirds to three-quarters of the forest was in the eastern third of the country.

There was no doubt that the area of forest had diminished significantly since first European settlement. In 1907 William Greeley calculated that a mere 515 million acres of commercial forest and 65 million acres of inferior noncommercial forest were left, and during the following year E. A. Zeigler put the figures a little lower at 495 and 50 million acres, respectively. Although later estimates varied considerably, there was little doubt that the forest had diminished in size by between 300 and 350 million acres by the opening years of the first decade of this century. In time the calculations became more precise, the commercial forest being divided into mature forest and two other types of forests – those growing sawtimber and those growing less valuable cordwood – the details of these divisions being shown in Table 13.4 for various times between 1907 and 1977.

With the accumulation of such information, the national balance sheet of forest loss gradually began to take shape. The magnitude of the loss was not unexpected, but the publication of the figures, however tentative, focused attention on the changes that had taken place in the past and, perhaps more importantly, that were likely to take place in the future. In 1907 Greeley pinpointed some of these fears when he calculated that the

Table 13.3. *Estimated area of commercial and noncommercial forest, and total standing sawtimber volume*

	Commercial forest (million acres)	Noncommercial forest (million acres)	Standing sawtimber volume (bill. b.f.)
1630	850	100	7,625
"Original"	850	150	5,200
1895	562	—	2,300
1902	c. 494	—	2,000
1907	515	65	—
1908	550	—	2,500
1920	464	150	2,215
1930	495	120	1,668
1944	461	163	1,601
1952	499	276	2,412
1962	509	250	2,430
1970	496	254	2,421
1977	483	254	2,569

Sources: Based on Marion Clawson, "Forests in the Long Sweep of American History," 2, table 1; and USFS, *An Analysis of the Timber Situation in the United States, 1952–2030*, 149.

annual drain on the forest then was 9.4 million acres, new farms taking just over 1 million acres, cutover land not restocking taking 2.7 million acres, the remaining 5.6 million acres restocking with second growth. When he projected his figures forward to 1950 on the basis of the expected population increase to midcentury and the amount of cropland and lumber needed to service that population increase, he came to the conclusion that another 100 million acres of forest would disappear for farms and that the total of unproductive cutover would reach 182 million acres, leaving a mere 298 million acres of forest.[21] It was an alarming prospect that struck at the very heart of the Americans' image of the boundless resources of their country; the nation might be reduced to the status of some impoverished, denuded Mediterranean country.

The situation was not helped by Greeley's professional but almost obsessive concern about the mature forest, or virgin forest, as he called it repeatedly.[22] At this time the virgin trees for him seemed to represent the major and, perhaps, the only true resource remaining in the forest. Though it is true that the old, mature trees were the source of the large merchantable timbers, they did not represent a source of growth for the future, and in that sense they resembled the rightly despised cutovers. In his writing at this time, Greeley did not seem to recognize that distinction, and he stoked the fires of concern about depletion by compiling and publishing three maps of the declining

Table 13.4. *Area of commercial and noncommercial forest, and growth potential of commercial forest lands, 1907–77 (millions of acres)*

	1907 Greeley	1908 Zeigler	1919 Forest Service	1919 Committee for Applicat.	1920 Capper Report	1923 Senate Report	1933 Copeland Report	1952 Forest Service	1962 Forest Service	1970 Forest Service	1977 Forest Service
Total forest											
Noncommercial	65	50	143	—	150	150	120	176	250	254	254
Commercial	515	495	507	500	463	469	495	499	509	496	483
Commercial forest only											
A. No net growth											
Mature timber	270	188	147	150	137	138	99	—	—	—	—
Cutover or not restocking	65	82	80	100	81	81	83	—	—	—	—
Total not growing	335	270	227	250	218	219	182	—	—	—	—
B. Net growth											
Sawtimber }					112	114	90	—	—	—	—
Cordwood }	180	225	280	250	133	136	121				
Fair restocking	—	—	—	—	—	—	102				
Total growing	180	225	280	250	245	250	313	—	—	—	—
Growth/acre (cu. ft.)											
120+	—	—	—	—	—	—	—	—	43	—	47
85–120	—	—	—	—	—	—	—	—	117	—	99
50–85	—	—	—	—	—	—	—	—	232	—	200
25–50	—	—	—	—	—	—	—	—	117	—	136
Total	—	—	—	—	—	—	—	—	509	—	483

Source: See text.

acreage of the virgin forest in 1620, 1850, and 1920, which were alarming in their stark and seemingly simple message of denudation (Fig. 13.1). These maps were, and often still are, interpreted as showing the diminution of the total forest cover rather than the diminution of an important but small part of it.[23] Of the 137 million acres of virgin timber left, 56 percent was in the Rocky Mountains and Pacific Coast States. Admittedly, there were another 245 million acres of forest growing timber, but in Greeley's estimation only half of that could grow sawtimber, the rest merely cordwood. The depletion of the eastern and midwestern forests had reached such a point that there was a net deficiency of softwood timber, and although the southern forests showed a surplus of production over consumption, they were declining rapidly (Fig. 13.2). The Northwest, therefore, was the last "great reservoir" of timber for the country.

The timber stand and its rate of growth

The preoccupation of the early surveys with the distribution and acreage of forest provided basic and essential information, but in truth it told people little about the more important attributes of the forest, the stock of timber present and its rate of growth. Calculations of the stand of timber in the forests were fraught with difficulties. Unlike acres, which were a constant measure, the timber worth measuring and the measure to be used changed over time. For example, with changing technology and supply-and-demand relationships, trees that were thought to be too young and too small, species that were once thought valueless (e.g. eastern hemlock in the Lake States), and parts of logs that were once thought waste became salable and usable. The measure of cubic feet of timber was also different from the measure of board feet of sawtimber, and scaling procedures varied over time. Add to these complications the deliberate underestimates by large landholders to evade taxation or competitive conditions or to "scare" the public into supporting them, and one can see why the calculation of early timber inventories was unreliable. There was, in addition, what Defebaugh called "the personal factor," which could color the calculations, for "with the best of intentions the pessimist and alarmist will underestimate the amount of standing timber, and so exaggerate the seriousness of the exigency. On the other hand, the optimist is likely to magnify the favourable facts and minimize the unfavourable ones."[24] During the early part of the century, the mood of pessimism certainly prevailed, and forest resources were constantly underestimated.

Our best estimate of the volume of all timber standing in the original forests of the country is 5,200 billion b.f. or 1,139 billion cu. ft. (using a conversion factor of 219 cu. ft. of forest per 1,000 b.f., allowing for usual waste), which by the end of the nineteenth century had been reduced to somewhere between 2,000 billion b.f. (439 billion cu. ft.) and 2,800 billion b.f. (613 billion cu. ft.) by commercial lumbering, fuel getting, and, above all, agricultural clearing. The pre–European settlement situation of the East had perhaps double the volume of timber of the West, but continual clearing had reduced that predominance, so that by 1920 the East had less than one-half the volume of the West.[25]

Like acreages, the estimates of the stock or volume of timber varied enormously at the

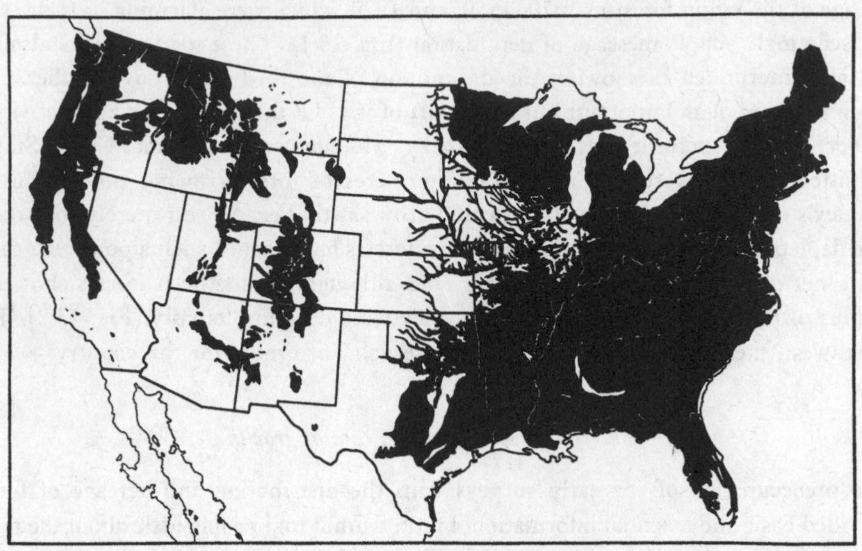

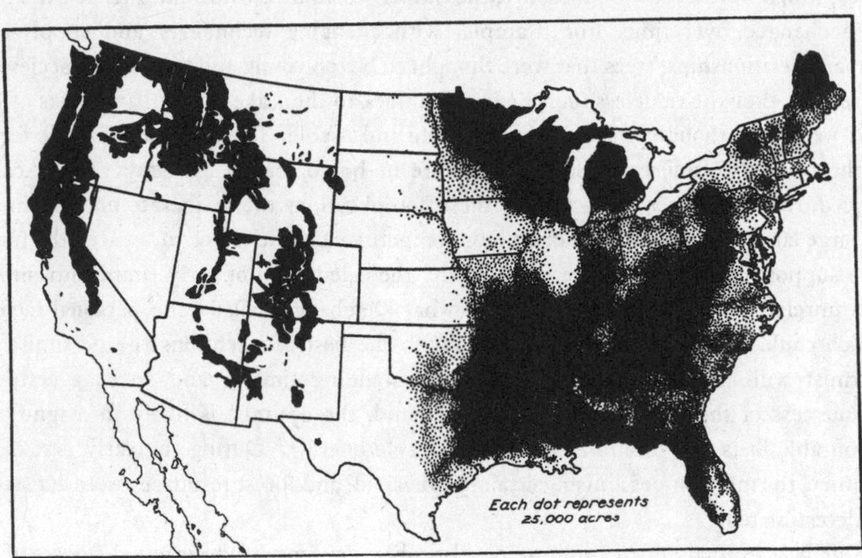

Figure 13.1. "Area of virgin forest": *top*, 1620; *bottom*, 1850; and *above right*, 1920. (W. Greeley, "The Relation of Geography to Timber Supply," 4, 5.)

turn of the century, and in 1907 Zeigler made a significant contribution to the debate by pointing out something that had been known intuitively by most people but was rarely appreciated: that whereas harvest could not exceed net timber growth indefinitely, neither could growth exceed harvest for very long because the standing timber accumulated to the level where no further net growth could take place.[26] It was all very

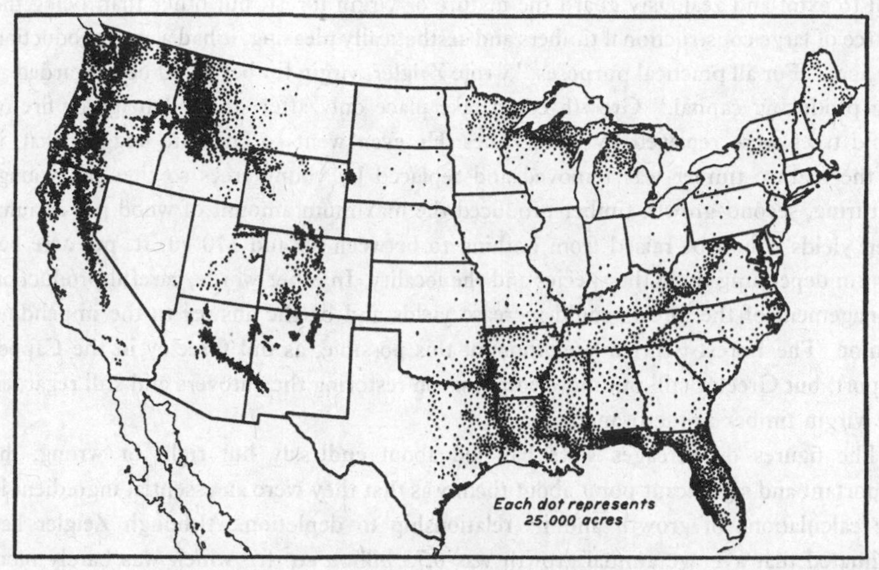

Figure 13.1. (*cont.*).

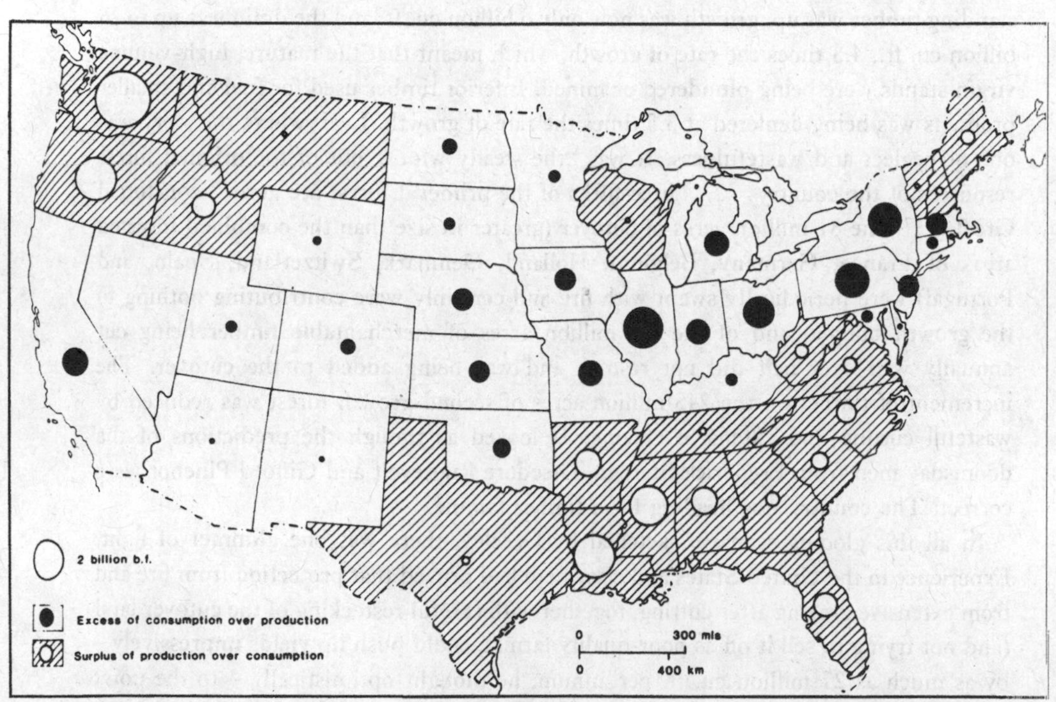

Figure 13.2. Lumber surpluses and shortages, 1920. (Adapted from National Archives GRG 95, Map Series 94, A–4, A–7.)

well to extol and zealously guard the mature or virgin forest, but other than being the source of large constructional timbers and aesthetically pleasing, it had a zero production per acre. "For all practical purposes," wrote Zeigler, virgin timber could be "regarded as non-producing capital." Growth could take place only after storm damage or fire or if old trees were replaced by young ones. He even went so far as to suggest that, if all the mature timber was removed and replaced by young trees so that the young, maturing, second-growth timber produced the maximum amount of wood per annum, then yields would be raised from nothing to between 30 and 110 cu. ft. per acre per annum depending upon the species and the locality. In other words, careful production management of the forests could increase yields and be the answer to the impending famine. The Forest Service later thought this possible, as did Greeley in the Capper Report, but Greeley still put more emphasis on restoring the cutovers and still regarded the virgin timber as sacrosanct.

The figures on acreages were bandied about endlessly but right or wrong, the important and significant point about them was that they were an essential ingredient in the calculations of growth and its relationship to depletion. Although Zeigler had estimated that average annual growth was 6.75 billion cu. ft., which was barely more than a quarter of the annual drain, it remained for Greeley in the 1920 Capper Report (with the assistance of the newly formed research section of the Forest Service under Earle Clapp) to refine the ideas further (Table 13.5). The result was, if anything, more alarming than the estimates of the number of acres cleared. Although the volume of all standing timber was up, growth was now only 6 billion cu. ft. and the drain was up to 26 billion cu. ft., 4.3 times the rate of growth, which meant that the mature, high-volume virgin stands were being plundered or mined. Inferior timber used for fuel and smaller products was being depleted at 3.5 times the rate of growth. Everywhere the story was one of neglect and wastefulness. It was "the steady wiping out of the original forest resources of the country. . . . Three-fifths of the primeval forest are gone," concluded Greeley.[27] The 81 million acres of cutover (greater in size than the combined forested areas of France, Germany, Belgium, Holland, Denmark, Switzerland, Spain, and Portugal) were periodically swept with fire and certainly were contributing nothing to the growth budget, and of the 5.5 million acres of merchantable timber being cut annually well over half did not restock and was being added to the cutover. The increment of timber on the 245 million acres of second-growth forest was reduced by wasteful cutting and extensive grazing. It looked as though the predictions of the doomsday men from Frederick Starr to Theodore Roosevelt and Gifford Pinchot were correct: The country was heading for a timber famine.

In all this gloomy scenario sketched by Greeley, there was one glimmer of light. Experience in the United States and elsewhere had proved that protection from fire and from extensive grazing after cutting, together with careful restocking of the cutover land (and not trying to sell it off as poor-quality farms), could push up yields impressively – by as much as 27 million cu. ft. per annum, he thought optimistically – to the point where it would exceed present consumption. But that sort of fire control and planting program needed the enactment of political and legislative provisions that were beyond the powers of any government or department of the day, and their implementation

Table 13.5. *Estimated volume and annual growth and drain, 1909–77 (billion cu. ft.)*

	1909 Zeigler	1920 Greeley	1933 Copeland	1950 Copeland Forecast I	1950 Copeland Forecast II	1952 Forest Service	1962 Forest Service	1970 Forest Service	1977 Forest Service
Volume, all standing timber	c. 548	746	487	634	634	?	699	?	792
Volume, growing timber only	n.a.	?	n.a.	n.a.	n.a.	603	648	680	711
Growth per annum									
Actual	6.75	6.0	8.9	—	—	13.9	16.7	19.8	21.7
Potential cutovers restocked		19.5	—						
Virgin timber cut	30 to 120	8.2	—	10.6	21.4	27.5	?	?	c. 36.0
All forest well stocked and managed		27.7	—						
Drain per annum									
Firewood and lumber	23.0	24.3	14.5	15.3	15.3	11.8	12.0	14.0	14.2
Fire and insects	c. 2.0	1.7	1.8	1.2	1.2	3.9	4.3	4.0	3.9
Total	25.0	26.0	16.3	16.5	16.5	15.7	16.3	18.0	18.1
Ratio growth to drain	1:3.5	1:4.3	1:1.8	1:1.5	1:0.77	1:1.13	1:0.97	1:0.90	1:0.83

Source: See text.

needed some degree of either coercion of or cooperation from the lumberman in order to succeed. Indeed, forestry in the United States during the 1920s was to become dominated by these opposing and deeply held views about the management of the stands. On the one hand, Pinchot thought that coercion of private forestry was necessary; on the other hand, Greeley thought that cooperation and example, as pursued since Fernow's time (and even by Pinchot himself while in office), were preferable. Ultimately, Greeley's view prevailed.[28]

By the time of the publication of the Copeland Report in 1933, the area devastated and poorly stocked still stood at 83 million acres, but the annual growth had risen significantly to 8.9 million cu. ft. whereas consumption had dropped to 16.3 million cu. ft. (largely due to the depression). Consumption was still greater than growth, but the gap between the two was significantly smaller than it had been before.[29] The greater growth was never explained at the time, but with hindsight we can say that more accurate measurement had something to do with it, as had natural regeneration. Few people realized the capacity of the forest lands to regenerate, even without fire suppression and replanting, nor did they appreciate fully that massive areas of farmland were being abandoned to regrowing forests, particularly in the eastern half of the country, and that this was leading to a great increase in the amount of timber. In just over a decade, between about 1935 and 1945, the whole idea of a timber famine was going to be reversed as timber growth slowly began to catch up with timber drain. The compilers of the Copeland Report had not been unmindful of the potential of the forest to increase its stock, and in two projections of growth to 1950, based on models of varying mixes of extensive and intensive management of the forest stand (Forecasts I and II, Table 13.5), they had estimated that growth could be pushed upwards to either 10.6 billion or 21.4 billion cu. ft., depending upon the protective measures and degree of replanting undertaken.[30] The upper estimate was optimistic, but by the early 1950s the lower prediction of 20 years before had been surpassed. In 1952 the actual annual growth was 13.9 billion cu. ft.; by 1962 it was 16.7 billion cu. ft. and was set to rise even further (Table 13.5). At last it looked as though the forest famine and the specter of dearth were over, and the rebirth of the forest had begun.

The political and administrative reaction

There is no doubt that the specter of the coming timber famine was actively sponsored by foresters, public servants, and politicians as well as by popular writers. The scare of a famine was an obvious lever to get changes in public opinion about the necessity to control forest destruction, preserve the forest, and manage it sensibly. It is not fanciful, either, to suggest that the lumbermen themselves played up the famine scare so that reserves would be protected and the existing privately held stock would be valued more highly. This alarmist concern reached its peak between 1900 and 1929 when lumber consumption and production reached heights (particularly in the year 1906) never attained before or since. With great enthusiasm and showmanship, Pinchot and Roosevelt used the "coming timber famine" issue skillfully again and again to further their wider aims in conservation, the creation of the national forests and the Forest

Service, Pinchot talking about it so much that he was even credited, erroneously, with coining the phrase "timber famine."[31] At the great American Forestry Congress in Washington, D.C., in January 1905, Roosevelt set out the issue in plain and uncompromising language:

> Our country . . . is only at the beginning of its growth. Unless the forests of the United States can be made ready to meet the vast demands which this growth will inevitably bring, commercial disaster, that means disaster to the whole country, is inevitable. The railroads must have ties . . . the miner must have timber . . . the farmer . . . must have timber . . . the stockman must have fence posts. If the present rate of forest destruction is allowed to continue, with nothing to offset it, a timber famine in the future is inevitable.[32]

The likelihood is that the outline, if not the very words, of Roosevelt's address was written by Gifford Pinchot, who had engineered and staged the whole congress and who, Roosevelt admitted freely, ghost-wrote many of his speeches and provided him with ideas on forestry and conservation. Nonetheless, the fact that the words were uttered by the president lent them much weight.

During the succeeding years, Pinchot's staff turned out forecasts that confirmed the pessimistic trend. Royal Kellogg, assistant to Pinchot, told the National Conservation Commission in 1909 that the forests contained about 2,500 billion b.f., that they were being cut "three times as fast as they were growing," and that, although the country might never reach "absolute timber exhaustion," serious disruption to life as it had been known would ensue unless the actions and aims of the federal government, the states, and individual owners came together in some way. The pressure built up. The Bureau of Corporations, probably a creation of Pinchot through Roosevelt, confirmed in a general way that depletion was proceeding rapidly and that reserves of all timber were only 2,826 billion b.f.[33] Several times Pinchot predicted that the supply of timber would end in 30 to 35 years' time. His most eloquent statement was in his little, but influential, book, *The Fight for Conservation*, published in 1910 just before he was dismissed as chief of the Forest Service, in which he reviewed the whole field and concept of conservation as interpreted by him. He brought the forest issue down from the world of commerce into the individual home by saying that it was "certain that the United States has already crossed the verge of a timber famine so severe that its blighting effects will be felt in every household in the land." The rising price of timber during the opening years of the century was only a presage of what was to come, so that the country "must necessarily suffer from the scarcity of timber long before our supplies are completely exhausted." Such dire warnings were becoming commonplace now, but coming from the man who had been chief forester, confidant of the president, and well known in his own right, his assertion that the famine was already present was alarming. "In a word," he concluded, "when the forests fail the daily life of the average citizen will inevitably feel the pinch on every side. And the forests have already begun to fail as a direct result of the suicidal policy of forest destruction which the people of the United States have allowed themselves to pursue."[34]

Pinchot's dismissal early in 1910 and the debate over its circumstances brought a temporary halt to the timber-famine question. One would have expected, however, that it would have been picked up again fairly quickly by Pinchot's successor, Henry S. Graves, who had been both close associate and friend of Pinchot throughout their professional lives. As a young man during the 1890s Graves had worked with Pinchot as a forest consultant to owners of spruce forests in the Adirondacks. Then, like Pinchot, he went to Germany to further his studies and was taught by Sir Dietrich Brandis. Between 1898 and 1900 he was superintendent of management in the Forestry Division and worked in close cooperation with Pinchot, moving on to become the first occupant of the Pinchot Chair of Forestry at Yale, an act of direct patronage by the Pinchot family.[35]

In many ways Graves seemed to be Pinchot's handpicked righthand man; yet on assuming office as chief he did not turn out to be Pinchot's puppet. He had little opportunity or liking for indulging in high politics and scaremongering, and he certainly had his work cut out in rebuilding the shattered morale of the service and combating three major internal threats to the national forests: the wholesale opening up of the land to homestead entry; the efforts of the western states to get back the forests and "unlock" the resources (water, minerals, power sites, grazing areas, as well as the land and timber); and the pressure to step up timber sales to increase revenue. Graves solved these three issues satisfactorily, and the forests were preserved almost totally intact. He also promoted the Forest Products Laboratory at Madison during his first years in office, encouraged fire protection of the watersheds of the navigable streams under the Weeks Act of 1911 – thus establishing the federal forests in the East – and created a research section in the Forest Service in 1915, all of which signaled a rethinking of Pinchot's priorities. Compared with Pinchot, he was described as "capable but less dramatic in his leadership," which was probably fortunate, and he had more in common with Fernow in his concern for scientific research, forestry education, service administration, and the custodial role of the service than with his flamboyant predecessor and mentor, Pinchot.[36]

Though Graves did not play down the timber famine and was aware of the possibility,[37] he did not heed the alarmist forecasts of the Bureau of Corporations and did little to further the scare actively, as he was painfully aware that since 1907, and certainly during the war years, the lumber industry had gone into a prolonged and downward depression and could therefore give little thought to the conservation of timber stands while prices were so low. Graves, with his head of timber management, William Greeley, was strongly influenced in this opinion by the publication in 1916 of Wilson Compton's *Organization of the Lumber Industry*, which analyzed the timber situation from the lumberman's point of view. Compton argued that large tracts of forest had been bought at low cost in the expectation of rising prices that had never eventuated; hence, even the standing or carrying costs in the huge timber tracts often could not be met unless cutting and selling continued regardless of prices, thus driving prices even lower than before. Moreover, substitutes to lumber, such as steel, concrete, and brick, had undermined the market so that per capita consumption had dipped steeply after the peak years of 1907–9, to fall by the time of the Great Depression to a mere quarter of what it had been. Greeley, particularly in his service report, *Some Public and Economic Aspects of the Lumber Industry*, of 1917, sympathized with the view that the lumberman was caught up in a vicious circle of overcapacity, high interest rates, and low prices, a situation made worse by the burden of timberland investments in the past and by the glutting of the eastern market with low-cost western timber, which the railroads were shipping east in search of markets because freight rates were based on weight and not value. Together, Graves and Greeley encouraged cooperation, not confrontation, with the lumber industry. They realized that the frontier era of "cut out and get out" was over and that the lumber business of the twentieth century needed a "more stable and rational economic order" where future prospects could be "predicted with a greater degree of certainty."[38]

But if Graves and his deputy Greeley were convinced that the lumber industry needed help and not censure, Pinchot and his associates were not. Pinchot was orchestrating his political comeback through the new Progressive party and the Bull Moose movement, and he cavalierly dismissed Greeley's Forest Service report of 1917 as "a whitewash of destructive lumbering."[39] Still concentrating on destruction and conservation of the forests, major planks in his political platform, his emphasis now changed from merely warning about the coming timber famine to fixing the blame wholly on the private forest holders, all of whom, he claimed, misused the land so grossly that they had to be regulated in their cutting, farm woodlots excluded, as usual. Cable yarding was the curse; all the young trees were swept off in getting the mature growth, regeneration was impossible, and forest fires in the debris finished off what the logger had left. Pinchot whipped up public opinion to believe that all lumbermen and industrial forest users were wicked, incompetent, greedy, and basically antisocial in their actions and motives and that the Forest Service was their dupe. He had a point: There were grave abuses; the lumber companies were powerful; and the Forest Service did seem to be taking the lumberman's point of view. His major, if only, weapon was invective, but whether tactics were politic was another matter.

These opinions were not new, and Pinchot was not alone in voicing them. Samuel Trask Dana wrote a very forceful and persuasive pamphlet for the Forest Service in 1918 entitled *Forestry and Community Development*, in which he traced the economic and social effects of forest devastation – lumber migration, cutovers, abandoned farms and towns, local shortages – rather than the physical effects such as erosion and reduced stream flow. He advocated greater public control and ownership (federal as well as state) of forest lands to mitigate these evils and achieve stability and a sustained yield from the forest.[40] Indeed, many people were thinking along these lines. The depression in lumbering, rising prices, the exigencies of wartime with its focus on essential raw materials, all accompanied by the clear downward trend in supplies, were leading toward a consensus after 1918 that forestry had to become a national affair needing national solutions; that the forest lands were not being managed in accordance with sound forestry principles; that depletion was occurring; and that action could not be confined to the western reserves but had to encompass all forest lands, in the East as well as the West, private as well as public.

However, the essential difference between the various advocates of national action lay in whether the regulation of cutting should be compulsory and enforced by federal legislation or voluntary and based on industrial/state/federal cooperation. Pinchot backed the first option uncompromisingly; the Forest Service under Graves, and then Greeley, backed the second option, knowing that many lumbermen accepted controls provided they got fire protection in return. Pinchot alienated private lumber companies (but undoubtedly scared some of them into considering cooperation more actively); the Forest Service tried to work with them. Thus, during 1919, the issue of timber depletion shifted rapidly from one of exhaustion to one of cutting regulation, reforestation, and fire protection. Rather than concentrating on the effect (i.e., depletion), the focus of attention was switched to what was thought to be the cause.[41]

Graves's cooperative approach to improve cutting and prevent depletion was set out in his annual report and in his speeches to lumbermen's conventions during early 1919, where he concluded that "The general practice of forestry on privately owned lands . . . will not take place through unstimulated private initiative" and needed a federal policy.[42] He proposed a broad federal/state cooperative program of forest land aquisition and planting on the cutovers; forest protection against fire, insects, and disease; taxation reform; and cutting supervision. But within that program he still wanted private forest owners to "assume the full responsibility of properly caring for their timberlands including protection and forest renewal," and by the next month lumbermen had responded cautiously but positively by setting up a joint working party with the Forest Service.[43]

Pinchot, however, as an influential member of the Society of American Foresters and now an aspiring politician, had set up his own rival "Committee for the Application of Forestry," which was filled with many "old crusaders, lawyers and publicists," as Greeley commented later, or "lunatic fringe," as E. T. Allen called them.[44] The committee forecast acute lumber shortages, exhaustion of mature supplies by the 1970s, and rapidly rising prices. Pinchot and his associates became even more extreme. They wanted immediate action for complete mandatory federal regulation of private cutting. State control was to be swept aside, and even state boundaries were to be rearranged in the pursuit of efficient forestry practices. Pinchot had no patience with the argument that the lumber industry had fallen on hard times; its plea that it could not afford to grow trees until they could be sold for more than the cost of growing them seemed merely to be a delaying tactic to his zealous crusaders, whereas to the Forest Service it was a desperate plea related to the scarcity of capital. The committee's report, "Forest Devastation: A National Danger and a Plan to Meet It," was vintage Pinchot. It was vital, disturbing, calculated to agitate, but easy to read and comprehend with its epigrammatic turns of phrase: "The annual cut is nearly three times the annual growth." Prices would rise, and "within less than fifty years our present timber shortage will have become a blighting famine." "Forest devastation has long been an unmitigated evil; today it threatens our national safety and undermines our industrial welfare." In a style and approach reminiscent of Frederick Starr nearly 60 years before, Pinchot thought that, "without the products of the forest, civilization as we know it would stop." The report was published in the *Journal of Forestry*, accompanied by a short prefatory article by Pinchot entitled "The Lines Are Drawn." It was a belligerent and uncompromising diatribe against private lumber owners, which proudly heralded the coming "fight." It concluded: "The field is cleared for action and the lines are plainly drawn. He who is not for forestry is against it. The choice lies between the convenience of the lumbermen and the public good."[45]

The lines need not have been plainly drawn as, in essence, the positions of Graves and Pinchot were not so very far apart. Both realized that the lumber industry needed regulation; Graves wanted a strong state role and local responsibility, and Pinchot wanted a strong, centralized federal role. However, the common ground of policy was lost as issues became polarized behind personalities. Once more, Pinchot, the outsider, was getting the limelight and the credit for action, and Graves and the Forest Service were not. Once more, money and public relations manipulation (Pinchot agreed to pay personally for the printing and distribution of 15,000 copies of "Forest Devastation") seemed likely to be superior to experience and service. A clash between Graves, the pragmatic cooperationalist, and Pinchot, the coercionist, seemed to be as inevitable as it was bitter.

The depth of feelings is highlighted in Pinchot's correspondence at the end of 1919 with J. Girvin Peters and Greeley (then assistant chief). Peters had written to Pinchot asking for his cooperation as state forester of Pennsylvania in the allocation of federal subsidies for fire control and the prevention of destructive lumbering. Pinchot was scathing in his rejection: "your program is fundamentally wrong in principle, can never be put through Congress and if it could would be unworkable. I cannot support it." For Pinchot the question was "not control by the nation or by the state, but the question of National control or no control at all." He was obsessed by the supposed venality and corruption of the lumbermen, and the ineffectualness of the states in controlling them, and the emphasis on fire control seemed to disregard the problem of forest destruction. Greeley now took up the exchange, replied in conciliatory tones, and pointed out the problems of imposing federal regulations, however desirable, a practical point that Pinchot seemed to ignore blithely. He added that destruction of timber by fire far exceeded destruction by logging. But Pinchot would have none of it and vowed to use all his influence to oppose the plan. He accused Greeley of calling him an "obstructionist," which Greeley had not, but it would have been difficult to find a more apt word for Pinchot's opposition and arguments. Each side sent their letters to the state departments of forestry in order to whip up support. But Greeley, who had begun to emerge as the principal spokesman of the Forest Service and lumber industry, stood firm against the formidable influence of Pinchot, even in a number of public debates with him before the State Foresters Assocaition.[46]

Pinchot sought congressional support for his position and got it through Sen. Arthur Capper of Kansas, who, in February 1920, managed to get the Senate to require the secretary of agriculture to submit a report on *Timber Depletion, Lumber Prices, Lumber Exports, and Concentration of Timber Ownership*.[47] It looked as though the Pinchot faction had won when Graves resigned the next month – weary from overwork, now quite ill, and reluctant to face Pinchot as adversary and enemy. However, his place was taken by his deputy, William B. Greeley, a spirited character with a strong personality, who, although he admired Pinchot for all he had achieved in the past for the Forest Service, found himself in fundamental disagreement with his former chief over the conditions and motives of the lumber industry. Greeley was the first wholly American-trained forester that the country had appointed, and he had a good appreciation of what would work and what would not. Thus he encouraged various lumber and forest industry groups to form a united front, which they did as the National Forestry Program Committee (NFPC) under the chairmanship of Royal Kellogg (now in private industry). The committee acted as a legislative body and supported Greeley's approach, provided he emphasized fire protection in his proposals, a noncontroversial, nonpolitical course of action that incidentally coincided with Greeley's own interests and pragmatic convictions.[48]

Consequently, though the publication and appearance of the Capper Report in June 1920 (written largely by Earle Clapp under the direction of Greeley) highlighted the seriousness and extent of timber depletion, it recommended unreservedly a cooperation between federal and state organizations regarding two practical and recognizable objectives: fire control and the growing of timber on the cutover lands. It also recommended an extension and consolidation of federal forest holdings and a study of the vexing question of taxation of forest lands. The uncharacteristically slim, 73-page report was a milestone in the acquisition of knowledge about the forests. It was accurate in its information and measured in its conclusions, and it was crucial in the subsequent formulation of positive policies to manage and protect the forests. Moreover, it had the support of the lumber industry, which thought it fair.

Not surprisingly, Capper was not persuaded by the mild report that bore his name, as it did not recommend national control as advocated by Pinchot. Three times he introduced bills along those lines, but each time they foundered on constitutional grounds, from sheer lack of support for the idea of compulsory federal regulation, and because of the NFPC lobby and the expert testimony of Greeley, Compton, and others against the measure. In any case, the national mood was not for radical solutions. The election of President Harding in November 1920 and his much-heralded "return to normalcy" went deep in the public psyche, and it was said that Pinchot's style and persistence as he badgered the president and other senior administration officials annoyed Harding. Paradoxically, too, the changed mood worked against Greeley and his supporters in the lumber industry as a majority of Congress was now prepared to sanction a number of rival bills introduced by Sen. Bertrand Snell of New York, which aimed to advance the less controversial idea of state regulation of forestry. The bills had been instigated by and had had the almost unqualified support of the lumber industry through the NFPC, except in the South.

There is no doubt that Pinchot and his allies in Congress and in the administration, particularly the secretary for agriculture, Henry C. Wallace, opposed Snell's program vigorously, and they succeeded in killing the bills, demonstrating that although Pinchot's influence in Congress was not strong enough to smooth the passage of the Capper bills it was strong enough to block the Snell bills. Pinchot was still a force to be reckoned with.[49] Greeley, on the other hand, recognized that events were at an impasse; federal control and state regulation were dead. The only common ground between the two factions was cooperative fire control. In 1922 he proposed a compromise; surprisingly, Pinchot agreed, and the Clarke–McNary bill became law in June 1924. The way now seemed open for what seemed a modest improvement in the protection of the forests.

How should the forest be protected?

The controversy between Pinchot and Graves, and then between Pinchot and Greeley, had shifted the debate from ownership and depletion to regulation and protection, although, of course, the former considerations were never entirely absent. The arguments that were bandied about over the next decade and a half were not empty rhetoric about what might be done or what might not be done. Far from being a void, the years between the Capper Report of 1920 and the next major pronouncement on the forests, the Copeland Report of 1933, were full of positive, if unspectacular, policies that were eventually to lead to real progress in forest regulation, protection, and use. Quietly but firmly, Greeley pursued his policies of cooperation and protection without the fanfare and anguish that had distinguished the creation of the national forests and the Forest Service during Pinchot's era. For example, during his first couple of years of office he pioneered, and Congress passed in March 1922, the General Exchange Act, whereby national forests were consolidated into compact blocks, land being exchanged for parts of the 24 million acres of privately owned property that had become hemmed in as a result of the checkerboard railroad grants and the arbitrary withdrawals of land for national forests by presidents in earlier years. At one blow, Greeley increased administrative and working efficiency and removed one of the festering complaints of western landowners.[50] In addition, in 1923 Greeley successfully warded off the threatened transfer of the Alaskan forests to the Department of the Interior as a prelude by Secretary Albert Fall to the total transfer of Forestry from Agriculture to Interior. Greeley's stand against Fall was vindicated totally when the Teapot Dome scandal was made public.[51]

Greeley believed that the public had a right to expect an assured supply of timber in the future but that private landholders, the suppliers of the bulk of the nation's timber, had either to make money or to be compensated in some way in order to "make compliance a reasonable undertaking." Quietly, he rethought the goals of forest management, particularly as they might be applied practically to private forests, and with reasonably sympathetic congressional support he went about framing policies on fire control, taxation, cooperative management, reclamation of the cutovers, and recreation, policies that are the foundation of many federal programs today. It was, he reminisced later, "a slow and halting revolution – not the quick reform demanded by the revivalists and crusaders," and it was going to take "another quarter century and a second world war to make the change decisive."[52]

Fire suppression

There was one thing that Greeley, like all foresters and lumbermen, knew instinctively: Wildfire destroyed the forest (Plate 13.1). It depleted the lumber stand and eliminated the young saplings that were the obvious source of future growth; to eliminate fire was one obvious way to increase timber supply. But the stopping of all fire was not a simple matter. From Indian times onward the forest undergrowth and small trees had been set aflame in order to clear the land for temporary tillage and to improve pasturage, and also to drive game for hunting. For more extensive agricultural clearings the large trees were

Plate 13.1. A fairly confined fire in the Olympic Peninsula, western Washington. Mount Rainier appears in the top left of the photograph.

felled and burned on the spot. These practices were adopted on a large scale by the pioneer farmers so that fire was as much a tool for clearing and using the forest as was the ax; it was part of a hallowed folk practice and tradition and was regarded as a sign of progress and therefore difficult to eradicate. That a controlled or prescribed fire could have a positive role in the management of the forest and, additionally, that it might be a factor in the shaping of the actual character of the forest were rarely appreciated.

But some fires were more serious. Inevitably, some of the tens of thousands of small agriculturally and pastorally induced fires got out of control, coalesced, and devastated large areas of the forest, and some of the largest fires – often caused naturally by lightning strikes in remote and inaccessible areas well beyond human control – could rage undetected for weeks and even months before burning themselves out. To these almost "traditional" causes of fire, however, was added a new cause from at least the mid-nineteenth century onward. Large-scale and unsupervised logging left a thick debris of highly combustible slash on the northern forest floor and cutovers, and this was ready fuel for the sparks from locomotives or logging campfires and stoves. Lumbermen thought that these fires were an almost inevitable consequence of logging, but for many people concerned with forest conservation such slash fires, which culminated in devastating "crown" fires, were the epitome of man's carelessness and abuse of the vegetational cover, the inevitable result of the rapacious and ruthless exploitation of the trees, which invariably ended in a holocaust. Consequently, wildfire suppression quickly became a prime and easily definable objective of forest conservation, management, and timber supply throughout the decades from 1910 to 1930, and it was seized upon as an acceptable compromise agreed to by all factions.

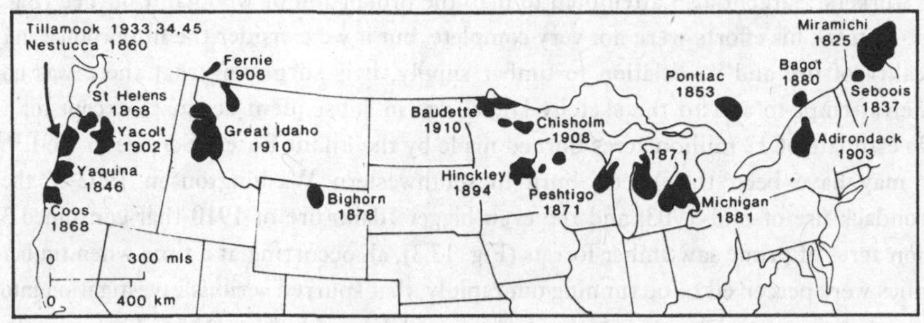

Figure 13.3. Location of some major forest fires in the United States since 1800. (Based on F. G. Plummer, *Forest Fires.*)

Accurate knowledge about the extent, location, and impact of the fires was sparse, so that the generally held concern about devastation was based on few facts and much emotion. The thousands of small fires that affected the forests, though universally deplored, were overshadowed by the massive and destructive wildfire conflagrations that burst into action from the small existing fires when the right combination of wind and related atmospheric conditions prevailed, to consume millions of acres and hundreds of lives (Fig.13.3). "Dark days," when day turned to a night produced by the drifting smoke from large fires somewhere in the interior, were first recorded in New England in May 1706, and the Black Friday of May 19, 1780, became part of folk mythology. But soon European settlers were to witness the fires at first hand. The Miramichi and Piscataquis fires that burned perhaps 3 million acres in Maine and New Brunswick in 1825 and claimed 160 lives were a dim but poignant memory by the late nineteenth century, but there were plenty of other fires from the late 1860s onward that kept the issue alive. For example, the Coos Bay and Mount St. Helens fires in Oregon and Washington in 1868 each burned about 300,000 acres; the terrible Peshtigo fire in northern Wisconsin in 1871 burned out 1.28 million acres and claimed 1,500 lives; and the Michigan fires of the same year consumed about 2.5 million acres. In 1881 the fires in the eastern Michigan "thumb" burned out 1 million acres and claimed about 160 lives; the Hinckley fire of east-central Minnesota burned hundreds of thousands of acres and caused 418 deaths; the northwestern Wisconsin fires of 1894 finally burned out several million acres; and there were many others. Most of these fires started in logging and small settled areas, and the correction and conclusion seemed inescapable.[53]

Despite Hough's county-by-county surveys of fire carried out for his *Report upon Forestry* and published in 1880, no overall view emerged, and consequently Charles Sargent set out to investigate systematically the extent of fire in his massive survey begun in the same year. As a result of his analysis of over 30,000 questionnaires, he thought that about 10.2 million acres were burned in that year, over 44 percent of the fires attributable to clearing for agriculture and improving grazing, 21 percent to hunters, and 17 percent to locomotives. The fact that logging seemed not to enter into the picture was lost on later workers. Sargent also attempted to map the proportion of woodland burned (Fig. 13.4). In total, his efforts were not very complete, but if we consider the importance and topicality of fire and its relation to timber supply, it is surprising that there was no further attempt to add to the sketchy knowledge in subsequent censuses except for a crude estimate of 12 million acres burned made by the infant Forest Service in 1891.[54]

It may have been the Yacolt burn in southwestern Washington in 1902–3, the Adirondack fire of mid-1903, and the even bigger Idaho fire of 1910 that consumed 3 million acres of prime sawtimber forests (Fig. 13.3), all occurring at a time when timber supplies were perceived to be running out rapidly, that spurred serious investigation into fires and their control by private lumbermen and Forest Service alike. For example, lumbermen in Washington had provided supplementary financial support for a state fire warden in 1903 and ultimately sponsored legislation through a newly formed Western Forestry and Conservation Association that required all private forestry owners in Washington to participate in fire control measures. The Forest Service began mapping the location of major fires – for example, in 1910 (Fig. 13.5) – and attempts were made to calculate the cost of the damage done by fires, the initial calculation being that, in round figures, at least 10 million acres were destroyed annually, causing $25 million in damage to trees and an equal amount of damage to stock, crops, buildings, and other improvements, and claiming 70 lives a year. Unquantifiable damage was done to the soil, to watercourses, and to wildlife. Above all, forest fires appeared to be burning all, if not more than, the annual growth of timber. It seemed as though moderately funded control measures would more than pay for themselves in the long run.[55]

Fire suppression did not seem to interest Pinchot much, but Graves was convinced that the disasters of 1910 pointed to the need for more forest trails, telephones, lookouts, and patrols and for a campaign to "soften" public opinion as to their need. In 1911 he successfully promoted and implemented the Weeks Act, the first federal initiative in fire suppression and the beginning of the federal forests in the East. In the beginning, the act applied particularly to the forest lands of the Appalachians and White Mountains where there had long been a movement to reserve forests for largely scenic reasons. Later, an appropriation of $11 million and an annual outlay of $200,000 were allotted for federal/state cooperation in protecting the forested watersheds of the navigable rivers, navigable river courses being the only constitutional excuse available to the federal government to transcend state boundaries. Erosion and silting were to be prevented and runoff and hence water levels equalized through the control of fires and the purchase of land in order to prevent lumbering.[56]

Much of the Forest Service's administration of the act was carried out by William Greeley, Graves's deputy, and Girvin Peters. Greeley had worked in Idaho when the great fire broke out in 1910, and the experience was so seared into his memory that fire suppression became a passion with him. With Peters he promoted conferences on fire protection, published bulletins on fire equipment and fire control methods, and publicized the crucial role of relative air humidity in the starting and spreading of fires. "We were evangelists out to get converts," confessed Greeley later, and he utilized new aids such as radios and aircraft, which he had seen in operation during wartime on the western front. Slowly the campaign began to achieve results as private owners began to

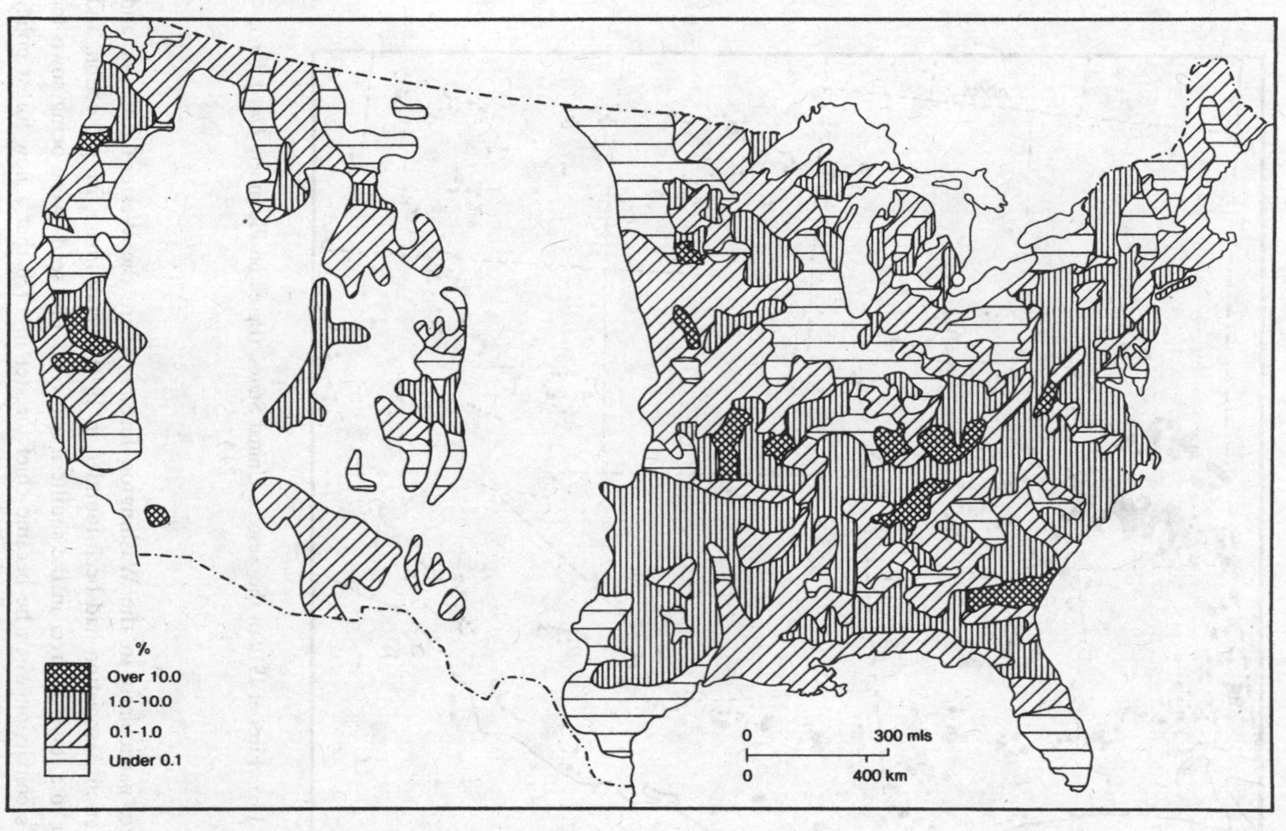

Figure 13.4. Proportion of woodland burned in the settled parts of the United States, 1880. (After Sargent, *Report on the Forests*, 491–2.)

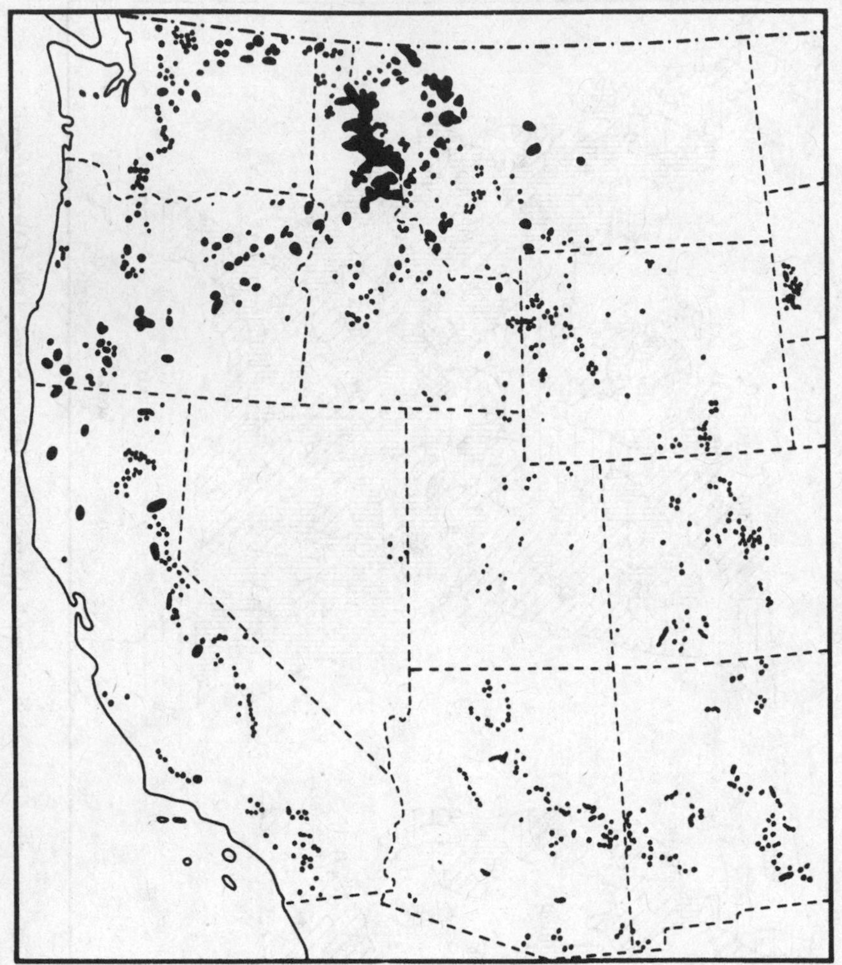

Figure 13.5. Fires of 1910 in the western United States. (Based on Plummer, *Forest Fires*, 24.)

cooperate, particularly in the Washington forests, where the Western Forestry and Conservation Association had been founded in 1910 to coordinate fire protection, and also, but to a lesser degree, in the southern cutovers. The seeds were being sown for Greeley's conviction when he became chief forester in 1920 that a new forest policy could be erected around cooperative fire protection, which would lead inevitably, so he thought, to better forest management and improved cutting practices. The cardinal point of his policy was reiterated many times: Success in fire suppression was the "yardstick of progress in American Forestry" or "The No 1 job of American foresters," which could be tackled by cooperation of federal, state, and private sectors of the industry.[57]

Greeley realized that considerable progress had been made already by the states under

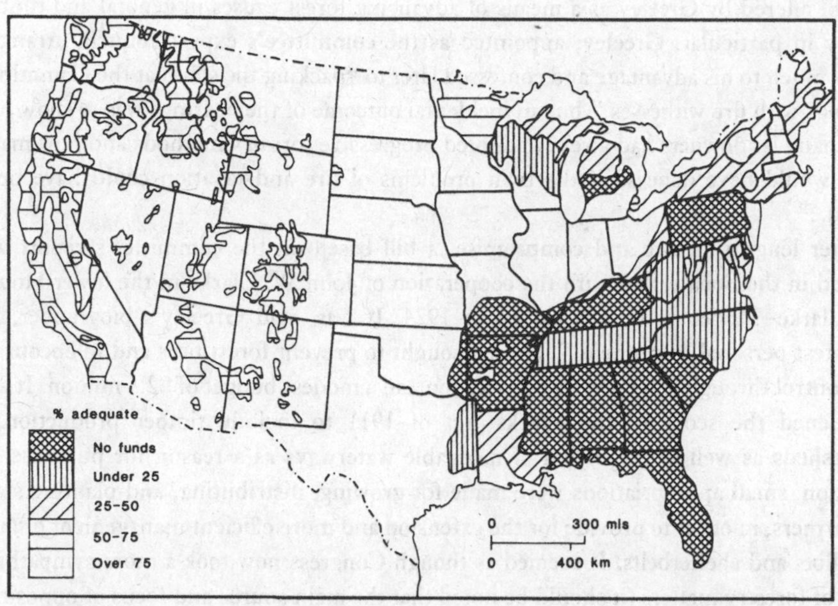

Figure 13.6. Organized protection against forest fires; funds spent in relation to funds received. (After W. Greeley et al., *Timber: Mine or Crop?*, 160.)

the guidance of Graves. In the national and state forests nearly 95 million acres were protected. In the private forests about 160 million acres were protected, mostly through the encouragement of fire protection in watersheds under the Weeks Act. Nearly all states had established forest departments, usually headed by trained foresters, and nearly all states had legislation providing for the control of fires. To be sure, there were still serious omissions: Eight southern states covering over 166 million acres of the most highly prized forest of the country had no protection, and other than Pennsylvania and Massachusetts all eastern states still left much to be desired in the coverage of their fire protection schemes (see Fig. 13.6). "The task," said Greeley, "is at present two-thirds done." Nevertheless, Greeley still wanted to build on what he saw as the solid progress already made and to cooperate with existing organizations as a way of advancing the cause of forest protection and hence securing national timber supply. His concentration on fire control was a means of deftly sidestepping contentious and difficult issues that might be resolved later.[58]

The failure of the Capper and Snell bills for forest protection and their many versions meant that the congressional battles came close to a stalemate. The way out came in January 1923 with the appointment of a Senate Select Committee on Reforestation, under the chairmanship of Sen. Charles L. McNary of Oregon, to consider various proposals to ensure the supply of timber in the future and to conduct hearings throughout the lumber areas of the country. Clearly, Greeley and McNary hit it off well; the difficulties of persuading the lumbermen of their responsibility to reforest the 81 million acres of cutover convinced McNary of the virtues of the compromise of fire control offered by Greeley as a means of advancing forest causes in general and timber supply in particular. Greeley, appointed as the committee's expert council, arranged affairs much to his advantage and confessed later to "packing the stand at the Committee hearings with fire witnesses," but an incidental outcome of the hearings was to show just how many landowners had already adopted progressive forestry methods and how many more would have done so if the twin problems of fire and taxation could have been solved.[59]

After lengthy debate and compromise, a bill based on the committee's report was framed in the Senate, and with the cooperation of John W. Clarke in the lower House, the Clarke–McNary Act was passed in 1924. It was, said Greeley's biographer, his "greatest personal monument." The act sought to prevent forest fires and to encourage fire control through federal/state cooperation and a modest budget of $2.5 million. It also broadened the scope of the Weeks Act of 1911 to include timber production in watersheds as well as protection of navigable waterways as a reason for purchase. In addition, small appropriations were made for growing, distributing, and planting stock for farmers, in order to provide for the extension and more efficient management of farm woodlots and shelterbelts. It seemed as though Congress now took a more sympathetic view of forestry matters (it should be noted that the main source and focus of opposition and obstruction, Gifford Pinchot, was diverted as he ran for and became governor of Pennsylvania) and, in what was quick succession compared to the inactivity of the previous decades, passed three more bills: the McSweeney–McNary Act and McNary–Woodruff Act of 1928 and the Knutson–Vandenberg Act of 1930. The details of these bills are numerous and complex, but the nub of them is as follows. The McSweeney–NcNary Act supported a $3.6 million research program into all matters relating to the forests: for example, more funding for the Forest Products Laboratory at Madison; more regional forest experimental stations; a survey of the forests and an inventory of standing timber, its growth and drain; in fact, all the things that Fernow had pressed for over 30 years before. The McNary–Woodruff Act provided money for the purchase of additional lands in watersheds, already authorized under the Clarke–McNary Act of 1924 but not provided for, with the stipulation that not more than 1 million acres could be purchased in any one state. Finally, the Knutson–Vandenberg Act of 1930 provided funds for the planting of trees in the national forests and in cutover areas.[60]

Taken together, the four acts had moved the forests from a neglected adjunct of public lands, or agriculture, or a host of other industries, to a joint federal, state, and industry responsibility. Federal coercion was a dead issue. The forest was a resource that needed nurturing for the future, and although new legislation altered this and that detail, or provided more or less funds as the case may be, the foundations of these policies had been laid in the 1920s. Many years later Greeley reminisced that "the turn in our forest economy from the timber *mine* to the timber *crop* came in the decade ending with 1930."[61] He was right and he had played a major part in that change of attitude. Perhaps the one issue he had not foreseen was that the forest had a value over and above that of timber supply, and during the years that followed the debate on the recreational and aesthetic value of the forest as playground and wilderness was going to consume about as much energy as all previous issues put together.

Taxation

The taxation of forest land was a complaint that had rumbled below the surface in all the lumbermens' utterances about the state of the industry since the middle of the nineteenth century, and it was claimed to be a deterrent to conservative, sustained yield in lumbering practices.

Taxation was a state matter, and many states wanted to boost local revenues in predominantly forested areas by taxing forest land in a manner similar to farmland, that is, on the ordinary annual general property tax assessed on the actual value of the product. But whereas a crop could be taken off farmland almost annually, timber could be taken off only after an interval of 50 to 80 years. Thus it was possible (and in fact happened) that the same tree could be taxed year after year although it could be cut and sold only once. In addition, the long-term nature of tree growing and the volatile nature of taxation meant that lumbermen were never sure of their capital investment. For example, would reforestation after logging attract a new high tax, which might just be acceptable in the long-term view? On the other hand, would a thinning after 10 years attract an even higher tax that would cripple the profitability of the reforestation?[62]

Faced with such uncertainties, many lumbermen realized their assets as quickly as possible, logged off thoroughly, disposed of the land quickly so as to avoid taxes and fire, and bought new standing timber. This pattern was very evident in the Lake States and the South where it was compounded by the abandonment of tax-delinquent land in the cutovers, the migration of the lumbermen to new areas, and a lack of reforestation, with the concomitant impoverishment of the local communities.[63]

Many of these issues were aired nationally in the report by Fred R. Fairchild to the National Conservation Commission in 1909. Although he concluded that the incidence of adverse taxation systems was probably not the cause of destructive logging, as many of the logged-off areas were often remote and not taxed, the fear of taxation, either actual or potential, and the compounding effect of tax-delinquent land on tax levels in the Lake States, in particular, were contributory causes to destructive logging and abandonment. By 1910, 26 states, and then another three immediately afterward, made adjustments in their property taxes in the form of rebates, exemptions, and bounties in recognition of these problems, and during 1920 some 10 states had passed "severance" or "yield" taxation laws, that is, a tax on the product only when it was harvested. Nevertheless, the problem of taxation still loomed large enough in the minds of the lumbermen and others for the Capper Report to recommend a thoroughgoing report on the problem.[64]

As the McNary committee moved from place to place across the country in 1924, many witnesses testified that taxation was as great a deterrent to conservative forest management as was fire, even though Greeley had "packed" the stand with "fire" witnesses. Eventually, the committee appointed Fairchild to produce a comprehensive report on taxation, and nearly eight years later his committee reported. Surprisingly, the massive 681-page report did not find that taxation forced rapid cutting and liquidation of holdings – it was overcapitalization in land and mills that did that – but it did conclude that unpredictable taxation deferred reforestation. It recommended a simple annual tax on the forest as bare land and a severance tax on the lumber when cut. But the recommendations of the Fairchild committee had little immediate effect (other than the notable exceptions of Washington and Oregon), possibly, suggests Harold Steen, because "it did not find property taxes to be the overwhelming villain as portrayed in the forest literature." Of the twin problems of how to manage the forest, fire was by far the more obvious and easier to understand when compared to the subtle complexities of the entrenched local and state interests in tax. Moreover, many forest owners, once they stopped generalizing about the problem and got down to specific cases of what the land would actually produce, found that taxes "were not the bogey they had previously pictured." An illustration of the changing nature of the problems is contained in the 1949 *Yearbook of Agriculture*, which was devoted entirely to the subject of "trees." One page only was enough for the topic of taxation, but fire could not be covered in less than 56 pages. Taxation was a dead issue.[65]

How should the forest be used?

The question of how the forest should be managed, and by implication how it should be used, had been an obvious adjunct to the debates about preservation from the 1890s onward. But just as preservation alone gave way to more sophisticated concepts of use, higher use and multiple use – conservation, no less – so conservation itself came under intense scrutiny, and in a sense the pendulum swung back to preservation. Arguments about the relative value and importance of the material and nonmaterial benefits of the forest, of, for example, timber supply versus wilderness preservation, or the "gospel of efficiency" versus the "gospel of beauty and pleasure," became polarized increasingly, and the two ways of thinking were personified in the ideas of Gifford Pinchot and John Muir. Their disagreement came to the fore in the celebrated inquiry of 1913 into the proposal to build a dam in the Hetch Hetchy Valley in the Yosemite National Park in order to supply water to San Francisco. The inquiry was the catalyst that activated the scattered sentiments about preserving wilderness into a major national movement.

Although the "users" won the day in the Hetch Hetchy episode, many people were aware that the success of the early movement to preserve the forest owed much to amenity arguments, which were put forward in the preservation of the Adirondacks, the White Mountains, and the Sierra. Now, with the creation of 180 million acres of national forest on the public lands in areas of sparse settlement and great natural beauty, it was perhaps reasonable that they wanted more land set aside purely for pleasure or recreational purposes. The advent and the impact of the automobile, which gathered momentum from the mid-1920s onward, added force to these demands, with more than 11 million people visiting the forests in 1924 (Plate 13.2). Naturally enough, it was the Forest Service, guardian and manager of the "people's" forests, that bore the brunt of these demands, and there is no doubt that the Forest Service adjusted to these demands slowly and uncomfortably. It saw recreation as a lesser use to that of supplying timber or even protecting watersheds or grazing, areas of activity in which all its early battles were fought. In its quest to show that it was efficient, that it could make a profit in some of its activities and was not, therefore, a total drain on the public purse, it was not disposed to

Plate 13.2. The automobile completely altered the concept of how the forest should be used. Automobile-led recreation was a crucial factor in the emerging debate between using the forest and preserving it, and in the concept of multiple use.

put the unquantifiable element of recreation on the same level as its other activities. In addition, the service was genuinely concerned that the recreational use of the forest increased the danger of fire and added to the pollution of water bodies.[66]

The question of putting parks and recreation high on the list of priorities of uses in the national forest management was constantly bedeviled by the fact that national parks were the province of the Department of the Interior, and they were constantly being enlarged and multiplied at public behest but always at the expense of national forest land. Despite Forest Service efforts to stop the creation of a separate parks bureau, the Bureau of Parks and its attendant National Park Service came into being in 1916, largely, suggests Nash, as an outcome of the passions raised during the Hetch Hetchy debate. Relations between the two were never easy despite deliberate efforts of both chiefs – Mather in Parks and Graves in Forest – to avoid anything that would create friction. Nevertheless, the overlapping and inherently competitive functions of the two services made for difficulties, and in addition the Forest Service was unduly sensitive to the newcomer in the bureaucratic ranks of Washington, D.C. Perhaps prompted by the creation of the new service, the Forest Service initiated in the next year an intensive study by Frank Waugh of its own recreational activities and potential. The study came out strongly in favor of a positive role for recreation, which was seen as a paying proposition. Fighting the public perception of the Forest Service as utilitarian and hostile to parks (a stigma that attached itself, unjustly, as it transpired, after Pinchot's testimony at the Hetch Hetchy inquiry), Graves issued a policy statement supporting national parks but accompanied by a plea for some stability in the situation and an end to the erosion of the forest reserves. But for all his sympathy and understanding, in the final analysis Graves could not accord recreation as high a priority as timber supply; the timber-famine controversy was always in the background. For example, Graves could not understand the critical public reaction to clear-cutting in the national parks. To him "the parks should comprise only areas which are not forested areas covered only with protective forest which would not ordinarily be cut."[67]

Greeley inherited these difficult problems, which were now compounded by an unlikely alliance between national park advocates and forest product users, who argued that any recreational program "was redundant and a misuse of the forests." Eventually in 1924, the issues were brought to a head by the newly formed National Conference on Outdoor Recreation (NCOR), which requested information on recreation in all federal lands. It was gradually becoming evident to anyone involved in recreation in the forested lands of the country that the concern was shifting from the provision of facilities for recreation – roads, cabins, hotels, which clearly gave an economic return – to the exclusion of facilities; in other words, the focus was now the preservation and maintenance of forest or wilderness areas. Simply, it was a shift from development to nondevelopment. To his credit, Greeley was perceptively sympathetic to the views of some of his staff, notably Arthur H. Carhart, and to Aldo Leopold, who was a forester in the Southwest and who proposed a "wilderness preserve" in the Gila National Forest, New Mexico. In 1924 Greeley converted nearly 600,000 acres of the Gila forest into the nation's first "wilderness area" and enunciated the policy that access should be only by foot, horse, or canoe.[68] But Greeley was treading a difficult path. Recreationists wanted roads and access, preservationists wanted untouched wilderness areas and roadside screens of uncut trees, and in Congress there were those who objected to the service's devoting so much time and money to these ends.

By the time of the Copeland Report in 1933, the ever-growing demand for recreation could be measured and was seen to be growing rapidly, with increased affluence, mobility, and leisure time (Fig. 13.7), although the enormous future demand was never foreseen. Robert Marshall, the energetic and brilliant articulator of the concept of wilderness as a source of peace, mental stability, and contemplation, and ultimately the founder of the Wilderness Society, wrote the recreation section of the report. He was convinced that private owners would never respond positively to recreational and wilderness needs, and he therefore recommended the setting aside of 45 million acres of land allotted to wilderness, primitive, and other recreational areas; for example, the 10 million acres of the proposed wilderness areas were described as areas with no permanent habitation or mechanical conveyance, and large enough for a person to walk "a week or two of travel without crossing his own tracks."[69] Undoubtedly, Marshall was ahead of his time in his crusading advocacy of wilderness, and although he did not have the satisfaction of seeing his recommendations implemented, he was made director of forestry in the U.S. Office of Indian Affairs and by 1937 had succeeded in establishing 16 wilderness areas on Indian reservations. In the same year he became head of the Forest Service's Division of Recreation and Lands, a position that allowed him to pursue his passion more effectively, aided by the real and imagined threats of a takeover by the Parks Service of recreational and scenic areas and of the whole Forest Service by the Department of the Interior.[70] Ultimately, in 1939, the U-regulations were promulgated, which established three categories of wilderness in the forest: wilderness, wild, and roadless. It was the beginning of an extended policy of setting aside wilderness areas, and to its credit, up until the

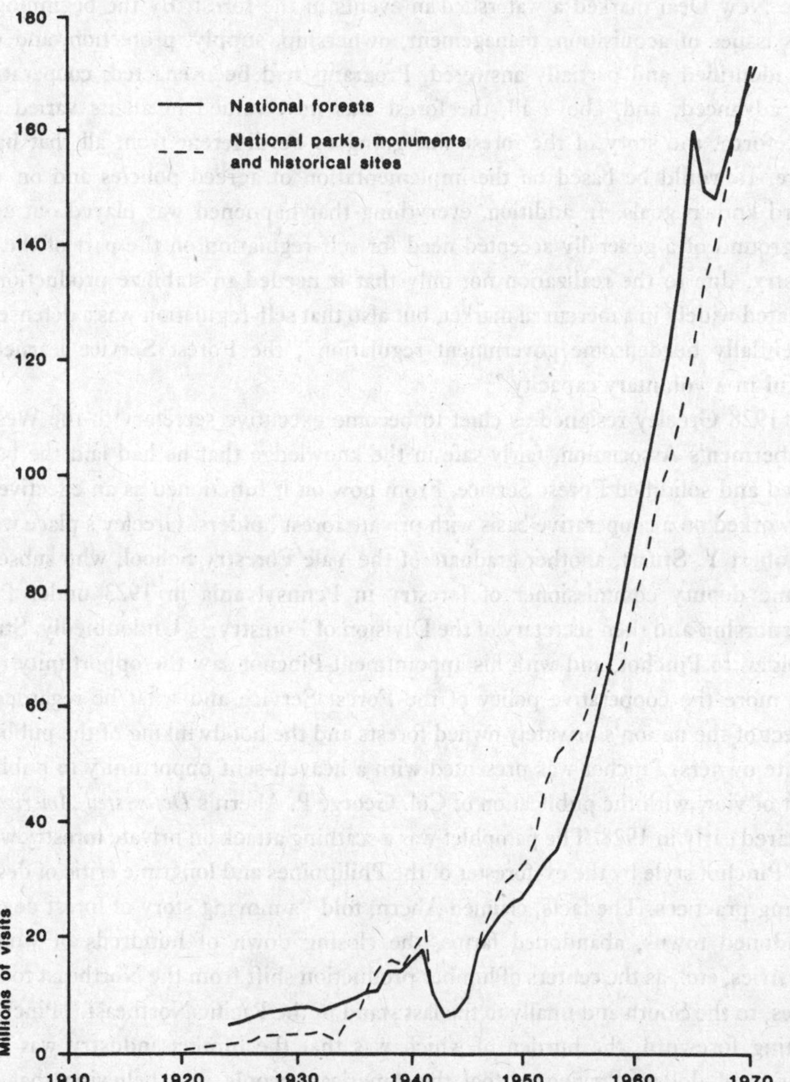

Figure 13.7. Visits to national forests, national parks, museums, and allied areas, 1923–70. (U.S. Bureau of the Census, *Historical Statistics*, tables 808, 829, 832.)

passing of the Wilderness Act of 1964, the Forest Service had set aside 14 million acres of its forests in this manner.[71]

Transition

As in so many other aspects of American life, the years between the time of the crash of 1929 and the beginning of the presidency of Franklin D. Roosevelt in 1933 and the start of the New Deal marked a watershed in events in the forest. By the beginning of the 1930s issues of acquisition, management, ownership, supply, protection, and use had been identified and partially answered. Programs had been enacted; cooperation had been advanced, and, above all, the forest was now valued in all its varied aspects. Henceforth, the story of the forest was going to be different from all that had gone before. It would be based on the implementation of agreed policies and on working toward known goals. In addition, everything that happened was played out against a background of a generally accepted need for self-regulation on the part of the lumber industry, due to the realization not only that it needed to stabilize production which oscillated widely in a mercurial market, but also that self-regulation was a defense against "potentially burdensome government regulation"; the Forest Service learned to be helpful in a voluntary capacity.[72]

In 1928 Greeley resigned as chief to become executive secretary to the West Coast Lumbermen's Association, fairly safe in the knowledge that he had laid the basis of a unified and solidified Forest Service. From now on it functioned as an effective agency that worked on a cooperative basis with private forest holders. Greeley's place was taken by Robert Y. Stuart, another graduate of the Yale Forestry School, who subsequently became deputy commissioner of forestry in Pennsylvania in 1923 under Pinchot's governorship and then secretary of the Division of Forestry.[73] Undoubtedly, Stuart was beholden to Pinchot, and with his appointment Pinchot saw the opportunity to attack once more the cooperative policy of the Forest Service and what he regarded as the neglect of the nation's privately owned forests and the hoodwinking of the public by the private owners. Pinchot was presented with a heaven-sent opportunity to publicize his point of view with the publication of Col. George P. Ahern's *Deforested America*, which appeared early in 1928. The pamphlet was a scathing attack on private forestry written in pure Pinchot style by the ex-forester of the Philippines and longtime critic of destructive logging practices. The facts, claimed Ahern, told "a moving story of forest devastation, abandoned towns, abandoned farms, the closing down of hundreds of wood using industries, etc., as the centers of lumber production shift from the Northeast to the Lake States, to the South and finally to the last stand in the Pacific Northeast." Pinchot wrote a biting foreword, the burden of which was that the lumber industry was spending millions of dollars "trying to fool the American people into believing that . . . the industry is regulating itself and has given up the practice of forest devastation. That is not true."[74]

Immediately, Pinchot enlisted the aid of Senator Capper, who asked for 10,000 copies to be printed by the Senate and then sent them to Pinchot who mailed them with a covering letter to influential persons and important organizations. Otherwise, he said, "the big press will kill it by silence" because of its "teamwork" with the lumber industry. He used all his considerable rhetorical skill in his attempt to stir opinion and concluded with a characteristic flourish: "We are still sowing in the wind, and the whirlwind is not far off."[75]

Although critical of the tardiness of the lumbermen in cooperating, and under great personal pressure form Pinchot, Stuart was not stampeded by his onetime boss and his host of influential friends. There was some evidence that the private companies were managing the forest on a long-term basis. Therefore he did not dismantle any previous legislation but embarked on the popular policy of increasing purchases of forest land for public ownership, an emphasis reflected later in the recommendations of the Copeland Report that the federal government should embark on an ambitious 20-year program of purchasing 225 million acres of cutover and tax-delinquent land. In recommending more public ownership, Stuart was taking a midway position between those of Greeley and Pinchot, and in reality it was the only sensible one because after the crash of October 1929 the forest industry was incapable of thinking of conservation practices as it reeled under the blows of financial failure and depression. Overall, production plummeted by two-thirds, from 39 billion b.f. in 1929 to a mere 13.5 billion b.f. in 1932. The number of mills reported as being in production also fell by about the same proportion, from 20,037 to 6,838. No state escaped the devastation of production (Fig. 13.8). Regionally, the Lake States, California/Nevada, the northern mountain states of Montana and Idaho, and the southern states of Arkansas, Alabama, and Mississippi were the hardest hit, and only the eastern and northern central states, together with the central mountain states of Wyoming, Utah, and Colorado, managed to maintain production at more than half the level of three years before. The impact on the economy was greatest in the biggest producing and most commercially oriented states, such as Washington, Oregon, Alabama, Mississippi, and Louisiana, for which the fall in production accounted for more than half (13.3 million b.f.) of the total national decline. It was a rout. The lumber industry, said Greeley, "degenerated into a struggle for survival."[76]

As the impact of the depression deepened, the industry sought relief, and the upshot was the appointment by President Hoover of the Timber Conservation Board in November 1930. The board was charged to attempt to equalize production and consumption, which were widely out of kilter. Among the many recommendations of the board, which included the usual pleas for more fire protection and land acquisition and less taxation, was a more controversial proposal, based on the ideas of lumberman David T. Mason, a member of the board, to coordinate public and private timber supplies in order to achieve a "sustained yield," an unconventional use of the term, which meant simply that over a long period of time the volume logged would not exceed the volume grown, any deficiencies of supplies being made up by the release of timber from the national forests. Stuart, for the Forest Service, was skeptical that production in the private sector could be either controlled or enforced to produce the desired result and that federal supplies were either sufficient or located strategically enough to influence overall production. He was less convinced of cooperative approaches than any of his predecessors. Knowing that the Timber Conservation Board could only recommend and not act, the Forest Service decided to use the legislation of the New Deal to achieve regulation and stabilize production. It invoked the National Industrial Recovery Act (NIRA) to establish a code of "Conservation and Sustained Production of Forest Resources," or Article X, as it became known, a package of minimum standards of logging practice and production worked out by the Forest Service in conjunction with the lumber industry, whose negotiators were Greeley and Mason. Ironically, no sooner had these voluntary but revolutionary codes of self-regulation been agreed upon and put into operation by about two-thirds of the private companies than the Supreme Court

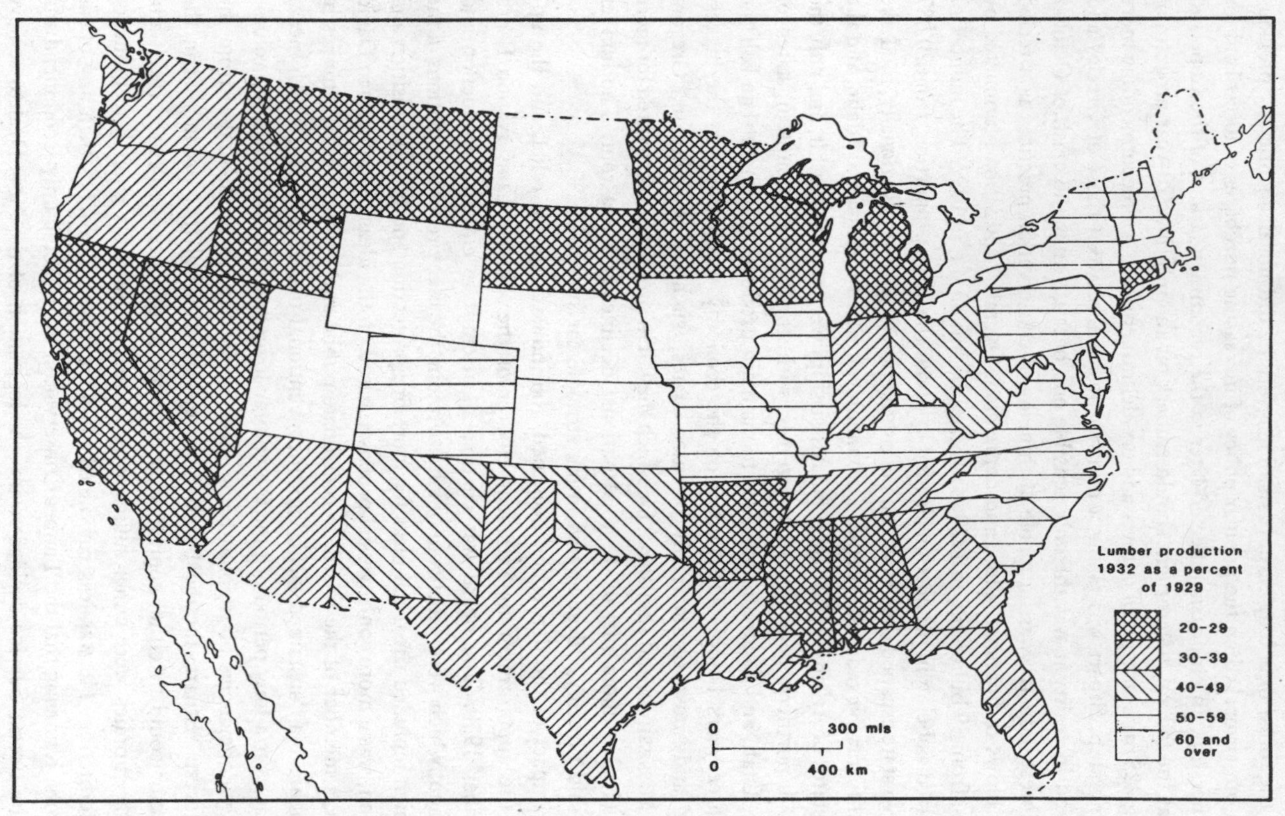

Figure 13.8. The decrease in lumber production in the United States, 1932, as a pecentage of 1929. (U.S. Census.)

ruled unanimously on May 27, 1935, that any industry-backed code involved an unconstitutional delegation of power and that, therefore, the whole of the NIRA program was unconstitutional and was to be dismantled. Its influence lingered on, however, in the general hope for regulation.[77]

The other great event in the story of the forests during the early 1930s was the publication of the provocative and massive 1,677-page Copeland Report, or *National Plan for American Forestry*. Commissioned in 1932 to reevaluate and update the Capper Report of 1920, it was produced in the incredibly short space of 12 months and was a tribute to the energy and editorial expertise of Earle Clapp, head of research in the Forest Service. Besides providing a massive inventory and review of the forest during the early 1930s, the report made new and positive recommendations about the future objectives of forest policy. Conceived at the time of the depression and in the mood of the New Deal, and at a time when the Forest Service was attempting to regulate private operators, the report was imbued with concepts of social welfare and practical management. It was highly critical of private forestry, which had been ineffective, it thought, in promoting national goals during the last 20 years when compared to public forestry. Although holding 90 percent of the growth capacity of the forests, the private owners had expended a mere 10 percent of "the total constructive effort" whereas the reverse figures were true for the public forests; this imbalance could not continue.

"The essence of the plan," said Clapp, is that "the public should in the shortest possible time take over at least half of the national enterprise forestry"; that is to say, it should purchase 225 million acres of private forests, including 32 million acres of abandoned farmland. Total multiple use of the forest was also envisaged, in which "the entire range of the management, protection, and administration of the timber watershed, recreational, wild life, forage, and other resources which make up the forest" would be undertaken in order to meet the objective of providing a "full economic and social service." In return for providing public aid and protection, the private sector would have to submit to regulation; it was the "quid pro quo in the public interest for concessions to private owners."[78]

The report and its recommendations were controversial and even "revolutionary" in a political sense, and despite extensive preparation by the Forest Service for a bill along the lines of the report, nothing happened. Congress found it too polemical and considered its recommendations disputable during times of economic hardship, and the intended bill was shelved. It was going to take a world war to shift Congress into action.

Nevertheless, other events were afoot during the early 1930s, under the innovative and reformist zeal of Franklin D. Roosevelt, that were to affect the forests directly by increasing their size and indirectly by raising public awareness of the value to the nation of its tree cover. With the creation and launching of the Civilian Conservation Corps (CCC) early in 1933, many of the unemployed youth of the country were put to constructive, long-term conservation work, of which tree planting and fire protection in the forests were a high priority. The creation of the Tennessee Valley Authority (TVA) resulted in a general acceptance of multipurpose river schemes in which reforestation played an important part, and the shelterbelt project, started in 1934 and designed to combat soil erosion and drift in the Dust Bowl, similarly resulted in millions of trees

Plate 13.3. Although the lumber industry moved into a period of prosperity in the 1940s with wartime demands, the legacy of past profligacy was still fresh in people's minds. It was, as the cartoonist J. N. "Ding" Darling said and drew, "Time to take an inventory of our Pantry."

being planted, in millions of acres being rehabilitated, and in a new public awareness of the role and utility of positive forestry practices.[79]

On October 23, 1933, Robert Y. Stuart, chief of the Forest Service, died after falling from the seventh floor of his office in Washington. The debate as to whether it was an accident or suicide was unresolved, but all were agreed that the pressure of work and the burdens of office during the complex and difficult years of the early 1930s had killed him.

In a sense, his death marked the end of an era begun in the early 1920s, an era in which all the elements and considerations that were to affect the story of the American forests until the present time had been laid bare and discussed. Goals had been defined and programs worked out. The basis of the new cooperative approach had been laid by the Clarke–McNary act of 1924, which had allotted federal funds for state forest fire programs, so that the 140 million acres protected in 1921 had risen to 223 million acres by 1933. The McSweeney–McNary Act of 1928 ensured the permanence of this approach. Contemporaneously, national forest lands were consolidated or expanded by acquisitions under the Clarke–McNary and the McNary–Woodruff acts of 1924 and 1928. Concentrating on watersheds and cutover areas, the government had purchased another 3 million acres by 1933, and the General Exchange Act of 1922 had tidied up forest holdings so that the net area of the federal forest had increased from 156.6 million acres to 162 million acres in 1933. In the private sector, although Article X had to be dismantled, the forest associations decided to adopt it as association policy so that by 1935 about 200 million acres were covered by minimum forestry regulations. Vast areas of the cutover lands still remained and much was still to be done, but an important start had been made, and new forces, new concepts, and a new acceptance of purposeful forestry practices as a means of rehabilitating the land had been set in motion with the CCC, the TVA, and the shelterbelt project.[80] All the elements and considerations that were to affect the story of the American forests during the years until the present had been laid bare and discussed by 1933 or 1934. The lumber industry survived, strengthened by the stimulus of the New Deal and then by wartime conditions and demands after 1941 (Plate 13.3).

14

The rebirth of the forest, 1933 and after

> The capacity of natural forest lands to regenerate timber stands and the capacity of timber to grow, even in the absence of man's help and often in spite of his wishes, tend to be overlooked or ignored.
>
> Marion Clawson, "Forests in the Long Sweep of American History" (1979), 4

By the early 1930s changes of major significance were underway in the American forest. A few discerning people might have read the signs of impending change, but for the vast majority what was happening went unnoticed and unappreciated. They were not to blame; the destruction and reduction of the forest had been the overriding experience for three-and-a-half centuries, and it was difficult to understand and recognize anything that was different from that common experience. Moreover, perceptions and policies had been notoriously short-term in nature – that had been the problem in the past – whereas forest growth and forestry were long-term matters. For example, few people realized the ultimate capacity of the forest to regenerate and of the timber stands to grow despite the destruction of previous decades. Therefore, seeing anything as having a significance beyond the contemporary present was alien to the forest users. Consequently, the characteristics of the age would appear only when they were measured against developments of later years.

Not all the trends and developments during the half-century from about the mid-1930s to about the present are or can be dealt with here. There are too many, and in a sense they are not altogether relevant to the main focus of this book, which is the forest as a geographical phenomenon that has changed through time in response to man's needs, wants, and ideas. Diminution of that forest cover was the theme of the first 300 years of occupation by Americans. It was only after about 1870 that the diminution was questioned seriously and efforts were made to inquire into ways of preserving, administering, and controlling the use of the forest for an expanding population for a longer time.

Since the mid-1930s inquiry and concern have continued unabated. Arguments over intensely and deeply felt issues have dominated: for example, the recurrent battles between the Departments of the Interior and Agriculture for the control and administration of the Forest Service and hence the national forests; the constant swing of opinion as to the virtues of coercion and cooperation between federal authorities and private lumber owners and millers over the regulation of logging amounts and logging practices; and the repeated conflict, particularly in the West, among timber, grazing, water, recreation, wildlife, and other interests in the allocation of uses in national and latterly private forested areas. These and many more issues have been dealt with in a host of publications.[1] Perhaps it is sufficient to say that the Forest Service has remained within Agriculture where it began; that the cooperative philosophy, begun in this century and given impetus by Greeley after the 1920s, yet so severely questioned during the 1930s, has dominated thinking during the 1950s and after; that a largely privately owned forestry industry has attempted to put its own house in order after 1941 with, among other things, the voluntary tree farm movement, which now has 39,000 sites and encompasses 80 million acres; that the conflicting claims of all forest users, in the national forests at least, have been catered to in a fashion by the Multiple Use–Sustained Yield Act of 1960; and that all federal lands are now subject to the regulations of the Forest and Rangeland Renewable Resources Planning Act of 1974.[2]

Many of these moves toward conservation have been the result of self-interest on the part of the lumber industry (with the acquiescence of successive administrations) to ensure survival amid the turbulent and chaotic market conditions of this century, conditions created by overproduction, economic fluctuations, and intense competition. Nevertheless, whatever the motives, since the Second World War some semblance of order and stability has prevailed where none did before, and this is a new era in forest politics and policies.[3]

The need in this final chapter, however, is to get back to the forest as geographical entity and to ask what is happening to it now. This means that the themes that most materially affect the extent, density, and appearance of the forest are those selected. In that way the story of the forest in the past may be balanced by the story of the forest in the present.

The rebirth of the forest

Whatever measure is taken, the conclusion seems unavoidable that the forest is building up today rather than declining. The trends of the last three-and-a-half centuries have been reversed so that the conclusion seems unavoidable that the forest is being reborn. The amount of land in commercial forest (see Table 13.3) has risen from a low point of about 461 million acres in 1944 to 483 million acres in 1977. The amount of land in noncommercial forest has increased even more, from a low point of 120 million acres in 1930 to 254 million acres in 1977. Standing timber has risen from 1,601 billion b.f. in 1944 to 2,569 billion b.f. in 1977. The annual drain on timber inventory is down from 23 billion to 18.1 billion cu. ft., whereas annual growth is up from somewhere near 6 billion cu. ft. in 1920 to a current estimate of about 36 billion cu. ft., an increase of about 350 percent (see Table 13.5). The evidence of one's eyes is better than statistics. Regrowth can be seen everywhere, and one is struck by the robustness of the forest.

The causes of the change from decline to growth, from death to rebirth, are many, but a few stand out for special mention, either because they are important in current forest management (and will be even more important in the future) or because they were the direct outcome of events, policies, and legislation that first became evident during the 1920s. One cause for the regrowth has been dealt with already, that is, the past cutting that has allowed the new harvest to grow. Not considered, however, is the abandonment of cropland and its reversion to forest, the suppression of fire and also the use of controlled fire, the regeneration and replanting of trees, better management of existing stands, and the falling demand for lumber and lumber-derived goods. When all these are looked at in detail, the conclusion is inescapable that the forest is entering into an era of increased application of scientific knowledge and careful management of the trees, all undreamed of in the "cut out and get out" days of previous decades and centuries.

The new harvest

The enormous growth of timber during the last 40 years has been possible only because the old forest has been removed. The old forest also grew, of course, but the yield of timber was minimal and could occur only as a result of mortality due to fire, storm, insects, disease, and old age. Therefore, without the clearing of the mature timber of the past during the last hundred years or so (whether done carefully or carelessly), there would not be the magnitude of timber growing at present to supply the needs of today. The abandonment of pasture and cropland to regrowth forest has also increased the opportunity for stock to grow, although often to species different from those of the original forest stand.

The power of the forest to regenerate is enormous (Plate 14.1). "Timber growth," as Clawson points out, "is a function of timber harvest," but other factors are important, too.[4] The major proviso to that relationship is that uncontrolled wildfires have to be suppressed in order that the young saplings and trees can colonize the cutover land, although there are some notable exceptions where some species, such as the longleaf pine in the South, respond to periodic controlled fires.

Farmland reversion

Even as Pinchot and his clique were proclaiming the coming timber famine during the early years of the century, the forest was making its comeback in the eastern half of the country. There always was a great deal of forest left in farming (Fig.14.1), but now farmland was being abandoned to forest. The suppression of forest fires allowed a greater growth of all existing trees and particularly of the young, low-standing trees in the first stages of recolonization of the abandoned fields. As early as 1840, farmland was being abandoned in New England and was gradually reverting to forest. After 1880 the process began in the Middle Atlantic States of New York, Pennsylvania, and New Jersey, and after 1910 in the east-central states of Ohio, Indiana, and West Virginia. In later years the incidence of abandonment spread to the South as old cotton and tobacco fields reverted to pine forest.[5] Of course, not all land reverted to forest. Some was purposefully converted to other uses. For example, in the vicinity of metropolitan locations, the expansion of industry, residences, and communications gobbled up the land, particularly after 1945. Again, the post–Second World War years saw major losses because of strip mining in eastern Ohio, Tennessee, western Pennsylvania, and West Virginia. But the bulk of the abandoned land did revert and always has reverted passively to forest.[6]

Plate 14.1. The ability of the forest to regenerate naturally after a time was always underestimated. *Above*, timber exploitation at Bear Gulch, in the redwood forest of northern California, in about 1888. *Below*, the same spot about 40 years later.

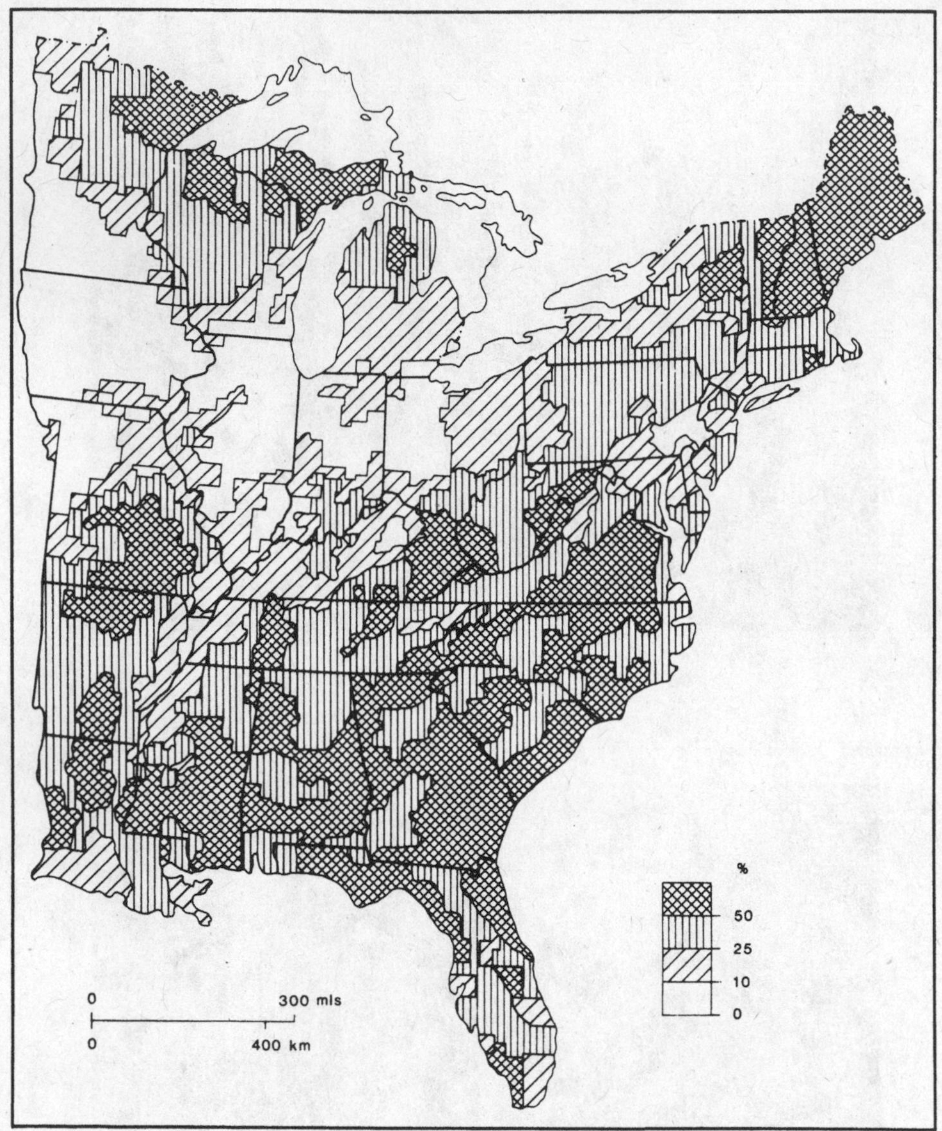

Figure 14.1. Woodland as a percentage of total farmland, 1959. (After J. F. Hart, "Loss and Abandonment of Cleared Farm Land in the Eastern United States," 419.)

Abandonment was a result of a complex set of forces that varied spatially and temporally across the continent. In New England and the Middle Atlantic States, land was often physically difficult to farm, being stony, steep, infertile, and eroded actively. Much of this land was also climatically marginal for many crops. Nearly all of it was created with the application of enormous pioneer effort. Economically, then, these lands were poor compared to the new lands that opened up in the Midwest and West, where climate and soils were favorable to agricultural production and where the generally level nature of the ground allowed the use of large-scale machinery. To these natural advantages could be added the ever-improving transport facilities that enabled the western farmers to ship their produce east at prices well below those the local eastern farmer could afford to take. As a consequence, many eastern farmers migrated to new farming areas as well as, of course, to the many industries growing up in the towns in the Northeast.

In the South the situation was not very different. There had long been a tradition of temporary land abandonment before the advent of fertilizer, as part of a strategy of combating soil exhaustion and/or erosion. Traditionally, "old" fields had been allowed to revert periodically to pine, but this had little effect on the total acreage of agricultural land, as "resting" in one area was usually compensated for by clearing in another area. What was different in the South was the long-term and often long-distance shifts of major crops from the piedmont of South Carolina, Georgia, and Alabama to the coastal plain and to the new, irrigated areas farther west. In the smallest farms throughout the South, these shifts of staple crop production encouraged the more widespread abandonment of all land in farms. For example, whereas there were 4.1 million acres of cotton in 88 counties in the piedmont in 1920, there were only 0.3 million acres left by 1959, but a total of 5.8 million acres of land had been abandoned in those same counties. In the piedmont of Virginia, the shift in production from the once-favored dark, fire-cured tobacco to the milder flue-cured type resulted in a drop in acreage in 24 major tobacco counties from 172,000 to 73,000 acres but in an overall abandonment of 1.3 million acres of cleared farmland.[7]

If we consider the importance of the question of reversion, it is surprising how difficult it is to find out how much forest has regrown and why reversion has happened in a particular location. Regrowth statistics do not appear in the agricultural census; they can be deduced only by comparing the amount of total farmland with the amount of farmland less farm woodland, and then only meaningfully in the predominantly wooded counties of the 31 eastern states of the country. Even then there are complications that make it difficult to be exact; for example, either the land cleared in one period reverted to forest and then was cleared again at a later time, or the abandoned land reverted to forest but was compensated for subsequently by clearing elsewhere in the original forest. One thing is certain, however: that aggregate statistics for the nation as a whole can be very misleading, as can statistics of the total acreage of farmland, because woodland already covers about one-quarter of the total area. Again, the causes of abandonment and the ultimate use of the land will vary, some abandoned land being taken over for urban and industrial and even strip-mining purposes and not allowed to revert to forest.

One important reason for our sparse knowledge about the gain in forest land that farm abandonment represents is that abandonment was ignored or even disapproved of. In a society imbued with the frontier ideals of development, progress, and the virtues of forest clearing, abandonment was retrogressive, difficult to comprehend, and even sinful to contemplate. It was something to be ignored tactfully rather than praised blatantly. Although there were hints and some preliminary analysis in the major reports on forestry in the pre-1940s era that reversion was happening, the implications for forest growth were never spelled out.

However, an analysis of the census data by J. F. Hart shows that in the 31 eastern states the net loss of cleared farmland between 1910 and 1959 was 43.8 million acres with 65.5 million acres having been abandoned but 21.7 million acres gained, mainly from the forests of Florida and coastal Alabama (Table 14.1). From 1959 to 1979 another 16.9 million acres net were lost to agriculture or gained from the forest. In other words, 865,000 acres of cleared farmland were lost from agriculture and added to the forest in every year between 1910 and 1959, and a slightly lesser amount, 845,000 acres, has been added per annum since then. The change for the period 1910 to 1959 is shown in Figure 14.2. The data for 1959–79 have not been mapped in detail, but preliminary results suggest a pattern for these latter 20 years that is similar in location and intensity to that of the previous 60 years.[8]

The magnitude of the changes taking place was rarely appreciated at the time. The Capper Report of 1920 barely recognized the reversion of farmland to forest. Rather, the "five Ds" – devastation, depletion, deterioration, decay, and disappearance – permeated its thinking and phraseology.[9] But, two years later, Greeley had modified his position slightly. Though still deploring the depletion of prime timber and the "idleness" of the cutovers, he did recognize that abandonment was occurring in the East and was affecting to some degree the growth of forest elsewhere. In New England, in particular, the forest area was some 13 percent greater than it had been 60 years before. Increasingly, it seemed that burned-over forest was not necessarily being taken over by agriculture.[10] By 1925 he thought that an equilibrium had been reached.

> There is small prospect that the area available for growing trees will be reduced materially, if at all, for many years to come. While the inroads of the farm are continuing here and there, the great tide of forest clearing for cultivation seems largely to have spent itself. For many years, indeed, the abandonment of farm land in forest growing regions of the old States has practically offset new clearing on the agricultural frontier.[11]

Recognition of this shift from a reduction of the forest area to some sort of equilibrium went a step further in the Copeland Report of 1933, in which farm abandonment was discussed in depth for the first time. C. I. Hendrickson, who wrote this section of the report, guessed that even equilibrium had been overtaken by growth and that for the first time more land was becoming available for forestry rather than less. He calculated that 2.3 million acres had been lost in New England between 1880 and 1910 alone, as well as 0.77 million acres in the Middle Atlantic States during the same period. He even calculated and mapped the percentage demise in agricultural land in the eastern states from its peak to 1930 (Fig. 14.3). In total, he thought that net abandonment between 1910 and 1930 might be on the order of 19.1 million acres (in fact, it was nearer 13 million), and he foresaw even greater changes in the future with the impact of mechanization and the migration of rural families to urban areas and to the more productive Great Plains; moreover, the decline of horses and mules had released so much land from producing animal feed that there was almost a surplus.[12] In all, Hendrickson was optimistic that an expanded and vigorous program of afforestation would not lack land. To the 9.1 million acres he added another 21.5 million acres of land designated as idle or fallow in the "first stage of abandonment," as well as some 29 million acres of

Table 14.1. *Cleared farmland in the United States, 1910–79 (millions of acres)*

		31 eastern states		
	Coterminous U.S. total farmland	Total farmland	Farm woodland	Cleared farmland
1979	?	351.1	72.5	278.6
1975	1,013.7	343.4	77.7	265.7
1969	1,059.6	369.4	78.2	291.2
1965	1,106.9	396.1	95.6	300.5
1959	1,120.2	415.4	113.2	302.2
1954	1,158.2	455.0	133.5	321.5
1950	1,161.4	470.5	135.9	334.6
1945	1,141.6	470.5	135.9	334.6
1940	1,065.1	459.9	106.1	353.8
1935	1,054.5	474.5	132.2	342.3
1930	990.1	443.4	110.1	333.3
1925	924.3	445.7	108.7	337.0
1920	958.7	478.0	130.4	347.6
1910	881.4	490.3	144.3	346.0

Sources: J. F. Hart, "Loss and Abandonment of Cleared Farm Land in the Eastern United States," 418, table 1; and U.S. Bureau of the Census, *Agricultural Census* (1965, 1969, 1975, 1979).

"pasture other than plowable or woodland pasture," much of it in "an advanced stage of abandonment." Therefore, even if only half these two estimates were added to the 19.1 million acres there would be nearly 40 million acres available for forestry immediately and possibly between 75 million and 80 million acres by 1950. It was these optimistic calculations that influenced the Copeland Report in its recommendations and encouraged its compilers to campaign vigorously for extensive state intervention and replanting.

In fact, the 1930s saw a slight gain nationally in the amount of forest cleared, although the picture varied regionally. In the South detailed surveys covering 283 million acres of forest revealed that abandonment was running at somewhat more than half the rate of clearing (5.3 million to 2.0 million acres) and that it dominated all areas except North Carolina, northern Alabama, and the Mississippi lowlands. Eventually, wartime emergencies caused a marked upswing in clearing so that between 1940 and 1945 more farmland was created out of the forest than at any other time in the nation's history, and with a marked destruction of much high-yielding timber (Table 14.1). Since then, however, the trend has been markedly downward with net losses of cleared land due to all causes totaling over 74 million acres since 1940.[13]

As was stressed earlier, not all abandoned land goes back to forest; other uses absorb a great deal of it, and even if it does revert to forest the stock may be poor and low-yielding. Evidence of this is provided by the 1982 U.S. Forest Service report, *An Analysis of the Timber Situation in the United States, 1952–2030*, which shows that the

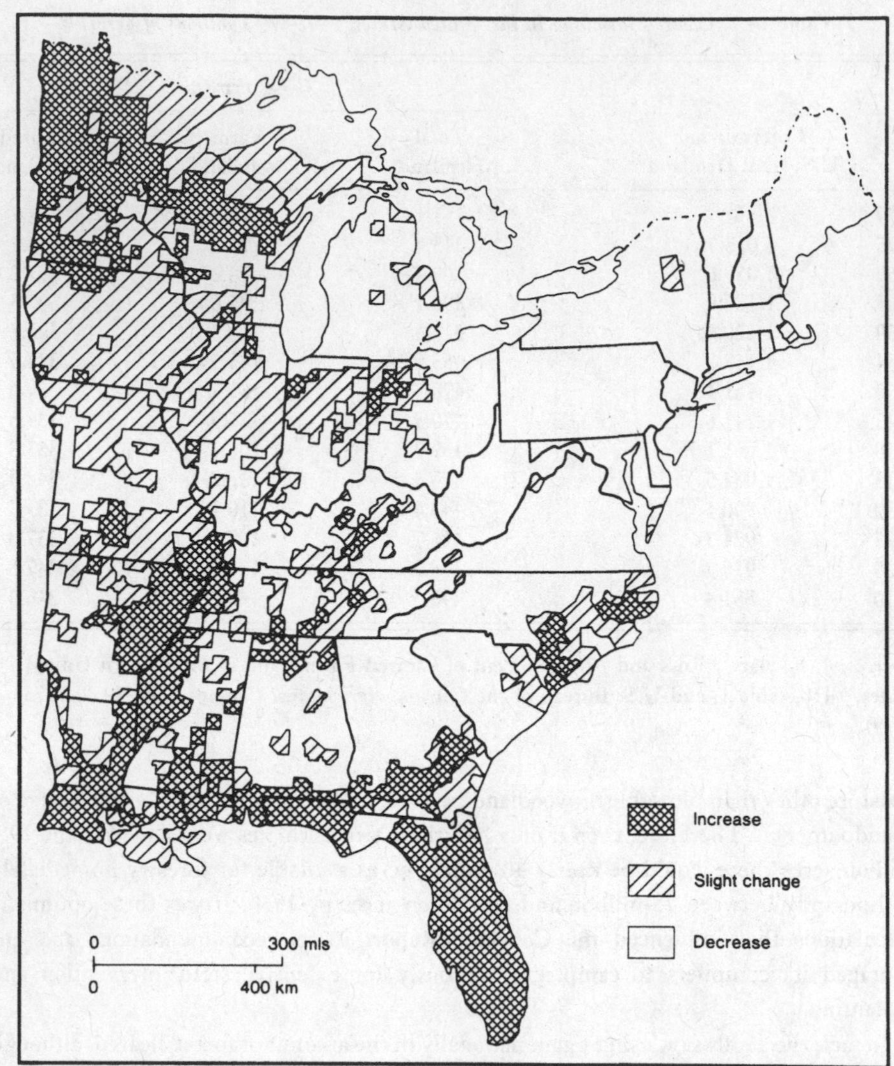

Figure 14.2. Change in cleared farmland acreage, 1910–59. Stability in cleared farmland was achieved if either the change was less than ±5 percent or the change amounted to 2.5 percent or less of the total area of the county. (After Hart, "Loss and Abandonment," 422.)

amount of land in commercial timber increased up to 1962 when it was 509.4 million acres but began to fall off markedly after that, the total dropping to 496.4 million acres in 1970 and to 482.5 million acres in 1977, a total fall of nearly 27 million acres (see Table 13.3). The fact that large acreages were lost in the Rocky Mountains and along the Pacific Coast was the result of additions to public forest lands, parks, and wilderness areas (i.e., it was still forest land, although withdrawn from commercial use). The decline in the South, however, was a result of clearing for soybean cultivation and pasture development.[14]

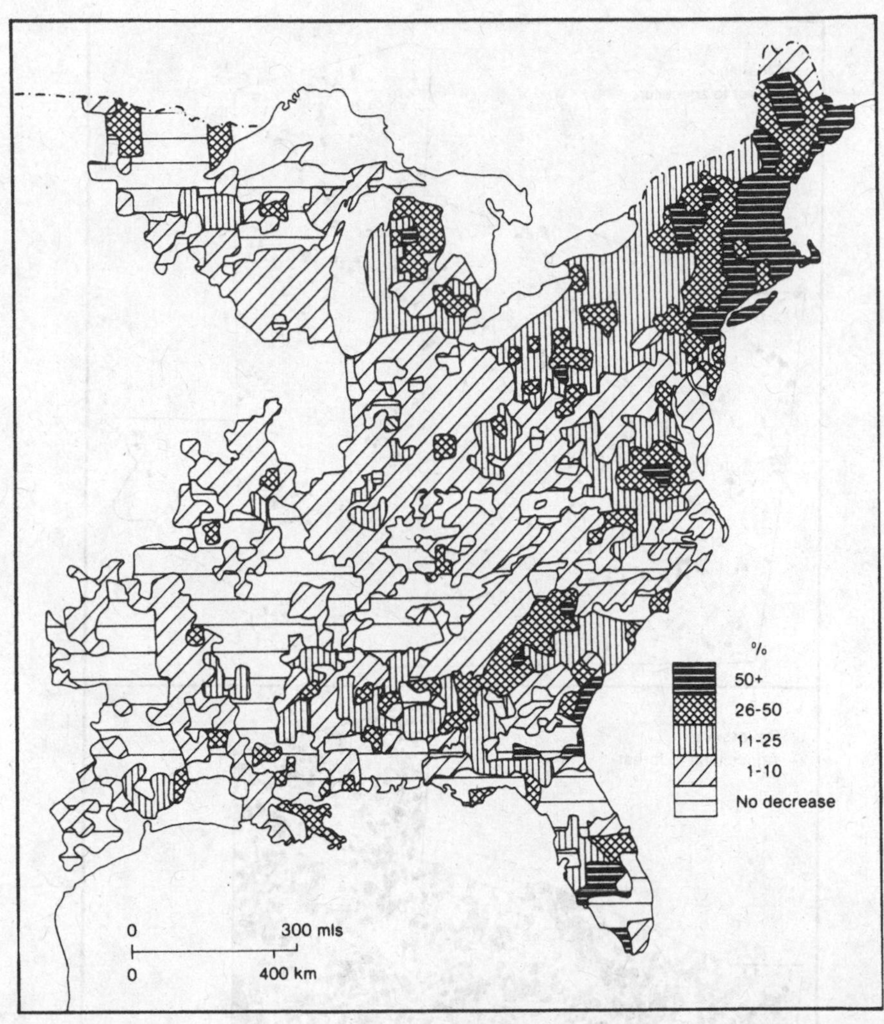

Figure 14.3. The percentage decrease in agricultural land in the eastern United States from its peak to 1930. (After C. I. Hendrickson, "The Agricultural Land Available for Forestry," 156.)

The magnitude of these statistics, the lack of comparability between different calculations, and the difficulties of reconciling gross and net figures of how much land actually reverts to forest make changes difficult to comprehend and appreciate. Perhaps the example of one county will help our understanding. Again, the work of J. F. Hart is instructive. Carroll County is located about 30 miles southwest of Atlanta, Georgia (Fig. 14.4). Whether it is typical or not and therefore whether its figures on land use can be extrapolated to other counties is beside the point; it is an example that illustrates processes and realities.[15] A comparison of aerial photos between 1937 and 1974 reveals the extent to which the land has been abandoned and the forests have taken over. The

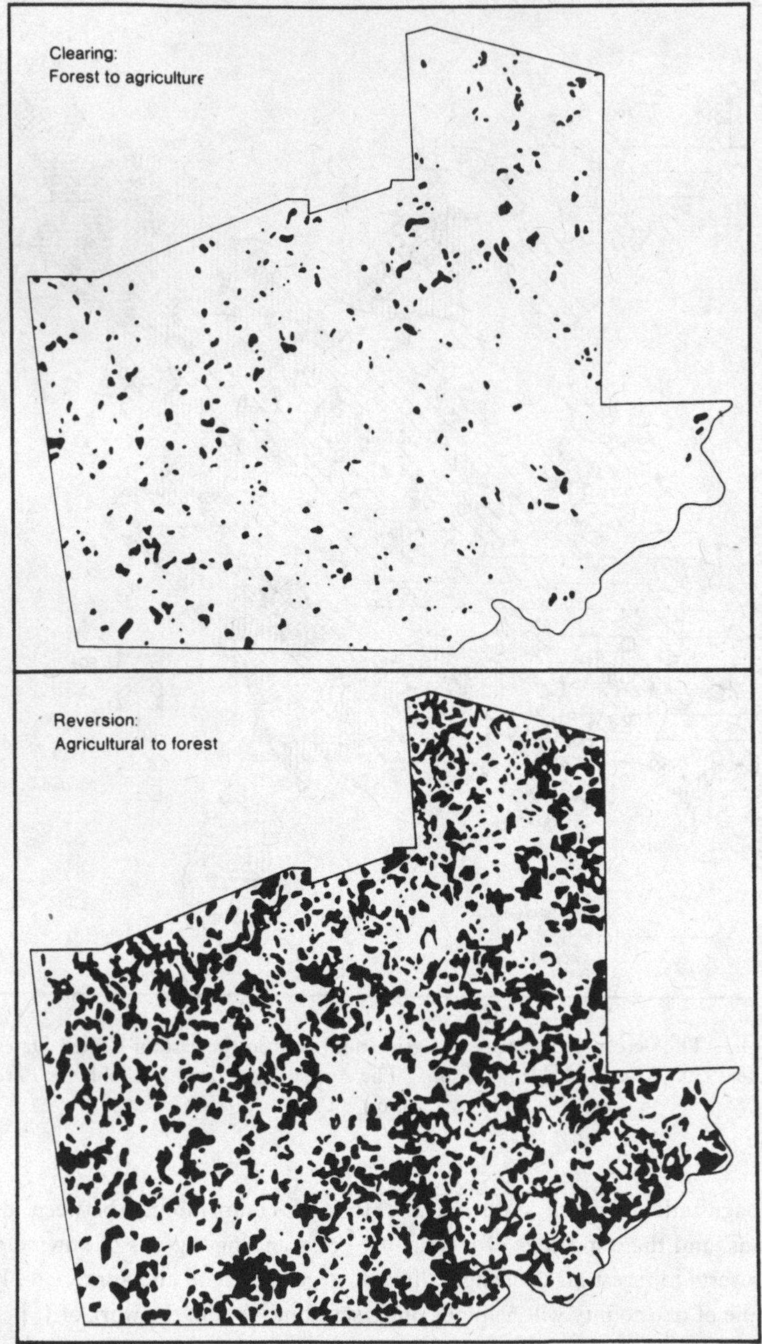

Figure 14.4. Conversion of forest to agricultural land and abandonment of agricultural land to forest, Carroll County, Georgia, 1937–74. (J. F. Hart, "Land Use Change in a Piedmont County," 515–16.)

Table 14.2. *Major land-use changes in Carroll County, Georgia, 1937–74*

	Acres
Agriculture to forest	90,807
Agriculture to built-up	6,171
Forest to built-up	515
Forest to agriculture	8,496

Source: J. F. Hart, "Land Use Change in a Piedmont County," 115–16.

change has been gradual rather than dramatic, and that tells us a lot about the incremental, scattered nature of abandonment by individuals that is so difficult to detect in the aggregate statistics. Like much of the South, there is a long tradition in Carroll County of turning livestock into the woods so that many pastures and fields do not have sharp boundaries or even fences but merge gradually into the woodland grazing zones. In the past, farmers have had to be constantly vigilant and expend a great deal of labor in keeping the brush and saplings from creeping out of the ravines, first onto the steeper slopes and then onto the gentler slopes. In some places land is being changed from forest to agricultural land, but that is about one-ninth of the total change (Table 14.2). Much of the pattern reflects the energies and inclinations of the individual farmers. Older men may be content to let the woods encroach slowly on land that was cleared when they were younger and more vigorous; younger men may want to extend their holdings. Abandonment and regrowth are just as individual and difficult to detect as was initial clearing in the forest during the last three centuries.

Regeneration

Unless a forest has been stripped completely bare of all trees, it will regenerate naturally in time. However, this natural process may be too slow and uncertain for modern forestry practice, and consequently the area planted to trees and seeded artificially has gone on at a greater rate than has perhaps been appreciated. The massive planting program of 50 million acres suggested in the Copeland Report of 1933 will probably never be fulfilled; nevertheless, a close look at the topic suggests that replanting and reseeding have been far greater than acknowledged, particularly if private industrial plantings are taken into account.

Under the Clarke–McNary Act of 1924 relatively small appropriations were made for the growing, distribution, and planting of stock for farmers in order to provide for the extension and more efficient management of woodlots and shelterbelts, and the Knutson–Vandenberg Act of 1930 provided funds for planting trees in national forests and in cutover areas.[16] With these acts and the various programs under the New Deal, the groundwork was laid for federal/state cooperation in the distribution of trees and

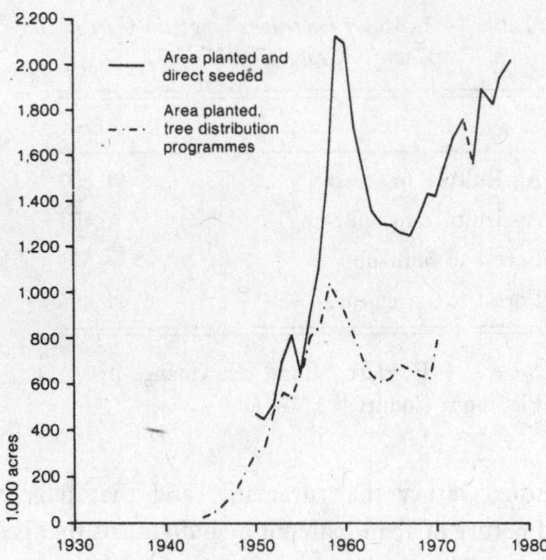

Figure 14.5. Direct planting and seeding of forest, 1943–78. (USFS, *Analysis of the Timber Situation*, 231; and U.S. Bureau of the Census, *Historical Statistics*, tables L33, L37.)

seedlings for forestry and windbreak purposes, culminating in the Cooperative Forest Management Act of 1950. In the 31 years between 1939 and 1970 at least 12.6 billion trees were distributed, 14.8 million acres planted, and 1.7 million woodland owners affected.

But this was the tip of the iceberg. A variety of other programs, such as the Agricultural Conservation Program (1930), the Conservation Reserve Soil Bank Program (1956), and its successor the Conservation Soil Bank Program (1960), added many more millions of acres (Fig. 14.5). Up until the 1950s, the main priorities were for shelterbelts on the Great Plains, forests on badly eroded land, forests on unstable sandhill country in Nebraska, Kansas, and South Dakota, and forests in silt-choked catchments.[17] After the mid-1950s the major part of the planting was on farms and other private forest ownerships in the South where payments were made under the soil-bank programs to farmers to retire land from crop use, and this continues still at nearly half a million acres a year. In more recent years the greatest amount of regeneration has come from the lumber industry, again in the South (Plate 14.2) and to a lesser extent in the Pacific Northwest. Most of these are minor plantings, and they would not affect materially the national timber supply, constituting as they do a mere one-half of 1 percent of commercial timberland. But collectively, with the other things that are happening in the forest, they do help to halt the cumulative decline of the nation's forest cover and the stock of growing trees.

A better and more complete picture of the extent and impact of replanting comes, however, from considering the impact of the lumber industry on the forests. The vast majority of lumber companies have long since tried to shed their image as ruthless woodland destroyers, and although they are clearly aiming to be profit-making concerns in the long-term and difficult business of growing trees, they have done a lot to put the

Plate 14.2. Cotton gives way to pine: reforestation in the South.

forest back. The situation has been the reverse of that which obtained in the nineteenth and early twentieth centuries. It is no longer cheaper to abandon a plant and move on to new stands (even if there were any); it is cheaper now to replant the forest, especially with improved fire control, and maintain the expensive capital equipment and its social and economic infrastructures. Lumber industry nurseries have multiplied since the 1930s, as have private commercial nurseries since the Second World War. Of the nearly 300 nurseries in existence now, 65 percent are industrial and/or commercial, and the remainder are federal- or state-owned.[18] Though the direct planting of artificially raised trees has been the normal method of regeneration, experiments at direct seeding have been carried out in the southern pine forest with some success.

One regional example of replanting must suffice to illustrate events elsewhere in the country. In the 13 southern states, including and south of Virginia, Kentucky, Arkansas, and Oklahoma, the graph of replanting and reseeding is impressive. In gross terms nearly 28.9 million acres of forest were replanted or reseeded to 1980; the forest industry contribution to that was over 53 percent and is now running at just under 1 million acres a year (Fig. 14.6). A word of caution is perhaps necessary in interpreting this graph. The acreages planted are gross acreages, and there is no way of knowing exactly how many acres were replanted in order to come up with net acreages. For example, in the Yazoo–Little Tallahatchie Flood Prevention Program in Mississippi, where nearly 800,000 acres have been planted on blocks owned by small-scale owners, it has been necessary to replant 18 percent of the acreage during the last 30 years. Checks made when the plantations should have been 15 years old revealed that 13 percent of them no longer existed. Much of the loss of trees occurs because the seeds are eaten by rodents, and therefore the extensive use of pesticides has been resorted to, but with increasing protest from concerned conservationists.[19]

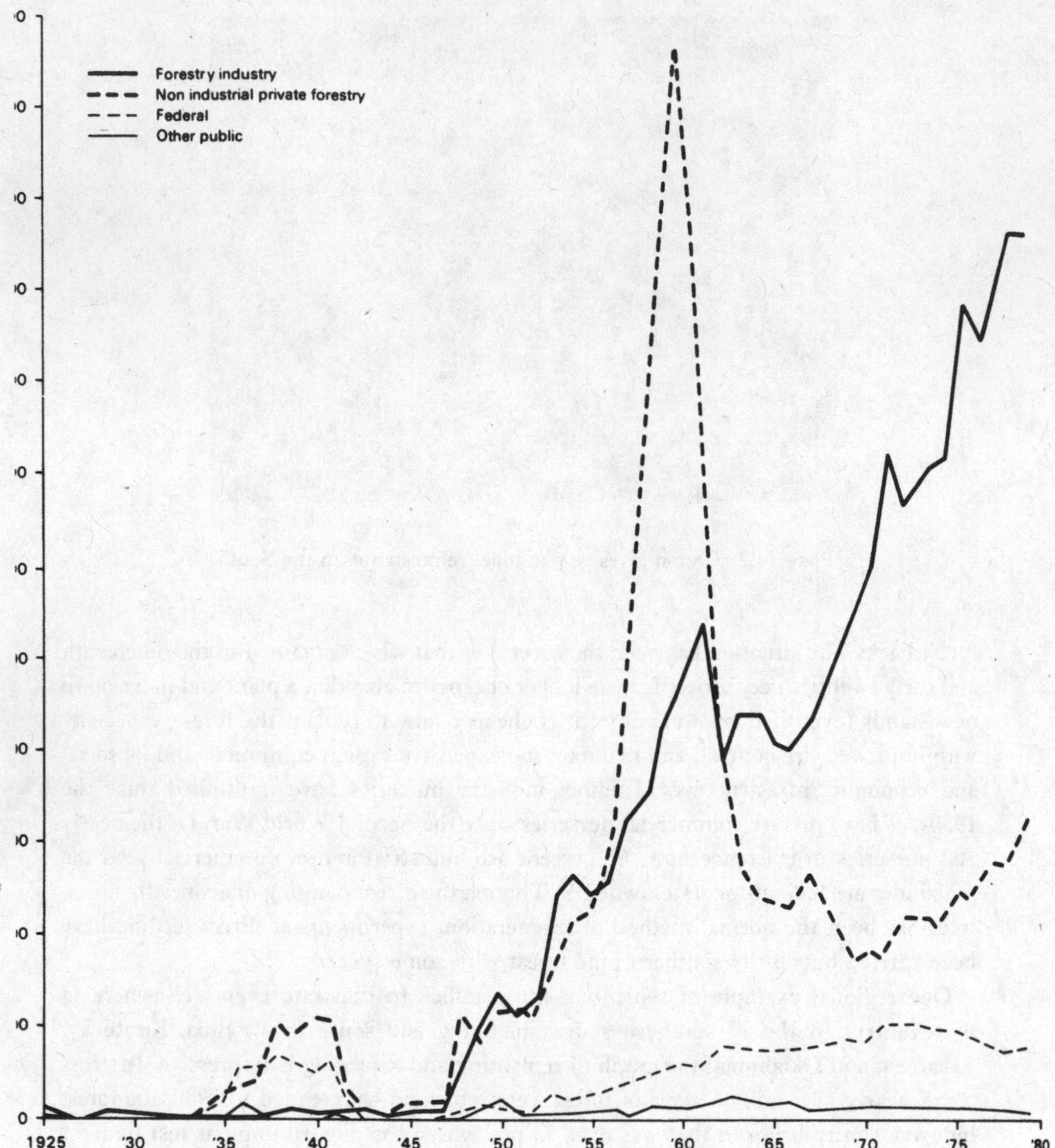

Figure 14.6. Replanting and reseeding of land to forest in 13 southern states, 1925–80. (After H. L. Williston, *A Statistical History of Tree Planting in the South, 1925–79*, app.)

Intermediate stand treatment and insect and disease control

The pleas of those in the past, such as Fernow, that the forest should be treated more as a crop to be tended carefully than as a mine to be plundered are finally being heard. Not only are forests being reseeded and replanted, but they are also being given what has come to be known as intermediate stand treatments – that is, measures to increase growth between the establishment and the harvest of the stand. At present just under 1.5 million acres of the forest are affected annually, mainly in the South and the Pacific Northwest and in the national forests in the Rocky Mountains.[20] The silvicultural measures include deadening less desirable trees; thinning young stands to remove excess or lower-quality trees; weeding out competing vegetation (or "release," as it is sometimes known), especially in the South where scrubby hardwood invasions into newly logged areas have reduced the pine stand by nearly 4 million acres between 1970 and 1976 alone; pruning lower limbs; and sometimes cutting out larger commercial timbers in order to allow small, shaded timber to grow. Fertilizer is also used in large quantities to establish and encourage young growths.[21]

Like crops, forests also suffer from diseases. Rarely is the toll a constant one as pests and insects are present all the time, but from time to time the balance is disrupted and they appear in great numbers over wide areas where they can cause serious damage. In the past, up to 1 billion cu. ft. of timber was destroyed by such insects, diseases, and parasites as spruce budworm, Dutch elm disease, pine bark beetle, Douglas-fir tussock moth, and mistletoe. Increased expenditure on detection and prevention up until the mid-1960s and more effective pesticides have kept losses low, but they will never be eliminated entirely, and, as with other aspects of the use of artificial agents in the forests, there are many opponents.[22]

The suppression and control of fires

Nothing has contributed more to the supply side of the timber in the forest than the suppression of wildfires and the careful use of controlled fires. Fire management is the most widespread form of forest management and has proved to be the most effective.

The suppression of wildfires. It was William Greeley who pursued the idea of cooperative fire control legislation during the 1920s in order to stop the devastation that swept away millions of acres of good timber annually and prevented new growth from taking hold. Greeley had been district officer in Montana at the time of the great Idaho fire of 1910 that wiped out over 2 million acres of prime timber forest, and the experience of that fire was seared into his soul so that achievement of fire control became a passion with him. He believed that the suppression of fire, which everyone could see as an obvious destroyer of timber, stock, property, and even lives, would be a practical beginning to the implementation of better forestry practices and regulation and would encourage the lumbermen to be more careful in leaving thick debris of highly combustible slash on the forest floor and the cutovers after lumbering. Greeley used fire suppression as his trump card during the McNary hearings of 1923, and it was offered as a compromise and part of a wider strategy to get cooperative action among the lumber industry, the states, and the Forest Service.[23]

Statistics of fires began to be collected, at first spasmodically, from about 1910 onward and regularly after 1930. They showed a dramatic increase in the amount of forest burned, from an average of 8.5 million acres per annum in the decade to 1920, to 23.7 million acres per annum in the next decade, to an astronomically high 39.1 million

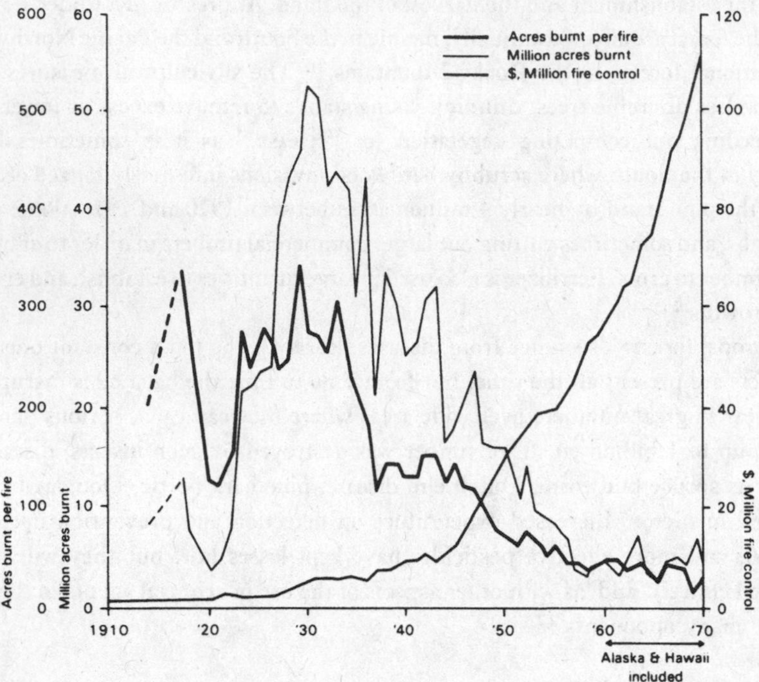

Figure 14.7. Annual extent of forest fires in the United States, amounts spent on fire control (in millions of dollars), and average size of area burned per fire, 1910–70. (U.S. Bureau of the Census, *Historical Statistics*, tables L48, L49; and USFS, *Analysis of the Timber Situation*, 234, 239.)

acres during the 1930s, 53 million acres being burned in 1931 alone (Fig. 14.7).[24] The increase was both apparent and real. It was apparent in that the fires that had gone undetected in the past were now recorded but real in that depressed prices for lumber during the years leading up to the depression had caused lumbermen to cut costs and corners, so that they left a legacy of litter that was to burn fiercely. The replacement of animal power in the forests by high-speed cable and high-level yarding led to the total destruction of all the timber in a wide radius around the donkey engine. Wood-fed locomotives and stationary hauling engines sent sparks flying everywhere in the mangled and torn forest and onto the slash on the forest floor, so that fires became more numerous.

Greeley campaigned for the measures that ultimately became incorporated in the Clarke–McNary Act of 1924, which, among other things, provided funds for a co-operative federal/state fire control system and strengthened the well-intentioned but largely ineffective provisions of the Weeks Act of 1911. The initially modest expenditures (later increased), but particularly the new awareness of the need for fire control that they engendered, began to bite into the fire problem after 1930, as did the provision of new techniques and equipment, such as watchtowers, telephone lines, spotter aircraft, and forest trackways, and the sharpening of public awareness of fire prevention through advertising campaigns, although the most successful, the creation of Smokey the Bear, did not occur until 1945.

National and state forests grew in size when huge areas of abandoned farms, tax-delinquent lands, and cutover in the southern states and Lake States were acquired during the 1930s. Often these new areas were remote from the systematically and easily protected lands. The common practice of letting fires in these remote, low-value, "backcountry" areas burn themselves out changed with the great Tillamook burn of 1933 and the Oregon and Stanley burns in the northern Rockies in 1934, which effectively put an end to such a "lax" policy. It was replaced by the "10 a.m. policy," which meant simply that control of the fire was to be achieved by 10 a.m. of the morning following the fire or, failing that, by 10 a.m. the next day and so on, a program of complete suppression that was made possible by the investment in New Deal programs and by the creation and use of the Civilian Conservation Corps, which supplied huge amounts of labor for fire fighting.[25]

After 1930 the number of fires and the area burned began to fall steadily as more land became protected, so that the average annual burn during the decade of the 1960s was 4.6 million acres and during the 1970s was a mere 3.2 million acres, below which it will probably not fall markedly because of the greater risks associated with improved access and greater public use of forests for recreation, and also because of the increased accumulation of fuel in specially protected areas, such as wilderness areas (Fig. 14.7). There is no doubt that the reduction in the area burned bears a close relationship to the amount of money spent on forest fire protection, expenditures having tripled between 1950 and 1977, rising from $97 million to $318 million in constant 1972 dollars. However, the really significant statistic that can be teased out of the data is that of the average size of the fire. This has dropped from 253 acres per fire per annum in the 1920s to 38 acres in the 1960s. Simply, detection and fire-fighting organization are quicker and more effective than before, and fires do not get the chance to spread.[26]

Another major contributory factor to limitation has been the realization since at least 1945 that prescribed light burning reduces the fuel buildup in the forest and hence reduces the incidence of disastrous fire. The age-old adage, "Fight fire with fire," has proved to be true. The cumulative effect of this reduction in damaging wildfire has been to allow the young stock to grow vigorously and without interference, which has contributed to the massive increase in the inventory of standing stock that is now evident. It has also contributed to an expanded program of replanting and reseeding, both of which are now feasible economic alternatives to cutting out and getting out. Investments no longer go up in smoke.

Fire control. The logic of fire suppression and its obvious benefits in enabling the forest to grow vigorously were not matched initially by the logic that controlled or prescribed fire is also an aid to growth. Gradually, it has become clear that fire is a part of nature, part of the ecological makeup of many forest types, which are in a symbiotic relationship with fire so that fire produces fire-adapted vegetation, which in turn thrives on regular fire. Fire can control the sort of forest that grows by preventing the "typical" forest stand from deteriorating, as, for example, in the southern longleaf pine forests, or the western ponderosa and lodgepole pine and sequoia forests; it can in some cases retard or promote successional patterns; it can release essential nutrients locked up in the forest litter or duff; it produces a mosaic of vegetational types and mixture of age classes. Fire can even favor wildlife and modify the impact of insects and diseases; and, above all, it can reduce the accumulation of hazardous fuel on the forest floor that may be the cause of a much more disastrous and destructive fire in future years.

The acceptance of fire as a positive rather than a negative tool in forest management took a long time to materialize. The Indian use of fire to manage the land was dismissed as primitive, and the disastrous and spectacular fires of the late nineteenth and early twentieth centuries were well known and their recurrence feared. Consequently, it is not surprising that the observations and investigations of those who suggested that light and controlled burning might actually assist tree growth in some ecosystems and/or prevent even more serious wildfire damage in the future were ignored, treated with abuse, and even suppressed. The controversy was most pronounced in two regions of the country: on the West Coast, where the issue was mainly, though not exclusively, fire prevention, and in the South, where it was mainly, though by no means solely, forest regeneration.[27]

From as early as the 1880s settlers and lumbermen on the West Coast had advocated light burning as a means of encouraging pasture growth on the range and of reducing the hazardous buildup of fuel in the forests. The virtues of managing the forest "the Indian way" were extolled by John Wesley Powell, who cited Paiute fire practices, much to the disgust of Fernow and Pinchot.[28] Nevertheless, by the turn of the century fire protection dominated thinking, and the advocates of prescribed burning were in a minority. Experiments in light burning during the first decade of the century, undertaken by S. B. Show of the Forest Service in the Douglas-fir forests near Mount Shasta, seemed to demonstrate that management by fire was a more expensive and less efficient method of control than fire elimination.[29] At the same time, Coert du Bois, a state forester in California, had been conducting many other experiments, and though his favorable conclusions about light firing were suppressed, his *Systematic Fire Protection in the California Forests* (1914) put the technique of fire protection on a new, quantified, and "scientific" footing. It was becoming increasingly clear that it would be difficult to show conclusively that the hazy notions about the benefits of light burning could compete with the hard facts about the economy, efficiency, and systematic logic that fire protection seemed to offer. The debate dragged on, but a further report in 1924 by Show and Kotok, *The Role of Fire in the California Pine Forests*, was effectively the culmination of the controversy in the West.[30] Attitudes had already hardened, the Forest Service took a fairly uncompromising stance, and light burning was seen as "a political threat and not as a management technique."[31] Fire had to be excluded from cutover, brushland, and forest, and the debate was not really reopened until the mid-1950s, after changes in the South had pointed the way to a radical reassessment of fire as a tool of management.

It was, perhaps, in the prized longleaf pine forest of the South that the issue of light or prescribed burning was most pressing and contentious because of the region's potential as a lumber-producing district and because of its folk tradition of forest firing. As early as 1889, Mrs. Ellen Long had observed that, although the suppression of wildfires was essential, the annual burning of the woods was desirable because it "is the prime cause and preserver of the *Pinus palustris* [longleaf] to be found there; but for the effects of these burnings . . . the maritime pine belt would soon disappear and give place to a jungle of hardwood and deciduous trees."[32]

When the astute observations of the amateur were supported by the experiments in controlled burning by the professional forester, namely, Herman Haupt Chapman of the Yale Forestry School, the case for prescribed fire management became stronger. Between about 1907 and 1944, Chapman published more than 20 papers dealing with the southern pines and their relation to fire. He showed that periodic winter fires did not kill all longleaf pine seedlings; rather, they cleared the seedbed of debris and helped the pine to establish itself. Fire also suppressed other pine (e.g., slash) and hardwood (e.g., scrub oak) competitors, reduced hazardous fuel accumulation, and even controlled brown-spot disease. Chapman recommended the use of periodic burning in the longleaf forest at intervals of about three years. His assertion that the protective measures suitable for the northern pines would lead ultimately to the destruction of the southern pines was too provocative for contemporary forestry opinion and went unheeded.[33]

The newly fire-conscious crusaders of the Forest Service were not inclined to listen to the heresies of people like Chapman. For them, such experimental ideas were seen, particularly in the South, as a license to the careless lumbermen, who let duff fires burn unattended in a forest they still thought inexhaustible, and an encouragement to the "slovenly," even "immoral," practices of the southern farmers, who had fired the forest for centuries to clear land for agriculture, to provide pasture growth for hogs and cattle, to safeguard the turpentine trees, and to eliminate chiggers, ticks, and snakes.[34] After the great Idaho fire of 1910 the Forest Service tightened its control by expanding the area of protected lands, preaching the need for complete fire suppression in the coming era of timber famine, and quashing further heretical statements by employees that might dilute the forest protection message and threaten government conservation and forestry. By 1927 the service had interpreted the Clarke–McNary program to mean that funds would be withheld from states that tolerated controlled burning.[35]

Perhaps the first chink in the armor of abolition and exclusion of references to the beneficial results of controlled fire came with the publication in 1931 by S. W. Greene of "The Forest That Fire Made," an article that received wide public attention. It was significant that Greene was not a forester but a husbandman in the Bureau of Animal Husbandry at the Coastal Plains Experimental Station, McNeill, Mississippi, and was thus beyond the reach of censorship of state and federal forestry departments. He found that the quality of grass and legumes on burned lands was much higher than on unburned lands and that cattle grazed on such burned pasture made greater weight gains than those on unburned pastures. The advantages of burning were not confined to grazing; H. L. Stoddard of the U.S. Biological Survey showed that bobwhite quail numbers increased in forests subject to light burning. The scientific evidence for what was known instinctively and traditionally for centuries by the Indian and by the southern farmer, herdsman, and hunter was mounting.[36]

The reluctance of state and federal officials to recognize these demonstrable facts is illustrated in the Copeland Report of 1933. This otherwise radical document deftly sidestepped the issue by saying that no research had been done that could prove conclusively that fire was beneficial. Nevertheless, censorship was questioned and removed by 1935, and in the same year the Forest Service quietly removed its prohibition on prescribed burning under the Clarke–McNary Act.[37] By 1941, disbelief and opposition in the South had given way to concern and agitation for change. The longleaf forest had decreased in size by well over 25 percent through encroachment of other trees, but more importantly, the buildup of debris on the forest floor after 30 years of strict fire suppression was so great that it was feared that disastrous fires would break out everywhere, just as they had in 25,000 acres of the Osceola National Forest in Florida. The protracted drought of the winter of 1942–3 brought about further heavy damage, and as a result a program of prescribed burning began in 1943. Ironically, funds for this purpose were now made available under the Clarke–McNary Act.[38]

By 1946 the *volte-face* was complete, aided in the next year by a fire in Maine that destroyed a quarter of a million acres of forest, much of it occupied by suburban development and hence exceptionally disastrous. Illustrative of the change of attitude was a section in the 1949 volume of the USDA *Yearbook of Agriculture*, which was devoted to trees. It was entitled "Fire: Friend or Enemy?," a title that would have been unthinkable 10 years before. Over four-fifths of the 60 pages of the section dealt with fire suppression, but one chapter by Arthur Hartman, regional chief of fire control in the South, was entitled "Fire as a Tool in Southern Pines." He reported that, between 1943 and 1948, 803,500 acres of forest had been deliberately burned over in the southern states from Texas to North Carolina, but particularly in Florida in the Osceola National Forest, in order to reduce the potentially dangerous fuel buildup.

Moreover, burning seemed to be good business; 26,000 acres burned in southern Alabama produced a 90 percent satisfactory-to-heavy seeding at a mere 1 percent of the cost of planting. Chapman's contention of nearly 40 years earlier that the longleaf forest may be a nonclimax forest seemed correct.[39]

Up until about 1949, the debate had concentrated largely on the southern forests because of their unique fire-resistant nature, and prescribed burning was not accepted in any other region of the country. Yet evidence was growing that fire was an integral part of many forest ecosystems, particularly in the West where the protracted debate on light burning at the turn of the century was reopened. The case for burning out the dangerous fuel buildup in the chapparal in the vicinity of the urbanized parts of southern California remained a source of contention, but a major advance in thinking and application came elsewhere in the West with the work of Harold Weaver and Henry Kallander of the Bureau of Indian Affairs, who made scores of studies from about 1943 onward on controlled burning in the ponderosa pine on the Indian reservations of Fort Apache, San Carlos, and Hualpai in Arizona. These showed that in its natural state lightning struck the ponderosa pine every six to seven years and reduced the fuel level to produce low-intensity fires that pruned back the woody vegetation, prepared the soil to receive seed fall, thinned out the young trees, and consequently produced a thicker and more vigorously growing forest.[40] Farther north, a disastrous wildfire in Kings Canyon National Park in 1955 threatened the Grant Grove of giant sequoias (*Sequoiadendron*), and this prompted research into the effect of frequent low-intensity surface fires in the mixed sequoia–white fir forests. It was found that fire cleared the undergrowth and thus prevented more damaging crown fires, actually helped in seeding by increasing soil/water penetration, removing shade-tolerant and fire-susceptible species like the white fir, and providing sunlit openings, thus allowing the sequoia seedling to grow. Wildlife increased, and disease and insects decreased after burning.[41]

Subsequently, the story of experiments could be repeated for nearly every other forest type as the beneficial role of fire became clearer. Douglas-fir, western white pine, longleaf pine, and parts of the lodgepole pine and ponderosa pine are subclimax fire types, and fire is crucial to maintain their state. Red and eastern white pine, and pure ponderosa, lodgepole, and Engelmann spruce, though not fire types, do need fire to hasten and ensure reproduction. At the other end of the scale, the extensive areas of aspen and jack pine varieties throughout the Lake States are not natural types and have been successful only because of major fire disasters in the past. Similarly, the low-value hardwoods that have invaded the productive southern pine forests have been the result of fire control and partial cutting in areas where wildfires had previously kept hardwoods at bay. Throughout the country, 60 years of fire suppression had caused a marked deterioration in the composition of the nation's forests.[42]

The results of many of the experiments were disseminated through the medium of the Tall Timbers Fire Ecology Conferences held between 1962 and 1976 and the National Fire Effects Workshop of 1978 sponsored by the Forest Service. The scientific evidence for the age-old dictum, "Fight fire with fire," and fire's beneficial effects in relation to the regeneration and maintenance of forest conditions led the National Park Service in 1968 to abandon total fire suppression, letting natural fires burn and encouraging prescribed fires, particularly after extensive experiments in the Everglades, and Kings Canyon and Sequoia National Parks. Slowly and cautiously, the Forest Service followed suit after 1971. With increased knowledge of the physics of fire and methods of control, ever more land is being burned by wildfire. Fire is now regarded as an important tool and technique in forest management and a significant factor in the improvement of forest growth. Its future use is limited, perhaps, only by the objections of those who do not understand the more complicated message of prescribed burning and by the threat of air pollution created by excessive smoke.[43]

Per capita consumption of timber products

The quite spectacular increase in the supply of wood that has boosted the timber inventory has been matched by an equally spectacular decrease in the demand for wood, initially in absolute terms and latterly in per capita terms. (Fig. 14.8).

The early alarmists had based their predictions about the ultimate death of the forest largely on the per capita consumption of timber products. Their assumptions of a coming famine were reasonable at the time as per capita consumption was rising steadily (157 cu. ft. for all products in 1900) and the population itself was increasing at a rapid rate. Per capita consumption of lumber was highest in 1906 with a figure of 82 cu. ft., but with the gradual introduction of substitute material such as steel, aluminum, concrete, and later plastics, and a less prodigal attitude toward wood use in general, consumption dropped steadily to about 30 cu. ft. in the late 1970s, a little more than a third of the previous total, and it has stayed fairly stable ever since. Fuel wood consumption dropped even more dramatically, from 63 cu. ft. in 1900 to 2.5 cu. ft. in the late 1970s with the conversion of home heating and mechanical power to natural gas, electricity, and oil, particularly from 1930 onward. Some of the decline in timber and fuel has been offset by

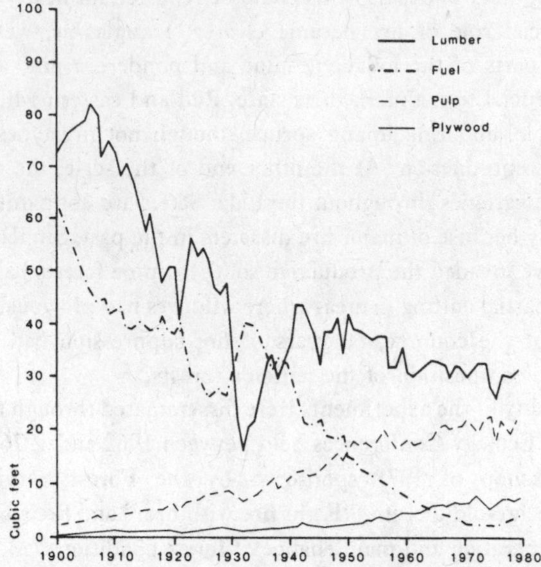

Figure 14.8. Per capita consumption of timber products, 1900–80. (U.S. Bureau of the Census, *Historical Statistics*, tables L87–L97; and M. Clawson, personal communication.)

the increase in the use of plywood, a substitute for timber in many cases (0.1 to 6.5 cu. ft.), and, of course, by pulpwood, which is converted to paper and board (2.2–21.5 cu. ft.) and which now bulks as the second-largest user of wood in the United States, far outstripping fuel and nearly equaling lumber.[44]

The remarkable thing is that when all these different types of per capita consumption are added together the total consumption in roundwood equivalent terms in 1980 was still one-sixth less than that for 1900, although the population of the country had tripled from 75.9 million to 226.5 million in the meantime. Simply, less wood use means more forest growth.

Taken together, these causes of regrowth have led to an expanding forest that at present is about 483 million acres in extent, the greatest it has been for over half a century. It probably will not change much in size from now on. The great land use changes of the nineteenth and early twentieth centuries can never be repeated, and the moving frontier of cultivation is over. There is no new forest potential cropland left, and improved forestry techniques and practices have absorbed much of the slack that existed in production. However, demands on the existing forest will undoubtedly increase, first for lumber and forest-related products from an expanding population, and second, for purposes other than for producing lumber, for example, recreation, protected areas, and more complex wilderness, scenic, and other multiple-use demands from an increasingly affluent and mobile population. In that sense, the United States has come of age. The resource is finite, although renewable; therefore, the country has to make the best use, if not better use, of what is left of that great mantle of trees that once covered over half the continent.

Table 14.3. *Ownership of commercial forests, 1977 (to nearest million acres)*

	Area (million acres)	Proportion (%)
Private		
Farmers	116	24.0
Forest industry	69	14.3
Other private	162	33.5
Total	347	71.8
Public		
National forest	89	18.4
Bureau of Land Management	6	1.2
Other federal	5	1.0
Indian	6	1.2
States	23	4.8
County and municipal	7	1.5
Total	136	28.1
All ownerships	483	99.9

Source: USFS, *Analysis of the Timber Situation in the United States, 1952–2030.*

With 347 million acres or nearly three-quarters of the productive capacity of the forests in private ownership (Table 14.3), constructive cooperation among federal, state, and private owners must continue to prevail.[45] The old controversies about private versus public ownership should not be allowed to dominate again. The forest of the future is too valuable to be subjected to the political bickering and rhetoric that debilitated positive management in the past. Nevertheless, the new element of a public that is now intensely interested in the forest rather than indifferent, as in the past, will generate new controversies over questions of ownership, use, and management.

Epilogue: gains and losses

The question that impinged on the author's mind when he first visited the United States over 10 years ago was, simply: What happened to the forest that once covered so much of the country? The answer to that question necessitated an investigation into the major impacts on the forest by settlement, agriculture, industry, transport, lumbering, and other means through nearly four centuries, and the answers gave rise to this book. It has been a story of fairly relentless modification and destruction of the biotic cover, tempered by a greater regard for conservation after 1870 and culminating in a rebirth of some of the forest cover after about 1940. It is a story of a complex interweaving of humankind and its actions with the forest environment, as it was variously and successively despised, exploited, and valued. The story could be projected forward by hazarding some guesses about the likely future forest – a not too difficult task given the long-term nature of tree growth – but more valuable than creating future scenarios would be an attempt to bring together some of the themes developed so far and a brief assessment of what has been achieved over 400 years and what has been lost.

The most obvious manifestation of change to which the clearing of the forest has contributed significantly has been the emergence of a dynamic, buoyant nation of over 240 million people who enjoy a greater affluence, probably have greater individual freedoms, and certainly exercise more power on a global scale than the people of any country at any other time in history. The attributes of wealth, prestige, and power go on unabated. Nearly a million people move to the nation's shores annually, its prosperity bolsters the economy of many parts of the world, and its military power is awesome. There are many reasons for this prosperity and power, but it would not be incorrect to say that the foundations were firmly laid with the exploitation and use of the abundant natural resources of the continent, of which the forest was one, if not the most important. At the most conservative estimate, more than 300 million acres of forest land had been cleared by the beginning of the twentieth century to create some of the most productive agricultural land known. Much of the timber released by this clearing was simply burned on the spot, but much of it entered into all parts of daily life for housing, general construction, fuel, transport, and industry and provided an abundance of energy and material that paralleled the abundance of new farmland created. Unlike Europe, where industrialization was partially a response to the scarcity of natural reserves of wood and land and the substitution of coal for wood was the critical step needed for industrialization, industrialization in America rose on the sheer abundance of wood. The forest was recognized as the basis of agricultural expansion, industrialization, and material progress. Frederick Starr got it just about right in 1860 when he said that the things that were most fundamental to a nation's growth and prosperity were "cheap bread, cheap house, cheap fuel and cheap transportation for passengers and freight," and wood and the forests entered largely and consistently into these "four great departments of industry and living."[46]

Another great gain must surely be the emerging awareness of the need to conserve resources, predicated admittedly in the first instance on the widespread destruction of the most valuable resource the nation possessed. Given the undoubted importance of the forests in the daily and national life of America, it is not surprising that people such as George Perkins Marsh in 1864 and Increase Lapham in 1867, among many, were beginning to question the wholesale destruction of such large areas of woodland by individual settlers and by the large companies of the industrial capitalists. Additionally, the forest and the backwoodsman were almost symbols of America; the forest environment and the type of person and qualities it was thought to produce were disappearing. Supplies of lumber were clearly deficient in some areas, and the subtle relationships between the forest and runoff, floods, rainfall, soil formation, erosion, micro- and macroclimatic occurrences, to say nothing of the disappearance of the fauna and flora, were being hotly debated, particularly with the advance into the Plains, which appeared to need only trees to make them an integral part of "The Garden of the World" that Americans saw emerging in the Midwest. Coupled with these economic and environmental causes was the fact that some people were even beginning to appreciate the aesthetic beauty and recreational potential of the forests. The first stirrings of the conservation movement as we know it in the Western world had their roots in the forests of the United States. Admittedly, the readiness of the lumber industry to embrace conservation was influenced largely by its desire to bring order out of the chaos of glut, overcutting, and fluctuating profits that had resulted in an unstable industry. Nonetheless, forest conservation was a contribution and a gain that was to have worldwide significance.

But these achievements, though undoubtedly real and beneficial to many millions of people, had been wrought at some considerable cost to the environment and its original occupants – human, animal, and vegetable. As pointed out already, the Indians had been eliminated or relegated to the back blocks of the country and the back blocks of the city; their habitation was to become the reservation or the slum. Many animal species were eliminated as their habitats were changed irreversibly or restricted to the densest and most distant forest remnants in, for example, Maine, the Appalachians, or the northern Lake States. Many animals, insects, birds, and fish are reduced in numbers, and some are so near extinction that special reserves must be set aside for their preservation. Original grasses are rare, and, above all, the original forest is diminished in size by more than half, its species impoverished and its ecology disrupted by, among other things, the absence of fire. Soils, runoff, and local climates have all changed. Too often progress was a sort of crusade of extirpation by the pioneer settlers and a quest for more than adequate profits by the lumbermen. Few people seemed capable of appreciating the simple truth of what George Perkins Marsh had said: "Man is a disturbing agent," and wherever he sets his foot "the harmonies of nature are set to discords."[47]

But what of the forest of today? Has the much-desired state of equilibrium and balance and sustained yield so earnestly advocated and sought after since the middle of the nineteenth century been achieved?

Since 1946, the Forest Service has made five major studies of timber growth and consumption and projected them into the future on the basis of assumptions about the income of the population, economic activity, and the growth of disposable income.[48] Up to the present time forecast and reality have not always coincided (Fig. 14.9). Growth has been underestimated, and demand has been overestimated. The latest exercise in 1982, however, suggests that there is a change in the essential parameters of demand, and the forecast is consequently less optimistic, so that the abundance of which the forest industry now boasts might well turn into a dearth as the end of the century approaches and the population of the country reaches nearly 300 million.

The reasons for the unsuspected growth have all been outlined earlier in this chapter, but the one put forward as the major factor is that timber growth is a function of past timber harvest. Though undoubtedly true, it raises many grave questions about forest management. For example, is the reverse proposition true? Does an assured supply of timber in the future mean the inevitable harvesting destruction of the remaining mature forest on the scale of the past (admittedly more controlled and less rapacious, but destruction all the same)? Forests have been growing for millions of years without human help or management, and harvesting is not a necessity for the maintenance of their timber yield. What harvesting does, however, is to accelerate growth. The danger could be that as the demand for timber rises the temptation will be to cut more

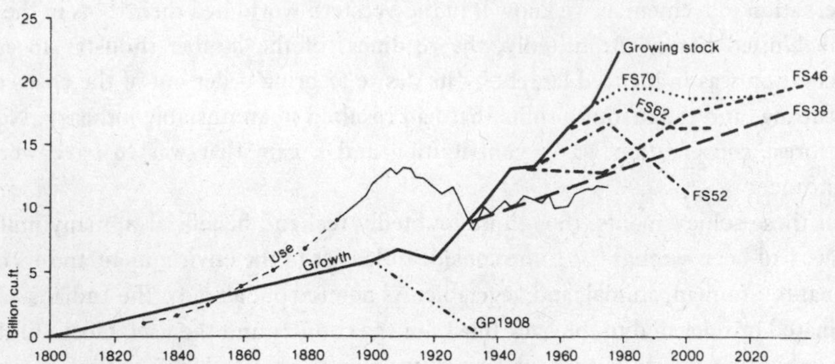

Figure 14.9. Annual net growth and use, 1800–1977, with projections of growth and demand to c. 2030. FS = Forest Service; GP = Gifford Pinchot. (Based on M. Clawson, "Forests in the Long Sweep of American History," 4; and USFS, *Analysis of the Timber Situation*, 69, table 3.16).

frequently and more vigorously to meet the demands, with the added justification that yet more growth will be brought about for the future.

Second, and closely related to the harvest/growth argument, are the oft-repeated aims of those of the past, from Fernow onward, that the forest be regarded as "a crop rather than a mine or quarry from which we take what is useful and then abandon it." In Greeley's view, that change came about after 1930, and it heralded a new chapter in the use and management of the forest. But farming, like farming everywhere in the developed world, is falling into disrepute. In many cases it is no longer a finely balanced activity in tune with the ecology of the countryside; it is a business like any other, which upsets "the harmonies of nature" in order to ensure maximum and continuous production in ways undreamed of in the past. In forestry, farming means careful and purposeful selection of species, experimentation with genetically improved species, weeding out of undesirable trees, pruning of the lower limbs of growing trees, careful ground preparation, substitution of productive for nonproductive trees, disease and insect control by pesticides, weed and fungal control, seed enhancement by herbicides and hormones, and bigger and better harvesting machinery. Yields can be increased by carefully regulating the timing of the harvest. For example, few hardwoods would be allowed to grow longer than 50 years, after which time net annual growth decreases.[49] Clear-cutting in the Douglas-fir region of the Pacific Northwest and in the pines of the reforested cutovers of the South in order to promote even-aged regeneration and minimize harvesting costs poses threats to the appearance and productivity of these forests, and the use of pesticides and herbicides has raised controversy ever since the publication in 1962 of Rachel Carson's seminal book, *Silent Spring* (Plates 14.3–14.4). Thus the danger is that the much-lauded aims of sustained yield and forest agriculture will impoverish the existing forest and harm the total forest environment. When some of the more radical and startling forecasts of the outcome of genetic engineering are added to these changes, Americans in the future may well look in vain for the forest as it was 150 years ago, or even for the forest of today. These matters are not pursued in detail, but they are presented, nonetheless, and they have their proponents and their critics.[50]

Plate 14.3. New forestry practices become environmental issues: Douglas-fir trees do not regenerate easily in the shade; therefore, block cutting, which is felt to be aesthetically undesirable, is used to promote growth and even-aged stands.

Plate 14.4. Another environmental issue: aerial spraying of herbicides and pesticides over inaccessible forest areas is perceived as a hazard to human health.

These concerns will have their manifestation in the appearance of the forest, but there are others that are not so clearly seen. These relate to the organization of the exploitation of the forest. The increasing centralization of power in large corporations such as the Weyerhaeuser Company, Georgia-Pacific, or International Paper and the elimination of the smaller independent operators through mergers and consolidation since 1945 raise specters of the past, as does the concentration of stumpage ownership in the hands of a few. The trend to vertical integration from stump to final product, encouraged by taxation legislation, also fans the fires of concern over monopoly and the power of large corporations as they strive for self-sufficiency, although perhaps unreasonably, as no corporation holds more than 4 percent of the total capacity of the industry.[51] Underlying all these concerns is the swell of public feeling about the environment in general and the conservation of resources.

Whatever the outcome of these vexing and difficult questions, one thing is certain. The saga of the relationship of Americans with their forest is by no means over. It is a continuous process, to which there is no end. In a decade, perhaps, another chapter will have to be written outlining and analyzing how Americans came to terms with the new advances and ideas and used the remaining forest – one of their greatest resources – during the closing years of the twentieth century.

Notes

1. The forest in American life

1 For two excellent recent studies of the American landscape, see John R. Stilgoe, *The Common Landscape of America, 1580 to 1845*, and W. Cronon, *Changes in the Land: Indians, Colonists, and the Ecology of New England*.

2 USFS, *Timber Depletion, Lumber Prices, Lumber Exports, and Concentration of Timber Ownership*, 32; USFS, *An Analysis of the Timber Situation in the United States, 1952–2030*, 149.

3 Alexis de Tocqueville, "A Fortnight in the Wilds," *Journey to America*, 329.

4 Perhaps Richard G. Lillard's popular account, *The Great Forest*, comes nearest to presenting an overall view of the multiple uses of, and impacts on, the forest over the centuries.

5 James Hall, *Statistics of the West*, 100–1.

6 Based on U.S. Bureau of the Census, *Census of Manufactures for the United States*, 1850–1920.

7 Brooke Hindle (ed.), *America's Wooden Age: Aspects of Its Early Technology*, 3; see also Charles Van Ravenswaay, "America's Age of Wood," and Brooke Hindle (ed.), *Material Culture of the Wooden Age*; J. Hall, *Statistics of the West*.

8 USFS, *Analysis of the Timber Situation*, 2–4; USFS, *The Economic Importance of Timber in the United States*; and USFS, *Timber in the United States Economy, 1963, 1967, and 1972*, 7–20.

9 For a discussion of the computation of these figures on clearing, see Chapter 5, herein.

10 Marion Clawson, *Man, Land, and the Forest Environment*, 27–33.

11 Hindle, *America's Wooden Age*, 12; Lewis Mumford, *Technics and Civilization*, 109–10.

12 This interpretation of attitudes toward the forests owes much to Roderick Nash's *Wilderness and the American Mind* and to Peter N. Carroll's *Puritanism and the Wilderness: The Intellectual Significance of the New England Frontier, 1629–1700*. Other intellectual histories of importance in understanding man's relationship with the forest are Henry Nash Smith, *The Virgin Land: The American West as Symbol and Myth*; Charles Leroy Sanford, *The Quest for Paradise: Europe and the American Moral Imagination*; and Leo Marx, *The Machine in the Garden: Technology and the Pastoral Ideal in America*.

13 Alexander Porteous, *Forest Folklore, Mythology, and Romance*, 84–148. In addition to the symbolism of the forest, there is the symbolism of trees, which enter Western thought in many ways, from Christmas trees to genealogical trees. The tree symbolizes man himself: Union "bears fruit"; untimely death is being "cut down in one's prime," so that death is, symbolically, an ax man. The tree also figures largely in Christian thought, for God "planted a garden eastward of Eden" where he made "to grow every tree that is pleasant to the sight, and good for food; the tree of life also in the midst of the garden, and the tree of knowledge of good and evil." The Koran's account of the beginning of the world is very little different from that of the Bible.

14 See Clarence J. Glacken, *Traces on the Rhodian Shore: Nature and Culture in Western Thought*

from Ancient Times to the End of the Eighteenth Century, 288–351, for an excellent review of medieval forest-clearing activity.

15 Bohumil Shimek, "The Pioneer and the Forest," 97–9; Gilbert Chinard, "The American Philosophical Society and the Early History of Forestry in America," 444.

16 Tocqueville, "A Fortnight in the Wilds," *Journey to America*, 335; Alexis de Tocqueville, *Democracy in America*, 2:74.

17 Carl Bridenbaugh, *Fat Mutton and the Liberty of Conscience*, 5–15, passim; Increase Mather, *A Brief History of the War with the Indians in New England*. See also Alan Heimert, "Puritanism, the Wilderness, and the Frontier," 361–82, for an extended discussion of attitudes and their interplay with the environment.

18 Nash, *Wilderness*, 28–30, 34–7; Michael Wigglesworth, "God's Controversy with New-England" (1662), 83–4; Glacken, *Traces on the Rhodian Shore*, 680–1. See also Roy Harvey Pearce, *The Savage of America: A Study of the Indian and the Idea of Civilization*; Charles F. Carroll, *The Timber Economy of Puritan New England*, 57, quoting Josh. 17:1.

19 J. Hector St. John de Crèvecoeur, *Letters from an American Farmer*, 55–7, 271.

20 Oscar Handlin, *Race and Nationality in American Life*, 114–15.

21 "Wilderness evil" has been identified as a major theme in American culture by Constance M. Rourke, *The Roots of American Culture and Other Essays*, 20–1.

22 Francis Parkman, "The Forests and the Census," 836; Frederick Jackson Turner, "Pioneer Ideals and the State University," in *The Frontier in American History*, 269–70.

23 Crèvecoeur, *Letters from an American Farmer*, 2, 39.

24 William Cooper, *A Guide in the Wilderness*, 6.

25 Sanford, *Quest for Paradise*, 119; Benjamin Franklin, in Albert H. Smyth (ed.), *The Writings of Benjamin Franklin*, 3: 72–3; Andrew Jackson, "Second Annual Message," in James D. Richardson (ed.), *A Compilation of the Messages and Papers of the Presidents*, 3:1084.

26 Merle Curti, *The Growth of American Thought*, 150–5; Christopher Hussey, *The Picturesque: Studies in a Point of View*, 20–37.

27 Walter J. Hipple, *The Beautiful, the Sublime, and the Picturesque in Eighteenth Century British Aesthetics*; Marjorie Hope Nicolson, *Mountain Gloom and Mountain Glory: The Development of the Aesthetics of the Infinite*.

28 Curti, *Growth of American Thought*, 239–43; Perry Miller, *Errand into Wilderness*, 209; Hoxie Neal Fairchild, *The Romantic Quest*, 123–40, 351–72. For examples of romantic poetry, see, particularly, Lord Byron, *Childe Harold's Pilgrimage*, canto 4, clxxvii: "There is a pleasure in the pathless woods . . ."; and William Wordsworth, *The Tables Turned*: "One impulse from a vernal wood . . ."

29 Vicomte de Chateaubriand, *Recollections of Italy, England, and America on Various Subjects*, 138–9; Curti, *Growth of American Thought*, 243–4.

30 Nash, *Wilderness*, 55–6; Philip M. Marsh (ed.), *The Prose of Philip Freneau*, esp. 196–202; George W. Corner (ed.), *An Autobiography of Benjamin Rush*, 72.

31 Mason Wade (ed.), *The Journals of Francis Parkman*, 1:3. For an example of the melancholy genre, see Philip H. Gosse, *Letters from Alabama, Chiefly Relating to Natural History*, 25.

32 Marx, *Machine in the Garden*, 73–144; H. Smith, *The Virgin Land*, 138–64; William Penn, *Some Fruites of Solitude* (1663), pt. 1, no. 220.

33 Thomas Jefferson, *Notes on the State of Virginia*, ed. William Peden, query 19, 164–5; Franklin, *Writings*, 1:245; Washington Irving, *Rip Van Winkle, and The Legend of Sleepy Hollow*, 63–4.

34 Curti, *Growth of American Thought*, 141–8, 238–44; Nash, *Wilderness*, 78–83; Chateaubriand, *Travels in America and Italy*, 98. For the works of Bryant, see Parke Godwin (ed.), *The Poetical Works of William Cullen Bryant*, 1:23–4, 130–4; for Paulding, see Samuel Kettell (ed.), *Specimens of American Poetry, with Critical and Biographical Notices*, 2:180–4.

35 See Roland Van Zandt, "The Catskills and the Rise of American Landscape Painting," 257–81; Mary E. Woolley, "The Development of the Love of Romantic Scenery in

America," 56–66; Charles L. Sanford, "The Concept of the Sublime in the Works of Thomas Cole and William Cullen Bryant," 434–48; Sanford, *Quest for Paradise*, 136–54.

36 Ralph Waldo Emerson, *The Collected Works of Ralph Waldo Emerson*, "Nature, Addresses, and Lectures": "Nature," 1:10; "The Young American," 1:221–7.

37 Arthur A. Ekirch, Jr., *Man and Nature in America*, 47–69; Nash, *Wilderness*, 84–95; Henry David Thoreau, "Walking," in "Nature Essays," *The Works of Thoreau*, 672.

38 Turner, "Frontier in American History," 266, 216, 293; Richard Hofstadter, *The Progressive Historians: Turner, Beard, Parrington*, 120–5, esp. 121.

39 Nash, *Wilderness*, 95; and Henry David Thoreau, *The Maine Woods*, 155–6.

40 Benjamin Franklin, *Franklin's Works*, "An Account of the New Invented Pennsylvania Fireplace," 2:392; Franklin B. Hough, *Report upon Forestry*, 1:446. For a more extended treatment of fuel and lumber supply problems, see Chapters 3 and 10, herein.

41 See esp. Chapter 12, herein.

42 For example, Benjamin Rush, "An Enquiry into the Cause of the Increase of Bilious and Intermitting Fevers in Pennsylvania"; Thomas Wright, "On the Mode Most Easily and Effectually Practicable for Drying Up the Marshes of the Maritime Parts of North America": Jared Eliot, *Essays upon Field Husbandry in New England, 1748–1762*; and F. C. Volney, *A View of the Soil and Climate of the United States of America*. For a penetrating review of these works, see Glacken, *Traces on the Rhodian Shore*, 687–98.

43 George Perkins Marsh, "The Woods," in *Man and Nature; or, Physical Geography as Modified by Human Action*, 113–280.

44 George Perkins Marsh, *Address Delivered before the Agricultural Society of Rutland County*, 17–19, quoted in David Lowenthal, "On the Author of *Man and Nature*," 10.

45 See esp. Chapter 11, herein.

46 Thoreau, *The Maine Woods*, 156; Marsh, *Man and Nature*, 327.

47 Morton White and Lucia White, *The Intellectual versus the City: From Thomas Jefferson to Frank Lloyd Wright*, 54–116.

48 H. D. Thoreau, *Journal*, quoted in Nash, *Wilderness*, 90; Richard W. B. Lewis, *The American Adam: Innocence, Tragedy, and Tradition in the Nineteenth Century*; James O. Robertson, *American Myth, American Reality*.

49 Anthony Trollope, *North America*, 136. For the descriptions of pioneers in the Genessee country of New York, see Orsamus Turner, *A Pioneer History of the Holland Purchase of Western New York*.

50 For an extensive appraisal of Boone and Cooper, see H. N. Smith, *Virgin Land*, 54–76; Nash, *Wilderness*, 75–8; and Nelson Van Valen, "James Fenimore Cooper and the Conservation Schism," 289–306.

51 H. N. Smith, *Virgin Land*, 88–125; Wilson O. Clough, *The Necessary Earth: Nature and Solitude in American Literature*, 64–74.

52 For a good factual account of the lumberjack's life, though with some eulogistic overtones, see Stewart H. Holbrook, *Holy Old Mackinaw: A Natural History of the American Lumberjack*; see also George B. Engberg, "Who Were the Lumberjacks?," 238–46.

53 D. C. Everest, "A Reappraisal of the Lumber Barons," 17–22; Charles E. Twining, "Plunder and Progress: The Lumber Industry in Perspective," 116–24.

54 William G. Robbins, *Lumberjacks and Legislators: Political Economy of the U.S. Lumber Industry, 1890–1941*, 8 et seq.

55 Marx, *Machine in the Garden*. For the destruction of the forest by lumbering, see esp. Chapters 7, 8, and 9, herein.

2. The forest and the Indian

1 See Glacken, *Traces on the Rhodian Shore*, 257–63, for an extended discussion of the significance of the new discoveries.

2 Gilbert Chinard, "Eighteenth Century Theories on America as a Human Habitat," 27.

3 John Smith, *A Description of New England*, in *The Travels of Captain John Smith*, 207; and Thomas Morton, *The New English Canaan*, 182–7. For other examples, see Virginia S. Eifert, *Tall Trees and Far Horizons: Adventures and Discoveries of Early Botanists in America*; Robert Elman, *First in the Field; America's Pioneering Naturalists*; Joseph A. Ewan (ed.), *A Short History of Botany in the United States*; Henry Savage, *Discovering America, 1700–1875*.

4 Pehr Kalm, *Travels into N. America*, vols. 1 and 2, passim.

5 François André Michaux and Thomas Nuttall, *The North American Sylva*. See also Jacob Richard Schramm, "Influence – Past and Present – of François André Michaux on Forestry and Forest Research in America"; Rodney H. True, "François André Michaux, the Botanist and Explorer."

6 Richard G. Beidleman, "Some Biological Sidelights on Thomas Nuttall, 1786–1859"; Francis W. Pennell, "Travels and Scientific Collections of Thomas Nuttall."

7 Volney, *Soil and Climate of the United States*, 8–9.

8 Ralph H. Brown, *The Historical Geography of the United States*, 195–201, 206–11.

9 Joseph Henry, "Meteorology and Its Connection with Agriculture," plate VII, and p. 437.

10 James G. Cooper, "On the Distribution of the Forests and Trees of North America," 250–72, and "The Forests and Trees of North America, as Connected with Climate and Agriculture," esp. 436–45. For other information about Cooper quoted here, see Eugene Coan, "James Graham Cooper, Pioneer Naturalist and Forest Conservationist," 126–9.

11 William H. Brewer, "The Woodland and Forest Systems of the United States."

12 Charles Sprague Sargent, *Report on the Forests of North America*, 485, and *Folio Atlas of Forest Trees of North America*.

13 For a good introduction to this, see Frank N. Egerton (ed.), *American Plant Ecology, 1897–1917: An Origimal Anthology*. For other useful information, see Robert H. Whittaker, "Classification of Natural Communities."

14 Fredric E. Clements, "Plant Formations and Forest Types"; Henry C. Cowles, "The Ecological Relations of the Vegetation on the Sand Dunes of Lake Michigan." For further discussion of the distinctive emphases of Clements and Cowles, see Ronald C. Tobey, *Saving the Prairies: The Life Cycle of the Founding School of American Plant Ecology, 1895–1955*; and for the influence of both workers on understanding the forest, see Susan L. Flader, "Ecological Science and the Expansion of Our Forest Heritage.

15 Henry David Thoreau, "The Natural History of Massachusetts," 127–62, and "The Succession of Forest Trees," 229, 231–2, 236.

16 Hugh M. Raup, "Some Problems in Ecological Theory and Their Relation to Conservation." For further discussion of Cowles's approach to the "science of communities" and its sociological overtones and Clements's approach to plant succession and its overtones of Social Darwinism, see Donald Worster, *Nature's Economy: The Roots of Ecology*, 212–15.

17 Herbert A. Gleason, "The Individualistic Concept of the Plant Association"; Arthur G. Tansley, "The Use and Abuse of Vegetational Concepts and Terms."

18 Homer L. Schantz and Raphael Zon, "Natural Vegetation," USDA, *Atlas of American Agriculture*, p. 1.

19 A. W. Küchler, *Manual to Accompany the Map of Potential Natural Vegetation of the Coterminous United States*.

20 USFS, *Analysis of the Timber Situation*, 120–4, 496–7.

21 For example, Nathaniel S. Shaler, *Nature and Man in North America*, 195; Francis Parkman, *A Half Century of Conflict: France and England in North America*, 1:32; William Fox, *History of the Lumber Industry in the State of New York*, 8; Ulysses P. Hedrick, *A History of Agriculture in the State of New York*, 329; Lillard, *The Great Forest*, 4. For other examples, see J. T. Adams, *The Epic of America*, 4; John Bakeless, *The Eyes of Discovery: The Pageant of North America as Seen by the First Explorers*, 407.

22 For accounts of Indian cultivation practices, and of the crops and their identification by Europeans, see Lyman Carrier, *The Beginnings of Agriculture in America*.

23 James Mooney, "The Aboriginal Population of America North of Mexico," 33; A. L. Kroeber, *Cultural and Natural Areas of Native North America*, 131–57; Angel A. Rosenblat, *La población indígena de América desde 1492 hasta la actualidad*, 92, and *La población indígena y el mestizaje en América*, vol. 1, *La población indígena, 1492–1950*, 102. Rosenblat defends his estimates in another book, *La población de América en 1492: Viejos y nuevos calculos*, 1–16.

24 Henry F. Dobyns, "Estimating Aboriginal American Population: An Appraisal of Techniques with a New Hemispheric Estimate," 395–416. See H. Paul Thompson, "Estimating Aboriginal American Population: A Technique Using Anthropological and Biological Data," 417–24, with appended critiques (425–49) of Thompson and Dobyns from 20 scholars; and Sherburne F. Cook and Woodrow Borah, *Essays in Population History: Mexico and the Caribbean*, vol. 1, passim.

25 Wilbur R. Jacobs, "The Tip of the Iceberg: Pre-Columbian Indian Demography and Some Implications for Revisionism," 123–33, and "The Indian and the Frontier in American History: A Need for Revision," 50–6; William W. Newcomb, *North Americans: An Anthropological Perspective*, 15–17.

26 Harold E. Driver, *The Indians of North America*, 63–5, and map 6.

27 Carl O. Sauer, *Sixteenth Century North America: The Land and the People as Seen by the Europeans*, 303.

28 The best overall accounts of the Indian impact on the forest are Gordon M. Day, "The Indian as an Ecological Factor in the Northeastern Forest," 344–6; and, though about the Ontario Indians, Conrad Heidenreich, *Huronia*, 107–218. See also S. W. Bromley, "The Original Forest Types of Southern New England," 63–89. Quotation is from William Byrd, *William Byrd's Natural History of Virginia*, 592–3.

29 Hugh Jones, *The Present State of Virginia*, 55.

30 Jo Françoise Lafitau, *Customs of the American Indians Compared with Those of the Customs of Primitive Times*, 39; Day, "The Indian as an Ecological Factor," 330.

31 J. N. B. Hewitt, "The Iroquoian Cosmology," 461–2; Edmund B. O'Callaghan (ed.), *Documentary History of the State of New York*, 1:27–41.

32 Benjamin Hawkins, *A Sketch of the Creek Country in the Years 1789 and 1799*, 30.

33 J. R. Swanton, *Early History of the Creek Indians and Their Neighbours*, 403–11.

34 John Smith, *The Generall Historie of Virginia, New England, and the Summer Isles*, 2:363, in *The Travels and Works of Captain John Smith*.

35 William Strachey, *The Historie of Travaile into Virginia Britannia*.

36 Quoted in Hu Maxwell, "The Use and Abuse of the Forests by the Virginia Indians," 81.

37 William Byrd, *History of the Dividing Line between Virginia and Other Tracts, 1728–1736*, 2:15–17, "Journey to the Land of Eden"; and John Lederer, *The Discoveries of John Lederer in Three Marches from Virginia to the West of Carolina . . . 1669 and 1670*, 24.

38 Maxwell, "Use and Abuse of the Forests," 85–6.

39 Alexander Young, *Chronicles of the Pilgrim Fathers of the Colony of Plymouth from 1602 to 1625, 206–7.* For other examples, see Day, "The Indian as an Ecological Factor," 330–4.

40 Francis H. Higginson, *New-England's Plantation*, 9. For other accounts of New England, see Morton, *New English Canaan*, 172; and William Wood, *New England's Prospect*, 40.

41 Sauer, *Sixteenth Century North America*, 88; Brown, *Historical Geography*, 17; and Day, "The Indian as an Ecological Factor," 332.

42 U.S. Congress, House, *Final Report of the United States De Soto Expedition Commission*, 82, 89, 98.

43 Hawkins, *Sketch of the Creek Country*, 22–35, passim; William Bartram, *Travels through North and South Carolina*, 374–401, passim. Erhard Rostlund, in "The Myth of a Natural Prairie Belt in Alabama: An Interpretation of Historical Records," 397–402, has many more examples.

44 Jones, *The Present State of Virginia*, 74.

45 F. Cook (ed.), *Journals of the Military Expedition of General John Sullivan against the Six*

Nations of Indians in 1779, 301; other descriptions in Martin Richard Kaatz, "The Black Swamp: A Study in Historical Geography," 14–18, 23, 25.

46 For other uses of fire, see Stephen J. Pyne, *Fire in America: A Cultural History of Wild Land and Rural Fire*, 72; H. Maxwell, "Use and Abuse of the Forests," 87–8; Omer C. Stewart, "Burning and Natural Vegetation in the United States," 317–18.

47 Pyne, *Fire in America*, 74.

48 Carl O. Sauer, "The Agency of Man on Earth," 54–6.

49 For example, see Day, "The Indian as an Ecological Factor," 334; Calvin Martin, "Fire and Forest Structure in the Aboriginal Eastern Forest," 40. Stewart, in "Burning and Natural Vegetation," said that "An undoubtedly incomplete review of the literature brought to light more than 200 references to Indians setting fire to vegetation in aboriginal times" (319).

50 Morton, *New English Canaan*, 172–3.

51 Andrew White, *A Narrative of the Voyage to Maryland*, 18.

52 Edward Johnson, *Johnson's Wonder-Working Providence of Sion's Saviour in New England, 1628–1681*, 85; quoted in Day, "The Indian as an Ecological Factor," 335; Adam Hodgson, *Remarks during a Journey through North America, in 1819, 1820, and 1821*, 1:273.

53 David Lowenthal and Hugh Prince, "English Landscape Tastes," 198–200.

54 For the prairies, see Douglas R. McManis, "The Initial Evaluation and Utilization of the Illinois Prairies, 1815–1840," 30–44; and Chapter 5, herein.

55 For Virginia, see accounts in H. Maxwell, "Use and Abuse of the Forests," 94–6; for Pennsylvania, see Harold J. Lutz, "The Vegetation of Heart's Content: A Virgin Forest in North Western Pennsylvania"; for Kentucky, see Shaler, *Nature and Man in America*, 186, and François André Michaux, *Travels to the West of the Allegheny Mountains* (1805), 221–2, 268.

56 T. Bigelow, *Journal of a Tour to Niagara Falls in the Year 1805*, 49–51.

57 Rostlund, "Myth of a Natural Prairie Belt," 392–411; J. Baird, *View of the Valley of the Mississippi*, 204.

58 Carl O. Sauer, "A Geographical Sketch of Early Man in America," 543–5, 551–4; Pyne, *Fire in America*, 76; Douglas Branch, *The Hunting of the Buffalo*, 52–63; and Frank Gilbert Roe, *The North American Buffalo: A Critical Study of the Species in the Wild State*, 26–68, 228–82.

59 Omer C. Stewart, "Fire as the First Great Force Employed by Man," 125–7; Sauer, "Agency of Man on Earth," 55.

60 E. J. Dyksterhius, "The Vegetation of the Western Cross Timbers," 332–3; Herbert Gleason, "Vegetational History of the Middle West," 80–5; Elizabeth Chavannes, "Written Records of Forest Succession," 76–85.

61 Bayard Taylor, *Colorado: A Summer Trip*, 5, 179.

62 Inter alia, Stewart, "Burning and Natural Vegetation," 318; Pyne, *Fire in America*, 79–80; Dyksterhius, "Vegetation of the Western Cross Timbers," 333.

63 K. H. Garren, "Effects of Fire on Vegetation of the Southeastern United States," 620–35; Elsie Quarterman and Catherine Keever, "Southern Mixed Hardwood Forest: Climax in the Southeastern Coastal Plain, USA," 168–9.

64 The numerous papers of Herman Haupt Chapman on the behavior of the southern pines in relation to fire are crucial; for example, H. H. Chapman, "Is the Longleaf Type a Climax?" 328–34, and "Fire and Pines: A Realistic Appraisal of the Role of Fire in Reproducing and Growing Southern Pines," 62–4, 91–3.

65 For a more extended discussion, see Arthur A. Brown and Kenneth P. Davis, *Forest Fire: Control and Use*, 31–44; Bruce M. Kilgore, "From Fire Control to Fire Management: An Ecological Basis for Policies"; Theodore T. Kozlowski and C. E. Ahlgren (eds.), *Fire and Ecosystems*, 195–320; Küchler, *Potential Natural Vegetation*, 2–4.

66 Kroeber, *Cultural and Natural Areas*, 131, 166; Heidenreich, *Huronia*, 195–200.

67 H. Maxwell, "Use and Abuse of the Forests," 81.

68 Carl O. Sauer, "Grassland: Climax, Fire, and Man," 16–21; Kozlowski and Ahlgren, *Fire and Ecosystems*, 139–94.

69 Omer C. Stewart, "Barriers to Understanding the Influence of the Use of Fire by Aborigines on Vegetation."

3. The forest and pioneer life, 1600–1810

1 Tocqueville, *Journey to America*, "A Fortnight in the Wilds," 329.
2 Marquis de Chastellux, *Travels in North America in the Years 1780, 1781, and 1782*, 2:44.
3 Paul W. Gates, "Problems of Agricultural History, 1790–1840," 34.
4 For interesting commentary on ideas of self-sufficiency, see Andrew Hill Clark, "Suggestions for the Geographical Study of Agricultural Change in the United States, 1790–1840," 165–72; and Rodney, C. Loehr, "Self-sufficiency on the Farm." For detailed accounts of pioneer farming during these years, see Howard S. Russell, *A Long Deep Furrow: Three Centuries of Farming in New England*, 2–107; Philip L. White, *Beekmantown, New York: Forest Frontier to Farm Community*, 53–92; P. D. Bidwell and J. I. Falconer, *History of Agriculture in the Northern United States, 1620–1860*, 1–146.
5 Horace Greeley, *Recollections of a Busy Life*, 60.
6 Timothy Dwight, *Travels in New-England and New-York in 1821*, Letter 13, 2:321–2; P. White, *Beekmantown*, 56.
7 Samuel Symonds, "Letter," 118–20, quoted in Stilgoe, *Common Landscape*, 172.
8 John Smith, *Generall Historie*, 1:165; Strachey, *Historie of Travaile*, xxxii.
9 O'Callaghan, *Documentary History of New York*, 4:23–4, "New Netherlands in 1644," a letter by Father Isaac Jogues.
10 Andriaes Van der Donck, "A Description of the New Netherlands."
11 Herbert Baxter Adams, "Village Communities of Cape Anne and Salem," 25.
12 Peter Force, *Tracts and Other Papers*, "A Planter's Plea; or, The Grounds for Plantation Examined" (1630) 12, item 3:14, of the tract, and "Virginia Richly Valued" (1609) 4; item 1:13, of the tract.
13 For a general view of New England settlement, see Charles E. Clark, *The Eastern Frontier: The Settlement of Northern New England, 1610–1763*; Stilgoe, *Common Landscape*, 99–107.
14 Turner, *Pioneer History of the Holland Purchase*, 322.
15 In emphasizing here the generalities of the material and locational aspects of early settlement, one should not lose sight of the importance of a host of nonmaterial facets of pioneer life, such as attitudes toward status, capital accumulation, individual versus community action, etc. See, in particular, James T. Lemon, *The Best Poor Man's Country: A Geographical Study of Early Southeastern Pennsylvania*; and "The Weakness of Place and Community in Early Pennsylvania"; James Henretta, "Families and Farms: *Mentalité* in Pre-industrial America"; James T. Lemon, "Early Americans and Their Social Environment."
16 See Rush, "Causes of the Increase of Fevers," 207; Erwin H. Ackerknecht, *Malaria in the Upper Mississippi Valley, 1760–1900*, 27–50.
17 See W. Cooper, *A Guide in the Wilderness*, 22, 34–5, for an extended discussion; see also Adam Hodgson, *Letters from North America Written during a Tour in the United States and Canada*, 1:33; Stilgoe, *Common Landscape*, 143–8.
18 Lillard, *The Great Forest*, 67–8.
19 Sylvester Judd, *History of Hadley, Massachusetts*, 432; Eliot, *Essays upon Field Husbandry*, 1; Johann David Schoepf, *Travels in the Confederation* (*1783–1784*), 1:264.
20 Dwight, *Travels*, Letter 11, 2:125–6; Jeremy Belknap, *The History of New Hampshire*, 3:131–7; and Carl Bridenbaugh, "Yankee Use and Abuse of the Forest in the Building of New England, 1620–1660," 3–35.
21 Benjamin Rush, *An Account of the Manners of the German Inhabitants of Pennsylvania*, 58; and O'Callaghan, *Documentary History of New York*, 4:30.
22 Richard Frame, "A Short Description of Pennsilvania by Richard Frame," 303.
23 David M. Ellis, *Landlords and Farmers in the Hudson–Mohawk Region, 1790–1850*, 73, passim.

24 Belknap, *History of New Hampshire*, 3:132–3. Another good example is in Dwight, *Travels*, Letter 14, 2:325–6.

25 Thomas Cooper, *Some Information Respecting America*, 117–18. In Pennsylvania in the late eighteenth century a stump grubber could earn 20–24 shillings plus victuals for every acre cleared and could clear one every three to four days, according to Schoepf, *Travels in the Confederation*, 208–9.

26 Based on O'Callaghan, *Documentary History of New York*, 3:1148; Adam Hodgson, *Letters from America*, 1:22; James Stuart, *Three Years in North America*, 1:260–6; Belknap, *History of New Hampshire*, 3:133–7.

27 For high initial yields, see Belknap, *History of New Hampshire*, 3:135; Adam Hodgson, *Letters from America*, 2:306–7.

28 John Smith, "Advertisements for the Inexperienced Planters of New England or Anywhere," 38.

29 Chastellux, *Travels in North America*, 1:45–6. Other excellent descriptions of clearing are in La Rochefoucauld-Liancourt, *Travels through the United States of North America*, 1:293–5; Dwight, *Travels*, Letter 11, 2:113–14; Tench Coxe, *A View of the United States of America*, 452; William Strickland, *Journal of a Tour in the United States of America, 1794–95*, 95–6; Henry Beaufoy, *Tour through Parts of the United States and Canada by a British Subject*, 80–2; Thomas Anburey, *Travels through the Interior Parts of America*, 2:323–4.

30 Thomas F. Gordon, *A Gazetteer of the State of Pennsylvania*, 32.

31 Based on Turner, *Pioneer History of the Holland Purchase*, 565; Gabriel Thomas, *An Historical and Geographical Account of the Province and Country of Pensilvania*, 319; Orsamus Turner, *History of the Pioneer Settlement of Phelps and Gorham's Purchase and Morris' Reserve*, 176; La Rochefoucauld-Liancourt, *Travels*, 1:294; Henry Wansey, *Journal of an Excursion in the United States in Summer of 1794*, 191; Adam Hodgson, *Letters from America*, 1:40; P. White, *Beekmantown*, 54.

32 La Rochefoucauld-Liancourt, *Travels*, 1:294; Beaufoy, *Tour*, 81–2.

33 For a good description of a logging and cabin-raising bee, see Crèvecoeur, *Letters from an American Farmer*, 114–15. Quotation from Chastellux, *Travels in North America*, 1: 46–7.

34 Tocqueville, *Journey to America*, "A Fortnight in the Wilds," 337; W. Cooper, *A Guide in the Wilderness*, 38–41.

35 Byrd, *Natural History of Virginia*, 93.

36 Hugh Jones, *The Present State of Virginia*, 77.

37 *Calendar of State Papers: America and the West Indies, 1696–97*, 642–3.

38 Edmund Ruffin, *An Essay on Calcareous Manures*, 12; Avery O. Craven, *Soil Exhaustion as a Factor in the Agricultural History of Virginia and Maryland, 1606–1860*; Melvin G. Herndon, "The Significance of the Forest to the Tobacco Plantation Economy of Antebellum Virginia," 230–9.

39 See John Clayton, "A letter from Mr. John Clayton . . . to the Royal Society, May 12, 1686," 21; he says, "every three or four years they must be for clearing a new piece of ground out the Woods, which requires much Labour and Toil." See also Schoepf, *Travels in the Confederation*, 48; Adam Hodgson, *Letters from America*, 1:33, 294.

40 Anburey, *Travels*, 2:322; see also the remarks by John Clayton in "A Letter to the Royal Society," 21, on the problem of "solitary and unsociable" living resulting from the wide spacing of the plantations.

41 J. G. W. De Brahm, *Philosophico-Historico-Hydrogeography of South Carolina, Georgia, and East Florida*, 197–8.

42 Thomas Nairne, *A Letter from South Carolina*, 49, 52–4.

43 Chastellux, *Travels in North America*, 1:47.

44 Dwight, *Travels*, Letter 13, 321–2.

45 Sam Bass Hilliard, *Hogmeat and Hoecake: Food Supply in the Old South, 1840–1860*, 28–56.

46 H. Clifford Darby, *The Domesday Geography of Eastern England*, 124–9, 179–80, 235–9, 362–3.

47 Rupert B. Vance, *The Human Geography of the South: A Study in Regional Resources and Human Adequacy*, 145–6; U.S. Congress, House, *Final Report of the De Soto Expedition Commission*, 82–98.

48 Based on Samuel Wilson, "An Account of the Province of Carolina in America," 171; Nairne, *Letter from South Carolina*, 131; Vance, *Human Geography of the South*, 145–60.

49 Lewis C. Gray and Esther K. Thompson, *History of Agriculture in the Southern United States to 1860*, 1:138–9; De Brahm, *Philosophico-Historico-Hydrogeography of South Carolina*, 200.

50 Based on Gray and Thompson, Agriculture in the Southern United States, 1:139–43.

51 Bidwell and Falconer, *Agriculture in the Northern United States*, 119–20.

52 William W. Hening (ed.), *Virginia: The Statutes at Large*, 1:199; Gray and Thompson, *Agriculture in the Southern United States*, 145–6; Bidwell and Falconer, *Agriculture in the Northern United States*, 21–2. For fencing in New England see C. Carroll, *Timber Economy*, 63–4.

53 Earl W. Hayter, "Livestock-Fencing Conflicts in Rural America," 10–20, and *The Troubled Farmer, 1850–1900*, 104–13.

54 William Oliver, *Eight Months in Illinois*, 239. For good contemporary descriptions of Virginia fencing, see T. Cooper, *Some Information*, 127–8; Beaufoy, *Tour*, 53–4.

55 Gray and Thompson, *Agriculture in the Southern United States*, 553; Russell, *Long Deep Furrow*, 104–7.

56 Clarence H. Danhof, "The Fencing Problem in the Eighteen-fifties," 168–86; Martin L. Primack, "Farm Formed Capital in American Agriculture, 1850–1910," 287–91; Hayter, *The Troubled Farmer*, 97–103.

57 Gray and Thompson, *Agriculture in the Southern United States*, 553–4; De Brahm, *Philosophico-Historico-Hydrogeography of South Carolina*, 199; Clarence H. Danhof, *Change in Agriculture: The Northern United States, 1820–1870*, 118.

58 Hough, *Report upon Forestry*, 3:292.

59 Kalm, *Travels into N. America*, 2:55–6; Michaux, *Travels to the West*, 136–7; La Rochefoucauld-Liancourt, *Travels*, 1:47.

60 H. R. Shurtleff, *The Log-Cabin Myth: A Study of the Early Dwellings of the English Colonists in North America*.

61 Based on Michaux, *Travels to the West*, 136–7; La Rochefoucauld-Liancourt, *Travels*, 1:295–7; Dwight, *Travels*, Letter 11, 2:114–15. Wansey, in his *Excursion in the United States*, 175, said that a cabin was commonly erected in a day. For farm utensils in the houses and implements on the farm, see Wayne D. Rasmussen, "Wood on the Farm," 22–35.

62 Everett Dick, *The Dixie Frontier*, 80–3. There is a vast literature about regional variations in folk housing in America. The following is a selection of the most important and/or accessible works: Henry Glassie, *Folk Housing in Middle Virginia*; Mary Gould, *The Early American House*; Henry J. Kaufman, *The American Farmhouse*; Fred B. Kniffen, "Folk Housing: Key to Diffusion"; Fred Kniffen and Henry Glassie, "Building in Wood in the Eastern United States: A Time–Place Perspective"; Peirce Lewis, "The Geography of Old Houses"; Carl W. Condit, *American Building: Materials and Techniques from the First Colonial Settlements to the Present*.

63 C. Clark, *Eastern Frontier*, 197–225. For an interesting appraisal of traditional timber frame housing, see Dell Upton, "Traditional Timber Framing," 35–96.

64 Timothy Flint, *A Condensed Geography and History of the Western States or the Mississippi Valley*, 2:75–6; *New York State Census* (1855), 245–6, (1865), 271–2; *United States Census* (1840).

65 Coxe, *View of the United States*, 451.

66 See Robert P. Multhauf, "Potash"; William I. Roberts, "American Potash Manufacture before the American Revolution"; Harry Miller, "Potash from Wood Ashes: Frontier

Technology in Canada and the United States," 187–208. Also useful for historical and chemical details are C. A. Browne, "Historical Notes on the Domestic Potash Industry in Early Colonial and Later Times"; Theodore J. Kreps, "Viccissitudes of the American Potash Industry."

67 Cooper, *Some Information*, 119; Coxe, *A View of the United States*, 452–3. See also Patrick Campbell, *Travels in the Interior Inhabited Parts of North America in the Years 1791 and 1792*, 222.

68 White, *Beekmantown*, 5.

69 Benjamin Franklin, *An Account of the New Invented Pennsylvania Fire-place*, 392; Rev. Jonas Michäelius quoted in Alfred P. Muntz, "The Changing Geography of the New Jersey Woodlands, 1600–1900," 52.

70 De Brahm, *Philosophico-Historico-Hydrogeography of South Carolina*, 198.

71 Robert Beverly, *The History of Virginia* (1720), 108, quoted in Howard N. Eavenson, *The First Century and a Quarter of the American Coal Industry*, 30.

72 Robert V. Reynolds and Albert H. Pierson, *Fuel Wood Used in the United States, 1630–1930*, 6–8.

73 Gray, *Agriculture in the Southern United States*, 1:554.

74 J. Hector St. John de Crèvecoeur, *Sketches of Eighteenth Century America*, 144.

75 Benjamin Franklin, *Observations on Smoking Chimneys; Their Causes and Cure*; Joseph B. Felt, *History of Ipswich, Essex, and Hamilton*, 25–6; Darret B. Rutman, *The Husbandmen of Plymouth: Farming and Village in the Old Colony, 1620–1692*, 43; and Reynolds and Pierson, *Fuel Wood in the United States*, 8–11.

76 Based on Charles Paullin (ed.), *Atlas of the Historical Geography of the United States*, 42, and plates 61A–D; Carl Bridenbaugh, *Cities in the Wilderness: The First Century of Urban Life in America, 1625–1742*, 143, 303–4; Evarts G. Green and Virginia D. Harrington, *American Population before the Federal Census of 1790*, 19, 97, 117, 177, passim; W. S. Rossiter, *A Century of Population Growth: From the First Census to the Twelfth Census of the United States, 1790–1900*, 11–15.

77 Bridenbaugh, *Cities in the Wilderness*, 11–12, 151–2; Kalm, *Travels into N. America*, 1:93–4; Schoepf, *Travels in the Confederation*, 1:5, 60–1.

78 Bridenbaugh, *Cities in the Wilderness*, 311–13; Eavenson, *American Coal Industry*, 35. For the severe winter of 1759–60 in New York, see Arthur H. Cole, "The Mystery of Fuel Wood Marketing in the United States," 346.

79 Benjamin Franklin, "*Account of the Pennsylvania Fire-place*," 392.

80 Clark, *Eastern Frontier*, 116.

81 Muntz, "Changing New Jersey Woodlands," 127–69.

82 Strickland, *Journal*, 75, 160, 203, 212; and William Strickland, *Observations on the Agriculture of the United States*, 6.

83 Coxe, *View of the United States*, 10, 70–1; *American Husbandry* 1 (1837), 83–4.

84 Reynolds and Pierson, *Fuel Wood in the United States*, 12–20.

4. Two centuries of change: the commercial uses of the forest

1 P. White, *Beekmantown*, 29; Russell, *Long Deep Furrow*, 93; Clark, *Eastern Frontier*, chap. 1; C. Carroll, *Timber Economy*.

2 There are many works on the topic of forests and sea power, but the most important for this discussion are Robert G. Albion, *Forests and Sea Power: The Timber Problem of the Royal Navy, 1642–1852*, 200–315; C. M. Andrews, *The Colonial Period of American History*; Joseph J. Malone, *Pine Trees and Politics: the Naval Stores and Forest Policy in Colonial New England, 1691–1775*; Eleanor L. Lord, *Industrial Experiments in the British Colonies of North America*; Justin Williams; "England's Colonial Naval Stores Policy, 1588–1776."

3 H. Clifford Darby, "The Changing English Landscape," 377–98; see also Darby, "The Clearing of the English Woodlands."
4 Albion, *Forests and Sea Power.*
5 Gray, *Agriculture in the Southern United States*, 45, 152.
6 Ibid., 153–5.
7 J. Williams, "England's Colonial Naval Stores Policy," 42; E. Lord, *Industrial Experiments in the British Colonies*, App. B.
8 Gray, *Agriculture in the Southern United States*, 159–60; *One Hundred Years of Progress in the United States*. For an excellent contemporary description of the manufacture of naval stores, see Schoepf, *Travels in the Confederation*, 140–3.
9 Victor S. Clark, *History of Manufactures in the United States, 1607–1914*, 1:164–5.
10 Malone, *Pine Trees and Politics*, 28–46.
11 Bernard Bailyn and Lotte Bailyn, *Massachusetts Shipping, 1697–1714: A Statistical Study*, 42, and tables XVII, XVIII; Samuel Eliot Morison, *Maritime History of Massachusetts, 1783–1860*, has much additional information.
12 Malone, *Pine Trees and Politics*, 35–8; Harry R. Merrens, *Colonial North Carolina in the Eighteenth Century: A Study in Historical Geography*, 88–92; Justin Williams, "English Mercantilism and Carolina Naval Stores, 1705–1776," 88–92; Public Record Office (hereafter PRO,), London, Customs, 16/1; Melvin Herndon, "Naval Stores in Colonial Georgia," 426–33; Sinclair Snow, "Naval Stores in Colonial Virginia," 75–93.
13 Timothy Pitkin, *A Statistical View of the Commerce of the United States: Its Connection with Agriculture and Manufacturers*, 47–9, and tables XIV, XV; see also Terry G. Sharrar, "The Search for a Naval Policy, 1783–1812," 28.
14 Albion, *Forests and Sea Power*, 283.
15 Leonard B. Chapman, "The Mast Industry of Old Falmouth," 392–3; Wiliam R. Carlton, "New England Masts and the King's Navy," 4–18.
16 Albion, *Forests and Sea Power*, 236–7.
17 Clark, *Eastern Frontier*, 54; Samuel Sewell, *The Diary of Samuel Sewell, 1674–1729*, 1:149. For a later description of mast getting, see Belknap, *History of New Hampshire*, 3:103–7. See also William B. Weeden, *Economic and Social History of New England, 1620–1789*, 2:765–7, for many additional details.
18 Jay P. Kinney, *Forest Legislation in America Prior to March 4, 1789*; Lillian M. Willson, *Forest Conservation in Colonial Times.*
19 Albion, *Forests and Sea Power*, App. C.
20 Carroll, *Timber Economy*, 69, 131; Joseph A. Goldenberg, "With Saw and Axe and Auger: Three Centuries of American Shipbuilding," 98–101.
21 Carroll, *Timber Economy*, 131–2; Bailyn and Bailyn, *Massachusetts Shipping*, 42, and tables XVII, XVIII, 102–9.
22 Herndon, "Forest Products of Georgia," 134; C. Carroll, *Timber Economy*, 133; Weeden, *New England, 1620–1789*, 1:252–5.
23 Goldenberg, "Saw and Axe and Auger," 117; Coxe, *View of the United States*, 99.
24 See James E. Defebaugh, *History of the Lumber Industry of America*, 2:307–12; and Fox, *Lumber Industry in New York*, 11–13.
25 Quoted in Defebaugh, *Lumber Industry of America*, 2:272–3; see also 2:8–9, 67–8.
26 Turner, *Phelps and Gorham's Purchase*, 528–9, 347.
27 The discussion of early sawmills and lumbering practice is based on a variety of sources, including Defebaugh, *Lumber Industry of America*, 2:271–3; Richard G. Wood, *A History of Lumbering in Maine, 1820–1861*, 150–65; Fox, *Lumber Industry in New York*, 11–16; Ralph Adams Brown, "The Lumber Industry in the State of New York, 1790–1830," esp. 75–101; Harry B. Weiss and Grace M. Weiss, *The Early Sawmills of New Jersey*, 12–36; and the contemporary description of a mill in T. Cooper, *Some Information*, 132–3.

28 Evelyn M. Dinsdale, "Spatial Patterns of Technological Change: The Lumber Industry of Northern New York," 258–63; William T. Langhorne, Jr. "Mill Based Settlement Patterns in Schoharie County, New York: A Regional Study," 73–92.
29 Terry A. McNealy, "Rafting on the Delaware: New Light from Old Documents," 27–9; William Heidt, Jr., *History of Rafting on the Delaware*; J. Herbert Walker (ed.), *Rafting Days in Pennsylvania*, 53–6.
30 Anne [McVickar] Grant, *Memoirs of an American Lady*, 2:258–9.
31 Wansey, *Excursion in the United States*, 107–8.
32 Defebaugh, *Lumber Industry of America*, 2:314–18.
33 Walker, *Rafting Days*, 39.
34 David F. Magee, "Rafting on the Susquehanna," 196.
35 Grant, *Memoirs of an American Lady*, 2:259; Evelyn M. Dinsdale, "The Lumber Industry of Northern New York," 66.
36 Heidt, *History of Rafting on the Delaware*, 9. See also Thomas R. Cox, "Transition in the Woods: Log Drivers, Raftsmen, and the Emergence of Modern Lumbering in Pennsylvania," 345–64, for details of rafting on the Susquehanna and other Pennsylvania rivers.
37 Carroll, *Timber Economy*, 79–97.
38 Thomas Hutchinson (ed.), "Copy of a Narrative of the Commissioners from England, about New England," 2:152; Clark, *Eastern Frontier*, 55–6, 64–5; Weeden, *New England, 1620–1789*, 2:503; and Defebaugh, *Lumber Industry of America*, 2:64.
39 Malone, *Pine Trees and Politics*, 55; Bernard Bailyn, *The New England Merchants in the Seventeenth Century*, 94–111, 132–5.
40 PRO London, Customs, 16/1. These documents comprise 267 pages of customs returns of vessels entering and clearing ports in North America, including imports, exports, and duties paid. The 134 pages on exports are examined here; timber exports, such as furniture, masts, topsails, booms, hoops, clapboards, cedar boards, mahogany and walnut logs and boards, etc., are not dealt with. For an appraisal of these returns, see Merrens, *Colonial North Carolina*, 201–2, and particularly the thorough study of James F. Shepherd and Gary M. Walton, *Shipping, Maritime Trade, and the Economic Development of Colonial North America*, app. 4, 204–36, for summary tables and a note on the validity of the returns. See also James F. Shepherd, *Commodity Exports from the British North America Colonies to Overseas Areas, 1768–1772*. For some near contemporary comments on this timber trade, see Coxe, *View of the United States*, 90.
41 Merrens, *Colonial North Carolina*, 97–9.
42 Ibid., 101. For an interesting and detailed study of the trading activities of one merchant between New England and North Carolina, see Virginia B. Platt, "Tar, Staves, and New England Rum: The Trade of Aaron Lopez of Newport, Rhode Island, with Colonial North Carolina."
43 Wood, *Lumbering in Maine*, 27–9.
44 Fox, *Lumber Industry in New York*, 14.
45 Louis C. Hunter, "The Heavy Industries before 1860," 172–89; James M. Swank, *The History of the Manufacture of Iron in All Ages and Particularly in the United States from Colonial Times to 1891*; V. Clark, *History of Manufactures*, 1:1607–1860.
46 These and other figures of iron production are based on Peter Temin, *Iron and Steel in Nineteenth Century America: An Economic Inquiry*, esp. table C1, 264–5.
47 Quoted in Charles S. Boyer, *Early Forges and Furnaces in New Jersey*, 128–9.
48 For the technology of iron making, see Hunter, "Heavy Industries before 1860," 173–4; Arthur C. Bining, *Pennsylvania Iron Manufacture in the Eighteenth Century*, 67–94; John B. Pearse, *A Concise History of the Iron Manufacture of the American Colonies up to the Revolution and of Pennsylvania until the Present Time*.
49 Bining, *Pennsylvania Iron Manufacture*, 71–6.
50 Kemper Jackson, *American Charcoal Making in the Era of the Cold Blast Furnace*.

51 Pearse, *History of Iron Manufacture*, 12, 69, 250–1.
52 Temin, *Iron and Steel*, 83.
53 Bining, *Pennsylvania Iron Manufacture*, 75–6, quotes the example of the Oley furnace in the Schuylkill Valley, which used the timber from one acre of woodland daily to every two tons of pig iron produced in 1787.
54 James M. Ransom, *The Vanishing Iron Works of the Ramapos: The Story of the Forges, Furnaces, and Mines of the New Jersey–New York Border Area*, 186–7.
55 See Bining, *Pennsylvania Iron Manufacture*, 30–1, and Boyer, *Early Forges and Furnaces*, 1–2.
56 Boyer, *Early Forges and Furnaces*, 124; Ransom, *Vanishing Iron Works of the Ramapos*, 116, 198.
57 Quoted in Swank, *Manufacture of Iron, 1889–9*; see also Arthur C. Bining, "The Iron Plantations of Early Pennsylvania," 117–39.
58 Swank, *Manufacture of Iron*, 189.
59 Schoepf, *Travels in the Confederation*, 5, 36–7.
60 Pearse, *History of Iron Manufacture*, 30; Ransom, *Vanishing Iron Works of the Ramapos*, 102; and Boyer, *Early Forges and Furnaces*, 48–9. Much more information can be found in Theodore W. Kury, "Historical Geography of the Iron Industry in the New York–New Jersey Highlands, 1700–1900."
61 Temin, *Iron and Steel*, 14–15; J. Peter Lesley, *The Iron Manufacturer's Guide to the Furnaces, Forges, and Rolling Mills of the United States*, 747.

5. The quickening pace: agricultural clearing, 1810–1860

1 Malcolm J. Rohrbough, *The Trans-Appalachian Frontier: People, Societies, and Institutions, 1775–1850*, 15. See R. Brown, *Historical Geography*, 173–287, for a detailed account of these population movements.
2 Michael Williams, "Clearing the United States Forests: The Pivotal Years, 1810–1860," 12–13.
3 See Bidwell and Falconer, *Agriculture in the Northern United States*, 149–51, esp. the population density maps, figs. 6–8.
4 J. Hall, *Statistics of the West*, 70.
5 Tocqueville, *Democracy in America* (1966 ed.), 2:166.
6 Benjamin Rush, "Information to Europeans Who Are Disposed to Migrate to the United States," 191; Dwight, *Travels*, Letter 13, 2:321–2; John Taylor, "Journal of Rev. John Taylor, Missionary on Tour through the Mohawk and Black River Counties in 1802," 1148; Stuart, *Three Years in North America*, 1:259.
7 Stuart, *Three Years in North America*, 1:254; see also Henry Tudor, *Narrative of a Tour in North America*, 2:437, for comments on the relative costs of cleared and uncleared lands.
8 Tocqueville, *Journey to America*, "A Fortnight in the Wilds," 343; see also comments on p. 209. For similar comments about the Pennsylvania backwoodsman, see Lorain, *Nature and Reason Harmonized*, 134.
9 Quoted in Danhof, *Change in Agriculture*, 114.
10 U.S. Patent Office, *Annual Report* (1852), 293–4, quoted in Danhof, *Change in Agriculture*, 115–16.
11 Based on David E. Schob, *Hired Hands and Plowboys: Farm Labor in the Mid West, 1815–1860*, 5–14; see also Neil McNall, *An Agricultural History o the Genessee Valley, 1790–1860*, 88–9, for labor costs in upstate New York.
12 Belknap, *History of New Hampshire*, 3:95.
13 Beaufoy, *Tour*, 80–2; Belknap, *History of New Hampshire*, 3: 131–7; Dwight, *Travels*, Letter 11, 2:113–14, and Letter 14, 2:325–6; Adam Hodgson, *Letters from America* 1:22, 270; Stuart, *Three Years in North America*, 1:134–5, 256–7; Tudor, *Narrative of a Tour*,

1:218–19; Lewis D. Stillwell, in his "Migration from Vermont, 1776–1860," 68–9, has details of clearing.

14 *Cultivator and Country Gentleman* 9:3, 25.

15 *Michigan Farmer* 9 (1851), 70–1.

16 Martin L. Primack, "Land Clearing under Nineteenth Century Techniques: Some Preliminary Calculations," 485–6.

17 Massachusetts State Agricultural Society, *Abstracts from the Returns of the Agricultural Societies in Massachussetts*, 32.

18 Primack, "Farm Formed Capital," 68–79.

19 See Rodney C. Loehr, "Moving Back from the Atlantic Seaboard," 90–6; Solon Buck, "Making a Farm on the Frontier: Extracts from the Diaries of Mitchell Young Jackson," 92–6.

20 Gates, "Problems of Agricultural History," 35.

21 McNall, *Agricultural History of the Genessee Valley*, 85.

22 Based on Primack, "Farm Formed Capital," tables 3 and 4, 157–61, with modifications.

23 Stuart, *Three Years in North America*, 1:272.

24 James Madison, Address to Agricultural Society [Charlottesville, Va.], in *Letters and Other Writings*, 3:94–5; see also "Woods and Woodlands," 3:197.

25 Adam Hodgson, *Letters from America*, 1:173.

26 Andrew Burnaby, *Travels through the Middle Settlements in North America in the Years 1759 and 1760*, 88.

27 Basil Hall, *Travels in North America in the Years 1827 and 1828*, 2:135, and *Forty Etchings from Sketches Made with the Camera Lucida in North America in 1827 and 1828.*

28 Tudor, *Narrative of a Tour*, 1:217–18.

29 Some of these ideas are suggested in Chinard, "American Philosophical Society and Forestry," 463. See also Jane L. Mesick, *The English Traveller in America, 1785–1835*; Max Berger, *The British Traveller in America, 1836–1860*, passim.

30 David Lowenthal, "The American Scene," 75–6; William Nowlin, *The Bark Covered House*, passim; Ulysses P. Hedrick, *Land of the Crooked Tree.*

31 Bigelow, *Tour to Niagara Falls*, 14.

32 Tocqueville, *Journey to America*, "Pocket Note Book," 128–9; and "Journey to Lake Oneida," 322.

33 Isaac Weld, *Travels through the States of North America and Provinces of Upper and Lower Canada during the Years 1795, 1796, and 1799*, 1:31–41, 231–3.

34 Adam Hodgson, *Letters from America*, 1:396–7.

35 See Chapter 1, herein.

36 Turner, *Pioneer History of the Holland Purchase*, plates opp. 562, 564, 565, 566. See also George Harvey, *Harvey's Scenes of the Primeval Forests of America at the Four Periods of the Year*; scene 1, "Spring," shows clearing and girdling and has a description very like Turner's.

37 Crèvecoeur, "What Is an American?" in *Letters from an American Farmer*, 49–53; Benjamin Rush, "An Account of the Progress of Population, Agriculture, Manners, and Government in Pennsylvania in a Letter to a Friend in England," 213–21; Dwight, *Travels*, Letter 13, 2:321–3.

38 Tocqueville, *Journey to America*, "Fortnight in the Wilds," 333; Adam Hodgson, *Letters from America*, 1:339–40.

39 Adam Hodgson, *Letters from America*, 2:318–19. The added fascination of this passage is its similarity to the ideas of Frederick Jackson Turner in his celebrated *Frontier in American History*, though Hodgson prefigures Turner by 60 years.

40 H. Clifford Darby, "The Clearing of the Woodland in Europe," 183.

41 For example, Walter Prescott Webb, *The Great Plains*; Terry Jordan, "Between the Forest and the Prairie," 205–16; McManis, "Initial Evaluation of the Illinois Prairies, 1815–1840," 30–88; and Brian P. Birch, "Initial Perception of Prairie: An English Settlement in Illinois."

42 J. Baird, *View of the Valley of the Mississippi*, 204–5.

43 Inter alia, J. Hall, *Statistics of the West*, 70, 72; Benjamin Harding, *A Tour through the Western Country, A.D. 1818 and 1819*, 8; George Flower, *The Errors of Emigrants*, 21; B. Hall, *Travels in North America*, 2:313.
44 Charles Dickens, *American Notes*, "A Jaunt to the Looking-Glass Prairie and Back," 384–5.
45 J. Hall, *Statistics of the West*, 70.
46 Harding, *Tour through the Western Country*, 8–9; J. Silk Buckingham, *The Eastern and Western States of America*, 3:216–17; Flint, *Condensed Geography and History*, 2:116. See also William V. Pooley, *The Settlement of Illinois from 1830 to 1850*, 95, 112, 154, for other accounts of settlers clinging initially to woodland locations.
47 John Bradbury, *Travels in the Interior of America in the Years 1809, 1810, and 1811*, 308; see also Harding, *Tour through the Western Country*, 9; Dwight, *Travels*, 4:63, where he says that the experience of heavier crops in the oak plains was a fact that had "rapidly raised the plains in the public esteem."
48 Patrick Sherriff, *A Tour through North America, Together with a Comprehensive View of the Canadas and the United States*, 249.
49 For a detailed evaluation of the effects of the canal on prairie settlement, see Harlan H. Barrows, *Geography of the Middle Illinois Valley*, 92–8; Carl O. Sauer, "Geography of the Upper Illinois Valley and History of Development," 167–73.
50 Timothy Flint, *The History and Geography of the Mississippi Valley*, 318; Paul W. Gates, *The Farmers' Age: Agriculture, 1815–1860*, 9. For aspects of fire on the prairie, see Pyne, *Fire in America*, 71–99.
51 McManis, "Initial Evaluation of the Illinois Prairie," 50–7; Primack, "Farm Formed Capital," 22–4.
52 See Schob, *Hired Hands and Plowboys*, 21–42, for copious details about the labor requirements of sod busting; and J. Hall, *Statistics of the West*, 103.
53 Reynolds and Pierson, "Fuel Wood in the United States." For other calculations, see Sam H. Schurr and Bruce C. Netschert, *Energy in the American Economy, 1850–1975: An Economic Study of Its History and Prospects*, 50, n. 6.
54 *Michigan Farmer* 14 (1956), 100. For other examples, see Bayrd Still, "Milwaukee in 1833 and 1849," 297; Tyrone Power, *Impressions of America, during the Years 1833, 1834, and 1835*, 1:92.
55 Gates, "Problems of Agricultural History," 34–6.
56 Cole, "Mystery of Fuel Wood Marketing," 339–40.
57 For a more extended discussion, see Michael Williams, "Products of the Forest: Mapping the Census of 1840," 6–11.
58 "Charlemagne Tower: His Journey through Maine in the Summer of 1829," 47, as quoted in Gates, "Problems of Agricultural History," 46.
59 Good accounts of the coastal trade may be seen in Muntz, "Changing New Jersey Woodlands," 127–69.
60 Cole, "Mystery of Fuel Wood Marketing," 342.
61 Herbert G. Schmidt, *Rural Hunterdon: An Agricultural History*, 69–70.
62 A. William Hoglund, "Forest Conservation and Stove Inventors, 1789–1850," 5–6.
63 Based on Marcus Bull, *Experiments to Determine the Comparative Value of the Principal Varieties of Fuel* (1827), and on calculations in Schurr and Netschert, *Energy in the American Economy*, 50–2.
64 Quoted in Eavenson, *American Coal Industry*, 294.
65 John Bernard, *Retrospections of America, 1797–1811*, 1:182.
66 Zadoc Cramer, *The Navigator, Containing Directions for Navigating the Monongahela, Allegheny, Ohio, and Mississippi Rivers*, 68–78; Michaux, *Travels to the West*, 64.
67 Hoglund, "Forest Conservation and Stove Inventors," 5–6.
68 Schoepf, *Travels in the Confederation*, 60–1. The rudimentary stoves were known as "Franklins."
69 Reynolds and Pierson, "Fuel Wood in the United States."

70 See Chapter 10, herein.
71 M. Williams, "Products of the Forest," 12–17.
72 Pitkin, *Statistical View of Commerce*, 47–9, 82–8.
73 Tench Coxe, in his *View of the United States*, 454, said that the expense of clearing an acre of land was "fully and completely reimbursed" by the sale of pot and pearl ashes.
74 See Hough, *Report upon Forestry*, 2:377–9, for exports during 1855–79.
75 Ibid., 375–6.
76 Ibid., 374–5.
77 Madison, "Address to Agricultural Society," 93.
78 Marsh, *Address before the Agricultural Society of Rutland*," 17–19, quoted in Lowenthal, "Author of *Man and Nature*," 10.
79 The following paragraphs, except where otherwise noted, are based on the ideas in Glacken, *Traces on the Rhodian Shore*, 685–98.
80 See Chapter 3, herein.
81 Volney, *Climate and Soil of the United States*, 213–16.
82 Lorain, *Nature and Reason Harmonized*, 25–7; for the descriptions of clearing, see pp. 333–9; quotation on pp. 335–6.
83 Quoted in Lowenthal, "Author of *Man and Nature*," 10.

6. The quickening pace: the industrial impact, 1810–1860

1 Walt W. Rostow suggests that the takeoff occurred in the United States during the 1840s, but Robert W. Fogel, in his *Railroads and American Economic Growth: Essays in Econometric History*, 128, 234–5, suggests that there are grounds for considering the 1820s (even 1807) as the critical period. For a thorough discussion, see Douglass C. North, *The Economic Growth of the United States, 1790–1860*, 165–6, 189–90; and Simon S. Kuznets, "Notes on the Take-off," 40.
2 Mumford, *Technics and Civilization*, 109–10.
3 J. Hall, *Statistics of the West*, 100–1.
4 T. S. Ashton, *Iron and Steel in the Industrial Revolution*, 89–90.
5 Louis C. Hunter, "The Influence of the Market upon Technique in the Iron Industry of Western Pennsylvania up to 1860," 241–81; Temin, *Iron and Steel*, 55–7; Richard H. Schallenberg, "Evolution, Adaption, and Survival: The Very Slow Death of the American Charcoal Iron Industry," 342–7.
6 Railroad mileage rose from 2,808 miles in 1840 to 8,924 miles in 1850 and 30,626 miles in 1860. See Poor's estimates in E. R. Wicker, "Railroad Investment before the Civil War," 503–45. Hunter, in "Heavy Industries before 1860," 179, argues that the railroads were soon taking half of the iron consumed during and after the Civil War. However, Temin, in *Iron and Steel*, 21, argues that railroad use should not be overstressed because British manufacturers supplied three-quarters of needs in the 1850s, as they could produce cheaper rails after the British railway boom ended in the 1850s while American technology lagged. For the quotation in reference to nail production, see Fogel, *Railroads and Economic Growth*, 135.
7 Swank, *Manufacture of Iron*, 187–9.
8 Schallenberg, "Evolution, Adaption, and Survival," 342–7; Temin, *Iron and Steel*, 58–62.
9 Schallenberg, "Evolution, Adaption, and Survival," 353.
10 Kenneth Warren, *The American Steel Industry, 1850–1970: A Geographical Interpretation*, 22. In 1842 Pennsylvania anthracite furnaces produced 1,250 tons on average, and charcoal furnaces 470 tons. The ratio became even more pronounced in 1856, when the figures were 3,300 and 670 tons, respectively.
11 Temin, *Iron and Steel*, 63–7; and U.S. Bureau of the Census, *Special Report on the Manufacture of Coke*, 22–3.

12 Lesley, *The Iron Manufacturer's Guide*, 747–57.

13 Ibid., passim; H. Phillips, *On the Bituminous Coalfield of Pennsylvania*, 96.

14 U.S. Congress, Senate, *Letter from the Secretary of the Treasury Transmitting Information on Steam Engines.*

15 G. B. Catlin, "Early Travels on the Ohio and Its Tributaries, 1818," 155.

16 Eavenson, *American Coal Industry*, 305, 310, quoting Committee of Kentucky Legislative Council, 1837, and the *Pittsburgh Daily Gazette*, 17 Oct. 1844, respectively.

17 See John H. Morrison, *The History of American Steam Navigation*, 229, 232; and, in particular, Erik F. Haites, James Mak, and Gary M. Walton, *Western River Transportation: The Era of Early Internal Development, 1810–1860*, 144–7; Erik F. Haites and James Mak, "Steam Boating on the Mississippi, 1810–1869: A Purely Competitive Industry," 56.

18 For an extended discussion on fuel for boats and the social aspects of woodhawks, see David E. Schob, "Woodhawks and Cordwood: Steamboat Fuel on the Ohio and Mississippi Rivers, 1820–1860," 124–32; and Schob, *Hired Hands and Plowboys*, 151–68. See also the reminiscences of Stillman Carter Larkin in *The Pioneer History of Meigs County*, 67: "More than one farm was paid for by the cord wood cut and sold to the steam boats for fuel."

19 B. Hall, *Travels in North America*, 3:352, 354. See also Power, *Impressions of America*, 2:118.

20 Buckingham, *Slave States of America*, 1:159; Louis C. Hunter, *Steam Boats on the Western Rivers: An Economic and Technological History*, 266; Stuart, *Three Years in North America*, 293.

21 Williams, "Products of the Forest," 8–10.

22 James Hall, *The West: Its Commerce and Navigation*, 129–30.

23 *Indiana Agricultural Report*, 512.

24 Hunter, *Steamboats on the Western Rivers*, 269–70.

25 Based on Albert Fishlow, *American Railroads and the Transformation of the Ante-bellum Economy*, 125–9.

26 Based on Dewhurst and Associates, *America's Needs and Resources: A New Survey*, app. 25.3, tables J and K, 114–15; Schurr and Netschert, *Energy in the American Economy*, 52–3; and Sargent, *Report on the Forests*, 489. The railroads also used little lumber. Fogel, in his *Railroads and Economic Growth*, 138, 233–4, calculates that between 1840 and 1860 railroads used only 0.9 percent of lumber production, but 5.4 percent if ax-hewn ties are included. Either way, railroad consumption at this time was modest and emphasizes the enormous consumption by other sectors of the economy.

27 Thomas Gamble (ed.), *Naval Stores: History, Production, Distribution, and Consumption*, 6, 13–24, 31–6, 59–82; and Hough, *Report upon Forestry*, 1:138–43, 2:331–61.

28 See M. Williams, "Products of the Forest," 11–12; Merrens, *Colonial North Carolina*, 85–92.

29 Gamble, *Naval Stores*, 29–30.

30 For a discussion of some of the cultural, social, and technical aspects of the movement, see David C. Smith, "The Logging Frontier," 96.

31 Dinsdale, "The Lumber Industry of Northern New York: A Geographical Examination of Its History and Technology," 152.

32 M. Williams, "Products of the Forest," 17–20.

33 The sources of information on technical improvements and changes in sawmills are numerous, but the following have been used and are the most important: Joseph S. Illick, "The Story of the American Lumbering Industry," 150–98; R. Wood, *Lumbering in Maine*, 162–3; Defebaugh, *Lumber Industry of America*, 2:8, 53; Ralph C. Bryant, *Lumber: Its Manufacture and Distribution*, 3–23; Alfred J. Van Tassel and David W. Bluestone, *Mechanization in the Lumber Industry: A Study of Technology in Relation to Resources and Employment Opportunity*, 8–11, 194; P. C. Bardin, "The Outline History of the Sawmill," 35–6, 46; and Bryan Latham, *Timber, Its Development and Distribution: A Historical Survey*, 207–23.

34 *Niles' Weekly Register*, 27 Mar. 1819, 28 Sept. 1833.
35 Hough, *Report upon Forestry*, 1:422, 465.
36 Nathan Rosenberg, "America's Rise to Woodworking Leadership," 48–50, and *Technology and American Economic Growth*, 26–8.
37 U.S. Congress, House, *Patent Office, Annual Report, 1847*, 758–60.
38 "Report of the Committee on the Machinery of the United States," quoted in Nathan Rosenberg, *The American System of Manufacture*, 171.
39 Ibid., 344.
40 Illick, "Story of the Lumbering Industry," 51; R. Wood, *Lumbering in Maine*, 182; and U.S. Congress, Senate, *Letter from the Secretary of the Treasury on Steam Engines.*
41 Hough, *Report upon Forestry*, 1:389; U.S. Bureau of the Census, *Manufactures* (1880), 2:15, "Power Used in Manufactures."
42 Ralph C. Bryant, *Logging: The Principles and General Methods of Operation in the United States*, 121; and William G. Rector, *Log Transportation in the Lake States Lumber Industry, 1840–1918: The Movement of Logs and Its Relationship to Land Settlement, Waterway Development, Railroad Construction, Lumber Production, and Prices*, 15–41.
43 R. Wood, *Lumbering in Maine*, 121.
44 For a colorful description of the start of a long drive, see Thoreau, *The Maine Woods*, 41–2.
45 Based on Fox, *Lumber Industry in New York*, 23–9; R. Wood, *Lumbering in Maine*, 105–7; John S. Springer, *Forest Life and Forest Trees*, which is full of firsthand accounts of life in the woods of Maine.
46 See Fox, *Lumber Industry in New York*, 23–4, for lists of rivers affected in that state; and R. Brown, "Lumber Industry in New York," 84–5, for details of early regulations on the Delaware and Susquehanna.
47 R. Wood, *Lumbering in Maine*, 114–16.
48 Hough, *Report upon Forestry*, 1:438–9, for New York, and 465–6, for Pennsylvania; Wood, *Lumbering in Maine*, 118–20, for Bangor; Robert F. Fries, *Empire in Pine: The Story of Lumbering in Wisconsin, 1830–1900*, 49–55, for Wisconsin. For an excellent description of the Glens Falls boom, see Benson J. Lossing, *The Hudson from the Wilderness to the Sea*, 65–6.
49 Rector, *Log Transportation*, 29–30.
50 Ibid., 28–9.
51 Based on Hough, *Report upon Forestry*, 1:446–53; Defebaugh, *Lumber Industry of America*, 2:317–18, 407, 413; Dinsdale, "Lumber Industry of Northern New York," 70 ff.
52 Defebaugh, *Lumber Industry of America*, 2:413.
53 Hough, *Report upon Forestry*, 1:447–8; Defebaugh, *Lumber Industry of America*, 1:325–6; Fox, *Lumber Industry in New York*, 44–5.
54 Hough, *Report upon Forestry*, 1:448.
55 Defebaugh, *Lumber Industry of America*, 1:413–14.
56 U.S. Congress, Senate, *The Trade and Commerce of the British North American Colonies upon the Great Lakes and Rivers....*
57 Defebaugh, *Lumber Industry of America*, 2:455–6; Fox, *The Lumber Industry in New York*, 44–5; Fries, *Empire in Pine*, 70–1.
58 Rector, *Log Transportation*, 57–8; Fries, *Empire in Pine*, 14–77.
59 Rector, *Log Transportation*, 57–8; Harlan H. Barrows, *Geography of the Middle Illinois Valley*, 92–8; Sauer, *Geography of the Upper Illinois Valley*, 167–73.
60 James Parton, "Chicago," 33–4. See also George W. Hotchkiss, *History of the Lumber and Forest Industry of the Northwest*, 661–85; Barbara E. Benson, "Logs and Lumber: The Development of the Lumber Industry in Michigan's Lower Peninsula," 165–9, 182–3; George Barclay, Chicago: The Lumber Hub," 177, on the importance of Chicago as a distribution center for Great Lakes lumber.
61 Hough, *Report upon Forestry*, 1:463, quoting Samuel P. Johnson, *Engle's Illustrated History of Pennsylvania* (1876), 1135.

62 Quoted in Fries, *Empire in Pine*, 22.

63 Based on ibid., 51–2, 55–6, 66–7; Hotchkiss, *Lumber and Forest Industry*, 350–60; Walter A. Blair, *A Raft Pilot's Log: A History of the Great Rafting Industry on the Upper Mississippi, 1840–1915*.

64 John G. Gregory (ed.), *West Central Wisconsin: A History*, 1: 219. For further accounts of the raftsman's life, see David E. Schob, *Hired Hands and Plowboys*, 165–6.

65 Fries, *Empire of Pine*, 75–9.

66 For the downriver trade, see George W. Sieber, "Lumbermen at Clinton: Nineteenth Century Sawmill Center," 779–802.

7. The lumberman's assault on the forests of the Lake States, 1860–1890

1 For the spread of lumbering westward, see Henry B. Steer, *Lumber Production in the United States, 1799–1946*; Robert V. Reynolds and Albert H. Pierson, "Tracking the Sawmill Westward: The Story of the Lumber Industry in the United States as Unfolded by Its Trail across the Continent," 643–8; Nelson C. Brown, *Lumber, Manufacture, Conditioning, Grading, Distribution, and Use*, 1–4.

2 Mumford, *Technics and Civilization*, 109–10; Bernard A. Weisberger, *The New Industrial Society*, 12–21; U.S. Bureau of Corporations, *The Lumber Industry*, 1, pt. 1, "Standing Timber": 11–12; David Noble, *America by Design: Science, Technology, and the Rise of Corporate Capitalism*, 15–17.

3 See R. H. Brown, *Historical Geography*, 291–310; Rolland H. Maybee, "Michigan's White Pine Era 1840–1900," 385–98, 408–10; Lawrence Martin, *The Physical Geography of Wisconsin*, 3–33.

4 For estimates of forest resources, see Sargent, *Report on the Forests*, 551–9; Filibert Roth, *On the Forestry Conditions of Northern Wisconsin*, 14–16; Hotchkiss, *Lumber and Forest Industry*, 22, 724–51, for intermediate estimates; Ray H. Whitbeck, "The Industries of Wisconsin and Their Geographical Basis," 59–60; William N. Sparhawk and Warren D. Brush, "The Economic Aspects of Forest Destruction in Northern Michigan," 4–6; Charles S. Wheeler, *A Sketch of the Original Distribution of the White Pine in the Lower Peninsula of Michigan*," 2–22. U.S. Bureau of the Census, Twelfth Census (1900), 9:842–3, estimated that at least 13 billion b.f. remained uncut.

5 Mary Dopp, "The Geographical Influences in the Development of Wisconsin," 401–12, 585–609; Roth, *Forestry Conditions of Northern Wisconsin*, 21–42.

6 On the New England contribution to lumbering in the Northwest, see Hotchkiss, *Lumber and Forest Industry*, 300 ff., 485 ff.; D. Smith, "The Logging Frontier," 96–106; Theodore L. Blegen, *Minnesota: A History of a State*, 320–1; Frederick W. Kohlmeyer, "Northern Pine Lumbermen: A Study in Origins and Migrations," 530; Benson, "Logs and Lumber," 53–9.

7 Hough, *Report upon Forestry*, 1:525; and U.S. Bureau of the Census, Thirteenth Census (1909), 10:502, *Special Statistics on Manufactures*.

8 Hough, *Report upon Forestry*, 1:542; Rodney C. Loehr, "Saving the Kerf: The Introduction of the Band Saw Mill," 168–72.

9 Illick, "Story of the Lumbering Industry," 194; Latham, *Timber*, 215–17; Allen J. Davis, "The First Log Band Saw: An Historical Monograph," 33–4; Van Tassel and Bluestone, *Mechanization in the Lumber Industry*, 1–10; Hough, *Report upon Forestry*, 1:542–3, for location of different types of saws. Further details in U.S. Bureau of the Census, Ninth Census (1870), 3:612, and Thirteenth Census (1909), 10:502.

10 Fries, *Empire in Pine*, 61–4; Maybee, "Michigan's White Pine Era," 428; U.S. Bureau of the Census, Fourteenth Census (1920), 10:419–40, *Manufactures: Reports for Selected Industries*.

11 Rector, *Log Transportation*, 99–109; U.S. Bureau of the Census, Tenth Census (1880), *Water Power*, pt. 2, "Report on the Water Power of the Mississippi River and Some of Its

Tributaries," 17:2, 77, 82, 86, and "Report on the Water Power of the Northwest," 17:65–7, both by James L. Greenleaf; Fries, *Empire in Pine*, 46–50.

12 Rector, *Log Transportation*, 111–13: Elizabeth Bachmann, "Minnesota Log Marks," 126–37; Clifford Allen (ed.), *Michigan Log Marks: Their Function and Use during the Great Michigan Pine Harvest*; James E. Lundsted, "Log Marks: Forgotten Lore of the Logging Era," 44–6.

13 Fries, *Empire in Pine*, 49–55; Rector, *Log Transportation*, 116–31; Benson, "Logs and Lumber," 200–7.

14 For various aspects of the Beef Slough company and its operations, see Rector, *Log Transportation*, 132–46; Fries, *Empire in Pine*, 141–60; Paul W. Gates, *The Wisconsin Pine Lands of Cornell University: A Study in Land Policy and Absentee Ownership*, 125–33; Bernhardt J. Kleven, "The Mississippi River Logging Company," 190–202; Robert F. Fries, "The Mississippi River Logging Company and the Struggle for the Free Navigation of Logs, 1865–1900," 429–48; Paul W. Gates, "Weyerhaeuser and Chippewa Logging Industry," 50–64. For the downstream Iowa milling and distribution system, see the detailed studies of Sieber, "Lumbermen at Clinton," 779–802; Lyda C. Belthuis, "The Lumber Industry in Eastern Iowa," 115–55; George B. Hartman, "The Iowa Sawmill Industry," 52–93.

15 There is an extensive literature on rafting, much of which is summarized in Rector, *Log Transportation*, 151–66, 305–8, and Fries, *Empire in Pine*, 66–75. For firsthand and/or detailed accounts, see Blair, *A Raft Pilot's Log*; Edward W. Durant, "Lumbering and Steamboating on the St. Croix River," 645–77; Wilbur H. Glover, "Lumber Rafting on the Wisconsin River," 155–77, 308–24; William J. Petersen, "Rafting on the Mississippi: Prologue to Prosperity," 289–320; Arthur R. Reynolds, *The Daniel Shaw Lumber Company: A Case Study of the Wisconsin Lumbering Frontier*.

16 Rector, *Log Transportation*, 163–70; Robert C. Johnson, "Logs for Saginaw: The Development of Raft-Towing on Lake Huron," 37–41, 83–90; U.S. Congress, House, *Regulating Management of Lumber Rafts on the Great Lakes*.

17 Rector, *Log Transportation*, 187–8. For a case study of river, lake, and railway competition, see Carl E. Krog, "Lumber Ports of Marinette–Menominee in the Nineteenth Century," 272–80.

18 Rector, *Log Transportation*, 192–3.

19 Bryant, *Logging*, 178–9; Illick, "Story of the Lumber Industry," 163–7; Rector, *Log Transportation*, 208–9.

20 Bryant, *Logging*, 172–7; Illick, "Story of the Lumbering Industry," 167–71; Rector, *Log Transportation*, 208–14, 242–3; Agnes M. Larson, *History of the White Pine Industry in Minnesota*, 362–4.

21 Hough, *Report upon Forestry*, 1:551, 557; and Bryant, *Logging*, 242–7.

22 *Muskegon Chronicle*, 15 Mar. 1878, p. 2, quoted in Hudson Keenan, "America's First Successful Logging Railroad," 298.

23 Rector, *Log Transportation*, 200–2, 220–2; and Illick, "Story of the Lumbering Industry," 190–1.

24 The previous three paragraphs are based on a variety of sources, but mainly see Hough, *Report upon Forestry*, 1:546–7; Maybee, "Michigan's White Pine Era," 417–22; Fries, *Empire in Pine*, 84–92, 93–9, for railroads and the lumber industry. For the rate battles between Chicago and Wisconsin, see Rector, *Log Transportation*, 193, 202–5, 222–5, 231–5; J. C. Ryan, "Minnesota Logging Railroads," 303–8.

25 Fries, *Empire in Pine*, 64–5, 84–91.

26 Rand E. Rohe, "The Landscape and the Era of Lumbering in Northeastern Wisconsin," 1–26.

27 Illick, "Story of the Lumbering Industry," 179–81; Rector, *Log Transportation*, 244–5; and Bryant, *Logging*, 196–221.

28 J. Willard Hurst, *Law and Economic Growth: The Legal History of the Lumber Industry in*

Wisconsin, 1836–1915, 107–42. See also his "Institutional Environment of the Logging Era in Wisconsin," 137–55.

29 For an excellent analysis of the acquisition of pine lands by the "pine land ring," see Gates, *Wisconsin Pine Lands*, 102–20. For Minnesota, see Blegen, *Minnesota*, 322–3.

30 For a good example of "cruising," see Maybee, "Michigan's White Pine Era," 401.

31 U.S. Bureau of Corporations, *The Lumber Industry*, 1, pt. 1, "Standing timber," xviii–xix, 33–9, 243; Larson, *White Pine Industry*, 51–6; Benson, "Logs and Lumber," 31–54.

32 U.S. Bureau of Corporations, *The Lumber Industry*, 2, pt. 3, "Land Holdings of Large Timber Owners," 188–216; 1, pt. 1, xx, xxi, for acquisitions in the Lake States.

33 Fries, *Empire in Pine*, 247; John E. Nelligan, *The Life of a Lumberman*.

34 See Bryant, *Logging*, 56, 203–4, for examples of the growing specialization and subdivision of jobs.

35 The example of Burt's mill is quoted in Carl A. Leach, "Paul Bunyan's Land and the First Sawmills of Michigan," 83; and the example of McGraw's mill is quoted in Maybee, "Michigan's White Pine Era," 427, from *New York Tribune*, reprinted in *Lumberman's Gazette*, 3 Nov. 1877.

36 Rector, *Log Transportation*, 32; Defebaugh, *Lumber Industry of America*, 2:319, 407; Hotchkiss, *Lumber and Forest Industry*, 123–6, for Knapp Stout. For other major business organizations, see Issac Stephenson, *Recollections of a Long Life, 1829–1915*; Charles E. Twining, *Downriver: Orrin H. Ingram and the Empire Lumber Company*; Ralph W. Hidy, Frank E. Hill, and Allan Nevins, *Timber and Men: The Weyerhaeuser Story*, 28–206.

37 Rector, *Log Transportation*, 33–5.

38 Fries, *Empire in Pine*, 131–9; Twining, *Downriver*.

39 U.S. Bureau of Corporations, *The Lumber Industry*, pt. 1, "Standing Timber," 98–100; Fries, *Empire in Pine*, 137–60, for the operations of the Mississippi River Logging Company; see also n. 14 to this chapter.

40 Leach, "Bunyan's Land," 69–89; Fries, *Empire in Pine*, 8–23.

41 Hough, *Report upon Forestry*, 1:517, 545–9.

42 Hotchkiss, *Lumber and Forestry Industry*, 733; Sargent, *Report on the Forests*, 550.

43 Rector, *Log Transportation*, 226–7. For descriptions of the cutovers by the end of the century, see USDA, Forestry Division, *Annual Report* (1888), 629; Sparhawk and Brush, *Economic Aspects of Forest Destruction*, 8–9, citing W. B. Mershon, *Recollections of My Fifty Years Hunting and Fishing* (Boston, 1923).

44 Fries, *Empire in Pine*, 239–40; Increase A. Lapham, J. G. Knapp, and H. Crocker; *Report on the Disastrous Effects of the Destruction of Forest Trees Now Going On So Rapidly in the State of Wisconsin* (1867), 18–30; and Roth, *Forestry Conditions of Northern Wisconsin*, 49.

45 Hough, *Report upon Forestry*, 1:513; and for quotation see Sargent, *Report on the Forests*, 552.

46 Steer, *Lumber Production*; Rector, *Log Transportation*, 287–8; J. S. Bradshaw, "Grand Rapids Furniture Beginnings," 279–98, and "Grand Rapids, 1870–1880: Furniture City Emerges," 321–42.

47 Fries, *Empire in Pine*, 241; Paul W. Gates, *The Illinois Central Railroad and Its Colonization Work*, 303–32.

48 Rector, *Log Transportation*, 237; USDA, *Handbook*, Bull. no. 718 (1918), 44; Blegen, *Minnesota*, 329.

49 Fries, *Empire in Pine*, 248; Franklin B. Hough, "On the Duty of Governments in the Preservation of Forests," 1–22.

50 *Report of the Forestry Commission of Wisconsin* (1897–8), 15–16, quoted in Fries, *Empire in Pine*, 243; Whitbeck, "Industries of Wisconsin," 58–61.

51 D. L. Gibson, *Socio-economic Evolution in a Timbered Area in Northern Michigan*, 8; Rollin Lynde Hartt, "Notes on a Michigan Lumber Town," 107. For a more recent analysis of the fortunes of various lumber towns in Wisconsin and Michigan between 1880 and 1910, see

James B. Smith, "Lumbertowns in the Cutover: A Comparative Study of the Stage Hypothesis of Urban Growth."

52 The classic reviews are Stewart H. Holbrook, *Burning an Empire: The Story of American Forest Fires*, and, more recently, Stephen J. Pyne, in his voluminous *Fire in America*, esp. 199–218, in which he deals with the Peshtigo and Hinckley fires at some length. For individual fires, see Joseph Schafer, "Great Fires of Seventy-one," 96–106; Josephine Sawyer, "Personal Reminiscences of the Big Fire of 1871," 422–3; J. Alfred Mitchell, "Accomplishments in Fire Protection in the Lake States," 748–50. For an analysis of the causes of fires in 1880, when 10.274 million acres were burned and damage to property amounted to $25,462,250, see Sargent, *Report on the Forests*, 491–2.

53 The best general account of the physical environment of the cutover areas is W. A. Hartman and J. D. Black, *Economic Aspects of Land Settlement in the Cut-over Region of the Great Lakes States*, 2–21.

54 This and the following paragraphs are based on Arlan C. Helgeson, "Nineteenth-Century Land Colonization in Northern Wisconsin," 115–21; James I. Clark, *Farming the Cutover: The Settlement of Northern Wisconsin*; Lucile Kane, "Selling Cut-over Lands in Wisconsin," 236–47; James I. Clark, *Cutover Problems: Colonization, Depression, Reforestation*, 1–20. For a more extended and thorough treatment, see Arlan C. Helgeson, *Farms in the Cutover: Agricultural Settlement in Northern Wisconsin*; Vernon R. Carstenson, *Farms or Forests: Evolution of a State Land Policy for Northern Wisconsin, 1850–1932*. For the contemporary booster view, see William A. Henry, *Northern Wisconsin: A Handbook for the Homeseeker*, 6–18.

55 A good case history is in Leo Alilunas, "Michigan's Cut-over 'Canaan,'" 188–201. The quotation is from *Publicity Bulletin*, no. 21 (8 Jan. 1907), West Michigan Development Bureau, Traverse City, Mich.

56 Gates, *Wisconsin Pine Lands*, 240.

57 Helgeson, "Nineteenth-Century Land Colonization," 118–21; Hartman and Black, *Economic Aspects of Land Settlement*, 52.

58 Erling D. Solberg, *New Laws for New Forests: Wisconsin's Forest-Fire, Tax, Zoning, and County-Forest Laws in Operation*, on conditions left when logging operations ceased. For the rehabilitation of the cutovers, see George S. Wehrwein, "A Social and Economic Program for the Sub-marginal Areas of the Lake States," 915–24.

59 U.S. Congress, Senate. *Report on Investigation into Reforestation*, statement of Filibert Roth, Grand Rapids, Mich., p. 459; quoted in Rector, *Log Transportation*, 61–2.

8. The lumberman's assault on the southern forest, 1880–1920

1 Based on USFS, *Timber Depletion, Lumber Prices*, 20, 32.

2 For general accounts of the antebellum South, see Vance, *Human Geography of the South*, 109–44; C. Vann Woodward, *The Origins of the New South, 1877–1913*; A. E. Parkins, *The South: Its Economic-Geographic Development*; Calvin B. Hoover and B. U. Ratchford, *Economic Resources and Policies of the South*.

3 *De Bow's Review*, 29 Dec. 1860, 771–2.

4 See John A. Eisterhold, "Savannah: Lumber Center of the South Atlantic," 537, and "Charleston: Lumber and Trade in a Declining Southern Port," 67–9.

5 Lillard, *The Great Forest*, 183.

6 *Northwestern Lumberman*, 19 July 1890, 3, quoted in Paul W. Gates, "Federal Land Policy in the South, 1866–1888," 314.

7 Unless otherwise stated, the following paragraphs on land purchase are based on Paul W. Gates's seminal articles, "Federal Land Policy in the South" (1940) and "Federal Land Policies in the Southern Public Land States" (1979), with additional local information from Hamilton P. Easton, "The History of the Texas Lumbering Industry," 208–23; Donald J.

Millet, "The Lumber Industry of 'Imperial' Calcasieu, 1865–1900," 58–9; Nollie W. Hickman, *Mississippi Harvest: Lumbering in the Longleaf Pine Belt, 1840–1915*, 83–4; Robert S. Maxwell, "Researching Forest History in the Gulf Southwest: The Unity of the Sabine Valley," 111–15; and Thomas D. Clark, "The Impact of the Timber Industry on the South," 141–64.

8 U.S. Bureau of Corporations, *The Lumber Industry*, 1, pt. 1, 184–5, quoting James D. Lacey before the Committee on Ways and Means of the House of Representatives, 20 Nov. 1908.

9 T. Frederick Davis, "The Disston Land Purchase," 200–10; William D. Scroggs, "Federal Swamp Act Grants of 1849–50," 159–64. See also *Southern Lumberman*, 1 July 1883, 9, and 15 Nov. 1883, 10.

10 Reuben McKittrick, *The Public Land System of Texas, 1823–1910*, 40–61. See also Woodward, *Origins of the New South*, 118; Hickman, *Mississippi Harvest*, 40–61.

11 *Southern Lumberman*, 1 Apr. 1883, 10, and 1 July 1883, 9.

12 Thomas D. Clark, "Early Lumbering Activities in Kentucky," 14–15, 42–3, and "Kentucky Logmen," 144–57; Steven A. Schulman, "The Lumber Industry of the Upper Cumberland River Valley"; Sargent, *Report on the Forests*, 511–15.

13 Ulrich B. Phillips, *Plantation and Frontier, Documents, 1649–1863*, 1:115–22; Nollie W. Hickman, "Logging and Rafting Timber in South Mississippi, 1840–1910," 170. See also Chapter 4, "Timber Products and Lumber."

14 Eisterhold, "Charleston," 65–8, and "Savannah," 526–43.

15 Sargent, *Report on the Forests*, 515, 518–19.

16 In addition to the Mississippi-borne supplies, New Orleans was supplied by schooners plying their way around the coast from places such as Apalachicola and Pensacola. They also brought boxed wood from the old turpentine farms in Alabama, which was highly prized as a fuel. See Sargent, *Report on the Forests*, 526–7; John A. Eisterhold, "Lumber and Trade in the Lower Mississippi Valley and New Orleans, 1800–1860," 71–91.

17 Emory Fiske Skinner, *Reminiscences*, 14–15; Maxwell, "Researching Forest History," 113; George A. Stokes, "Log-Rafting in Louisiana," 81–9; Millet, "Lumber Industry of Calcasieu," 51–69.

18 Hickman, "Logging and Rafting," 163, 167, 172, and *Mississippi Harvest*.

19 For a full account of the botanical nature of baldcypress and its exploitation, see Wilbur R. Mattoon, *The Southern Cypress*, and for a somewhat historical account of the industry, see Rachel E. Norgress, "The History of the Cypress Lumber Industry in Louisiana," 979–1059. The best account is in Ervin Mancil, "An Historical Geography of Industrial Cypress Lumbering in Louisiana," portions of which are published in Mancil's "Some Historical and Geographical Notes on the Cypress Lumbering Industry" and, more recently, in his "Pullboat Logging." For studies of individual mill owners, see John Hebron Moore, *Andrew Brown and Cypress Lumbering in the Old Southwest*; and Anna M. C. Burns, "Frank B. Williams, Cypress Lumber King."

20 Sargent, *Report on the Forests*, 525–6.

21 For fuller details of methods of exploiting the baldcypress, see Bryant, *Logging*, 196–214, 233–4; Mancil, "Historical Geography," 90–3; Norgress, "Cypress Lumber Industry," 1002; Stanley F. Horn, *This Fascinating Lumber Business*, 120–2.

22 Illick, "Story of the Lumbering Industry," 166–7; Bryant, *Logging*, 178–84; Richard W. Massey, "A History of the Lumber Industry in Alabama and West Florida, 1880–1914," 62–6; *Southern Lumberman*, 1 Jan. 1904, 42.

23 John S. Spratt, *The Road to Spindletop: Economic Change in East Texas, 1875–1901*, 257–8.

24 Details of these early skidders and loaders are based on Hickman, *Mississippi Harvest*, 165; Illick, "Story of the Lumbering Industry," 187; Easton, "Texas Lumbering Industry," 243; *Southern Lumberman*, 17 Oct. 1908, 61.

25 For details of the many forms of power skidding and their comparative advantages, see Bryant, *Logging*, 205–9.

26 Quoted in Robert S. Maxwell, "One Man's Legacy: W. Goodrich Jones and Texas Conservation," 307.
27 For a preliminary exploration of some of these themes, see T. Clark, "Impact of the Timber Industry," 141–5.
28 Marx, *Machine in the Garden*, 227–80.
29 William Faulkner, "The Bear," in "*Go Down Moses" and Other Stories*, 227–9.
30 The following details of main lines are based on Hickman, *Mississippi Harvest*, 212–32; Gerald L. Collier, "The Evolving East Texas Woodland," 162–274; Carlton J. Corliss, *Main Line of Mid-America: The Story of the Illinois Central*, 170–83, 232–44, 375–84; Robert C. Black, *The Railroads of the Confederacy*, 29–35, passim; John F. Stover, *The Railroads of the South, 1865–1900: A Study in Finance and Control*, 155–85, 186–209; James H. Lemly, *The Gulf, Mobile, and Ohio: A Railroad that Had to Expand or Expire*, 276–7, 289–97.
31 Robert S. Maxwell, *Whistle in the Piney Woods: Paul Bremond and the Houston, East and West Texas Railway*, and "Researching Forest History," 115–16.
32 M. B. Hillyard, *The New South: A Description of the Southern States, Noting Each State Separately*, 247.
33 Hickman, *Mississippi Harvest*, 215–16.
34 Jerome Swineford, "Lumber Industry of Texas," 242.
35 Richard W. Massey, "Logging Railroads in Alabama, 1880–1914," 41–50; Nollie W. Hickman, "Mississippi Lumber Industry from 1840 to 1950," 133.
36 For descriptions of southern logging railroads and their construction, see Bryant, *Logging*, 422–47; Ruth A. Allen, *East Texas Lumber Workers: An Economic and Social Picture, 1870–1950*, 25; Massey, "Lumber Industry in Alabama," 91–110.
37 Quoted in Massey, "Lumber Industry in Alabama," 108–9.
38 U.S. Bureau of Corporations, *The Lumber Industry*, 1, pt. 1, 178.
39 For details of the peculiar technology of building swamp railroads, see Bryant, *Logging*, 311, 319; Mattoon, *The Southern Cypress*, 5–16; Mancil, "Pullboat Logging," 135–41.
40 Van Tassel and Bluestone, *Mechanization in the Lumber Industry*, 82–3, 101–2, 104–5.
41 Details of early mills are culled from John A. Eisterhold, "Colonial Beginnings in the South's Lumber Industry, 1607–1800," 150–3; Hickman, "Mississippi Lumber Industry," 133; Horn, *This Fascinating Business*, 194; Fred B. Kniffen, *Louisiana: Its Life and People*, 75; Mancil, "Historical Geography," 68–9; Moore, *Andrew Brown*, 11–15.
42 Collier, "East Texas Woodland," 183–5.
43 Bryant, *Lumber*, 251.
44 Easton, "Texas Lumbering Industry," 213; 231; Hickman, *Mississippi Harvest*, 159–60; U.S. Bureau of the Census, Twelfth Census (1900), 10, pt. 3, "Special Reports," 826; Fourteenth Census (1920), 10:437–8, "Reports for Selected Industries."
45 Michael Curtis, "Early Development of Operations of the Great Southern Lumber Company," 347–68; Easton, "Texas Lumbering Industry," 123–231.
46 Vance, *Human Geography of the South*, 128–9; Van Tassel and Bluestone, *Mechanization in the Lumber Industry*, 79–81.
47 M. B. Hillyard, *The New South*, 40.
48 U.S. Bureau of Corporations, *The Lumber Industry*, 2, pt. 2, 8–12, 132–4.
49 Based on Easton, "Texas Lumbering Industry," 218–23; John O. King, *The Early History of the Houston Oil Company of Texas, 1901–1908*; John Henry Kirby, "The Timber Resources of East Texas: Their Recognition and Development," 43–78; Mary Lasswell, *John Henry Kirby, Prince of the Pines*.
50 George A. Stokes, "Lumbering and Western Louisiana Cultural Landscapes," 250–66; George Creel, "The Feudal Towns of Texas," 76–8. See also G. M. Hudson, "A Study of a Permanent Alabama Lumber Town," 310.
51 There is a large, growing literature on labor conditions in the South, of which the following are useful: R. Allen, *East Texas Lumber Workers*, 15–60; Jerrell N. Shofner, "Negro Laborers

and the Forest Industries in Reconstruction Florida," 180–91; Hickman, *Mississippi Harvest*, 139–52, 233–52; Richard W. Massey, "Labor Conditions in the Lumber Industry in Alabama, 1880–1914," 172–81; Grady McWhiney, "Louisiana Socialists in the Early Twentieth Century: A Study in Rustic Radicalism," 315–16; James R. Green, "The Brotherhood of Timber Workers, 1910–1913: A Radical Response to Industrial Capitalism in the Southern U.S.A.," 161–200; Vernon H. Jensen, *Lumber and Labor*.

52 *Southern Lumberman*, 1 May 1883, 13.

53 Ibid., 1 Apr. 1883, 13.

54 U.S. Bureau of Corporations, *The Lumber Industry*, 1, pt. 1, 40; Charles A. Gillett, "Citizens and Trade Associations Dealing with Forestry," 285.

55 Based on James Boyd, "Fifty Years in the Southern Pine Industry," 63–4; Anthony Bruce, "Ninety Years On," 76–8; Stanley F. Horn and Charles W. Crawford, "Perspectives on Southern Forestry: The *Southern Lumberman*, Industrial Forestry, and Trade Associations," 19–30; Lee M. James, "Restrictive Agreements and Practices in the Lumber Industry, 1880–1930," 115–25; Massey, "Lumber Industry in Alabama," 181–7; U.S. Bureau of Corporations, *The Lumber Industry*, 3, pt. 4; Hickman, *Mississippi Harvest*, 191–211; Robbins, *Lumberjacks and Legislators*, 39–40.

56 Vance, *Human Geography of the South*, 127, citing R. W. Woolley, "Lumbering around Mobile, Alabama," 191–2.

57 Sargent, *Report on the forests*, 489.

58 Charles Mohr, *The Timber Pines of the Southern United States*, 31–47.

59 For regional studies of lumbering, see Collier, "East Texas Woodland"; Easton, "Texas Lumbering Industry"; Mancil, "Historical Geography"; Massey, "Lumber Industry in Alabama"; Millet, "Lumber Industry of Calcasieu"; Norgress, "Cypress Lumber Industry"; Hickman, *Mississippi Harvest*; Boyd, "Fifty Years," Vance, *Human Geography of the South*.

60 Merchants' Association of New York, *The Natural Resources of Texas*.

61 U.S. Bureau of the Census, Twelfth Census (1900) 9, pt. 3, 806.

62 Parkins, *The South*, 394.

63 Hough, *Report upon Forestry*, 4:105, 108, 193.

64 Boyd, "Fifty Years"; U.S. Bureau of Corporations, *The Lumber Industry*, 1, pt. 1, 44–5.

65 Hough, *Report upon Forestry*, 4:194, 108; Charles Mohr, "The Interest of the Individual in Forestry in View of the Present Condition of the Lumber Industry," 36.

66 Hickman, *Mississippi Harvest*, 259–61; Norgress, "Cypress Lumber Industry," 1047–54; Boyd, "Fifty Years," 66.

67 Hillyard, *The New South*, 41. For a description of one of the new mill towns, see "Meridian as a Lumber Center," *Southern Lumberman*, 8 Aug. 1908, 31–8.

68 Curtis, "Early Development," 347–68.

69 Based on Reginald D. Forbes, "The Passing of the Piney Woods," 131–6, 185.

70 Georgia State Board of Forestry, Report to the Georgia General Assembly on Forestry in Georgia, 9–10.

71 For the fluctuating fortunes of a town in Louisiana, see Stokes, "Lumbering and Western Louisiana," 250–66; Herman H. Chapman, "Why the Town of McNary Moved," 589–92, 615–16; Nollie W. Hickman, "The Yellow Pine Industries in St. Tammany, Tangipahoa, and Washington Parishes, 1840–1915," 75–88; Grace Ulmer, "Economic and Social Development of Calcasieu Parish, Lousiana, 1840–1912," 519–630. For a contrast, see Hudson, "Permanent Alabama Lumber Town," 310.

72 Mancil, "Historical Geography," 171–86, 256–9.

73 Forbes, "Passing of the Piney Woods," 133. For a discussion of the shortsightedness of cutting the future merchantable second growth, see David T. Mason, *Forests for the Future: The Story of Sustained Yield as Told in the Diaries and Papers of David T. Mason, 1907–1950*, 54, 264.

74 Easton, "Texas Lumbering Industry," 233–7; Helene King, "The Economic History of the Long-Bell Lumber Company," 14–17.
75 R. L. Bennett, "Settlement of Cut-over Lands," 37; E. O. Wild, "The Cut-over Empire of Louisiana," 137–43. For additional information, see Corliss, *Main Line of Mid-America*, 414–22; F. V. Emerson, "The Southern Long-leaf Pine Belt," 81–90.
76 W. T. Penfound, "Plant Distribution in Relation to the Geology of Louisiana," 32–5.
77 Emerson, "Southern Long-Leaf Pine Belt," 81–90; Forbes, "Passing of the Piney Woods," 85–90.
78 U.S. Congress, Senate, *Report of the National Conservation Commission*, 2:639; USFS, *Timber Depletion, Lumber Prices*, 20.
79 Norgress, "Cypress Lumber Industry," 1047–56; Robert K. Winters (ed.), *Fifty Years of Forestry in the U.S.A.*, 248–50. The baldcypress manufacturers had already formed a Committee for the Utilization of the Cypress Cut-overs as early as 1908, which concluded that natural regeneration was too slow a solution and that the cutovers should be either drained and farmed or planted to quicker-growing willow and tupelo.
80 Charles Mohr, quoted in Boyd, "Fifty Years," 66.
81 Frank Heyward, "Austin Cary: Yankee Peddler in Forestry," 29–30, 43–4; Roy R. White, "Austin Cary, the Father of Southern Forestry," 2–5.
82 Anna M. C. Burns, *A History of the Louisiana Forestry Commission*, 13–20; Herman H. Chapman, "The Initiation and Early Stages of Research on Natural Reforestation of Longleaf Pine," 505–10; Elwood L. Demmon, "Henry E. Hardtner," 885–6; Anna M. C. Burns, "Henry E. Hardtner: Louisiana's First Conservationist," 78–83. For incidental details, see Horn and Crawford, "Perspectives on Southern Forestry," 19–30.
83 John A. Galloway, "John Barber White and the Conservation Dilemma," 9–16. Others are dealt with in Robert S. Maxwell, "The Impact of Forestry on the Gulf South," 33–4.
84 Maxwell, "One Man's Legacy," 355–80.
85 Curtis, "Early Development," 358–68; Philip C. Wakeley, "F. O. ('Red') Bateman: Pioneer Silviculturalist," 91–9; Frank Heyward, "Planting the Largest Man-made Forest in the World," 62.
86 See Chapter 13, "How should the forest be protected?" for subsequent events.
87 I. James Pikl, Jr., "Pulp and Paper and Georgia: The Newsprint Paradox," 6–19; Jack P. Oden, "Charles Holmes Herty and the Birth of the Southern Newsprint Paper Industry, 1927–1940," 77–89. For background to the development of the pulp and paper industry, see David C. Smith, *History of Papermaking in the United States (1691–1969)*; Jack P. Oden, "Development of the Southern Pulp and Paper Industry, 1900–1970."

9. The last lumber frontier: the rise of the Pacific Northwest, 1880–1940

1 Donald W. Meinig, "The Growth of Agricultural Regions in the Far West, 1850–1910," 229–32.
2 For some idea of the methods and costs of clearing (which could be between $50 and $150 per acre, most commonly $100), see Harry Thompson, *The Cost of Clearing Logged-off Land for Farming in the Pacific Northwest*, 3–16; Byron Hunter and Harry Thompson, *The Utilization of Logged-off Land for Pasture in Western Oregon and Western Washington*, 8–10; Harry Thompson, *Costs and Methods of Clearing Land in Western Washington*, 7–58.
3 Harold K. Steen, *The U.S. Forest Service: A History*, 174.
4 This theme is taken up by Thomas R. Cox in "Trade, Development, and Environmental Change: The Utilization of North America's Pacific Coast Forests to 1914 and Its Consequences," 14–29.
5 For early descriptions of the forests, see Sargent, *Report on the Forests*, 571–80; Nathaniel H. Egleston (ed.), *Report upon Forestry*, 4:251–4. See also Raphael Zon and William N. Sparhawk, *The Forest Resources of the World*, 2:499–502, 525–6; Küchler, *Potential Natural*

Vegetation, map and manual to accompany map; Richard M. Highsmith, Jr., *Atlas of the Pacific Northwest: Resources and Development*, 49–60.

6 For early lumbering activity, see Sherwood D. Burgess, "Lumbering in Hispanic California," 237–48, and "The Forgotten Redwoods of the East Bay," 1–11. For the emerging Pacific trade, see Thomas R. Cox, *Mills and Markets: A History of the Pacific Coast Lumber Industry to 1900*, 71–100, and "The Passage to India Revisited: Asian Trade and the Development of the Far West, 1850–1900," 85–103.

7 T. Cox, *Mills and Markets*, 161–83, quotation on p. 71; Karl Kortum and Roger Olmstead, ". . . It Is a Dangerous Looking Place: Sailing Days on the Redwood Coast," 43–58.

8 Thomas R. Cox, "Lumber and Ships: The Business Empire of Asa Mead Simpson," 16–26.

9 Iva L. Buchanan, "Lumbering and Logging in the Puget Sound Region in Territorial Days," 34–5; quotation from Cox, *Mills and Markets*, 116.

10 These paragraphs are based on Edwin T. Coman and Helen M. Gibbs, *Time, Tide, and Timber: A Century of Pope & Talbot*, esp. 174–209; T. Cox, *Mills and Markets*, 115–20; Elwood R. Maunder, "Building on Sawdust," 57–60.

11 T. Cox, "Lumber and Ships," 18–23; Stephen D. Beckham, "Asa Mead Simpson: Lumberman and Shipbuilder," 259–73; T. Cox, *Mills and Markets*, 171–7; and Egleston, *Report upon Forestry*, 4:252, for the Puget Sound concentration.

12 Hyman Palais, "Pioneer Redwood Logging in Humboldt County," 18–27; John R. Finger, "Seattle's First Sawmill, 1853–1869: A Study of Frontier Enterprise," 24–31; Hyman Palais and Earl Roberts, "The History of the Lumber Industry in Humboldt County," 1–16.

13 Howard Brett Melendy, "Two Men and a Mill: John Dolbeer, William Carson, and the Redwood Lumber Industry in California," 59–71. For his broader study, see Howard B. Melendy, "One Hundred Years of the Redwood Lumber Industry, 1850–1950."

14 Based on U.S. Bureau of the Census, Eighth, Ninth, and Tenth Censuses 1860, 1870, 1890. Confirmation of these values, based on a sample of statistics for sawed lumber and shingles produced, 1874–6, is given in Hough, *Report upon Forestry*, 1:608.

15 Based on T. Cox, *Mills and Markets*, 304, table 2.

16 Ibid., 201–2; William H. Kensel, "The Early Spokane Lumber Industry, 1871–1910," 25–31.

17 For the Pacific markets, see Cox, *Mills and Markets*, 214–22. For the freight rate battles, see Robert W. Vinnedge, "The Genesis of the Pacific Northwest Lumber Industry," 30–1; John H. Cox, "Organizations of the Lumber Industry in the Pacific Northwest, 1889–1914," 6–70, 137–65; Hidy, Hill, and Nevins, *Timber and Men*, 212–13.

18 J. Cox, "Organizations of the Lumber Industry," passim; Oscar O. Winter, *The Great Northwest*, 272–3.

19 A. J. Dufur, *Statistics of the State of Oregon*, 121.

20 Sargent, *Report on the Forests*, 490.

21 For the movements from Lake States, see Defebaugh, *Lumber Industry of America*, 1:452–3; Hidy, Hill, and Nevins, *Timber and Men*, 207–27; Kenneth A. Erickson, "The Morphology of Lumber Settlements in Western Oregon and Washington," 11–12; Melendy, "One Hundred Years of the Redwood Industry," 23–4. Other details in Hough, *Report upon Forestry*, 1:602–3.

22 Thomas R. Cox, "Lower Columbia Lumber Industry, 1880–1893," 160–78.

23 The lack of success in 1890 is illustrated by the plight of William Kyle, a small operator working out of an inferior harbor. See Thomas R. Cox, "William Kyle and the Pacific Lumber Trade: A Study in Marginality," 4–14.

24 Allen H. Hodgson, *Logging Waste in the Douglas Fir Region*, 25.

25 Sargent, *Report on the forests*, 574.

26 T. Cox, *Mills and Markets*, 227–8.

27 For traditional methods of moving logs, see, inter alia, Melendy, "One Hundred Years of the Redwood Industry," 231–2, 245–6.

28 Thomas R. Cox, "Pacific Log Rafts in Economic Perspective," 21–9.

29 Inter alia, Edmond S. Meany, "A History of the Lumber Industry in the Pacific Northwest to 1917," 246–8; Van Tassel and Bluestone, *Mechanization in the Lumber Industry*, 10; Asa S. Williams, "Logging by Steam," 1–33; Bryant, *Logging*; 84–94; Peter J. Rutledge and Richard H. Tooker, "Steam Power for Loggers: Two Views of the Dolbeer Donkey," 18–29.

30 Nelson C. Brown, *Logging Transportation: The Principles and Methods of Log Transportation in the United States and Canada*, 203–5, 211–14; Meany, "Lumber Industry in the Pacific Northwest," 260; Van Tassel and Bluestone, *Mechanization in the Lumber Industry*, 9–12.

31 For general principles of flume construction, see N. Brown, *Logging Transportation*, 269–73. For particular examples of spectacular and complex flumes, see William H. Hutchinson, "The Sierra Flume & Lumber Company of California, 1875–1878," 14–20; Robert J. Moser, ". . . And Then There Was One," 21, 43–4, on the Broughton Lumber Company flume at Underwood, Wash., said to be the longest ever built.

32 Loehr, "Saving the Kerf," 168–72; Van Tassel and Bluestone, *Mechanization in the Lumber Industry*, 7–9; Henry Disston and Sons, *The Saw in History*, 13–14, 30–3, and the discussion in Chapter 6, "The lumber industry."

33 *West Shore* 10 (1884), 136, cited in T. Cox, *Mills and Markets*, 238.

34 Hidy, Hill, and Nevins, *Timber and Men*, 211–16; T. Cox, *Mills and Markets*, 212–42.

35 T. Cox, *Mills and Markets*, 245–54, and "Kyle and the Lumber Trade," 4–14.

36 For the disposal of public lands, see Frederick J. Yonce, "Lumbering and the Public Timberlands in Washington: The Era of Disposal," 4–17; for cases of subversion and fraud, see Stephen A. D. Puter and H. Stevens, *Looters of the Public Domain*, 22–35, passim. The crusade for family farms is covered in John Minto, "From Youth to Age as an American," 140–2, 154, 164–72, 375–87; and in Thomas R. Cox, "The Conservationist as Reactionary: John Minto and American Forest Policy."

37 Discussion is based on the 900-odd pages of the Professional Papers of the United States Geological Survey, 1902–3. For a general view, see Henry Gannett, *The Forests of Oregon*, and *The Forests of Washington: A Revision of Estimates*. For more detailed regional studies, see Fred G. Plummer, *Forest Conditions in the Cascade Ranges, Washington, between the Washington and Mount Rainier Forest Reserves*; Arthur Dodwell and Theodore F. Rixon, *Forest Conditions in the Olympic Forest Reserve, Washington*; John B. Leiberg, *Forest Conditions in the Northern Sierra Nevada, California*; H. D. Langille et al., *Forest Conditions in the Cascade Range Forest Reserve, Oregon*. The 1912–13 maps are part of the Forest Service Collection, National Archives, Washington, D.C., GRG 95.

38 See esp. H. Thompson, *Cost of Clearing Logged-off Land*, 1–16.

39 For example, Robert White, *Land Use, Environment, and Social Change: The Shaping of Island County, Washington*, 113–30, deals with the cutovers near the sound; and Robert E. Ficken, *Lumber and Politics: The Career of Mark E. Reed*, 62–7, deals with Washington more generally. The quotation is from Sargent, *Report on the Forests*, 574.

40 Leiberg, *Forest Conditions in the Sierra Nevada*, 17.

41 Conrad Bahre, "Effects of Historic Fuelwood Cutting on the Semidesert Woodlands of Arizona-Sonora Borderlands," 101–10.

42 For grazing in the uplands, see Harold K. Steen, "Grazing and the Environment: A History of Forest Service Stock-Reduction Policy," 238–42; William Voigt, Jr., *Public Grazing Lands: Use and Misuse by Industry and Government*, 43–203; Ronald F. Lockmann, *Guarding the Forests of Southern California: Evolving Attitudes toward Conservation of Watersheds, Woodlands, and Wilderness*, 91–112, 135–47.

43 See U.S. Congress, Senate, *Report of the Committee Appointed by the National Academy of Sciences upon the Inauguration of a Forest Policy*. See also Gifford Pinchot, *Breaking New Ground*, 105–7, for the committee's activities, and Linnie Marsh Wolfe, *Son of the Wilderness*, 270–2, for Muir's role.

44 The best summary of Minto's ideas is in John Minto, *A Paper on Forestry Interests*; his

opposition to the conservationists' arguments is in E. A. Carman, H. A. Heath, and John Minto, *Special Report on the History and Present Condition of the Sheep Industry of the United States*, 961–83. For an analysis, see Lawrence Rakestraw, "Sheep Grazing in the Cascade Range: John Minto vs. John Muir," 371–82. Rakestraw's thesis, "A History of Forest Conservation in the Pacific Northwest, 1891–1913," contains much more detail, esp. 96–124.

45 Rakestraw, "Sheep Grazing in the Cascade Range," 371–82, and "History of Forest Conservation," 106–12; Frederick V. Coville, *Forest Growth and Sheep Grazing in the Cascade Mountains of Oregon*.

46 Will C. Barnes, *Apaches and Longhorns: The Reminiscences of Will C. Barnes*, 202.

47 Linnie Marsh Wolfe, *Son of the Wilderness: The Life of John Muir*, 275–6; Nash, *Wilderness*, 161–7 ff.

48 Steen, *U.S. Forest Service*, 65–7; William D. Rowley, "Privilege vs. Right: Livestock Grazing in U.S. Government Forests," 61–7.

49 Production statistics are based on Steer, *Lumber Production*.

50 Hidy, Hill, and Nevins, *Timber and Men*, 211–15; U.S. Bureau of Corporations, *The Lumber Industry*, 1, pt. 1, 208; Roy E. Appleman, "Timber Empire from the Public Domain," 193 ff.

51 Hidy, Hill, and Nevins, *Timber and Men*, 105–8; Ross R. Cotroneo, "Western Land Marketing by the Northern Pacific Railway," 299–320. For other comments, see Appleman, "Timber Empire," 205.

52 See Hidy, Hill, and Nevins, *Timber and Men*, 222–4, for the creation of block holdings. Charges of shady dealing were made against the Weyerhaeuser Company; see U.S. Bureau of Corporations, *The Lumber Industry*, 1, pt. 1, 19; Appleman, "Timber Empire," 201–3.

53 U.S. Bureau of Corporations, *The Lumber Industry*, 3, pt. 3, 173–6; 2, pt. 2, 5; 1, pt. 1, 20.

54 Hidy, Hill, and Nevins, *Timber and Men*, 237; for a review of criticisms as they affected the Weyerhaeuser Company, see pp. 290–304.

55 Ibid., 221; Robert E. Ficken, "Weyerhaeuser and the Pacific Northwest Timber Industry, 1899–1903," 147, 148–51.

56 Hidy, Hill, and Nevins, *Timber and Men*, 231–2; U.S. Bureau of Corporations, *The Lumber Industry*, 1, introduction by Herbert Knox Smith, vii. For the later mills, see Hidy, Hill, and Nevins, *Timber and Men*, 274–6, 305–9.

57 Mumford, *Technics and Civilization*, 109–10.

58 Based on, inter alia, "Power Saws Come of Age: Chronicle of Heartaches and Ultimate Triumph," 150–63; Ellis J. Lucia, "A Lesson from Nature: Joe Cox and His Revolutionary Chain Saw," 158–65; Thomas R. Cox, "Logging Technology and Tools," 350; Nelson C. Brown, *Logging: Principles and Practices in the United States and Canada*, 73–8, 84, 124–8; Nelson C. Brown, *Logging: The Principles and General Methods of Harvesting Timber in the United States and Canada*, 117–18.

59 See Chapter 7, herein.

60 For high-lead logging, see Van Tassel and Bluestone, *Mechanization in the Lumber Industry*, 12–14, 16; W. S. Taylor, "Different Stages in the Evolution of Overhead System of Logging," 30–1; Newell L. Wright, "Logging the Pacific Slopes," 695–701; N. Brown, *Logging Transportation*, 82–3.

61 Based on, inter alia, Van Tassel and Bluestone, *Mechanization in the Lumber Industry*, 17–18, 37–9; N. Brown, *Logging Transportation*, 33–59, 70–7; William K. Cox, "Forty Years of Tractor Logging," 118–21; F. Hal Higgins, "Logging with Tractors in the '80s," 68, 116. Charles H. Holt founded the Holt Manufacturing Company, which became the Caterpillar Tractor Company in time.

62 Based on a variety of sources, including Van Tassel and Bluestone, *Mechanization in the Lumber Industry*, 42–5; Rector, *Log Transportation*, 361; "Billion Feet of Logs by Motor Truck," 172–80, 228; N. Brown, *Logging Transportation*, 176–80, 186–93, 211–12 (the costs of railroad construction and operation are dealt with on pp. 211–12, and railroad versus truck transport on pp. 191–3).

63 Van Tassel and Bluestone, *Mechanization in the Lumber Industry*, 41–2, 49–50.
64 Erickson, "Morphology of Lumber Settlements," 41–2; Van Tassel and Bluestone, *Mechanization in the Lumber Industry*, 49–50; N. Brown, *Logging: Principles and Methods*, 307–18.
65 Margaret E. Felt, *Gyppo Logger*, 30–9; Edward B. Mittelman, "The Gyppo System," 840–51.
66 Bryant, *Lumber*, 161–2; Van Tassel and Bluestone, *Mechanization in the Lumber Industry*, 26–7.
67 The best general accounts are in Holbrook, *Burning an Empire*, 112–20; William F. Cox, "Recent Forest Fires in Oregon and Washington," 462–9; Fred. G. Plummer, *Forest Fires: Their Causes, Extent, and Effects*, 462–9. For a general appraisal of the effects (beneficial or otherwise) of burning vegetation in the West, see Harold Weaver, "Effects of Fire on Temperate Forests: Western United States," 279–319.
68 George T. Morgan, Jr., "Conflagration as Catalyst: Western Lumbermen and American Forest Policy," 167–9; Pyne, *Fire in America*, 328.
69 Holbrook, *Burning an Empire*, 108–11; Pyne, *Fire in America*, 336–8; Morgan, "Conflagration as Catalyst," 169; Egleston, *Report upon Forestry*, 4:253. See also George T. Morgan, Jr., "The Fight against Fire: Development of Cooperative Forestry in the Pacific Northwest," 20–30.
70 Morgan, "Conflagration as Catalyst," 171, 175–6; Hidy, Hill, and Nevins, *Timber and Men*, 229–30; Ficken, "Weyerhaeuser and the Timber Industry," 153.
71 Morgan, "Conflagration as Catalyst," 172–3, 176–7; Pyne, *Fire in America*, 233.
72 See Eloise Hamilton, *Forty Years of Western Forestry: A History of the Movement to Conserve Forest Resources by Cooperative Effort, 1909–1949*; Clyde S. Martin, "History and Influence of the Western Forestry and Conservation Association on Cooperative Forestry in the West," 167–70; E. T. F. Wohlenberg, "Western Forestry and the Conservation Association," 505–6; Steen, *U.S. Forest Service*, 174; Pyne, *Fire in America*, 315; Robbins, *Lumberjacks and Legislators*, 60–1.
73 For a fuller treatment, see Chapter 13, "How should the forest be protected?"
74 For general accounts of the fires, see Holbrook, *Burning an Empire*, 134–6; Pyne, *Fire in America*, 329–32. For more detailed accounts, see Lynn F. Cronemiller, "Oregon's Forest Fire Tragedy," 487–90, 531, and "The Tillamook Burn," 29–31, 51–2. For the effects on, and the restoration of, vegetation, see Bonita J. Neiland, "Forest and Adjacent Burn in the Tillamook Burn Area of Northwestern Oregon," 660–71; William B. Greeley, "Oregon Restores a Green Tillamook," 13–14, 30, 43; Richard M. Highsmith, Jr., and John C. Beh, "Tillamook Burn: The Regeneration of a Forest," 139–48.
75 USFS, *Timber Depletion, Lumber Prices*, 23, 24, 32; U.S. Congress, Senate, *A National Plan for American Forestry*, 1:177; George W. Peavy, *Oregon's Commercial Forests*; Vinnedge, "Genesis of the Lumber Industry," 33.
76 USFS, *Timber Depletion, Lumber Prices*, 24, 25, 37; H. Thompson, *Cost of Clearing Logged-off Land*, 4–5; Erickson, "Morphology of Lumber Settlements," 104–8.
77 For a general outline of the situation, see T. Cox, "Trade, Development, and Change," 28; Marion Clawson, *Decision Making in Timber Production, Harvest, and Marketing*; U.S. Congress, Senate, *National Plan for Forestry*, 1:177.
78 Woodrow R. Clevinger, "Locational Change in the Douglas Fir Lumber Industry," 23–31. There are similar maps in Erickson, "Morphology of Lumber Settlements," 90–5. For a description of the declining small towns, see Erickson, "Morphology of Lumber Settlements," 284–326.
79 Willis B. Merriam, "Forest Situation in the Pacific Northwest," 104–5.
80 Thornton T. Munger, *Timber Growing and Logging Practice in the Douglas-Fir Region*, 13; George P. Ahern, *Deforested America: Statement of the Present Forest Situation in the United States*, 4–8, for many graphic descriptions.

81 U.S. Congress, Senate, *National Plan for Forestry*, 178; Merriam, "Forest Situation," 105; Marion Clawson and Roger Sedjo, "History of Sustained-Yield Concept and Its Application to Developing Countries," 7–8.

82 Bernhard E. Fernow, "What Are We After?," 3, and "Address," *Proceedings of the American Forestry Association* 10 (1889), 143, quoted in Herbert D. Kirkland, "The American Forests, 1864–1898: A Trend towards Conservation"; William B. Greeley et al., "Timber: Mine or Crop?" For Long, see Charles E. Twining, "Weyerhaeuser and the Clemons Tree Farm: Experimenting with a Theory," 34; for Hardtner and Jones, see Chapter 8, "Reforestation."

83 For a general appraisal of Mason's ideas, see Mason, *Forests for the Future*. For the details of the National Industrial Recovery Act (NIRA), see Steen, *U.S. Forest Service*, 225–8. For the origin and management of the O&C lands, see Richard C. Ellis, "The Oregon and California Railroad Grant, 1866–1945," 253–83; David T. Mason, "The Effect of O&C Management on the Economy of Oregon," 55–67.

84 Hidy, Hill, and Nevins, *Timber and Men*, 281–9, 500–6, 508; Twining, "Weyerhaeuser and the Clemons Tree Farm," 36–40; Paul F. Sharp, "The Tree Farm Movement: Its Origin and Development," 41–5; Richard Lewis, "Tree Farms," 654–6.

10. Industrial impacts on the forest, 1860–1920

1 Personal communication and figures supplied to the author by Dr. Marion Clawson, Resources for the Future, Inc., Baltimore, 23 Oct. 1980. It must always be borne in mind that data on fuel use are estimates, and totals could include fuel for mechanical energy generation.

2 Frederick Starr, "American Forests: Their Destruction and Preservation," 223; USDA, *Annual Report* (1883), 452–3.

3 Schurr and Netschert, *Energy in the American Economy*, 36–47.

4 Reynolds and Pierson, *Fuel Wood in the United States*, 8–10.

5 U.S. Bureau of the Census, Tenth Census (1880), 3:251–327.

6 Sargent, *Report on the Forests*, 489.

7 Schurr and Netschert, *Energy in the American Economy*, 49; Sargent, *Report on the Forests*, 489. In 1919 Henry S. Graves, in *The Use of Wood for Fuel*, 2–6, calculated that consumption was 12.6 cords "per farm."

8 John J. Thomas, "Culture and Management of Forest Trees," 43.

9 Hoglund, "Forest Conservation," 6–8.

10 Cole, "Mystery of Fuel Wood Marketing," 348.

11 Adolph Leue (ed.), *First Annual Report of the Ohio State Forestry Bureau: For the Year 1885*, 166.

12 Sargent, *Report on the Forests*, 489.

13 Reynolds and Pierson, "Fuel Wood in the United States," 17–18.

14 Graves, "Use of Wood for Fuel," 1:17–21, 33–7.

15 C. R. Tillotson, "The Woodlot: Its Present Problems and Probable Future in the United States," 198–208.

16 Earl H. Frothingham, *The Status and Value of Farm Woodlots in the Eastern United States*.

17 Ibid., 1–3, 11–15, 29–30, 35–43. For later figures, see Schurr and Netschert, *Energy in the American Economy*, 57; USFS, *The Friday Newsletter*, 19 Mar. 1982.

18 Lesley, *The Iron Manufacturer's Guide*, 44; Chapters 7–8, herein.

19 Temin, *Iron and Steel*, 82–3, 266–7.

20 Richard H. Schallenberg and David A. Ault, "Raw Material Supply and Technological Change in the American Charcoal Iron Industry," 438–43.

21 Alfred D. Chandler, "Anthracite Coal and the Beginnings of the Industrial Revolution in the United States"; Temin, *Iron and Steel*, 65; Hunter, "Influence of the Market," 262–3.

22 Edward H. French and R. James Withrow, *The Hardwood Distillation Industry of America*, passim.

23 Schallenberg, "Evolution, Adaption, and Survival," 344–5. For a good review of many of these points, see Richard H. Schallenberg, "Charcoal Iron: The Coal Mines of the Forest," 271–99.

24 See Temin, *Iron and Steel*, 82–3; Hunter, "Influence of the Market," 261–2.

25 Schallenberg and Ault, "Raw Material Supply," 445–6, 450; John D. Tyler, "The Charcoal Iron Industry in Decline, 1855–1925," 24–9.

26 *Journal of the United States Association of Charcoal Iron Workers* 6 (1885), 117–21.

27 Janice C. Beatley, "The Primary Forests of Vinton and Jackson Counties, Ohio," 96–108; Leue, *Report, Ohio Forestry Bureau*, 218. For other statistics on rotations, see Hough, *Report upon Forestry*, 3:65. They ranged from 18 to 40 years, with a median value of 25 years for the 75 cases examined.

28 Hough, *Report upon Forestry*, 1:215–27. See also N. W. Lord, "Iron Manufacture in Ohio," 491–5.

29 U.S. Bureau of the Census, Tenth Census (1880), 2:3, *Manufactures*, by J. M. Swank. Note that the capacity of a bushel varied between states; see Hough, *Report upon Forestry*, 3:62–3.

30 Lord, "Iron Manufacture in Ohio," 483.

31 Marx, *Machine in the Garden*.

32 E. F. Palmer, "Westward the Course of Empire Takes Its Way" (1868), Print Collection, Bancroft Library, Berkeley, Calif.

33 Andrew S. Fuller, *The Forest Tree Culturist: A Treatise on the Cultivation of American Forest Trees*, 12.

34 Starr, "American Forests," 213.

35 Hough, *Report upon Forestry*, 1:112–15.

36 For statistics on track mileage, see U.S. Bureau of the Census, *Historical Statistics of the United States from Colonial Times to 1970*, 2:732, tables Q 323, 288; E. E. Russell Tratman, *Railway Track and Trackwork*.

37 Daniel Millikin, "The Best Practical Means of Preserving and Restoring the Forests of Ohio," 325–6; John H. Klippart, "Condition of Agriculture in Ohio in 1876," 507.

38 Leue, *Report, Ohio Forestry Bureau*, 138, 153, 166, 96, 101, respectively.

39 Hough, *Report upon Forestry*, 1:115–16.

40 USDA, *Annual Report* (1883), 445, and *Annual Report* (1885), 185. Bernhard E. Fernow, in his *Consumption of Forest Supplies by Rail Roads and Practicable Economy in Their Use*, 13–14, estimates that between 10 and 15 million acres of forest were needed to feed the railroads. In addition, see Hough in Egleston, *Report upon Forestry*, 4:119–73, "Report on Kinds and Quantities of Timber Used in Railroad Ties."

41 M. G. Kern, *The Relation of Railroads to Forest Supplies and Forestry*, 14.

42 Quoted in ibid., 15–16.

43 Sherry H. Olson, *The Depletion Myth: A History of Railroad Use of Timber*, 28 ff.: Kern, *Relation of Railroads to Forest Supplies*, 19–20, 16.

44 S. Olson, *Depletion Myth*, 25; Fernow, *Consumption of Forest Supplies*, 14; William H. Wroten, "The Railroad Tie Industry in the Central Rocky Mountain Region, 1867–1900."

45 Andrew D. Rodgers, *Bernhard Eduard Fernow: A Story of North American Forestry*, 109; Fernow, *Consumption of Forest Supplies*, 14.

46 Hough, *Report upon Forestry*, 1:116–17; Kern, *Relation of Railroads to Forest Supplies*, 20; S. Olson, *Depletion Myth*, 57, 63–8, 104–9.

47 See Chapter 2 herein.

48 Fuller, *Forest Tree Culturist*, 12; Richard C. Overton, *Burlington West: A Colonization History*, 103–5, passim.

49 Kern, *Relation of Railroads to Forest Supplies*, 24; Hough, "Tree Planting by Railroads," in *Report upon Forestry*, 1:118–22; William L. Hall and Hermann von Schrenk, *The Hardy Catalpa*.

50 S. Olson, *Depletion Myth*, 93–5.

51 Fernow, "Consumption of Forest Supplies," 15–16; Bernhard E. Fernow, *Report upon the Forestry Investigations of the United States Department of Agriculture, 1877–1898*, 22; Rodgers, *Bernhard Fernow*, 174–83, 187–91, 239–40. See also E. E. Russell Tratman, *Report on Substitution of Metal for Wood in Railroad Ties*; Charles A. Nelson, *A History of the U.S. Forest Products Laboratory (1910–1963)*, 25–36.
52 S. Olson, *Depletion Myth*, 12–13; U.S. Bureau of the Census, *Historical Statistics of the United States*, 2:727.
53 S. Olson, *Depletion Myth*, quoting Howard Miller in *The Forester* 3 (1 Jan. 1897), 6.

11. Agricultural impacts on the forest, 1860–1920

1 Millikin, "Best Means of Preserving the Forest," 319.
2 Calculations of land cleared based on U.S. Bureau of the Census, county statistics of land improved for each census in predominantly forested and nonforested areas.
3 For this and other calculations on the time expended in clearing, see Primack, "Land Clearing," 485–6.
4 Horace Greeley, *What I Know of Farming: A Series of Brief and Plain Expositions of Practical Agriculture as an Art Based on Science*, 23–4.
5 The literature on clearing and stump removal in the logged-off areas of the Pacific Northwest and the Lake States is copious between 1909 and 1924, and some of it is repetitive; nevertheless, all references are cited, as they do add up to an impressive body of information on this important subject. See Harry Thompson, *Cost of Clearing Logged-off Land* (1909), and *Costs of Clearing Land in Western Washington* (1912); A. J. McQuire, *Land Clearing* (1913); Harry Thompson and Earl D. Strait, *Costs and Methods of Clearing Land in the Lake States* (1914); Mark J. Thompson, *Investigation in Costs and Methods of Clearing Land* (1916); F. M. White and E. R. Jones, *Getting Rid of the Stumps* (1918); Mark J. Thompson and A. J. Schwantes, *Investigations in Stump and Stone Removal* (1924). For early Pennsylvania, see Berton E. Beck, "Stump-Pulling," 20–31.
6 *American Agriculturist* 34 (1875), 16; 36 (1877), 19.
7 See esp. Thompson, *Costs of Clearing Land in Washington*, 13–59; White and Jones, *Getting Rid of the Stumps*, 5–31.
8 Based on Primack, "Farm Formed Capital," app., table 5, and note, p. 161; see also pp. 26–7.
9 For the probable distribution of the early vegetation in Ohio, see Paul B. Sears, "The Natural Vegetation of Ohio," pt. 1, 139–49; pt. 2, 128–46, 213–32. See also Edgar Transeau and H. Sampson, "Map of the Primary Vegetation Areas of Ohio," 18.
10 R. Brown, *Historical Geography*, 214–15.
11 These and all other statistics of improved and unimproved land, number and size of farms, etc., are based on the Agricultural Statistics in the *United States Census* for the appropriate years.
12 Quoted in R. Brown, *Historical Geography*, 27. For other early descriptions of pioneer life in the forests of Ohio, see William C. Howells, *Recollections of Life in Ohio from 1813 to 1840*; Julia Perkins Cutler, *The Life and Times of Ephraim Cutler, Prepared from His Journals and Correspondence*, 29–49, 87–90.
13 For a detailed study of the draining, clearing, and settlement of the Black Swamp, with many details about the early vegetation, see Kaatz, "The Black Swamp," 1–35.
14 Millikin, "Best Means of Preserving the Forests," 319–33; Klippart, "Agriculture in Ohio in 1876," 486–538; Franklin B. Hough, "The Decrease of Woodlands in Ohio," 174–80.
15 Leue, *Report, Ohio Forestry Bureau* (1885). This report of 314 pages was a detailed survey of the past and present condition of the forests of the state and of the consequences of cutting.
16 Wilbur Stout, "The Charcoal Iron Industry of the Hanging Rock Iron District: Its Influence on the Early Development of the Ohio Valley," 72–104; Beatley, "Primary Forests," 9–18, 19–123.

17 Robert L. Jones, "The Beef Cattle Industry," 176–83; Paul C. Henlein, "Cattle Driving from the Ohio Country, 1800–1850," 90–101.
18 These and the following quotations are from Leue, *Report, Ohio Forestry Bureau*, and can be found under the appropriate county heading.
19 See Kaatz, "The Black Swamp," 32–5.
20 John T. Curtis, "The Modification of Mid-latitude Grasslands and Forests by Man," 725–8.
21 J. Thomas, "Culture and Management of Forest Trees," 43.
22 For an authoritative and fascinating biography of Marsh, see David Lowenthal, *George Perkins Marsh: Versatile Vermonter*, esp. 246–76, see also David Lowenthal, "George Perkins Marsh and the American Geographical Tradition," 207–13.
23 Marsh, *Man and Nature*, viii.
24 Glacken, "Changing Ideas," 81.
25 William L. Thomas (ed.), *Man's Role in Changing the Face of the Earth*, "Introduction," xxx.
26 Starr, "American Forests," 210.
27 Ibid., 211, 218.
28 Herbert A. Smith, "The Early Forestry Movement in the United States," 326–32.
29 Starr, "American Forests," 210.
30 Lapham, Knapp, and Crocker, *Disastrous Effects of Destruction* (1867). For Lapham's very early concern about forest clearance, see Graham Hawks, "Increase A. Lapham, Wisconsin's First Scientist"; for the background to the 1867 report, see Carstenson, *Farms or Forests*, 6–9.
31 Lapham, Knapp, and Crocker, *Disastrous Effects of Destruction*, 33.
32 Quoted in Carstenson, *Farms or Forests*, 13–14, and referred to in Susan L. Flader, "Scientific Resource Management: An Historical Perspective," 21–2.
33 Lapham, Knapp, and Crocker, *Disastrous Effects of Destruction*, 3.
34 Bernhard E. Fernow, *A Brief History of Forestry in Europe, the United States and Other Countries.*
35 USDA, *Annual Report* (1872), 45–6, "The Forests of the United States."
36 USDA, *Annual Report* (1875), 245, "Statistics on Forestry." In this and the 1872 report, the data on forest area within farms are mapped for each state by county, and copious details are printed about local resources of wood and timber, the most abundant species, the condition of the forests and rate of growth, the home prices of wood and lumber, and "other practical points" (249).
37 Brewer, "Woodland and Forest Systems," Plates III and IV, and accompanying commentary.
38 Millikin, "Best Means of Preserving the Forest," 319–20.
39 George Pinney, *An Essay upon the Culture and Management of Forest Trees*, 3–15.
40 For a comprehensive account of the Adirondack movement, see Roger C. Thompson, "Politics in the Wilderness: New York's Adirondack Forest Preserve," 14–23, together with details in Arthur S. Hopkins, "Within and without the Blue Line," 10–11. For the original proposals concerning clearing and watersheds, see Verplanck Colvin, *Report on a Topographical Survey of the Adirondack Wilderness of New York* (1873). Colvin produced another three reports on the topic in 1874, 1879, and 1884.
41 Klippart, "Agriculture in Ohio in 1876," 508.
42 For details about Hough's career, see Edna L. Jacobsen, "Franklin B. Hough: A Pioneer in Scientific Forestry in America"; Ralph S. Hosmer, "Franklin B. Hough: Father of American Forestry."
43 Hough, "On the Duty of Government." For the association's memorial, see U.S. Congress, Senate, *Memorial from the American Association for the Advancement of Science upon the Cultivation of Timber and the Preservation of Forests*, 43rd Cong., 1st sess., S. Ex. Doc. 28; and U.S. Congress, House, *Report on the American Association for the Advancement of Science Memorial on the Cultivation of Timber and the Preservation of Forests*, 43rd Cong., 1st sess., H. R. 259. See Hough, *Report upon Forestry*, 49–60, for a history of these events.

44 The story of the formation and progress of the Division of Forestry is outlined in USDA, *Annual Report* (1877), 20–2; (1878), 27–32; (1880), 653–6.

45 Hough, *Report upon Forestry*, 1 (1878), 2 (1880), 3 (1882); Egleston, *Report upon Forestry*, 4 (1894).

46 USDA, *Annual Report* (1878), 27–8.

47 Stephanne B. Sutton, *Charles Sprague Sargent and the Arnold Arboretum*, 81–9; see also Carl Schurz, *Reminiscences*, 231–3.

48 Sargent, *Report on the Forests*, 485. For reactions to the *Report*, see Parkman, "The Forests and the Census," 835–9.

49 Hough, *Report upon Forestry*, 1:386; Sargent, *Report on the Forests*, 547–8, for example.

50 USDA, *Annual Report* (1885), 191.

51 Egleston, *Report upon Forestry*, 4:174–80.

52 Leue, *Report, Ohio Forestry Bureau*, 107, 290, 283.

53 H. Smith, *The Virgin Land*, 202. Smith's book is the seminal work on the yeoman myth.

54 For example, Elliott Coues (ed.), *The Expeditions of Zebulon Montgomery Pike*, 2:525; *Account of an Expedition from Pittsburgh to the Rocky Mountains Performed in the Years 1819, 1820* (1823), printed in Reuben G. Thwaites (ed.), *Early Western Travels*, 17:191, 147–8. See also Ralph C. Morris, "The Notion of a Great American Desert East of the Rockies," 190–200.

55 Particularly good basic accounts exist in H. Smith, *The Virgin Land*, 201–13, with a more extensive view in his "Rain Follows the Plow: The Notion of Increased Rainfall for the Great Plains, 1844–1880." Another treatment is in David M. Emmons, "American Myth, Desert to Eden: Theories of Increased Rainfall and the Timber Culture Act of 1873," 6–14, and in his *Garden in the Grasslands: Boomer Literature of the Central Great Plains*, 128–61. Walter M. Kollmorgen and Johanna Kollmorgen, "Landscape Meteorology in the Plains Area," 424–41, covers all the rainmaking theories.

56 James G. Cooper, "The Forests and Trees of North America, as Connected with Climate and Agriculture," 428; B. Taylor, *Colorado*, 5, 179.

57 Comte de Buffon, *Natural History, General and Particular: The History of Man and Quadrupeds*, 2:341. See also Glacken, *Traces on the Rhodian Shore*, 668–70.

58 Josiah Gregg, *Commerce of the Prairies; or, The Journal of a Santa Fe Trader*, 2:202–3.

59 Daniel Lee, "Agricultural Meteorology," 40–1.

60 Emmons, "American Myth," 8–9.

61 G. Marsh, *Man and Nature*. Marsh never asserted that the West could be watered by afforestation. Most of the evidence led him to believe that "more rain falls in wooded than in open country," but "we cannot positively affirm that the total annual quantity of rain is diminished or increased by the destruction of the woods" (196–8). He was far surer of the influence of forests on vegetation, soils, and water, and it was this certainty that was often misconstrued by the rainmakers.

62 U.S. Department of the Interior, *Annual Report* (1866), 34.

63 Joseph Wilson, "Observations ... on Forest Culture," 197.

64 Ferdinand V. Hayden, "The Geology of Nebraska," in "Report of the Commissioner of the General Land Office," 1867, 159–60.

65 H. Smith, *The Virgin Land*, 210.

66 J. Sterling Morton, address to the Horticultural Society, Lincoln, 4 January 1872, quoted in James C. Olson, "Arbor Day: A Pioneer Expression of Concern for the Environment," 9. For a more extended treatment of Morton, see Olson's *J. Sterling Morton*; for Morton's own account, see "Arbor Day: Its History and Growth," 22–7. Another major work by another tree propagator, fellow Nebraskan, and contemporary of Morton is Robert W. Furnas (ed.), *Arbor Day*.

67 Walter M. Kollmorgen, "The Woodsman's Assault on the Domain of the Cattleman," 215.

68 Thomas R. Wessel, "Prologue to the Shelterbelt, 1870–1934," 121, H. Greeley, *What I Know of Farming*, 44.

69 USDA, *Annual Report* (1869), 601; Ulysses Grant, Fourth Annual Message, 2 Dec. 1872, in James D. Richardson (ed.), *A Compilation of the Messages and Papers of the Presidents*, 6:4158.
70 U.S. Congress, *Congressional Globe*, 42nd Cong., 2nd sess., 10 June 1872, pt. 5, 4464.
71 U.S. Department of the Interior, *Annual Report* (1883), 7–8.
72 Benjamin H. Hibbard, *A History of the Public Land Policies*, 411–23; quotation on 421. For more detailed local studies of the application of the act in the Plains, see Kollmorgen, "The Woodsman's Assault," 221–3; William F. Raney, "The Timber Culture Acts"; C. Barron McIntosh, "Use and Abuse of the Timber Culture Acts."
73 Samuel Aughey, "The Geology of Nebraska: A Lecture Given in the Representative Hall," 14.
74 Samuel Aughey, *Sketches of the Physical Geography and Geology of Nebraska*, 41–7.
75 Charles Dana Wilber, *The Great Valleys and Prairies of Nebraska and the Northwest*, 69–70.
76 For example, Samuel Aughey and Charles D. Wilber, *Agriculture beyond the 100th Meridian; or, A Review of the United States Public Land Commission.* For reviews of the work and activities of the Nebraskans and Kansans, see Kollmorgen, "The Woodsman's Assault," 225–6; Kollmorgen and Kollmorgen, "Landscape Meteorology," 431, 434–6; H. Smith, "Rain Follows the Plow"; Charles R. Kutzleb, "Rain Follows the Plow: The History of an Idea"; Richard A. Overfield, "Trees for the Plains: Charles E. Bessey and Forestry," 18–28.
77 Kollmorgen, "The Woodsman's Assault," 227.
78 See n. 43 to this chapter.
79 Hough, *Report upon Forestry*, 1:221–33; Franklin B. Hough, *The Elements of Forestry.*
80 Franklin B. Hough, Sketch and Remarks accompanying "A Report on the Range and Ranch Cattle Traffic . . . ," 130–1, in U.S. Congress, House, *Report on the Internal Commerce of the United States.*
81 Nathaniel H. Egleston, in USDA, *Annual Report* (1883), 453–5, 457; (1884), 157–8; (1885), 184, 186, 192–3, 196; Nathaniel H. Egleston, "The Value of Our Forests," 176–86.
82 Henry J. Gannett, "The Influence of Forests on the Quantity and Frequency of Rainfall," 242–4. For Fernow's answer, see USDA, *Annual Report* (1888), 603–18.
83 Bernhard E. Fernow, "Forest Planting in the Plains," 140. For the background on Fernow and his attitudes toward this issue, see Rodgers, *Bernhard Fernow*, 146–9.
84 Charles R. Kutzleb, "American Myth, Descent to Eden; Can Forests Bring Rain to the Plains?" 21.
85 USDA, Forestry Division, *Forest Influences*, 9–22, 23–187, passim, 187–91.
86 J. Thomas, "Culture and Management of Forest Trees," 44–5; Lapham, Knapp, and Crocker, *Disastrous Effects of Destruction*, 6–7, 17–22, 40–3; Daniel Millikin, "Best Means of Preserving the Forest," 324. See also Hough, *Report upon Forestry*, 1:23, for comments on planting in belts rather than clumps in Kansas in 1875.
87 Samuel Edwards, "Timber on the Prairies," 496. For a compilation of material about shelterbelts, see Wilmon H. Droze, *Trees, Prairies, and People: Tree Planting in the Plains States.*
88 Danhof, "The Fencing Problem," 183; Hayter, *The Troubled Farmer*, 97–101.
89 Joseph G. Konvalinka, *How to Improve the Climate of the United States and Other Countries*, 8–10.
90 Wessel, "Prologue to the Shelterbelt," 127.
91 Paul H. Carlson, "Forest Conservation on the South Dakota Prairies," 28; Wessel, "Prologue to the Shelterbelt," 126–9; Charles A. Keffer, *Experimental Tree Planting in the Plains*, 7, 18, 35–6; John H. Hatton, "A Review of Early Tree Planting Activities on the Plains Region," in USFS, *Possibilities of Shelterbelt Planting in the Plains Region*, 51–5. In 1890 Fernow and Bessey collaborated in an experiment to plant trees in the Nebraska Sand Hills, which was successful enough by the early 1900s for both state and federal authorities to start planting the Nebraska National Forest and the Garden City Reserve in Kansas. See Carlos G. Bates and G. R. Pierce, *Forestation of the Sand Hills of Nebraska and Kansas*; Overfield, "Trees for the Plains," 24–31.

92 Raphael Zon, *Forests and Water in the Light of Scientific Investigation*, 51–5. For an excellent account of Zon's career and an appraisal of his contribution to research in forestry, see Norman J. Schmaltz, "Raphael Zon: Forest Researcher," 25–39, 86–97.

93 Raphael Zon, "The Relation of Forests in the Atlantic Plain to the Humidity of the Central States and Prairie Region," 139–53. By 1945, in his "Forests in Relation to Soil and Water," Zon concluded that "Forests *may not* increase the total precipitation but they certainly help to dispose of it more economically and fruitfully than land cleared of forests" (401; italics added). I am indebted to Norman Schmaltz for this and other information about Zon.

94 Schmaltz, "Raphael Zon," 30–7.

95 Raphael Zon, "What the Study Discloses," in USFS, *Possibilities of Shelterbelt Planting*, 5. The basis for the assertion that Zon influenced the administration is documented in Edgar B. Dixon (ed.), *Franklin D. Roosevelt and Conservation, 1911–45*, vol. 1, item 72, 200–3, and is repeated in Wilmon H. Droze, "The New Deal's Shelterbelt Project, 1934–1942," 33–4. Schmaltz suggests that, although the direct link cannot be proven, it would not be unreasonable (personal communication).

96 USFS, *Possibilities of Shelterbelt Planting*, 7–9.

97 Ralph A. Read, *The Great Plains Shelterbelt in 1954: A Re-evaluation of the Field Windbreaks Planted between 1935 and 1942 and a Suggested Research Program*, 8–10, 116–225, and map facing p. 11; Schmaltz, "Raphael Zon," 35. The Clarke–McNary Act of 1924, which promoted federal/state cooperation in the establishment of windbreaks and woodlots, resulted in the distribution of over 200 million trees by 1934. It was complementary to the shelterbelt project. See U.S. Bureau of the Census, *Historical Statistics of the United States*, tables L32–47.

12. Preservation and management, 1870–1910

1 J. Thomas, "Culture and Management of Forest Trees," 43.

2 Starr, "American Forests," 219. Fuller, *Forest Tree Culturist*, 5.

3 Roy M. Robbins, *Our Landed Heritage: The Public Domain, 1776–1936*, 302.

4 USDA, *Annual Report* (1872), 45; (1878), 245.

5 R. Robbins, *Our Landed Heritage*, 301.

6 John Ise, *The United States Forest Policy*, 61.

7 Thomas C. Donaldson, *The Public Domain: Its History, with Statistics*, 543.

8 Ise, *Forest Policy*, 83–5; Edward A. Bowers, "The Present Condition of the Forests on the Public Lands," 67–70.

9 Ise, *Forest Policy*, 81–2. US Department of the Interior, *Annual Report* (1874), xvi.

10 For an evaluation of Schurz's role in forest conservation, see Hildegard B. Johnson, "Carl Schurz and Conservation," 4; Jeannie S. Peyton, "Forestry Movement of the Seventies, in the Interior Department, under Schurz," 392–4.

11 U.S. Department of the Interior, *Annual Report* (1877), 21–2, Report of the Commissioner of the General Land Office (1877), xvi, xx.

12 U.S. Congress, Senate, "Bill Providing for the Disposition of Public Timber and Timbered Lands in the United States" (Plumb bill). See also Peyton, "Forestry Movement," 393–8.

13 20 *U.S. Statutes at Large* (1878), 88, often called the Timber Cutting Act.

14 Ise, *Forest Policy*, 62, 64.

15 Ibid., 65, 70; U.S. Department of the Interior, *Annual Report* (1887), 566–7; Report of the Commissioner of the General Land Office, (1888), 54.

16 20 *U.S. Statutes at Large* (1878), 89; Ise, *Forest Policy*, 70, 74–5; Bowers, "Present Condition of the Forests," 72–3.

17 U.S. Department of the Interior, *Annual Report* (1878), xiii–xiv. See also U.S. Department of the Interior, *Annual Report* (1885), 22.

18 Ise, *Forest Policy*, 89; Congressional Record, 46th Cong., 2nd sess., 20 May 1880, 10, pt.4:3579.

19 Ise, *Forest Policy*, 78–9. U.S. Congress, House, *Report of the Public Lands Commission*. A more elaborate final report appeared the next year: U.S. Congress, House, *Final Report of the Public Lands Commission*. Jenks Cameron, *The Development of Governmental Forest Control in the United States*, 172–5; Carl Schurz, *Speeches, Correspondence, and Political Papers*, 5:77, "The Need for a Rational Forest Policy"; Frank E. Smith, *The Politics of Conservation*, 63–4.

20 For a lengthy discussion, see Ise, *Forest Policy*, 109, 114–18.

21 Hough, "On the Duty of Governments," 1–22. See also Hough, *Report upon Forestry*, 3:49–58, for background details.

22 U.S. Congress, House, "A Bill for the Appointment of a Commission to Enquire into the Destruction of Forests." For background, see Cameron, *Development of Forest Control*, 189–98; USDA, Annual Report (1889), 90.

23 USDA, *Annual Report* (1884), 7.

24 Fernow, *Brief History of Forestry*, 482, quoted in Rodgers, *Bernhard Fernow*, 108.

25 USDA, *Annual Report* (1883), 456.

26 USDA, *Annual Report* (1880), 656; (1883), 455–7; (1884), 15–16; (1880), 656.

27 For background information about Fernow, see Rodgers, *Bernhard Fernow*, esp. 13–34; Fernow, *Brief History of Forestry*.

28 USDA, *Annual Report* (1886), 166; Cameron, *Development of Forest Control*, 198.

29 For a good review of these experiments, see Nelson, *A History of the US Forest Products Laboratory*, 6–11; Rodgers, *Bernhard Fennow*, 235.

30 William N. Sparhawk, "The History of Forestry in America," 708.

31 Pinchot, *Breaking New Ground*, 28, 38, and "How Conservation Began in the United States," 257–8.

32 USDA, *Annual Report* (1886), 162–8, 187; (1887), 611–13.

33 For example, see James B. Trefethen, *Crusade for Wildlife: Highlights in Conservation Progress*, 51; and Donald C. Swain, *Federal Conservation Policy, 1921–1933*, 6, 10, who base their assessments of Fernow on Pinchot's interpretation in his *Breaking New Ground*, 28–31.

34 Fernow, *Forestry Investigations*, "What Are We After?," 3, and Fernow, "Address," in *Proceedings of the American Forestry Association*, 10:143.

35 Based on *Proceedings of the American Forestry Congress* (Washington, D.C., 1883), 33–5; John A. Warder, "Report on Forestry"; "Records Relating to the First Convention of the American Forestry Congress, 1882," National Archives, Washington, D.C., R. G. 95.

36 *Proceedings of the American Forestry Congress*, 1:11–13; *Proceedings of the American Forestry Association*, 10:173–5; Fernow, *Forestry Investigations*, 171; Samuel T. Dana, *Forest and Range Policy: Its Development in the United States*, 2nd ed., 42–3.

37 Based upon *Proceedings of the American Forestry Congresses* (1882–9); *Proceedings of the American Forestry Association* (1890–7); USDA, *Annual Reports* (1885–97).

38 Ise, *Forest Policy*, 148–9; Cameron, *Development of Forest Control*, 440.

39 Nash, *Wilderness*, 96. For a general appreciation of Thoreau's view of the forest wilderness, see ibid., 84–95; for the quotation from Thoreau, see Thoreau, *The Maine Woods*, 212–13.

40 For the literary and artistic background, see Nash, *Wilderness*, 67–83; Horace Greeley, *Glances at Europe: In a Series of Letters . . . during the Summer of 1851*, 39.

41 USDA, *Annual Report* (1860), 445.

42 Linnie Marsh Wolfe, *John of the Mountains: The Unpublished Journals of John Muir*, 313. For a biographical detail of Muir, see William F. Badé, *The Life and Letters of John Muir*. Nash, *Wilderness*, 122–9, has a good appreciation of Muir.

43 John Muir, "God's First Temples: How Shall We Preserve Our Forests? . . . (Communicated to the Record-Union)," p. 8, cols. 6–7; Nash, *Wilderness*, 130–3.

44 USDA, *Annual Report* (1860), 445. As early as 1832 the Arkansas Hot Springs had been set aside as a national reservation; John Ise, *Our National Park Policy: A Critical History*, 13.

45 For the creation of Yosemite, see Hans Huth, "Yosemite: The Story of an Idea"; Ise, *Our*

National Park Policy, 52–5; Holway R. Jones, *John Muir and the Sierra Club: The Battle for Yosemite*, 25 ff. For Yellowstone Park, see Merrill D. Beal, *The Story of Man in Yellowstone*; and for the 1886 railroad proposal, see Nash, *Wilderness*, 114–15.

46 Donald W. Meinig, "The Geography of Expansion," 140–71.

47 For the writing of Hammond and May, see Nash, *Wilderness*, 103–4, 116. For the recreational impact generally, see Alfred L. Donaldson, *A History of the Adirondacks*, 1:190–201; Colvin, *Seventh Annual Report*, 10. For the best overall view of the Adirondack forest preservation movement, see Roger C. Thompson, "The Doctrine of Wilderness: A Study of the Policy and Politics of the Adirondack Preserve-Park," and his "Politics in the Wilderness."

48 Dinsdale, "Lumber Industry of New York," 109–10; Fernow, *Brief History of Forestry*, 491–2; Edna Jacobsen, "Franklin B. Hough," 316–17.

49 New York State Forest Commission, *Annual Report* (1890), 85; Fernow, *Brief History of Forestry*, 491–2; Rodgers, *Bernhard Fernow*, 86, 96–9. For the long-drawn-out campaign of the sportsmen, see articles in *Forest and Stream* as far apart as "The Adirondack Park," 1 (11 Sept. 1873), 73, and "The Adirondacks in 1898," 51 (1 Oct. 1898), 262–3. Nash, *Wilderness*, 118–21, reviews the Adirondack movement.

50 For the act of 1885, see *New York Laws* (1885), chap. 238, p. 482, and Hopkins, "Within and without the Blue Line," 10. For the 1892 act, see *New York Laws* (1892), chap. 709, p. 1459, and Nash, *Wilderness*, 120. Art. VII, sec. 7, of the New York State Constitution, quoted in Clifford J. Hynning, *State Conservation of Resources*, 13.

51 For the involvement of Sargent and Parkman, see Charles S. Sargent, "A Suggestion," and Francis Parkman, "The Forests of the White Mountains" and "Editorial." For a good coverage of the topic, see Charles D. Smith, "The Movement for Eastern National Forests, 1899–1911." Some of Smith's findings have been published in article form: "The Appalachian National Park Movement" and "The Mountain Lover Mourns: Origins of the Movement for a White Mountains National Forest, 1880–1903."

52 Rudyard Kipling, *From Sea to Sea*, 2:155; Robbins, *Lumberjacks and Legislators*, 10–12, 18–19, passim.

53 For a list of these measures, see *Proceedings of the American Forestry Association*, 12:41–58; Ise, *Forest Policy*, 110–14.

54 USDA, Forestry Division Bull. no. 2, "A Report on the Forest Conditions of the Rocky Mountains"; Rodgers, *Bernhard Fernow*, 131–2.

55 U.S. Congress, House, *Plan for the Management and Disposition of the Public Timber Lands*.

56 For Fernow's original proposals, see *Proceedings of the American Forestry Association*, 6: 9–13. For the Hale bill, see U.S. Congress, Senate, "Bill for the Protection and Administration of the Forests on the Public Domain." For discussion of the bill, see *Congressional Record*, 50th Cong., 1st sess., 1888, 19, pt. 1:787–8. Rodgers, *Bernhard Fernow*, 131. For a detailed evaluation and description of this and other legislation up to the 1891 Forest Reserve Act, see Kirkland, "The American Forests," 140–86.

57 *Proceedings of the American Forestry Association*, 8: 19–20; USDA, *Annual Report* (1889), 289–90.

58 For Dunnell's bill, see U.S. Congress, House, "Bill for the Reservation and Preservation of the Forest Lands"; for Fernow's remarks, see *Proceedings of the American Forestry Association*, 8:17. For discussion of the bill, see *Congressional Record*, 51st Cong., 1st sess., 1890, 21, pt.2: 1428, 21; pt.3: 2537–8.

59 U.S. Congress, House, "A Bill to Repeal the Timber Culture Laws and for Other Purposes." The progress of this bill can be followed in the *Congressional Record* from 51st Cong., 1st sess., 19 Feb. 1890, 21, pt. 2, to 51st Cong., 2nd sess., 4 Mar. 1891, 22, pt. 4.

60 Pinchot, *Breaking New Ground*, 85; Ise, *Forest Policy*, 117–18, for consequences.

61 U.S. Congress, Senate, *Message from the President of the United States Transmitting a Report Relative to the Preservation of Forests on the Public Domain*, pp. 1–4; Robbins, *Our Landed*

Heritage, 304. For a partial account of affairs in Congress, see Ise, *Forest Policy*, 115–16, 187, but a better view of the proceedings comes from reading *Congressional Record*, 51st Cong., 2nd sess., 28 Feb. 1890, 21, pt.4:3545–7.

62 For a list and description of these reserves, see *Proceedings of the American Forestry Association*, 10:118–21.

63 There are a number of biographies of Pinchot, esp. Martin N. McGeary, *Gifford Pinchot, Forester-Politician* (1960); Martin L. Fausold, *Gifford Pinchot, Bull Moose Progressive* (1961), for his career as leading spokesman for the Progressive party, 1910–17; and, more recently, Harold T. Pinkett, *Gifford Pinchot, Private and Public Forester* (1970).

64 Pinchot, *Breaking New Ground*, 33, 47–51, and "How Conservation Began," 261. For his career at Biltmore, see Harley E. Jolley, "Biltmore Forest Fair, 1908"; Harold T. Pinkett, "Gifford Pinchot at Biltmore."

65 U.S. Department of the Interior, *Annual Report* (1891), 224.

66 USDA, Forestry Division, *Annual Report* (1891), 224; (1893), 31; (1894), 133.

67 "A Plan to Save the Forests: Forest Preservation by Military Control," 631, for Muir's views.

68 Charles S. Sargent, "The Protection of Forests," 400–1. "A Plan to Save the Forests," 626, for Sargent's views. Sutton's *Sargent and the Arnold Arboretum* covers Sargent's life very fully.

69 Pinchot Diary, 16 Dec. 1894, and 4 Sept. 1895; box 3319, Pinchot Papers, Library of Congress. See also Ise, *Forest Policy*, 129.

70 Robbins, Our *Landed Heritage*, 311–13.

71 Pinchot Diary, 26 Aug. 1896, box 3319, Pinchot Papers; Nash, *Wilderness*, 135.

72 Pinchot, *Breaking New Ground*, 105. Pinchot Diary, 24 Oct. 1896 and 9 Jan. 1897; C. S. Sargent to Walcott Gibbs, 1 Feb. 1897; Gibbs to D. R. Francis, 1 Feb. 1897, Box 3319, Pinchot Papers. U.S. Congress, Senate, *Report of the Committee Appointed by the National Academy of Sciences* (1897), 8, 18–27.

73 See Robbins, *Our Landed Heritage*, 314–21, for an extensive account of the opposition. See also Rodgers, *Bernhard Fernow*, 223–4.

74 Pinchot, *Breaking New Ground*, 108.

75 Fernow, *Proceedings of American Forestry Association*, 12:31; Frederick H. Newell, "The National Forest Reserves," 186–7. U.S. Congress, Senate, *Report of the Committee Appointed by the National Academy of Sciences.*

76 Robbins, *Our Landed Heritage*, 320–4; and U.S. 30, *Statutes at Large*, 2:34–6.

77 Louis E. Peffer, *The Closing of the Public Domain: Disposal and Reservation Policies, 1900–1950*, 14–17; Samuel P. Hays, *Conservation and the Gospel of Efficiency: The Progressive Conservation Movement, 1890–1920*, 36–7, 44; Ise, *Forest Policy*, 119–42; Pinchot, *Breaking New Ground*, 86–104.

78 U.S. Bureau of Corporations, *The Lumber Industry*, 1, pt.1, 230; Robbins, *Lumberjacks and Legislators*, 24.

79 Rodgers, *Bernhard Fernow*, 241, 227. Martin Fausold, in his *Gifford Pinchot*, 12–13, suggests that Secretary of Agriculture Wilson ousted Fernow to provide a place for Pinchot.

80 Pinchot, *Breaking New Ground*, 38, and "How Conservation Began," 257–8. See, e.g., Hays, *Gospel of Efficiency*, 29, for the conventional interpretation of Fernow.

81 The total consumption for all wood products reached its highest point at 13.38 billion cu. ft. in 1907. The total for 1906 was the third-highest ever, the second-highest being in 1910.

82 William B. Greeley, *Forests and Men*, 64; Robbins, *Our Landed Heritage*, 337; and McGeary, *Gifford Pinchot*, 54–5.

83 Hays, *Gospel of Efficiency*, 39. See Cameron, *Development of Forest Control*, 220–2, for the interim arrangements in the Department of the Interior with Filibert Roth in charge.

84 Pinchot, *Breaking New Ground*, 167–77. Pinchot was not an unbiased witness, as chief of a rival bureau, but there was ample evidence of corruption. See Paul W. Gates, *History of Public Land Law Development*, 379, 591.
85 For Hitchcock's stance, see U.S. Department of the Interior, *Annual Report* (1901), 56; Ise, *Forest Policy*, 150–1; Cameron, *Development of Forest Control*, 232–3; Hays, *Gospel of Efficiency*, 7, 26.
86 Robbins, *Our Landed Heritage*, 345.
87 U.S. Congress, Senate, *Report of the Public Lands Commission*. The Division of Geography and Forestry of the Geological Survey also passed to Pinchot's control. The division had been formed in 1897 to map the reserves when the forest-lieu provisions were introduced, and it was probably the means by which the National Academy got funds for an underworked survey. Henry Gannett was geographer to the division and supervised the precise mapping of the 70 million acres of forest reserves.
88 Pinchot, *Breaking New Ground*, 254–6; *Proceedings of the American Forest Congress, 1905*, 4.
89 Robbins, *Our Landed Heritage*, 341; Peffer, *Closing of the Public Domain*, 57; Robbins, *Lumberjacks and Legislators*, 31, 44, 56.
90 For Muir, see Nash, *Wilderness*, 137; for Pinchot, see Hays, *Gospel of Efficiency*, 42.
91 Nash, *Wilderness*, 137; Pinchot quoted in Hays, *Gospel of Efficiency*, 41.
92 U.S. Congress, House, *First Message to Congress by the President*, xxvii; Peffer, Closing of the Public Domain, 30–1, 56–9; Hays, *Gospel of Efficiency*, 24–6; Robbins, *Our Landed Heritage*, 337–45.
93 Hays, *Gospel of Efficiency*, 1–4. Sparhawk, "History of Forestry," 711; Cameron, *Development of Forest Control*, 239–40.
94 Peffer, *Closing of the Public Domain*, 64; Nash, *Wilderness*, 163, for quote from Roosevelt. For an extended commentary on Pinchot's motives in elaborating his policy, see Hays, *Gospel of Efficiency*, 41–2.
95 Peffer, *Closing of the Public Domain*, 57; Cameron, *Development of Forest Control*, 240–5; Robbins, *Our Landed Heritage*, 349.
96 Robbins, *Our Landed Heritage*, 341, 349–51.
97 Ibid., 350–4, for a very full treatment.
98 For example, Gifford Pinchot, *The Fight for Conservation*, 79–80, "The Moral Issue"; see also Hays, *Gospel of Efficiency*, 122–7, for an excellent treatment.
99 Hays, *Gospel of Efficiency*, 135; Ise, *Forest Policy*, 51–2; *Proceedings, Conference of Governors, Washington, 1908*.
100 U.S. Congress, Senate, *Report of National Conservation Commission*, esp. vol. 2.
101 Ise, *Forest Policy*, 153.
102 Hays, *Gospel of Efficiency*, 148–51.
103 The Ballinger–Pinchot controversy and its context and implication must be among the most written about episodes in American conservation history. See esp. James L. Bates, "Fulfilling American Democracy: The Conservation Movement, 1907–1921"; Alpheus T. Mason, *Bureaucracy Convicts Itself: The Ballinger–Pinchot Controversy of 1910*; James L. Penick, *Progressive Politics and Conservation: The Ballinger–Pinchot Affair*; Pinkett, *Gifford Pinchot*; Rose M. Stahl, *The Ballinger–Pinchot Controversy*. For Pinchot's own account, see *Breaking New Ground*, 430–3.
104 Based on Hays, *Gospel of Efficiency*, 197–8; Peffer, *Closing of the Public Domain*, 332.

13. Ownership, supply, protection, and use, 1900–1933

1 *Congressional Record*, 59th Cong., 2nd sess., Dec. 1906, 41, pt.1:82; 18 Jan. 1907, 41, pt.1:1330–1.
2 Greeley, *Forests and Men*, 69, says categorically that the bureau's investigations were part of

the "Pinchot strategy," James R. Garfield, son of a president and commissioner of the bureau, being part of Roosevelt's "tennis cabinet" and directed to carry out the investigation. See also Robbins, *Lumberjacks and Legislators*, 49–50.

3 U.S. Congress, Senate, *Report of National Conservation Commission*, 2:188–91. For the background to the commission, see Ise, *Forest Policy*, 152–4.

4 These and the following paragraphs are based on U.S. Bureau of Corporations, *The Lumber Industry*, esp. vol. 2, pt.2, "Concentration of Timber Ownership in Important Selected Regions."

5 See Ficken, "Weyerhaeuser and the Timber Industry," for the background to the purchases; also Hidy, Hill, and Nevins, *Timber and Men*, 212–21, 496–500.

6 For details of ownership, see Ise, *Forest Policy*, 327–32; for maps of holdings, see U.S. Bureau of Corporations, *The Lumber Industry*, 2, pt.2, opp. p. 18, for Louisiana.

7 U.S. Bureau of Corporations, *The Lumber Industry*, 1, pt.1, 7; pt.3, 115, 182–3, 187.

8 See Ise, *Forest Policy*, 316 and note, on the bureau's prejudging of issues.

9 Wilson M. Compton, *The Organization of the Lumber Industry with Special References to the Influences Determining the Price of Lumber in the United States*, esp. chap. 6, "Influences that Have Affected Prices," 107–25.

10 USFS, *Timber Depletion, Lumber Prices*, 65.

11 U.S. Congress, Senate, *National Plan for Forestry*, 1:130–1.

12 Steen, *U.S. Forest Service*, 204.

13 USDA, *Annual Report* (1878), 245. See also Chapter 12 herein.

14 Defebaugh, *Lumber Industry of America*, 1:272.

15 Henry Gannett, "Forests of the United States," in U.S. Geological Survey *19th Annual, Report*, pt. 5, 1, 2, 18.

16 For Henry's map of 1858, see Joseph Henry, "Meteorology and Its Connection with Agriculture," plate VI; for Brewer's map of 1873, see Brewer, "Woodland and Forest Systems"; for Sargent's map, see Sargent, *Report on the Forests*, and accompanying *Folio Atlas of Forest Tress.*

17 Edgar T. Ensign, *Report on the Forestry Conditions of the Rocky Mountains and Other Papers.*

18 Gilbert Thompson et al., "Distribution of Woodland in Oklahoma" (1899).

19 Schantz and Zon, "Natural Vegetation."

20 The estimates of forest volume and area in Table 13.3 are a selection of those present in Marion Clawson's "Forests in the Long Sweep of American History"; however, there are many more estimates, as, e.g., in Defebaugh's *Lumber Industry of America*, 1:282–300, and particularly in the Forest Service Records in the National Archives, Washington, D.C.: e.g., GRG 95/144, "Estimated Forest Area of States" by G. Peters (1912), "Forest Statistics" (20 Nov. 1919), "Area of Logged-off Land" (1919), "Forest Area in 1920," R. E. Zon to unnamed congressman (27 Mar. 1930); GRG 95/241, "Outline for Work on Timber Supply Statistics" by W. D. Sterrett, "The Forest Problem of the United States" by Earle H. Clapp (31 May 1923).

21 William B. Greeley, "Reduction of the Timber Supply through Abandonment or Clearing of Forest Lands," 2:640–2; E. A. Zeigler, "Rate of Forest Growth," 2:203–69.

22 For example, ibid., 633; USFS, *Timber Depletion, Lumber Prices*, 31–2, 37–9; William B. Greeley, "The Relation of Geography to Timber Supply," 3–5, Greeley et al., "Timber: Mine or Crop?," 84–93.

23 The three maps were first produced in Greeley, "Relation of Geography to Timber Supply" (1925), and were subsequently included in Paullin, *Historical Geography of the United States*, plates 3A–C. In retrospect, Greeley (*Forests and Men*, 69) admitted that "None of us, foresters or lumbermen, had yet any conception of the reproductive power of the logged over forest, especially in the South, or how the growth rate was increasing as young trees replaced old timber." See also Greeley, *Forests and Men*, 226–9.

24 Defebaugh, *Lumber Industry of America*, 1:273.

25 See Table 13.3, based on Clawson's "Forests in the Long Sweep," table 1, p. 2. The conversion factor of 219 cu. ft. per 1,000 b.f. is based on Nelson C. Brown, *Forest Products: The Harvesting, Processing, and Marketing of Materials Other than Lumber*, 50.
26 Zeigler, "Rate of Forest Growth," 2:203–69.
27 USFS, *Timber Depletion, Lumber Prices*, 70.
28 For details, see following section.
29 USFS, *National Plan for Forestry*, 1:222–5.
30 Ibid., 242.
31 W. Greeley, *Forests and Men*, 69.
32 Theodore Roosevelt, "The Forest in the Life of the Nation," 8–9.
33 U.S. Congress, Senate, *Report of National Conservation Commission* (1909), 2:188–90; U.S. Bureau of Corporations, *The Lumber Industry*, 2, pt.2, 1–6.
34 Gifford Pinchot, *The Fight for Conservation*, 14–15, 74.
35 See W. Greeley, *Forests and Men*, 62; Rodgers, *Bernhard Fernow*, 218, 232–3, 269; Steen, *U.S. Forest Service*, 103–4.
36 For details of Graves's accomplishments, see Dana, *Forest and Range Policy*, 100–3, 119; Steen, *U.S. Forest Service*, 105–7, 133–5.
37 See, e.g., Henry S. Graves, "Private Forestry," in which he predicted the exhaustion of the southern yellow pine within 10 years.
38 Compton, *Organization of the Lumber Industry*, 5–6; W. Greeley, *Forests and Men*, 117, and *Some Public and Economic Aspects of the Lumber Industry in the United States*, 13–68; Steen, *U.S. Forest Service*, 112–13; Robbins, *Lumberjacks and Legislators*, 71–80.
39 See McGeary, *Gifford Pinchot*, for Pinchot's political career; also, W. Greeley, *Forests and Men*, 118.
40 Samuel T. Dana, *Forestry and Community Development*.
41 For a general appraisal of the opposing views, see Swain, *Federal Conservation Policy*, 11–15; Cameron, *Development of Forest Control*, 396–402; W. Greeley, *Forests and Men*, 102–3.
42 USFS, *Annual Report* (1919); Henry S. Graves, *A National Lumber and Forest Policy*. See also William B. Greeley in *The Timberman* 18 (18 Nov. 1916), 36, 38–9.
43 See Henry S. Graves, "A Policy of Forestry for the Nation"; Bernhard E. Fernow, "Editorial Comment." For additional details, see Steen, *U.S. Forest Service*, 176–9; Cameron, *Development of Forest Control*, 395–6.
44 Greeley, *Forests and Men*, 102; Edward T. Allen, "Men, Trees, and an Idea: The Genesis of a Great Fire Protective Plan," 529–32.
45 Gifford Pinchot, "Forest Devastation: A National Danger and a Plan to Meet It," 925, 914, 922, 915, and "The Lines Are Drawn," 900.
46 Pinchot Papers, container 2878, including Pinchot to Peters, 21 Sept. 1920; Greeley to Pinchot, 6 Oct. 1920; Pinchot to Peters, 21 Sept. 1920; Greeley to Pinchot, 6 Oct. 1920; Pinchot to Greeley, 22 Oct. 1920; Greeley to Pinchot, 26 Oct. 1920. For background material to this correspondence, see W. Greeley, *Forests and Men*, 104–6; Steen, *U.S. Forest Service*, 181.
47 USFS, *Timber Depletion, Lumber Prices*. The congressional background is covered in Dana, *Forest and Range Policy*, 125–6.
48 W. Greeley, *Forests and Men*, 102–6; Steen, *U.S. Forest Service*, 142–3; George T. Morgan, Jr., *William B. Greeley: A Practical Forester, 1879–1955*, 23–9; Robbins, *Lumberjacks and Legislators*, 90, 99–100.
49 See Dana, *Forest and Range Policy*, 125–6, and Steen, *U.S. Forest Service*, 182–6, for the complex background and details of this political activity. For Pinchot's campaign, see Pinchot Papers, container 556, copy of G. Pinchot, "National Forest Policy Now before Congress," from *Bulletin of National Conservation Association*, June 1921, which compares the merits of the Capper and Snell bills; and container 561, "Gifford Pinchot's interview with President Harding," 28 Nov. 1921, typescript.

50 USFS, *Annual Report* (1920–1), 6–8, (1921–2), 2–12; Dana, *Forest and Range Policy*, 121–2, 125–6.

51 For the attempted transfer, see W. Greeley, *Forests and Men*, 95–101; and an extensive treatment in Steen, *U.S. Forest Service*, 148–52. See also Burl Noggle, *Teapot Dome: Oil and Politics in the 1920s*, 8–30.

52 Greeley, *Forests and Men*, 119.

53 There is a vast literature, both scientific and popular, on the subject of forest fires. The best overall reviews are Stewart H. Holbrook's vivid account in *Burning an Empire* and, more recently, Stephen J. Pyne's voluminous *Fire in America*. Brown and Davis, *Forest Fire*, is an authoritative technical account. Examples of some readable and informative accounts of individual fires are to be found in Stewart H. Holbrook, "The Great Hinckley Fire"; Sawyer, "Personal Reminiscences"; Schafer, "Great Fires of Seventy-one"; Donald H. Clark, "The Yacolt Burn: Forest Graveyard"; Elers Koch, *When the Mountains Roared: Stories of the 1910 Fire*; Betty G. Spencer, *The Big Blowup: The Northwest's Great Fire*.

54 Hough, *Report upon Forestry*, 3:128–76; Sargent, *Report on the Forests*, 491–2.

55 See Morgan, "Fight against Fire"; Plummer, *Forest Fires*. For calculations of the amount of forest burned, see USFS Circulars 166 (1909), 171 (1909), which put losses at 2 billion b.f. per annum.

56 Ise, *Forest Policy*, 207–33; Cameron, *Development of Forest Control*, 168–77; C. Smith, "Movement for Eastern National Forests," 4–63; Steen, *U.S. Forest Service*, 129–31, 175.

57 W. Greeley, *Forests and Men*, 18, 24–6; Earl S. Pierce and William J. Stahl, *Cooperative Forest Fire Control*, for a factual overall view.

58 W. Greeley et al., "Timber: Mine or Crop?," 161 ff.

59 W. Greeley, *Forests and Men*, 107, passim. For the report, see U.S. Congress, Senate, *Investigation into Reforestation*, 1–15.

60 For a summary of the contents of the bills, see Dana, *Forest and Range Policy*, 126–31. William Greeley, in *Forests and Men*, 106–10, gives personal reminiscences of the passage of the bills and his part in the proceedings; Steen, *U.S. Forest Service*, 188, follows their passage through Congress.

61 W. Greeley, *Forests and Men*, 129.

62 Louis B. Murphy, "Forest Taxation Experience of the States and Conclusions Based on them"; John B. Woods, "Taxes and Tree Growing."

63 The literature on the effects of tax delinquency on the productive and settlement capacity of the land is immense, especially as it affects the Lake States. In particular, see George S. Wehrwein and Raleigh Barlowe, *The Forest Crop Law and Private Forest Taxation in Wisconsin*; Solberg, *New Laws for New Forests*; W. Deitz, "A Study in Michigan Forest Land Taxation"; James M. Lee, "Property Taxes and Alternatives for Michigan." For a general statement on the visual and landscape effects, see Dana, *Forestry and Community Development*.

64 Fred R. Fairchild, in U.S. Congress, Senate, *Report of National Conservation Commission*, 2:581–632; W. Greeley et al., "Timber: Mine or Crop?," 166; USFS, *Timber Depletion, Lumber Prices*, 72.

65 Fred R. Fairchild, *Forest Taxation in the United States*; Steen, *U.S. Forest Service*, 193; W. Greeley, *Forests and Men*, 113–14; USDA, *Trees: Yearbook of Agriculture, 1949*, 269–70, 477–532.

66 Nash, *Wilderness*, 161–81. For a general overview of the concept of conservation of wilderness, see Michael Frome, *Battle for the Wilderness*, 105–40.

67 For the national parks in general, see Ise, *Our National Park Policy*, esp. 27–50. For a detailed study of the Forest Service role, see Donald Cate, "Recreation and the U.S. Forest Service: A Study of Organizational Response to Changing Demands," 31–85; Frank A. Waugh, *Recreation Uses on the National Forests*, 10–12, 25–8.

68 Dana, *Forest and Range Policy*, 131–2; for the seminal role of Leopold and Cahart, see Nash, *Wilderness*, 182–99; for a more extended treatment of Leopold, see Susan L. Flader, *Thinking like a Mountain: Aldo Leopold and the Evolution of an Ecological Attitude toward Deer, Wolves, and Forests*, esp. 79–81. Steen, *U.S. Forest Service*, 155–6, summarizes the Forest Service reaction. See also Glen O. Robinson, *The Forest Service: A Study in Public Land Management*, 119–21, 154–7.

69 For Marshall, see Nash, *Wilderness*, 202–5; Robert Marshall, *The People's Forests*, 176–81, 187–8. For the Copeland Report, see U.S. Congress, Senate, *National Plan for Forestry*, 1:464–86, 154–6.

70 Dana, *Forest and Range Policy*, 157–8. The best account of the U-regulations is in James P. Gilligan, "The Development of Policy and Administration of Forest Service Primitive and Wilderness Areas in the Western United States," 199.

71 Dana, *Forest and Range Policy*, 157–8.

72 Robbins, *Lumberjacks and Legislators*, 131–4.

73 Swain, *Federal Conservation Policy*, 21–3.

74 Ahern, *Deforested America*, i, vi.

75 Pinchot Papers, container 980, circular letter by Pinchot, 28 Nov. 1928, and correspondence with Senator Capper, 4 Mar. 1929.

76 Swain, *Federal Conservation Policy*, 23–4; W. Greeley, *Forests and Men*, 131–2. Production statistics are from Steer, *Lumber Production*, 15–16. Robbins, *Lumberjacks and Legislators*, 133–53, deals exhaustively with the reactions of the lumber industry.

77 See Dana, *Forest and Range Policy*, 169–70, for NIRA and the lumber code; Mason, *Forests for the Future*, chap. 6–8, for sustained yield and the career of Mason; W. Greeley, *Forests and Men*, 133–4, 136–7. For a very full coverage of these events, see Steen, *U.S. Forest Service*, 223–7; Henry Clepper, *Professional Forestry in the United States*, 148–9; Robbins, *Lumberjacks and Legislators*, 181–4.

78 U.S. Congress, Senate, *National Plan for Forestry*, 1:76–8. For general background to the Copeland Report, see Dana, *Forest and Range Policy*, 168–9; Steen, *U.S. Forest Service*, 199–204. See also National Archives, GRG 95/120, Earle Clapp, "The Philosophy of the Copeland Report" (26 Oct. 1932).

79 For the CCC, see Perry H. Merrill, *Roosevelt's Forest Army: A History of the Civilian Conservation Corps, 1933–1942*; John A. Salmond, *The Civilian Conservation Corps, 1933–1942: A New Deal Case Study*. For the TVA, see the account of the first director, David E. Lilienthal, *TVA: Democracy on the March*; for the shelterbelt, see Chapter 11 herein.

80 Based on Swain, *Federal Conservation Policy*, 27–8; Steen, *U.S. Forest Service*, 196–7; W. Greeley, *Forests and Men*, 137–8.

14. The rebirth of the forest, 1933 and after

1 In particular, see Luther H. Gulick, *American Forest Policy: A Study of Government Administration and Economic Control*; Steen, *U.S. Forest Service*, 223–323; Dana, *Forest and Range Policy*, 142–348; Marion Clawson, *Forests for Whom and for What?*; Clepper, *Professional Forestry*; Robinson, *The Forest Service*; Frank E. Smith (ed.), *Conservation in the United States: A Documentary History*; Henry Clepper and Arthur B. Meyer (eds.), *American Forestry: Six Decades of Growth*.

2 Dana, *Forest and Range Policy*, 167, 323–7, 331–47; J. Michael McCloskey, Note and Comment, Natural Resources–National Forests: the Multiple Use–Sustained Yield Act of 1960"; Lewis, "Tree Farms"; Clawson, *Man, Land, and Forest*, 59–60.

3 See, e.g., Robbins, *Lumberjacks and Legislators*, 11–12; James Weinstein, *The Corporate Ideal in the Liberal State*, x.

4 Clawson, "Forests in the Long Sweep," 4.

5 For data on farm abandonment, see U.S. Bureau of the Census, Agricultural returns, esp. Fourteenth Census (1920), "Agriculture: General Report and Analytical Tables," 34–47; L. M. Vaughan, *Abandoned Farm Areas in New York*, 245–55; C. I. Hendrickson, "The Agricultural Land Available for Forestry," 151–69; and L. A. Reuss, H. H. Wooten, and F. J. Marschner, *Inventory of Major Land Uses, United States*, 26–49; Gates, "Problems of Agricultural History," 46–9; Harold F. Wilson, *The Hill Country of Northern New England: Its Social and Economic History, 1790–1830*; Stilwell, "Migration from Vermont."

6 John F. Hart, "Loss and Abandonment of Cleared Farm Land in the Eastern United States," 426–30.

7 For the changes in the South, see Vance, *Human Geography of the South*, 177–204; Hendrickson, "Agricultural Land Available," 161–2; Hart, "Loss and Abandonment," 430.

8 Reuss, Wooten, and Marschner, *Inventory of Land Uses*, 39–40; Clawson, "Forests in the Long Sweep," 2, and "Competitive Land Use in American Forestry and Agriculture," 225; Hart, "Loss and Abandonment," 418–21; and my own analysis of the *U.S. Agricultural Census* for 1965, 1969, 1975, and 1979, and the distribution of clearing and abandonment by counties.

9 USFS, *Timber Depletion, Lumber Prices*, 3.

10 W. Greeley et al., *Timber: Mine or Crop?*, 88, 92.

11 W. Greeley, "Relation of Geography to Timber Supply," 14.

12 Hendrickson, "Agricultural Land Available," 157–62, 164–5.

13 Reuss, Wooten, and Marschner, *Inventory of Land Uses*, 45–7; see also Robert W. Harrison, "Clearing Land in the Mississippi Alluvial Valley."

14 USFS, *Analysis of the Timber Situation*, 118–19.

15 John F. Hart, "Land Use Change in a Piedmont County," 514–16. For another example of reversion in a small area using the evidence of the Harvard Forest at Petersham, Mass., see Hugh M. Raup, "The View from John Sanderson's Farm: A Perspective for the Use of the Land," 2–11.

16 For a brief summary of the acts, see Dana, *Forest and Range Policy*, 126–9, 130–1.

17 Based on U.S. Bureau of the Census, *Historical Statistics*, tables L32–L43, and accompanying notes.

18 See Ronald S. Adams, "Tree Nurseries and Tree Seed Collection," 656–9; USFS, *Analysis of the Timber Situation*, 228–33.

19 Based on Hamlin L. Williston, *A Statistical History of Tree Planting in the South, 1925–79*. See also Thomas D. Clark, *The Greening of the South: The Recovery of Land and Forests*, 23, passim.

20 USFS, *Analysis of the Timber Situation*, 32.

21 George W. Bengtson, "Forest Fertilization in the United States: Progress and Outlook," 222–9.

22 For the history and characteristics of many of these insects, diseases, and parasites, see USDA, *Trees: Yearbook of Agriculture, 1949*, 407–76.

23 Greeley, *Forests and Men*, 15–29. See also Chapter 13, "How should the forest be protected?"

24 Based on U.S. Bureau of the Census, *Historical Statistics of the United States*, pt. 1, tables L48, L49.

25 Pyne, *Fire in America*, 261 271, 275–94.

26 Based on U.S. Bureau of the Census, *Historical Statistics of the United States*, and *Analysis of the Timber Situation*, 294, table 9.5; for money spent on fire protection, see p. 233, table 9.4.

27 The outstanding work on fire control is Pyne, *Fire in America*, 100–22, 143–60; but see also Ashley L. Schiff, *Fire and Water: Scientific Heresy in the Forest Service*, 2–115; Brown and Davis, *Forest Fire*, 31–44, 559–70; E. V. Komarek, Sr., "Comments on the History of

Controlled Burning in the Southern United States"; Kilgore, "Fire Control to Fire Management."

28 Rodgers, *Bernhard Fernow*, 154; Pinchot, *Breaking New Ground*, 24.

29 During these years perhaps only T. B. Walker, a lumberman near Mount Shasta, spoke up for prescribed burning. See Walker, in Clyde Leavitt, "Forest Fires," in U.S. Congress, Senate, *Report of National Conservation Commission*, 2:424–5. For the work of S. B. Show, see Pyne, *Fire in America*, 104–5; Steen, *U.S. Forest Service*, 135–6.

30 Coert du Bois, *Systematic Fire Protection in the California Forests*; Stuart B. Show and E. I. Kotok, *The Role of Fire in the California Pine Forests.*

31 Pyne, *Fire in America*, 101.

32 Ellen Long, "Forest Fires in Southern Pine," quoted in Schiff, *Fire and Water*, 19.

33 Inter alia, Herman H. Chapman, "A Method of Studying Growth and Yield of Longleaf Pine Applied to Tyler County, Texas" (1909), "Forest Fires and Forestry in the Southern States" (1912), "Factors Determining the Natural Reproduction of Longleaf Pine on Cutover Lands in Lasalle Parish, Louisiana" (1926), "Is the Longleaf Type a Climax?" (1932), and "Fire and Pines" (1944).

34 Schiff, *Fire and Water*, 17–18; Pyne, *Fire in America*, 143–51.

35 Pyne, *Fire in America*, 115.

36 S. W. Greene, "The Forest that Fire Made"; Herbert L. Stoddard, *The Bobwhite Quail: Its Habits, Preservation, and Increase.*

37 U.S. Congress, Senate, *National Plan for Forestry* (Copeland Report), 1:673; Schiff, *Fire and Water*, 60–2.

38 See, e.g., W. G. Wahlenberg, "Forest Fire Control in the Coastal Plains Section of the South"; USDA, *Trees: Yearbook of Agriculture, 1949*, 518–19, 522–3.

39 USDA, *Trees: Yearbook of Agriculture, 1949*, 477–532; Winters, *Fifty Years of Forestry.*

40 Harold Weaver, "Fire as an Ecological and Silvicultural Factor in the Ponderosa Pine Region of the Pacific Slope" (1943), and "Effects of Fire on Temperate Forests" (1974).

41 Kilgore, "Fire Control to Fire Management," 479–82, and "Restoring Fire to National Park Wilderness"; Richard J. Hartesveldt, "Fire Ecology of the Giant Sequoias: Controlled Fires May Be One Solution to the Survival of the Species."

42 Brown and Davis, *Forest Fire*, 31–44; Kilgore, "Fire Control to Fire Management," 477, 491; Kozlowski and Ahlgren, *Fire and Ecosystems.*

43 Much of the new knowledge has been contained in and disseminated through the *Proceedings of the Annual Fire Ecology Conferences* of the Tall Timbers Research Station, 1962 to date. See esp. E. V. Komarek, Sr., *A Quest for Ecological Understanding: The Secretary's Review.*

44 For trends in consumption, see U.S. Bureau of the Census, *Historical Statistics of the United States*, tables L87–L97; Clawson, "Forests in the Long Sweep," 5; USFS, *Timber Trends in the United States*, 43–5, 66–7, and *Analysis of the Timber Situation*, 293–9, tables 1.16–1.22.

45 USFS, *Analysis of the Timber Situation*, 119–20, 69–71, table 3.4.

46 Starr, "American Forests," 211–12.

47 G. Marsh, *Man and Nature*, 191.

48 USDA, *Forests and National Prosperity* (1958); USFS, *Timber Resources for America's Future* (1958), *Timber Trends* (1965), *The Outlook for Timber in the United States* (1973), and *Analysis of the Timber Situation* (1982).

49 Stephen G. Boyce, "How to Double the Harvest of Loblolly and Slash Pine Timber"; USFS, *Analysis of the Timber Situation*, 240–6.

50 For the clear-cutting controversy, see Nancy Wood, *Clearcut: The Deforestation of America*, 20–5, 79–93; Jack Shepherd, *The Forest Killers: The Destruction of the American Wilderness*, 113–70, 340–64. For the pesticides controversy, see Rachel Carson, *Silent Spring*; Thomas R. Dunlap, *DDT: Scientists, Citizens, and Public Policy*, 95–125; Paul W. Gates, "Pressure Groups and Recent American Land Policies," 110–19. The literature on general environ-

mental issues is enormous, but see esp. Samuel P. Hays, "The Structure of Environmental Politics since World War II"; Donald Fleming, "Roots of the New Conservation Movement."

51 See, e.g., Dennis C. Le Master, *Mergers among the Largest Forest Products Firms, 1950–1970*; Walter J. Mead, *Mergers and Economic Concentration in the Douglas-Fir Lumber Industry*; Robbins, *Lumberjacks and Legislators*, 244–6.

References

Archives

Forest History Society, Durham, N.C. Photographic Collection.
Library of Congress, Washington, D.C. Pinchot Papers.
National Archives, Washington, D.C. Research Compilation File.
Forest Service Papers. GRG 95.
Map Collection.
Public Record Office, London. Customs, 16/1.
State Historical Society of Wisconsin, Madison, Division of Archives and Manuscripts. Increase Lapham Papers.
William Freeman Vilas Papers.
Photographic and Map Collection.
University of California, Berkeley, Bancroft Library. Records of Regional Oral History Office.
University of Wisconsin, Madison. Forestry Department Records.

General references

Ackerknecht, Erwin H. *Malaria in the Upper Mississippi Valley, 1760–1900*. Supplement to *Bulletin of the History of Medicine*. Baltimore: Johns Hopkins University Press, 1945.

Adams, Herbert Baxter. "Village Communities of Cape Anne and Salem." *Johns Hopkins University Studies in History and Political Sciences*, 1st ser. 9–10. Baltimore: Johns Hopkins Press, 1883.

Adams, J. T. *The Epic of America*. Boston: Little, Brown, 1931.

Adams, Ronald S. "Tree Nurseries and Tree Seed Collection." In Davis, Richard (ed.), *Encyclopedia of American Forest and Conservation History*, 2: 656–9. New York: Macmillan, 1983.

Ahern, George P. *Deforested America: Statement of the Present Forest Situation in the United States*. With a Foreword by Gifford Pinchot. Printed privately, 1928. Later, the pamphlet without the Foreword appeared in U.S. Congress, Senate, 70th Cong., 2nd sess., S. Doc 216, 1929.

Albion, Robert Greenhaugh. *Forests and Sea Power: The Timber Problem of the Royal Navy, 1652–1862*. Cambridge, Mass.: Harvard University Press, 1926.

Alilunas, Leo. "Michigan's Cut-over 'Canaan.'" *Michigan History* 26 (1942), 188–201.

Allen, Clifford (ed.). *Michigan Log Marks: Their Formation and Use during the Great Michigan Pine Harvest*. Compiled by the workers of the WPA Writers' Program. Michigan Agric. Exper. Stat. Mem. Bull. 4, Lansing, 1942.

Allen, Edward T. "Men, Trees, and an Idea: The Genesis of a Great Fire Protective Plan." *American Forests and Forest Life* 32 (1926), 529–32.

Allen, Ruth A. *East Texas Lumber Workers: An Economic and Social Picture, 1870–1950*. Austin: University of Texas Press, 1961.

American Agriculturist [New York], 1875–7.

American Husbandry 1 (1837), 83–4.

Anburey, Thomas. *Travels through the Interior Parts of America, in a Series of Letters by an Officer.* 2 vols. London, 1791, 1798. Reprint, Boston: Houghton Mifflin, 1923.

Andrews, C. M. *The Colonial Period of American History.* 4 vols. New Haven: Yale University Press, 1934–8.

Appleman, Roy E. "Timber Empire from the Public Domain." *Mississippi Valley Historical Review* 26 (1939), 193–208.

Ashton, T. S. *Iron and Steel in the Industrial Revolution.* Manchester: Manchester University Press, 1924.

Aughey, Samuel. "The Geology of Nebraska: A Lecture Given in the Representative Hall." In Nebraska Agricultural Board, *Annual Report*, Lincoln, 1873, pp. 67–85.

Sketches of the Physical Geography and Geology of Nebraska. Omaha: Daily Republican Book and Job Office, 1880.

Aughey, Samuel, and Wilber, Charles D. *Agriculture beyond the 100th Meridian; or, A Review of the United States Public Land Commission.* Lincoln: Journal Company, 1880.

Bachmann, Elizabeth. "Minnesota Log Marks." *Minnesota History* 26 (1945), 126–37.

Badé, William Frederic. *The Life and Letters of John Muir.* 2 vols. Boston: Houghton Mifflin, 1924.

Bahre, Conrad. "Effects of Historic Fuelwood Cutting on the Semidesert Woodlands of Arizona–Sonora Borderlands." In Steen, Harold K. (ed.), *History of Sustained-Yield Forestry: A Symposium*, 101–10. Santa Cruz, Calif.: Forest History Society, 1984.

Bailyn, Bernard. *The New England Merchants in the Seventeenth Century.* Studies in Entrepeneurial History. Cambridge, Mass: Harvard University Press, 1955.

Bailyn, Bernard, and Bailyn, Lotte. *Massachusetts Shipping, 1697–1714: A Statistical Study.* Cambridge, Mass: Harvard University Press, 1959.

Baird, J. *View of the Valley of the Mississippi; or The Emigrants' and Travellers' Guide to the West.* Philadelphia: H. S. Tanner, 1832.

Bakeless, John. *The Eyes of Discovery: The Pageant of North America as Seen by the First Explorers.* Philadelphia: J. B. Lippincott, 1950.

Barclay, George. "Chicago: The Lumber Hub." *Southern Lumberman* 193 (1956), 177–178B.

Bardin, P. C. "The Outline History of the Sawmill." *Hardwood Record* 55, no. 8 (1923), 35–6, 46.

Barnes, Will C. *Apaches and Longhorns: The Reminiscences of Will C. Barnes.* Edited by Frank C. Lockwood. Los Angeles: Ward Ritchie Press, 1941.

Barrows, Harlan H. *Geography of the Middle Illinois Valley.* Illinois State Geological Survey bull. no. 15. Urbana: University of Illinois, 1910.

Bartram, William. *Travels through North and South Carolina, Georgia, East and West Florida, the Cherokee Country, the Extensive Territories of Muscogulges, or Creek Confederacy, and the Country of the Chactaws.* Philadelphia: James and Johnson, 1791; London: reprinted for J. Johnson in St. Paul's Church-yard, 1794.

Bates, Carlos G., and Pierce, G. R. *Forestation of the Sand Hills of Nebraska and Kansas.* USFS bull no. 121. Washington, D.C.: GPO, 1913.

Bates, James Leonard. "Fulfilling American Democracy: The Conservation Movement, 1907–1921." *Mississippi Valley Historical Review* 44 (June 1957), 29–57.

Beal, Merrill D. *The Story of Man in Yellowstone.* Caldwell, Idaho: Caxton Printers, 1949.

Beatley, Janice Carson. "The Primary Forests of Vinton and Jackson Counties, Ohio." Ph.D. diss., Ohio State University, 1953.

Beaufoy, Henry. *Tour through Parts of the United States and Canada by a British Subject.* London: Longman, Rees, Orme, Browne, and Green, 1828.

Beck, Berton E. "Stump-Pulling." *Pennsylvania Folk-Life* 16 (1960), 20–31.

Beckham, Stephen D. "Asa Mead Simpson: Lumberman and Shipbuilder." *Oregon Historical Quarterly* 68 (1967), 259–73.

Beidleman, Richard G. "Some Biological Sidelights on Thomas Nuttall, 1786–1859." *Proceedings of the American Philosophical Society* 104 (1960), 86–100.

Belknap, Jeremy. *The History of New Hampshire.* 3 vols. Boston: printed for the author, 1791–2.

Belthuis, Lyda C. "The Lumber Industry in Eastern Iowa." *Iowa Journal of History and Politics* 46 (April 1948), 115–55.

Bengtson, George W. "Forest Fertilization in the United States: Progress and Outlook," *Journal of Forestry* 78 (1979), 222–9.

Bennett, R. L. "Settlement of Cut-over Lands." *Southern Lumberman* (30 July 1910), 37.

Benson, Barbara E. "Logs and Lumber: The Development of the Lumber Industry in Michigan's Lower Peninsula, 1837–1870." Ph.D. diss., Indiana University, 1976.

Berger, Max. *The British Traveller in America, 1836–1860.* New York: Columbia University Libraries for Library of Congress, 1943.

Bernard, John. *Retrospections of America, 1797–1811.* Edited from a manuscript by Mrs. Bayle Bernard. New York: Harper and Bros., 1887.

Bidwell, Percey W., and Falconer, John I. *History of Agriculture in the Northern United States, 1620–1860.* Carnegie Institute Publication no. 358. 1925. Reprint, New York: P. S. Smith, 1941.

Bigelow, Thomas. *Journal of a Tour to Niagara Falls in the Year 1805.* Boston: J. Wilson, 1876.

"Billion Feet of Logs by Motor Truck." *Timberman* 50 (1949), 172–80, 228.

Bining, Arthur C. "The Iron Plantations of Early Pennsylvania." *Pennsylvania Magazine of History and Biography* 57 (1933), 117–37.

Pennsylvania Iron Manufacture in the Eighteenth Century. Pennsylvania Historical Commission Publication no. 4. Harrisburg, 1938.

Birch, Brian P. "Initial Perception of Prairie: An English Settlement in Illinois." In Ironside, R. G. (ed.), *Frontier Settlement,* 178–98. University of Alberta Studies in Geography, Monograph 1. Calgary, 1974.

Black, John D., and Gray, L. C. *Land Settlement and Colonization in the Great Lakes States.* USDA bull. no. 1295. Washington, D.C.: GPO, 1925.

Black, Robert C. *The Railroads of the Confederacy.* Chapel Hill: University of North Carolina Press, 1952.

Blair, Walter A. *A Raft Pilot's Log: A History of the Great Rafting Industry on the Upper Mississippi, 1840–1915.* Cleveland: Arthur H. Clark, 1930.

Blegen, Theodore L. *Minnesota: A History of a State.* Minneapolis: University of Minnesota Press, 1963.

Borah, Woodrow. "America as Model: The Demographic Impact of European Expansion on the Non-European World." In *Actas y. memorias, XXXV: Congreso International de Americanistas,* 3: 379–87. Mexico City, 1962.

Borah, Woodrow, and Cook, Sherburne F. "Conquest and Population: A Demographic Approach to Mexican History." *Proceedings of the American Philosophical Society* 113 (1969), 177–83.

Bormann, F. H., and Likens, G. E. *Pattern and Process in a Forested Ecosystem.* New York: Springer-Verlag, 1979.

Bowen, William A. *The Willamette Valley: Migration and Settlement on the Oregon Frontier.* Seattle: University of Washington Press, 1978.

Bowers, Edward A. "The Present Condition of the Forests on the Public Lands." *Publications of the American Economic Association* 6 (May 1891), 57–74.

Boyce, Stephen G. "How to Double the Harvest of Loblolly and Slash Pine Timber." *Journal of Forestry* 73 (1975), 761–6.

Boyd, James. "Fifty Years in the Southern Pine Industry." *Southern Lumberman* 144 (15 Dec. 1931), 59–67; 145 (1 Jan. 1932), 23–4.

Boyer, Charles Shimer. *Early Forges and Furnaces in New Jersey.* Philadelphia: University of Pennsylvania Press, 1931.

Bradbury, John. *Travels in the Interior of America in the Years 1809, 1810, and 1811.* London: Sherwood, Neeley, and Jones, 1817. Reprinted as *Bradbury's Travels in the Interior of North America.* Cleveland: Arthur H. Clark, 1904.

Bradshaw, J. S. "Grand Rapids, 1870–1880: Furniture City Emerges." *Michigan History* 55 (1968), 321–42.

"Grand Rapids Furniture Beginnings." *Michigan History* 52 (1968), 279–98.

Branch, Douglas. *The Hunting of the Buffalo*. Lincoln: University of Nebraska Press, 1962.

Brewer, William H. "The Woodland and Forest Systems of the United States." Plates I, III, and V, together with commentary. In Francis A. Walker (ed.), *A Statistical Atlas of the United States, Based on the Results of the Ninth Census*. Washington, D.C.: GPO, 1874. Commentary was also published in the USDA *Annual Report* (1875), 352–8.

Bridenbaugh, Carl. *Cities in the Wilderness: The First Century of Urban Life in America, 1625–1742*. New York: Rowland Press, 1938.

Fat Mutton and the Liberty of Conscience. Providence: Brown University Press, 1974.

"Yankee Use and Abuse of the Forest in the Building of New England, 1620–1660." *Proceedings of the Massachusetts Historical Society* 89 (1977), 3–35.

Bromley, S. W. "The Original Forest Types of Southern New England."*Ecological Monographs* 5 (1935), 61–89.

Brown, Arthur A., and Davis, Kenneth P. *Forest Fire: Control and Use*. 2nd ed. New York: McGraw-Hill, 1973.

Brown, Nelson Courtlandt. *Forest Products: The Harvesting, Processing, and Marketing of Materials Other than Lumber*. New York: John Wiley and Sons, 1950.

Logging: Principles and Practices in the United States and Canada. New York: John Wiley and Sons, 1934. Reissued and revised in 1949 as *Logging: The Principles and General Methods of Harvesting Timber in the United States and Canada*.

Logging-Transportation: The Principles and Methods of Log Transportation in the United States and Canada. New York: John Wiley and Sons, 1936.

Lumber, Manufacture, Conditioning, Grading, Distribution, and Use. New York: John Wiley and Sons, 1947.

Brown, Ralph Adams. "The Lumber Industry in the State of New York, 1790–1830." M.A. thesis, Faculty of Political Science, Columbia University, 1933.

Brown, Ralph H. *The Historical Geography of the United States*. New York: Harcourt Brace, 1948.

Browne, C. A. "Historical Notes on the Domestic Potash Industry in Early Colonial and Later Times." *Journal of Chemical Education* 3 (1926), 749–56.

Bruce, Anthony. "Ninety Years On." *Southern Lumberman* 223 (15 Dec. 1971), 76–8.

Bryant, Ralph Clement. *Logging: The Principles and General Methods of Operation in the United States*. New York: John Wiley and Sons, 1913; 2nd ed., 1923.

Lumber: Its Manufacture and Distribution. New York: John Wiley and Sons, 1922; and 2nd ed., 1934.

Buchanan, Iva L. "Lumbering and Logging in the Puget Sound Region in Territorial Days." *Pacific Northwest Quarterly* 27 (1936), 34–53.

Buck, Solon. "Making a Farm on the Frontier: Extracts from the Diaries of Mitchell Young Jackson." *Agricultural History* 4 (1930), 92–102.

Buckingham, James Silk. *The Eastern and Western States of America*. 3 vols. London: Fisher and Sons, 1842.

The Slave States of America. 2 vols. London: Fisher and Sons, 1842.

Buffon, comte de [Georges-Louis Leclerc]. *Histoire naturelle, générale et particulière*. 44 vols. Paris, 1749–1804. Another edition: William Wood (ed.), *Comte de Buffon, Natural History, General and Particular: The History of Man and Quadrupeds*. Translated with Notes and Observations by William Smellie. 10 vols. London: T. Cadell and W. Davies, 1812.

Bull, Marcus. *Experiments to Determine the Comparative Value of the Principal Varieties of Fuel*. Philadelphia, 1827.

Burgess, Sherwood D. "The Forgotten Redwoods of the East Bay." *California Historical Society Quarterly* 30 (1951), 1–14.

"Lumbering in Hispanic California." *California Historical Society Quarterly* 41 (1962), 237–48.

Burnaby, Andrew. *Travels through the Middle Settlements in North America in the Years 1759 and 1760*. London: T. Payne, 1765.

Burns, Anna M. C. "Frank B. Williams, Cypress Lumber King." *Journal of Forest History* 24 (1980), 127–34.

"Henry E. Hardtner: Louisiana's First Conservationist." *Journal of Forest History* 22 (1978), 79–85.

A History of the Louisiana Forestry Commission. Louisiana Studies Institute, monograph ser. no. 1. Natchitoches: Northwestern State College, 1968.

Butterick, P. L. "Forest Growth on Abandoned Agricultural Land." *Scientific Monthly* 5 (n.d.): 80–91.

Byrd, William. *History of the Dividing Line between Virginia and Other Tracts, 1728–1736: From the Papers of William Byrd, Virginia, Esq*. Edited by T. H. Wynne. Richmond, Va. 1866.

William Byrd's Natural History of Virginia; or, The Newly Discovered Eden. Edited and translated from a German version by Richmond C. Beatty and W. J. Mulloy. Richmond, Va.: Dietz Press, 1940.

Calendar of State Papers, Colonial Series: America and the West Indies, 1696–97. London: H.M.S.O., 1904.

Cameron, Jenks. *The Development of Governmental Forest Control in the United States*. Institute for Government Research, Studies in Administration. Baltimore: Johns Hopkins University Press, 1928. Reprint, New York: Da Capo Press, 1972.

Campbell, Patrick. *Travels in the Interior Inhabited Parts of North America in the Years 1791 and 1792*. Edinburgh, 1793. Reprinted in 1937 with Introduction by H. H. Langton, notes by H. H. Langton and W. F. Ganong. Toronto: Champlain Society, 1937.

Carlson, Paul H. "Forest Conservation on the South Dakota Prairies." *South Dakota History* 2 (1971), 23–45.

Carlton, William R. "New England Masts and the King's Navy." *New England Quarterly* 12 (1939), 4–18.

Carman, E. A., Heath, H. A., and Minto, J. *Special Report on the History and Present Conditions of the Sheep Industry of the United States*. Washington, D.C.: Bureau of Animal Husbandry, 1892.

Carrier, Lyman. *The Beginnings of Agriculture in America*. New York: McGraw-Hill, 1923.

Carroll, Charles F. *The Timber Economy of Puritan New England*. Providence: Brown University Press, 1973.

Carroll, Peter N. *Puritanism and the Wilderness: The Intellectual Significance of the New England Frontier, 1629–1700*. New York: Columbia University Press, 1969.

Carson, Rachel L. *Silent Spring*. Greenwich, Conn.: Crest Books, 1962.

Carstenson, Vernon. R. *Farms or Forests: Evolution of a State Land Policy for Northern Wisconsin, 1850–1932*. Madison: University of Wisconsin, College of Agriculture, 1958.

Castiglioni, Luigi. *Viaggio negli Stati Uniti dell' America Settentrionale fatto negli Anni 1785, 1786, e 1787*. 2 vols. Milan: Nella Stamperid di Giuseppe Marelli, 1790.

Cate, Donald. "Recreation and the U.S. Forest Service: A Study of Organizational Response to Changing Demands." Ph.D. diss., Stanford University, 1963.

Catlin, G. B. "Early Travel on the Ohio and Its Tributaries, 1818." *Michigan Historical Magazine* (1936), 150–62.

Chandler, Alfred D. "Anthracite Coal and the Beginnings of the Industrial Revolution in the United States." *Business History Review* 46 (1972), 141–81.

Chapman, Herman Haupt. "Factors Determining the Natural Reproduction of Longleaf Pine on Cut-over Lands in Lasalle Parish, Louisiana." Yale School of Forestry Bull. 16. New Haven, 1926.

"Fires and Pines: A Realistic Appraisal of the Role of Fire in Reproducing and Growing Southern Pines." *American Forests* 50 (1944), 62–4, 91–3.

"Forest Fires and Forestry in the Southern States." *American Forests* 18 (1912), 510–17.

"The Initiation and Early Stages of Research on Natural Reforestation of Longleaf Pine." *Journal of Forestry* 46 (1948), 505–10.

"Is the Longleaf Type a Climax?" *Ecology* 13 (1932), 328–34.

"A Method of Studying Growth and Yield of Longleaf Pine Applied to Tyler County, Texas."*Proceedings, Society of American Foresters* 4 (1909), 217–18.

"Why the Town of McNary Moved: A Tragedy of the Southern Pines and a Parallel Which Carries Its Own Lesson." *American Forests and Forest Life* 30 (1924), 589–92, 615–16, 626.

Chapman, Leonard B. "The Mast Industry of Old Falmouth." *Maine Historical Society Collections*, 2nd ser. 7 (1896), 390–405.

"Charlemagne Tower: His Journey through Maine in the Summer of 1829." *Down East* 16 (1970), 47–9.

Chastellux, François Jean, marquis de. *Travels in North America in the Years 1780, 1781, and 1782*. 2 vols. New York: White, Gallaher, and White, 1789 and 1827.

Chateaubriand, François Auguste René, vicomte de. *Recollections of Italy, England, and America on Various Subjects*. Philadelphia: M. Carey, 1816.

Travels in America and Italy. 2 vols. London: Henry Colburn, 1828.

Chavannes, Elizabeth. "Written Records of Forest Succession." *Scientific American* 53 (1941), 76–80.

Chinard, Gilbert. "The American Philosophical Society and the Early History of Forestry in America." *Proceedings, American Philosophical Society* 89 (1945), 444–88.

"Eighteenth Century Theories on America as a Human Habitat." *Proceedings, American Philosophical Society* 91 (1947), 27–57.

Clark, Andrew Hill. "Suggestions for the Geographical Study of Agricultural Change in the United States, 1790–1840." *Agricultural History* 46 (1972), 155–72.

Clark, Charles E. *The Eastern Frontier: The Settlement of Northern New England, 1610–1763*. New York: Alfred A. Knopf, 1970.

Clark, Donald H. "The Yacolt Burn: Forest Graveyard." *American Forests* 60 (1954), 18–20.

Clark, James I. *Cutover Problems: Colonization, Depression, Reforestation*. Chronicles of Wisconsin Series, no. 13. Madison: State Historical Society of Wisconsin, 1956.

Farming the Cutover: The Settlement of Northern Wisconsin. Chronicles of Wisconsin Series, no. 10. Madison: State Historical Society of Wisconsin, 1956.

Clark, Thomas D. "Early Lumbering Activities in Kentucky." *Northern Logger* 13 (1965), 14–15, 42–4.

The Greening of the South: The Recovery of Land and Forests. Lexington: University Press of Kentucky, 1984.

"The Impact of the Timber Industry on the South." *Mississippi Quarterly* 25 (1972), 141–64.

"Kentucky Logmen." *Journal of Forest History* 25 (1981), 144–57.

Clark, Victor S. *History of Manufactures in the United States, 1607–1914*. 3 vols. Washington, D.C. Carnegie Institute Publication no. 2158. 1929. Reprint, New York, 1949.

Clawson, Marion. "Competitive Land Use in American Forestry and Agriculture." *Journal of Forest History* 25 (1981), 222–7.

Decision Making in Timber Production, Harvest, and Marketing. Research Paper no. R-4. Washington D.C.: Resources for the Future, 1977.

Forests for Whom and for What? Baltimore: Johns Hopkins University Press, for Resources for the Future, 1975.

"Forests in the Long Sweep of American History." *Science* 204 (1979), 1168–74. Reissued as Reprint no.164, for Resources for the Future, Washington, D.C., 1980.

Man, Land, and the Forest Environment. Seattle: University of Washington Press, 1977.

"The National Forest: A Great National Asset Is Poorly Managed and Unproductive." *Science* 191 (20 Feb. 1976), 762–7.

Clawson, Marion, and Sedjo, Roger. "History of Sustained-Yield Concept and Its Application to

Developing Countries." In Steen, H. K. (ed.), *History of Sustained-Yield Forestry: A Symposium*, 3–15. Santa Cruz, Calif.: Forest History Society, 1984.

Clayton, John. "A Letter from Mr. John Clayton . . . to the Royal Society, May 12, 1686, Giving an Account of Several Observables in Virginia." Reprinted in Force, Peter (ed.), *Tracts and Other Papers Relating to the Origin, Settlement, and Progress of the Colonies of America*, 3:20–59. Washington, D.C., 1836–41.

Clements, Fredric E. "Plant Formations and Forest Types." *Proceedings, Society of American Foresters* 4 (1909), 50–63.

Clepper, Henry. *Professional Forestry in the United States*. Baltimore: Johns Hopkins University Press, for Resources for the Future, 1971.

Clepper, Henry, and Meyer, Arthur B. (eds.). *American Forestry: Six Decades of Growth*. Washington, D.C.: Society of American Foresters, 1960.

Clevinger, Woodrow R. "Locational Change in the Douglas Fir Lumber Industry." *Yearbook of the Association of Pacific Coast Geographers* 15 (1953), 23–31.

Clough, Wilson O. *The Necessary Earth: Nature and Solitude in American Literature*. Austin: University of Texas Press, 1964.

Coan, Eugene. "James Graham Cooper, Pioneer Naturalist and Forest Conservationist." *Journal of Forest History* 27 (1983), 126–9.

Cole, Arthur H. "The Mystery of Fuel Wood Marketing in the United States." *Business History Review* 44 (1970), 339–59.

Collier, Gerald L. "The Evolving East Texas Woodland." Ph.D. diss., University of Nebraska, Lincoln, 1964.

Colvin, Verplanck. *Report on a Topographical Survey of the Adirondack Wilderness of New York*. Albany: Argus, 1873.

Seventh Annual Report on the Progress of the Topographical Survey of the Adirondack Region of New York to the Year 1879. Albany: Weed and Parsons, 1880.

Coman, Edwin, T., and Gibbs, Helen M. *Time, Tide, and Timber: A Century of Pope & Talbot*. Stanford Business Series no. 7. Stanford, Calif: Stanford University Press, 1949.

Compton, Wilson M. *The Organization of the Lumber Industry, with Special References to the Influences Determining the Price of Lumber in the United States*. Chicago: American Lumberman, 1916.

Condit, Carl W. *American Building: Materials and Techniques from the First Colonial Settlements to the Present*. Chicago: University of Chicago Press, 1968.

Conway, Steve. *Logging Practices: Principles of Timber Harvesting Systems*. San Francisco: Miller Freeman Publications, c. 1976.

Cook, Frederick (ed.). *Journals of the Military Expedition of General John Sullivan against the Six Nations of Indians in 1779*. Auburn, N.Y.: Knapp, Peck, and Thompson, 1887.

Cook, Sherburne, F., and Borah, Woodrow. *Essays in Population History: Mexico and the Caribbean*. Vol. 1. Berkeley: University of California Press, 1972.

Cooper, James G. "The Forests and Trees of North America, as Connected with Climate and Agriculture." *Report of the Commissioner of Patents for 1860*. U.S. Congress, House, 36th Cong., 2nd sess., 1861 H. Ex. Doc. 48 (serial no. 1099), 416–45. For other information see Cooper's earlier paper, "On the Distribution of the Forests and Trees of North America with Notes on Its Physical Geography." U.S. Congress, House, *Annual Report of the Board of Regents, Smithsonian Institution*. 35th Cong., 2nd sess., 1859, Misc. Doc. 49, 246–80.

Cooper, Thomas. *Some Information Respecting America, Collected by Thomas Cooper*. Dublin: P. Wogan, 1774. 2nd ed., London: J. Johnston, 1794.

Cooper, William. *A Guide in the Wilderness; or, The History of the First Settlements in the Western Counties of New York, with Useful Instructions to Future Settlers*. Dublin: Gilbert and Hodges, 1810. Reprint, Rochester, N.Y.: G. P. Humphrey, 1897.

Corliss, Carlton John. *Main Line of Mid-America: The Story of the Illinois Central*. New York: Creative Age Press, 1950.

Corner, George W. (ed.). *An Autobiography of Benjamin Rush*. Princeton, N.J.: Princeton University Press, 1948.

Cotroneo, Ross R. "Western Land Marketing by the Northern Pacific Railway." *Pacific Historical Review* 27 (1968), 299–320.

Coues, Elliott (ed.). *The Expeditions of Zebulon Montgomery Pike*. 3 vols. New York: Francis P. Harper, 1895.

Coville, Frederick V. *Forest Growth and Sheep Grazing in the Cascade Mountains of Oregon*. USDA, Division of Forestry, Bull. no. 15, Washington, D.C.: GPO, 1898.

Cowles, Henry C. "The Ecological Relations of the Vegetation on the Sand Dunes of Lake Michigan." *Botanical Gazette* 27 (1899), 95–100.

Cox, John H. "Organizations of the Lumber Industry in the Pacific Northwest, 1889–1914." Ph.D. diss., University of California, Berkeley, 1937.

"Trade Associations in the Lumber Industry of the Pacific Northwest, 1899–1914." *Pacific Northwest Quarterly* 41 (1950), 285–311.

Cox, Thomas R. "The Conservationist as Reactionary: John Minto and American Forest Policy." *Pacific Northwest Quarterly* 74 (1983), 146–55.

"Logging Technology and Tools." In Davis, Richard (ed.), *Encyclopedia of American Forest and Conservation History*, 347–54. New York: Macmillan, 1983.

"Lower Columbia Lumber Industry, 1880–1893." *Oregon Historical Quarterly* 67 (1966), 160–78.

"Lumber and Ships: The Business Empire of Asa Mead Simpson." *Forest History* 14 (1970), 16–26.

Mills and Markets: A History of the Pacific Coast Lumber Industry to 1900. Seattle: University of Washington Press, 1974.

"Pacific Log Rafts in Economic Perspective." *Journal of Forest History* 15 (1972), 21–9.

"The Passage to India Revisited: Asian Trade and the Development of the Far West, 1850–1900." In Carroll, John A. (ed.), *Reflections of Western Historians*, 85–103. Tucson: University of Arizona Press, 1969.

"A Pennsylvania Lumber Raftsman's Year: The Diary of William Langdon for 1855." *Journal of Forest History* 26 (1982), 124–39.

"Trade, Development, and Environmental Change: The Utilization of North America's Pacific Coast Forests to 1914 and Its Consequences." In Tucker, Richard P., and Richards, John F. (eds.), *Global Deforestation and the Nineteenth Century World Economy*, 14–29. Durham, N.C.: Duke University Press, 1983.

"Transition in the Woods: Log Drivers, Raftsmen, and the Emergence of Modern Lumbering in Pennsylvania." *Pennsylvania Magazine of History and Biography* 104 (1980), 345–64.

Cox, William F. "Recent Forest Fires in Oregon and Washington." *Forestry and Irrigation* 8 (1902), 426–9.

Cox, William K. "Forty Years of Tractor Logging." *Southern Lumberman* 144 (15 Dec. 1931), 118–21.

Coxe, Tench. *A View of the United States of America, in a Series of Papers Written at Various Times between the Years 1787 and 1794*. Philadelphia: William Hall and Wrigley and Berriman, 1794. Reprinted in Economic Classics. New York: Augustus M. Kelley, 1965.

Cramer, Zadoc. *The Navigator, Containing Directions for Navigating the Monongahela, Allegheny, Ohio, and Mississippi Rivers*. Philadelphia, 1811–12.

Craven, Avery O. *Soil Exhaustion as a Factor in the Agricultural History of Virginia and Maryland, 1606–1860*. Urbana: University of Illinois Press, 1926.

Creel, George. "The Feudal Towns of Texas." *Harper's Weekly* 60 (23 Jan. 1915), 76–8.

Crèvecoeur, J. Hector St. John de. *Letters from an American Farmer*. London: Thomas Davies, 1782. Another edition edited by L. Lewishon. New York: Fox Duffield, 1904.

Sketches of Eighteenth Century America (c. 1770–4). Edited by Henri L. Bourdin. New Haven: Yale University Press, 1925.

Croker, Thomas C., Jr. "The Longleaf Pine Story." *Journal of Forest History* 23 (1979), 32–43.
Cronemiller, Lynn F. "Oregon's Forest Fire Tragedy." *American Forests* 40 (1934), 487–90, 531.
"The Tillamook Burn." *American Forests* 67 (1961), 29–31, 51–2.
Cronon, W. *Changes in the Land: Indians, Colonists, and the Ecology of New England*. New York: Hill and Wang, 1983.
Cultivator and Country Gentleman. Albany, New York, and Philadelphia. Various titles, 1853–1920.
Curti, Merle. *The Growth of American Thought*. New York: Harper and Row, 1943.
Curtis, John T. "The Modification of Mid-latitude Grasslands and Forests by Man." In Thomas, William (ed.), *Man's Role in Changing the Face of the Earth*, 721–36. Chicago: University of Chicago Press, 1956.
Curtis, Michael. "Early Development of Operations of the Great Southern Lumber Company." *Louisiana History* 14 (1973), 347–68.
Cutler, Julia Perkins. *The Life and Times of Ephraim Cutler, Prepared from His Journals and Correspondence*. Cincinnati: Robert Clarke, 1890. Reprint, New York: Arno Press, 1971.
Dana, Samuel Trask. *Forest and Range Policy: Its Development in the United States*. American Forestry Series. New York: McGraw-Hill, 1956; 2nd ed., 1980, revised and enlarged by Sally K. Fairfax.
Forestry and Community Development. USDA, Departmental Bull. no. 638. Washington, D.C.: GPO, 1918.
Danhof, Clarence H. *Change in Agriculture: The Northern United States, 1820–1870*. Cambridge, Mass.: Harvard University Press, 1969.
"The Fencing Problem in the Eighteen-fifties." *Agricultural History* 18 (1944), 168–86.
Darby, H. Clifford. "The Changing English Landscape." *Geographical Journal* 117 (1951), 377–98.
"The Clearing of the English Woodlands." *Geography* 36 (1951), 71–83.
"The Clearing of the Woodland in Europe." In Thomas, William L. (ed.), *Man's Role in Changing the Face of the Earth*, 183–216. Chicago: University of Chicago Press, 1956.
The Domesday Geography of Eastern England. Cambridge: Cambridge University Press, 1952.
Davis, Allen J. "The First Log Band Saw: An Historical Monograph." *Lumber World Review* 39 (1920), 33–4.
Davis, Richard (ed.). *Encyclopedia of American Forest and Conservation History*. 2 vols. New York: Macmillan, 1983.
North American Forest History: A Guide to Archives and Manuscripts in the United States and Canada. Santa Barbara, Calif.: ABC–Clio Books, in conjunction with Forest History Society, 1977.
Davis, T. Frederick. "The Disston Land Purchase." *Florida Historical Quarterly* 27 (1939), 200–10.
Day, Gordon M. "The Indian as an Ecological Factor in the Northeastern Forest." *Ecology* 34 (1953), 329–46.
De Bow, J. D. B. (ed.). *De Bow's Review of the Southern and Western States*. New Orleans, 1846–80.
De Brahm, John Gerar William. *Philosophico-Historico-Hydrogeography of South Carolina, Georgia, and East Florida*. In Weston, P. C. J. (ed.), *Documents Connected with the History of South Carolina*, 155–227. London, 1856.
Defebaugh, James Elliot. *History of the Lumber Industry of America*. 2 vols. Chicago: American Lumberman, 1906–7.
Deitz, W. "A Study in Michigan Forest Land Taxation." M.A. thesis, Michigan State University, 1954.
Demmon, Elwood L. "Henry E. Hardtner." *Journal of Forestry* 33 (1935), 885–6.
Denevan, William M. *The Native Population of the Americas in 1492*. Madison: University of Wisconsin Press, 1976.

Dewhurst and Associates. *America's Needs and Resources: A New Survey*. New York: Twentieth Century Fund, 1955.

Dick, Everett. *The Dixie Frontier*. New York: Alfred A. Knopf, 1948.

Dickens, Charles. *American Notes* (1867). Reprint, London: Hazell, Watson and Viney, 1930.

Dinsdale, Evelyn M. "The Lumber Industry of Northern New York: A Geographical Examination of Its History and Technology." Ph.D. diss., Syracuse University, 1963.

"Spatial Patterns of Technological Change: The Lumber Industry of Northern New York." *Economic Geography* 41 (1965), 258–65.

Disston, Henry, and Sons. *The Saw in History: A Comprehensive Description of This Most Useful of Tools from the Earliest Times to the Present Day*. 2nd ed. Philadelphia: Henry Disston and Sons, 1916.

Dixon, Edgar B. (ed.). *Franklin D. Roosevelt and Conservation, 1911–1945*. 2 vols. Hyde Park, N.Y.: Franklin D. Roosevelt Library, 1957.

Dobyns, Henry F. "Estimating Aboriginal American Population: An Appraisal of Techniques with a New Hemispheric Estimate." *Current Anthropology* 7 (1966), 395–416.

Dodwell, Arthur, and Rixon, Theodore F. *Forest Conditions in the Olympic Forest Reserve, Washington*. USGS Professional Paper no. 7. Washington, D.C.: GPO, 1902.

Donaldson, Alfred Lee. *A History of the Adirondacks*. 2 vols. New York: Century, 1921.

Donaldson, Thomas Corwin. *The Public Domain: Its History, with Statistics*. 1880. 3rd ed. Washington, D.C.: GPO, 1884.

Dopp, Mary. "The Geographical Influences in the Development of Wisconsin. V: The Lumber Industry." *Bulletin, American Geographical Society* 45 (1913), 401–12, 490–9, 585–609, 653–63, 736–49, 831–46, 902–20.

Driver, Harold E. *The Indians of North America*. Chicago: University of Chicago Press, 1961; 2nd rev. ed., 1970.

Droze, Wilmon Henry. "The New Deal's Shelterbelt Project, 1934–1942." In Hollingsworth, Harold M., and Holmes, William F. (eds.), *Essays of the New Deal*, 23–48. Austin: University of Texas Press, 1969.

Trees, Prairies, and People: Tree Planting in the Plains States. Denton: Texas Women's University, 1977.

du Bois, Coert. *Systematic Fire Protection in the California Forests*. USDA, Forest Service. Washington, D.C.: GPO, 1914.

Dufur, A. J. *Statistics of the State of Oregon*. Salem: Willamette Farmers Office, 1869.

Dunlap, Thomas R. *DDT: Scientists, Citizens, and Public Policy*. Princeton, N.J.: Princeton University Press, 1981.

Durant, Edward W. "Lumbering and Steamboating on the St. Croix River." *Collections of the Minnesota Historical Society* 10, pt. 2 (1905), 645–75.

Dwight, Timothy. *Travels in New-England and New-York in 1821*. 4 vols. New Haven: T. Dwight, 1821–2.

Dyksterhius, E. J. "The Vegetation of the Western Cross Timbers." *Ecological Monographs* 18 (1948), 326–76.

Easton, Hamilton Pratt. "The History of the Texas Lumbering Industry." Ph.D. diss., University of Texas, 1947.

Eavenson, Howard N. *The First Century and a Quarter of the American Coal Industry*. Pittsburgh, 1942.

Edwards, Samuel. "Timber on the Prairies." In USDA, *Annual Report, 1862*, 495–8. Washington, D.C.: GPO, 1863.

Egerton, Frank N. (ed.). *American Plant Ecology, 1897–1917: An Original Anthology*. New York: Arno Press, 1977.

Egleston, Nathaniel H. (ed.). *Arbor Day: Its History and Observance*. USDA Report no. 56. Washington, D.C.: GPO, 1896.

"The Value of Our Forests." *Popular Science Monthly* 19 (June 1881), 176–86.

Report upon Forestry. Vol. 4. Washington, D.C.: GPO, 1884.

Eifert, Virginia S. *Tall Trees and Far Horizons: Adventures and Discoveries of Early Botanists in America*. New York: Dodd, Mead, 1965.

Eisterhold, John A. "Charleston: Lumber and Trade in a Declining Southern Port." *South Carolina Historical Magazine* 74 (1973), 61–73.

"Colonial Beginnings in the South's Lumber Industry, 1607–1800." *Southern Lumberman* 223 (15 Dec. 1971), 150–3.

"Lumber and Trade in the Lower Mississippi Valley and New Orleans, 1800–1860." *Louisiana History* 13 (1972), 71–91.

"Savannah: Lumber Center of the South Atlantic." *Georgia Historical Quarterly* 57 (1973), 526–45.

Ekirch, Arthur A., Jr. *Man and Nature in America*. New York: Columbia University Press, 1963.

Eliot, Jared. *Essays upon Field Husbandry in New England [1760] and Other Papers, 1748–1762*. Edited by Henry J. Carver and Rexford Tugwell. New York: Columbia University Press, 1934.

Ellicott, Joseph. *Reports of Joseph Ellicott as Chief of Survey (1797–1800) and as Agent (1800–1821) of the Holland Land Company's Purchases in Western New York*. Edited by Robert W. Bingham. Buffalo, N.Y.: Buffalo Historical Society, 1937–41.

Ellis, David Maldwyn. *Landlords and Farmers in the Hudson–Mohawk Region, 1790–1850*. Ithaca, N.Y.: Cornell University Press, 1946.

Ellis, Richard C. "The Oregon and California Railroad Land Grant, 1866–1945." *Pacific Northwestern Quarterly* 39 (1948), 253–83.

Elman, Robert. *First in the Field: America's Pioneering Naturalists*. New York: Mason-Charter, 1977.

Emerson, F. V. "The Southern Long-leaf Pine Belt." *Geographical Review* 7 (1919), 81–90.

Emerson, Ralph Waldo. *The Collected Works of Ralph Waldo Emerson*. Edited by A. R. Ferguson. Vol. 1, *Nature: Addresses and Lectures*. Cambridge, Mass.: Harvard University Press (Belknap Press), 1902.

Emmons, David M. "American Myth, Desert to Eden: Theories of Increased Rainfall and the Timber Culture Act of 1873." *Forest History* 15 (1971), 6–14.

Garden in the Grasslands: Boomer Literature of the Central Great Plains. Lincoln: University of Nebraska Press, 1971.

Engberg, George B. "Who Were the Lumberjacks?" *Michigan History* 32 (1948), 238–46.

Ensign, Edgar T. *Report on the Forestry Conditions of the Rocky Mountains and Other Papers*. USDA, Forestry Division, Bull. no. 2. Washington, D.C.: GPO, 1889.

Erickson, Kenneth A. "The Morphology of Lumber Settlements in Western Oregon and Washington." Ph.D. diss., University of California, Berkeley, 1965.

Everest, D. C. "A Reappraisal of the Lumber Barons." *Wisconsin Magazine of History* 36 (1952), 17–22.

Ewan, Joseph Andorfer (ed.). *A Short History of Botany in the United States*. New York: Macmillan (Hafner Press), 1969.

Fahl, Ronald J. *North American Forest and Conservation History: A Bibliography*. Santa Barbara, Calif.: ABC and Clio Books, 1977.

Fairchild, Fred Rogers. "Taxation of Timber Lands." In U.S. Congress, Senate, *National Conservation Commission Report* 2: 581–626. Washington, D.C.: GPO, 1909.

Fairchild, Fred Rogers, and Associates. *Forest Taxation in the United States*. USDA, Miscellaneous Publication no. 218. Washington, D.C.: GPO, 1935.

Fairchild, Hoxie Neal. *The Romantic Quest*. New York: Columbia University Press, 1931.

Faulkner, William. *"Go Down Moses" and Other Stories*. London: Chatto and Windus, 1942.

Fausold, Martin L. *Gifford Pinchot, Bull Moose Progressive*. Syracuse, N.Y.: Syracuse University Press, 1961.

Felt, Joseph Barlow. *History of Ipswich, Essex, and Hamilton*. Cambridge, Mass.: C. Folsom, 1834.

Felt, Margaret E. *Gyppo Logger*. Caldwell, Idaho: Caxton Printers, 1963.

Fernow, Bernhard E. "Address." *Proceedings of the American Forestry Association* 10 (1889), 143.

A Brief History of Forestry in Europe, the United States, and Other Countries. Toronto: Toronto University Press, 1907, 1911.

Consumption of Forest Supplies by Rail Roads and Practicable Economy in Their Use. USDA, Forestry Division, Bull. no. 4. Washington, D.C.: GPO, 1890.

"Editorial Comment." *Journal of Forestry* 17 (1919), 880–3.

"Forest Planting in the Plains." *Annual Report, Nebraska State Board of Agriculture, 1890*, 139–50. Lincoln: Nebraska State Journal, 1891.

Forestry in the U.S. Department of Agriculture, 1877–1898: Resume. Washington, D.C.: GPO, 1899. Reprinted from U.S. Congress, House, 55th Cong., 3rd sess., 1899, H. Doc. 181.

"The Influence of Forests on Water Supplies." In USDA, *Annual Report, 1889*, 297–330. Reissued with extensive alterations as *Relations of Forests to Water Supply*. USDA, Forestry Division, Bull. no. 7, 123–70. Washington, D.C.: GPO, 1893.

Report upon the Forestry Investigations of the United States Department of Agriculture, 1877–1898. Washington, D.C.: GPO, 1899.

"What Are We After?" *New Jersey Forester* 1 (1895), 2–3.

Ficken, Robert E. *Lumber and Politics: The Career of Mark E. Reed*. Seattle: University of Washington Press, 1979.

"Weyerhaeuser and the Pacific Northwest Timber Industry, 1899–1903." *Pacific Northwest Quarterly* 70 (1979), 146–54.

Fickle, James E. *The New South and the "New Competition": Trade Association Development in the Southern Pine Industry*. Champaign: University of Illinois Press, 1980.

Finger, John R. "Seattle's First Sawmill, 1853–1869: A Study of Frontier Enterprises." *Forest History* 15 (1972), 24–31.

Fishlow, Albert. *American Railroads and the Transformation of the Ante-bellum Economy*. Harvard Economic Studies 127. Cambridge, Mass.: Harvard University Press, 1965.

Flader, Susan L. "Ecological Science and the Expansion of Our Forest Heritage." In *America's Renewable Resource Potential: The Turning Point*, 108–20. Proceedings, Annual Conference of Society of American Foresters. Washington, D.C., 1976.

"Scientific Resource Management: An Historical Perspective." *Transactions, 41st North American and Natural Resources Conference*, pp. 17–30. Washington, D.C., 1976.

Thinking like a Mountain: Aldo Leopold and the Evolution of an Ecological Attitude toward Deer, Wolves, and Forests. Columbia: University of Missouri Press, 1974.

Flader, Susan L. (ed.). *The Great Lakes Forest: An Environmental and Social History*. Minneapolis: University of Minnesota Press, 1983.

Fleming, Donald. "Roots of the New Conservation Movement." *Perspectives in American History* 6 (1972), 7–34.

Flint, Timothy. *A Condensed Geography and History of the Western States or the Mississippi Valley*. 2 vols. Cincinnati: W. M. Farnsworth, 1828.

The History and Geography of the Mississippi Valley, to Which Is Appended a Condensed Physical Geography of the Atlantic United States and the Whole of the American Continent. 2nd ed. Cincinnati: E. H. Flint and L. R. Lincoln, 1832.

Flower, George. *The Errors of Emigrants*. London: Cleave, 1841.

Fogel, Robert W. *Railroads and American Economic Growth: Essays in Econometric History*. Baltimore: Johns Hopkins University Press, 1964.

Forbes, Reginald D. "The Passing of the Piney Woods." *American Forestry* 29 (Mar. 1923), 131–36, 185.

Force, Peter (ed.). *Tracts and Other Papers Relating to the Origin, Settlement, and Progress of the Colonies of America*. 4 vols. Washington, D.C.: P. Force, 1836–41.

Fowler, John. *Journal of a Tour of New York in the Year 1810, with Remarks on Agriculture*. London: Whittaker, Treacher, and Arnot, 1831.

Fox, William, F. *History of the Lumber Industry in the State of New York*. USDA, Bureau of Forestry, Bull. no. 34. Washington, D.C.: GPO, 1902.

Frame, Richard. "A Short Description of Pennsilvania by Richard Frame" (1692). In Myers, Albert Cook (ed.), *Original Narratives of Early American History: Narratives of Early Pennsylvania, West New Jersey, and Delaware, 1630–1709*. New York: Charles Scribner and Sons, 1912.

Franklin, Benjamin. *An Account of the New-Invented Pennsylvania Fire-place: Wherein Their Construction and Manner of Operation, Particularly Explained*. Philadelphia: B. Franklin, 1744.

Franklin's Works. 20 vols. Philadelphia: McCarty and Davis, 1834.

Observations on Smoking Chimneys, Their Causes and Cure: With Some Considerations on Fuel and Stoves. London: I. and J. Taylor, 1793.

The Writings of Benjamin Franklin. 10 vols. Collected and edited with a life and introduction by Albert Henry Smyth. New York: Macmillan, 1905–7.

French, Edward H., and Withrow, R. James. *The Hardwood Distillation Industry of America*. Ohio State University Bull. 19, no. 17. Columbus, 1905.

Fries, Robert F. *Empire in Pine: The Story of Lumbering in Wisconsin, 1830–1900*. Madison: State Historical Society of Wisconsin, 1951.

"The Mississippi River Logging Company and the Struggle for the Free Navigation of Logs, 1865–1900." *Mississippi Valley Historical Review* 35 (1948), 429–48.

Friis, Herman Ralph. "A Series of Population Maps of the Colonies of the United States, 1625–1790." *Geographical Review* 30 (1940), 463–70.

Frome, Michael. *Battle for the Wilderness*. New York: Praeger Publishers, in cooperation with the Wilderness Society, 1974.

Frothingham, Earl H. *The Status and Value of Farm Woodlots in the Eastern United States*. USDA, Bull. no. 481. Washington, D.C.: GPO, 1917.

Fuller, Andrew S. *The Forest Tree Culturist: A Treatise on the Cultivation of American Forest Trees*. New York: Geo. and F. W. Woodward, 1866.

Furnas, Robert Wilkinson (ed.). *Arbor Day*. Lincoln: Nebraska State Journal, 1888.

Galloway, John A. "John Barber White and the Conservation Dilemma." *Forest History* 5 (1962), 9–16.

Gamble, Thomas (ed.). *Naval Stores: History, Production, Distribution, and Consumption*. Savannah, Ga.: Review Publishing and Printing, 1921.

Gannett, Henry. *The Forests of Oregon*. USGS, Professional Paper no. 4. Washington, D.C.: GPO, 1902.

Forests of the United States. Washington, D.C.: GPO, 1899. Also printed in U.S. Geological Survey, *19th Annual Report* (1897–8), pt. 5.

The Forests of Washington: A Revision of Estimates. USGS, Professional Paper no. 5. Washington, D.C.: GPO, 1902.

"The Influence of Forests on the Quantity and Frequency of Rainfall." *Science* 12 (23 Nov. 1888), 242–4.

Garren, K. H. "Effects of Fire on Vegetation of the Southeastern United States." *Botanical Review* 9 (1943), 617–54.

Gates, Paul W. *The Farmers' Age: Agriculture, 1815–1860*. Vol. 3 of *An Economic History of the United States*. New York: Holt, Rinehart and Winston, 1960.

"Federal Land Policies in the Southern Public Land States." *Agricultural History* 53 (1979), 206–27.

"Federal Land Policy in the South, 1866–1888." *Journal of Southern History* 6 (1940), 303–30.

The Illinois Central Railroad and Its Colonization Work. Harvard Economic Studies, vol. 42. Cambridge, Mass.: Harvard University Press, 1934.

"Pressure Groups and Recent American Land Policies." *Agricultural History* 55 (1981), 103–27.

"Problems of Agricultural History, 1790–1840." *Agricultural History* 46 (1972), 33–51.

"Weyerhaeuser and Chippewa Logging Industry." In Ander, P. Fritiof (ed.), *The John H.*

Hauberg Historical Essays, 50–64. Augustana Library Publications, no. 26. Rock Island, Ill.: Denkmann Memorial Library, Augustana College, 1954.

The Wisconsin Pine Lands of Cornell University: A Study in Land Policy and Absentee Ownership. Ithaca, N.Y.: Cornell University Press, 1943.

Gates, Paul W., with Swenson, Robert W. *History of Public Land Law Development*. Written for the Public Land Law Review Commission. Washington, D.C.: GPO, 1968.

Georgia State Board of Forestry. *Report to the Georgia General Assembly on Forestry in Georgia*, 9–10. 1922.

Gibson, D. L. *Socio-economic Evolution in a Timbered Area in Northern Michigan*. Michigan Agricultural Experiment Station, Tech. Bull. no. 193. Lansing, 1944.

Gillett, Charles A. "Citizens and Trade Associations Dealing with Forestry." In Winters, Robert K. (ed.), *Fifty Years of Forestry in the U.S.A.* Washington, D.C.: Society of American Foresters, 1950.

Gilligan, James P. "The Development of Policy and Administration of Forest Service Primitive and Wilderness Areas in the Western United States," Ph.D. diss., University of Michigan, 1954.

Glacken, Clarence J. "Changing Ideas of the Habitable World." In Thomas, William L. (ed.), *Man's Role in Changing the Face of the Earth*, 70–92. Chicago: University of Chicago Press, 1956.

Traces on the Rhodian Shore: Nature and Culture in Western Thought from Ancient Times to the End of the Eighteenth Century. Berkeley: University of California Press, 1967.

Glassie, Henry. *Folk Housing in Middle Virginia*. Knoxville: University of Tennessee Press, 1975.

Gleason, Herbert A. "The Individualistic Concept to the Plant Association."*Bulletin of Torrey Botanical Club* 53 (1926), 7–26.

"Vegetational History of the Middle West." *Annals, Association of American Geographers* 12 (1922), 78–85.

Glover, Wilbur H. "Lumber Rafting on the Wisconsin River." *Wisconsin Magazine of History* 25 (1941–2), 155–78, 308–29.

Godwin, Parke (ed.). *The Poetical Works of William Cullen Bryant*. 2 vols. New York: Russell and Russell, 1883.

Goldenberg, Joseph A. "With Saw and Axe and Auger: Three Centuries of American Shipbuilding." In Hindle, Brook (ed.), *Material Culture of the Wooden Age*, 97–128. Tarrytown, N.Y.: Sleepy Hollow Press, 1981.

Gordon, Thomas F. *A Gazetteer of the State of Pennsylvania*. Philadelphia: Belknap Press, 1832.

Gosse, Philip H. *Letters from Alabama, Chiefly Relating to Natural History*. London: Morgan and Chase, 1859.

Gould, Mary. *The Early American House*. Rutland, Vt.: Tuttle, 1965.

Graham, Frank. *Since Silent Spring*. London: Hamish, 1970.

Grant, Anne [McVickar]. *Memoirs of an American Lady: With Sketches of Manners and Scenes in America as They Existed Previous to the Revolution* (1808). 2nd ed. New York: D. Appleton, 1846.

Graves, Henry S. *A National Lumber and Forest Policy*. Address before American Lumber Company, Chicago, 16 Apr. 1919. USDA, Circular 134. Washington, D.C.: GPO, 1919.

"A Policy of Forestry for the Nation." *Journal of Forestry* 17 (1919), 901–10.

Practical Forestry in the Adirondacks. USDA, Forestry Division, Bull. no. 26. Washington, D.C.: GPO, 1899.

"Private Forestry." *Journal of Forestry* 17 (1919), 113–21.

The Use of Wood for Fuel. USDA, Bull. no. 753. Washington, D.C.: GPO, 1919.

Gray, Lewis C., and Thompson, Esther K. *History of Agriculture in the Southern United States to 1860*. 2 vols. Carnegie Institute Publication no. 430. Washington, D.C., 1933.

Greeley, Horace. *Glances at Europe: In a Series of Letters from Great Britain, France, Italy, Switzerland, etc., during the Summer of 1851*. New York: Dewitt and Davenport, 1851.

Recollections of a Busy Life: Including Reminiscences of American Politics and Politicians, to Which Is Added Miscellanies. New York: J. B. Ford, 1868.

What I Know of Farming: A Series of Brief and Plain Expositions of Practical Agriculture as an Art Based on Science. New York: G. W. Carleton, 1871.

Greeley, William B. *Forests and Men.* Garden City, N.Y.: Doubleday, 1951.

"Oregon Restores a Green Tillamook." *American Forests* 59 (1953), 12–14, 30, 43.

"Reduction of the Timber Supply through Abandonment or Clearing of Forest Lands." In U.S. Congress, Senate, *Report of the National Conservation Commission,* 2: 633–44. Washington, D.C.: GPO, 1909.

"The Relation of Geography to Timber Supply." *Economic Geography* 1 (1925), 1–11.

Some Public and Economic Aspects of the Lumber Industry in the United States. Studies in the Lumber Industry, pt. 1. USDA, Forest Service, Rep. no. 114. Washington, D.C.: GPO, 1917.

Greeley, William B.; Clapp, Earl; Smith, Herbert A.; Zon, Raphael; Sparhawk, W. N.; Shepard, Ward; and Kitteridge, J. "Timber: Mine or Crop?" In USDA, *Yearbook, 1922,* 83–183. Washington, D.C.: GPO, 1923.

Green, Evarts G., and Harrington, Virginia D. *American Population before the Federal Census of 1790.* New York: Columbia University Press, 1932.

Green, James, R. "The Brotherhood of Timber Workers, 1910–1913: A Radical Response to Industrial Capitalism in the Southern U.S.A." *Past and Present* 60 (1973), 161–200.

Greene, S. W. "The Forest that Fire Made." *American Forests* 37 (1931), 583–4, 618.

Gregg, Josiah. *Commerce of the Prairies; or, The Journal of a Santa Fe Trader.* 2 vols. New York, 1844. Another edition with an Introduction by Milo Milton Quaife, Lincoln: University of Nebraska Press, 1967. Also published in Thwaites, Reuben (ed.), *Early Western Travels, 1748–1846,* vols. 19, 20.

Gregory, John Goadby (ed.). *West Central Wisconsin: A History.* 4 vols. Indianapolis: S. G. Clarke, 1933.

Gulick, Luther H. *American Forest Policy: A Study of Government Administration and Economic Control.* New York: Duell, Sloan and Pearce, for the Institute of Public Administration, 1951.

Haites, Erik F., and Mak, James. "Steam Boating on the Mississippi, 1810–1860: A Purely Competitive Industry." *Business History Review* 45 (1971), 52–79.

Haites, Erik F.; Mak, James; and Walton, Gary M. *Western River Transportation: The Era of Early Internal Development, 1810–1860.* Baltimore: Johns Hopkins University Press, 1975.

Hall, Basil. *Forty Etchings from Sketches Made with the Camera Lucida in North America in 1827 and 1828.* Edinburgh: Cadell, 1829.

Travels in North America in the Years 1827 and 1828. 3 vols. Philadelphia: Cary, Lea, and Carey, 1829.

Hall, James. *Statistics of the West.* Cincinnati: J. A. James, 1836.

The West: Its Commerce and Navigation. Cincinnati: H. W. Derby, 1848.

Hall, William L., and Schrenk, Hermann von. *The Hardy Catalpa.* USDA, Forest Service, Bull. no. 37. Washington, D.C.: GPO, 1902.

Hamilton, Eloise. *Forty Years of Western Forestry: A History of the Movement to Conserve Forest Resources by Cooperative Effect, 1909–1949.* Portland, Ore.: Western Forestry and Conservation Association, 1949.

Handlin, Oscar. *Race and Nationality in American Life.* Garden City, N.Y.: Doubleday, 1957.

Harding, Benjamin. *A Tour through the Western Country, A.D. 1818 and 1819.* New London, Conn.: Samuel Green, 1819.

Harrison, Robert W. "Clearing Land in the Mississippi Alluvial Valley." *Arkansas Historical Quarterly* 13 (1954), 352–71.

Hart, John Fraser. "Land Use Change in a Piedmont County." *Annals, Association of American Geographers* 70 (1980), 492–527.

"Loss and Abandonment of Cleared Farm Land in the Eastern United States." *Annals, Association of American Geographers* 58 (1968), 417–40.

Hartesveldt, Richard J. "Fire Ecology of the Giant Sequoias: Controlled Fires May Be One Solution to the Survival of the Species." *Natural History Magazine* 73 (1964), 12–19.

Hartman, George B. "The Iowa Sawmill Industry." *Iowa Journal of History and Politics* 40 (Jan. 1942), 52–93.

Hartman, W. A., and Black, J. D. *Economic Aspects of Land Settlement in the Cut-over Region of the Great Lakes States.* USDA, Circular 160. 1931.

Hartt, Rollin Lynde. "Notes on a Michigan Lumber Town." *Atlantic Monthly* 85 (Jan. 1900), 101–9.

Harvey, George. *Harvey's Scenes of the Primeval Forests of America at the Four Periods of the Year.* London: G. Harvey, 1841. Reprint, Tarrytown, N.Y.: W. Abbott, 1925.

Hawkins, Benjamin. *A Sketch of the Creek Country in the Years 1798 and 1799.* Reprinted in *Collections of the Georgia Historical Society*, vol. 3, pt.1. Savannah, 1848.

Hawks, Graham. "Increase A. Lapham, Wisconsin's First Scientist." Ph.D. diss., University of Wisconsin, Madison, 1960.

Hayden, Ferdinand V. "The Geology of Nebraska." "Report of the Commission of the General Land Office," 1867, in U.S. Department of the Interior Annual Report, 1867, 40th Cong., 3rd sess., H. Ex. Doc. 1 (Serial no. 1326), 159–60. Washington, D.C.: GPO, 1867.

Hays, Samuel P. *Conservation and the Gospel of Efficiency: The Progressive Conservation Movement, 1890–1920.* Harvard Historical Monographs, no. 40. Cambridge, Mass.: Harvard University Press, 1959.

"The Structure of Environmental Politics since World War II." *Journal of Social History* 14 (1981), 719–38.

Hayter, Earl W. "Livestock-Fencing Conflicts in Rural America." *Agricultural History* 37 (1963), 10–20.

The Troubled Farmer, 1850–1900: Rural Adjustment to Industrialization. Dekalb: Northern Illinois University Press, 1968.

Hedrick, Ulysses Prentiss. *A History of Agriculture in the State of New York.* New York: J. B. Lyon for the New York State Agricultural Society, 1933.

Land of the Crooked Tree. New York: Oxford University Press, 1948.

Heidenreich, Conrad. *Huronia: A History and Geography of the Huron Indians, 1600–1650.* Ontario: McClelland and Stewart, 1971.

Heidt, William, Jr. *History of Rafting on the Delaware.* Paper read to meeting of the Minisink Valley Historical Society. Port Jervis: New York Gazette Printers, 1922.

Heimert, Alan. "Puritanism, the Wilderness, and the Frontier." *New England Quarterly* 26 (1953), 361–82.

Helgeson, Arlan C. *Farms in the Cutover: Agricultural Settlement in Northern Wisconsin.* Madison: State Historical Society, for the Department of History, University of Wisconsin, 1962.

"Nineteenth-Century Land Colonization in Northern Wisconsin." *Wisconsin Magazine of History* 36 (1953), 115–21.

Hendrickson, C. I. "The Agricultural Land Available for Forestry." In U.S. Congress, Senate, *A National Plan for American Forestry* [Copeland Report], 1:151–69, 1933.

Hening, William Walter (ed.). *Virginia: The Statutes at Large; Being a Collection of All the Laws of Virginia from the First Session of the Legislature, in the Year 1619.* 13 vols. New York: printed for the editor, 1819–23; Charlottesville: University Press of Virginia, for the Jamestown Foundation of the Commonwealth of Virginia, 1969.

Henlein, Paul C. "Cattle Driving from the Ohio Country, 1800–1850." *Agricultural History* 28 (1954), 90–101.

Henretta, James. "Families and Farms: *Mentalité* in Pre-industrial America." *William and Mary Quarterly*, 3rd ser. 35 (1978), 3–32.

Henry, Joseph. "Meteorology and Its Connection with Agriculture." In U.S. Congress, House,

Report of the Commissioner of Patents for 1858, 35th Cong., 1st sess., 1858, H. Ex. Doc. 32 (serial no. 954), 429–93 and frontispiece.

Henry, William Arnon. *Northern Wisconsin: A Handbook for the Homeseeker*. Madison, Wisc.: Democrat Printing, State Printer, 1896.

Herndon, Melvin G. "The Forest Products of Colonial Georgia." *Journal of Forest History* 23 (1979), 130–5.

"Naval Stores in Colonial Georgia." *Georgia Historical Quarterly* 52 (1968), 426–33.

"The Significance of the Forest to the Tobacco Plantation Economy of Antebellum Virginia." *Plantation Society* 2 (1981), 230–9.

Hewitt, J. N. B. "The Iroquoian Cosmology." In *43rd Annual Report of the Bureau of American Ethnology, 1925*, pt. 2, 449–63. Washington, D.C.: GPO, 1926.

Heyward, Frank. "austin Cary: Yankee Peddler in Forestry." *American Forests* 61 (May 1955), 29–30, 43–4, 61; (June 1955), 28–9, 52–3.

"Planting the Largest Man-Made Forest in the World." In *Forest and People*, Golden Anniversary Issue, 1963.

Hibbard, Benjamin H. *A History of the Public Land Policies*. New York: Macmillan, 1924. Reprinted with a foreword by Paul W. Gates. Madison: University of Wisconsin Press, 1965.

Hickman, Nollie W. "Logging and Rafting Timber in South Mississippi, 1840–1910." *Journal of Mississippi History* 19 (1952), 154–72.

Mississippi Harvest: Lumbering in the Longleaf Pine Belt, 1840–1915. University: University of Mississippi, 1962.

"Mississippi Lumber Industry from 1840 to 1950." *Southern Lumberman* 193 (15 Dec. 1956), 132–7.

"The Yellow Pine Industries in St. Tammany, Tangipahoa, and Washington Parishes, 1840–1915." *Louisiana Studies* 5 (1966), 75–88.

Hidy, Ralph W.; Hill, Frank Ernest; and Nevins, Allan. *Timber and Men: The Weyerhaeuser Story*. New York: Macmillan, 1963.

Higgins, F. Hal. "Logging with Tractors in the '80s." *Timberman* 48 (1947), 68, 116.

Higginson, Francis H. *New-England's Plantation: or, A Short and True Description of the Commodities and Discommodities of That Country*. London: T.C. and R.C., for M. Sparke, 1630. Reprinted in Force, P. (ed.), *Tracts, and Other Papers*, vol. 1, no. 12. Washington, D.C.: P. Force, 1836–41.

Highsmith, Richard M., Jr. *Atlas of the Pacific Northwest: Resources and Development*. 4th ed. Corvallis, Ore.: State University Press, 1968.

Highsmith, Richard M., Jr., and Beh, John C. "Tillamook Burn: The Regeneration of a Forest." *Scientific Monthly* 75 (1952), 139–48.

Hilliard, Sam Bass. *Hogmeat and Hoecake: Food Supply in the Old South, 1840–1860*. Carbondale: Southern Illinois University Press, 1972.

Hillyard, M. B. *The New South: A Description of the Southern States, Noting Each State Separately*. Baltimore: Manufacturers' Record Company, 1887.

Hindle, Brooke (ed.). *America's Wooden Age: Aspects of Its Early Technology*. Tarrytown, N.Y.: Sleepy Hollow Restorations, 1975.

Material Culture of the Wooden Age. Tarrytown, N.Y.: Sleepy Hollow Press, 1981.

Hipple, Walter J. *The Beautiful, the Sublime, and the Picturesque in Eighteenth Century British Aesthetics*. Carbondale: Southern Illinois University Press, 1957.

Hodgson, Adam. *Letters from North America Written during a Tour in the United States and Canada*. 2 vols. London: Hurst, Robinson, 1824.

Remarks during a Journey through North America, in 1819, 1820, and 1821. New York: Samuel Whiting, 1823.

Hodgson, Allen H. *Logging Waste in the Douglas Fir Region*. Portland, Ore.: West Coast Lumberman, 1930.

Hofstadter, Richard. *The Progressive Historians: Turner, Beard, Parrington*. New York: Alfred A. Knopf, 1968.

Hoglund, A. William. "Forest Conservation and Stove Inventors, 1789–1850." *Forest History* 5 (1962), 2–8.

Holbrook, Stewart H. *Burning an Empire. The Story of American Forest Fires*. New York: Macmillan, 1943.

"The Great Hinckley Fire." *American Mercury* 57 (Sept. 1943), 348–55.

Holy Old Mackinaw: A Natural History of the American Lumberjack. New York: Macmillan, 1938. Reprinted as *The American Lumberjack*, 1962.

Hoover, Calvin B., and Ratchford, B. U. *Economic Resources and Policies of the South*. New York: Macmillan, 1951.

Hopkins, Arthur S. "Within and without the Blue Line." *New York State Conservationist*, Apr.–May 1951, 10–11.

Horn, Stanley F. *This Fascinating Lumber Business* (1943). 2nd ed. Indianapolis: Bobbs-Merrill, 1951.

Horn, Stanley F., and Crawford, Charles W. "Perspectives on Southern Forestry: The *Southern Lumberman*, Industrial Forestry, and Trade Association." *Journal of Forest History* 21 (1977), 18–30.

Hosmer, Ralph S. "Franklin B. Hough: Father of American Forestry." *North Country Life* 6 (1952), 16–20.

Hotchkiss, George W. *History of the Lumber and Forest Industry of the Northwest*. Chicago: George W. Hotchkiss, 1898.

Industrial Chicago: The Lumber Interests. Chicago: Goodspeed Publishing, 1894.

Hough, Franklin B. "The Decrease of Woodlands in Ohio." In Egleston, Nathaniel H. (ed.), *Report upon Forestry*, 4:174–80. Washington, D.C.: GPO, 1884.

The Elements of Forestry. Cincinnati: Robert Clarke, 1882.

"On the Duty of Governments in the Preservation of Forests." *Proceedings of the American Association for the Advancement of Science*, 1–22. Portland, Me., 1873 (1874). Also printed in U.S. Congress, House, *Cultivation of Timber and the Preservation of Forests*, 43rd Cong, 1st sess., 1874, H. R. 259 (serial no. 1623).

"Report on Kinds and Quantities of Timber Used for Railroad Ties." In Egleston, Nathaniel H. (ed.), *Report upon Forestry*, 4:119–73, Washington, D.C.: GPO, 1884.

Report upon Forestry. Submitted to Congress by the Commissioner of Agriculture. Vol. 1 (1878); vol. 2 (1880); and vol. 3 (1882). Vol. 4 (1884) was prepared by Nathaniel H. Egleston. Washington, D.C.: GPO.

Howells, William C. *Recollections of Life in Ohio from 1813 to 1840*. Cincinnati: Robert Clarke, 1895.

Hudson, G. M. "A Study of a Permanent Alabama Lumber Town." *Journal of Geography* 36 (1937), 310–15.

Huizinga, Johan. "The Task of Cultural History." In *Men and Ideas: History, the Middle Ages, the Renaissance: Essays*. Translated by James S. Holmes and Hans Van Marle. New York: Meridian Books, 1959.

Hunter, Byron, and Thompson, Harry. *The Utilization of Logged-off Land for Pasture in Western Oregon and Western Washington*. USDA, Farmers' Bull. no. 462. Washington, D.C.: GPO, 1911.

Hunter, Louis C. "The Heavy Industries before 1860." In Williamson, Harold F. (ed.), *The Growth of the American Economy*, 172–89. New York: Prentice-Hall, 1944; 2nd ed., 1951.

"The Influence of the Market upon Technique in the Iron Industry of Western Pennsylvania up to 1860." *Journal of Economic and Business History* 1 (1929), 241–81.

Steamboats on the Western Rivers: An Economic and Technological History. Cambridge, Mass.: Harvard University Press, 1949.

Hurst, James Willard. "The Institutional Environment of the Logging Era in Wisconsin." In

Flader, Susan L. (ed.), *The Great Lakes Forest: An Environmental and Social History*, 137–55. Minneapolis: University of Minnesota Press, 1983.

Law and Economic Growth: The Legal History of the Lumber Industry in Wisconsin, 1836–1915. Cambridge, Mass.: Harvard University Press (Belknap Press), 1964.

Hussey, Christopher. *The Picturesque: Studies in a Point of View* (1927). 2nd ed. London: Frank Cass, 1967.

Hutchinson, Thomas (ed.). "Copy of a Narrative of the Commissioners from England, about New England" (1665). In *A Collection of Original Papers Relative to the History of the Colony of Massachusetts Bay*. Boston: Thomas and John Fleet, 1769; reprint, Albany, 1865.

Hutchinson, William H. "The Sierra Flume & Lumber Company of California, 1875–1878." *Forest History* 17 (1973), 14–20.

Huth, Hans. "Yosemite: The Story of an Idea." *Sierra Club Bulletin* 33 (Mar. 1948), 47–78.

Hynning, Clifford J. *State Conservation of Resources*. Prepared for the National Resources Committee. Washington, D.C.: GPO, 1939.

Illick, Joseph S. "The Story of the American Lumbering Industry." In Kaempffert, Waldemar (ed.), *A Popular History of American Invention*, 2:150–98. New York: Charles Scribner and Sons, 1924.

Indiana Agricultural Report. Agricultural Experimental Station, Bloomington, 1888– .

Irving, Washington. *Rip Van Winkle, and The Legend of Sleepy Hollow* (1820). 1848 ed. reproduced, Tarrytown, N.Y.: Sleepy Hollow Restorations, 1974.

Ise, John. *Our National Park Policy: A Critical History*. Baltimore: Johns Hopkins University Press, for Resources for the Future, 1961.

The United States Forest Policy. New Haven: Yale University Press, 1920.

Jackson, Kemper. *American Charcoal Making in the Era of the Cold Blast Furnace*. National Park Service, Population Studies Series Hist., no. 14. Washington, D.C.: GPO, 1941.

Jacobs, Wilbur R. "The Indian and the Frontier in American History: A Need for Revision." *Western Historical Quarterly* 4 (1973), 50–6.

"The Tip of the Iceberg: Pre-Columbian Indian Demography and Some Implications for Revisionism." *William and Mary Quarterly* 31 (1974), 123–33.

Jacobsen, Edna L. "Franklin B. Hough: A Pioneer in Scientific Forestry in America." *New York History* 15 (1934), 311–25.

James, Lee M. "Restrictive Agreements and Practices in the Lumber Industry, 1880–1939." *Southern Economic Journal* 13 (1946), 115–25.

Jefferson, Thomas. *Notes on the State of Virginia*. Edited by William Peden. Chapel Hill: University of North Carolina Press, 1955.

Jensen, Vernon H. *Lumber and Labor*. New York: Farrar and Rinehart, 1945.

Johnson, Edward. *Johnson's Wonder-Working Providence of Sion's Saviour in New England, 1628–1681*. Edited by John F. Jameson. Original Documents of Early American History. New York: Charles Scribner and Sons, 1910.

Johnson, Hildegard B. "Carl Schurz and Conservation." *American-German Review* 23 (Oct.–Nov. 1956), 4–8.

Johnson, Robert C. "Logs for Saginaw: The Development of Raft-Towing on Lake Huron." *Inland Seas* 5 (1949), 37–41, 83–90.

Jolley, Harley E. "Biltmore Forest Fair, 1908." *Forest History* 14 (1970), 6–17.

Jones, Holway R. *John Muir and the Sierra Club: The Battle for Yosemite*. San Francisco: Sierra Club, 1965.

Jones, Hugh. *The Present State of Virginia, from Whence Is Inferred a Short View of Maryland and North America* (1753). Edited by R. L. Morton. Chapel Hill: University of North Carolina Press, for the Virginia Historical Society, 1956.

Jones, Robert L. "The Beef Cattle Industry." *Ohio Archaeological and Historical Quarterly* 64 (1955), 170–92.

Jordan, Terry. "Between the Forest and the Prairie." *Agricultural History* 38 (1964), 205–16.

Judd, Sylvester. *History of Hadley, Massachusetts*. Springfield, Mass.: H. R. Hunting, 1863; 2nd enlarged ed., 1905.

Kaatz, Martin Richard. "The Black Swamp: A Study in Historical Geography." *Annals, Association of American Geographers* 45 (1955), 1–35.

Kalm, Pehr. *Travels into N. America, Containing Its Natural History and a Circumstantial Account of Its Plantations and Agriculture*. Translated by John Reinhold Forester. 3 vols. London: William Eyres, 1770–1.

Kane, Lucile. "Selling Cut-over Lands in Wisconsin." *Business History Review* 28 (1954), 236–47.

Kaufman, Henry J. *The American Farmhouse*. New York: Hawthorne, 1975.

Keenan, Hudson. "America's First Successful Logging Railroad." *Michigan History* 44 (1960), 292–302.

Keffer, Charles A. *Experimental Tree Planting in the Plains*. USDA, Forestry Division, Bull. no. 16. Washington, D.C.: GPO, 1898.

Kellogg, Royal S. *The Timber Supply of the United States*. USFS, Circular 166. Washington, D.C.: GPO, 1909.

Kensel, William H. "The Early Spokane Lumber Industry, 1871–1910." *Idaho Yesterdays* 12 (1968), 25–31.

Kern, M. G. *The Relation of Railroads to Forest Supplies and Forestry*. USDA, Forestry Division, Bull. no. 1. Washington, D.C.: GPO, 1887.

Kettell, Samuel (ed.). *Specimens of American Poetry, with Critical and Biographical Notices*. 3 vols. New York: Benjamin Blom, 1967.

Kilgore, Bruce M. "From Fire Control to Fire Management: An Ecological Basis for Policies." *Transactions: 41st North American Wildlife and Natural Resources Conference*, 477–93. Washington, D.C.: Wildlife Management Institute, 1976.

"Restoring Fire to National Park Wilderness." *American Forests* 81 (Mar. 1975), 16–19, 57–9.

King, Helene. "The Economic History of the Long-Bell Lumber Company." M.A. thesis, Louisiana State University, 1936.

King, John O. *The Early History of the Houston Oil Company of Texas, 1901–1908*. Vol. 3, no. 1. Houston: Texas Gulf Coast Historical Association Publications, 1959.

Kinney, Jay P. *Forest Legislation in America Prior to March 4, 1789*. Cornell University, Agricultural Experimental Station of the New York College of Agriculture, Department of Forestry, Bull. no. 370, 359–405. Ithaca, N.Y., 1916.

Kipling, Rudyard. *From Sea to Sea*. New York: Doubleday and McClure, 1899.

Kirby, John H. "The Timber Resources of East Texas. Their Recognition and Development." *American Lumberman* 68 (22 Nov. 1902), 43–78.

Kirkland, Herbert Donald. "The American Forests, 1864–1898: A Trend towards Conservation." Ph.D. diss., Florida State University, 1971.

Kleven, Bernhardt J. "The Mississippi River Logging Company." *Minnesota History* 27 (1946), 190–202.

Klippart, John H. "Condition of Agriculture in Ohio in 1876." *Ohio Agricultural Report*, 2nd ser. (1876), 486–538.

Kniffen, Fred B. "Folk Housing: Key to Diffusion." *Annals, Association of American Geographers* 55 (1965), 549–77.

Louisiana: Its Life and People. Baton Rouge: University of Louisiana Press, 1968.

Kniffen, Fred, and Glassie, Henry. "Building in Wood in the Eastern United States: A Time–Place Perspective." *Geographical Review* 56 (1966), 40–66.

Koch, Elers. *When the Mountains Roared: Stories of the 1910 Fire*. Coeur d'Alene, Idaho: USFS, Coeur d'Alene National Forest, 1942.

Kohlmeyer, Frederick W. "Northern Pine Lumbermen: A Study in Origins and Migrations." *Journal of Economic History* 16 (1956), 529–38.

Kollmorgen, Walter M. "The Woodsman's Assault on the Domain of the Cattleman." *Annals, Association of American Geographers* 59 (1969), 215–39.

Kollmorgen, Walter M., and Kollmorgen, Johanna. "Landscape Meteorology in the Plains Area." *Annals, Association of American Geographers* 63 (1973), 424–41.

Komarek, E. V., Sr. "Comments on the History of Controlled Burning in the Southern United States." *Proceedings, 17th Annual Watershed Symposium* 17 (1973), 11–17.

A Quest for Ecological Understanding: The Secretary's Review. Tall Timbers Research Station, Publication no. 5. Tallahassee, Fla., 1977.

Konvalinka, Joseph G. *How to Improve the Climate of the United States and Other Countries*. Long Island City, N.Y., 1889.

Kortum, Karl, and Olmstead, Roger. ". . . It Is a Dangerous Looking Place: Sailing Days on the Redwood Coast." *California Historical Quarterly* 50 (Mar. 1971), 43–58.

Kozlowski, Theodore T., and Ahlgren, C. E. (eds.). *Fire and Ecosystems*. New York: Academic Press, 1974.

Kreps, Theodore J. "Vicissitudes of the American Potash Industry." *Journal of Economic and Business History* 3 (1931), 630–66.

Kroeber, A. L. *Cultural and Natural Areas of Native North America*. University of California Publications in American Archaeology and Ethnology, 38. Berkeley, 1939.

Krog, Carl E. "Lumber Ports of Marinette–Menominee in the Nineteenth Century." *Inland Seas* 28 (1972), 272–80.

Küchler, A. W. *Manual to Accompany the Map of Potential Natural Vegetation of the Coterminous United States*. American Geographical Society, Special Publication no. 36. New York, 1964. A condensed commentary and smaller-scale map appear in *The National Atlas of the United States of America*, plate. 89. U.S. Department of the Interior, Geological Survey. Washington, D.C.: GPO, 1970.

Kury, Theodore William. "Historical Geography of the Iron Industry in the New York–New Jersey Highlands, 1700–1900." Ph.D. diss., Louisiana State University, 1968.

Kutzleb, Charles R. "American Myth, Desert to Eden: Can Forests Bring Rain to the Plains?" *Forest History* 15 (1971), 14–21.

"Rain Follows the Plow: The History of an Idea." Ph.D. diss., University of Colorado, 1968.

Kuznets, Simon Smith. "Notes on the Take-off." In Rostow, Walt W. (ed.), *The Economics of Take-off into Sustained Growth*. Proceedings of Conference held by the International Economic Association. New York: St. Martin's Press, 1963.

Lafitau, Jo Françoise. *Customs of the American Indians Compared with Those of the Customs of Primitive Times*. Edited by William N. Fenton and Elizabeth L. Moore. 2 vols. Toronto: Champlain Society, 1974. Based on *Moeurs, des Sauvages Ameriquains, compares aux moeurs des premier temps*. Paris, 1724.

Langhorne, William T., Jr. "Mill Based Settlement Patterns in Schoharie County, New York: A Regional Study." *Historical Archaeology* 10 (1976), 73–92.

Langille, H. D.; Plummer, Fred; Dodwell, Arthur; Rixon, Theodore F.; and Leiberg, John B. *Forest Conditions on the Cascade Range Forest Reserve, Oregon: With Introduction by Henry Gannett*. USGS, Professional Paper no. 9. Washington, D.C.: GPO, 1903.

Lapham, Increase A.; Knapp, J. G.; and Crocker, H. *Report on the Disastrous Effects of the Destruction of Forest Trees Now Going On So Rapidly in the State of Wisconsin*. Madison: Atwood and Rublee, State Printers, 1867; reprint, Madison: State Historical Society of Wisconsin, 1967.

Larkin, Oliver W. *Art and Life in America*. New York: Rinehart, 1949.

Larkin, Stillman Carter. *The Pioneer History of Meigs County*. Columbus: Berlin Printing, 1908.

La Rochefoucauld-Liancourt, duc de. *Travels through the United States of North America and the Country of the Iroquois and Upper Canada in the Years 1795, 1796, and 1797*. 2 vols. London: R. Phillips, 1799. Another edition, 4 vols., London, 1810.

Larson, Agnes M. *History of the White Pine Industry in Minnesota*. Minneapolis: University of Minnesota Press, 1949.

Lasswell, Mary. *John Henry Kirby, Prince of the Pines*. Austin: Encino Press, 1967.

Lathan, Bryan. *Timber, Its Development and Distribution: A Historical Survey*. London: George G. Harrap, 1957.

Leach, Carl A. "Paul Bunyan's Land and the First Sawmills of Michigan." *Michigan History* 20 (1936), 69–89.

Lederer, John. *The Discoveries of John Lederer in Three Marches from Virginia to the West of Carolina and Other Parts of the Continent, 1669 and 1670*. Collected and translated by Sir William Talbot. London, 1672; reprint, Rochester, N.Y., 1902.

Lee, Daniel. "Agricultural Meteorology." In U.S. Congress, House, *Report of the Commissioner of Patents for 1850*, 31st Cong., 1st sess., 1850, H. Ex. Doc. 243 (serial no. 630), pt. 2, pp. 40–41.

Lee, James M. "Property Taxes and Alternatives for Michigan." *Journal of Forestry* 58 (1960), 86–92.

Leiberg, John B. *Forest Conditions in the Northern Sierra Nevada, California*. U.S. Geological Survey, Professional Paper no. 8. Washington, D.C.: GPO, 1902.

Le Master, Dennis C. *Mergers among the Largest Forest Products Firms, 1950–1970*. Washington State University, College of Agricultural Research Center, Bull. no. 854, Pullman, Wash., 1977.

Lemly, James Hutton. *The Gulf, Mobile, and Ohio: A Railroad that Had to Expand or Expire*. Homewood, Ill.: R. D. Irwin, 1953.

Lemon, James T. *The Best Poor Man's Country: A Geographical Study of Early Southeastern Pennsylvania*. Baltimore: Johns Hopkins University Press, 1972.

"Early Americans and Their Social Environment." *Journal of Historical Geography* 6 (1980), 115–32.

"The Weakness of Place and Community in Early Pennsylvania." In Gibson, James R. (ed.), *European Settlement and Development in North America: Essays on Geographical Change in Honour and Memory of Andrew Hill Clark*, 190–207. Toronto: Toronto University Press, 1978.

Lesley, J. Peter. *The Iron Manufacturer's Guide to the Furnaces, Forges, and Rolling Mills of the United States*. New York: John Wiley, 1859.

Leue, Adolph (ed.). *First Annual Report of the Ohio State Forestry Bureau: For the Year 1885*. Columbus: Westbote, 1886.

Lewis, Peirce. "The Geography of Old Houses." *Earth and Mineral Sciences* 39 (1970), 33–7.

Lewis, Richard. "Tree Farms." In Davis, Richard (ed.), *Encyclopaedia of America Forest and Conservation History*, 2: 654–6. New York: Macmillan, 1983.

Lewis, Richard W. B. *The American Adam: Innocence, Tragedy, and Tradition in the Nineteenth Century*. Chicago: University of Chicago Press, 1955.

Lilienthal, David E. *TVA: Democracy on the March*. New York: Harper and Brothers, 1944.

Lillard, Richard G. *The Great Forest*. New York: Alfred A. Knopf, 1948.

Lockmann, Ronald F. *Guarding the Forests of Southern California: Evolving Attitudes toward Conservation of Watersheds, Woodlands, and Wilderness*. Glendale, Calif.: Arthur H. Clarke, 1981.

Loehr, Rodney C. "Moving Back from the Atlantic Seaboard." *Agricultural History* 17 (1943), 90–6.

"Saving the Kerf: The Introduction of the Band Saw Mill." *Agricultural History* 23 (1949), 168–72.

"Self-sufficiency on the Farm." *Agricultural History* 26 (1952), 37–41.

Long, Ellen. "Forest Fires in Southern Pine." *Forest Leaves* 2 (1889), 29.

Lorain, John. *Nature and Reason Harmonized in the Practice of Husbandry*. Philadelphia: H. C. Carey and I. Lea, 1825.

Lord, Eleanor L. *Industrial Experiments in the British Colonies of North America*. Johns Hopkins University Studies in Historical and Political Science, extra vol. no. 17. Baltimore: Johns Hopkins University Press, 1898.

Lord, N. W. "Iron Manufacture in Ohio." *Report of the Geological Survey of Ohio Economic Geology* 5 (1884), 438–554.

Lossing, Benson J. *The Hudson from the Wilderness to the Sea.* New York: Virtue and Yorston, 1866.

Lowenthal, David. "The American Scene." *Geographical Review* 58 (1968), 61–88.

"George Perkins Marsh and the American Geographical Tradition." *Geographical Review* 43 (1953), 207–13.

George Perkins Marsh: Versatile Vermonter. New York: Columbia University Press, 1958.

"On the Author of *Man and Nature.*" *American Forests* 83 (Sept. 1977), 6, 8–11, 44–8.

Lowenthal, David, and Prince, Hugh. "English Landscape Tastes." *Geographical Review* 55 (1965), 186–22.

Lucia, Ellis J. "A Lesson from Nature: Joe Cox and His Revolutionary Chain Saw." *Journal of Forest History* 25 (1981), 158–65.

Lundsted, James E. "Log Marks: Forgotten Lore of the Logging Era." *Wisconsin Magazine of History* 39 (1955), 44–6.

Lutz, Harold J. "The Vegetation of Heart's Content: A Virgin Forest in North Western Pennsylvania." *Ecology* 11 (1930), 1–29.

McCloskey, J. Michael. "Note and Comment. Natural Resources–National Forests: The Multiple Use–Sustained Yield Act of 1960." *Oregon Law Review* 41 (Dec. 1961), 49–78.

McGeary, Martin N. *Gifford Pinchot, Forester-Politician.* Princeton, N.J.: Princeton University Press, 1960.

McIntosh, C. Barron. "Use and Abuse of the Timber Culture Acts." *Annals, Association of American Geographers* 65 (1975), 347–62.

McKittrick, Reuben. *The Public Land System of Texas, 1823–1910.* Madison: University of Wisconsin Press, 1918. Also published as *Bulletin of the University of Wisconsin,* no. 905, Economics and Political Sciences Series, vol. 9, no. 1.

McLeod, W. C. "Fuel and Early Civilization." *American Anthropologist,* n.s. 27 (1925), 344–6.

McManis, Douglas R. "The Initial Evaluation and Utilization of the Illinois Prairies, 1815–1840." University of Chicago, Department of Geography Research Paper no. 94. Chicago, 1964.

McNall, Neil A. *The Agricultural History of the Genessee Valley, 1790–1860.* Philadelphia: University of Pennsylvania Press, 1952.

McNealy, Terry A. "Rafting on the Delaware: New Light from Old Documents." *Bucks County Historical Journal* (1977), 27–9.

McQuire, A. J. *Land Clearing.* University of Minnesota Agricultural Experimental Station Bull. no. 134. St. Paul, 1913.

McWhiney, Grady. "Louisiana Socialists in the Early Twentieth Century: A Study in Rustic Radicalism." *Journal of Southern History* 20 (1954), 315–36.

Madison, James. *Letters and Other Writings of James Madison Ordered to Be Printed by Congress.* 4 vols. Philadelphia: J. B. Lippincott, 1865.

Magee, David F. "Rafting on the Susquehanna." *Papers of the Lancaster [Pa.] County Historical Society* 24 (Nov. 1920), 193–202.

Malone, Joseph J. *Pine Trees and Politics: The Naval Stores and Forest Policy in Colonial New England, 1691–1775.* Seattle: University of Washington Press, 1964.

Mancil, Ervin. "An Historical Geography of Industrial Cypress Lumbering in Louisiana." Ph.D. diss., Louisiana State University, 1972.

"Pullboat Logging." *Journal of Forest History* 24 (1980), 135–41.

"Some Historical and Geographical Notes on the Cypress Lumbering Industry of Louisiana." *Louisiana Studies* 8 (1969), 14–25.

Marsh, George Perkins. *Address Delivered before the Agricultural Society of Rutland County, Sept. 30th, 1847.* Rutland, Vt.: Rutland Herald, 1848.

Man and Nature: Or Physical Geography as Modified by Human Action. New York: Scribner,

1864. Another edition, with introduction by David Lowenthal, Cambridge, Mass.: Harvard University Press (Belknap Press), 1965.

Marsh, Philip M. (ed.). *The Prose of Philip Freneau*. New Brunswick, N.J.: Scarecrow Press, 1955.

Marshall, Robert. *The People's Forests*. New York: Harrison Smith and Robert Haas, 1933.

Martin, Calvin. "Fire and Forest Structure in the Aboriginal Eastern Forest." *Indian Historian* 6 (1973), 23-26, 38–42, 54.

Martin, Clyde S. "History and Influence of the Western Forestry and Conservation Association on Cooperative Forestry in the West." *Journal of Forestry* 43 (Mar. 1945), 167–70.

Martin, Lawrence. *The Physical Geography of Wisconsin*. Wisconsin Geological and Natural History Society, Bull. no. 36. Madison, 1910.

Marx, Leo. *The Machine in the Garden: Technology and the Pastoral Ideal in America*. New York: Oxford University Press, 1964.

Mason, Alpheus T. *Bureaucracy Convicts Itself: The Ballinger–Pinchot Controversy of 1910*. New York: Viking Press, 1941.

Mason, David T. "The Effect of O&C Management on the Economy of Oregon." *Oregon Historical Quarterly* 64 (Mar. 1963), 55–67.

Forests for the Future: The Story of Sustained Yield as Told in the Diaries and Papers of David T. Mason, 1907–1950. Edited with introduction and notes by Rodney C. Loehr. Forest Products History Foundation Series, no. 5. St. Paul: Forest Products History Foundation and Minnesota Historical Society, 1952.

Massachusetts State Agricultural Society. *Abstract from the Returns of the Agricultural Societies in Massachusetts*. 1846–47.

Massey, Richard Walter. "A History of the Lumber Industry in Alabama and West Florida, 1880–1914." Ph.D. diss., Vanderbilt University, 1960.

"Labor Conditions in the Lumber Industry in Alabama, 1880–1914." *Journal of the Alabama Academy of Science* 37 (1966), 172–81.

"Logging Railroads in Alabama, 1880–1914." *Alabama Review* 14 (1961), 41–50.

Mather, Increase. *A Brief History of the War with the Indians in New England*. Boston: John Foster, 1676.

Mattoon, Wilbur R. *The Southern Cypress*. USDA, Bull. no. 272. Washington, D.C.: GPO, 1915.

Maunder, Elwood R. "Building on Sawdust." *Pacific Northwest Quarterly* 51 (1960), 57–62.

Maxwell, Hu. "The Use and Abuse of the Forests by the Virginia Indians." *William and Mary College Quarterly* 19 (1910), 73–104.

Maxwell, Robert S. "The Impact of Forestry on the Gulf South." *Forest History* 17 (1973), 30–5.

"One Man's Legacy: W. Goodrich Jones and Texas Conservation." *Southwestern Historical Quarterly* 77 (1974), 355–80.

"Researching Forest History in the Gulf Southwest: The Unity of the Sabine Valley." *Louisiana Studies* 10 (1971), 109–22.

Whistle in the Piney Woods: Paul Bremond and the Houston, East and West Texas Railway. Texas Gulf Coast Historical Association Publication Series, vol. 7, no. 2. Houston, 1963.

Maybee, Rolland H. "Michigan's White Pine Era 1840–1900." *Michigan History* 34 (1959), 385–431.

Mead, Walter J. *Mergers and Economic Concentration in the Douglas-Fir Lumber Industry*. Portland, Ore.: USFS, Pacific Northwest Forest and Range Experimental Station, 1964.

Meany, Edmond S. "A History of the Lumber Industry in the Pacific Northwest to 1917." Ph.D. diss., Harvard University, 1935.

Meining, Donald W. "The Geography of Expansion." In Thompson, John W. (ed.), *The Geography of New York State*, 140–71. Syracuse, N.Y.: Syracuse University Press, 1966.

"The Growth of Agricultural Regions in the Far West, 1850–1910." *Journal of Geography* 54 (1955), 221–32.

Melendy, Howard Brett. "One Hundred Years of the Redwood Lumber Industry, 1850–1950." Ph.D. diss., Stanford University, 1953.

"Two Men and a Mill: John Dolbeer, William Carson, and the Redwood Lumber Industry in California." *California Historical Society Quarterly* 38 (1959), 59–71.

Merchants' Association of New York. *The Natural Resources of Texas.* Report of an examination made by a special committee of the Merchants' Association of New York, by invitation of the governor and legislature of Texas. Buffalo, 1901.

Merrens, Harry R. *Colonial North Carolina in the Eighteenth Century: A Study in Historical Geography.* Chapel Hill: North Carolina University Press, 1964.

Merriam, Willis B. "Forest Situation in the Pacific Northwest." *Economic Geography* (1938), 103–8.

Merrill, Perry H. *Roosevelt's Forest Army: A History of the Civilian Conservation Corps, 1933–1942.* Montpelier, Vt.: P. H. Merrill, 1981.

Mesick, Jane L. *The English Traveller in America, 1785–1835.* New York: Columbia University Press, 1922.

Michaux, André. *Flora Boreali-Americana.* 2 vols. Paris: Frates Levrault, 1803.

Historie des chênes de l'Amérique Septentrionale. Paris, 1801. Translated by Walter Wade as *Quercus, or, Oaks, from the French of Michaux.* Dublin: Graisberry and Campbell, 1809.

Michaux, François André. *Travels to the West of the Allegheny Mountains.* London, 1805. Reprinted in Thwaites, Reuben G. (ed.), *Early Western Travels, 1748–1846,* vol. 3. Cleveland: Arthur H. Clark, 1904.

Michaux, François André, and Nuttall, Thomas. *The North American Sylva; or, A Description of the Forest Trees of the United States, Canada, and Nova Scotia.* Translated from the French. 3 vols. Paris: C. D'hautel, 1819; reprinted 1841, 1857, 1865.

Miller, Harry. "Potash from Wood Ashes: Frontier Technology in Canada and the United States." *Technology and Culture* 21 (1980), 187–208.

Miller, Perry. *Errand into Wilderness.* Cambridge, Mass.: Harvard University Press (Belknap Press), 1956.

Millet, Donald J. "The Lumber Industry of 'Imperial' Calcasieu, 1865–1900." *Louisiana History* 7 (1966), 51–64.

Millikin, Daniel. "The Best Practical Means of Preserving and Restoring the Forests of Ohio." *Ohio Agricultural Report,* 2nd ser. (1871), 319–33.

Minto, John. "From Youth to Age as an American." *Oregon Historical Quarterly* 9 (1908), 140–2, 154, 164–72, 375–87.

A Paper on Forestry Interests. Salem, Ore.: W. H. Leeds, State Printer, 1898.

Mitchell, Alfred J. "Accomplishments in Fire Protection in the Lake States." *Journal of Forestry* 37 (1939), 748–50.

Mittelman, Edward B. "The Gyppo System." *Journal of Political Economy* 31 (Dec. 1923), 840–51.

Mohr, Charles. "The Interest of the Individual in Forestry in View of the Present Condition of the Lumber Industry." *Proceedings of the American Forestry Congress* (1889), 36–38.

The Timber Pines of the Southern United States. USDA, Forestry Division, Bull. no. 13. Washington, D.C.: GPO, 1897.

Mooney, James. "The Aboriginal Population of America North of Mexico." In Swanton, J. R. (ed.), *Smithsonian Institution Miscellaneous Collections,* vol. 80, no. 7. Washington, D.C.: GPO, 1928.

Moore, John Hebron. *Andrew Brown and Cypress Lumbering in the Old Southwest.* Baton Rouge: Louisiana State University Press, 1967.

"William H. Mason: Southern Industrialist." *Journal of Southern History* 27 (1961), 169–83.

Morgan, George T., Jr. "Conflagration as Catalyst: Western Lumbermen and American Forest Policy." *Pacific Historical Review* 47 (1978), 167–87.

"The Fight against Fire: Development of Cooperative Forestry in the Pacific Northwest." *Idaho Yesterdays* 6 (1962), 20–30.

William B. Greeley: A Practical Forester, 1879–1955. St. Paul: Forest History Society, 1961.

Morison, Samuel Eliot. *Maritime History of Massachusetts, 1783–1860*. Boston: Houghton Mifflin, 1941.

Morris, Ralph C. "The Notion of a Great American Desert East of the Rockies." *Mississippi Valley Historical Review* 13 (1926), 190–200.

Morrison, John H. *The History of American Steam Navigation*. New York: Stephen Daye Press, 1958.

Morton, J. Sterling, "Arbor Day: Its History and Growth." In Egleston, Nathaniel H. (ed.), *Arbor Day: Its History and Observance*, 22–7. USDA Report no. 56. Washington, D.C.: GPO, 1896.

Morton, Thomas. *The New English Canaan of Thomas Morton* (1642). Edited by Charles Adams, Jr. Publications of the Prince Society. Boston: John Wilson and Son, 1883.

Moser, Robert J. ". . . And Then There Was One." *American Forests* 65 (Dec. 1959), 21, 43–4.

Muir, John. "God's First Temples: How Shall We Preserve Our Forests? The Question Considered by John Muir, the California Geologist – The Views of a Practical Man and a Scientific Observer – A Profoundly Interesting Article (Communicated to the Record-Union)." Sacramento *Daily Union*, Feb. 5, 1876, p. 8, cols. 6–7. Reprinted in Sacramento *Semi-Weekly Record Union*, Feb. 9, 1876, ed.

Multhauf, Robert P. "Potash." In Hindle, Brook (ed.), *Material Culture of the Wooden Age*, 227–40. Tarrytown, N.Y.: Sleepy Hollow Press, 1981.

Mumford, Lewis. *The Brown Decades: A Study of the Arts in America, 1865–1895*. New York: Harcourt Brace, 1931.

Technics and Civilization. New York: Harcourt Brace, 1934.

Munger, Thornton T. *Timber Growing and Logging Practice in the Douglas Fir Region*. USDA Bull. no. 1493. Washington, D.C.: GPO, 1927.

Muntz, Alfred P. "The Changing Geography of the New Jersey Woodlands, 1600–1900." Ph.D. diss., University of Wisconsin, Madison, 1959.

Murphy, Louis B. "Forest Taxation Experience of the States and Conclusions Based on Them." *Journal of Forestry* 22 (1924), 453–63.

Nairne, Thomas. *A Letter from South Carolina: Giving an Account of the Soil, Air, Products, Trade, Government, Laws, Religion, People, Military Strength of that Province*. London: A. Baldwin, 1710.

Nash, Roderick. *Wilderness and the American Mind*. New Haven: Yale University Press, 1967.

Neiland, Bonita J. "Forest and Adjacent Burn in the Tillamook Burn Area of Northwestern Oregon." *Ecology* 39 (1958), 660–71.

Nelligan, John E. *The Life of a Lumberman*. Edited by Charles M. Sheridan. Madison, Wisc., 1929.

Nelson, Charles A. *A History of the U.S. Forest Products Laboratory (1910–1963)*. Madison, Wisc.: USFS, Forest Products Laboratory, 1971.

Newcomb, William W. *North Americans: An Anthropological Perspective*. Pacific Palisades, Calif.: Goodyear Publishing, 1974.

Newell, Frederick H. "The National Forest Reserves." *National Geographic Magazine* 8 (1897), 177–87.

New York State Agricultural Society. *Transactions*. Albany, 1851.

New York State Census. 1855, 1865.

New York State Forest Commission. *Annual Report, 1886–1895*. Albany: Argus. Succeeded by the Reports of the Forest, Fish, and Game Commission.

Nicolson, Marjorie Hope. *Mountain Gloom and Mountain Glory: The Development of the Aesthetics of the Infinite*. Ithaca, N.Y.: Cornell University Press, 1959.

Niles' Weekly Register, 7 Sept. 1811–28 Sept. 1849. Philadelphia. Published under various titles: Sept. 1811–Aug. 1814, as *The Weekly Register*; Sept. 1814–Aug. 1837, as *Niles' Weekly Register*; and thereafter as *Niles' National Register*.

Noble, David. *America by Design: Science, Technology, and the Rise of Corporate Capitalism*. New York: Oxford University Press, 1977.

Noggle, Burl L. *Teapot Dome: Oil and Politics in the 1920s.* Baton Rouge: Louisiana State University Press, 1962.

Norgress, Rachel E. "The History of the Cypress Lumber Industry in Louisiana." *Louisiana Historical Quarterly* 30 (1947), 979–1059.

North, Douglass C. *The Economic Growth of the United States, 1790–1860.* Englewood Cliffs, N.J.: Prentice-Hall, 1961.

Nowlin, William. *The Bark Covered House; or, Back in the Woods Again* (1876). Edited by Milo M. Quaife. Chicago: Lakeside Press, R. R. Donnelley and Sons, 1937.

O'Callaghan, Edmund B. (ed.). *Documentary History of the State of New York.* 4 vols. Albany: Weed, Parsons, 1849–50 (vols. 1–3). C. van Benthuysen, 1851 (vol. 4). This is a shortened version of O'Callaghan, Edmund B., and Fernow, Berthold (eds.), *Documents Relative to the Colonial History of the State of New York.* 15 vols. Albany: Weed, Parsons, 1853–87.

Oden, Jack P. "Charles Holmes Herty and the Birth of the Southern Newsprint Paper Industry, 1927–1940." *Journal of Forest History* 21 (1977), 77–89.

"Development of the Southern Pulp and Paper Industry, 1900-1970." Ph.D. diss., Mississippi State University, 1973.

Ohio Forests: Proceedings of the Ohio State Forestry Association Meeting, Columbus, March 28th, 1884, Together with a Report upon the Forest Condition of Ohio. Columbus: G. J. Brand, 1884.

Oliver, William. *Eight Months in Illinois, with Information to Immigrants.* Newcastle upon Tyne, M. A. Mitchell, 1843. Reprint, Chicago: Walter M. Hill, 1924.

Olson, James C. "Arbor Day: A Pioneer Expression of Concern for the Environment." *Nebraska History* 53 (1972), 1–13.

J. Sterling Morton. Lincoln: University of Nebraska Press, 1942. Reprint, Lincoln: Nebraska State Historical Society Foundation, 1972.

Olson, Sherry H. *The Depletion Myth: A History of Railroad Use of Timber.* Cambridge, Mass.: Harvard University Press, 1971.

One Hundred Years of Progress in the United States. Hartford: L. Stebbens, 1870.

Overfield, Richard A. "Trees for the Plains: Charles E. Bessey and Forestry." *Journal of Forest History* 23 (1979), 18–31.

Overton, Richard C. *Burlington West: A Colonization History.* Cambridge, Mass.: Harvard University Press, 1941.

Palais, Hyman. "Pioneer Redwood Logging in Humboldt County." *Forest History* 17 (1974), 18–27.

Palais, Hyman, and Roberts, Earl. "The History of the Lumber Industry in Humboldt County." *Pacific Historical Review* 19 (1950), 1–16.

Parkins, A. E. *The South: Its Economic-Geographic Development.* New York: Chapman and Hall, 1938.

Parkman, Francis. "Editorial." *Garden and Forest* 1 (12 Dec. 1888), 494.

"The Forests and the Census." *Atlantic Monthly* 55 (June 1885), 835–9.

"The Forests of the White Mountains." *Garden and Forest* 1 (29 Feb. 1888), 2.

A Half Century of Conflict: France and England in North America. London: Macmillan, 1892.

Parton, James. "Chicago." *Atlantic Monthly* 19 (1867), 325–45.

Paullin, Charles (ed.). *Atlas of the Historical Geography of the United States.* Washington, D.C., 1932.

Pearce, Roy Harvey.*The Savages of America: A Study of the Indian and the Idea of Civilization.* Baltimore: Johns Hopkins University Press, 1953; rev. ed., 1965.

Pearse, John B. *A Concise History of the Iron Manufacture of the American Colonies up to the Revolution and of Pennsylvania until the Present Time.* Philadelphia: Allen, Lane, and Scott, 1876.

Peavy, George W. *Oregon's Commercial Forests.* Oregon State Board of Forestry, Bull. no. 2. Salem: State Printing Department, 1922.

Peffer, E. Louis. *The Closing of the Public Domain: Disposal and Reservation Policies, 1900–1950.* Stanford, Calif.: Stanford University Press, 1951.

Penfound, W. T. "Plant Distribution in Relation to the Geology of Louisiana." *Proceedings of the Louisiana Academy of Science* 8 (1943), 32–50.

Penick, James L. *Progressive Politics and Conservation: The Ballinger–Pinchot Affair.* Chicago: University of Chicago Press, 1968.

Penn, William. *Some Fruites of Solitude.* Philadelphia, 1663.

Pennell, Francis W. "Travels and Scientific Collections of Thomas Nuttall." *Bartonia* 18 (1936), 1–51.

Perry, Percival. "The Naval-Stores Industry in the Old South, 1790–1860." *Journal of Southern History* 34 (1968), 509–26.

Petersen, William J. "Rafting on the Mississippi: Prologue to Prosperity." *Iowa Journal of History* 58 (1960), 289–320.

Peyton, Jeannie S. "Forestry Movement of the Seventies, in the Interior Department, under Schurz." *Journal of Forestry* 18 (1920), 391–405.

Phillips, H. *On the Bituminous Coalfield of Pennsylvania.* Reports of the British Association. London, 1837.

Phillips, Ulrich Bonnell. *Plantation and Frontier Documents, 1649–1863.* 2 vols. Cleveland: Arthur H. Clark, 1929.

Pierce, Earl S., and Stahl, William J. (comp.). *Cooperative Forest Fire Control: A History of Its Origin and Development under the Weeks and Clarke–McNary Acts.* Washington, D.C.: USFS, 1964.

Pikl, Ignatz James, Jr. "Pulp and Paper and Georgia: The Newsprint Paradox." *Forest History* 12 (1968), 6–19.

Pinchot, Gifford. *Breaking New Ground.* New York: Harcourt Brace, 1947.

The Fight for Conservation. New York: Doubleday, Page, 1910. Reprint, Seattle: University of Washington Press, 1967.

"Forest Devastation: A National Danger and a Plan to Meet It." *Journal of Forestry* 17 (1919), 911–45.

"Now Conservation Began in the United States." *Agricultural History* 11 (1937), 255–65.

"The Lines Are Drawn." *Journal of Forestry* 17 (1919), 899–900.

Pinkett, Harold T. "Gifford Pinchot at Biltmore." *North Carolina Historical Review* 34 (1957), 346–57.

Gifford Pinchot, Private and Public Forester. Urbana: University of Illinois Press, 1970.

Pinney, George. *An Essay upon the Culture and Management of Forest Trees and Natural Evergreens, Exhibiting the Vast Amount of Timber Being Consumed Here, the Various Profits and Advantages of Forest Tree, Culture and Directions for Planting and Cultivating the Same.* Galena, Ill.: D. W. Scott, 1869.

Pitkin, Timothy. *A Statistical View of the Commerce of the United States: Its Connection with Agriculture and Manufacturers.* Hartford: Charles Hosmer, 1816. Reprint, New York: Augustus M. Kelly, 1967.

"A Plan to Save the Forests: Forest Preservation by Military Control." *Century Magazine* 49 (Feb. 1895), 626–34.

Platt, Virginia B. "Tar, Staves, and New England Rum: The Trade of Aaron Lopez of Newport, Rhode Island, with Colonial North Carolina." *North Carolina Historical Review* 48 (1971), 2–22.

Plummer, Fred G. *Forest Conditions in the Cascade Range, Washington, between the Washington and Mount Rainier Forest Reserves.* U.S. Geological Survey, Professional Paper no. 6. Washington, D.C.: GPO, 1902.

Forest Fires: Their Causes, Extent, and Effects, with a Summary of Recorded Destructions and Loss. USFS, Bull. no. 117. Washington, D.C.: GPO, 1912.

Plummer, George H.; Plummer, F. G.; and Raynor, J. H. *Map of Washington Showing Classification of Land.* U.S. Geological Survey, Professional Paper, not numbered and no commentary. Washington, D.C.: GPO, 1902.

Pooley, William Vipond. *The Settlement of Illinois from 1830 to 1850.* Bulletin of the University of Wisconsin, no. 220. Madison, 1908.

Porteous, Alexander. *Forest Folklore, Mythology, and Romance.* London: George Allen and Unwin, 1928.

"Power Saws Come of Age: Chronicles of Heartaches and Ultimate Triumph." *Timberman* 50 (1949), 150–63.

Power, Tyrone. *Impressions of America, during the Years 1833, 1834, and 1835.* 2 vols. London: Richard Bentley, 1836.

Primack, Martin L. "Farm Formed Capital in American Agriculture, 1850–1910." Ph.D. diss., University of North Carolina, Chapel Hill, 1963.

"Land Clearing under Nineteenth Century Techniques: Some Preliminary Calculations." *Journal of Economic History* 22 (1962), 485–96.

Prince, Hugh C. "The Real, Imagined, and Abstract Worlds of the Past." *Progress in Geography* 3 (1971), 3–86.

Proceedings of the American Forestry Congress, 1882–9.

Proceedings of the American Forest Congress Held in Washington, D.C., Jan. 2–6, 1905. Washington, D.C.: H. M. Suter, 1905.

Proceedings, Conference of Governors, Washington, 1908. Washington, D.C.: GPO, 1908.

Puter, Stephen A. D., and Stevens, H. *Looters of the Public Domain by S. A. D. Puter, King of the Oregon Land Fraud Ring.* Portland, Ore.: Portland Printing House, 1908. Reprint, New York: Arno Press, 1972.

Pyne, Stephen J. *Fire in America: A Cultural History of Wild Land and Rural Fire.* Princeton, N.J.: Princeton University Press, 1982.

Quarterman, Elsie, and Keever, Catherine. "Southern Mixed Hardwood Forest: Climax in the Southeastern Coastal Plain, USA." *Ecological Monographs* 32 (1962), 167–85.

Rakestraw, Lawrence. "A History of Forest Conservation in the Pacific Northwest, 1891–1913." Ph.D. diss., University of Washington, 1955.

"Sheep Grazing in the Cascade Range: John Minto vs. John Muir." *Pacific Historical Review* 27 (1958), 371–82.

Raney, William F. "The Timber Culture Acts." *Mississippi Valley Historical Association Proceedings* 10 (1919–20), 219–29.

Ransom, James Maxwell, *The Vanishing Ironworks of the Ramapos: The Story of the Forges, Furnaces, and Mines of the New Jersey–New York Border Area.* New Brunswick, N.J.: Rutgers University Press, 1966.

Rasmussen, Wayne D. "Wood on the Farm." In Hindle, Brooke (ed.), *Material Culture of the Wooden Age*, 3–34. Tarrytown, N.Y.: Sleepy Hollow Press, 1981.

Raup, Hugh M. "Some Problems in Ecological Theory and Their Relation to Conservation." *Journal of Ecology* 52 supp. (1964), 19–28.

"The View from John Sanderson's Farm: A Perspective for the Use of the Land." *Forest History* 10 (1966), 2–11.

Read, Ralph A. *The Great Plains Shelterbelt in 1954. A Re-evaluation of the Field Windbreaks Planted between 1935 and 1942 and a Suggested Research Program.* Lincoln: Experimental Research Station, University of Nebraska, 1958.

Rector, William G. *Log Transportation in the Lake States Lumber Industry, 1840–1918: The Movement of Logs and Its Relationship to Land Settlement, Waterway Development, Railroad Construction, Lumber Production, and Prices.* American Waterways Series no. 4. Glendale, Calif.: Arthur H. Clark, 1953.

Reuss, L. A.; Wooten H. H.; and Marschner, F.J. *Inventory of Major Land Uses, United States.* USDA, Miscellaneous Publication no. 663. Washington, D.C.: GPO, 1949.

Reynolds, Arthur R. *The Daniel Shaw Lumber Company: A Case Study of the Wisconsin Lumbering Frontier.* New York University, Graduate School of Business Administration, Business History Series no. 5. New York: New York University Press, 1957.

Reynolds, Robert V., and Pierson, Albert H. *Fuel Wood Used in the United States, 1630–1930.* USDA Circular 641. Washington, D.C.: GPO, 1942.

"Tracking the Sawmill Westward: The Story of the Lumber Industry in the United States as Unfolded by Its Trail across the Continent." *American Forests* 31 (Nov. 1925), 643–8.

Richardson, James D. (ed.). *A Compilation of the Messages and Papers of the Presidents.* 10 vols. Washington, D.C.: Bureau of National Literature, 1897.

Robbins, Roy M. *Our Landed Heritage: The Public Domain, 1776–1936.* Princeton, N.J.: Princeton University Press, 1942.

Robbins, William G. *Lumberjacks and Legislators: Political Economy of the U.S. Lumber Industry, 1890–1941.* College Station: Texas A&M University Press, 1982.

Roberts, William I. "American Potash Manufacture before the American Revolution." *Proceedings, American Philosophical Society* 116 (1972), 383–95.

Robertson, James Oliver. *American Myth, American Reality.* New York: Hill and Wang, 1980.

Robinson, Glen O. *The Forest Service: A Study in Public Land Management.* Baltimore: Johns Hopkins University Press, for Resources for the Future, 1975.

Rodgers, Andrew Denny. *Bernhard Eduard Fernow: A Story of North American Forestry.* Princeton, N.J.: Princeton University Press, 1951.

Roe, Frank Gilbert. *The North American Buffalo: A Critical Study of the Species in the Wild State.* Toronto: Toronto University Press, 1951.

Rohe, Rand E. "The Landscape and the Era of Lumbering in Northeastern Wisconsin." *Geographical Bulletin* 4 (Apr. 1972), 1–27.

Rohrbough, Malcolm J. *The Trans-Appalachian Frontier: People, Societies, and Institutions, 1775–1850.* New York: Oxford University Press, 1978.

Roosevelt, Theodore. "The Forest in the Life of the Nation." In *Proceedings of the American Forest Congress*, 3–12. Washington, D.C.: H. M. Suter, 1905.

Rosenberg, Nathan. *The American System of Manufacture: The Report of the Committee on the Machinery of the United States, 1855, and the Special Report of George Wallis and Joseph Wentworth, 1854.* Edinburgh: Edinburgh University Press, 1969.

"America's Rise to Woodworking Leadership." In Hindle, Brooke (ed.), *America's Wooden Age: Aspects of Its Early Technology*, 36–62. Tarrytown, N.Y.: Sleepy Hollow Press, 1975.

Technology and American Economic Growth. New York: Harper and Row (Harper Torchbooks), 1972.

Rosenblat, Angel. *La población de América en 1492: Viejos y nuevos calculos.* Mexico City, D.F., 1967.

La población indígena de América desde 1492 hasta la actualidad. Buenos Aires, 1954.

La población indígena y el mestizaje en América. 2 vols. Vol. 1: *La población indígena, 1492–1950.* 3rd ed. Buenos Aires, 1954.

Rossiter, W. S. *A Century of Population Growth: From the First Census to the Twelfth Census of the United States, 1790–1900.* Washington, D.C.: GPO, 1909.

Rostlund, Erhard. "The Myth of a Natural Prairie Belt in Alabama: An Interpretation of Historical Records." *Annals, Association of American Geographers* 47 (1957), 392–411.

Rostow, Walt Whitman. *The Process of Economic Growth.* Oxford: Oxford University Press, 1953.

Roth, Filibert. *On the Forestry Conditions of Northern Wisconsin.* Geological and Natural History Survey of Wisconsin, Bull. no. 1. Madison, 1898. Reprinted as *Forestry Conditions and Interests of Wisconsin* (with an introduction by B. E. Fernow). USDA, Forestry Division, Bull. no. 16. Washington, D.C.: GPO, 1898.

Rourke, Constance Mayfield. *The Roots of American Culture and Other Essays.* Edited with preface by Van Wyck Brooks. New York: Harcourt Brace, 1942.

Rowley, William D. "Privilege vs. Right: Livestock Grazing in U.S. Government Forests." In Steen, Harold K. (ed.), *History of Sustained-Yield Forestry: A Symposium*, 61–7. Santa Cruz, Calif.: Forest History Society, 1984.

Ruffin, Edmund. *An Essay on Calcareous Manures*. 2nd ed. Shellbanks, Va.: Farmers' Register, 1835.

Rush, Benjamin. *An Account of the Manners of the German Inhabitants of Pennsylvania Written in 1789 by Benjamin Rush, M.D.* Edited by I. Daniel Rupp. Philadelphia: S. P. Tour, 1875.

"An Account of the Progress of Population, Agriculture, Manners, and Government in Pennsylvania in a Letter to a Friend in England." In *Essays, Literary, Moral, and Philosophical*, 23–5. Philadelphia: T. and W. Bradford, 1798.

"An Enquiry into the Cause of the Increase of Bilious and Intermitting Fevers in Pennsylvania, with Hints for Preventing Them." *Transactions, American Philosophical Society* 2, no. 25 (1786), 206–12. Also printed in his *Medical Inquiries and Observations*, 2: Philadelphia, 1799. 265–76.

"Information to Europeans Who Are Disposed to Migrate to the United States." In *Essays, Literary, Moral, and Philosophical*, 189–212. 2nd ed. Philadelphia: T. and W. Bradford, 1798.

Russell, Howard S. *A Long Deep Furrow: Three Centuries of Farming in New England*. Hanover, N. H.: University Press of New England, 1976.

Rutledge, Peter J., and Tooker, Richard H. "Steam Power for Loggers: Two Views of the Dolbeer Donkey." *Forest History* 14 (1970), 19–29.

Rutman, Darret B. *The Husbandmen of Plymouth: Farming and Village in the Old Colony, 1620–1692*. Boston: Beacon Press, 1968.

Ryan, J. C. "Minnesota Logging Railroads." *Minnesota History* 27 (1946), 300–8.

Salmond, John A. *The Civilian Conservation Corps, 1933–1942: A New Deal Case Study*. Durham, N.C.: Duke University Press, 1967.

Sanford, Charles Leroy. "The Concept of the Sublime in the Works of Thomas Cole and William Cullen Bryant." *American Literature* 28 (1957), 434–48.

The Quest for Paradise: Europe and the American Moral Imagination. Urbana, Ill.: University of Illinois Press, 1961.

Sargent, Charles Sprague. "The Protection of Forests." *North American Review* 135 (1882), 386–401.

Report on the Forests of North America (Exclusive of Mexico), vol. 9 of *The Tenth Census of the United States* (1880). Accompanied by a *Folio Atlas of Forest Trees of North America* (1880). Washington, D.C.: GPO, 1884.

"A Suggestion." *Garden and Forest* 5 (13 July 1892), 325.

Sauer, Carl O. "The Agency of Man on Earth." In Thomas, William, *Man's Role in Changing the Face of the Earth*, 49–69. Chicago: University of Chicago Press, 1956.

"A Geographical Sketch of Early Man in America." *Geographical Review* 34 (1944), 529–73.

Geography of the Upper Illinois Valley and History of Development. Illinois State Geological Survey, Bull. no. 27, Urbana, Illinois, 1916.

"Grassland: Climax, Fire, and Man." *Journal of Range Management* 3 (1950), 16–21.

Sixteenth Century North America: The Land and the People as Seen by the Europeans. Berkeley: University of California Press, 1971.

Savage, Henry. *Discovering America, 1700–1875*. New York: Harper and Row, 1979.

Sawyer, Josephine. "Personal Reminiscences of the Big Fire of 1871." *Michigan History Magazine* 16 (1932), 422–30.

Schafer, Joseph. "Great Fires of Seventy-one." *Wisconsin Magazine of History* 11 (1927), 96–106.

Schallenberg, Richard H. "Charcoal Iron: The Coal Mines of the Forest." In Hindle, Brooke (ed.), *Material Culture of the Wooden Age*, 271–99. Tarrytown, N.Y.: Sleepy Hollow Press, 1981.

"Evolution, Adaption, and Survival: The Very Slow Death of the American Charcoal Iron Industry." *Annals of Science* 32 (1975), 341–59.

Schallenberg, Richard H., and Ault, David A. "Raw Material Supply and Technological Change in the American Charcoal Iron Industry." *Technology and Culture* 18 (1977), 436–66.

Schantz, Homer L., and Zon, Raphael. "Natural Vegetation." In USDA, *Atlas of American*

Agriculture. Pt. 1: "The Physical Basis of Agriculture," sec. E. Washington, D.C.: GPO, 1924.

Schiff, Ashley Leo. *Fire and Water: Scientific Heresy in the Forest Service*. Cambridge, Mass.: Harvard University Press, 1962.

Schmaltz, Norman J. "Raphael Zon: Forest Researcher." *Journal of Forest History*, pt. 1, 24 (1980), 25–39; pt. 2, 24 (1980), 86–97.

Schmidt, Herbert G. *Rural Hunterdon: An Agricultural History*. New Brunswick, N.J.: Rutgers University Press, 1946.

Schob, David E. *Hired Hands and Plowboys: Farm Labor in the Mid West, 1815–1860*. Urbana: University of Illinois Press, 1975.

"Woodhawks and Cordwood: Steamboat Fuel on the Ohio and Mississippi Rivers, 1820–1860." *Journal of Forest History* 21 (1977), 124–32.

Schoepf, Johann David. *Travels in the Confederation (1783–1784)*. Reproduced as *Schoepf's Travels in the Confederation, 1783–84*. Translated and edited by Alfred J. Morrison. Philadelphia: W. J. Campbell, 1911.

Schramm, Jacob Richard. "Influence – Past and Present – of François André Michaux on Forestry and Forest Research in America." *Proceedings, American Philosophical Society* 101 (1957), 336–43.

Schulman, Steven A. "The Lumber Industry of the Upper Cumberland River Valley." *Tennessee Historical Quarterly* 32 (1973), 255–64.

Schurr, Sam H., and Netschert, Bruce C. *Energy in the American Economy, 1850–1975: An Economic Study of Its History and Prospects*. Baltimore: Johns Hopkins University Press, for Resources for the Furture, 1960.

Schurz, Carl. *Reminiscences*. 3 vols. New York: McClure, 1907–8.

Speeches, Correspondence, and Political Papers of Carl Schurz. Selected and edited by Frederic Bancroft. 6 vols. New York: G. P. Putnam's Sons, 1913. Rpt., Westport, Conn.: Greenwood Press, 1974.

Scroggs, William D. "Federal Swamp Act Grants of 1849–50." *Louisiana State University Quarterly* 6 (1911), 159–64.

Sears, Paul B. "The Natural Vegetation of Ohio." *Ohio Journal of Science*, pt. 1, "Forest History," 25 (1925), 139–49; pt. 2, "Plant Succession," 26 (1926), 128–46; pt. 3, "The Prairies," 26 (1926), 213–32.

Sewell, Samuel. *The Diary of Samuel Sewell, 1674–1729*. Edited by M. Halsey Thomas. 2 vols. New York: Farrar, Strauss and Giroux, 1973.

Shaler, Nathaniel S. *Nature and Man in North America*. New York: Charles Scribner and Sons, 1899.

Sharp, Paul F. "The Tree Farm Movement: Its Origin and Development." *Agricultural History* 23 (Jan. 1949), 41–5.

Sharrar, Terry G. "The Search for a Naval Policy, 1783–1812." In Hagan, Kenneth J. (ed.), *In Peace and War: Interpretation of American Naval History, 1775–1978*. Westport, Conn.: Greenwood Press, 1978.

Shepherd, Jack. *The Forest Killers: The Destruction of the American Wilderness*. New York: Weybright and Talley, 1975.

Shepherd, James F. *Commodity Exports from the British North American Colonies to Overseas Areas, 1768–1772: Magnitude and Patterns of Trade*. Herman C. Krannert Graduate School of Industrial Administration, Purdue University, Paper no. 258. Lafayette, Ind., 1969.

Shepherd, James F., and Walton, Gary M. *Shipping, Maritime Trade, and the Economic Development of Colonial North America*. Cambridge: Cambridge University Press, 1972.

Sherriff, Patrick. *A Tour through North America, Together with a Comprehensive View of the Canadas and the United States*. Edinburgh: Oliver and Boyd, 1835.

Shimek, Bohumil. "The Pioneer and the Forest." *Proceedings, Mississippi Valley Historical Association* 3 (1911), 96–105.

Shofner, Jerrell N. "Negro Laborers and the Forest Industries in Reconstruction Florida." *Journal of Forest History* 19 (1975), 180–91.

Show, Stuart B., and Kotok, E. I. *The Role of Fire in the California Pine Forests*. USDA, Bull. no. 1294. Washington, D.C.: GPO, 1924.

Shurtleff, H. R. *The Log-Cabin Myth: A Study of the Early Dwellings of the English Colonists in North America*. Cambridge, Mass.: Harvard University Press, 1939.

Sieber, George W. "Lumbermen at Clinton: Nineteenth Century Sawmill Center." *Annals of Iowa* 41 (1971), 779–802.

Sim, Robert J., and Weiss, Harry B. *Charcoal Burning in New Jersey from Early Times to the Present*. Trenton, N.J.: Jersey Agricultural Society, 1955.

Skinner, Emory Fiske. *Reminiscences*. Chicago: Vestal Printing, 1908.

Smith, Charles D. "The Appalachian National Park Movement, 1885–1901." *North Carolina Historical Review* 37 (1960), 38–65.

"The Mountain Lover Mourns: Origins of the Movement for a White Mountains National Forest, 1880–1903." *New England Quarterly* 33 (Mar. 1960), 37–56.

"The Movement for Eastern National Forests – 1899–1911." Ph.D. diss., Harvard University, 1956.

Smith, David C. *A History of Lumbering in Maine, 1861–1960*. University of Maine Studies no. 93. Orono: University of Maine Press, 1972.

History of Papermaking in the United States (1691–1969). New York: Lockwood, 1970.

"The Logging Frontier." *Journal of Forest History* 18 (1974), 96–106.

Smith, Frank Ellis. *The Politics of Conservation*. New York: Random House (Pantheon Books), 1966.

Smith, Frank E. (ed.). *Conservation in the United States: A Documentary History*. 5 vols. New York: Chelsea House Publishers, in association with Van Nostrand Reinhold, 1971.

Smith, Henry Nash. "Rain Follows the Plow: The Notion of Increased Rainfall for the Great Plains, 1844–1880." *Huntingdon Library Quarterly* 10 (Feb. 1947), 169–93.

The Virgin Land: The American West as Symbol and Myth. Cambridge, Mass.: Harvard University Press, 1950. Another edition, New York: Alfred A. Knopf, 1961.

Smith, Herbert A. "The Early Forestry Movement in the United States." *Agricultural History* 12 (1938), 326–46.

Smith, James B. "Lumbertowns in the Cutover: A Comparative Study of the Stage Hypothesis of Urban Growth." Ph.D. diss., University of Wisconsin, Madison, 1973.

Smith, John. "Advertisements for the Inexperienced Planters of New England or Anywhere." *Massachusetts Historical Society Collections*, 3rd ser. 3 (1833), 1–54.

The Generall Historie of Virginia, New England, and the Summer Isles (1624). Vol. 1 of *The Travels of Captain John Smith*. 2 vols. Glasgow: James Maclehose and Sons, 1897.

Travels and Works of Captain John Smith. 2 vols. Edited by Edward Arber. Birmingham, 1884. New edition by A. G. Bradley, Burt Franklin Research and Source Works Series, no. 130.

Snow, Sinclair. "Naval Stores in Colonial Virginia." *Virginia Magazine of History and Biography* 72 (1964), 75–93.

Solberg, Erling D. *New Laws for New Forests: Wisconsin's Forest-Fire, Tax, Zoning, and County-Forest Laws in Operation*. Madison: University of Wisconsin Press, 1961.

Southern Lumberman. 1881–90.

Sparhawk, William N. "The History of Forestry in America." In USDA, *Trees: Yearbook of Agriculture, 1949*, 702–14. Washington, D.C.: GPO, 1949.

Sparhawk, William N., and Brush, Warren D. *The Economic Aspects of Forest Destruction in Northern Michigan*. USDA Tech. Bull. no. 92. Washington, D.C.: GPO, 1929.

Spencer, Betty G. *The Big Blowup: The Northwest's Great Fire*. Caldwell, Idaho: Caxton Printers, 1956.

Spratt, John S. *The Road to Spindletop: Economic Change in East Texas, 1875–1901*. Dallas: Southern Methodist University Press, 1955.

Springer, John S. *Forest Life and Forest Trees: Comprising Winter Camp-Life among the Loggers and Wild-Wood Adventure, with Descriptions of Lumbering Operations on Various Rivers of Maine and New Brunswick*. New York: Harper and Brothers, 1851. Rpt., Somersworth, N. H.: New Hampshire Publishing, 1971.

Stahl, Rose M. *The Ballinger–Pinchot Controversy*. Smith College Studies in History, vol. 11, pt. 2. Northampton, Mass.: Smith College, 1926.

Starr, Frederick. "American Forests: Their Destruction and Preservation." USDA, *Annual Report* (1865), 210–34.

Steen, Harold K. "Grazing and the Environment: A History of Forest Service Stock-Reduction Policy." *Agricultural History* 49 (1975), 238–42.

The U.S. Forest Service: A History. Seattle: University of Washington Press, 1977.

Steen, Harold K. (ed.). *History of Sustained-Yield Forestry: A Symposium*. Santa Cruz, Calif.: Forest History Society, 1984.

Steer, Henry B. *Lumber Production in the United States, 1799–1946*. USDA, Miscellaneous Publication no. 669. Washington, D.C.: GPO, 1948.

Stephenson, Isaac. *Recollections of a Long Life, 1829–1915*. Chicago, 1915.

Stewart, Omer C. "Barriers to Understanding the Influence of the Use of Fire by Aborigines on Vegetation." *Proceedings, 2nd Annual Tall Timbers Fire Ecology Conference*, 117–26. Tallahassee, Fla., 1963.

"Burning and Natural Vegetation in the United States." *Geographical Review* 41 (1951), 317–20.

"Fire as the First Great Force Employed by Man." In Thomas, William, *Man's Role in Changing the Face of the Earth*, 115–33. Chicago: University of Chicago Press, 1956.

Stilgoe, John R. *The Common Landscape of America, 1580 to 1845*. New Haven: Yale University Press, 1982.

Still, Bayrd. "Milwaukee in 1833 and 1849." *Wisconsin Magazine of History* 52 (1970), 297.

Stillwell, Lewis D. "Migration from Vermont, 1776–1860." *Proceedings, Vermont Historical Society*, 2 (1937), 62–245.

Stoddard, Herbert Lee. *The Bobwhite Quail: Its Habits, Preservation, and Increase*. New York: Charles Scribner's Sons, 1931.

Stokes, George A. "Log-Rafting in Louisiana." *Journal of Geography* 58 (1959), 81–9.

"Lumbering and Western Louisiana Cultural Landscapes." *Annals, Association of American Geographers* 47 (1957), 250–66.

Stout, Wilbur. "The Charcoal Iron Industry of the Hanging Rock Iron District: Its Influence on the Early Development of the Ohio Valley." *Ohio Archaeological and Historical Quarterly* 42 (1933), 72–104.

Stover, John F. *The Railroads of the South, 1865–1900: A Study in Finance and Control*. Chapel Hill: University of North Carolina Press, 1955.

Strachey, William. *The Historie of Travaile into Virginia Britannia*. Reprinted in Hakluyt Society Publication no. 5. London, 1849.

Strickland, William. *Observations on the Agriculture of the United States*. London: Bulmer, 1801. Reprinted as *Journal of a Tour in the United States of America, 1794–95*. Edited by J. E. Strickland. New York: New-York Historical Society, 1971.

Stuart, James. *Three Years in North America*. 2 vols. 2nd rev. ed. Edinburgh: R. Cadell, 1833.

Sutton, Stephanne B. *Charles Sprague Sargent and the Arnold Arboretum*. Cambridge, Mass.: Harvard University Press, 1970.

Swain, Donald C. *Federal Conservation Policy, 1921–1933*. Berkeley: University of California Press, 1963.

Swank, James M. *The History of the Manufacture of Iron in All Ages and Particularly in the United States from Colonial Times to 1891: Also a Short History of Early Coal Mining in the United States and a Full Account of the Influences Which Long Delayed the Development of all American Manufacturing Industries*. 2nd ed. Philadelphia: American Iron and Steel Association, 1892.

Swanton, J. R. *Early History of the Creek Indians and Their Neighbours.* Bureau of American Ethnology, Bull. no. 73. Washington, D.C.: GPO, 1922.

Swineford, Jerome. "Lumber Industry of Texas." In Raines, C. W. (ed.), *Year Book of Texas,* 239–43. Austin: Gammel Book Company, 1902.

Symonds, Samuel. "Letter" (1637). *Massachusetts Historical Society, Collections,* 4th ser. 8 (n.d.), 118–20.

Tansley, Arthur G. "The Use and Abuse of Vegetational Concepts and Terms." *Ecology* 16 (1935), 284–307.

Taylor, Bayard. *Colorado: A Summer Trip.* New York: G. P. Putnam and Sons, 1867.

Taylor, George Rogers. *The Transport Revolution, 1815–1860.* New York: Harper and Row, 1951. Also published as vol. 4 of *The Economic History of the United States.* New York: Harper and Row (Harper Torchbooks), 1968.

Taylor, John. "Journal of Rev. John Taylor, Missionary on Tour through the Mohawk and Black River Counties in 1802." In O'Callaghan, E. B. (ed.), *Documentary History of the State of New York,* 3:1107–50. Albany: Weed, Parsons, 1854.

Taylor, W. S. "Different Stages in the Evolution of the Overhead System of Logging." *Timberman* 15 (Jan. 1914), 30–1.

Temin, Peter. *Iron and Steel in Nineteenth Century America: An Economic Inquiry.* Cambridge, Mass.: MIT Press, 1964.

Thomas, Gabriel. *An Historical and Geographical Account of the Province and Country of Pensilvania; and of West-New-Jersey in America.* London: A. Baldwin, 1698. Reprinted in Myers, Albert C. (ed.), *Narratives of Early Pennsylvania, West New Jersey, and Delaware, 1630–1709,* 307–52. New York: Charles Scribner and Sons, 1912.

Thomas, John J. "Culture and Management of Forest Trees." USDA, *Annual Report* (1864), 43–7.

Thomas, William L., Jr., with the collaboration of Carl O. Sauer, Marston Bates, and Lewis Mumford. *Man's Role in Changing the Face of the Earth.* Chicago: University of Chicago Press, 1956.

Thompson, Gilbert; Matthes, F. E.; and Cudlipp, M. L. "Distribution of Woodland in Oklahoma." In U.S. Geological Survey, 21st Annual Report, pt. 5. Washington, D.C.: GPO, 1899.

Thompson, Harry. *The Cost of Clearing Logged-off Land for Farming in the Pacific Northwest.* USDA, Bureau of Plant Industry, Circular 25. Washington, D.C.: GPO, 1909.

Costs and Methods of Clearing Land in Western Washington. USDA, Bureau of Plant Industry, Bull. no. 239. Washington, D.C.: GPO, 1912.

Thompson, Harry, and Strait, Earl D. *Costs and Methods of Clearing Land in the Lake States.* USDA, Bull. no. 91. Washington, D.C.: GPO, 1914.

Thompson, H. Paul. "Estimating Aboriginal American Population: A Technique Using Anthropological and Biological Data." *Current Anthropology* 7 (1966), 417–24, see also 425–9 for comments by other scholars.

Thompson, Mark J. *Experimental Station Work.* USDA, Farmers' Bull. No. 381. Washington, D.C.: GPO, 1909.

Investigation in Costs and Methods of Clearing Land. University of Minnesota Agricultural Experimental Station, Bull. no. 163. St. Paul, 1916.

Thompson, Mark J., and Schwantes, A. J. *Investigations in Stump and Stone Removal.* University of Minnesota Agricultural Experimental Station, Bull. no. 208. St. Paul, 1924.

Thompson, Roger C. "The Doctrine of Wilderness: A Study of the Policy and Politics of the Adirondack Preserve-Park." Ph.D. diss., State University of New York, 1962.

"Politics in the Wilderness: New York's Adirondack Forest Preserve." *Forest History* 6 (1963), 14–23.

Thoreau, Henry David. *The Maine Woods* (1877). Edited by Joseph J. Moldenhauer. Princeton, N.J.: Princeton University Press, 1972. Also in *Excursions in the Writings of Henry David Thoreau,* 3:55. Boston: Houghton Mifflin (Riverside Press), 1895.

"The Natural History of Massachusetts" (1842), and "The Succession of Forest Trees" (1860). In *Excursions in the Writings of Henry David Thoreau*, 9:127–62, 225–50. Boston: Houghton Mifflin (Riverside Press), 1895.

"Walden." In *The Works of Thoreau*, 243–456. Edited by Henry Seidel Canby. Boston: Houghton Mifflin (Riverside Press), 1937.

Thwaites, Reuben G. (ed.). *Early Western Travels, 1748–1846*. 32 vols. Cleveland: Arthur H. Clark, 1904–7.

Tillotson, C. R. "The Woodlot: Its Present Problems and Probable Future in the United States." *Proceedings, Society of American Foresters* 11 (1916), 198–208.

Tobey, Ronald C. *Saving the Prairies: The Life Cycle of the Founding School of American Plant Ecology, 1895–1955*. Berkeley: University of California Press, 1981.

Tocqueville, Alexis de, *Democracy in America*. 3rd ed. 4 vols. Translated by Henry Reeves. London: Saundars and Otley, 1838.

Journey to America. Comprising 14 notebooks including "Pocket Note-Books," "Journey to Lake Oneida," and "A Fortnight in the Wilds." Translated by George Lawrence. Edited by J. P. Mayer. London: Faber and Faber, 1960.

Transeau, Edgar N. "The Prairie Peninsula." *Ecology* 16 (1935), 423–37.

Transeau, Edgar, and Sampson, H. "Map of the Primary Vegetation Areas of Ohio." In Sitterley, J. H., and Falconer, J., *Better Land Utilization for Ohio*. Ohio Agricultural Experimental Station, Department of Rural Economics, Mimeo Bull. no. 108. Columbus, 1938.

Tratman, E. E. Russell. *Railway Track and Trackwork*. 2nd ed. New York: Engineering News, 1901.

Report on Substitution of Metal for Wood in Railroad Ties, Together with a Discussion on Practicable Economies in the Use of Work for Railway Purposes by B. E. Fernow. USDA, Forestry Division, Bull. no. 4. Washington, D.C.: GPO, 1890.

Trefethen, James B. *Crusade for Wildlife: Highlights in Conservation Progress*. Boone and Crockett Club Book. Harrisburg: Stackpole, 1961.

Trollope, Anthony. *North America*. London: Chapman and Hall, 1862. Edited with notes by Donald Smalley and Bradford Allen Booth. New York: Alfred A. Knopf, 1951.

True, Rodney H. "François André Michaux, the Botanist and Explorer." *Proceedings, American Philosphical Society* 78 (1937), 313–27.

Tudor, Henry. *Narrative of a Tour in North America*. 2 vols. London: James Duncan, 1834.

Turner, Frederick Jackson. *The Frontier in American History*. New York: Holt, 1920.

Turner, Orsamus. *History of the Pioneer Settlement of Phelps and Gorham's Purchase and Morris' Reserve*. Rochester, N.Y.: W. Alling, 1851.

A Pioneer History of the Holland Purchase of Western New York. Buffalo: Jewett Thomas, 1849.

Twining, Charles E. *Downriver: Orrin H. Ingram and the Empire Lumber Company*. Madison: State Historical Society of Wisconsin, 1975.

"Plunder and Progress: The Lumber Industry in Perspective." *Wisconsin Magazine of History* 42 (1963), 116–24.

"Weyerhaeuser and the Clemons Tree Farm: Experimenting with a Theory." In Steen, Harold K. (ed.), *History of Sustained-Yield Forestry: A Symposium*, 33–42. Santa Cruz, Calif.: Forest History Society, 1984.

Tyler, John D. "The Charcoal Industry in Decline, 1855–1925." M.A. Thesis, University of Delaware, 1967.

Ulmer, Grace. "Economic and Social Development of Calcasieu Parish, Louisiana, 1840–1912." *Louisiana Historical Quarterly* 32 (1950), 519–630.

Upton, Dell. "Traditional Timber Farming." In Hindle, Brooke (ed.), *Material Culture of the Wooden Age*, 35–96. Tarrytown, N.Y.: Sleepy Hollow Press, 1981.

U.S. Association of Charcoal Ironworkers. "The Charcoal Iron Industries of the Hanging Rock Region," and "Transport of Charcoal." *U.S. Association of Charcoal Ironworkers Journal* 6 (1885), 117–21.

U.S. Bureau of the Census. *Historical Statistics of the United States from Colonial Times to 1957.* A statistical abstract supplement prepared by the Bureau of the Census with the cooperation of the Social Science Research Council. Washington, D.C.: GPO, 1960. This was revised, updated, and reprinted as 2 vols. in 1977.

Manufacturers: Special Reports on Selected Industries. "Lumbering," by Henry Gannett and J. E. Whelchel. Twelfth Census (1900), 9, pt. 3: 805–97.

Manufactures. "Iron and Steel," by William G. Gray. Twelfth Census (1900), 10, pt. 4: 1–95.

Manufactures. "Power Used in Manufacture." Tenth Census (1880), 2:13–18.

Special Report on the Manufacture of Coke, Jasper D. Weeks. Tenth Census (1883), vol. 10.

Special Statistics on Manufactures. Thirteenth Census (1909), vol. 10.

Statistics of the Iron and Steel Industry Production of the United States, by James Moore Swank. Tenth Census (1881), vol. 2.

Water Power: pt. 2, "Report on the Water Power of the Mississippi River and Some of Its Tributaries" and "Report on the Water Power of the Northwest," by James L. Greenleaf. Tenth Census (1880), 17: 7–118, 119–276.

U.S. Bureau of Corporations. *The Lumber Industry.* 3 vols. Vol. 1, pt. 1, "Standing Timber" (1913); vol. 2, pt. 2, "Concentration of Timber Ownership in Important Selected Regions," pt. 3, "Land Holdings of Large Timber Owners" (1914); and vol. 3, pt. 4, "Conditions in Production and Wholesale Distribution Including Wholesale Prices" (1914). Washington, D.C.: GPO, 1913–14. Also in U.S. Congress, Senate, 61st Cong., 3rd sess. S. Doc. 818 (serial no. 5943). Washington, D.C.: GPO, 1910–11.

U.S. Congress. *Congressional Globe.* 1850–72.

Congressional Record. 1873–1930.

U.S. Congress, House. "Bill for the Appointment of a Commission to Enquire into the Destruction of Forests." 43rd Cong., 1st sess., 1874, H. R. 2497.

"Bill for the Reservation and Preservation of the Forest Lands on the Public Domain and to Establish a Commission to Examine the Condition of the Said Lands . . ." (Dunnell bill). 51st Cong., 1st sess. 1890, H. Res. 7026.

"Bill to Repeal the Timber Culture Laws and for Other Purposes." 51st Cong., 1st sess., 1890, H. Res. 7254.

Final Report of the Public Lands Commission. 46th Cong., 3rd sess., H. Ex. Doc. 47 (serial nos. 1975, 1976). Washington, D.C.: GPO, 1881.

Final Report of the United States De Soto Expedition Commission. 76th Cong., 1st sess., H. Doc. 71 (serial no. 10104). Washington, D.C., 1939.

First Message to Congress by the President. 57th Cong., 1st sess., H. Doc. 1, xvi–xvii (serial no. 4268). Washington, D.C.: GPO, 1901.

Patent Office, Annual Report, 1847. 30th Cong., 1st sess., H. Ex. Doc. 54 (serial no. 519). Washington, D.C.: GPO, 1848.

Plan for the Management and Disposition of the Public Timber Lands. (Report of Edward A. Bowers.) 50th Cong., 1st sess., H. Ex. Doc. 242 (serial no. 2560). Washington, D.C.: GPO, 1888.

Regulating Management of Lumber Rafts on the Great Lakes. 51st Cong., 2nd sess., H. R. 3759 (serial no. 2888). Washington, D.C.: GPO, 1891.

Report of the Public Lands Commission. 46th Cong., 2nd sess., H. Ex. Doc. 46 (serial no. 1923). Washington, D.C.: GPO, 1880.

Report on the American Association for the Advancement of Science Memorial on the Cultivation of Timber and the Preservation of Forests. 43rd Cong., 1st sess. 1874, H. R. 259 (serial no. 1623). Washington, D.C.: GPO, 1874.

Report on the Internal Commerce of the United States. 48th Cong., 2nd sess., H. Ex. Doc. 267 (serial no. 2304). Washington, D.C.: GPO, 1885. Sketch and remarks by Franklin Hough in App. 14, "A Report on the Range and Ranch Cattle Traffic in the Western States," 130–1.

U.S. Congress, Senate. "Bill for the Protection and Administration of the Forests on the Public Domain" (Hale bill). 50th Cong., 1st sess., 1888, S. Bill 1779.

"Bill Providing for the Disposition of Public Timbered Lands in the United States" (Plumb bill). 46th Cong., 2nd sess., 1878, S. Bill 609.

Letter from the Secretary of the Treasury Transmitting Information on Steam Engines. 25th Cong., 3rd sess., S. Ex. Doc. 21 (serial no. 345). Washington, D.C.: GPO, 1838.

Memorial from the American Association for the Advancement of Science upon the Cultivation of Timber and the Preservation of Forests. 43rd Cong., 1st sess., S. Ex. Doc. 28 (serial no. 1580). Washington, D.C.: GPO, 1874.

Message from the President of the United States Transmitting a Report Relative to the Preservation of Forests on the Public Domain. 51st Cong., 1st sess., S. Ex. Doc. 36 (serial no. 2682). Washington, D.C.: GPO, 1890.

A National Plan for American Forestry: Letter from the Secretary of Agriculture, Transmitting in Response to Senate Resolution 175 the Report of the Forest Service . . . on the Forest Problem of the United States (Copeland Report). 2 vols. 73rd Cong., 1st sess., S. Doc. 12 (serial nos. 9740, 9741). Washington, D.C.: GPO, 1933.

Report of the Committee Appointed by the National Academy of Sciences upon the Inauguration of a Forest Policy . . . 55th Cong., 1st sess., S. Doc. 105 (serial no. 3562). Washington, D.C.: GPO, 1897.

Report of the National Conservation Commission. 3 vols. Edited by Henry Gannett. 60th Cong., 2nd sess., S. Doc. 676 (serial nos. 5397–9). Washington, D.C.: GPO, 1909.

Report of the Public Lands Commission. 58th Cong., 3rd sess., S. Doc. 189 (serial no. 4766). Washington, D.C.: GPO, 1905.

Report on Investigation into Reforestation. 68th Cong., 1st sess., S. Rep. 28 (serial no. 8222). Washington, D.C.: GPO, 1923.

The Trade and Commerce of the British North American Colonies upon the Great Lakes and Rivers . . . 32nd Cong., 1st sess., S. Ex. Doc. 112 (serial no. 622). Washington, D.C.; GPO, 1852.

USDA. *Annual Report of the Commissioner of Agriculture*. Washington, D.C.: GPO, 1862.

Forest Influences. Bull. no. 7. Forestry Division, Washington, D.C.: GPO, 1893.

Forests and National Prosperity (Reappraisal Report). Publication no. 668. Washington, D.C.: GPO, 1948.

Trees: Yearbook of Agriculture, 1949. Washington, D.C.: GPO, 1949.

Yearbook of Agriculture. Washington, D.C.: GPO, 1893.

U.S. Department of the Interior. *Annual Reports* (1865–1903). Includes the Report of the Commissioner of the General Land Office. Washington, D.C.: GPO.

USFS. *An Analysis of the Timber Situation in the United States, 1952–2030*. Forest Resources Rep. no. 23. Washington, D.C.: GPO, 1982.

The Economic Importance of Timber in the United States. Compiled by Dwight Hair. Miscellaneous Publication 941. Washington, D.C.: GPO, 1963.

The Friday Newsletter. 19 Mar. 1982.

Handbook. Bull. No. 718. Washington, DC.: GPO, 1918.

(Lake State Forest Experimental Station). *Possibilities of Shelterbelt Planting in the Plains Region: A Study of Tree Planting for Protective and Ameliorative Purposes as Recently Begun in the Shelterbelt Zone of North and South Dakota, Nebraska, Kansas, Oklahoma, and Texas by the Forest Service; Together with Information as to Climate, Soils, and Other Conditions Affecting Land Use and Tree Growth in the Region*. Washington, D.C.: GPO, 1935.

The Outlook for Timber in the United States. Forest Resource Rep. no. 20. Washington, D.C.: GPO, 1973.

Timber Depletion, Lumber Prices, Lumber Exports, and Concentration of Timber Ownership (Capper Report). U.S. Congress, Senate, 66th Cong., 2nd sess., report on S. Res. 311. Washington, D.C.: GPO, 1920.

Timber in the United States Economy, 1963, 1967, and 1972, by Robert B. Phelps. Gen. Tech. Rep. WO-21. Washington, D.C.: GPO, 1980.

Timber Resources for America's Future (Timber Review Report). Forest Resource Rep. no. 14. Washington, D.C.: GPO, 1958.

Timber Trends in the United States. Forest Resource Rep. no. 17. Washington, D.C.: GPO, 1965.

U.S. Patent Office. *Annual Reports.* 1845–70.

United States Statutes at Large. 1799–1900.

Vance, Rupert B. *The Human Geography of the South: A Study in Regional Resources and Human Adequacy.* Chapel Hill: University of North Carolina Press, 1932.

Van der Donck, Andriaes. "A Description of the New Netherlands" (1656). In *New York Historical Society Collections*, 2nd ser. 1 (1841), 148.

Van Ravenswaay, Charles. "América's Age of Wood." *Proceedings, American Antiquarian Society* 80 (1970), 49–66.

Van Tassel, Alfred J., and Bluestone, David W. *Mechanization in the Lumber Industry: A Study of Technology in Relation to Resources and Employment Opportunity.* National Research Project Rep. no. M-5, Philadelphia Work Projects Administration, Mar. 1940.

Van Valen, Nelson. "James Fenimore Cooper and the Conservation Schism." *New York History* 62 (1981), 289–306.

Van Zandt, Roland. "The Catskills and the Rise of American Landscape Painting." *New York Historical Society Quarterly* 49 (1965), 257–81.

Vaughan, L. M. *Abandoned Farm Areas in New York.* Cornell Agricultural Experimental Station Bull. no. 490. Ithaca, N.Y., 1928.

Vinnedge, Robert W. "The Genesis of the Pacific Northwest Lumber Industry." *Lumber World Review* 45 (25 Dec. 1923), 29–33.

Voigt, William, Jr. *Public Grazing Lands: Use and Misuse by Industry and Government.* New Brunswick, N.J.: Rutgers University Press, 1976.

Volney, Constantine François Chasseboeuf. *A View of the Climate and Soil of the United States of America.* London: J. Johnson, 1804.

Wade, Mason (ed.). *The Journals of Francis Parkman.* London: Eyre and Spottiswood, 1947.

Wahlenberg, W. G. "Forest Fire Control in the Coastal Plains Section of the South." *Journal of Forestry* 33 (1935), 360.

Wakeley, Philip C. "The Adolescence of Forestry Research in the South: A Reflective View." *Journal of Forest History* 22 (1978), 136–45.

"F. O. ('Red') Bateman: Pioneer Silviculturalist." *Journal of Forest History* 20 (1976), 91–9.

Walker, Francis A. (ed.). *Statistical Atlas of the United States, Based on the Results of the Ninth Census.* Washington, D.C.: GPO, 1874.

Walker, J. Herbert (ed.). *Rafting Days in Pennsylvania.* Altoona, Pa.: Times–Tribune Group, 1922.

Wansey, Henry. *Journal of an Excursion in the United States in the Summer of 1794.* London: G. T. Wilkie, 1796.

Warder, John A. "Report on Forestry." In *Reports of the Commissioners of the United States to the International Exposition.* U.S. Congress, House, 44th Cong., 1st sess., H. Ex. Doc 196, vol. 1, sec. D. (serial no. 1710). Washington, D.C.: GPO, 1876.

Warren, Kenneth. *The American Steel Industry, 1850–1970: A Geographical Interpretation.* Oxford: Oxford University Press (Clarendon Press), 1973.

Waugh, Frank A. *Recreation Uses on the National Forests.* Washington, D.C.: GPO, 1918.

Weaver, Harold. "Effects of Fire on Temperate Forests: Western United States." In Kozlowski, T. T., and Ahlgren, C. E. (eds.), *Fire and Ecosystems*, 279–319. New York: Academic Press, 1974.

"Fire as an Ecological and Silvicultural Factor in the Ponderosa Pine Region of the Pacific Slope." *Journal of Forestry* 41 (1943), 7–15.

Webb, Walter Prescott. *The Great Plains*. Boston: Ginn, 1931.

Weeden, William B. *Economic and Social History of New England, 1620–1789*. 2 vols. Cambridge, Mass.: Houghton Mifflin, 1891.

Wehrwein, George S. "A Social and Economic Program for the Sub-marginal Areas of the Lake States." *Journal of Forestry* 29 (1931), 915–24.

Wehrwein, George, and Barlowe, Raleigh. *The Forest Crop Law and Private Forest Taxation in Winsonsin*. Bull. no. 519. Madison: Wisconsin Conservation Commission, 1945.

Weinstein, James. *The Corporate Ideal in the Liberal State*. Boston: Beacon Press, 1968.

Weisberger, Bernard A. *The New Industrial Society*. New York: John Wiley, 1969.

Weiss, Harry B., and Weiss, Grace M. *The Early Sawmills of New Jersey*. Trenton: New Jersey Agricultural Society, 1968.

Weld, Isaac. *Travels through the States of North America and Provinces of Upper and Lower Canada during the Years 1795, 1796, and 1797*. 2 vols. 2nd ed. London: John Stockdale, 1799.

Weslager, C. A. *The Log Cabin in America: From Pioneer Days to the Present*. New Brunswick, N.J.: Rutgers University Press, 1969.

Wessel, Thomas R. "Prologue to the Shelterbelt, 1870–1934." *Journal of the West* 6 (1967), 119–34.

Wheeler, Charles S. *A Sketch of the Original Distribution of White Pine in the Lower Peninsula of Michigan*. Bull. no. 162. Lansing: Michigan Agricultural College. Experimental Station, 1898.

Whitbeck, Ray H. "The Industries of Wisconsin and Their Geographical Basis." *Annals, Association of American Geographers* 2 (1912), 55–64.

White, Andrew. "A Relation of the Colony of the Lord Baron of Baltimore in Maryland, Near Virginia," in *A Narrative of the Voyage to Maryland by Father Andrew White*. In Force, Peter (ed.), *Tracts and Other Papers*, vol. 4, no. 12. Washington, D.C.: P. Force, 1836.

White, F. M., and Jones, E. R. *Getting Rid of the Stumps*. University of Wisconsin Agricultural Experimental Station, Bull. no. 295. Madison, 1918.

White, Morton, and White, Lucia. *The Intellectual versus the City: From Thomas Jefferson to Frank Lloyd Wright*. Cambridge, Mass.: Harvard University Press and MIT Press, 1962.

White, Philip L. *Beekmantown, New York: Forest Frontier to Farm Community*. Austin: University of Texas Press, 1978.

White, Robert. *Land Use, Environment, and Social Change: The Shaping of Island County, Washington*. Seattle: University of Washington Press, 1980.

White, Roy R. "Austin Cary, the Father of Southern Forestry." *Forest History* 5 (1961), 2–5.

Whittaker, Robert H. "Classification of Natural Communities." *Botanical Review* 28 (1962), 1–239.

Wicker, E. R. "Railroad Investment before the Civil War." In *Trends in the American Economy in the Nineteenth Century: Studies in Income and Wealth*, 503–45. Princeton, N.J.: Princeton University Press, 1960.

Wiebe, Robert. *The Search for Order, 1877–1920*. New York: Hill and Wang, 1967.

Wigglesworth, Michael. "God's Controversy with New-England" (1662). *Proceedings, Massachusetts Historical Society* 12 (1871–3), 83–4.

Wilber, Charles Dana. *The Great Valleys and Prairies of Nebraska and the Northwest*. Omaha, Neb.: Daily Republican Printer, 1881.

Wild, E. O. "The Cut-over Empire of Louisiana." *Southern Lumberman* (16 Dec. 1916), 137–43.

Williams, Asa S. "Logging by Steam." *Forest Quarterly* 6 (Mar. 1908), 1–33.

Williams, Justin. "England's Colonial Naval Stores Policy, 1588–1776." *University of Iowa Studies in the Social Sciences* 10, no. 3 (1934), 32–45.

"English Mercantilism and Carolina Naval Stores, 1705–1776." *Journal of Southern History* 1 (1935), 169–85.

Williams, Michael. "Clearing the United States Forests: The Pivotal Years, 1810–1860." *Journal of Historical Geography* 8 (1982), 12–28.

"Products of the Forest: Mapping the Census of 1840." *Journal of Forest History* 24 (1980), 4–23.

Williston, Hamlin L. *A Statistical History of Tree Planting in the South, 1925–79.* USDA, Forest Service. Atlanta, 1979.

Willson, Lillian M. *Forest Conservation in Colonial Times.* Forest Products History Foundation Ser. No. 3, Minnesota Historical Society. S. Paul, 1948.

Wilson, Joseph. "Observations . . . on Forest Culture." Report of the Commissioner of the General Land Office, 1868, in U.S. Department of the Interior, *Annual Report for 1868*, 40th Cong., 3rd sess., H. Ex. Doc. 1 (serial no. 1366), 173–98. Washington, D.C.: GPO, 1868.

Wilson, Harold F. *The Hill Country of Northern New England: Its Social and Economic History, 1790–1830.* New York: Columbia University Press, 1936.

Wilson, Samuel. "An Account of the Province of Carolina in America, Together with an Abstract of the Patent" (London, 1682). In Salley, Alexander Samuel (ed.), *Narratives of Early Carolina, 1650–1708*, 161–76. New York: Charles Scribner's Sons, 1911.

Winter, Oscar O. *The Great Northwest.* 2nd ed. New York: Alfred A. Knopf, 1952.

Winters, Robert K. (ed.). *Fifty Years of Forestry in the U.S.A.* Washington, D.C.: Society of American Foresters, 1950.

Wohlenberg, E. T. F. "Western Forestry and Conservation Association." *Journal of Forestry* 44 (1946), 505–6.

Wolfe, Linnie Marsh. *John of the Mountains: The Unpublished Journals of John Muir.* Boston: Houghton Mifflin, 1938.

Son of the Wilderness. The Life of John Muir. New York: Alfred A. Knopf, 1945.

Wood, Nancy. *Clearcut: The Deforestation of America.* San Francisco: Sierra Club, 1971.

Wood, Richard G. *A History of Lumbering in Maine, 1820–1861.* University of Maine Studies in History and Government, 2nd ser., no. 33. Orono: University of Maine Press, 1935.

Wood, William. *New England's Prospect* (1634). Publications of the Prince Society. Boston: John Wilson and Sons, 1865.

"Woods and Woodlands." *New England Farmer* 24 (17 Dec. 1845), 197.

Woods, John B. "Taxes and Tree Growing." *American Forests* 54 (Aug. 1948), 354–5, 384.

Woodward, C. Vann. *The Origins of the New South, 1877–1913.* Vol. 9 of Stephenson, Wendell Holmes, and Coulter, E. Merton (eds.), *A History of the South.* Baton Rouge: Louisiana State University Press, 1951.

Woolley, Mary E. "The Development of the Love of Romantic Scenery in America." *American Historical Review* 3 (1897), 56–66.

Woolley, R. W. "Lumbering around Mobile, Alabama." *Review of Reviews* 33 (1906), 190–2.

Worster, Donald. *Nature's Economy: The Roots of Ecology.* Garden City, N.Y.: Doubleday (Anchor Books), 1979; 2nd ed., Cambridge: Cambridge University Press, 1985.

Wright, Newell L. "Logging the Pacific Slopes." In USDA, *Trees: Yearbook of Agriculture, 1949*, 695–701. Washington, D.C.: GPO, 1949.

Wright, Thomas. "On the Mode Most Easily and Effectually Practicable for Drying Up the Marshes of the Maritime Parts of North America." *Transactions, American Philosophical Society* 4, 29 (1799), 243–6.

Wroten, William Harvey. "The Railroad Tie Industry in the Central Rocky Mountain Region, 1867–1900." Ph.D. diss., University of Colorado, 1956.

Yonce, Frederick J. "Lumbering and the Public Timberlands in Washington: the Era of Disposal." *Journal of Forest History* 22 (1978), 4–17.

Young, Alexander. *Chronicles of the Pilgrim Fathers of the Colony of Plymouth from 1602 to 1625.* Boston: C. C. Little and J. Brown, 1841.

Zeigler, E. A. "Rate of Forest Growth." In U.S. Congress, Senate, *Report of the National Conservation Commission*, 2: 203–69. Washington, D.C.: GPO, 1909.

Zon, Raphael. *Forests and Water in the Light of Scientific Investigation.* Washington, D.C.: National Water Commission, 1912; reprinted with revised bibliography, 1927.

"Forests in Relation to Soil and Water." *Symposium on Forestry and the Public Welfare: Proceedings, American Philosophical Society* 89 (1945), 399–488.

"The Relation of Forests in the Atlantic Plain to the Humidity of the Central States and Prairie Region." *Proceedings, Society of American Foresters* 8 (1913), 139–53.

Zon, Raphael, and Sparhawk, William N. *The Forest Resources of the World*, 2 vols. New York: McGraw-Hill, 1923.

Index

Abbot, Henry, 414
Adirondack State Park, 375, 406–7, 456
Adirondacks, 27, 96, 141, 442
Agassiz, Alexander, 414
Agen, James, 236
Agriculture: "natural" sequence to lumbering, 356, 429–30, 447, 470–1, 579; virtue of, 24–5, 581–2, 589
Agriculture, Department of, *see* Department of Agriculture
Ahern, George P., 460
Aiken (S.C.), 245
Alabama, 27; cutovers and clearing, 283, 285, 471, 473; fuel wood, 337; Indian clearings, 39–40, 42, 45–6, 129; iron making, 339, 342; land purchases, 241, 243, 397; lumbering, 248, 255, 278, 279, 461; naval stores, 158–9; regrowth, 30, 728
Alabama River, 42, 247
Alaska: coal holdings case, 423; forest transfer, 447; furs, 141
Albany (N.Y.), 138, 145, 222; milling center, 104, 178–9, wholesale center, 176, 178–82, 193, 197, 271; *see also* Glens Falls
Albemarle (N.C.), 144
Albion, Robert G., 91
Alexandria (La.), 241, 280
Allegheny Mountains, 41, 56, 111, 132, 152, 186
Allegheny River (Pa.), 91, 99, 136, 150, 151, 152, 186
Allen, E. T., 444
Allen, R. T., 322
Altamaha River (Ga.), 245
American Association for the Advancement of Science, 400, 410, 411
American Forestry Association, 383, 403, 407, 410, 413, 416
American Forestry Conquress: first meeting, 351, 403; Fernow's promotion of, 401; as pressure group, 410, 411
American Forestry Congress (1905), 418, 419, 441
Anburey, Thomas, 66
Androscoggin River (N.H./Maine), 104, 174, 175
Annapolis (Md.), 121
Apalochicola River (Fla.), 380
Appalachians/Appalachia, 27, 56, 73, 111, 141, 152; fire control, 450; proposed national park, 407
Arbor Day, 350, 383; *see also* reforestation; shelter, shelterbelts
Arizona, 308, 398, 486
Arkansas, 42; land purchases, 243; lumbering, 262, 272, 278, 481; settlement, 111; woodlots, 337
Arkansas River, 334, 380,
Article X, 461, 463
Ashland (Wis.), 216
Atlantic coast, 271; lumbering near, 4, 244, 245, 247, 253, 277; wood dealers on, 334
Audubon, John J., 15
Aughey, Samuel, 384
Augusta (Ga.), 154
Ault, David, 341
Ausable (Mich.), 232
Ausable River (Mich.), 226
Ausable River (N.Y), 173
Australia, 291, 292, 298, 383
automobile: impact on national forests, 456
axes: as symbols, 19, 56; types of, 300, 315

backwoodsmen, 11–12, 19–21, 153–4, 490; *see also* pioneer life
Bailyn, Bernard, 93
Baird, John, 45, 130
Baker, F. P., 278, 279
Baker, R. Stannard, 236
baldcypress (*Taxodium distichum*), 102, 244, 248–50, 271, 282, 316
Ballinger–Pinchot controversy, 423
Baltic: masts, 91; naval stores and timber, 75, 83–4
Baltimore, 78, 141, 168, 341; as timber market, 99–100
Baltimore and Ohio Railroad, 155, 255, 360
Bangor (Maine), 104
Baptist, William, 249
Barnes, Will C., 308
barrens, *see* prairies and openings
Bartram, William and John, 15, 23, 42, 45
Bateman, F. O., 286
Baton Rouge (La.), 277, 281
Bay City (Mich.), 206, 207, 219, 224, 227, 241
Beaufoy, Henry, 115
Beaumont (Tex.), 277
Beef Slough, *see* Weyerhaeuser, Frederick
Beekmantown (N.Y.), 75
Belknap, Jeremy, 115
Bend (Oreg.), 325
Berkshire Mountains, 150
Bessey, Charles E., 384

Biloxi River (Miss.), 247
Biltmore Forestry School (N.C.): Pinchot at, 412; Schenck at, 404
Bining, Arthur C., 108
Birch, Brian P., 131
Bisbee (Ariz.), 307
Black River (Miss.), 241
Black River (Wis.), 186
blacks, 268; in lumber camps, 253; *see also* slaves/slavery
Black Swamp (Ohio), 362, 363, 367
Blodgett, Delos, 241, 280
Bodega Head (Calif.), 292
Bogalusa (La.), 255, 262, 280, 287
Bonita logging line (Wis.), 312
booms, 276; Bangor (Maine), 175; Beef Slough (Wis.), 204–5, 206, 221, 222, 228; Black River (Wis.), 204; Brunswick (Maine), 175; Choctawhatchee (Miss.), 247; Ferry Pass (Miss.), 247; Glens Falls (N.Y.), 175; Muskegon (Mich.), 204, 328; Saco (Maine), 175; St. Croix (Wis.), 186, 204; Tittabawassee (Mich.), 204, 206, 213, 228; West Newton (Minn.), 206, 228; Williamsport (Pa.), 175
Boone, Daniel, 20, 141
Boone and Crockett Club, 419
booster literature: Great Plains, 382; Lake States cutovers, 234–5
Boston, 169, 279; exports of forest products, 75, 89, 102, 141; fuel wood needs, 78–9, 134, 136; shipbuilding, 93
Boston and Providence Railroad, 156
Bowers, Edward A., 410
Bowie (La.), 282
Boyd, James, 272
Boyer, Charles S., 108
Brackenridge brothers, 241
Bradbury, John, 130
Bradley, N. B., 241
Brandis, Dietrich, 442
Brewer, William H., 648; map of "Woodland and Forest Systems," 25–7, 374, 376, 414, 431
Brewton (Ala.), 280
Britain, 101, 168, 169, 244, 365; coal exports, 74, 136; fuel costs, 78; iron making, 106, 110, 147; mercantilist policy, 84; naval stores trade, 83–9, 152–8; potash imports, 75; steam engines, 152–3
Britton, A. T., 399
broad arrow policy, 92, 102
Brooklyn (N.Y.), 216
Brotherhood of Timber Workers, 268
Brunswick (Ga.), 158, 277
Brunswick (Maine), 175
Brunswick (N.C.), 89, 102, 258
Bryant, William Cullen, 16, 125, 404
Buckingham, James Silk, 154
Buffalo (N.Y.), 178, 262–71; as wholesale lumber center, 176, 182–3, 197, 207, 208
buffalo (bison), 38, 46
Buffon, comte de, 380–1
building materials, 2; *see also* houses, pioneer
Bull Moose campaign, 422, 443; *see also* Pinchot, Gifford
Bumpo, Natty, 20; *see also* Cooper, James Fenimore
Bureau of Corporations, 194, 274, 426–9, 441, 442; on Lake states, 218; on Pacific Northwest, 313, 314, 426; on South, 263–8, 279
Bureau of Forestry, 286; *see also* Forest Service
Bureau of Parks, 457
Burlington (Iowa), 204
Burlington Railroad, 346
Burnaby, Andrew, 120
Butters, Horace, 216, 248, 259, 316
Byrd, William, 65
Byron, Lord, 14

Calcasieu basin (La.): declining towns, 281–2; land ownership, 264–7; land purchases, 241; railroad net, 256–8
Calcasieu Pine and Southern Timber Company, 265
Calcasieu River, 247
California, 3, 37, 46, 290, 298, 307; Butte Co., 295; central valley, 289; coastal lumber trade, 291–5; Eldorado Co., 295; fire control, 321, 484; gold rush, 291, 292, 293; grazing policy in forests, 476; Humboldt Co., 295, 398; Jefferson Co., 295; Klamath Co., 295; land ownership, 426; lumbering, 161, 291–5, 316–18, 461; Mendocino Co., 292, 295; Nevada Co., 295; Placer Co., 295; Santa Ana Co., 295; Santa Clara Co., 295, Santa Cruz Co., 295; Sierra Co., 295; Sonoma Co., 295; *see also* San Francisco
Canada, 89, 176, 207, 230, 280, 291; masts from, 92; pulp from, 287; shelterbelt planting, 387, 388
canals: Chemung, 136; Chenango, 136; Erie, 136, 178–82, 198, 375, 406; Genesee, 136; Hudson–Champlain, 178–9; Illinois–Michigan, 132, 185; Lehigh and Morris, 136; Oswego, 178; Richelieu, 99, 178; Union and Schuylkill, 136
Canandaigua (N.Y.), 122
Cannelton (Ind.), 153, 155
Canton (Iowa), 187, 204
Cape Fear Valley, 102
Cape Providence, 102
Capper, Arthur, 445, 446, 453, 460
Capper Report (1920), 325, 429, 438, 445, 447, 455, 463, 472
Carhart, Arthur H., 458
Caribbean, 101, 244; *see also* West Indies
Carnegie, Andrew, 221, 422
Carroll, Charles, 93
Cary, Austin, 286
Cascade Mountains, 291, 292, 297, 302, 304, 313, 318–19, 328; grazing in, 307–9
Casco Bay, 90, 101
Catlin, George, 404
Catskill Mountains, 16
Cedar Grove (Wis.), 220
Central Coal and Coke Company (4Cs), 262
Central Michigan Railroad, 214
Central Pacific Railroad, 346
Chapman, Herman Haupt, 281, 485, 486
charcoal: as fuel, 7, 133–8, 146; furnaces, 104–10,

147–52, 337–43; impact on forest, 148–52, 343–4; *see also* furnaces, charcoal iron
Champlain, Lake, 42, 75, 141, 178
Charleston (S.C.), 169; lumber center, 245, 254, 277, 280; naval stores, 89, 158; port, 79, 141; shipbuilding, 93
Charlestown (Mass.), 93
Chastellux, François Jean, marquis de, 55, 64, 67, 144, 370
Chateaubriand, François Auguste Réńe, vicomte de, 14, 16
Chattahoochee River (Ga./Ala.), 40, 42
Cheboygan (Mich.), 231–2
Cheboygan River, 226
Chehalis River (Wash.), 313
Chesapeake Bay, 123, 238
Chicago, 178, 179, 183, 198, 208, 234, 267, 345; crosstie market, 349; effect of railroad, 214, 222–3; timber center, 132, 176, 179, 183–9, 271–2, 298
Chicago and Northwestern Railroad, 212
Chicago, St. Paul, Minneapolis, and Omaha Railroad, 234
Chile, 291, 292, 298
China, 41, 141, 291, 298
chipboard, 238
Chippewa River (Wis.), 186, 202, 203, 204–6, 217, 221, 224, 226, 228
Choctawhatchee River (Fla.), 247
Christianity: and clearing, 10, 11, 14, 380; and tree planting, 383
Church, Frederick, 404
Cincinnati, 99, 138, 145, 155, 178, 202, 362, 383, first American Forestry Congress, 403
Civilian Conservation Corps (CCC), 323, 463, 465, 483
Civil War, 112, 144, 146, 148, 156, 157, 158, 187, 187, 188, 201, 204, 277, 337, 339, 379, 393; effect on South, 239–40, 243, 254, 371
Clapp, Earle, 438, 446, 463
Clarke–McNary Act (1924), 446, 454, 465; on fire prevention, 323, 482, 485, 486; on reforestation, 477
clear-cutting, repacious logging, 230, 252, 329; silvicultural practice, 492; *see also* clearing, agricultural; forest depletion
clearing, agricultural: after 1800, 112–18, 132, 353–7, 394; before 1800, 60–7; climatic effect, 145–6, 229, 352, 380; descriptions, 120–8; environmental effect, 368–73, 378–9; extent, 8, 118–20, 357–61, 368, 371–2, 490; Ohio, 361–8, 377–9; proportion, 373–5, 380, 394; Wisconsin, 367
Clements, Fredric E., 30
Cleveland, Grover, 383, 413, 414, 646
Cleveland, 182
climatic change, *see* clearing, agricultural; trees and rainfall
climax vegetation, 30; effect of fire, 43–4, 46–8, 483, 486–7; and succession, 30
Clinton, De Witt, 139
Clinton (Iowa), 187, 204
coal: for domestic heating, 77, 136–9, 334–5; in iron furnaces, 148–51; in river boats and locomotives, 155–6
Cohocton River (N.Y.), 99
Cole, Thomas, 16, 125, 404; *see also* Hudson River school
Collier, Gerald L., 272
Colorado, 3, 188, 398, 461; forest reserves, 421; Free Timber Act, 398; tree planting, 384
Columbia (Miss.), 255
Columbia River and basin, 289, 299, 300, 302, 313, 325
Colvin, Verplanck, 406
Compton, Wilson, 429, 442, 446
Comstock (Nev.), 398
Conaunga River (N.Y.), 99
Conecuh River (Ala.), 247
Conestigo River (N.Y.), 99
Connecticut, 68, 95, 105, 362
Connecticut River, 58
Connellsville (Pa.), 151
conservation, 18–19, 490–1; awareness of need for, 368–73, 376, 406; concept, 145; Pinchot's wise-use policy, 416, 418, 419–20, 422–4; *see also* environmental awareness; fire protection; Marsh, George Perkins; reforestation
controlled burning, 483–7
Coon Lake (Wis.), 214
Cooper, James Fenimore, 16, 20, 125, 139, 141, 404
Cooper, James G., 23, 25
Cooper, Thomas, 75
Cooper, William, 13
Cooperative Forest Management Act (1950), 478
coopers, 4
Coos Bay, 291, 294, 327–8
Coosa River (Ala.), 42
Copeland Report (1933), 325, 329, 447, 463; fire control, 485; holding size, 630; land purchases, 430, 461; reforestation, 477; timber regrowth, 440, 472–3
corduroy roads, 7
cordwood, *see* fuel wood
corn (maize), 33, 63, 234, 367
Cornell, Ezra, 217
cotton, 56–7, 59, 65, 245; migration of plantations, 245, 468, 471
Coville, Frederick, V., 308–9
cowboys, 20, 68, 354
Cowles, Henry C., 30
Cowlitz River (Wash.), 313
Cox, Thomas, 293
Coxe, Tench, 80, 94
Crèvecoeur, Hector St. John, 11, 12, 78, 125
Crockett, Davy, 20, 141
Crivitz (Wis.), 235
cross timbers (Okla./Tex.), 46
Crown Zellerbach Corporation, 286, 287
Culberton, Harvey, 384
cultural myths and symbols, 2, 9, 19, 20–1, 490; in European heritage, 10, 14; *see also* backwoodsmen; cowboys; frontier; pioneer life
Cumberland River (Tenn.), 349
Cummings, Benjamin, 168

Currie, William, 144
Curtis, George E., 386
Cushing, Caleb, 234
Cut-Over Conference of the South (1917), 285
cutovers: fire in, 448, 453–4; Lake States, 18, 233–7, 356; Pacific Northwest, 290, 305, 325, 328, 356; reforestation of, 477; South, 279, 283–5; United States, 355, 433, 438

Dana, Samuel Trask, 443
Darby, Abraham, 110
Darien (Ga.), 277, 280
Davenport (Iowa), 204
De Brahm, John G. W., 66
Defebaugh, James Elliot, 219, 430, 435
Defoe, Daniel, 14
Delaware, 163
Delaware River, 60, 72, 78, 104, 136, 150; rafting on, 98–100, 187
Denis River (Maine), 104
Denkmann, F. C. A., 204
Department of Agriculture, 222, 308, 462; beginnings of forestry in, 376, 400; disputes with Interior, 410, 416–17, 423, 447, 466–7; see also Department of the Interior, Forest Service; national forests
Department of the Interior: Division of Forestry in, 416, 473–4; national forests, 457; relations with Agriculture, 402, 410, 416–17, 423, 466; timber abuse, 327; see also Ballinger–Pinchot cortroversy; Department of Agriculture
De Soto expendition, 42, 67
Detroit, 182
Dingley tariff (1897), 208, 232
Dinsdale, Evelyn, 162
disease: humans, 53, 59, 132; trees, 30, 47, 481, 484, 486
Disston, Henry, 243
Division of Forestry (USDA), *see* Forest Service
Dobyns, Henry, 33, 34
Dolbeer and Carson Lumber Company, 294; engines, *see* skidders
Donaldson, Thomas, 399
Douglas, David, 23
Douglas-fir (*Pseudotsuga menziesii*), 48, 117, 229, 291, 294, 300, 301, 311, 318, 319, 323, 325, 329, 330, 429, 484, 487, 492
Downsville (N.Y.), 220
du Bois, Coert, 484; *see also* fire, controlled burning
Dubuque (Iowa), 187, 204, 220, 221
Dufur, A. J., 298
Dulany, D. M., 220
Durand, Asher, 404
Durell, Frank H., 207
Dust Bowl, 323, 463
Dwight, Timothy, 60, 113, 115, 125

eastern redcedar, 37, 71, 93, 102
eastern seaboard, 58, 59, 83, 101, 111
Eastman, Robert, 168
Easton, Hamilton P., 272
Eau Claire (Wis.), 186, 187, 198, 204, 210, 220, 221, 231
Eau Claire River, 224, 226
ecosystem, 30, 352, 371, 492
Edwards, Samuel, 386
Egg River (N.J.), 80
Egleston, Nathaniel H., 331, 377; as chief of Forestry Division, 385, 400–401; on railroad ties, 347, 349
Emerson, Ralph Waldo, 16, 17, 125, 370, 404, 405; *see also* transcendentalism
Ensign, Edgar T., 432
Enterprise (Miss.), 255
environmental awareness, 112, 144–5, 290, 352, 394; concern, 368–74, 378–9, 492, 494
Erie (Ohio), 202
Erie, Lake, 43, 178, 183, 347
erosion: Ohio, 378–9; soil, 8, 25, 112, 353, 388, 422, 450, 463, 478; wind, 387–8; *see also* Dust Bowl
Escambia River (Fla.), 7, 247, 272
Escatawa River (Miss.), 247
Eureka (Calif.), 294, 301, 399
Europe, 46, 188, 234, 269; comparison of landscapes with U.S., 16, 22, 45, 120–1, 405; cultural heritage, 16; log rafting, 170; trade, 56, 188
Europeans (settlers), 10, 32–3, 44, 49, 53, 144, 432
Evelyn, John, 90, 373
Everett (Wash.), 313, 314

Fairchild, Fred, R., 455–6
Fall, Albert, 447
Fall River (Mass.), 101
Falmouth, 90, 101, 102
farm abandonment, 128; Great Plains, 386; Lake States cutovers, 236–7; late nineteenth century, 357, 379, 431, 440; post-1945, 468–77; *see also* forest regrowth and rejuvenation
Faulkner, William, 253
Federal Timber Conservation Board, 330
fences and fencing, 7, 47, 69–72; lack of timber for on prairie, 128, 130, 132, 386–7; *see also* shelter, shelterbelts
Fernow, Bernhard E., 230, 279, 395, 409, 412, 424, 440, 442; in American Forest Congress, 401, 403; as chief of Forest Division, 400, 401–2, 413–14, 416, 418, 419; and congressional bills, 410, 413; on conservation, 329–30, 402–3, 420, 480, 492; at Cornell, 404, 416; on railroad crossties, 349, 351, 484; on rainfall, 385–6
fire, causes, 232–3, 447–8; *see also* fire use
fire, controlled burning 483–7
fire effects: on forest composition, 47–49, 483, 486–7; on grassland, 45–7, 68, 130, 300, 484; on wildlife, 484, 485; *see also* Chapman, Herman Haupt
fire, historic: Adirondack (1903), 450; Hinckley (1894), 449, 232–3, 499; Idaho/Montana (1910), 322, 450, 481, 485; Maine (1947), 486; Michigan (1881), 232, 449; Miramachi (1825), 449; N. Wisconsin (1885), 232, 449; Oceala (1943), 486;

Peshtigo (1871), 232, 449; Piscataquis (1825), 449; Tillamook (1939, 1945, 1951), 321, 323, 328, 483; Yacolt (1902–3), 321, 323, 450
fire protection: cooperative, 438, 445, 446, 455, 481; in Pacific Northwest, 321–3, 328; *see also* Clarke–McNary Act
fire suppression, 447–8, 454, 456, 481–3
fire use, 30, 51; Indian, 35, 38–9, 43–5, 130, 448, 484, 485; pioneer, 60–2, 63, 65, 115, 121–2, 448
firewood, *see* fuel wood
Fisher-Victoria tramway (La.), 257–9
Fishlow, Albert, 156
Flambeau River (Wis.) 186
Flint, Timothy, 15, 74, 131
Flint and Père Marquette Railroad, 211, 213, 219
flood control, 388, 394; *see also* runoff
Florida, 27, 39, 53, 111; agricultural clearing, 360, 472; cutting and depletion, 278, 279, 285, 360; land sales, 242–3; lumbering, 238, 245, 257, 262; naval stores, 89, 158; stock, 42, 67; timber ownership, 264, 426
flour and grist mills, 5, 95
flumes, 302
Fond du Lac (Wis.), 184
Fordyce (Ark.), 277
Forest and Rangeland Renewable Resources Planning Act (1974), 467
forest area, 3–4, 25–32, 431–5, 467, 474, 489, 490
forest (vegetation) associations, 27–31
forest depletion: general concern, 368–73, 393–6; in Lake States, 228–30; in Pacific Northwest, 323–6; in South, 252, 277–80, 283–5; *see also* clearing, agricultural; timber famine/shortage; *individual state entries*
forest genetics, 492
forest-lieu provisions, 310, 418, 426
forest management, *see* national forests; recreation
Forest Management Act (1897), 414–15, 416
forest ownership: Bureau of Corporations on, 425–9; Capper report on, 429; Copeland report on, 430; during 1980s, 489, 494; in Lake States, 217–18, 303; in Pacific Northwest, 303, 310–14, 426–9; in South, 406–9; *see also* stumpage
Forest Products Laboratory (Madison, Wis.), 351, 442, 459
forest regrowth and rejuvenation, 4, 368, 467, 471, 477; *see also* farm abandonment; timber growth and inventories
Forest Reserve Act (1891), 399, 415, 416, 419; origin, 410–11; proclamations under, 414, 420, 421
forest reserves, 409–11; *see also* national forests
Forest Service, 230, 308, 320, 330, 336, 443, 438, 445, 460, 466–7; cooperation with lumber industry, 409, 481; under Fernow, 351–2, 401–3; fire protection, 450; grazing, 307–8; under Greeley, 322–3, 445–6, 450–4; growth studies, 491–2; under Pinchot, 415–24; recreation, 456–8; under Stuart, 460–4; in U.S., 400–401
Forbes, R. D., 281
Fort Edward (N.Y.), 170
Fort Leavenworth (Kans.), 185
Fort Madison (Iowa), 220
Fort Wayne (Ind.), 202
Fort Worth (Tex.), 237
Fox brothers, 170
France, 23, 168, 252, 287; forest area, 438
Frankfort (Ky.), 145
Franklin, Benjamin, 13, 15; on fuel scarcity, 17, 75, 78, 79; on stoves, 137–8
Free Timber Act (1878), 398–9
French settlement in Gulf, 244, 248
Freneau, Philip, 15
frontier, 1, 12, 17, 19; *see also* stages of civilization
Forest, Robert, 69
Frost-Johnson Lumber Company, 265
fuel wood, 2, 7, 118, 331–3, 386; heat value of trees, 77 Indian use, 37; lack on prairie, 128, 130, 133–7, 333–7; nineteenth-century trade, 78–81; pioneer use, 56, 69, 72, 75–81, 154; twentieth century, 487; *see also* stoves, woodburning
Fuller, Thomas, 350, 393, 394
Fullerton (La.) 428
Fulton, Robert, 152
furnaces, charcoal iron: Center Furnace (Ky.), 342; Champlain Ore and Iron Co. (N.Y.), 343; Clinton Furnace (N.J.), 109; Cumberland Furnace (N.J.), 109; Hamilton Iron Works (Mo.), 342; Martic Furnace (Pa.), 108–9; Monmouth Furnace (N.J.), 109; Pompton Furnace (N.J.), 108; Sterling Ironworks (N.Y.), 107, 109; Union Furnace (N.J.), 109; Woodstock Iron Co. (Ala.), 342
furniture making, 7, 147, 198; at Buffalo, 182; at Grand Rapids, 229

Galveston (Tex.), 267
Gannett, Henry, 377, 431; mapping of western forests, 304; National Conservation Commission, 423; on rainfall, 386
Garden City (La.), 282
Garden of the World, 379–80, 383, 388, 496
Gardiner (Oreg.), 294
Garfield, James R., 422, 423
Garren, K. H., 47
Garrish, W. Scott, 211
Garyville (La.), 282
Gates, James Leslie, 360
Gates, Paul, 242
General Exchange Act (1922), 447, 465
Genesee (N.Y.), 42, 59, 95, 136
Geneva (N.Y.), 119
Georgia, 47, 68, 79, 104; Carroll Co., 475–6 (*see also* forest regrowth and rejuvenation); farm abandonment, 357, 471, 475–6; iron making, 339; lumbering, 237, 245, 252, 262, 272, 278, 285; naval stores, 89, 158; State Board of Forestry, 281; woodlots, 337
Georgia Pacific Corporation, 494
German training of American foresters, 372, 385, 399, 401
Germany, 329, 375; forest area, 438
Gibbs, Walcott, 414

Gila National Forest, 458
ginseng, 141, 157
girdling, *see* clearing, agricultural
Glacken, Clarence J., 144
Gleason, Henry A., 30, 46
Glens Falls (N.Y.), 104, 170, 175, 178, *see also* Albany
Glover, George T., 209
Goldie, William, 207
Goodyear brothers, 255, 262, 280, 286, 287
Governors' Conference on Conservation, 422
Grand Haven (Ohio), 182
Grand Rapids (Mich.), 373
Grand River (Mich.), 184, 224, 226
Grant, Anne McVickar, 99, 100
Grant Grove (General Grant National Park), 486, 487
Graves, Henry Solon, 27, 450, 453; as chief of Forestry Division, 442–5, cooperation with lumber industry, 442–3, 444–5; on recreation, 457–8
Gray, Asa, 145, 376, 381
Grays Harbor (Wash.), 291, 294, 304, 325, 327, 330
grazing: in National Forests, 307–8, 419, 456, 466; in Ohio, 560
Great American Desert, 380, 382, 384, 386
Great Egg Harbor (N.J.), 108
Great Lakes, 18, 136, 152, 198, 291; steamboats on, 155
Great Lakes Experimental Station, 388
Great Plains: settlement of, 3, 234, 289, 354, 379, 384, 472, 478; *see also* Dust Bowl: Garden of the World; Great American Desert; shelter, shelterbelts
Great Southern Lumber Company, 255, 262, 280
Greeley, Elam, 208
Greeley, William Buckhout, 330, 395, 401, 438, 467, 492; arguments with Pinchot, 445; cooperation with lumber industry, 440, 442–5, 467; fire supression policy, 447–8, 450–4, 481, 482; forest area, 432–3, 435, 438; Pacific Northwest, 324, 325; recreation, 458; West Coast Lumbermen's Association, 460–1
Greene, S. W., 485
Green River (Ky.), 344
Gregg, Josiah, 381
Gregory, William, 239, 244
Grinnell, C. Bird, 404
Gulf, Mobile, and Ohio Railroad, 255
Gulf and Ship Island Railroad, 409
Gulf Coast states, 242, 264, 271–2; lumbering, 245, 247; *see also individual state entries*
Gulf of Mexico, 45, 252; *see also* Gulf Coast States
Gulfport (Miss.), 255
gyppo loggers, 319, 328, 329; *see also* peckerwood mills

Hackley, Charles H., 241
Hague, Arnold, 414
Haites, Eric, 153
Hall, Basil, 121, 122, 154
Hall, James, 5, 7, 112, 130, 133, 147, 155
Hammond, S. H., 406
Hanging Rock (Ohio), 150, 342, 344, 364
Hannibal (Mo.), 204, 220
Harding, Warren G., 446
Hardtner, Henry E., 286, 287, 330
Harrington, Mark W., 386
Harrison, Benjamin, 411, 412
Hart, J. Fraser, 472, 475
Hartman, Arthur, 486
Hartt, Rollin L., 232
Hattiesburg (Miss.), 255, 267, 280
Havana (Cuba), 89, 267
Hawaii, 291, 292, 293
Hawesville (Ky.), 153
Hawkins, Benjamin, 39, 40, 42
Hawthorne, Nathaniel, 20
Hayden, V. Ferdinand, 382, 383
Hays, Samuel, 420
Hedrick, Ulysses P., 123
Heidenreich, Conrad, 37, 48
Hendrickson, C. I, 472
Henry, Joseph, 23–5, 381, 431
Henry, William A., 234–5
Hermann, Binger, 418
Herndon, Melvin G., 93
herbicides, *see* pesticides/herbicides
Herty, Charles H., 287–8
Hetch Hetchy dam controversy, 308, 424, 456; *see also* Muir, John
Hibbard, Benjamin H., 384
Hickman, Nollie W., 272
Higginson, Francis, 41
Hill, James J., 298, 310, 418, 422
Hindle, Brook, 9
Hines, Edward, 280
hired labor, 114–5
Hitchcock, Ethan, 418
Hitchcock, Phineas, 383
Hodgson, Adam, 44, 63, 66, 115, 124, 125, 128
Hoffman, Jacob, 202
Holmes, Joseph A., 423
Holt, Charles, 317
Homestead Act (1862), 240, 304, 354, 381; abuse of, 396–7
Hong Kong, 141
Hoover, Herbert, 461
Hotchkiss, George W., 198, 228, 279
Hough, Franklin, 226, 230, 395, 405, 409, 424; *The Duty of Governments*, 376–6, 385, 400; fire, 449; Forest Agent, 400–401; New York State Park Commission, 407; Ohio clearing, 363, 377–8, 400; rainfall, 385, *Reports upon Forestry*, 278, 376, 400; railroad impacts, 345, 349, 351
Housatonic River (Conn.), 109
houses, pioneer, 7, 72–4, 146, 147; descriptions, 125; lack of timber for in prairie, 128, 130, 132
Houston (Tex.), 255
Houston, East and West Texas Railroad, 256
Huckleberry Finn, 254
Hudson River, 57, 59, 104, 152, 178–9, 334; log drive on, 170, 173, 175; log rafts on, 99, 100, 178–9; water supply, 373, 406

Hudson River school, 16, 125; *see also* Cole, Thomas
Humboldt, Alexander von, 381
Humboldt Bay (Calif.), 291, 292, 293, 294, 295
Hunnewell, H. H., 350
Huron, Lake, 184, 207, 233
hydrology, 4

ice roads, 208–10
Idaho, 317, 398; fire of 1910, 322, 450, 481, 485; forest reserves, 421; Free Timber Act, 398; lumbering, 297, 298, 461; ownership concentration, 426
Illinois, 27, 377; fuel use, 334; openings and prairies, 23, 46, 60, 112, 128; settlement, 130–1, 133, 360, 379; St. Clair Co., 132; supply of timber, 184, 226, 386; tree planting, 380
Illinois Central Railroad, 229, 241, 242, 255, 256
Illinois River, 132, 185
improved land, 13, 188–9; *see also* clearing, agricultural
Indiana, 27, 43, 153, 198; clearing, 27, 360, 377; farm abandonment, 468; fuel use, 335; lumber from, 161, 182; openings, 23, 112, 128, 129
Indianapolis, 202, 267
Indians, 4, 14, 30, 53, 432, 491; Abnaki, 38; clearing by, 35–43, 53, 55–7, 129; Creek, 39; crops, 33, 37; fire use, 35, 39, 43–9, 130, 447, 484; Huron, 37, 48; Iroquois, 32, 38–9, 41, 42, 44, 59; lessons for pioneers, 57–8; Narragansett, 41; Pawhatan, 40; resistance, 11, 58, 72, 101, 111
industrialization and wood, 9, 146, 152, 240, 490
Ingram, Orrin, 220, 221
Inland Waterways Commission, 422
insects, 30, 481, 486
Interior, Department of the, *see* Department of the Interior
International Paper Company, 494
Iowa: fuel use, 334; openings and prairies, 45, 183, 379; timber supply, 226
iron manufacture, 5; *see also* charcoal; furnaces, charcoal iron
irrigation, 8, 420
Irving, Washington, 16
Ise, John, 399
Ishmael (*Moby Dick*), 254

Jackson (Miss.), 255, 280
Jacksonville (Fla.), 277, 280
Jacksonville (N.C.), 158
Jacquith, John, 168
James River (Va.), 89, 102, 136
Jamestown (Va.), 40, 57
Jefferson, Thomas, 15, 65
Johnson, Edward, 44, 123
Johnson, Robert Underwood, 410, 413, 414, 421
Jones, W. Goodrich, 252, 280, 330
Jordan, Terry, 131
Jordan River (Miss.), 247
journals, forestry, 404
Juniata Valley (Pa.), 150

Kalamazoo River (Mich.), 184
Kallender, Henry, 486
Kalm, Pierre, 15, 23, 72, 78
Kanawha River (W.V.), 41, 349
Kansas, 4, 18, 46, 185, 188, 271; fuel use, 334–5; planting, 478; settlement, 379, 380, 386; Timber Act abuse, 384
Kansas City (Mo.), 267
Kansas City, Fort Scott and Gulf Railroad, 350
Kansas City and Southern Railroad, 256
Kansas Pacific Railroad, 382
Kaul Lumber Company, 257
Kennebec River (Maine), 104, 134, 174, 175
Kellogg, Royal, 441, 446
Kentfield, Buhure, and Jones, 294, 295
Kentucky, 23, 45, 56, 58, 60, 66, 128, 129, 141, 357; bluegrass country, 45, 71, 111, 379; iron making, 342; lumbering, 244
kerf, 201–2, 303
Kern, M. G., 347, 349–50
kiln drying of timber, 202, 214, 219, 261, 303
Kings Canyor National Park, 486, 487
Kipling, Rudyard, 409, 411
Kirby, John Henry, 242, 264, 266–7; Kirby Lumber Company, 264; Kirby-Bonner Lumber Company, 266–7
Kitsap (Wash.), 295
Kittery (Maine), 101
Kittredge, Alfred B., 425
Klamath River (Calif.), 314
Klippart, John H., 363, 375, 378
Klondike, 309
Knapp Stout and Co., 220
Knutson–Vandenberg Act (1930), 454, 477
Kohlmeyer, Frederick M., 201
Kollmorgen, Walter M., 383
Konvalinka, Joseph G., 387–90
Kotok, E. I., 484
Küchler, A. W., 484

labor in lumber industry: convict, 268–9; unionization, 268; *see also* blacks
Lacey, James, 241, 262, 280
La Crosse (Wis.), 186, 187, 202, 204, 231
Lake Cadillac (Mich.), 256
Lake Charles (La.), 281
Lake George (Mich.), 211
Lakeshore, Michigan and Southern Railroad, 345
Lake States, 4, 48, 111, 178, 179, 182, 188–9, 243–4, 263, 271, 283, 290, 300, 303, 314, 320, 349, 402, 425; agricultural clearing, 355, 357; character of forest, 197–200; depopulation and cutovers, 230–7; land purchases, 396; lumbering and log transport, 201–16; lumber organization, 216–22; lumber production and depletion, 222–30, 289, 375, 461; migration to other regions, 229–35, 239–44, 262, 265–6, 279–80, 298–9, 303, 310, 321; tax delinquency, 236, 455; timber source, 128, 133, 161, 379, 435; woodlots, 336; *see also* Michigan, Minnesota, Wisconsin
Lambeville, Jean de, 38

laminated board, 7
land grants, abuse of, 217–18, 298, 330, 396–7
Land Office, 25, 382, 411; control of Forest Division, 416–18; transfer of forest reserves, 417–18
landscape, 2; aesthetics, 44–5, 121–5, 405; American qualities of, 404–5; change in, 55, 144–5; made landscape, 12–13, 14, 16, 55, 121, 373–5, 380; middle landscape, 15, 17; of prairie, 130–1
Lapham, Increase Allen, 372–3, 380, 386, 394, 490
La Rochefoucauld, 63, 72, 73
Las Vegas (Nev.), 350
Latin America, 247, 269, 292; *see also* South America
Laurel (Miss.) 251, 255, 277, 280
Lee, Daniel, 381
Leighton, Marshall O., 424
Leopold, Aldo, 458
Lesley, J. Peter, 150–1, 337
Lexington (Ky.), 145
Licking River (Ky.), 349
Lindsay, John, 251
Little Kazar River (Maine), 174
Little Rock (Ark.), 255
livestock in the forest, 62, 67–9, 385
loblolly pine (*P. taeda*), 30, 47, 238, 239
locomotives, 152, 155–7, 344–5
log marks, 173–4, 203–4
log and lumber transport, land: Dolbeer engines and skidders, 300–302, 315–16; high lead, 216, 316–7, 329; ice roads, 208–10; logging railroads, 211–16, 251–61, 301–2, 318–19; oxen, 96–7, 185, 208, 231–2, 251, 300–301; power skidding, 216, 251–2, 259; trucks and tractors, 261, 317–20
log and lumber transport, water: arks, 99–100; log drives, 98–9, 100, 170–5, 186, 203–5, 247; pull-boat logging, 216, 248–50, 254, 259–60, 316; rafting, 97–100, 175–9, 186–9, 206–8, 244–7; western coastal trade, 291–5, 300
log cabin, *see* houses, pioneer
logging railroads, 21; Lake States, 211–16; Pacific Northwest, 301–2, 318–19; South, 251–61; *see also* log and lumber transport, land
Lombard, Alvin O., 209, 317
London, 75, 137
Long, Ellen, 484
Long, George Smith, 310, 314, 321–2, 329
Long-Bell Lumber Company, 241, 265
Long Haven (Pa.), 168
Long Island (N.Y.), 78, 113
longleaf pine (*P. palustris*), 43, 47, 85, 88, 188, 237, 238, 239, 258, 267, 468, 484–5, 487
Longview (Wash.), 313, 318, 328
Lorain, John, 145, 370
Lord, N. W., 344
Los Angeles, 294
Louisiana, 42, 111, 129, 130; Bayou Par., 277; Beauregarde Par., 281; Calcasieu Par., 241; clearing, 360; land purchases, 241–3, 264–7, 426; lumbering, 255, 262, 268, 271–7, 278–9, 283, 285, 426, 461; Natchitoches Par., 258, 260; naval stores, 89, 158–60; Rapides Par., 241, 281; Sabine Par., 258; steam engines, 152; Tangipahoa Par., 260; Vernon Par., 241
Louisville (Ky.), 145, 153, 155
Ludington (Mich.), 216
Lufkin (Tex.), 287
lumber barons, 20, 222, 396, 418
lumber code, *see* National Industrial Recovery Act
lumber consumption, 161, 442, 487–9
lumber industry, 4, 6, 160–7, 216–17; capital and landownership, 425–26, 466, competition, 421–9; cooperation versus coercion, *see* Graves, Henry Solon; Greeley, William Buckout; Pinchot, Gifford; employment, 7, 163–4; interwar conditions, 443–6, 459–64; value, 5, 8
lumberjacks, 20–1, 268
lumber production: Lake States, 200, 222–8; mass production methods, 28, 30, 170, 309, 346; Pacific Northwest, 295–6, 304, 309–10, 323–9; South, 271–80; U.S., 100–104, 160–3, 166–7, 193–7, 200, 222–8, 461
lumber prices, 179, 211–12, 230, 298, 429; in prairies, 132, 185, 187
lumber towns and logging camps: decline of, 230–7, 280; Lake States, 216, 218–19; Pacific Northwest, 320; South, 268–9
lumber wholesale centers, 175–6, 179–89, 194–7, 216, 222, 263–71, 292–3, 298–9
Lutcher, Henry J. (Lutcher and Moore), 241–2, 262, 265
Lutcher (La.), 282

McCaslin, Ike, 253–4
McComb (Miss.), 256, 277
McElroy (La.), 282
McGee, WJ., 422, 423
McGraw, John, 219
Machias River (Maine), 104
McKinley, William, 417
McKittrick, Reuben, 241
McManis, Douglas R., 131
McNary (La. and Ariz.), 281
McNary–Woodruff Act (1928), 454, 465
McSweeney–McNary Act (1928), 454, 465
Madison, James, 119, 144
Mahoning River (Pa.), 151
Maine, 27, 35, 53, 58, 60, 90, 92, 111, 141, 175, 293, 375; fuel wood source, 79, 134; fuel wood use, 335, 336; Hancock Co., 134; Lincoln Co., 134; lumbering, 102, 142, 160–3, 169, 193, 198–201; migration to Lake States, 201; Waldo Co., 134
Mancil, Ervin, 272
Manhattan (N.Y.), 78
Manistee (Mich.), 209, 237
Manistee River, 184, 214, 224, 226, 231
maple sugar/syrup, 37, 56, 82, 139, 157
Marblehead (Mass.), 101, 102
Marsh, George Perkins, 18, 25, 144–5, 490, 491; on deleterious effects of clearing, 370–3, 385, 394; on trees and rainfall, 380–1, 302
Marshall, Robert, 458

Maryland, 45, 62, 66, 67, 89, 105, 150
Marx, Leo, 15, 21
Mason, David T., 330, 461
Mason, William H., 288
Massachusetts, 33, 46, 69, 79, 89, 93, 152, 169, 334, 375; clearing, 117; fire protection, 168–9, iron making, 104, 150; sawmills, 168–9
Massachusetts Bay Company, 92
Massey, Richard W., 272
masts and spars, 90–2, 193, 170
Mather, Increase, 11
Mather, Stephen T., 457
Maumee River (Ohio), 42
Maxwell, George H., 424
Maxwell, Hu, 48
meat packing and slaughtering, 6
Memphis (Tenn.), 255, 270, 277
Menominee (Mich.), 202, 220, 241
Menominee River (Wis.), 189, 203, 224, 228
mercantilism, *see* Britain
Meridian (Miss.) 255, 279, 280
Merrens, Harold R., 102
Merrimac River (N.H./Mass.), 58
Mexico, 267, 271, 291, 298
Michaux, André, 15, 23, 25
Michaux, François André, 23, 25, 72, 137
Michigan, 111, 128, 176, 177, 178–9, 182, 184; agricultural clearing, 11, 123, 134, 360; Clare Co., 211; fire, 232, 449; fuel wood use, 335; iron making, 152, 339, 341–2; Lake Co., 231; lumbering, 161–2, 169, 175, 193–237 passim, 240, 251 272; Manistee Co., 231; Mason Co., 231; Oceana Co., 231; timber to prairie, 132, 298
Michigan, Lake, 184, 224, 231
Midwest, 42, 178, 189, 221, 379, 470, 490
migration/migrants, 12, 53, 113, 146, 234; Dutch, 57; German; 56, 60, 95; Polish, 235; Scots-Irish, 56; Swedish, 61, 72, 234;
Millet, Donald J., 272
Millikin, Daniel, 346, 362, 374–5, 386
Milwaukee, 188, 202, 214, 235
Minneapolis, Sault St. Marie and Atlantic Railroad, 214
Minnesota, 27, 176, 179, 188, 272, 310; agricultural clearing, 357, 379, fire, 232, 449; lumbering, 161, 193–237 *passim*, 272; woodlots, 335–6
Minto, John, 304, 307, 308
Mississippi, 44, 73, 264, 485; Copiah Co., 279; fuel wood use, 336–7; Green Co., 241; Hines Co., 279; Jackson Co., 241; land purchases, 241, 243, 264; lumbering, 255–6, 267, 271, 272, 277, 278, 279, 461; Madison Co., 279; Marion Co., 241; naval stores, 158; Pearl River Co., 241, 280, Perry Co., 241; Rankin Co., 279; reforestation, 285, 479; taxation, 279–80; Yazoo Co., 279
Mississippi River, 23, 25, 27, 42, 46, 111, 128, 141, 220, 221, 238, 242, 247; rafting on, 132, 177, 185–9, 206–7, 228; river boats on, 152–5; timberland in valley, 310, 214, 473
Mississippi River Logging Company, 204–5, 221
Missouri, 45, 71, 111, 141, 152, 204, 349, 379, 380; iron making, 339; taxation, 287, timber to, 226
Missouri-Pacific Railroad, 255
Missouri River, 46, 155, 188
Mobile, 42, 89, 136, 158, 255, 267; as lumber center, 239, 242, 247, 255, 256, 277
Mohawk River, 178; valley, 123
Mohr, Charles, 248, 272, 279, 285, 400
Moline (Iowa), 187, 204
Monongahela River (Pa.), 136, 151
Monroe (Ohio), 182
Montana: forest reserves, 421; Free Timber Act, 398; lumber, 297, 461; tree planting, 387
Monterey (Calif.), 291
Montgomery (Ala.), 280
Montreal, 64
Moore, G. Bedell (Lutcher and Moore), 241–2, 262, 265
Moosehead Lake (Maine), 174
Morgan, J. Pierpont, 236
Morrison, John H., 153
Morton, J. Sterling, 382–3
Morton, Thomas, 22, 43
Moss Point (Miss.), 247–8
Mount Rainier National Park, 307, 310, 426
Muir, John, 395, 405, 413; grazing in the Sierras, 307–8; Hetch Hetchy controversy, 424, 456; and Pinchot, 414, 419; *see also* conservation; preservation
Multiple use: Copeland Report on, 463, 488; Pinchot on, 420
Multiple Use–Sustained Yield Act (1960), 467
Mumford, Lewis, 4, 146, 162, 315
Munger, Thornton, 328
Mullica River (N.J.), 80
Muscatine (Iowa), 187, 204
Muskegon (Mich.), 224, 241
Muskegon River, 184, 204, 211, 214, 224, 226

Narragansett (R.I.), 78
Nash, Roderick, 15, 407, 414, 457
Nashville (Tenn.), 277
Natchez (Miss.), 2, 44
National Academy of Science, 307, 414; *see also* National Forestry Commission
National Conference on Outdoor Recreation (NCOR), 458
National Conservation Commission, 422–3, 441, 455
National Forestry Commission, 414–15, 416
National Forestry Program Committee (NFPC), 446
national forests, 290, 420, 421, 424, 447, 466, 477; area, 425, 465; fire protection, 453, 465, 483; grazing, 307–8; proclamations, 412, 419; recreation, 456–8; sustained yield, 461; *see also* grazing; wilderness
National Industrial Recovery Act (NIRA), 330, 461, 465; Article X, 461, 465
national parks: beginnings, 405–6; fire policy, 487; and Forest Service, 423, 457; *see also*, King's Canyon National Park; Mount Rainier National Park; Sequoia National Park
National Resources Commission, 422

nature: organic wholeness, 18, 144–5, 405; subduing/altering, 12–13, 16; wonder at, 14
Nauvoo (Ill.), 186
naval stores, 82, 83–92, 157–60; labor, 268; *see also* broad arrow policy; masts and spars; shipbuilding
Nebraska, 4, 18, 30, 46, 478; dearth of timber, 188, 271; fuel use, 334; rainmaking, 382–3; settlement, 379, 380; tree planting, 384, 478
Neenah (Wis.), 216
Netschert, Bruce C., 341
Nevada, 398, 461
New Bern (N.C.), 158
New Deal programs, 323, 424, 460, 461, 463, 465, 477
Newell, Frederick, 415, 418, 420, 422, 423
New England, 4, 41, 44, 53, 57, 58, 83, 201, 203, 233, 289; agricultural clearing, 55–6, 60–5, 123, 145; farm abandonment, 336, 379, 431, 468–9, 472; fencing and stock, 68–9, 71; fuel wood, 81; houses, 73–4; lumbering, 73–4, 160–4, 193, 233; mast trade, 90–2; naval stores, 88–9; outmigration, 300, 362; potash, 139; rafting and log drives, 98, 174–5; timber economy, 82, 101–4, 176
New Hampshire, 58, 59, 60, 101, 407; fuel wood and woodlots, 79, 336; log transport, 170, 174
New Haven (Conn.), 101
New Jersey, 57, 92, 113, 152; Atlantic Co., 134; Burlington Co., 225; Cumberland Co., 134; farm abandonment, 468; fuel wood supply, 78–80, 134–6; Gloucester Co., 134; iron making, 105, 108, 109; lumbering, 98, 160, 163; Monmouth Co., 134
Newlands, Francis G., 424
New London (Conn.), 101
New Mexico, 398
New Orleans, 99, 111, 153, 186, 188, 248, 255, 278, 285; lumber center, 169, 239, 247, 267; naval stores, 158
New Orleans, Jackson and Great Northern Railroad, 255
New Orleans and Northeastern Railroad, 255
New Orleans Pacific Railroad, 426
Newport (R.I.), 78, 79
New York Centre Railroad, 345
New York City, 68, 138, 176, 267, 335; Chamber of Commerce, 407; fuel use, 78–9, 134, 136, 180, 333; lumber consumption and port, 89, 99, 102, 141
New York State, 41, 112, 129, 152, 216, 233, 262, 406; Allegheny Co., 178; Brown Co., 100; Chanago Co., 100; Chemung Co., 178; clearing, 13, 27, 121–2, 125, 371–2, 377; farm abandonment, 379, 468; fuel wood cutting, 134–6; impact on other lumbering regions, 201, 203, 207, 219; iron making, 105, 109, 150, 342, 343; Lewis Co., 134; lumbering, 60–3, 94–6, 98, 132, 160–3, 169, 174–5, 178–9, 182, 193; migration to other lumber regions, 222; Monroe Co., 136; Montgomery Co., 134; Oneida Co., 136; Onondaga Co., 136; Otsego Co., 136; potash, 75, 141; railroads, 211, 345; Seneca Co., 134; settlement and pioneer life, 58, 59–71, 73–4, 113–14; timber deficit, 17, 176, 182, 197, 220, 400; Warren Co., 170
New Zealand, 291, 293
Niagara (N.Y.), 16, 145, 183, 405
Nisqually River (Wash.), 313
Noble, John W., 410
noncommercial forest, 4, 432–3, 467
Norgress, Rachel E., 272
North Carolina, 41, 42, 56, 60, 66, 68, 141, 407, 486; Beaufort Co., 158; Bertie Co., 158; Bladen Co., 158; Carteret Co., 158; clearing, 102, 115, 360; Craven Co., 158; Duplin Co., 158; Edgecombe Co., 158; farm abandonment, 473; Greene Co., 158; Jones Co., 158; lumbering, 245, 249, 261, 277, 278; Martin Co., 158; naval stores, 85, 88–9, 158–60; New Hanover Co., 158; Pitt Co., 158
North Dakota, 398; tree planting, 384, 387
Northern Pacific Railway Co., 297, 298–9, 304, 310–13, 396, 426
Nowlin, William, 123
Nuttall, Thomas, 23

oaks, 30, 48, 60, 93, 101, 102, 117, 182, 190, 347–50
O&C lands, 330, 417
Ogeechee River (Ga.), 68
Ohio, 141, 186, 198; agricultural clearing, 27, 60, 354, 355, 361–8, 375; Beaumont Co., 346; Fairfield Co., 334, 346; farm abandonment, 361–7, 468; fuel wood, 334, 344; Geagu Co., 139; iron making, 152, 339, 342, 344, 364; Jackson Co., 342; lumbering, 161, 178, 182, 201; Muskingum Co., 346; openings, 23, 71, 128, 129; potash, 139; Scioto Co., 342; Trumbull Co., 139, 346; Vinton Co., 342
Ohio River, 41, 42, 111, 128, 141, 349, 366; rafts on, 99, 132; river boats on, 136, 152, 153–5; valley, 59
Okefinokee Swamp, 248
Oklahoma, 46; forest/prairie edge, 432
Oklahoma City, 267
Oliver, William, 71
Olson, Sherry, 352
Olympic Mountains National Park, 313, 320
Omaha, 46, 350
Onondaga (N.Y.), 38, 42
Ontario, 37, 48, 178, 182, 208
Ontario, Lake, 178
Openings, *see* prairies and openings
Oquawka (Iowa), 187, 204
Orange (Tex.), 242, 255, 277, 280
Oregon, 3, 262, 307, 311, 313, 314; Clackamas Co., 321; fire control, 321–2, forest resources, 281–9, 304–5; lumbering methods, 292, 298, 301, 316–18; Marion Co., 321; production, 295, 327–8, 461; recreational concerns, 307; reforestation, 295; taxation, 456–7; Tillamook Co., 321; timber holdings, 428
Oscoda (Mich.), 241
Oshkosh (Wis.), 231

Oswego River (N.J.), 80
Ouachita River (Ark.), 42
Ozarks, 349

Pacific Coast, 289; timber resources of, 394, 396, 431, 435, 474
Pacific Northwest, 355, 419, 421, 425; early lumbering, 291–6, 299, 300, 304; fire protection, 320, 328–9; grazing, 304–9; land purchases, 309–15, 426–7; log and lumber handling, 300–304, 315, 320; migration of lumbermen to, 298–9; planting, 478, 481; production, 304, 309–10, 323–9, 327–8; setting, 289–90, 421; sustained yield, 329–30; transcontinental rail links, 297–99
Pacific Ocean/basin, 291, 292, 293, 294, 298, 304
Pallisades, 79
Palmer, E. F., 344
Panama Canal, 298, 314
paper and pulp, 7, 238, 287, 406, 488
Parkins, A. E., 227
parklike vegetation, 32–3, 44; *see also* prairies and openings
parks, *see* national parks
Patterson (La.), 282
Patuxent River (Md.), 102
Payson, J., 411
Peavey, George, 324
peckerwood mills, 261, 262–3, 282; *see also* gyppo loggers
Peffer, E. Louis, 424
Penn, William, 15
Pennsylvania, 46, 58, 111, 198, 362, 468; Clarion Co., 152; clearing, 59, 60–1, 113, 145; coal, 136, 153, 334–5; fuel wood, 155, 345; iron making, 105, 108–9, 150–1, 152, 339, 342, 364; kerosene, 158; lumbering, 160–2, 174, 178, 186, 193; Mercer Co., 152; migration to other lumber regions, 201, 222, 262; openings, 45, 57; pioneer life, 72, 73; potash, 75; Venango Co., 152
Penobscot River (Maine), 100, 102, 175
Pensacola (Fla.), 89, 158, 256; lumber center, 242, 247, 272, 277
Peoria (Ill.), 132, 185
Peru, 291, 298
Peshtigo River (Wis.), 184, 244
pesticides/herbicides, 315, 481, 492
Peters, J. Girvin, 445, 450
petroleum, 5
Philadelphia, 99, 105, 121, 123, 176; centennial exhibition, 168, 202; exports, 75, 102; fuel wood supplies, 78–9; shipbuilding, 93, 99; stove manufacture, 138
Piedmont, 57, 471
Pierson, Albert H., 78
Pike, Zebulon, 380
Pinchot, Gifford, 322, 395, 411, 426, 438, 440, 450, 484; American Forestry Congress (1905), 418, 441; association with Theodore Roosevelt, 417, 420, 421, 422; background, 412–13, 424; Ballinger controversy/dismissal, 421, 423–4; Committee for the Application of Forestry, 443–6; conservation claims, 402; forester and governor of Pennsylvania, 423, 445, 454, 460; Hetch Hetchy, 424, 456, 457; National Forestry Committee, 414, 415; Stuart controversy, 460–1; timber famine, 438, 440–1, 444; wise-use policy, 416–22, 424; Western Ranges National Committee, 308; *see also* Ahern, George P.; Bull-Moose campaign; Capper, Arthur; Graves, Henry Solon; Greeley, William Buckout; Roosevelt, Theodore
Pine Barrens (N.J.): fuel source, 79–80, 134, 136; iron making, 105
pioneer life, 10–11, 15, 55–60, 112–13; part time lumbering, 82, 94–5, 163–7; *see also* clearing, agricultural; fences and fencing; fuel wood; houses, pioneer; potash/pearl ash
Piscataqua River (Maine/N.H.): exports, 102; mast trade, 90, 92, 170; shipbuilding, 93, 101
pitch, *see* naval stores
Pittsburgh, 99, 153, 169; coal use, 136–8; iron making, 150, 152; stove manufacture, 138
Pittsfield (Mass.), 95, 109
plantations, 84
Platte river (Nebr.), 334, 380
Plumb, Preston B., 397
plyboard/wood, 7, 488
Plymouth Colony, 57, 92
Poa pratensis (bluegrass), 45
pole roads, 256
ponderosa pine, 291, 314, 325, 483, 487
Pontiac (Mich.), 113
Pope and Talbot, 293, 299, 314
population: internal movements, 198, 230–1, 236; U.S., 13, 53, 54, 111, 490; *see also* migration/migrants
Port Gamble (Wash.), 293
Port Huron (Mich.), 178
Portland (Oreg.), 297, 307, 327, 330
Port Ludlow (Wash.), 293
Portsmouth (N.H.), 90, 101
Portugal, 89, 92, 101
potash/pearl ash, 7, 53, 56, 61, 69, 74–5, 83, 133, 139, 157, 290
Potlach (Idaho), 317
Potomac River, 44, 45
Potter, Albert F., 308
Poughkeepsie (N.Y.), 168
Powell, John Wesley, 399, 484
power skidding, *see* log and lumber transport, land
Prairie du Chien (Wis.), 204
Prairie Farm (Wis.), 220
prairies and openings: cost of "clearing," 132–3, 356; encounter with, 23, 44, 45–7, 49, 111–12, 128–9; lack of timber, 71, 128, 130, 131–2, 183–4, 197; settlement of edge, 130–3, 177, 197, 354, 356, 375, 432
Preemption Act, 396
preservation, 394, 396–9, 411; American Forestry Association, 403–4; Division of Forestry, 399–403; forest reserves, 409–11; lumberman's lobby, 407–9; wilderness enthusiasts, 404–7
Presumpscot River (Maine), 102
Price, Overton, 423

Providence (R.I.), 101, 138
Public Land Office, 240
public lands, 290, 396–9
Public Lands Commission (1879), 399
Puget Sound (Wash.), 291, 292, 293, 294, 295, 298, 299, 300, 302, 304, 313, 321
pulp, *see* paper and pulp
Putnam, Henry C., 198

Quarterman, Elsie, 47
Queensbury (N.Y.), 173
Quincy (Ill.), 187, 204

rafting, 97–100, 175–9, 186–9, 206–7, 222–3, 244–7
railroad impact on: forests, 253, 344–52; linking the West, 297–99; logging practice, 212–16, 253; marketing, 182, 186, 188; mill location, 214–15, 255; river and lake traffic, 182, 183, 185; *see also* locomotives; log and lumber transport, land; ties, railroad
"rain follows the plow," 384
Rainier, Mount, 313
Rappahannock River (Va.), 45
Raquette River (N.Y.), 174
Reconstruction, 239–40, 247
recreation, 2, 18–19, 394; Adirondacks, 406; fire risk of, 483; in national forests, 454, 456–9; protecting Sierras and Cascades, 307; use conflicts, 46; *see also* Muir, John; wilderness
Rector, William, 220
Red River (La.), 247, 281
Reed River (Miss.), 247
Redwood (*Sequoia semperivens*), 117, 229, 299, 300, 318
Reed, Edward, 243
reforestation: in Pacific Northwest, 329–30; relationship to taxation, 455; since 1930, 463, 477–81; in South, 252, 285–8, 438, 479; *see also* Knutson–Vandenberg Act
Reitbrock, Frederick, 235
Report upon Forestry, *see* Hough, Franklin
Report on the Forests of North America, *see* Sargent, Charles Sprague
Rice Lake (Wis.), 220, 221
Reynolds, Robert V., 78
Richards, William A., 418
Richelieu River (Canada), 99
Richmond (Va.), 57
Rio Grande, 33
Roanoke (N.C.), 89
Roanoke Valley, 41
Robbins, William, 409
Rochester (N.Y.), 42, 122
Rockefeller, John D., Jr., 221, 299
Rockford (Ill.), 184
Rock Island (Ill.), 204
Rock River (Ill.), 184
Rocky Mountains, 3, 290, 297, 382, 387, 481; forest extent, 409–10, 432, 435, 474; Free Timber Act, 398; untouched resources, 349, 394, 396
Rodgers, Andrew Denny, 416
romantic movement, 14–15
Roosevelt, Franklin Delano, 388, 463, 459–60
Roosevelt, Theodore, 395; American Forestry Congress, 418, 441; Bureau of Corporations, 426; friendship with Pinchot, 412, 417, 420, 421, 422; Inland Waterways Commission, 422; proclamation of forest reserves, 420–1; timber famine, 482; watershed management, 420
Rostlund, Erhard, 45
Rostow, Walt W., 146
Roth, Filibert, 198, 229, 237
Rousseau, Jean-Jacques, 14
runoff, 4, 8, 25, 112, 144, 145, 353, 388, 394, 450, 491; Ohio, 378–9; western ranges, 307–8; *see also* erosion; Weeks Act
rural idealization, 15–16, 17, 253; *see also* Jefferson, Thomas; Marx, Leo; Penn, William
Rush, Benjamin, 15, 112, 125, 144, 370

Sabine River (Tex.), 247, 281
Sacadanga River (N.Y.), 178
Saco (Maine), 101
Saco River, 102, 170, 175
Sacramento (Calif.), 168, 293
Sage, H. W., 235
Saginaw, 178, 206, 207, 214, 224, 228, 272
Saginaw River, 204, 206, 213, 224, 226, 227, 228
St. Croix (Wis.), 234, 235
St. Croix River (Maine), 104, 174
St. Croix River (Wis.), 186, 203, 204, 206, 213, 221, 228, 247
St. Helens, Mount, 313, 328, 449
St. Lawrence River, 99, 178–9
St. Louis, 45, 178, 267; cross-tie market, 349; fur center, 141; lumber center, 176, 185, 186–7, 188, 204, 220
St. Marys (Ga.), 277
St. Paul, 226, 298, 310
St. Paul and Pacific Railroad, 350
St. Paul and Tacoma Lumber Co., 299
Salem (Mass.), 41, 93, 101, 102
Salmon Falls (N.H.), 92
Salmon River (N.Y.), 100
San Diego, 293, 294, 300
Sandusky (Ohio), 178, 182
Sandy Hill (N.Y.), 170
San Francisco, 291, 307, 456; earthquake, 314; exports, 141, 293; lumber hub, 292, 294, 295, 299, 303
San Jose (Calif.), 292
Santa Barbara (Calif.), 291
Santa Cruz (Calif.), 291, 292
Santa Fe Railroad, 298, 346, 349, 350
Sargent, Charles Sprague, 27, 198, 227, 230, 375, 407, 412; on fire, 449; on forest density, 431; on fuel wood, 334, 346; *Garden and the Forest*, 404, 413; on Lake States, 198, 227, 230, 277; and National Forestry Commission, 414–15, 449–50; on Pacific Northwest, 298, 300, 305; *Report on the Forests*, 376–7; on South, 241, 245, 254, 271, 277

Savannah (Ga.), exports, 102; lumbering, 239, 245, 254, 277, 280; naval stores, 158; river, 42, 68, 154; shipbuilding, 93
savannas, *see* prairies and openings
sawmills: early, 73, 90, 95–6, 101; effect of railroads, 214–16, 253–6; handling devices in, 202–3, 261; mill size and holding size, 428–9; nineteenth-century, 163–9; size increase, 219–20, 224–8; steam power in, 96, 146, 194, 201–3, 261–2, 303; twentieth-century, 261–2, 302–3, 319–20; water power in, 96, 152, 163, 245
saws: band, 168, 201, 202, 303; chain/power, 65, 103, 315; circular, 168, 201, 303; crosscut, 300, 315; gang, 90, 96, 168, 201; muley, 96, 201, 303, planing machinery, 168–9, 202; up-and-down, 96
Schallenburg, Richard H., 341
Schenck, Carl A., 404
Schenectady (N.Y.), 186, 267
Schoepf, Johann D., 65, 78, 109, 138
Schroon (N.Y.), 96
Schroon River, 170, 178
Schurr, Sam H., 334
Schurz, Carl: on timber abuse, 397–8, 399, 402, 407, 427; as secretary of Interior, 376
Schuylkill River (Pa.), 150
Scituate (Mass.), 93
Seattle, 309
Senate Select Committee on Reforestation (McNary hearings), 323, 453, 455, 481
sequoia (*Sequoiadendron*), 291, 483
Sequoia National Park, 486–7
Sewell, Judge Samuel, 92
Shantz, Homer, 30
Shay, E. E., 212
shelter, shelterbelts, 25, 336, 386–8, 454, 477, 478; project, 388–90, 463
Shenandoah River (Va.), 45
Shenango River (Pa.), 151
shingles, 83, 94, 102
shipbuilding, 5, 90, 93–4, 101
shortleaf pine (*P. echinata*), 238, 239, 285
Show, S. D., 484
Shreveport (La.), 255, 280, 394
Sierra Club, 308, 405
Sierra Nevada, 291, 292, 295, 302, 304, 456; grazing, 307–9; Muir, 405
Simpson, Asa Mead, 292, 294
skidders, 21, 251–2, 260, 279; Bernhardt, 252; Dolbeer, 300–302, 315–16, 317; McGiffert, 252; *see also* log and lumber transport, land
skid roads, 466
skins and furs, 141, 157
slaves/slavery, 239, 240, 268; clearing, 66–7; collecting fuel, 77; fencing, 71, 72; naval stores, 85, 157
Smith, John, 22, 40, 44, 57, 63
Smith, Hoke, 414
Smithsonian Institute, 23, 376, 381
Smokey Bear, 482
Snell, Bertrand, 446, 453
Snow, Frank H., 384
Society of American Foresters, 388, 444
soil: fertility and exhaustion, 59–60, 66, 71, 471; Lake States cutover, 430; prairie, 220; South cutover, 430
Soil Bank and conservation programs, 478
South, 42, 59, 82, 87, 108, 128, 289, 290, 292, 298, 303, 319, 320; clearing, 65–7, 73, 77, 360–1; depopulation and cutovers, 280–5; farm abandonment, 468–71, 472, 473, 475–7; fire control, 484–5; Indian clearing, 39, 42, 44–6; lumbering and log transport, 102, 244–60, 492; lumber production and depletion, 271–80, 289; northern purchases, 239–44, 390; reforestation, 285, 478–9, 481; stock in forest, 67–8
South America, 244, 267; *see also* Latin America
South Carolina, 85, 88–9; early clearing 57, 58, 63, 66, 68; exports, 90; lumbering, 245, 272, 278; naval stores, 158–60; woodlots, 336–7
South Dakota, 386, 398; Black Hills, 414; tree planting, 385, 387, 478
Southern Homestead Act (1866), 240–41
Southern Lumber Co., 257
Southern Lumberman, 243, 269
Southern Pacific Railroad, 297, 311, 313, 426
Spain, 92, 101, 244
Spaulding, N. W., 168, 303
Spaulding, V. M., 402
splash dams, 174, 201, 203, 300
stages of civilization, 11–12, 125–8
Starr, Frederick, 331, 345, 371–2, 393, 394, 438, 444, 490
staves, 83, 94, 102
steam engines: mills, 146, 168–9, 194, 202–3; motive power, 152–6; *see also*, locomotives; sawmills, skidders; steam navigation
steam navigation: riverboats, 7, 146, 152–5; towboats on lakes and rivers, 183, 187, 201–7
Steen, Harold K., 456
Steuben, Baron Richard von, 403
Stillwater (Wis.), 183
Stockbridge (N.Y.), 109
Stoddard, H. L., 485
Stoney River (Conn.), 95
stoves, woodburning, 28, 78, 80, 136, 137–8, 334; *see also* fuel wood
Strachey, John, 40
Strickland, John, 80, 91
Stuart, James, 112, 119
Stuart, Robert Young, 460–5
stumpage: Lake States, 204, 217–18, 221–2; large holdings, 425–30; Pacific Northwest, 290, 303, 309–15, 328–9; South, 239–43, 263–9, 279; *see also* forest ownership
stumps, 37, 58, 62, 65, 132; gunpowder and, 198, 356; in Lake States, 283; in Pacific Northwest, 304–7; in South, 283; winches, 115, 356; *see also* cutovers
Sullivan, Daniel F., 242
Superior, Lake, 184, 207
Susquehanna River, 100, 123, 134; anthracite transport, 150; arks, 100; log rafting, 98, 99, 100, 187
sustained yield, 309, 315, 329, 419, 461, 467, 491;

sustained yield (*cont.*)
Pinchot at Biltmore, 412; *see also* Mason, David T.
swampland: baldcypress, 248–50; grants, 243–4
Swank, James M., 109
Syracuse (N.Y.), 38, 42

Tacoma, 297, 313
Taft, William, H., 423
Tallahassee (Fla.), 39, 42
Tallapoosa River (Ala.), 39, 42
Tall Timbers Fire Ecology Conferences, 487
tannin, 82
Tansley, Arthur G., 30
tar, *see* naval stores
Taunton River (Mass.), 41
tax problems: Lake States, 233, 236, 455, 461, 485, 494; Pacific, 305; solutions, 455–6; South, 279–80, 285–6, 287, 455, 483
Taylor, Bayard, 46
Taylor, John, 113
Teapot Dome scandal, 447
Temin, Peter, 107
Tennessee, 42, 56, 58, 66, 141; clearing, 60; farm abandonment, 357, 468; iron making, 339; openings, 71, 128, 129
Tennessee River, 349
Tennessee Valley Authority (TVA), 463, 465
Tensas River (La.), 241
Terre Haute (Ind.), 45
Texarkana (Tex.), 255, 277
Texas, 45, 46, 111, 188, 366, 388, 486; clearing, 360; controlled fire, 486; land purchasing, 243; lumbering, 242, 252, 255–6, 257, 261–2, 264, 267, 271–2, 278; naval stores, 158; reforestation, 252
Thomas, John: on clearing, 369; on fuel wood 344, 393, 404; on shelterbelts, 386
Thompson, Gilbert, 432
Thompson, Mark J., 325
Thoreau, Henry David, 17, 18, 49, 125, 255, 370, 404, 405
ties, railroad, 7, 345–52
Tillamook Bay, 291
Tillamoot fire, *see* fire, historic
timber and forest bills, 409; Dunnell bill, 410; Hale bill, 410; McRae bill, 413–14, 415, 416; Paddock bill, 413
Timber and Stone Act (1878), 304, 398–9
Timber Conservation Board, 461
timber cruising, 198–9, 218, 241, 262, 435
Timber Culture Act (1873), 350, 383–4, 385; abuse of, 396; repeal, 410–11
timber famine/shortage, 320, 393–4, 407, 425, 439, 440, 442, 487; Pinchot and Roosevelt, 441, 443–4, 468
timber growth and inventories, 435–40, 467–8; forecasts, 491–2; *see also* timber cruising
timber "physics," 251, 402, 416
timber theft, 395
Tioga (N.Y.), 211
Tioga River, 99
tobacco, 33, 56–7, 59, 65–6, 71, 84, 336, 365, 468, 471
Tocqueville, Alexis de, 5, 11, 55, 112, 113, 123, 125
Toledo (Ohio), 178, 182
Tombigbee River (Miss.) 42, 247
Tombstone (Ariz.), 307
Tonawanda (N.Y.), 176, 182–3; *see also* Buffalo
Tookabatchee River (Ala.) 39
tractors: bulldozer (crawler), 319; diesel, 317–19; steam, 209–11; *see also* Lombard, Alvin O.
trade associations and alliances, 220–2, 269–71, 322–3
transcendentalism, 17, 125, 404, 405; *see also* Emerson, Ralph Waldo; Thoreau, Henry David
tree farms, 467; Clemons, 330
tree planting, *see* Arbor Day; reforestation; shelter, shelterbelts
trees, *see individual names of main species*
trees and rainfall, 18, 25, 112, 353; climate, 144–5, 387; planting, 350, 380, 386, 387–8; *see also* Arbor Day; Marsh, George Perkins; shelter, shelterbelts
trees as indicators of fertility, 60, 130, 132, 379
Trollope, Anthony, 19
trucks, 260–1, 318–19
Tudor, Henry, 115, 122
Turner, Frederick Jackson, 12, 17; *see also* frontier
Turner, Orsamus, 125
turpentine, *see* naval stores

Umpqua River (Oreg.), 294
Union Pacific Railroad, 297, 346, 350, 381
Upham, F. W., Lumber Company, 214
urbanization: antiurbanism, 15, 19, 125, 377; growth of towns, 78, 146, 177–8, 468; rural–urban transition, 112, 146
U regulations, 458
Utah, 461; Free Timber Act, 398
Utilitarianism, *see* Pinchot, Gifford
Utslady (Wash.), 293

Vance, Rupert B, 272
Van Rensselaer family, 180
Van Tassel, Alfred J., 319
vegetation mapping/knowledge, 22–32, 431–2
Vermont, 18, 42, 58, 59, 111, 144; clearing, 60; Marsh, 370; migration of lumber industry, 308; woodlots, 336
Vinnedge, Robert W., 324
virgin forest, 325, 433, 435
Virginia: clearing and agriculture, 56, 57, 58, 63, 66, 67, 72; coal, 77, 79, 80, 136, 320, 336; farm abandonment, 357, 471; Indian clearing, 37, 41, 48; iron making, 341; lumbering, 161, 163, 272; naval stores, 84–9
Virginia fence, *see* fences and fencing
Volney, Count, 23, 144, 370

Wabash (Minn.), 204
Walker, Francis, 25
Warder, John A., 351, 403

Ware, Walter M., 373
Washington, George, 41
Washington (D.C.), 308, 397; bureaus and bureaucracy, 422, 423; bureau rivalry, 402, 457
Washington (N.C.), 158
Washington State, 3, 25, 262, 272, 397, 422–3, 456; Clark Co., 321; fire control, 321–7, 449–50; King Co., 295, 325; lumbering, 292–301, 316–19; lumber resources, 298–301, 304–5; production, 161, 295, 304, 309–10, 325–8, 418, 461; reforestation, 330; Skamania Co., 321; Snohomish Co., 295, 325; timber holdings, 310–11, 428
Wansey, Henry, 63, 99
Warrior River (Ala.), 136
waste elimination, 232, 261, 303; *see also* kerf
water power, 146, 148; impact of railroad on sites, 214–15; sawmills, 163, 169, 226–7, 245; steam versus, 152, 201
watershed management, 308, 394, 415, 453, 456, 467; multiple use management, 422, *see also*, Adirondacks, Weeks Act
Waugh, Frank, 457
Weaver, Harold, 486
Weeks Act (1911), 308, 322, 407, 442, 450, 453, 454, 482
Weld, Isaac, 123
Wells, Charles, 255
Wells, Samuel, 144
Wells (Maine), 101
Weisberger, Bernard A., 193
West, the, 290, 334–5, 406, 407; as concept, 354–5, 397–9; resentment of East, 414, 415, 420–2, 422, 466; *see also* national parks; national forests; public lands
West Coast, 290, 297; lumbering on, 292, 293, 294, 299; lumber migration to, 299; physical nature, 296; railroads to, 297–8, 381
West Coast Lumbermen's Association, 460
Western Forestry and Conservation Association, 322, 450, 452
West Indies, 68, 84, 89, 93, 101, 188; *see also* Caribbean
West Troy (N.Y.), 178
West Virginia, 56, 141; farm abandonment, 357, 468; lumbering, 244
Weyerhaeuser, Frederick, 322, 428; Beef Slough operation, 205–6, 222, 228; Lake States interests, 220, 221; Pacific Nothwest interests, 229, 235–6, 286, 299, 303, 310–4, 321, 421; Southern interests, 229, 242, 265–6, 270; Washington Forestry Congress, 418
Weyerhaeuser, R. M., 270
Weyerhaeuser Company, 303, 311–14, 318, 321, 329–30, 426–7
Wheeling, 169
White, James Barber, 436
White Mountains (N.H.), 407, 456; fire control, 450
white pine (*P. strobus*), 93, 102, 104, 117, 132, 170, 176–7, 178–9, 182, 183, 198, 198–9, 207, 222, 228–9, 299, 402
Wilber, Charles Dana, 384
wilderness: areas, 458, 474; concept, 10, 17, 19, 308, 344, 454, 488; fire 483; movement, 403, 404–7; work, 123
Wilderness Act (1964), 459
Wilderness Society, 458
wildlife, 4, 8, 11, 467; effect of fire on, 484, 485, 486; elimination, 491; game, 38, 46
Willamete Valley, 289, 295, 304
Willapa Bay, 313
Williamson, J. A., 397
Williamsport (Pa.), 168, 175, 242
Wilmington (N.C.), 158, 245, 255, 277, 280
Wilson, James, 381, 382, 383, 420
wind storm damage, 30
Wine Islands, 84, 101
Winona (Minn.), 187, 204, 220
Wisconsin, 19, 27, 45, 46, 111, 176, 184, 242, 272, 397; Cadiz Co., 367; clearing, 357, 367, 373; fire, 232–3, 449; iron making, 152, 339; lumbering, 179, 188, 193–237 passim, 272; Marinette Co., 235; openings, 379; rafting and prairie lumber supplies, 186–8; *see also* Lake States; prairies and openings; rafting
Wisconsin Central Railroad, 216, 234
Wolf River (Miss.), 247
Wolf River (Wis.), 184, 212, 214, 224, 226
woodlots, 118, 119, 120, 125, 333, 336–7, 386, 428, 454, 477
Woodruff, George, 423
Woodstock (Vt.), 144
woodworking machinery, 168–9, 202, 303
Wordsworth, William (N.Y.), 168
Wordsworth, William (U.K.), 14
Wyoming, 3, 461; forest reverves, 421; Free Timber Act, 398; tree planting, 605; *see also* Yellowstone National Park

Yale Forestry School, 442, 460, 485
Yawkey, William C., 241
Yazoolands, 243, 246; Yazoo–Little Tallahatchie Flood Prevention Program, 479
Yellowstone National Park, 405, 406
Yoknapatawpha County, 253
York (Maine), 101
Yosemite National Park, 308, 405–6, 411, 456
Youghiogheny River (Pa.), 151
Young River (Pa.), 136

Zeck, Heronymous, 235
Zeigler, E. A.: on forest area, 432; on timber growth, 436, 438
Zon, Raphael, 30, 432; on tree planting, 387–8

Printed in the United States
By Bookmasters